Agaves of Continental North America

Agave shawii in flower at Peñasco Lobera.
Photograph by Reid V. Moran.

Agaves of Continental North America

Howard Scott Gentry

THE UNIVERSITY OF ARIZONA PRESS
Tucson, Arizona

About the Author . . .

Howard Scott Gentry made his first field trip into the Sierra Madre Occidental of Mexico in 1933. Most of the next twenty years were spent in exploring the plant life of northwestern Mexico. As an agricultural explorer for the U.S. Department of Agriculture from 1950 to 1971, he searched for germ plasm through North America and parts of India, western Europe, and Africa. In the process of working with useful plants he became an economic botanist. In 1971 he affiliated with the Desert Botanical Garden in Phoenix, Arizona, which supported his work in agave research. His principal written works include *Rio Mayo Plants of Sonora-Chihuahua; The Warihio Indians of Sonora-Chihuahua, an Ethnographic Survey; Los Pastizales de Durango; The Agave Family in Sonora;* and *The Agaves of Baja California.*

The Desert Botanical Garden in Phoenix, Arizona, has supported the work of the author with concern and practical assistance for more than a decade.

THE UNIVERSITY OF ARIZONA PRESS

Copyright © 1982
The Arizona Board of Regents
All Rights Reserved

This book was set in 10/11 V-I-P Times Roman
Manufactured in the U. S. A.

Library of Congress Cataloging in Publication Data

Gentry, Howard Scott.
    Agaves of continental North America.

    Bibliography: p.
    Includes index.
    1. Agave. 2. Botany—North America. 3. Botany,
Economic—North America. 4. Ethnobotany—North America.
I. Title.
QK495.A26G453   584'.43   82-6896

ISBN 0-8165-0775-9   AACR2

Dedicated to my father and mother
Irvin and Elizabeth Gentry
of the past and for the future
1875 to 1965

Contents

Part I. Background to Agaves

1. The Man-Agave Symbiosis   3

2. Taxonomic History and Morphology   25

3. Geographic Guide to Species and the Exsiccatae   49

Part II. Systematic Account of Genus and Species: Subgenus **Littaea**

Part III. Systematic Account of Genus and Species: Subgenus **Agave**

Part IV. Culture of Agaves and Other Addenda

Tables

Preface

This book is a detailed account of the North American agaves, 136 species in twenty generic groups. The summary of *Agave* taxa (below) gives particulars for all 197 taxa in the two subgenera, **Littaea** and **Agave**. The generic groups bring related species together and assist in taxonomic orientation and identification of species. I have followed Trelease and Berger in groupings to a large extent but have not hesitated to redefine and make new groupings whenever new evidence (specimens) recommended changes. These groups are tantamount to sections in more formal taxonomy. Many boundaries between groups are not sharp, as also obtains among many species, because variation in *Agave* is mostly of a gradual or clinal type; one form or character changes to another by degrees, a condition which also characterizes variations in other genera of the Agavaceae. It appears that the family has been slowly evolving for many millions of years. The groups are given in alphabetical order under each subgenus.

Agaves are unique in the whole plant world, not merely because of their succulent character but because of a special role they have played in the indigenous civilizations in North America. Hence, as introduction, I have given a thoughtful description of the useful role of agaves and how they have been employed by New World men. Following that section the taxonomic history and general morphology are conventional in scope but written in as nontechnical a way as I thought would appeal to succulent plant fanciers and biologists in general. There are far more of them with special interests in agaves than there are professional taxonomists.

The main corpus of this work is the 136 essays on each species recognized. These essays vary in length according to what is known about the subject plants. In general my procedure has been to give (1) a brief taxonomic description; (2) notes on the taxonomic position and pointed comment on its main distinguishing features; (3) distribution and habitat; (4) uses, past and present; (5) vernacular names; (6) horticultural notes. Many of the species are represented by line drawings, many by photos of the plants in natural habitats; there are also close-up photos of leaf and/or inflorescence details.

One of the chief aids in the identification of agaves, whether herbarium or garden specimen or a wildling, is knowledge of their original location. It is easier to select the correct morphological type from a few candidates than from many candidates. In Chapter

Summary of *Agave* Taxa

SECTION	SPECIES	SUBSPECIES	VARIETIES	FORMAE	TOTAL
Amolae	8		1		9
Choritepalae	3				3
Filiferae	8				8
Marginatae	21			7	28
Parviflorae	4	2	1		7
Polycephalae	5		2		7
Striatae	3	1			4
Urceolatae	2	1	2		5
Total Subgenus **Littaea**	54	4	6	7	71
Americanae	6	5	8		19
Campaniflorae	3				3
Deserticolae	10	11			21
Crenatae	6	1	1		8
Ditepalae	10	2			12
Hiemiflorae	12				12
Marmoratae	4				4
Parryanae	6		4		10
Rigidae	12		7		19
Salmianae	5	1	3		9
Sisalanae	6				6
Umbelliflorae	2	1			3
Total Subgenus **Agave**	82	21	23		126
TOTAL	136	25	29	7	197

3, as a further aid to identification of agaves, I have prepared some synopses of species according to their geographical regions.

In all too many cases, the species dimensions remain indeterminate; populations with fuzzy edges, grading fleetingly or strongly into others. However, by stressing the variability of species, I hope I have dispelled the clonal image of species portrayed by the narrow binomials of earlier botanists. At least my names are suitable for population markers, if not as ideal species that include all interbreeding varieties and forms. I leave the next agave taxonomist a good opportunity for improvement.

H. S. G.

Acknowledgments

There are many people and institutions to thank, to cite, and to commend for making this book possible. The appropriation of funds, the support given me for both personal and professional living and work over many years, has amounted to hundreds of thousands of dollars. I mention this matter of money not to boast my position but to indicate the seriousness with which our botanical community has viewed the need for an illuminating work on agaves. I came to feel a heavy responsibility in accepting funds and positions, which in themselves expressed confidence and faith in my abilities and intentions. The agave project became for me a community duty to produce a quality product.

The taxonomic investigation of agaves was initiated by the New Crops Research Branch, Agricultural Research Service, Beltsville, Maryland, in October 1957, as Line Project CR il-15. I was designated as project leader and the estimated duration was five years at the rate of 0.4 man-year per year.

The agave project was continued in 1972 under the auspices of the Desert Botanical Garden, which I joined as research botanist in 1971. The Garden has furnished an office and herbarium headquarters, funds for routine operations, and part-time assistance from staff as needed.

The National Science Foundation supported the agave project with fund grants (BMS 74-24553, 75-14087) from 1972 to 1978.

The California Academy of Sciences published at their expense two botanical papers on the *Agaves of Baja California* (1978, 1978a). In this connection publication was particularly enhanced by colored illustrations with funds generously provided by Mr. E. R. LeRoy of San Francisco.

The Huntington Botanical Gardens, San Marino, California, probably has the most complete collection of living agaves in the world. Myron Kimnach and others there, including the founding father William Hertrich, provided special privileges for study of the living agaves, many of which I collected for them myself. Since 1951 it has been a rich asset for my agave studies.

Many herbaria have provided specimens for critical study or for identification. A list of these institutions is provided under Catalogue of Specimens. The *Agave* collections at the Missouri Botanical Gardens Herbarium and the U.S. National Herbarium are particularly rich in materials, and I was provided working space and special attentions for repeated visits at both these institutions. I am grateful for the consideration shown me by the curators and others involved at all these marvelous institutions.

Clarification of agave characters by art work has been ably done by Regina Hughes, former staff artist for Agricultural Research Center, Beltsville, Maryland; by Lucretia Hamilton of Tucson, Arizona; and finally by Wendy Hodgson, graduate student of botany, Arizona State University, Tempe, Arizona.

My field trips and collections of agaves were greatly benefitted by Juan Arguelles of San Bernardo, Sonora; by Arthur Barclay of ARS, Beltsville, Maryland; Tony Burgess of Texas Tech University, Lubbock, Texas; Frank Cech of San Diego, California; Rodney Engard, Lyle McGill, and John Henry Weber of the Desert Botanical Garden staff; Dudley Gold of Cuernavaca, Morelos; Francisco Gonzales Medrano and H. Sanchez Mejorada of the National University, México, D. F.; William Fox, Charles L. Gilly, Eugene C. Ogden, formerly with New Crops Research Branch, Beltsville, Maryland, who worked with me on agaves during the early 1950s.

Special contributions with photographs in the field came from my brother Bruce Gentry, Arthur Barclay, John McClure, Reid Moran, and, in the photo laboratory, from Walter Hodge of Cornell University.

Professional botanical assistance and advice has been given generously by Richard Felger of Arizona-Sonora Desert Museum; Frederick Hermann, retired USDA; Thomas Howell and George Lindsay, California Academy of Sciences; David Kiel, San Luis Obispo College; Rogers McVaugh, University of Michigan, Ann Arbor; Donald Pinkava, Arizona State University, Tempe, Arizona; Jane Reese Sauck, University of Arizona, Tucson, Arizona.

Final typist has been Regina DeRose of Arizona State University, Tempe. Staff members have helped in many special ways: Wendy Hodgson with errands and the volunteer Joyce Baldwin with the onerous task of alphabetizing the exsiccatae.

I am grateful to the University of Arizona Press, especially to my editor, Marie L. Webner, and to the other members of the staff for their competent work in making this a handsome book.

Finally, I accord my respects and gratitude to my wife, Marie Ann Cech Gentry, who assisted me on several field trips, and although she would have nothing to do with agaves, no doubt kept me from going completely monomaniacal over agaves on many occasions.

H. S. G.

PART I

Background to Agaves

Metl plant, one of the earliest illustrations of agaves, as depicted by Hernandez in 1651. In the beginning man's image of agave was crude.

1.

The Man-Agave Symbiosis

The uses of agaves are as many as the arts of man have found it convenient to devise. At least two races of man have invaded Agaveland during the last ten to fifteen thousand years, where, with the help of agaves, they contrived several successive civilizations. The region of greatest use development is Mesoamerica (Fig. 1.1). Here the great genetic diversity in a genus rich in use potential came into the hands of several peoples who developed the main agricultural center of the Americas. Perhaps, as the Aztec legends suggest (Goncalves, 1956: 72), it was the animals that first showed man the edibility of agave. Evolution in use ranges all the way from the coincidental and spurious, through tool and food-drink subsistence with mystical overlay, to the practical specialties of modern industry and art. The historic period of agave will be outlined here as briefly as that complicated development will allow.

The Historical Perspective

The diffusion of agave cultivation from its original nucleus in the Mesoamerican highlands occurred rapidly after the conquest. When the Spaniards began colonization of more northern regions, like Durango and Saltillo, they took Nahuatl people with them as interpreters, laborers, and farmers. The farmers took maguey with them and established the pulque culture which still persists as the northern fringe of the pulque complex. Other agaves, for ornamental and fiber uses, were apparently first carried overseas by both Spaniards and Portuguese: *Agave americana* to the Azores and Canary Islands; *A. angustifolia, A. cantala,* and others to Asia and Africa. By the eighteenth century *A. americana, A. lurida,* and others were established along the Mediterranean coasts. The spread of the genus to the Old World reached its height in the nineteenth century, when agaves became popular throughout Europe as ornamental succulents in both private and public gardens. In northern Europe their culture was generally limited, because of the cold winters, to pot and greenhouse culture. Agave fiber industries were developed in the nineteenth century by colonial interests in Indonesia and the Philippines, and in East Africa in the twentieth century with *A. sisalana.* Methods of culture, fiber harvest, and selection of varying forms have been developed in different regions, according to the regional environments and

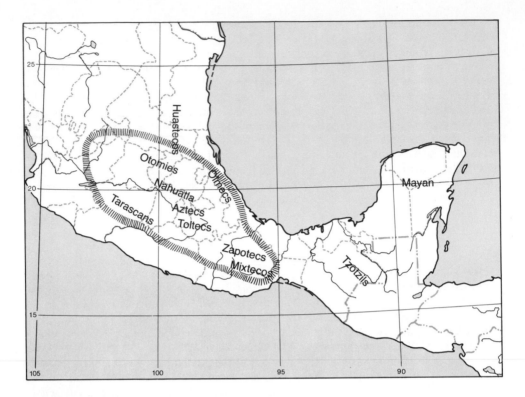

Fig. 1.1. Region of intense agave culture with some of the principal Indian nations. Olmecs were extinct before the Spaniards appeared.

available working resources. Additional observations are given in the section dealing with fiber, pp. 16–20, and under individual species in the taxonomy part of this work.

Overseas, as in America, the man-agave relationship has required mutual adaptation of both organisms for cosurvival. In the modern, competing, complex, organic world, both ornamental and fiber culturists of agaves have waxed and waned individually through the centuries, according to how interest or economic motive was served. Today in the Old World agaves are carried on, or, in localities favorable to them, they persist by themselves spontaneously. If we consider the genes as causal factors, then which of the two organisms is carrying the other? Perhaps mutualism is just a fortuitous alliance of two dissimilar gene pools, which operate better than either would do alone. Agave hosted man in the New World; man transported and hosted agave in the Old World.

The aesthetic nature of man has conserved agave as ornamentals, while his intellectual character in devising uses has fostered its evolution, but his hunger and greed have also destroyed agaves. Since both the aesthetic and intellectual character of man are dependent upon the satisfaction of his hunger, the man-agave relationship may continue for as long as mankind is not too hungry or too crowded for space.

In North America, agave perhaps had as much to do in fostering the beginnings of agriculture as any other genus of plants. In Agaveland anyone can plant and grow agaves. All that is needed is to dig up or pull up a young offset and bury its base in moist or dry soil, with or without roots, wherever it is wanted. If it does not strike root and grow the first season, the chances are good that it will the next. Sauer (1965) has made a strong case that such transplants were the primary agricultural subjects of the Amerindians. Compared with seeds, the shift of useful plants from the open wild site to camp or village was more obvious and direct with transplants, and their care, protection, and culture were simpler.

The hunting and gathering tribes had good reason to regard agaves with special attention, because agaves supplied them with food, fiber, drink, shelter, and miscellaneous natural products. Protection may have been one use, for when planted around a cottage, the larger species make armed fences, a common practice in modern Mexico. While much about the first beginnings of agriculture will always remain obscure, there is a great deal now known about the history of man-agave relationship. A review of this history, especially in archaeological, historical, and ethnological literature, is perhaps the most interesting way to show the development of this indigenous American complex. The many uses of agave will become more meaningful if the man-agave mutualism is stressed rather than the uses simply enumerated. Castetter, Bell, and Grove (1938) have provided an excellent review of agave use among the Indians in the southwestern United States. This account will focus on Mesoamerica, the center of agricultural origin of the genus, and will thus provide conceptual background to agave taxonomy.

Agave as Food

In Mesoamerica man has chewed agave for at least 9,000 years. Callen (1965) published irrefutable evidence for this statement, after having examined several hundred coprolites (mummified human feces) furnished him by McNeish, leader of the Peabody Foundation archaeological expeditions to Mexico during the 1950s and early 60s. Callen summarized the principal items in the diet from 7000 B.C. to A.D. 1500. With *Agave* these were *Setaria, Ceiba, Cactus, Cucurbita,* bean, *Capsicum, Amaranthus, Diospyros,* bone, meat, and *Zea*—from about 5000 B.C. Agave was found throughout the time scale in 25–60 percent of the coprolites. Two of Callen's summary charts are reproduced here (Figs. 1.2, 1.3).

Callen explains that fragments of consumed plant materials pass through the alimentary tract sufficiently intact for microscopic identification: "Agave epidermis with cells arranged in a diamond shape around the stoma, and the characteristic type of crystals in its tissues." The time scale was established by carbon-14 datings. Because the coprolites represent cave sites on rocky mountain terrain, the materials do not reflect fairly the

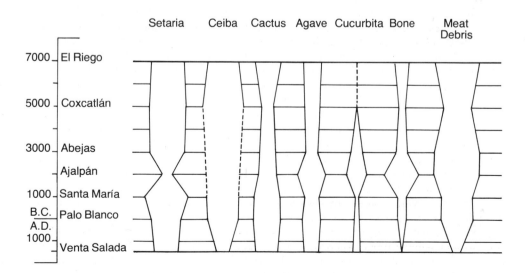

Fig. 1.2. Food sources as found in coprolites in caves and other culture sites in Tamaulipas. (Percentage occurrence.)

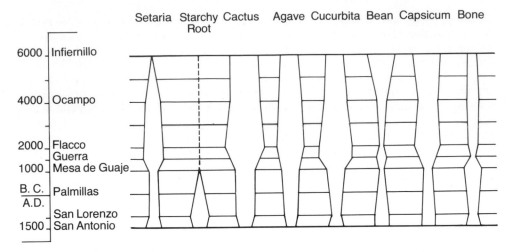

Fig. 1.3. Food sources as found in coprolites in caves and other culture sites in the Valley of Tehuacán, Puebla. (Percentage occurrence.)

cultivated agricultural plant products, which were developing from 5000 B.C. onward, with the establishment of villages and later the Mexican city states. "It should be clearly understood right away, however, that this is a cave diet, and not a city diet." Use of the caves in village and city times was attributed to seasonal visits by hunting or gathering parties. However, the presence of cooked portions of agave in modern Mexican markets testifies to the use of agave as food through the late twentieth century.

From 7000 B.C. onward the use of agave is documented also by archaeological specimens of quids (chewed fiber rejects), by artifacts made of agave fiber, and the tools used in their manufacture. The records of such uses are widely scattered in numerous reports of travelers and archaeologists (Byers et al., 1967). In Mesoamerica the many evolving varieties and forms of *Agave* species were selected by man, moved from place to place with him, and inadvertently crossed. As man lived with these varietal eventualities through the centuries, he was provided with new genetic combinations that he could check empirically for yield and quality of fiber, food, beverage, and other special products. As he specialized with civilization, he specialized agave, selecting characteristics according to his wants. Even though he had no concept of genetics, he quite innocently fostered an explosive evolution in agave diversification.

The main source of food in agave is the soft starchy white meristem in the short stem and the bases of leaves, excluding the green portion. As the plant matures the starch and sugar content of these organs increases, as does their palatability. Some species and varieties are more palatable than others; those with high sapogenin content and other toxic compounds were generally known and avoided. The young, turgid, tender flowering shoot of most species is edible, as are the flowers of many. The early agriculturist doubtless selected only the sweet sorts for cultivation. The sapogenous species were not domesticated. Since merely supplying heat converts the starches to sugars, the Indian cooked the softer parts by direct fire or with hot water. In earliest times the cooking of agave was crude. "Cooking appears to have been largely of the roasting type, with the outside frequently charred, and the interior still raw. This is true of such plants as *Ceiba, Agave,* and *Opuntia,* though they appear to have been eaten raw almost as frequently" (Callen, 1965). Charring of agave flowering shoots by laying them in the fire or in hot coals and ashes overnight was still observed among the backcountry Mexicans in the 1970s, especially to appease hunger on longer journeys. From the time the Mexicans had pots, these flowering shoots were probably boiled, a practice extended to modern times in Mexico. A more sophisticated or communal method for cooking agave was *pit baking*, which became universal, at least north of Mesoamerica, and which has been mapped and fully discussed by Castetter et al. (1938).

Agave pit baking was a family or group effort, generally with men and boys collecting the wild *mescal heads* (''cabezas''), the women and girls gathering firewood and cooking. A pit large enough to hold many heads was dug and lined on sides and bottom with stones. A large fire was burned in the pit to heat the rocks and form coals. When the fire burned down, the Agave heads were pitched in the pit on the hot stones, or sometimes on a layer of green grass, palm leaves, or other green leaves. The heads were covered over with leaves or grass and sufficient earth on top to prevent steam from escaping. The mass was allowed to steam-cook for one or two days. The time necessary for cooking depended on the amount to be cooked, size of the heads, tribal customs and ceremonies. The cooked heads were cut up into chunks, after separation from the leaf butts and both could be eaten at once or could be stored indefinitely for furture use. They were also pressed into flat cakes that could be easily carried and were bartered in trade between tribes. The expressed juice from the cooked chunks was rendered into syrup. Candies were also prepared with this syrup. There is considerable fiber in these chunks of agave, which when chewed are rejected and, if left in the litter of dry camp sites and caves, become the quids of archaeology.

This description barely outlines the practices of agave pit baking as we know them from northern Mexico and southwestern United States. I have seen mescal pits as far south as northern Sinaloa and Durango and have seen baked agave in the markets of central Mexico. However I have seen no account of the practice in Mesoamerica. Miguel del Barco wrote an excellent account of the practice in Baja California (see Gentry, 1978: 5).

Among other food uses of agave is the boiling of the flowers of some species or, more commonly, scrambling them with eggs, as reported about Tehuacán. Among the Mixe Indians in the mountains of Oaxaca, the cuticle of *Agave atrovirens* and other species is put to a singular culinary use. The cuticle is peeled from the leaf and employed as a wrapper for tortilla sandwiches and other foods carried as lunch to the field. It forms a translucent sheet looking like an archaic forerunner of modern polyethylene plastic. This practice was drawn to my attention by the conspicuous rectangles appearing on the broad leaves of agaves, growing about the Mixe fields and houses, as a result of skinning off sections of the cuticular wrapper. In Saltillo, Coahuila, bread is still made with pulque, which gives the bread the distinctive flavor of pulque. Vinegar is easily made from aguamiel and alcohol has been distilled from pulque.

In northeastern Mexico, agave leaves are fed to livestock. In San Luis Potosí in 1963 I observed *Agave salmiana* being carted daily to the diary herds supplying milk to that city. The fresh green leaves amounted to several thousand tons annually. Along with *Opuntia* pads, this agave constitutes an important animal food resource in that desert country. Extensive young plantings of agave on hillsides 8 to 12 miles east of the city of San Luis Potosí appeared to be destined for animal forage. In Baja California the flowering panicles of *Agave shawii* and *A. s. goldmaniana* are cut and trucked to range cattle. On open cattle ranges the flowering shoots of all species of the smaller agaves are commonly cropped by the animals. However, the sharp terminal spines on the leaves of the larger species are apparently effective in foiling cattle away from the flowering shoots.

Agave as cattle forage is a recent development reflecting the pressure of human population on the available plant resources. Local ingenuity is extending milk production in a desert region with desert plants, far removed from the region of origin of animal industry. I have seen no definitive account of agave as cattle forage, but a studied comparison of this resource with conventional forage crops would be most interesting. What would be the production of meat and milk from one hectare of agaves? Is *Agave salmiana* the most suitable species for cattle feed? Have others been tried? Are there other peoples in other desert regions of the world that would benefit by the introduction of forage agaves to their countries? These are some of the obvious questions that appear to recommend a serious evaluation of agave as a forage resource. For a recent work on food values in agave, see Marroquin (1979).

The Beverages of Agaves

The agaves produce two distinct kinds of beverages. The first is drawn as sap from the living plants and consumed fresh as *aguamiel* or fermented as *pulque* (a Caribbean word; see Goncalves de Lima, 1956). The second kind is the distilled liquors known as *mescal (mezcal)* or *tequila,* which were developed subsequent to European occupancy of Mexico. The Native American did not know distillation. The distilled liquors now comprise a modern industry, rapidly expanding and of increasing economic importance. Pulque, however, has been produced since prehistoric times, has remained local or endemic in use, but is more important as a product that stimulated and nourished the cultivation of agave and became basic to ancient Mexican civilizations.

The first historical records of agaves are the Mexican pictographs on ruins and in the codices. These have been made intelligible by the chronicles of the Spanish conquest and by both Indian and European scholars following the conquest. Goncalves de Lima, in his "El Maguey y el pulque en los codices Mexicanos" (1956), has provided an excellent résumé of history pertaining to agave. I recommend it to all who can read Spanish as an account that will illuminate the fascinating subject of man-agave relationship in this period. The historical accounts show the several centuries preceding the conquest as a heyday of agave culture.

Aguamiel and Pulque

A specialized agave cultivation became habitual to both rural and urban societies among the various tribes in Mesoamerica. In addition to fiber, food, and the sweet-juice drink aguamiel, a fermented drink was developed in the product now called pulque (*octli* to the Aztecs). Its mild alcoholic effects were extended with infusions of drug plants and incorporated into religious and mystical ceremonies. Such liquors were used in the rites of human sacrifice, for example. Associated with maguey were several gods, but the ascendant one seems to have been *Mayahuel,* the Aztec goddess of pulque (Fig. 1.4). She is depicted in various ways in the codices but is identifiable by the leaves, by the stylized inflorescence of the maguey, and by foaming pulque in her hair or by pots of pulque in her hand or nearby.

According to the Mexican codices, those of Boturini being particularly helpful here, the Aztecs during their migration to the Valley of Mexico, discovered maguey and "invented" pulque between A.D. 1172 and 1291. This can only be considered as meaning they discovered the magueys and pulque for themselves, because the archaeological records show that earlier peoples of this region had been using maguey for thousands of years previously. Although there is no direct archaeological evidence for when agave sap and pulque were first drunk, we can be reasonably sure it was consumed much earlier than the Aztec period, considering the worldwide use of plant juices, both fresh and fermented, by relatively simple societies. Such legendary inaccuracy does not detract from the obvious success of the Mesoamericans in the development of the magueys.

From the "relaciones" of early history we learn that the Aztecs and other Nahuatl peoples did much to refine, elaborate, and extend the cultivation of maguey and its products. We can also see the results in the array of living plants in central Mexico today. New varieties would have been introduced from wild sources and from chance crosses already in cultivation, while new tools and techniques were devised for tapping, brewing, drinking, and in conceptualizing social behavior with the stimulus of drink. The whole

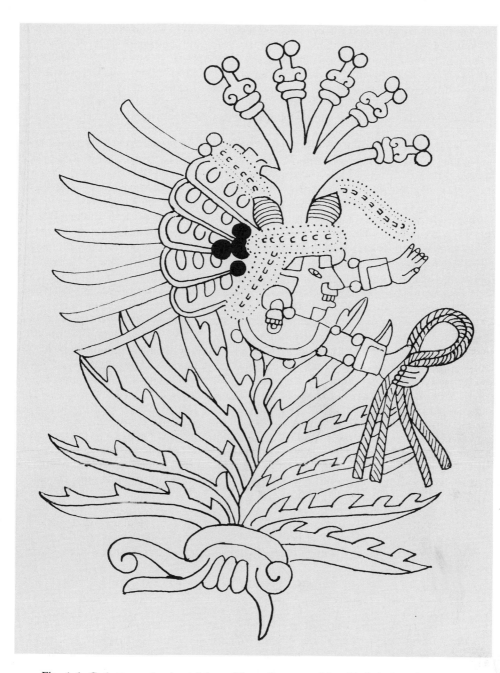

Fig. 1.4. Codex reproduction of the goddess of maguey, Mayahuel, in the Borbonica collection (from Goncalves de Lima, 1956).

industry developed specialized labor from the farming peasants to the consuming intellectual priesthood, and social layering would intensify the civilization. As the new varieties of maguey would have been tested in different soils and climates, their productivity compared with older varieties, so would have been tried and evaluated their varying flavors of aguamiel and the different methods of brewing pulque.

About thirty different microbes have been identified in aguamiel and pulque in the following genera: *Bacterium, Lactobacillus, Leuconostoc, Sarcina, Pseudomonas, Streptococcus, Diplobacter, Bacillus, Hansenia, Saccharomyces, Picchia, Torulopsis, Rhodotorula, Mycoderma* (Goncalves, 1956: 21). Not all of these microbes are involved in pulque brewing, and little appears to have been published regarding the roles of those genera in the fermenting of pulque. Goncalves de Lima (1956: 23) found that *Pseudomonas lindneri* was an important agent in the initial fermentation of aguamiel. The Indian, of course, did not know of these organisms, but he could refine his selections of brew according to the conditions in which fermentation proceeded, as temperature, type of container, and the critical stage in brewing or timing. The Indian brewer elaborated empirically.

Also, like the beer brewers of Europe, the Mexicans flavored pulque with herbs, roots, and barks. The Aztecs, according to Clavigero (1807), added the roots of an herb they called *ocpatli* to facilitate fermentation and add force to the drink. The identity of this plant is not known. Such fortifiers probably carried inoculums for fermentation. Roots and barks of many plants have been and still are used around the world in making fermented drinks. However that may be, the arts of the brewer were enjoined with those of the agriculturists in domesticating the pulque agaves.

Overlaying and infusing all the specialized activities was the developing priesthood. They became the ruling class who, while they propitiated the gods with pulque, laid out detailed rules governing not only the use of pulque but all activities of their social organization. They possessed pulque with seeming avidity, appropriating its use especially in their ceremonies, when inebriation was not only permissible but desirable. The story apparent in Figure 1.5 shows the rigorous discipline of the Aztecs, which was also a prescription for preconditioning to pain and perhaps rationalized as the punishment of the gods. The Puritans bedeviled alcohol, the Mexicans deified it, but both practiced rigid moral standards. Peterson (1961) informs us about the Aztecs: "Sale of pulque, an intoxicant, was allowed only under certain restrictions. To the sick, especially privileged, and to people who had passed their fifty-second birthday three cups were allowed daily. But there were festivals when everyone might drink. If a person became drunk illegally, his hair was cut off the first time, on the second occasion his house was demolished and he lost his employment; and the third time brought death to him as an incorrigible offender." This was in the disciplined state of the city.

Fig. 1.5. Use of agave spines in disciplining boys and of agave thongs in punishing a runaway slave (after Peterson, 1961).

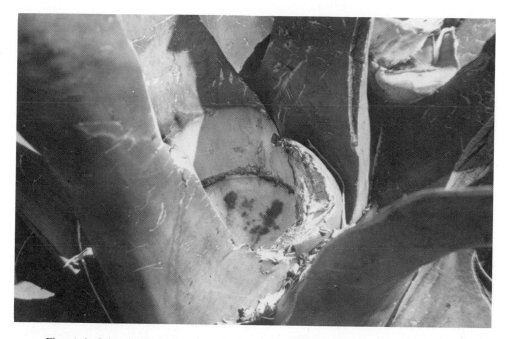

Fig. 1.6. Juice basin in large maguey plant, formed by cutting out the terminal leaf bud.

Maguey culture was coincident with warm and arid climates. The upwelling sap, held like a live spring in the heart of the huge maguey plants, was a ready source of sweet water for the sweating peasant and the weary traveler where potable water was scarce or absent. "Es el pulque la bebida fundamental de los habitantes de enormes extensiones de terrenos desprovistos de aguas potable, y claro que es con ventaja preferible a las aguas contaminadas"* (Martin del Campo, 1938: 12). Fermentation will occur in the maguey basins (Figs. 1.6, 1.7), if not collected fresh daily and is comparable as intoxicant to our hard cider. The nourishing sap of maguey was available in its varying forms to all country people in the region of its culture. Corn and beans provided a substantial basic diet of carbohydrates and proteins and were well established in Mesoamerica by 4000 B.C., as in the Coxcatlan phase of Tehuacán Valley (MacNiesh, 1967). As in other agricultural nuclei, diversification of agriculture in nonbasic crops would have followed the basic; thus, maguey was probably burgeoning before the rise of such city states as those of the Otomis, Zapotecs, and Mayans. In addition to water, pulque supplied such dietary needs as minerals, amino acids, and vitamins. Pulque helps to balance the nearly meatless diet of many of the poor Indians who consume it regularly today (Tables 1.1, 1.2).

There is little doubt that alcoholic pulque had much to do with the esoteric and exotic development of Mesoamerican culture. Martin del Campo (1938:7), in his historical review of pulque, remarks on the dual nature of pulque: "Armas de dos filos, las bebidas embriagantes han tenido importante participación en la evolución de las costumbres, las creencias, las artes, etc., de los pueblos. . . . En un sentido, el de su uso mesurado, avivan dichas bebidas la imaginación, aguzan el ingenio, despertan facultades estéticas latentes, estimulan el ánimo y provocan agradable euforia; en el otro, el de su abuso, lejos de

---

*"Pulque is an essential drink for the inhabitants over enormous extensions of land without potable water. Obviously, this is preferable to contaminated water."

Fig. 1.7. A *pulquero,* who collects the juice in the maguey basins and transports it by burro to the *pulquería,* where the juice is brewed.

estimular, abaten toda manifestación superior de la inteligencia y del sentimiento humanos, con excepciones que, si fueran la regla, bien valdria la pena de tomarlas como ejemplo.''*

Alcohol, the physiologists tell us, is a depressant, but its effects on the psyche of man are quite the reverse, because it releases individual responses generally inhibited by society. Quiet men become noisy, stupid men become brutish, hidden grudges and secret hopes become manifest, weakness or strength may be revealed in individuals where these characters were not hitherto suspected. Among many, alcohol is a palliative because it permits mental and psychic exercises not otherwise functional.

Through psychological release it may stimulate sexual intercourse, becoming a positive agent for reproduction. Among the Tarahumara and Warihio Indians of Chihuahua, sexual contacts are seldom made by the shy youths unless they are under the influence of their brew ''tesqüino.'' Carl Lumholtz (1973, vol. 1:352) in his famous study of the Tarahumara made a social judgment: ''Incredible as it may sound, yet, after prolonged and careful research into this interesting psychological problem, I do not hesitate to state that in the ordinary course of his existence the uncivilized Tarahumare is too bashful and modest to enforce his matrimonial rights and privileges; and that by means of tesqüino chiefly the race is kept alive and increasing.'' Most likely pulque functioned as a breeding stimulant among Mesoamericans also. Had not those thousands of Aztec codices been destroyed by the horrified Spanish priests, we might have found among them not only the maguey gods but also an agave cupid.

Alcohol promotes fellowship and communication and many a bargain or arrangement is made today in the social office of the ''cantina'' or club. As a social catalyst, pulque

---

*''The intoxicating liquors have played an important part in the evolution of the customs, the beliefs, the arts, and so on of the pueblos....They are like weapons with two edges; in one sense, with moderate use, such beverages enliven the imagination, sharpen the mind, waken the latent aesthetic faculties, stimulate the soul and provoke an agreeable euphoria; on the other hand, when their use is abused, they are far from noble stimulants and they repress all the superior human qualities of intelligence and sentiments. If the exceptions were the rule, it would be well worth the trouble to take them as an example.''

Table 1.1. Amino Acids of Pulque

NITROGEN G/100 ML	AMINO ACIDS: AMOUNTS IN MG/100 ML								
	Lysine	Tryptophan	Histidine	Phenylalanine	Leucine	Tyrosine	Methionine	Valine	Arginine
0.14	16.2	2.7	4.7	11.2	10.5	6.4	0.7	6.6	10.9

SOURCE: Massieu et al. (1948).

appears to have fostered mental gyrations in a fanatical ruling priesthood with a pulque-based culture. Mayahuel became the surrogate that nourished the body, slaked the parched throat, relieved the duty pressures, exalted the spirit, provided at least temporary surcease from hard life, and, being god-like, protected the home. These effects constitute another contribution of agave to man during the centuries of the symbiosis.

The principal sources of pulque in Mexico (as I have observed them 1940–70) are the agave species listed below. Each of these species has selected varieties and minor forms, which are most numerous in the first and last listed. Because of crossing they are frequently hard to identify to species, although the *pulqueros* who live with their varieties know them intimately and call them all by name. (See P. and I. Blasquez's ''Tratado del Maguey,'' published in Puebla, if you can find it!)

1. *Agave salmiana* Otto ex Salm.

The most abundantly cultivated maguey. It dominates the fields about Puebla, Tlaxcala, and the plains of Apam; in less profusion in Michoacán, Aguascalientes, and San Luis Potosí. It grows spontaneously in northern Puebla, Hidalgo, and San Luis Potosí. Such wild populations, generally of smaller individuals, I regard as progenitors to the cultivates. This is the *Agave atrovirens* of authors, not the original of Karwinsky and Salm (see taxonomic text, pp. 468).

2. *Agave mapisaga* Trel.

The long-leaved giant frequently grown with *salmiana* (above); eastern Michoacán, state of Mexico, Morelos, Puebla, Michoacán, and sporadically to Zacatecas.

Table 1.2. Composition of Aguamiel, Tlachique, and Pulque
(amounts in 100 g)

COMMON NAME	Water, g	Ash, g	Protein, g	Non-nitrogenous Extract, g	Calcium, mg	Phosphorus, mg	Iron, mg	Carotene	Thiamine, mg	Riboflavin, mg	Niacin, mg	Ascorbic Acid, mg
Aguamiel	87.8				10	20	0.40	0.00	0.10	0.01	0.50	11.3
Aguamiel	94.0	0.4	0.30	5.30	20	9		0.00	0.02	0.03	0.40	6.7
Pulque	97.0		0.44		10	10	0.70	0.00	0.02	0.02	0.30	6.2
Pulque	98.3	0.2	0.37	1.13	11	6	0.70	0.00	0.02	0.03	0.35	5.1
Tlachique	97.3	0.2	0.20	2.50	10	5		0.00	0.03	0.02	0.15	4.6

SOURCE: Cravioto et al. (1951).

3. *Agave atrovirens* Karw. ex Salm.

Sierra Madre Oriental in the moist cool mountains from the Puebla-Veracruz border south into Oaxaca; also in the Sierra Madre del Sur. It is little used and poorly developed as a pulque source in Oaxaca. This is not the *A. atrovirens* of Trelease and other modern authors.

4. *Agave ferox* Koch

Southern Puebla and northern Oaxaca; intergrades with *A. salmiana*.

5. *Agave hookeri* Jacobi

Highlands of Michoacán.

6. *Agave americana* L.

More arid regions from Neuvo León and Durango to Michoacán and Oaxaca. Both wild and cultivated varieties are numerous; the former occur mainly along the slopes of northern Sierra Madre Oriental and were doubtless the sources from which the early cultivators selected their varieties.

Several other species are used locally, even wild individuals being tapped, but their use is relatively minor. More specific details can be found in the taxonomic annotation of species in the latter part of this work. The map (Fig. 1.1) outlines the region of pulque culture and the inhabiting Indian tribes that use or used the magueys.

Mescal and Tequila

Until recently ''mescal'' was to ''tequila'' what moonshine is to whiskey. Both Mexican products are distilled liquors fabricated from the meristem and leaf base of agave species. When an agave plant is mature and ready to flower, the short broad stem with the attached white leaf bases are cut and carried to the distillery. These globose vegetables, weighing from 60 to 120 pounds (25–50 kg) called ''cabezas'' or heads, are a familiar sight along the roads in the mescal- and tequila-producing regions of Oaxaca and Jalisco, respectively (Fig. 20.25).

In the distilleries these heads are first cooked for 30 to 48 hours in large steam-producing ovens and the raw starches of the meristem are converted to sugars. They are next macerated and allowed to ferment in large vats until the sugars are bacterially changed to alcohol. The alcoholic juice is then expressed from the cooked and fermented pulp and distilled. This manufacture has been carried on in the state of Jalisco for more than 150 years by well-established companies. They are all located in or near Tequila, Jalisco, and the product takes its official name from that town. Each company, however, markets its various grades of tequila under such registered brand names, as ''Herradura, Cuervo, Viuda de Martinez, Sauza.'' All of the product is derived from cultivated varieties of *Agave tequilana* Weber, which still exists wild in the adjacent region of western Jalisco. ''Tequila azul'' has in the past several decades become the favorite variety, and its glaucous blue color dominates the extensive fields between Tequila and Guadalajara and beyond. This appears to be the most productive variety out of many that have been practicably tried during the preceding half century. Diguet (1902) listed the varieties ''manolargo, chato, bermejo, azul, pata de mula, siquin, zopilote'' as in cultivation in Jalisco. Trelease (1920) treated several of these as species, which I have listed as synonyms under *A. tequilana*.

Between 1950 and 1970 a thriving mescal industry developed in the Oaxaca Valley. Here, a bluish variety of an agave species, probably *Agave angustifolia*, came to dominate the agave fields, displacing other species previously cultivated for mescal and pulque.

New plantings in 1967 were observed almost halfway to Tehuantepec long the main highway. Here, also, brand names are being established by companies becoming sophisticated in the distiller's art and marketing methods. "Mescal with its own worm" is a famous curiosity caller to the increasing tourist trade.

Previous to the development of mescal in Oaxaca and the respect its commerce is engendering, mescal was a maverick product of either wild agaves or of small plot plantings. It was usually produced in primitive stills, by poor distillers dodging state supervision and taxes, quite similar to the "hill-billy moonshiners" of southeastern United States. Their product was generally obtained from wild agaves, suitable or unsuitable for liquor, frequently poorly distilled, unaged, and drunk entirely at the consumer's own risk. The mescal hunter with his ragged clothing and machete may still be seen frequently upon the mountain trails and slopes of hinterland Mexico.

One of the more famous mescals of Mexico is "mescal bacanora" of Sonora, which takes its name from the locality where it is produced. Bacanora is no more than a rancho in eastern Sonora, not far from Sahuaripa. The principal species employed by the earthy still there is wild variety of *Agave angustifolia*, earlier accounted as *Agave pacifica* Trel. (Gentry, 1972: 147). Most of the heads are still carried in from the surrounding mountainous slopes by mules and burros. Whether owing to this particular variety, or to the minerals in the soil in which it grows, or to the particular bacteria of fermentation, or to other unknown factors, this mescal has an outstanding flavor, even when tasted fresh from the still. For further notes, see pp. 559–63.

The cultivated variety "azul" of *Agave tequilana* requires about eight years to mature from offsets. During this time each plant produces a "cabeza" weighing 70 to 100 pounds (27–44 kg), containing enough starches to produce, on the average, about 5 liters of tequila, or 7 "fifths" per plant. At $2 per fifth this would yield a gross value of $14 per plant, or $8,400 per acre (600 plants per acre) over an eight-year period. One could presume that Mexico would soon have a competing industry in the United States or elsewhere. Mexico has already, as an industry-protecting move, put an embargo on the export of propagation stocks of *Agave tequilana*.

The product value of distilled agave now exceeds the original Indian pulque. Mescal and tequila are regularly exported to the United States. In 1967 total import of tequila into the United States was over 581,000 gallons, in 1968, 698,700 gallons, and a level of over 1,000,000 gallons annually was predicted by 1971 (Brief of the National Association of Alcoholic Beverages, Inc., Washington, D.C., 1969, mimeograph). The same source provides the following figures for tequila production for six consecutive years, by an estimated 80 to 100 million plants of *Agave tequilana*, "blue variety," in the Jalisco region.

Year	Millions of Liters	Increase Over Previous Year
1963	19.7	5%
1964	20.0	1%
1965	20.5	3%
1966	22.1	8%
1967	23.5	6%
1968	25.0 (est.)	7%

These data indicate that this product, nutritious, nonessential, deleterious, or otherwise, is costing the United States consumer considerably more than 30 million dollars annually. Much of the recent increase in U.S. imports is due to the cocktail "margarita" which has become popular in the bar habitat.

Among the rather many wild species of agave which have been used for making mescal, the following have been recommended to me by the various mescal distillers and mescaleros:

Agave angustifolia Haw., vars.   *Agave palmeri* Engelm.
Agave rhodacantha Trel.   *Agave zebra* Gentry
Agave tequilana Weber, vars.   *Agave asperrima* Jacobi
Agave shrevei Gentry   *Agave potatorum* Zucc.
Agave wocomahi Gentry   *Agave weberi* Cels
Agave durangensis Gentry

These can be regarded as species with relatively high starch or sugar content. The first two belong to the section Rigidae, while four, *A. shrevei*, *A. wocomahi*, *A. durangensis*, *A. palmeri* are in the section Ditepalae. All belong to the subgenus **Agave.** Some other species recommended, as *A. schidigera* and *A. polianthiflora* of the subgenus **Littaea,** although among the sweeter kinds, are small in size and exist in small or inaccessible stands on cliffs not suitable for exploitation. Species high in sapogenins, as the Amolae group and the astringent members of the Crenatae, are not fit for consumption.

As with all hard liquors, mescal and tequila are frequently abused by individual consumers. In Mexico, state and local laws are periodically applied to control the nuisance of intoxication on city streets or its demoralization of family life. However, the product provides income to business and to a broad array of people in liquor factories, on agave plantations, and to the hinterland mescaleros. Agave is hosting man in a kind of social eroticism. The moral problem it poses as a hard liquor has been argued for centuries, but were any case brought against agave itself, I am sure the court would find agave innocent. Man's problems are his own.

Agave Fibers

Emigrant man in America must have found the fibers of agave useful as soon as he began to live among them. Just as the pliable stems of grasses, of the Malvaceae, and the strippings of palm leaves were naturally ready at hand for trussing firewood and captured game, so would have been the fibers of agaves. At least, he would have recognized the strength and use of agave fibers when he began eating them, because he had to cut the fibrous leaves from the nutritious fecula of leaf base and hearty stem. No doubt, he was using the fiber by 7000 B.C., the earliest date archaeology has found it as food (Callen, 1965).

Indian Fiber Products

Castetter, Bell, and Grove (1938) found in the southwestern United States, where agaves are native, the use of agave fiber was widespread: for cordage, nets, bags, basketry, mats, blankets, clothing (particularly aprons), sandals, pottery rests, headrings, braids and other miscellaneous woven objects, hair brushes, paint brushes, needle and thread, fish stringers, armor, lances, fire hearths, musical instruments, and ceremonial objects. Figure 1.8 shows an Apache stringed instrument collected late in the last century by Edward Palmer, the indefatigable plant explorer. This list covers a wide range of uses; many other special uses developed with the city-state cultures of Mesoamerica during the last 3,000 years.

At the end of the last century the botanist J. N. Rose composed his "Notes on Useful Plants of Mexico" (1899). Included are his observations made during previous years of travel on the uses of agave and some excellent photos of the cottage fiber industry and its specialized tools. Finer fabrics were woven from the more pliable fibers in the conal buds

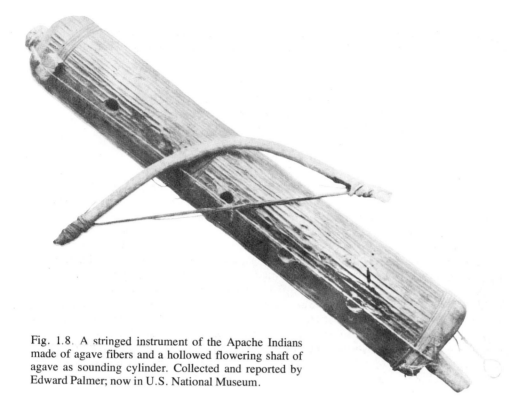

Fig. 1.8. A stringed instrument of the Apache Indians
made of agave fibers and a hollowed flowering shaft of
agave as sounding cylinder. Collected and reported by
Edward Palmer; now in U.S. National Museum.

of agaves; as finely decorated bags for various uses, tablecloths, blouses, and other
articles of dress. Some of the villages among tribes specialized in the production of these
goods. Rojas Gonzalez (1939) has provided an interesting account of the local fiber
industry carried on by the Otomis in the Valle de Mezquital in Hidalgo. Although his
account reveals adaptation to modern conditions, the Indian dedication to fiber work was
still in 1939 among old prehispanic traditions. "Ayate" is the indigenous term for the
loose woven goods of twine, nets, carrying cloths, bags, and related fabrics made from
the conal buds of the magueys, which are cut from the pulque plants to form the
basin in which the aguamiel is secreted. It is, therefore, a byproduct of the pulque industry. Figure 1.9 shows an Indian woman at work over a simple hanging loom in the Valle
de Mezquital.

Rojas found that the majority of the Otomis work at the fiber industry on a part-time
basis. Not only are the pulque agaves used, so also are the wild stands of *Agave
lechuguilla* and *A. striata,* called "tzita," on the rocky heights around the valley. The
latter yields a short, coarse yellow fiber, "jarcia," collected also in the bud stage. The
industry is specialized. Both men and women work at the various stages. Men usually
separate the fibers from the leaf buds, while the women do the weaving. The sons help the
fathers, the girls help the mothers, and the children carry water, frequently long distances,
to the houses where the main work is done. The extracted fibers must be washed. The pulp
of *Agave lechuguilla,* called "shite," has a large demand in all markets for washing and
removing spots from fine clothing. There are private plantings where the extracted fiber is
sold at a price per kilogram. There are men who sell the "ixtle" (from Aztec, fiber) from
the maguey. There are women who peddle the products in the markets and women who
tend the market stores featuring the fiber products.

Fig. 1.9. Indian woman weaving cloth with bud fibers of *Agave salmiana*. Valley of Mezquital, Hidalgo.

Rojas' (1939) comment aptly portrays the hard necessity-disciplined life of these Indians:

> The spinning of maguey fiber in the over-worked life of the Otomi, can be considered not only as a constant occupation, but more as an indispensable monomania whose product is money to round out the limited budget of poverty. It is a notable sight to see the Indian woman with her baby slung on her back and carrying a heavy load of wood on her head while she herds a flock of goats, and, as though these occupations were not sufficient to cope with the daily problem, the spit-moistened hands still also spin the tireless spindle.*

The introduction of horses, burros and mules by the Spaniards further stimulated the Mesoamerican fiber industry. The handicrafters learned quickly to provide harness of ropes, hackamores, bridles, saddle and packsaddle pads, and ornamental trappings as well. Braided webbing was also made for furniture, as for European beds and chairs. One has only to visit the various fiber product departments in any of the large markets in central Mexico to see the results of this highly crafted American industry. The importance of agave to man in America is again made abundantly clear.

---

*"El torcido del *santhé* puede considerarse dentro de la atareada vida otomí, más que una ocupación constante, una monomanía imprescindible, con cuyo producto en dinero se redondea el paupérrimo presupuesto doméstico. Es notable observar a algunas mujeres indias con el crío cargado a la espalda y llevando sobre la cabeza un pesado tercio de leña, mientras apacienta una manada de cabras...como si estas ocupaciones no fueran suficientes para resolver su diario problema, todavía entre sus dedos ensalivados el malacate gira incansable.''

Henequen and Sisal

With the development of planation farming and harvesting machinery, agave fiber has grown into a world industry (see Figs. 20.18, 20.19). Henequen fiber production from *Agave fourcroydes* is still largely confined to Mexico and adjacent Caribbean islands, while *A. sisalana,* source of sisal fiber, was introduced via Florida to several Old World countries in Africa and Asia and in Brazil. *A. sisalana* now supplies about 70 percent and *A. fourcroydes* about 15 percent of the world's hard long fibers, as twine, rope, and bags. About 90 percent of the henequen is grown in Mexico. Both of these cultigens or cultivars originated in Mexico, but details are obscure. They appear to be sterile clones, derived by chance hybridization, which were found by American Indians and cultivated by vegetative offsets about their houses.

There is no room in this book to discuss this large agroeconomic subject in detail. Additional notes pertaining to these economic plants can be found under individual species in the taxonomic part of this work. Extensive information can be found in the world's literature on agave fibers and their cultivation (see bibliography in *Der Pflanzer,* pp. 294–310). Two that I currently find useful are Lock's *Sisal* (1969) and especially Wienk's more recent comprehensive article (1969). Tables 1.3, 1.4, and 1.5 show the relative positions of henequen and sisal in world fiber trade.

Agave fibers are usually in a weak competitive position with other principal long fibers, represented by abaca and the synthetics of rayon and nylon, all of which have

Table 1.3. Estimated Production of Henequen Fiber for 10-Year Period
The bulk is from Mexico, but Cuba is also included.

Year	Metric Tons	Maximum Price, $ U.S.	Minimum Price, $ U.S.
1930	108,600	.08¼	.04½
1931	76,300	.04¼	.02⅜
1932	94,000	.02½	.02¼
1933	99,000	.03¼	.02½
1934	102,000	.03⅞	.03
1935	96,000	.06⅛	.02¾
1936	102,000	.06⅜	.06⅛
1937	107,000	.06	.05⅞
1938	91,000	(no quotations)	
1939	95,000	(no quotations)	

Source: Dewey (1941).

Table 1.4. Estimated World Production of Sisal Fiber for an 8-Year Period

Year	Metric Tons	Maximum Price, $ U.S.	Minimum Price, $ U.S.
1930	131,000	.08⅞	.04¾
1931	149,000	.04¼	.02½
1932	161,000	.02¾	.02⅜
1933	193,000	.04⅜	.02½
1934	201,000	.04½	.03⅜
1935	224,000	.06⅜	.03½
1936	227,000	.06½	.05⅞
1937	228,000	.06½	.05¾

Source: Dewey (1941).

Table 1.5. Estimated 1963 World Output of Hard Fibers in Metric Tons

FIBER	PRODUCING COUNTRY	PRODUCTION	AGAVE FIBERS, %	HARD FIBERS, %
Sisal	Tanganyika	210,800	29.3	25.3
	Mozambique and Angola	91,250	12.7	11.0
	Kenya and Uganda	69,400	9.6	8.3
	Former French Africa and Madagascar	24,100	3.3	2.9
	Brazil	157,400	21.9	18.9
	Haiti	19,700	2.7	2.4
	Others	7,200	1.0	0.9
	Total	579,850	80.5	69.7
Henequen	Mexico	127,900	17.8	15.3
	Cuba	9,800	1.3	1.2
	Others	2,600	0.4	0.3
	Total	140,300	19.5	16.8
Others		112,500	—	13.5
World Total of Hard Fibers		832,650	—	100.0

SOURCE: Hard Fibres No. 57 (1965).

captured more of the world trade. Agave fibers can only be produced where hand labor is cheap, which actually reflects an economic malaise for any developing country. The comparative estimate made in this study of sisal, tequila, and steroids, pp. 21–23, strongly suggests that agave fiber agriculturists should seriously consider switching to cultivation of *Agave tequilana* and *A. vilmoriniana* or other steroid producers.

Agave as a Paper Source

The Nahua tribes, e.g. Otomis, Tlaxcalans, Aztecs, and others, have historically been credited with making paper from maguey fibers or the pulp of the leaves of maguey. Von Hagen (1943) questioned this, concluding that the Mayan codices were made from bast fibers of "amate," *Ficus* spp. Lenz (1950), after examining forty-four preconquest manuscripts microscopically, attributed thirty-six to wild *Ficus* species, four to linen or hemp, and four to maguey. Christenson (1963) recently observed paper-making among living Otomis and Aztecs in northwestern Veracruz. He found them using the bast fibers of the mulberry trees, *Morus niger,* introduced from the Old World, and of the wild figs, *Ficus petiolaris, F. padifolia, F. involuta, F. tecolutensis, F. continifolia,* and *F. elastica.* He did not find agave being used, and it appears likely that agave paper has been abandoned. I suggest that, if maguey fibers were used, they were the bud fibers, which are softer, more spinable and workable than the hard fibers in the mature leaf. The finer fabrics found today in the markets of central Mexico, as in the finer twines, carrying bags, nets, and tablecloths, are made from the conic buds of the magueys (Fig. 1.10). Considering the strength, general durability, and length of fiber cells, the paper-making potentialities of the various kinds of agave fibers should be thoroughly investigated by modern methods. It should be noted in passing that in some species it is only the immature leaf fiber that is employed, as in *Agave lechuguilla, A. funkiana,* and the magueys, whereas in the modern fiber industries of sisal and henequen it is the mature leaf fiber that is used.

Medicinal and Miscellaneous Uses

The leaf chemistry of agaves is complex and only a few of the leaf constituents have been identified. The substance in the leaves which causes dermatitis is not known, for instance, nor is the pain-causing exudate on the surface of the spines. The Indians used the leaves as poultices for itches, sores, bruises, and wounds. Pennington (1958) and Bye et al. (1975) reported that several agaves are employed by the Tarahumara Indians as fish poisons. The juice of *Agave lechuguilla* was used to poison arrow tips (Palmer notes). The leaves of several species, or the leaf juice, is still widely used in Mexico as soap. We do know the leaves contain sapogenins, sterols, terpenes, and vitamins (Wall et al., 1955). Hecogenin, one of the sapogenins, occurs in the plantation fiber agaves. As a byproduct of fiber operations in Mexico and East Africa, it has been extracted and used in the manufacture of steroid drugs. Because sapogenins are suitable starting compounds, from which cortisone and the chemically related sex hormones can be synthesized, many agaves were quantitatively assayed during the early 1950s by several drug companies and especially by the U.S. Department of Agriculture (USDA). The following list gives those species in

Fig. 1.10. The white conal bud of the maguey *Agave salmiana*. The bud is cut out to leave a basin of juice.

which the leaves contained up to 1 percent or more of sapogenins, based on USDA analyses (Wall et al., 1954–57).

1.	*vilmoriniana*	0.5–4.4%	12.	*lophantha*	0–1.5
2.	*capensis*	0.3–3.6	13.	*funkiana*	0.5–1.4
3.	*promontorii*	0.2–3.4	14.	*ocahui*	0.5–1.4
4.	*cerulata*	0.1–3.4	15.	*sobria*	0.6–1.3
5.	*nelsonii*	0.5–2.5	16.	*goldmaniana*	0.1–1.3
6.	"*maguey comiteca*"	0–2.2	17.	*pedunculifera*	0–1.2
7.	*roseana*	1.3–2.2	18.	*schottii*	0.2–1.2
8.	*toumeyana*	0.5–1.8	19.	*felgeri*	0.8–1.2
9.	*aurea*	0.3–1.6	20.	*colimani*	0–1.0
10.	*chrysoglossa*	0.1–1.6	21.	*schidigera*	0–1.0
11.	*lechuguilla*	0.3–1.6			

All of these species showed higher sapogenin content than the plantation fiber species; *Agave sisalana, A. fourcroydes, A. angustifolia, A. deweyana,* the analyses of which ranged from 0 to 0.5 percent. It is of economic interest to compare the resource value of sapogenous species with the two leading producers of fiber and liquor. *Agave vilmoriniana,* according to our series of analyses and observations of growth habits, is the leading prospect as a sapogenin resource. It compares well as a cultivate in several respects with *A. sisalana* and *A. tequilana.* It also reproduces vegetatively, is about the same size and would, therefore, require about the same planting space, and requires a similar maturation period of eight to ten years. Table 1.6 compares gross product value of the three species per hectare according to current fiber and tequila wholesale prices and the price of smilagenin at an estimated price of $20/kg. to drug manufacturers. The 3 percent smilagenin is considered a conservative estimate for *A. vilmoriniana* after it has flowered and dried. In the green growing stage the leaves contain only 0.5 to 1.0 percent smilogenin.

These evaluations when converted to our familiar acre unit are: *A. sisalana,* $2,000 per acre or $333 per acre per year; *A. tequilana,* $6,000 per acre or $600 per acre per year. *A. vilmorinana,* $7,500 per acre or $750 per acre per year. For cultivation in the United States, *A. vilmoriniana* has two other advantages. It is more cold hardy, withstanding short freezes down to 20°F (7°C), and can be grown easily from the Rio Grande Valley of Texas to California. Unlike the fiber and tequila agaves, *A. vilmoriniana* does not require hand-pruning of leaves, and so the harvest could be entirely mechanized.

Table 1.6. Relative Productions of Fiber, Liquor, and Sapogenin on One Hectare, with Their Respective Wholesale Market Values

Agave Species & Product	No. of Plants	Weight, Tons		Total Product	Price of Product US $*	Gross Value US $*	Term Years
		Lf. or Pl.	Dry				
Sisalana (fiber)	5000	500	50	22.6 tons	220/ton	4,972	8
Tequilana (tequila)	5000	175		25,000 liters	.60/liter	15,000	10
Vilmoriniana (smilagenin)	5000	300	30	900 kg	20/kg	18,000	10

SOURCE: Wienk, 1969.

*In 1960.

Neither our Department of Agriculture nor our drug firms have seen fit to develop this resource. Perhaps one of our more enlightened and constructive entrepreneurs, who has more money than he can use but who would nevertheless like to turn another pretty dollar, and who is sympathetic to the diversification of agriculture, will take up the challenge. As a source of sex hormones and birth control, smilagenin should be produced as cheaply as possible, and, considering the critical overpopulation of humans on our limited earth, as soon as possible.

In recent decades the National University of Mexico and Mexican government agencies have carried on extensive chemical and nutritional investigations of the agaves. As ensilage the leaves have been found suitable for livestock. The juice is rich in sugars and when properly treated can be employed in refreshment drinks. A pasteurized form of the juice, "miel de maguey," has been exported to the United States. A form of the juice carries fructose sugars, which is acceptable to diabetics and may, therefore, have formulative possibilities as a health food for diabetics. Some of the nutritional studies have been reviewed and published by Sanchez Marroquin (1966).

There are many other local uses of the agaves. The tall flowering shafts when dried out are used as poles in house construction, as fishing poles, harvest poles for cactus fruits, for fences, and props for fruit trees in Baja California. Recently, the dry candalabra-like panicles have been used for Christmas trees and other decorations. The old dry rosettes, found scattered over the deserts of Baja California, make wonderful evening bonfires in camp. The Tlaxcalans use the dried butts of maguey leaves for cooking fuel (Wilken, 1970). The leaves are still employed as shingles on roofs of temporary field shelters in northern Oaxaca. Palmer noted (unpublished notes in National Herbarium), among the Indians of the southwestern United States, that the leaves were used as poultices for rheumatism and lumbago; "leaf mixed in a mortar, very acid, placed between – – – – – and applied locally." The quids were used as wads over gunpowder when guns were loaded. Notes on uses are also given under respective species in the taxonomic part of this work.

The Native Names of Agaves

The plant nomenclature of the Native American is an informative study in itself. Like our own, it is essentially binomial, reflecting that associative ploy of universal value for mental filing and recall. In northwestern Mexico the most prevalent term for Agave is "mezcal," also spelled "mescal," of Indian origin. It is basically generic in context, as it is modified according to species or variety, e.g., "mescal pelon" (*Agave pelona*), "mescal lechuguilla" (*A. bovicornuta* and *A. lechuguilla*), "mescal ceniza" (*A. shrevei),* and "mescal de maguey" or "mescal de Bacanora" (*A. angustifolia*). Vernacular binomials can be cited in numerous examples, as for species of *Bursera* "torote blanco" (*B. stenophylla*), "torote papelio" (*B. laxiflora*), "torote prieto" (*B. inopinnata*) (Gentry, 1942: 160).

In central Mexico with the Spanish conquest, the common term for agave became "maguey," which scholars trace (Palmer, Von Martins, Jaramillo, Cortez; see Goncalves, 1956: 10–13) to the Antilles. The Spaniards first encountered agave on the Caribbean Islands. The Nahuatl name for agave was "metl," and this generic term is still used by the Indians in central Mexico. For instance, near Tehuacán, I was given the name "pitzometl" for *Agave marmorata* and "papalometl" for *Agave potatorum.* Martin del Campo lists similar names, "metometl" for *A. lechuguilla* and fourteen others he ascribes to Hernandez's records (1649). A study of such names among the relict Mesoamerican tribes, like the Otomi, Aztec, Zapotec, Mixe, and others, would no doubt add a great deal of ethnic information about agaves. It is, of course, understood that such unsophisticated name systems are not botanically precise but are still useful for local communications. The Aztecs, for instance, applied the name *metl* to some genera of the

Bromeliaceae, which superficially resemble *Agave*. Whatever their system of names was in *Agave*, it could hardly have been more inept than the ones we have employed for the last 100 years.

Summary of the Symbiosis

The ancient Mexicans, while they propitiated the gods of maguey, cultivated and coddled maguey incessantly. They cleared the wild land and put agave into it. They opened up a new and nurturing environment with varying habitats and ecologic niches for the random variants of the gene-rich agave genus. The cultivators made agave a home on the deep productive soils and in time provided water and manure. They protected the plants from weedy trees. They selected the genetic deviates of high production by planting the vegetative offsets. Agave species multiplied into more varieties than man has been able to characterize and count. Generally, that is what man did for agave in this Mesoamerican symbiosis.

In return agave has nurtured man. During the several thousand years that man and agave have lived together, agave has been a renewable resource for food, drink, and artifact. As man settled into communities, agave became fences marking territories, protecting crops, providing security, and ornamenting the home. Agave fostered in man the settled habit, attention to cultivation, and the steadfast purpose through years and life spans, all virtues required by civilization. As civilization and religion increased, the nurturing agave became a symbol, until with its stimulating juice man made it into a god. The religion and the god have gone, but agave still stands as a donor species of the first water. Among the world's crops, are there others that have played a more useful and as bizarre a role? If we are to ask more of agave, we must give it attention and growing room.

2.

Taxonomic History and Morphology

In 1871 Sir Joseph Hooker wrote, "Of all cultivated plants none are more difficult to name accurately than the species of *Agave,* partly because of the imperfection of the published descriptions, and more from the impossibility of fixing their characters by words." Hooker was no stranger to agaves, as he had the genus under observation in the Royal Botanic Gardens at Kew at least from 1856 when he named *Agave celsii.* He also had at hand Jacobi's then recent enumeration of species (1864–67). Bafflement among botanists in understanding species in *Agave* continues to the present, in spite of the monographic works of Trelease (1909–20) and Berger (1915). Few are the taxonomists who will undertake to identify agaves, except in limited regional areas, and none is competent to name them on a worldwide basis. Modern genetics and the fugitive concept of species appear to complicate rather than resolve the *Agave* species problem. The following is a brief review of the taxonomy and the characters of the genus. Particular attention is paid to the problems of variation and the breeding habits.

Agave Collectors

Portage of agaves from the New World to the Old are thought to have begun with the early voyages of the Spaniards and Portuguese. Introductions to Europe as ornamental novelties were particularly numerous during the early and middle nineteenth century. Most of the early collectors of agaves satisfied their purposes with seeds or capsules and young live plants. No records or herbarium specimens were published. To trace the routes of the nineteenth century collectors would require a great deal of library research. In the course of my readings, I have incidentally noted the areas where some of the collectors of agaves have traveled, Hemsley being the best general historical source (1887). These are indicated on the accompanying map (Fig. 2.1) with the inclusive years of their travel.

A favorite point of contact for collectors in the New World was Sartorius's hacienda El Mirador near Huatusco, Veracruz, a tropical coffee-growing district. Sartorius was a hospitable plant enthusiast himself and frequently shipped plants to European importers. All were billed as from Mirador, e.g. *A. miradorensis, A. sartorii,* but they actually came from many miles away. This became very clear to me when I visited El Mirador in the

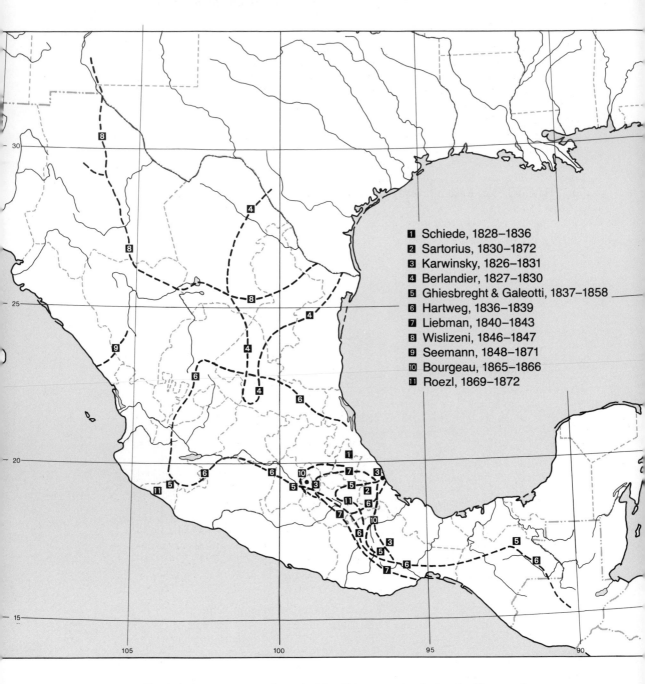

1 Schiede, 1828–1836
2 Sartorius, 1830–1872
3 Karwinsky, 1826–1831
4 Berlandier, 1827–1830
5 Ghiesbreght & Galeotti, 1837–1858
6 Hartweg, 1836–1839
7 Liebman, 1840–1843
8 Wislizeni, 1846–1847
9 Seemann, 1848–1871
10 Bourgeau, 1865–1866
11 Roezl, 1869–1872

Fig. 2.1. Areas or routes of travel of the 19th-century horticultural collectors of agaves on the American mainland.

1950s, looking for the sources of old collections. There were no agaves growing at El Mirador; the humid, tropical, forested habitat thereabouts is not suitable for agaves. *Agave sartorius* (= *A. pendula*), however, was located many years later on the sides of a canyon several miles north of El Mirador. *A. miradorensis* origin has never been found, but it could still be located perhaps on one of the many steep barrancas descending from the high Veracruz Sierra Madre.

The precise origins of the earliest named agaves are very obscure, their authors not knowing themselves from whence came their subjects, whether from North America or South America. Examples of some of these are: *Agave americana* L. introduced to the Azore Islands and thence to the Mediterranean Region; *A. cantala* Roxb. and *A. angustifolia* Haw. to Asia. Recent investigations have located several hitherto unknown origins as *A. geminiflora* Ker-Gawl, ascribed by him as perhaps from South America but recently rediscovered in Nayarit, Mexico (Gentry, 1968). Most of the limited herbarium materials from original localities were made between 1870 and 1910 by the American collectors, Palmer, Pringle, Rose, Nelson, and Goldman. However, the beginnings of taxonomy in the genus were based on the cultivated introductions to Europe, and most of these, as indicated in Fig. 2.1, came from central Mexico.

Agave Taxonomists

Linnaeus recognized four species in *Agave* when he founded the genus in 1753. One of them was removed to *Furcraea* by Ventenat in 1793 and another to *Manfreda* by Salisbury in 1866. This narrowed the concept of genus, but scarcely affected the concept of species. Specific additions during the latter part of the eighteenth and through the nineteenth centuries were mainly for horticultural convenience. Importations of agaves to Europe as ornamental novelties were particularly numerous during the early and middle nineteenth century. Many of the specific names were published in lists or brief descriptions of garden plants, as Otto's list of nomen nudum in 1842.

Salm-Dyck in 1834 described or listed thirty-four species growing in his garden at Dusseldorf. His later work in 1859 constitutes a generic revision describing forty-five species, including fourteen new ones. He divided the genus into five sections: *Macranthae, Heteracanthae, Micranthae, Inermis,* and *Herbaceae.* He included also "Species dubiosae, amplius inquirendae." The most prolific name raiser was General von Jacobi, who described seventy-eight species and many varieties (1864–67). He based his descriptions on living plants observed in various European gardens, many of which were immature potted plants. Jacobi's position commanded respect, and he was regarded as the authority on agaves, an influence lasting on through Berger's work forty-five years later, apparently outweighing the rules of custom in Berger's mind, as he often honored Jacobi's names over earlier ones. Neither Salm-Dyck nor Jacobi used inflorescence in diagnosing their species. They provided no illustrations, cited no preserved specimens, and our *types* were unknown in those times. In reference to *A. macroacantha,* Trelease (1907: 236) stated that Hooker's illustration of *A. besseriana* in 1871, "seems to be the first flowering record relating to any of this group of Agaves." The flowers were first described by Zuccarini in 1833. The binomials of these authors were founded on the diagnoses of leaves grown under conditions that varied from the pots of northern greenhouses to the sunny outdoor gardens of the Mediterranean. As will be shown later, root crowding, shade, and immaturity stunt, distort, and misrepresent the normal leaves of the helio-xerophytic agaves. Most of these taxa are unrecognizable on the basis of written descriptions alone. Nevertheless, in the interest of nomenclatorial stability as maintained by priority, these early names must be seriously considered. The mid-nineteenth century accounts also lead back to still earlier and even more obscure names.

The identities of some early taxa have been solved by revisits to the origin of collection. Schiede in 1829 very briefly described *A. lophantha*. Trelease was able to visit the locality where Schiede collected and, by obtaining new specimens, to reestablish the species. His identification was quite different from that of Berger, who misinterpreted the species on the basis of horticultural and bibliographical history. Not all doubtful cases can be solved in this way, as the nineteenth century botanists usually failed to give any specific geographic origin of the plants they described.

The insular botanists were almost as active as the continentals; through the nineteenth century, they added more than thirty binomials. Principal among these men were Hooker (five species) and Baker (twenty-six species). Nearly all of these were illustrated, many of them in the colored plates of Curtis Botanical Magazine and Saunders Refugium Botanicum. Herbarium specimens deposited at Kew may be accepted as the types. Baker was more fortunate than the earlier Germans, as he observed and lightly described many flowering agaves. He failed, however, to use floral morphology significantly. His final summation in 1888 accounts 138 species, including the Manfredas. Although his organization recognized three subgenera and sections, the characters of the inflorescence were scantily used.

Among European botanists it became Berger (1915) who finally organized the European botany of *Agave* into a monograph describing 274 species. He recognized three subgenera, Manfreda, Littaea, and Euagave. For Littaea he devised seven "sektionen" and for Euagave eighteen "reihen." Some of these sektionen or reihen were subdivided into subsektionen or other undesignated subdivisions. He was fortunate in having the rich living collection established at Mórtola by Hanbury in northwestern Italy in a climate that fostered more normal growth and flowering. He made use of the inflorescence in his separation of sektionen and reihen but made little use of it on the species level. Traditionally, vegetative variation guided his concept of species, rather than floral morphology. Berger, like other European botanists before him, cited no type specimens, but left a rich legacy of specimens prepared from Mórtola cultivates, some of which are designated as types. These are now in the U. S. National Herbarium.

Americans eventually were in a better position for studying agaves but made little use of it. The first serious student of the group was Engelmann (1875–1911). However, he was at St. Louis, nearly as removed from Agaveland in those times as the Europeans, and depended on collectors for study material. Of European extraction, he followed European tradition. By emphasizing a larger cognizance of characters (1875), he broadened the concept of species; this was reflected in his disciple, Elizabeth Mulford, who published the first American revision of agaves in 1896 but limited it virtually to the United States.

The apogee of *Agave* taxonomic flights was William Trelease, who became a successor to Engelmann at the Henry Shaw Botanical Garden in St. Louis. His works from 1907 to 1920 show him as contemporary to Alwin Berger, and there was correspondence between them. However, World War I interrupted their contact, 1912–18, and Trelease was unaware of Berger's 1915 monograph until about 1920, when he composed his "Agaves of Mexico" (1920) for Standley, and Berger was apparently unaware of Trelease's *Agaveae of Guatemala* (1915). Trelease's taxonomy in *Agave* was partly brilliant and generally uninhibited in describing species. Except for Mulford's visit to the western United States, he was the first active taxonomist to observe the wild, but mainly cultivated, populations. His published collection records (1913) show that he visited the islands of Barbados, Jamaica, and Cuba in 1907. From 1901 to 1911 he made several trips through eastern and central Mexico as far south as Mitla in Oaxaca. He visited Guatemala in 1915 and wrote an account of the agaves he found there (1915). He cited some herbarium specimens and types, some of which are not locatable. Some of his citations were ephemeral, as, "type cultivated in hedges at Quezaltenango" (1915). Altogether, he collected very few herbarium specimens himself, apparently relying on photographs and living specimens he carried or sent to the Missouri Botanical Garden for culture and on the herbarium collections and photographs of others.

The nomenclatorial labyrinth is particularly burdensome in the *Agave,* not because of the large numbers of names so much as the difficulty of applying them for lack of specimens. Applied binomials number over 400, and there are numerous varietal names. Trelease was adept in nomenclature and obviously was informed about horticultural events in Europe during the previous or nineteenth century. He composed two monographs, several revisions, and various articles on the agaves (1907–20). His concept of species was narrow, frequently based on vegetative characters of fragmentary materials, which in certain instances appear trivial; e.g., his account of the *Zapupae* agaves of Veracruz (1909), where the terminal spine is emphasized as the principal character for separating species. In other groups his acumen in recognizing taxa was astute, considering the fragmentary materials available. He recognized groups of agaves as the *Applanatae* (1912), but his scanty use of floral characters makes the natural relations of some species in such groups unnatural. Berger showed a better grasp of floral relationships, perhaps because he had more flowering material under observation. Trelease never fulfilled the broader species concept indicated but never matured by Engelmann; he continued the traditional out-of-difference-comes-species taxonomy of the nineteenth century. Since his last agave publication in 1924, other botanists have proposed thirty-five additional species to the 310 recognized by Trelease and Berger.

The agave taxonomists did not state their concepts of species, Trelease excepted (1910), so we can judge only by the way they separated their species. The criteria remain obscure. The general practice has become for regional floristic taxonomists to interpret species according to the morphological agreement or disagreement with established known species. Johnston's (1924) cursorial revision of part of the peninsular Californian Agaves is exceptional in trying to relate species according to floral and population characteristics. He exposed the weakness in Trelease's close splitting, but lacked the long experience with agave that their taxonomy appears to require.

Genetics, the most significant aid to taxonomy of the twentieth century, has had but little application to the problem of species in agave. Granick in 1944 published her chromosome patterns in the Agavaceae. Her thesis considered the position and nature of this family as proposed by Hutchinson and contributed only incidentally to the species problem. Cave (1964) has recently reported additional chromosomes. Sharma and Bhattacharyya (1962), in support of their theory of "speciation through vegetative means," reported on karyotypes of agaves grown in the Imperial Nursery in Calcutta; they stated that varieties are identifiable by the chromosome type. As all the agave names reported by these cytologists were for garden materials, which are frequently misapplied, the monographer can only wonder what taxa they may represent. However, the karyotypes and polyploid series they established are particularly useful to the taxonomist.

In summary, we have a plethora of names for agaves, rather than a system of classification according to natural relations. Species have generally been erected on vegetative criteria as a mental convenience to horticultural interests. Demonstrable difference called forth new binomials. The unvarying generations perpetuated in horticultural clones have become common reference series, tending to accentuate clonal differences, whether interclonal or interspecific in scope. Jacobsen's (1960) tabulation shows us nomina conservanda by repetition. According to modern precepts of systematics, these species or taxa should be studied and revised according to both classical and modern techniques; the former builds on comparative morphology, the latter on observation and analyses of populations, ecology, cytology, and all other branches of our science that may be brought to bear on the study of speciation. Before we review the prospects of such applications in detail, some general characteristics of the agave plant are briefly reviewed below.

Characters of Agave

The Rosette

Agaves may be regarded as rosette perennials since they require several to many years to grow and flower. However, if we disregard the years required for maturation, they appear like giant herbs that grow up and flower but once like an annual. They can be visualized as gross succulent herbs, tree-like in the time needed to mature and in the size of many of the inflorescences. Growth and the accumulation of reserves proceed together for periods of 8 to 20 years, like a boiler building up a head of steam until enough pressure is developed to open the valve or the apical meristem and send high an ephemeral superstructure, the inflorescence. Flowering drains out the leaves which wither and dry as the seeds and/or bulbils mature. If the hazardous transition from flowering to seedling is successful, the genetic code repeats itself, exactly if by bulbil, approximately if by seed.

Those that flower but once from the apex of the axis are monocarpic rosettes or multiannuals. Those that flower repeatedly from the leaf axils, excentric to the axis, are polycarpic rosettes or perennials. Others with branching stems must be thought of as polycarpic plants, or arborescent perennials, producing several to many monocarpic rosettes, the individual rosettes dying after flowering. Some are truly arborescent and develop trunks, as *A. karwinskyi* Zucc. and *A. goldmaniana* Trel. The latter buds and branches from the leaf axils and reaches several meters in length. With age the trunk reclines upon the ground and new roots form with soil contact. The older rosettes flower and dry, and, as old sections of stem deteriorate, the younger rosettes continue to grow, to branch, and root until a fragmented reclining "tree" or clone is formed. Large clones may cover several hundred square yards with the appearance of many individual plants (see Fig. 2.2). Such clones appear to be over a hundred years old. Most of the cultivated commercial agaves are clones, reproducing by rhizomatous suckers and by bulbils of the inflorescence. Some of them appear to be sterile polyploids that rarely or never set viable seed. *A fourcroydes* Lem. and *A. tequilana* Weber are among these, and plantation men in Mexico call the bulbils and suckers seeds ("semillas"), as they are the only propagules known to them. Rhizomatous suckers develop above the roots at the base of the rosette in many species. Other species rarely or never proliferate by suckers. Some species develop suckers only when the rosettes are young, while others may develop them throughout their life, and still others will sucker only after final maturation with inflorescence. However, most wild agave populations form seed, sometimes in conjunction with bulbils or with suckers, so that both sexual and asexual reproduction is available for generation. The ornithophilous and protandrous flowers indicate outbreeding, but some plants have demonstrated self-fertility, as *A. funkiana* Koch and Bouche.

Except in the few arborescent forms, the stem of agave is a thick abbreviated shoot. Many agaves appear as sessile or acaulescent rosettes. Others may be acaulescent or caulescent depending upon environmental conditions. Where floral maturation is delayed, as in gardens foreign to the natural habitat, leaf and stem growth may continue for many years, resulting in an unnaturally elongated plant. The stem is thick, water-storing, and contains a large terminal portion of central meristem in and below the large clonal bud of new leaves. This, together with the spirally imbricated leaf bases, forms a mass of fibrous white meristematic tissue, the carbohydrates of which are easily converted to sugars by heat. When trimmed from the roots and the green parts of leaves, the stem becomes the "cabezas" or heads widely employed by the Native Americans for food and fermented drinks and later by the immigrant Occidentals for the distilled liquors of "mescal" and "tequila."

The rosette of agave is also a defensive form. The stiff sharp spines and teeth ward off the larger animals, which is especially important for protecting the soft palatable flower-

Fig. 2.2. A clone of *Agave shawii goldmaniana* showing the decumbent arborescent habit with old desiccated trunks radiating away from the old vanished mother rosette.

ing stalks. Man employs many of the larger forms about his yards and fields as fence plants. Cattle will eat some agave leaves when pressed by hunger. Some of the smaller agaves have toxic constituents in the leaves, as *A. lechuguilla,* which in Texas has poisoned hungry goats. However, the agaves are to be encouraged in such cases, not the goats! This suggests that agaves might be planted to advantage in other regions where overgrazing has reached chronic proportions.

The Leaves

The leaves of *Agave* are borne spirally. They are generally thick and succulent, but there are a few hard-leaved species, as *A. striata*. After the seedling stage the leaves require several years for forming in the bud. During the bud time they are tightly overlapped in a long tapering cone, each maturing leaf at unfolding leaving its marginal impressions upon the following leaf within; what I have called *bud-printing*. Matured and spreading in the rosette, the leaves endure for several to many years until final maturation of the plant or rosette. They are the principal source of stored nutrients moving eventually into the prolific inflorescence. They are grossly thickened and specialized with spongy parenchyma for holding water within a waxy epidermis. In most species a copious system of strong vascular strands traverses the length of leaf to the spine and shortens progressively to the contacts with the margins or teeth. Some of these fibers are economically important resources: *A. sisalana, A. fourcroydes*. As long as the leaf remains turgid, it remains erect or ascending to horizontal in age. Under drought the tissues shrink, the conduplicate shape, or "guttered" form, becomes accentuated, and with extreme drought the leaves collapse, as in *A. americana* L., and lie prone with the terminal bud incongruously erect in the center. When given water, most of the leaves regain their turgor and fullsome shape. Flowering causes the final decline of the leaves, and during the maturation of the fruits and seeds, usually several months, the leaves become dry and hard, the position in drying varying in different species. In some species (*A. palmeri, A. shawii*) the leaves turn yellow, or red, or vermillion at floral time. The great longevity of the leaf is perhaps its most singular character; twelve to fifteen years is a great age for a leaf.

The radial arrangement of leaves covers the plant's area of occupancy, the leaves collecting rain like imbricated troughs and directing the water inward around the base of the stem (Fig. 2.3). During severe drought the stem shrinks, drawing away from the confining soil and rocks, giving the water subsurface ingress around the roots. The soil here is shaded, with consequent reduction of evaporation. This physical arrangement would be more advantageous in climates having short irregular rains, as are characteristic of the arid regions inhabited by agaves. The imbricated radial pattern of the leaves appears to be an important adaptation for survival.

The outline form, the length, the thickness and color, and the epidermis of the leaves have all been noted in describing species of agaves (see Fig. 3.1), but attention has been drawn more to the varying armature. This consists of a terminal spine, almost universal in *Agave,* and a lateral hard cuticular border, continuous or discontinuous, sometimes lacking. The edges of the leaf frequently have lateral prickles, which, following Trelease, have come to be called "teeth." Teeth or serrations may be lacking. Lack of teeth may be a specific character, or some species may show forms in which teeth are lacking, as in *A. pendunculifera* Trel. of **Littaea** and *A. deserti* Engelm. and *A. cerulata* Trel. of subgenus **Agave.** The clonal species *A. sisalana* is usually toothless but produces individuals occasionally with irregular teeth. The number, sizes, and forms of teeth are numerous (Fig. 2.4). Color of teeth has also been employed in characterizing species, but color in some varies according to age of leaf and perhaps with season or with physiological stages of growth. The pink zones on leaves of *A. colorata* Gentry and the crimson of *A. palmeri* Engelm. add to the beauty of the plants but are given by soil and post-maturity, respectively. The length, color, form, thickness, and amount of dorsal grooving of the terminal spine (hereafter referred to as spine as distinct from teeth) have also been employed in specific diagnoses. When the cuticular structure of the spine continues down the edge of the leaf, the spine is said to be decurrent. Many of the varying teeth and color forms on leaves were used to delimit specific standing by early garden taxonomists, but population studies indicate they are often but forms of single species.

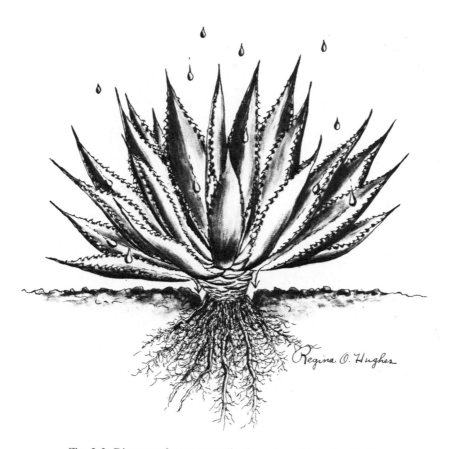

Fig. 2.3. Diagram of an agave collecting rain and irrigating itself.

Some diverse forms of agave leaves are shown in Figures 2.5 and 2.6; details of spines and teeth forms in Figure 2.4.

Leaf variation will be discussed under two headings; ontological leaf variation and individual leaf variation.

Ontological Leaf Variation

The size and form of *Agave* leaves and armature usually vary a great deal according to the age of the plant (Fig. 2.5). From the small early leaves they gradually increase in size with plant maturation and then relatively suddenly decrease in shape and size as they graduate into bracts borne on the base of the peduncle. The bracts up the peduncle may continue the diminishing transition gradually or the transition may be abrupt. The whole series eventually ends in the ultimate bracteoles of the pedicels. In their final form, the bracts have little similarity to leaves. The progression of leaf to bract is shown in Figure 2.5, where the growth of the successive leaves and bracts is depicted as a natural curve representing the growth, maturation, and end of a generation. In many other groups of plants the gradual transition from leaf to bract is not apparent, as the two organs are very unlike in form and size. In the latter case, the size progression has short-circuited the lineage, while in *Agave* the ''missing links'' are present. It is an energy pattern as graceful as a skyrocket. An evolution so perfect may go unnoticed.

Fig. 2.4. Variation in spines and teeth in the subgenus **Littaea,** A–H, and the subgenus **Agave,** J–S. A, *striata falcata;* B, *filifera;* C, *celsii;* D, E, *obscura;* F, *lophantha;* G, *kerchovei;* H, *victoriae-reginae;* J, *scabra;* K, *parryi;* L, *peacockii;* M, *marmorata;* N, *applanata;* O, *avellanidens;* P, *ferox,* Q, *inaequidens;* R, *americana;* S, *shawii goldmaniana.* All teeth from the mid-blades.

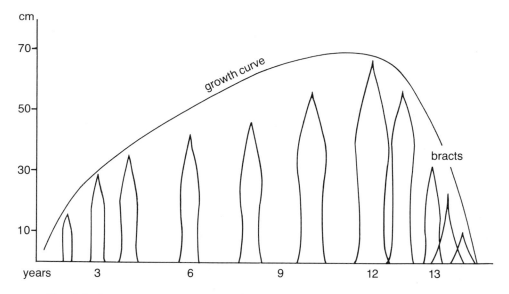

Fig. 2.5. Ontogeny of leaf on a single plant of *Agave murpheyi;* from seedling to maturity and sudden decline to bracts with flowering. Ages are approximate with measured lengths.

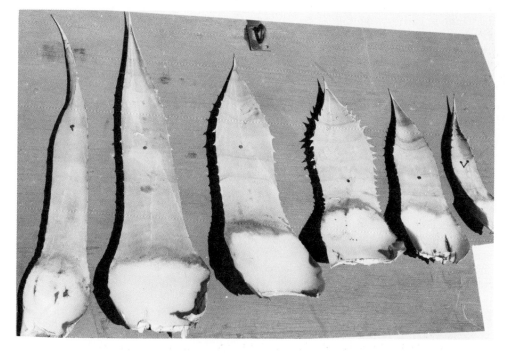

Fig. 2.6. Leaves of individual plants in a population of *Agave deserti* on Sierra Santa Rosa, California.

Individual Leaf Variation

This refers to the differences between the leaves as borne by separate plants within a population. Now that we are aware of ontological variation, it might be supposed that perception of individual variation would be relatively simple. It is not always so. A population deployed over several acres may actually be a single clone grown vegetatively and the differences observed attributable to those engendered physiologically by site, by insect attacks, or soil effects. If the population is partly asexual and partly cross-pollinated and free-seeding, there will be a mixture of physiological and genetic variants. Error of interpretation is easy in such cases, as, for instance, a recurring clonal member can be mistaken for a Mendelian segregate or as a distinct species, since no morphological intergrades are apparent. Such puzzling mixtures of clonal fidelity and rampant individual genetic variation are amply illustrated in the massive population of *Agave cerrulata* Trel. of the Baja California desert, where generation is both by suckers and by seed. I once counted more than twenty forms separable on leaf characters alone. A better case in point is *A. deserti*. In Figure 2.6 are leaves of plants in one population on the sierra above Palm Desert, California. A taxonomist with only two leaf specimens in this series, without other organs, could regard them as two species. Trelease did make a second species (*A. consociata*) of the smaller forms on the right. However, the uniform inflorescence of this population assures its conspecific nature.

Having duplicated this experience with several other variable populations, it appeared to the author that one could relegate many a specific leaf-oriented species to synonomy and obscurity. However, in a section of the subgenus **Littaea** it subsequently appeared that floral difference could be accompanied by very similar leaves (Fig. 2.7a). The floral characters establish such similar leaves as belonging in distinct taxa. The reverse of this leaf-flower alignment obtains in a section of the subgenus **Agave.** Here quite distinctive leaves are accompanied by similar flowers; *A palmeri* and *A. durangensis* are compared in Figure 2.7b. Reevaluation of the taxonomic evidence caused a reversal of opinion in the case represented by Figure 2.7a and a more sophisticated appraisal of relationships in Figure 2.7b. It is obvious from such experiences that taxonomic judgments in *Agave* are inept without complete knowledge of ontogeny of the plants and of the meaning of variability in its population context. One erects species with confidence on the basis of both floral and leaf distinctions, but the species based on leaf differences alone may evoke taxonomic unease.

Indument

Indument on *Agave* is nearly lacking. The leaves are all glabrous, but the waxy cuticle may appear glaucous. A pruinose or waxy bloom also develops on some flowers. Some species develop minute epidermal tubercles making them scabrous, as in *Agave marmorata* and *A. scabra*. Short papillae are apparently universal on the hooded tips of the tepals, rather conspicuous in some species and variable in extent or pattern. Glandular papillae cover the stigmas. Presumably, the waxy exudate composing the bulk of the cuticle would be protective to the leaf and moderate the extremes of temperature, wind, sand abrasion, and other environmental stresses. Some studies have indicated protective function for waxy exudates (McClendon, 1908), but others failed to find positive correlations between the various degrees of exudation and aridity (Gentry and Sauck, 1978).

Inflorescence

Inflorescence is a spectacular climax to years of rosette growth. It has two distinct forms; the spicate or racemose form of the subgenus **Littaea** and the paniculate form of the subgenus **Agave** (Fig. 2.8). The smallest spikes do not reach 2 meters in length, e.g., *A. parviflora,* while the larger panicles reach to 9 or 10 meters above ground

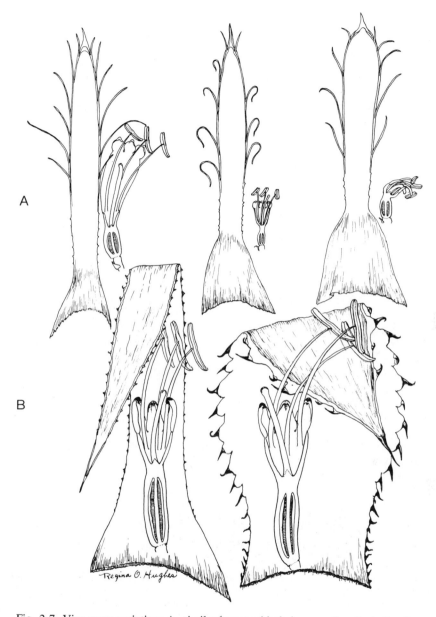

Fig. 2.7. Vice versa variation: A, similar leaves with their respective dissimilar flowers; B, dissimilar leaves with their respective similar flowers.

in two to four months of growth. The leaves of the upper part of the rosette graduate into peduncular bracts and continue in ever smaller forms upward as the bractlets and bracteoles of the lateral peduncles and terminal pedicels. The foliar sequence is commonly one of gradual increase followed by decrease, but there may also be abrupt change in size and in form as in *A. pumila*. When taken as a whole or in part, the leaves may constitute a character to correlate with species. *A. shawii* for instance, is characterized by unusually large thick bracts enclosing the umbellate clusters of buds.

Generally the spicate and paniculate types of inflorescences are distinct (Fig. 2.8), but there are forms in the spicate *A. lechuguilla* Torr. and the paniculate *A. potatorum*

Fig. 2.8. Forms of inflorescence in agaves. A, *Agave obscura,* as typical of subgenus **Littaea;** B, *A. americana,* as typical of subgenus **Agave.** C, D, E are intermediate forms: C, *A. lechuguilla;* D, *A. glomeruliflora;* E, *A. potatorum.*

Zucc. which combine or bridge the two (Figs. 2.8c and 2.8d, respectively). The racemose form of the *A. potatorum* inflorescence appears to become paniculate by the simple elongation of the primary branches or laterals. As the spicate form is more universal among the monocotyledons than the unique paniculate umbels of subgenus **Agave,** it is inferrable that the spicate **Littaea** represents the phylogenetic and geologically older form. However, it does not follow that all spicate species are older or more primitive than all paniculate species. In cases like *A. potatorum,* the narrow racemose form of inflorescence may be a regression from the branched umbellate form by simple inhibition of the growth of the lateral peduncle. In fact, the incidence of pedunculate or subsessile umbels in some populations suggest multiple genetic factors in Mendelian segregations. There may be two or more lines of evolution presently contemporary in time and space. It is clear from these cases that the two subgenera are not completely distinct according to key criteria for separating subgenera. However, the main matrices of the two are distinct enough— umbellate versus spicate inflorescences.

Some of the forms of agave flowers are depicted in Figure 2.9. The variability is remarkable within the strict determinance of the generic limitations. All agave flowers are perfect with a three-celled inferior ovary of many ovules, six equal to unequal tepals in one or two series usually united below in a shallow to deep tube. The perianth may appear as a single series, but the overlap by the outer tepals in the bud is in some species, as *A. palmeri,* maintained conspicuously in the opened flower by larger and overlapping outer tepals. Floral diversity is greater in the **Littaea,** but petal dimorphy has advanced especially in subgenus **Agave.** Six filiform, exserted filaments are variously inserted in the tube or upon the tepal base and have two-celled, versatile, longitudinally dehiscent anthers. Where outer tepals are clearly larger than the inner, their opposite filaments are sometimes inserted higher than the alternate filaments. Insertion on the tube is therefore unequal or biserial. Between the apex of the ovary cells and the tube there is ovarian tissue usually forming a narrow "neck" ("hals" fid. Berger), or the neck may be virtually lacking. Rarely, the ovary apex protrudes into the bottom of the tube, as in *A. striata* Zucc. and related species (Fig. 10.5). This suggests a morphological relic of the inferior ovary in the putative ancestors of the Agavaceae.

The relation and proportions of the tube to tepals are highly variable, when the genus as a whole is studied. The tube may be broad, shallow, or narrow and deep. It is usually lined, deeply or shallowly, by grooves descending from the tepal sinuses. The tepals may be thick and succulent or thin, planar or keeled on the back, of various colors, narrow or broad in outline form, short or long, recurving outward or bent inward or falcate at anthesis, wilting before anthesis and crimping outward and downward (Rigidae section), or remaining erect throughout anthesis (Ditepalae). The apex is usually glandular-tipped, simply obtuse or markedly hooded, or with a peaked tip over the hood and variously papillate pubescent.

Additional genetic expression is carried in the forms and colors of the filaments and anthers. The former are doubled in the bud, with the "knee" high enough to clear the petals as the filaments extend. They vary in thickness, length, and upward alignment. The stamens are all versatile, but they may be markedly excentric (filaments attached off-center), as in some of the sword-leaved group. Apiculations of the connective occur irregularly. The insertion of the filaments in the tube sometimes forms a thickened and rather undulate ring, as about the orifice of the tube in the section Filiferae. Nearly all agave flowers are protandrous. One exception appears to be *A. polianthiflora.*

Ideographs

A diagrammatic cross section of the agave flower is shown in Figure 2.10. It has been found useful to take measurements of the respective organs. While the whole flower and its parts vary in size on the same inflorescence and between individual plants, the ratios between the respective organs remain more constant. A series of measurements are helpful in analyzing variability and for correlating with leaf variations. However, any flower varies in measurements according to its stage of development; from bud stage to opening and on to extension of filaments. The pistil and ovary continue to elongate after the anthers dehisce, as the anthers and the tepals are withering. In order to assure uniformity in comparisons, therefore, flowers should be measured at or about the time of anthesis, which occurs in a single day. Fortunately, the agave inflorescence continues flowering for several weeks from the base upward, so that after the first few days the whole presents a sequence of flowers in all stages and one can select flowers just reaching anthesis for measurements. From such measurements representative ideographs can be devised as a

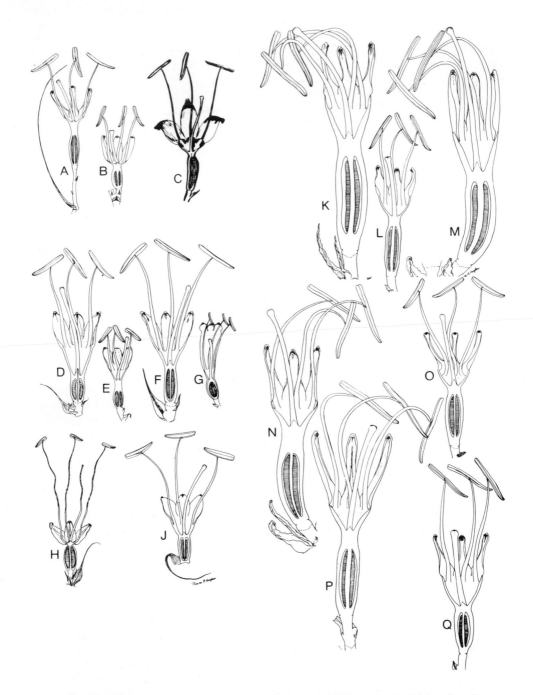

Fig. 2.9. Diversity in agave flowers. Subgenus **Littaea:** A, *A. lophantha;* B, *lechuguilla;* C, *pelona;* D, *celsii;* E, *schottii;* F, *polyacantha;* G, *polianthiflora;* H, *bracteosa;* J, *dasylirioides.* Subgenus **Agave:** K, *salmiana;* L, *aurea;* M, *mapisaga;* N, *goldmaniana;* O, *angustifolia;* P, *inaequidens;* Q, *sisalana.* About half size, filaments bent to conserve space.

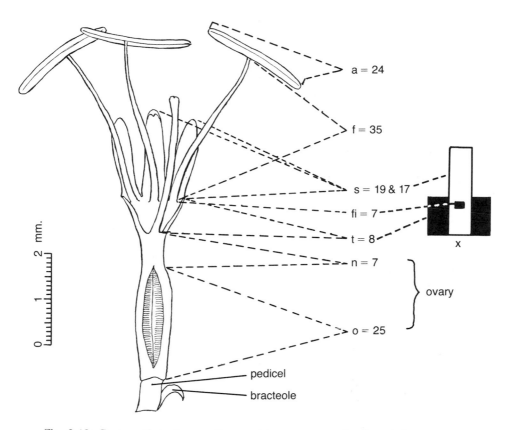

Fig. 2.10. Cross section of agave flower with parts measured and a tube/tepal ideogram, X. The white column represents the tepal, the black the tube, and the black square the insertion in the tube. o, ovary body length; n, neck of ovary length; t, tube length; fi, filament insertion (measured to bottom of tube); s, tepal lengths; f, filament length; a, anther length.

form of systematic shorthand, as the one shown in Figure 2.10-x. While this procedure assists in forming objective judgments, it by no means portrays many of the subtle variations expressed in other details of the flowers, as the form of tubes, color patterns, and the minute variations of the tepal tips and their pubescent patterns. In the taxonomic accounts of species a table of flower measurements is organized for each Group or Section. In the interest of space and printing costs, it was necessary to eliminate many of the measurements actually made. Those reported in the tables are intended as representative of the various series made.

The upper structure of the agave flower is a tube with a two-ranked proliferation into stamens and tepals. Whether the vascular traces of these organs leave the tube high or low would seem to be of small ontogenological or systematic moment. Nevertheless, study of a large series of flowers shows that the insertion of filament on tube is fixed within limits, reflecting sectional and specific alignments. The characters of filament insertion and tepal/tube ratio appear, therefore, as barometers reflecting the relative stability of populations and morphology. Where irregularities occur, other morphological characters and environmental situations indicate hybridization or vagrant genes from introgressive populations. They will be elaborated and discussed below.

Teratological variation is not uncommon in agave inflorescences. Observed are broomy branch types in *A. tequilana* Weber at Huntington Botanical Garden and

A. goldmaniana Trel. in Baja California. Seven and more tepal lobes have been observed in *A. shawii* Engelm., ten tepals, ten stamens in *A. lechuguilla,* and two fused pistils in other species. The fusion of a single stamen to style has also been observed. Wiggins has reported a crestate inflorescence of *A. shawii*.

The pistil is large and still elongating during anthesis; the stigma not maturing or receptive until after pollen release. In post-anthesis the style becomes a moist long tube, barely or actually open between the three stigmatic lobes; a structure that appears to foster the growth of innumerable pollen tubes necessary for fertilization of the numerous ovules. The stigmata are covered with glandular hairs and a syrupy excretion, and the stylar canal has been observed filled with a colorless gel. It may be viewed as the nourishing race track of the gametes where the lucky and the stronger survive to unite with the ovarian nucleus and carry on their part of genetic duplication.

The fruits of *Agave* are all inferior, three-celled and with axile or central placenta. They mature as loculicidally dehiscent capsules with six columns of thin, disciform, numerous seeds, black and shiny when fertile, dull and whitish when sterile. The capsules vary greatly in size, in thickness of walls and pericarp, and relative succulence and color when immature. They may be stipitate or sessile, beaked or rounded, and range in shape from oblong to ovoid and obovoid. Specific differences are frequently insignificant, but sometimes exhibit correlative taxonomically significant characters. As used by Trelease, capsule characters can be misleading, as capsules from a single plant may show differences in size and texture that he employed for separating species, e.g., *A. goldmaniana* and *A. pachyacantha* (1912); see also Figure 19.4.

Seeds are produced in abundance, gradually being shaken out of the cracks of the splitting erect capsules by animals and wind. Their winglike form appears to be suited for wind lift out of the erect capsules. Most of them fall near the parent plant, but others in strong wind may be blown several hundred feet. A partial count of the seeds in one spike of *A. chrysoglossa* gave an estimated number of 720,000 seeds. In morphology the seeds of *Agave* are rather uniform, the greatest variation being in size, while minor differences appear in shape and minute sculpturing. As with capsules, closely related species are hardly separable by their seeds, but sectional differences are notable in some cases, especially in size. Some capsules and seeds of the two subgenera are shown in Figure 2.11.

Pollination and Fertilization

The flowers of many agaves are particularly well structured for producing and containing nectar. The deep-tubed flowers held geotropically erect in umbellate clusters appear designed for this, while the shallow-tubed, nongeotropic flowers of many of the **Littaea** are not designed to hold accumulating nectar.

Bats are perhaps the most important visitors to agave flowers, as dramatically shown in Fig. 2.12. Recent studies have established that bats are regular and important pollinators of agaves, especially *Agave palmeri*. The bats of the genus *Leptonycteris*, subfamily Glossophaginae, are anatomically structured for nectar-lapping and pollen-feeding. They are of migratory habit and have been observed to feed in small flocks during the seasonal flowering of *A. palmeri* (Howell, 1979). Anthers dehisce during night hours. These bats also feed on flowers other than agave, but certain structures of the latter are notably co-adaptive with bats; e.g., abundance of nectar in a strongly scented mass in individual cuplets held erect by geotropic flowers. The tough short leathery tepals of the Ditepalae appear unusually well structured to support the clambering bats and protective of the nectar-holding tubes. Such structures and functions in disparate organisms can develop only over long periods of time and indicate adaptive co-evolution. Geologic time is an appropriate term in the case of agaves, where generation spans require 15 to 20 or more years. If you ever asked yourself, how and when did agaves get this way?, this co-adaptation is one clue

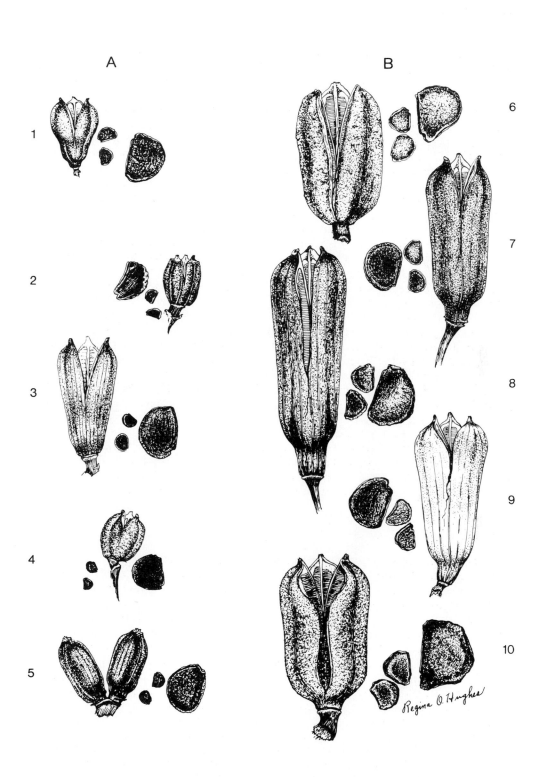

Fig. 2.11. Capsules and seeds in *Agave*: 1, *A. lechuguilla*; 2, *A. striata*; 3, *A. potrerana*; 4, *A. parviflora*; 5, *A. geminiflora*; 6, *A. palmeri*; 7, *A. shawii*; 8, *A. mapisaga*; 9, *A. cerulata*; 10, *A. angustifolia*.

Fig. 2.12. Bat feeding on the flowers of *Agave schottii*. Photo by George Olin.

to consider. I would call this bio-relationship another case of symbiosis, rather than a ''syndrome.'' There is nothing psychopathic about it: nectar-feeding, not nectar-phobia.

Presumably, other members of the Ditepalae participate in the bat-agave symbiosis, because their flowers are structurally similar. The flowering seasons in Figure 16.1 (Ditepalae) indicate there is a wave of nectar flow from spring to winter, north to south, making a nectar flyway for migrating bats; as discussed in chapter 16 under Ditepalae, *A. palmeri*.

Many other animals visit the agave flowers, obviously attracted by the masses of color, scent, and the nectar abundantly produced in the tube. The nectar is sweet and syrupy indicating a nutritious liquid attractive to animals in an open environment, frequently arid. The following have been observed to visit agave flowers, but it is not to be inferred that they all carry or transfer pollen from one flower to another.

Hummingbirds habitually visit the flowers, their long beaks inserted into the tubes as they hover (Fig. 2.13). Their breasts, heads, and necks brush against anthers and pistils. The horizontal branches of the high panicles provide excellent perches for them. Many other birds have been observed on the thick clusters of flowers and the inflorescence branches, but little is known of their roles as pollinators. The author has seen flickers,

Fig. 2.13. Hummingbird visiting the flowers of *Agave durangensis*.

doves, pigeons, wrens, ravens, caracaras, and Mexican parrots, the latter apparently chewing up the fresh flowers. Flower mutilation is also caused by rodents such as ground squirrels and pack rats that have climbing habits. Range cattle frequently eat and destroy the young tender flowering shoots of the smaller agaves. No doubt the wild game animals do the same, as probably did the numerous fossil ungulates and other herbivores through geologic time.

Many insects are found on agave flowers. Those observed include the diurnal hymenoptera: wild bees, honey bees, wasps, as well as various flies, beetles, and others. These alight and clamber over the flowers, many missing the exerted stamens and going into the tube below. Hawk moths hover like hummingbirds, inserting their long uncurling tongues into the tubes to suck in the copious nectar.

There seems little doubt that the elevated structure of the agave inflorescence with its colorful sented lures and nutrients provides a valuable life asset for the kinds of animals cited above. Not only do some of these animals serve as pollinators, but the larger may also carry about fruits and disseminate seeds. No doubt the association of these animals has had significant effects on agave populations and their evolution through geologic time and they are therefore to be considered in taxonomic studies. The recent study on pollinators by the Schaffers (1977) illuminates the evolutionary significance of animal-plant associations.

I have found no cytological account of pollen tubes and fertilization in *Agave,* but Newman gives an account of pollination and embryology in the distantly related *Doryanthes* of the Australian *Agavaceae. Doryanthes* has a protandrous flower like agave and structures are similar. Newman wrote: "The cleft of the stigma is well provided with glandular hairs and together with the stylar canal—lined with glandular cells—is filled with fluid for the nourishment of the numerous (pollen) tubes which pass down to the cavity through the micropyles across the parietal (wall) tissues and push past the synergids to the eggs. The well-nourished cells of the inner integument and of the parietal tissue, and the vigorous synergids all serve to facilitate passage of the tube by providing an abundant food supply."

Altogether, there is a wonderful array of taxonomic characters in all the major organs of *Agave* throughout the genus. Any widely representative living collection is a striking exposition of variability within the strict limits of generic organization; basically the plants are all alike, but all species differ in detail. As collection increases, variation becomes more and more a series of gradual sequences: in the growth of individual plants, in the variability patterns of populations, and in what we may devise as sectional or subgeneric groups. There is even a sequence, if incomplete, in the inflorescence characters dividing the two subgenera. However, here and there are more striking deviations in certain organs, as toothless leaf margins, tortuous spines, bicuspid teeth, ventral leaf streak, and tepal lobing. Some of such appear as specific key characters and still others as homologous variations as discussed elsewhere.

The situation of morphological gradation is familiar to all monographic taxonomists. The more specimens that are studied, the more difficult it may become to delimit satisfactory taxa. The old boys, we may say, had it easy because they had little material and hence clear breaks in variability for separating species. Lately, we have devised new techniques, as statistics and graphs, for handling large series of variables and are eventually able to clarify with reasonable systems. The unique deterrent in *Agave* appears to be the difficulties in procuring sufficient specimens and other relevant evidence. Because of their nature they do not lend themselves to easy specimen collections, to simple field measurements, to quick observation of progeny, or to experimental genetic proof by manipulation of generations. However, their arrays of sequential variability, together with their insular patterns of distribution, qualify them admirably as subjects for studies in evolution.

The factors that effect and maintain isolation between populations may support or indicate where specific lines can be drawn. Geographic isolation and sexual isolation are therefore of primary importance. Are sympatric species to be inferred because of differing flowering seasons, when morphologic separation is not certain? Many such problems arise with the study of agaves and call for more information than classical comparative morphology can supply. Many related disciplines will bear application; geology, ecology, genetics, embryology, mathematics, and even human history for man had already modified agaves before western science began. Taxonomy in its widest scope employs many sciences, and any system devised should relate the life forms to one another and each to its specific world. As Edgar Anderson wrote (letter, 1964), "I kept looking forward to monographs which would not only discriminate between species but would illuminate them as well."

Methods for Preparing Herbarium Specimens

The coarse and succulent nature of agaves makes them difficult to collect for the herbarium. The teeth, spines, and caustic juices are wounding and repellent. The leaves will rot, mold, and ferment in the plant press unless special care is taken to clean out internal tissues and juices. Generally, when the botanist does collect them, he makes short work with fragments of leaves, and the mold-prone flowers shrivel into unrecognizable shapes. One can collect twenty or forty other species of plants while preparing one agave for the press, so they are generally ignored. Only rarely does the botanist find plants with inflorescence, as many natural populations do not flower every year. Some populations are virtually limited to inaccessible cliffs.

There is no herbarium with adequate study material of the genus. If all the agave material of all the world's herbaria were combined, this would still be true. Hooker and Baker considered it not feasible to make representative herbarium specimens and, like other nineteenth century botanists, used live garden plants and illustrations. However, herbarium specimens as basic factual records are indispensable. The lack of them is partly responsible for the retarded state of *Agave* taxonomy. The botanist with time, patience, and a little special equipment can prepare good herbarium specimens. The writer's experience suggests the following procedure.

Specimen leaves should be selected from the middle portion of mature rosettes. They should be cut entire at the base where the green meets the white of the clasping butt. The lower or abaxial side of the leaf should be removed, leaving the margins. The internal tissue of fiber and pulp should be pulled and scraped out down until a thin but substantial layer of fibers and the epidermis of the top of the leaf is left (Fig. 2.14). Wiped off with toweling, the entire leaf can then be easily doubled to fit the herbarium sheet dimensions and put in the press. Even the gigantic leaves of the "maguey" agaves can be made to

Fig. 2.14. Skinning an agave leaf for the plant press in Baja California, using an old butt of a fallen agave shaft for work bench. Photo by Bruce Gentry.

conform in this way, although the widest ones may have to be halved longitudinally and mounted on more than one sheet. Extra blotters inserted between the leaf folds and replaced dry daily for a short term will assure proper curing of the thicker specimens. Figure 7.4 shows a well-prepared and mounted *Agave* specimen. In mounting, only the lower dissected fiber side of the first fold should be attached (by glue or thread) to the herbarium sheet. The leaf can then be unfolded for measurement or outline scanning.

Segments of the large inflorescences are sufficient as specimens when they represent the essential characters of the whole. The flowers cure much better when dipped in a preservative solution of formalin and alcohol, or kerosene before pressing. It has become my practice also to preserve flowers in jars of 6-3-1 solution (6 parts water, 3 parts alcohol, 1 part formalin), as many characters in the shape and proportions of the flower are significant. Pint jars with resinoid tops resistant to corrosion are suitable. Ordinary canning jar tops will be corroded by formalin in a few years. These with a shallow pan for dipping flowers to press, a pair of tweezers, pruning shears, machete, a lasso rope, and a good agave skinning knife are essential equipment.

Photographs are an excellent supplement for recording habitat, the entire inflorescence, population variants, distribution records, and the innumerable shapes and colors not recorded in dried specimens. A good set of field binoculars are invaluable for locating distant stands, flowering individuals, or for preventing needless and arduous mountain climbing.

Documentation of specimens is essential and should be done in the field, at the time and place of collection. The field label copy below shows a form used by the author for years.

Field No. _18414_ .... Latin name _Agave aff. scabra_ ....

Local name _Maguey_ .... P. I. No. ....

Locality data _4 miles N of route 83 on Farm Road 649 (Rosita) Starr Co., Texas. Sandy open chaparral desert._

Plant description _Large, suckering, few-lvd, big-headed, ashy glaucous to yellowish green rosettes 1.5 X 1.5 m. lvs. coarse, guttered, plicate, with deflexed teats & strong teeth; spine dark; panicles 6-7 m tall,_

Special notes _Laterals 16-14 in terminal ⅓ of shaft._ Wild? ✓

Photos, 4 live for prop. .... Cult? ....

Collector _Gentry & Barclay_ .... Date _March 16, 1960_

PLANT INTRODUCTION SECTION, HORTICULTURAL CROPS RESEARCH BRANCH, AGRICULTURAL RESEARCH SERVICE
UNITED STATES DEPARTMENT OF AGRICULTURE

3.

Geographic Guide to Species and the Exsiccatae

The following lists of agave species are composed according to the geographic regions in which they occur; Arizona (12 species), Central Mexico (31), Chichuahuan Desert (24), Jaliscan Plateau Region (22), Texas (9). They include both wild species and species that are cultivated for economic purposes; not horticultural collections in gardens. To facilitate identification, the lists have been put into selective keys based on synoptic morphology. Many students and travelers are concerned with plants in limited regions. Therefore, this arrangement should facilitate their acquaintance with regional agaves.

For Agaves in Baja California, see Gentry (1978).
For Agaves in Central America, see Group Hiemiflorae in this work.
For Agaves in Sonora, see Gentry (1972).
For Agaves in Southern Mexico, see Group Hiemiflorae in this work.

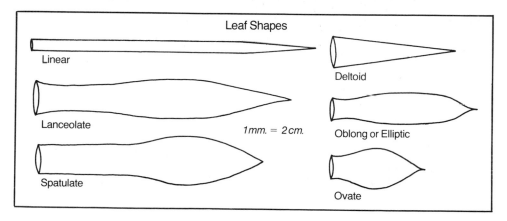

Fig. 3.1. Leaf shape terms, as generally used in plant descriptions and keys.

Synopsis of Agaves in Arizona

1. Inflorescence spicate or racemose, the peduncles slender, less than 3 cm diameter at rosette height; flowers small, less than 40 mm long   2
1. Inflorescence paniculate, the flowers on lateral branches in umbellate clusters, the peduncles stouter, 5–20 cm diameter; flowers 50–80 mm long   6
2. Leaves filiferous   3
2. Leaves not filiferous   5
3. Plants very small; leaves 6–12 cm long; flowers not exceeding 12 mm length, greenish to pink
          parviflora p. 200
3. Plants larger, leaves usually 20–40 cm long; flowers larger, 15–40 mm long, yellow   4
4. Leaves narrowly linear, 8–10 mm wide; flowers 30–40 mm long, tube 9–14 mm deep   *schottii*, p. 205
4. Leaves thicker, broader, 15–20 mm wide; flowers 22–28 mm long, tube only 3–4 mm deep *toumeyana*, p. 209
5. Leaves with horny margin extending nearly to base; flowers not markedly urceolate, tepals erect-ascending; filaments inserted in mid-tube
          arizonica, p. 254
5. Leaves without horny margin except for pine decurrency, not reaching to mid-blade; tepals connivent; filaments inserted near bottom of tube
          utahensis, p. 257
6. Plants large, to 1.8 m tall; leaves 1 m or more long, broadly lanceolate
          americana, p. 278
6. Plants smaller, to 1 m tall; leaves generally less than 1 m long   7

7. Leaves linear, or nearly so; spines conical, short, to 2 cm long; panicles regularly bulbiferous; flowers 70 mm long
          murpheyi, p. 440
7. Leaves lanceolate; spines acicular, 2–6 cm long; panicles normally not bulbiferous; flowers 30–60 mm long 8
8. Flower tube deep, frequently equaling or exceeding tepals in length; filaments inserted below orifice, frequently in mid-tube   9
8. Flower tube shallow, much shorter than tepals; filaments inserted at base of tepals or on rim of tube   11
9. Rosettes compact; leaves short, ovate, 20–40 mm long, with largest teeth toward apex; tepals much longer than tube; filaments usually inserted on one level   *parryi*, p. 538
9. Rosettes less compact with fewer leaves; largest teeth not limited to apex but distributed along mid-blade; tepals dimorphic, shorter than or equaling tube; filaments inserted on two levels   10
10. Leaf edge relatively straight with closely set teeth, mostly 5 mm apart; tepals conspicuously tinged with red or reddish brown   *palmeri*, p. 443
10. Leaf edge undulate; larger teeth mostly 1–3 cm apart; tepals yellow
          chrysantha, p. 426
11. Leaves 18–30 x 3–5 cm; teeth firmly attached; flowers 30–40 mm long
          mckelveyana, p. 390
11. Leaves larger, mostly 25–40 x 5–10 cm; teeth feebly attached; flowers 40–60 mm long   *deserti*, p. 376

Synopsis of Agaves in Central Mexico

Subgenus **Littaea**

1. Leaves without regular teeth, the margin filiferous or denticulate   2
1. Leaves generally armed with teeth, the margin corneous   7
2. Leaves filiferous, white bud-printed
   filifera, p. 110
2. Leaves not filiferous   3
3. Leaves narrow, linear, 1–3 cm wide, hard, always striate   4
3. Leaves broad, soft, ovate to oblong, finely serrulate or entire, not striate   6
4. Leaves pliable, 1.5–3 cm wide
   dasylirioides, p. 237
4. Leaves rigid, narrow, 1–1.5 cm wide   5
5. Flower tube much longer than the segments   *striata,* p. 242
5. Flower tube about equaling segments
   stricta, p. 248
6. Leaves broadly lanceolate, acute, light glaucous gray; rosettes developing long stems; flowers yellow
   attenuata, p. 66
6. Leaves ovate to oblong, serrulate, green to pale green; rosettes stemless; flowers lavender to reddish
   celsii, p. 220
7. Leaves narrow, 10 to 20 times longer than wide, mostly 2–6 cm wide in mid-blade; spikes more laxly flowered   8
7. Leaves wider, usually 4 to 10 times longer than wide, mostly 5–15 cm wide in mid-blade; spikes densely flowered except *A. angustiarum* and *A. xylonacantha*   10
8. Leaves generally 2.5–4 cm wide, the margin straight, detachable; teeth directed downward *lechuguilla,* p. 154
8. Leaves wider, generally 4–6 cm wide in mid-blade, the margin straight to undulate   9
9. Teeth coarse, gray, irregular in size and spacing, sometimes toothless
   difformis, p. 135
9. Teeth slender, brown, regular in size and spacing   *funkiana,* p. 139
10. Plants generally smaller, compact, cespitose except in *horrida* and *obscura*; leaves mostly 30–80 cm long, deltoid to ovate   11

10. Plants larger, more open, single; leaves mostly 70–140 cm long, lanceolate, long-acuminate   15
11. Leaves 40–80 cm long; teeth remote, mostly 3–6 cm or more apart; spines with open or narrow groove
    kerchovei, p. 149
11. Leaves 20–40 cm long; teeth mostly proximal, 1–3 cm apart; spines grooved or hollowed or flat above   12
12. Leaves mottled olivaceous in color, rarely green; straight, deltoid, with few straight or curved teeth or without teeth; spines grooved above
    triangularis, p. 181
12. Leaves green, lanceolate, with more numerous teeth, variously curved; spines grooved or ungrooved   13
13. Leaves widest in mid-blade, narrowed toward base; teeth usually moderate in size and nearly straight, more remote or reduced or lacking toward leaf apex; spines roundly grooved, 2–3.5 cm long   *ghiesbreghtii,* p. 141
13. Leaves more elliptic, little narrowed toward base; teeth generally large, frequently confluent, highly flexed, continuing to near base of spine; spines flat above, 2.5–4 (-5) cm long   14
14. Plants single; leaves generally ca. 80 in a rosette; spikes relatively small, laxly flowered, 2–3 m tall; pedicels 4–10 mm long. Morelos and Puebla
    horrida, p. 144
14. Plants single or cespitose; leaves more numerous, 100 or more in a rosette; spikes large, more densely flowered, 3–5 m tall; pedicels 1–2 mm long
    obscura, p. 161
15. Inflorescence racemose or subumbellate, the flowers clustered on short lateral branches; teeth very large, irregular   *peacockii,* p. 165
15. Inflorescence regularly spicate   16
16. Leaf margin nearly straight; teeth reduced, sometimes lacking on upper leaf margin   *angustiarum,* p. 134
16. Leaf margin sinuate; teeth large, 2–3-cuspidate   *xylonacantha,* p. 187

Subgenus **Agave**

1. Rosettes large, 1.5 m or more tall   2
1. Rosettes smaller, seldom exceeding 1 m tall   8
2. Inflorescence massive, the panicles pyramidal in outline with largest branches near base; branches subtended by large fleshy bracts   3
2. Inflorescence less massive, the panicles deeply oval in outline, with longest branches in mid-panicles; bracts smaller, chartaceous   4
3. Leaves lanceolate, 1–2 m long, the apex noticeably sigmoid; larger teeth in mid-blade 5–6 mm long
    salmiana, p. 605
3. Leaves linear, 2–2.5 m long, apex not noticeably sigmoid; teeth small, the larger 3–4 mm long
    mapisaga, p. 602
4. Flowers in dense spherical clusters with large bracteoles; flushed reddish to purple   *atrovirens,* p. 468
4. Flowers in spreading decompound umbels with small sparse bracteoles; yellow   5
5. Leaves glaucous gray; spines narrowly channeled   *americana,* p. 278
5. Leaves green; spines openly grooved   6
6. Plants developing a long trunk, 1.5–2 m tall; leaves less than 1 m long
    karwinskii, p. 577
6. Plants acaulescent; leaves over 1 m long   7
7. Leaves bright green, shiny, long-acuminate; apex with teeth to within 5–10 cm of spine. Wild
    inaequidens, p. 340
7. Leaves pruinose grayish green, the apex without teeth for 10–20 cm below spine. Cultivated   *hookeri,* p. 338
8. Leaves narrow, mostly linear, 10 to 20 times longer than wide, patulous, stiff to rigid, not thickly succulent   9
8. Leaves broader, ovate to lanceolate, thickly succulent   11
9. Leaves grayish blue glaucous, less than 30 cm long
    macroacantha, p. 579
9. Leaves not grayish blue glaucous, 50–120 cm long   10
10. Leaves lanceolate, rigid, heavily armed with large teeth to apex; spines 3–7 cm long   *applanata,* p. 421
10. Leaves linear, more flexible, teeth small; spines 2–3 cm long
    angustifolia, p. 559
11. Panicles short, compact, subtended by large succulent bracts; leaves very broad at base, deltoid
    macroculmis, p. 598
11. Panicles elongate, subtended by chartaceous bracts; leaves not broad at base, ovate to lanceolate   12
12. Leaves scabrous, mottled or zoned grayish or pale green; spines short, conic, 1.5–3 cm long; flowers small, 40–50 mm long   *marmorata,* p. 512
12. Leaves not scabrous, green or glaucous gray; spines 3–6 cm long; flowers larger, 60–80 mm long   13
13. Leaves bright shiny green, long-acuminate, lanceolate, 70–140 cm long   *inaequidens,* p. 340
13. Leaves light glaucous gray to pale green, ovate, short-acuminate, 25–50 cm long   *potatorum,* p. 490

Synopsis of Agaves in the Chihuahuan Desert

Subgenus **Littaea**

1. Leaves without teeth and without corneous margins, merely serrulate of filiferous   2
1. Leaves with corneous margins, usually armed with teeth   5
2. Leaves without terminal spines; flowers virtually tubeless, white or pale yellow   *bracteosa,* p. 91
2. Leaves with terminal spines; flowers tubular, yellow or variously colored   3
3. Plants frequently cespitose by axillary branching; leaves with finely serrulate margins, striate, hard-fibrous   4
3. Plants single, rarely rhizomatous; leaves with filiferous margins, white bud-printed, relatively soft fleshy, not striate   *schidigera,* p. 119
4. Leaves 1 cm or less wide above base, if wider, then 40–80 cm long and generally recurving; flowers 25–35 cm long   *striata,* p. 242
4. Leaves 1–2 cm wide above base, frequently falcate but not recurving, thick, stiff; flowers 35–45 mm long   *striata falcata,* p. 245
5. Leaves without marginal teeth, usually white-marked with bud-printing   *victoriae-reginae,* p. 183
5. Corneous margins of the leaves with teeth, but sometimes reduced in size and number; leaves not white bud-printed   6
6. Plants freely suckering by rhizomes   7
6. Plants single, not (or rarely) suckering   9
7. Corneous margins of leaves undulate to crenate, firmly attached; teeth proximal, mostly 1–2 cm apart, sometimes 2 on a teat, curved up or down   *lophantha,* p. 157
7. Corneous margins of leaves nearly straight, loosely attached; teeth remote, generally 2–4 cm apart, retrorse   8
8. Leaves 2–4 cm wide; inflorescence usually slender, simple, spicate; flowers yellow or red   *lechuguilla,* p. 154
8. Leaves 5–10 cm wide; inflorescence racemose, the flowers clustered on short lateral peduncles; flowers yellow   *glomeruliflora,* p. 142
9. Teeth smaller (mostly 2–5 mm long), frequently reduced or lacking on terminal half of leaves   *potrerana,* p. 172
9. Teeth larger (to 10 mm long), not reduced along upper half of leaves and continuing to near apex   *obscura,* p. 161

Subgenus **Agave**

1. Mature rosettes large, 1–3 m tall, if shorter, then leaves 20–30 cm wide at base; leaves large, thick fleshy; inflorescence large with broad (1–2 m), spreading diffusive umbels of large flowers; buds usually not red-flushed   2
1. Mature rosettes smaller, usually less than 1 m tall; leaves shorter, less than 20 cm wide at base, more compact, fleshy but more rigid; inflorescence more compact, if broadly spreading, the umbels more compact; flowers generally smaller, the buds frequently flushed with red   8
2. Leaves regularly well-armed on margins   3
2. Leaves with small reduced teeth, 1–2 mm long, the upper mature leaves frequently toothless except near base. Cultivated   *weberi,* p. 631
3. Leaves light gray or nearly white to pale green; bracts on peduncles not closely imbricate, quickly drying chartaceous; panicles deeply oval in outline   4
3. Leaves green; bracts of the peduncles closely imbricate, congested at base of panicle; panicles pyramidal to broadly oval in outline   5

4. Leaves smooth, 1 m or more long; teeth straight or variously curved
americana, p. 278

4. Leaves scabrous, usually less than 1 m long; teeth along lower part of leaf retrorse *scabra,* p. 296

5. Mature plants massive with thick stems; leaves linear or lanceolate, 1–2 m long; flowers larger, 90–110 mm long 6

5. Mature plants shorter, frequently less than 1 m tall; leaves deltoid or ovate, 20–30 cm wide at base, rigid, less than 1 m long *macroculmis,* p. 598

6. Leaves broadly lanceolate, broadest in the mid-blade, 1–2 m long; teeth larger, 5–8 mm long in mid-blade; spines 5–10 cm long 7

6. Leaves linear lanceolate, 1.7–3 m long; teeth smaller, 3–5 mm long; spines 3–5 cm long. Cultivated
mapisaga, p. 602

7. Plants larger; leaves 1.5–2 m long. Cultivated *salmiana,* p. 605

7. Plants smaller; leaves 1 m or less long. Wild *salmiana crassispina,* p. 609

8. Leaves smooth to slightly asperous; teeth below the mid-leaf not consistently down-flexed 9

8. Leaves scabrous; teeth below mid-leaf characteristically down-flexed
scabra, p. 296

9. Rosettes compact, often with closely imbricate leaves; leaves with largest teeth toward leaf apex; flower tube shorter than or about equaling tepals in length; filaments usually inserted on one level 10

9. Rosettes less compact with fewer leaves; largest teeth not limited to leaf apex but distributed along mid-leaf; flower tube longer than tepals; filaments inserted on two levels within tube *flexispina,* p. 436

10. Bracts of the peduncle chartaceous, not congested at base of panicle; leaves more acuminate 11

10. Bracts of the peduncle broad, succulent, congested near base of panicle; leaf apex truncate to short-acuminate
parrasana, p. 537

11. Leaves relatively short, broad, mostly 2–3.5 times longer than wide; panicles relatively wide; flowers large, mostly 60–90 mm long 12

11. Leaves relatively slender, more acuminate, mostly 3.5–6 times longer than wide; panicles slender; flowers smaller, mostly 40–60 mm long 13

12. Rosettes frequently globose, suckering copiously with long rhizomes; mature leaves 20–40 cm long; panicles deep with 20–40 umbels; flowers mostly 60–75 mm long *parryi,* p. 538

12. Rosettes more openly flat-topped with few suckers; mature leaves 30–70 cm long; panicles rather wide, open, with 10–20 large umbels; flowers larger, mostly 70–90 mm long
havardiana, p. 531

13. Flowers larger, mostly 55–67 mm long, with deeper tubes (10–14 mm); spring flowering *neomexicana,* p. 535

13. Flowers smaller, mostly 40–55 mm long, with shallow tubes (5–7 mm); fall flowering *gracilipes,* p. 526

Synopsis of Agaves in Jaliscan Plateau Region

Subgenus **Littaea**

1. Leaves without teeth, the margins filiferous or minutely serrate   2
1. Leaves with teeth on a corneous margin   8
2. Leaf margins filiferous   3
2. Leaf margins smooth or minutely serrulate   6
3. Leaves very narrow, less than 1 cm wide   4
3. Leaves wider, 1–4 cm wide in mid-leaf   5
4. Stem large with symmetrical rosettes and numerous leaves; not surculose   *geminiflora*, p. 112
4. Stem small with asymmetrical rosettes and few leaves; surculose   *ornithobroma*, p. 117
5. Leaves elongate, 40–90 cm long; flower tube deep and slender, 10–17 x 5–9 mm   *colimana*, p. 102

5. Leaves shorter, 25–40 cm long; flower tube broader, 8–10 x 7–12 mm   *schidigera*, p. 119
6. Leaves broadly lanceolate, light glaucous gray, plane to hollowed   7
6. Leaves linear lanceolate, broadest near the base, green, deeply guttered   *vilmoriniana*, p. 82
7. Plants developing long stems   *attenuata*, p. 66
7. Plants acaulescent   *pedunculifera*, p. 79
8. Leaves green to faintly glaucous, faintly bud-printed; teeth reduced in number, 1–3 cm apart, lacking toward apex   *angustiarum*, p. 134
8. Leaves yellowish green, boldly bud-printed white; teeth more numerous, 1 cm or less apart, continuing through to apex   *impressa*, p. 146

Subgenus **Agave**

1. Leaves ensiform, 10 to 20 times longer than wide   2
1. Leaves not ensiform, 2 to 10 times longer than wide   4
2. Plants smaller, shorter-stemmed; leaves generally less than 10 cm wide; inflorescence less profuse. Wild   *angustifolia*, p. 559
2. Plants larger with larger stems; leaves generally more than 10 cm wide, 12–20 dm long; inflorescences more profuse   3
3. Leaves 1.5–2 m long, green; teeth usually large, 5–10 mm long, closely set; capsules long-stipitate   *rhodacantha*, p. 580
3. Leaves 10–14 dm long, grayish to bluish gray glaucous; teeth variable, mostly small, 4–6 mm long; capsules unknown. Cultivated   *tequilana*, p. 582
4. Plants very large, rosettes 1.5–2.5 m tall   5
4. Plants medium to small, rosettes less than 1.3 m tall   7

5. Axis of inflorescence with large fleshy bracts; panicles pyramidal in outline with widest branches below; leaves massive, generally green   *salmiana*, p. 605
5. Axis of inflorescence with small chartaceous bracts; panicles deeply oval in outline; leaves light gray to glaucous green   6
6. Plants suckering profusely; leaves light gray glaucous   *americana*, p. 278
6. Plants not surculose; leaves light pruinose green   *hookeri*, p. 338
7. Leaves with short spines, 1–2.5 cm long; flowers small, 30–50 mm long, in small flat-topped umbels   8
7. Leaves with longer spines, 3–6 cm long; flowers larger, 50–80 mm long, in larger diffuse umbels   9
8. Leaves weak, brittle, grayish, scabrous, conduplicate, the margins closely mammillate *gypsophila*, p. 510
8. Leaves not weak and brittle, green, nearly smooth; margins straight to undulate   *nayaritensis*, p. 515

9. Plants not surculose; flowers with shallow tubes, the tepals 2 to 4 times longer than the tubes   10
9. Plants often sparingly surculose; flowers with deep tubes, the tepals about equaling the tubes   11
10. Leaves 1 m or more long, bright green, long-acuminate   *inaequidens*, p. 340
10. Leaves less than 1 m long, usually pruinose glaucous, short-acuminate
maximiliana, p. 346
11. Leaves glaucous gray, the margins

crenate with mammillate teeth   12
11. Leaves green, the margins undulate, the teeth generally not set on high teats
wocomahi, p. 456
12. Plants small, compact, the leaves 25–30 cm long; flowers unknown
guadalajarana, p. 531
12. Plants larger, more spreading, the leaves 40–70 (–100) cm long; flower tube very deep, exceeding the short, leathery tepals   *shrevei*, p. 447

Synopsis of Agaves in Texas

1. Inflorescence spicate or racemose; flower tube shallow and open   2
1. Inflorescence paniculate, flowers borne in umbellate clusters on lateral branches; flower tube deep   4
2. Leaves with a firm, sinuate, horny border; teeth 20–30 per side or more; spines weak, 1–1.5 cm long
lophantha, p. 157
2. Leaves with detachable, nearly straight, horny border; teeth 10–20 per side; spines strong, 2–4.5 cm long   3
3. Leaves 2–3 cm wide, light green to yellowish, the lower surface (when fresh) checked with green lines
lechuguilla, p. 154
3. Leaves 5–7 cm wide, dark green to glaucous green, lower surface not checked with green lines
glomeruliflora, p. 142
4. Leaves large, 12–18 cm broad; panicles large and broad with lateral peduncles sigmoid   5
4. Leaves small, usually less than 10 cm broad; panicles smaller, narrower, the lateral peduncles straight   8

5. Leaves without teeth, or few small teeth irregularly along margins, mostly below mid-blade. Cultivated
weberi, p. 631
5. Leaves always armed with prominent teeth along margins   6
6. Leaves asperous; teeth large (1 cm or more) and generally deflexed along lower leaf edges   *scabra*, p. 296
6. Leaves smooth; teeth usually smaller, less than 1 cm long, straight or flexed
7
7. Leaves lanceolate, 10–17 dm long, rarely shorter, 7 to 9 times longer than wide. Cultivated and wild
americana, p. 278
7. Leaves short, broad, deltoid, 4–7.5 dm long, only 3 to 4 times longer than wide. Wild   *havardiana*, p. 531
8. Flowers larger, mostly 55–67 mm long, with deeper tubes (10–14 mm); spring flowering
neomexicana, p. 535
8. Flowers smaller, mostly 40–55 mm long, with shallow tubes (5–7 mm); fall flowering   *gracilipes*, p. 526

Catalogue of Specimens or the Exsiccatae

Specimens that document the taxonomy, the distribution maps, and illustrations are enumerated in the index to Exsiccatae that concludes each chapter in the Systematic Account of Genus and Species. Taxa, state names, and collectors' names are listed in alphabetical order. To facilitate these data for ready reference, the collectors' last names with their collection numbers are listed first in italics, followed by the containing herbaria in capital letters (see abbreviations below). The localities with dates and notes in brackets follow. Type collections and the herbaria containing holotypes are given in boldface. Curators having duplicates of these listed specimens can, therefore, easily identify their materials by using this reference. I have been unable to annotate all specimens in all herbaria, and some of my earlier identifications have been changed. Hence, this list of specimens will bring identifications up-to-date, provided curators use these listings.

It is not usual to include dates of collection with exsiccatae citations, but I have done this to document flowering times. The plant parts composing the collections are listed as abbreviations (see table below) in parentheses, as well as my comments on some collections. Locality citations are edited copies from collectors' labels; their appended notes are frequently given in quotes. Altogether, the data represent a record of the living agave populations and should be useful to future students.

Abbreviations For Herbaria

*(Gentry Herbarium specimens are presently on deposit in the Desert
Botanical Garden Herbarium and are included in that listing, DES.)*

A	Arnold Arboretum, Harvard University, Cambridge, Massachusetts.
AHFH	Allan Hancock Foundation, University of Southern California, Los Angeles, California.
ARIZ	University of Arizona, Tucson, Arizona.
ASU	Arizona State University, Tempe, Arizona.
B	Botanischer Garten und Botanisches Museum Berlin-Dahlam, Berlin, Germany.
BH	Bailey Hortorium, Cornell University, Ithaca, New York.
CAS	California Academy of Sciences, Golden Gate Park, San Francisco, California.
CSLA	California State University, Los Angeles, California.
DES	Desert Botanical Garden, Phoenix, Arizona.
DS	Dudley Herbarium, now in CAS.
ENCB	Escuela Nacional de Ciencias Biológicas, México, D.F.
GH	Gray Herbarium, Harvard University, Cambridge, Massachusetts.
HNT	Huntington Botanical Gardens, San Marino, California.
ILL	University of Illinois, Urbana, Illinois.
INIF	Institución Nacional de Investigaciones Forestales, Tacubaya, México, D.F.
K	Royal Botanic Gardens, Kew, England.
M	München: Botanische Staatssammlung, Menzingerstrasse 67, D-8 München 19, Germany.
MEXU	Instituto de Biología, Universidad Nacional Autónoma de México.
MICH	University of Michigan, Ann Arbor, Michigan.
MO	Missouri Botanical Garden, St. Louis, Missouri.
NA	National Arboretum Herbarium, Washington, D.C.
NY	New York Botanical Garden, Bronx, New York.

OSH	University of Wisconsin, Oshkosh, Wisconsin.
POM	Pomona College, Claremont, California.
SBBG	Santa Barbara Botanic Garden, Santa Barbara, California.
SD	San Diego Museum of Natural History, San Diego, California.
TEX	University of Texas, Austin, Texas.
UC	University of California, Berkeley, California.
UNLV	University of Nevada, Las Vegas, Nevada.
US	National Herbarium, Natural History Museum, Washington, D.C.
WIS	University of Wisconsin, Madison, Wisconsin.

Parenthetical Abbreviations

br	bract	f	flower	l	leaf
bul	bulbil	infl	inflorescence	photo	photograph
cap	capsule	infr	infructescence	s	seed

List of Agave States by Countries
(with abbreviations)

Central America

Belice (British Honduras), Bel.
Costa Rica, C.R.
Guatemala, Guat.
Honduras, Hond.

Nicaragua, Nic.
Panama, Pan.
El Salvador, Sal.

Mexico

Aguascalientes, Agsc.
Baja California Norte, B.C.N.
Baja California Sur, B.C.S.
Campeche, Cam.
Coahuila, Coah.
Colima, Col.
Chiapas, Chis.
Chihuahua, Chih.
Distrito Federal, D.F.
Durango, Dur.
Guanajuato, Guan.
Guerrero, Gro.
Hidalgo, Hgo.
Jalisco, Jal.
México, Mex.
Michoacán, Mich.

Morelos, Mor.
Nayarit, Nay.
Nuevo León, N.L.
Oaxaca, Oax.
Puebla, Pue.
Querétaro, Quer.
Quintana Roo, Q.R.
San Luis Potosí, S.L.P.
Sinaloa, Sin.
Sonora, Son.
Tabasco, Tab.
Tamaulipas, Tamp.
Tlaxcala, Tlax.
Veracruz, Ver.
Yucatán, Yuc.
Zacatecas, Zac.

United States

Arizona, Ariz.
California, Cal.
Florida, Flor.
Nevada, Nev.

New Mexico, N.M.
Texas, Tex.
Utah, Utah

PART II

Systematic Account of Genus and Species

Subgenus **Littaea**

AGAVE L., Sp. Pl. 323, 1753

Succulent rosettes, monocarpic or polycarpic, perennials or multiannuals with long-lived leaves, frequently suckering at base and occasionally bulbiferous in the inflorescence; roots hard fibrous, radiately and shallowly deployed; stems thick, very short, usually shorter than the terminal bud, simple or branched; leaves large, generally succulent, spine-tipped, margin armed or unarmed with teeth; inflorescence tall, bracteate, scapose, spicate, racemose, or paniculate with flowers in umbellate clusters; flowers mostly large, generally protandrous; perianth tubular to shallowly funnel-form, the six segments erect to variously curved similar or dimorphic, imbricate in the bud; stamens six, exserted; filaments long, inserted in tube or on tepal bases; anthers versatile; ovary inferior, three-celled, succulent, thick-walled with numerous axile ovules in two rows per cell; pistil elongate, filiform, tubular; stigma three-lobate, papillate glandular; fruit a dehiscent, loculicidal capsule; seeds flattened, black.

Type of genus = *Agave americana* L., Sp. Pl. 323, 1753.

Flowers spicate in pairs or clusters or more rarely racemose in small distinct clusters
Subgenus **Littaea,** p. 61
Flowers paniculate in large umbellate clusters on lateral peduncles
Subgenus **Agave,** p. 267

Subgenus **Littaea** (Tagliabue) Baker

Baker J. G. 1888. Handbook of the Amaryllideae 1: 164.
Tagliabue, 1816. Biblioteca Italiana 1: 100. (non vidi).
Type species = *Agave geminiflora* (Tagl.) Ker-Gawl.

Key to Groups of Subgenus **Littaea**

1. The weak-armed group; leaves generally with smooth or serrulate or filiferous margins and weak spines or spineless   2
1. The strong-armed group; leaves generally with large teeth (except some Polycephalae and some Marginatae), strong spines, and frequently with corneous margins   6
2. Leaves with firm margins, not filiferous, not white bud-printed   3
2. Leaves with margins splitting off into white threads, the surfaces white bud-printed   5

3. Leaves striate, not soft succulent, with fine serrulate margins; ovary neckless, intruding a well-developed tube
Striatae, p. 235

3. Leaves not striate, soft succulent, margins smooth or irregularly serrulate; ovary with neck; tube generally short or lacking   4

4. Flower tube present, short to medium; filaments inserted on rim of tube; leaves with end spine   Amolae, p. 63

4. Flower tube lacking or receptacular; filament insertion not elevated; leaves without end spine (except *A. guiengola*)   Choritepalae, p. 89

5. Rosettes small, only a long-leaved form of *A. toumeyana* exceeding 30 cm; flowers small, tubular; tubes 6–30 mm long, usually much exceeding the short tepals   Parviflorae, p. 195

5. Rosettes mostly medium-sized, 30–90 cm tall; flowers larger, companulate, 30–55 mm long, the tube usually much shorter than the tepals
Filiferae, p. 101

6. Plants mainly polycarpic with branching stems; leaves broad, soft succulent, usually with close-set teeth; tepals equaling to twice as long as the deep tube   Polycephalae, p. 216

6. Plants mainly monocarpic with simple stems, often surculose; leaves various, generally firm or rigid with conspicuous corneous margins and large teeth; tepals 2 to 6 times as long as the shallow tube (except *A. pelona*)   7

7. Tepals 5 to 6 times as long as the shallow cup-like tube, frequently involute around filaments; leaves generally with conspicuous margins and teeth
Marginatae, p. 124

7. Tepals 2 to 3 times as long as the urceolate tube, not clasping filaments; leaves with or without corneous margins   Urceolatae, p. 251

Sectional List of Species

Amolae (9 taxa)
attenuata
bakeri
chrysoglossa
nizandensis
ocahui
pedunculifera
vilmoriniana
yuccaefolia

Choritepalae (3 taxa)
bracteosa
ellemeetiana
guiengola

Filiferae (7 taxa)
colimana
felgeri
filifera
geminiflora
multifilifera
ornithobroma
schidigera

Marginatae (28 taxa)
albomarginata
angustiarum
difformis
ensifera
funkiana
ghiesbreghtii
glomeruliflora
horrida
impressa
kerchovei
lechuguilla
lophantha
obscura
peacockii
pelona
potrerana
pumila
titanota
triangularis
victoriae-reginae
xylonacantha

Parviflorae (7 taxa)
parviflora
polianthiflora
schottii
toumeyana

Polycephalae (7 taxa)
celsii
chiapensis
pendula
polyacantha
warelliana

Striatae (4 taxa)
dasylirioides
striata
stricta

Urceolatae (5 taxa)
arizonica
utahensis

4.

Group Amolae

Plants small to medium-sized, simple or more rarely surculose, monocarpic short-stemmed, or perennial and long-stemmed, rosettes with smooth, soft, pliant leaves with unarmed margins and small terminal spines or spine lacking. Inflorescence a densely flowered spike with geminate or clustered flowers, bulbiferous and/or seed-bearing; flowers slender, small, yellow to greenish; tepals equal, the inner wider, ascending to partly outcurving, frequently clasping filaments in post-anthesis; tube generally shallow, much shorter than tepals; filaments inserted near or with tepals on the orifice of tube. Sierra Madre Occidental and central Mexico.

Typical species: *Agave attenuata* Salm, Hortus Dyckensis, 1834: 3.

Group Relationships

Most plants in this group have been very incompletely know to all previous students of the genus. *Agave yuccaefolia* was the first named, by De Candolle in 1812, but the other species did not appear in literature until the present century. Berger (1915) recognized three species. *A bakeri* and *A. yuccaefolia* he placed in his Section Anacamptagave, characterized mainly by polycarpic, mound-forming cespitose plants with relatively deep tubes. He placed *A. vilmoriniana* in his mixed Section Anoplagave, but he did not know its flowers. With the addition of considerable specimens since his work of 1915, the species described below appear to constitute a related group, which clarifies their inter-relationships and their relation to the rest of the subgenus **Littaea.**

The Amolae is a heterogeneous group. Some of the species have extra-sectional characteristics, such as the relatively deep flower tubes of *A. bakeri* and *A. yuccaefolia*. The diminutive *A. nizandensis* is especially anomalous because of its small umbellate inflorescence on a Littaea-like leaf rosette. It does not fit well into any section of subgenus **Agave** or **Littaea** and is placed here as a provisional convenience. On the whole, the Amolae is a distinctive group, showing diversity in both leaf and flower evolution. The small thin denticulate leaves of *A. yuccaefolia* are very different from other members. There is diversity in flower tube structure and the tepals may be erect and filament-clasping, as in the Marginatae, or recurving as in some other groups. Only *A. vilmoriniana* is strongly bulbiferous. The others, so far as known, appear to be more fertile when cross-pollinated. The flowers are visited by hummingbirds, perhaps by bats and other night-flying animals. The flowering habit of *A. yuccaefolia* is quite unknown to me. Except for *A. chrysoglossa* and *A. vilmoriniana,* the species are easily distinguished.

Distribution and Habitat

The group is wholly Mexican. There is a northwestern segment consisting of *A. chrysoglossa* and *A. ocahui* limited to Sonora, and *A. vilmoriniana,* which extends all along the highly dissected western slopes of the Sierra Madre Occidental from Sonora to Jalisco. They occur from sea level to 1,500-m elevations, are restricted to rocky sites and particularly cliffs. *A. nizandensis* is a local endemic of central Oaxaca, while the natural habitats of *A. bakeri* and *A. yuccaefolia* are unknown. The latter apparently originated in or near the state of Hidalgo. The climatic vegetation types include the Sonoran desert shrub, the tropical drought-deciduous short-tree forest, and the sierran oak-pine forests, all having the convectional summer rainfall regime. The rainfall of each species is indicated in Fig. 4.1. The plant community associates are too many to enumerate here. The known distributions are mapped in Figs. 4.4 and 4.8. Further particulars are given in the accounts of each species following.

As Ornamentals

The unarmed Amolae permit more intimacy in the garden than most agaves. Only *Agave ocahui* with its short stiff leaves with pungent tips provokes avoidance. All of them are distinctly ornamental with individually varied form of grace and symmetry, according to site and species. They have obviously relied more on their unpalatable juices and

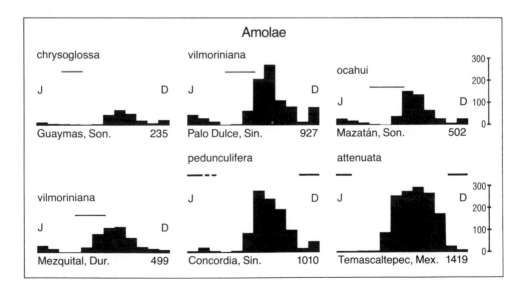

Fig. 4.1. Rainfall (silhouettes) and flowering perimeters (bars) of some species of Amolae. Relevant meteorological stations with average annual rainfall in millimeters. Data from Atlas Climatológico de México (1939) and Hastings et al. (1965). Flowering periods are based on herbarium specimens and field observations, supplemented by plants in cultivation.

inaccessible cliff sites for survival than on defensive armor. The northern taxa, *A. chrysoglossa*, *A. ocahui*, and *A. vilmoriniana*, have demonstrated frost resistance to 20°F ($\approx$6.7°C) during the last 30 years of their culture in southwestern U.S. gardens. No doubt they will increase in popularity as garden plants as they become more widely known.

Chemistry

During the 1950s, the U.S. Department of Agriculture screened a great many plants for sapogenin content in a search for the natural precursors of steroid drugs. The chemical assays were made at the Eastern Regional Laboratory in Philadelphia and were reported in various publications (Wall et al., 1954–61). The writer collected many of the samples and has collated his notes with the chemical reports.

Table 4.1 shows that some of the Amolae agaves carry a relatively high content of sapogenins (whence the name amole). As renewable sources of steroid products, some species merit further investigations for development in the United States. Combined with other potential products, such as fiber, detergent, and fertilizers, they might prove economically viable in agriculture, especially if the present sources of such products should become inadequate to our needs. The reader will find additional notes on uses under the species headings.

Table 4.1. Sapogenin Content in the *Amolae*
(given in percentages on dry weight basis)

COLL. No.	SOURCE LOCALITY	MONTH COLL.	PLANT PART	TOTAL	SMIL.	YUC.	GIT.
chrysoglossa							
11349	San Pedro Nolasco Is.	Dec.	leaf	1.6	100		
16611	Bacanora, Son.	Feb.	leaf†	0.7	92	8	
16613	Bacanora, Son.	Feb.	leaf†	4.4	84	16	
16623	Bacanora, Son.	Feb.	leaf	1.4	63	37	
ocahui							
16637	Guasabás, Son.	Feb.	leaf†	1.0			44
16639	Guasabás, Son.	Feb.	fruit	1.4			
16639	Guasabás, Son.	Feb.	peduncle	0			
ocahui var. *longifolia*							
11610	Sierra de Matapé, Son.	Feb.	leaf	0			
16603	Aguaje de Pescado, Son.	Feb.	leaf†	1.0	95	5	
vilmoriniana							
11099	San Bernardo, Son.	Aug.	leaf	0			
	Hunt. Bot. Gard., Cal.	Oct.	leaf	0.3			
11405	San Bernardo, Son.	Dec.	leaf	1.7	100		
12431	Hunt. Bot. Gard., Cal.	Dec.	leaf†	3.2	100		
12432	Hunt. Bot. Gard., Cal.	Dec.	stem†	1.5			
12551	San Bernardo, Son.	Mar.	leaf†	3.3	80		
16497	San Bernardo, Son.	May	leaf†	0			
16523-26	San Bernardo, Son.	May	leaf & stem	0.2–0.5	(Pfeizer)		

* Smil. = smilogenin; Yuc. = yuccagenin; Git. = gitogenin.
† Part from plant flowered and dried.

The Flowers of the Amolae

The flowers of the Amolae, as in most groups of agaves, are essential for judging the relationships and perimeters of species. Unfortunately, fresh or pickled flowers are not available for three species in this group, but long sections and ideographs have been prepared from dried specimens (Fig. 4.2). The flowers show a general relationship, quite close except for the deep tubes of *A. bakeri* and *A. yuccaefolia*. The measurement of flowers and the devising of their ideographs have been described in detail in the Introduction, p. 39, and are mentioned again here in this first section for the reader's orientation. Representative measurements of Amolae flowers are given in Table 4.2. (See p. 41.)

Key to Species of Amolae

1. Leaves broad, ovate lanceolate, widest in middle, narrowed near the base   2
1. Leaves narrow, linear lanceolate, broadest near base   4
2. Flower tube deep, 10–12 mm long   *bakeri*, p. 71
2. Flower tube shallow, 2–7 mm long   3
3. Plants developing long stems   *attenuata*, p. 66
3. Plants acaulescent   *pedunculifera*, p. 79
4. Leaves with smooth margins, green, not spotted or striped; plants generally not surculose   5
4. Leaves with fine serrulate margin, frequently red-purplish-spotted or with pale mid-stripe; plants surculose   7
5. Leaves thick, 1–2 m long, 4.5–12 cm wide; inflorescence with or without bulbils; flowers long-pedicellate (8–20 mm)   6

5. Leaves thin, 20–80 cm long, straight; inflorescence not bulbiferous; tepals ovate lanceolate, the inner 8 mm wide   *ocahui*, p. 75
6. Leaves plane, at least below mid-blade, ca. 1 m long, straight; inflorescence not bulbiferous; tepals ovate lanceolate, the inner 8 mm wide   *chrysoglossa*, p. 71
6. Leaves deeply guttered, arching, 1–1.8 m long; inflorescence frequently bulbiferous; tepals linear lanceolate, 4–6 mm wide   *vilmoriniana*, p. 82
7. Leaves 50–65 cm long, deeply concave above, red-purplish-spotted, without mid-stripe; flower tube about ½ as long as the recurving tepals   *yuccaefolia*, p. 85
7. Leaves 20–30 cm long, plane above, with pale mid-stripe; flower tube only ¼ as long as the ascending tepals   *nizandensis*, p. 75

Agave attenuata
(Figs. 4.1, 4.2, 4.3, 4.4; Table 4.2)

Agave attenuata Salm, Hortus Dyckensis 1834: 3.
  Agave glaucescens Hook., Curtis Bot. Mag. Tab 5333, 1862.
  Agave cernua Berger, Agaven 1915: 122.
  Agave pruinosa Lem. ex Jacobi, Hamb. Gart. u. Blumenz. 21: 499, 1865.

Perennial plants with 1 to several stems, usually ascending-bent, 5–15 dm long from the base, becoming naked in age, and with indeterminate number of relatively short-lived leaves. Leaves ovate-acuminate, 50–70 x 12–16 cm, broadest in middle, soft succulent, plane to concave, light glaucous gray to pale yellowish green, the margin smooth or

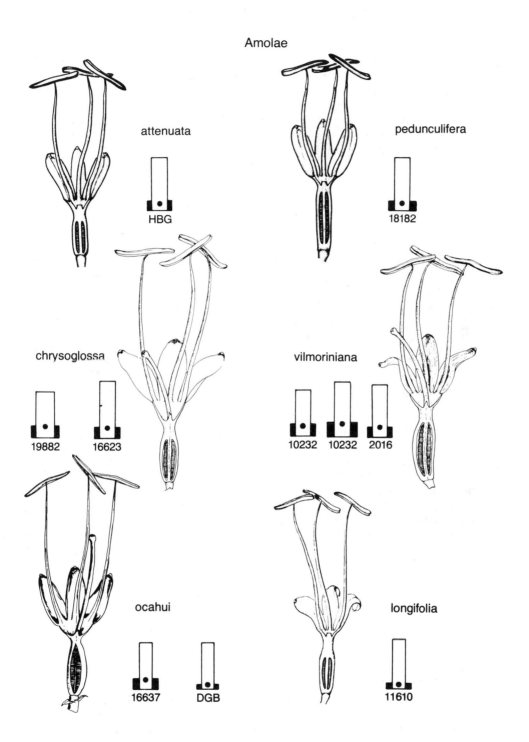

Amolae

attenuata

HBG

pedunculifera

18182

chrysoglossa

19882   16623

vilmoriniana

10232   10232   2016

ocahui

16637   DGB

longifolia

11610

Fig. 4.2. Long sections of some Amolae flowers with their respective flower ideographs depicting relative proportions of tubes (black) to outer tepals (white column) and filament insertion in tube. *Longifolia* is a variety of *ocahui*.

Fig. 4.3. *Agave attenuata* in Huntington Botanical Gardens. The smooth arching lines contrast strongly with the hard prickly lines of cactus. Photo from Huntington staff.

serrulate, apex spineless but finely tapered, soon fraying; inflorescence a dense raceme, thickly flowered, 2–3.5 m long; flowers 35–50 mm long, greenish yellow, in short pedicellate fascicles of 3–8, in axils of chartaceous bracts; ovary 15–25 mm long, fusiform, green, with constricted neck; tube 3–5 mm long, shallow-funnelform; tepals equal, 16–24 mm long, thin, linear oblong, recurving, mucronate with white-floccose tips, the inner frequently wider, with broad low keel; filaments 35–45 mm long, inserted on rim of tube, whitish, flattened, tapered, slender; anthers yellow, 15–20 mm long, greenish yellow, centric; capsules oblong, 2–3 x 1 cm, or smaller, thin-walled, stipitate, short-beaked, freely seeding; seeds 3–3.5 x 2–2.5 mm, lunate to deltoid with low marginal wing.

Neotype: Tab. 5333, Curtis Botanical Magazine, III, 18, 1862.

Kew received their plants from an unspecified locality in central Mexico collected by the botanical explorer Galeotti in 1834.

Table 4.2. Flower Measurements in the Amolae (in mm)

Taxon & Locality	Ovary	Tube	Tepal	Filament Insertion & Length		Anther	Total Length	Coll. No.
attenuata								
Hunt. Bot. Gard., Cal.	15	3×8	17×5&17	3–2	33	15	35†	7
	16	3×8	17×5&17	3–2	40	16	36†	7
	16	3×7	16×4.5	3–2	44	15	34†	7
bakeri								
Kew Bot. Gard.	20	11–12	20	11	45	14	53*	type
chrysoglossa								
Bahía San Carlos, Son.	18	4.5×9	15×6	4–4.5	40	15	37†	19882
	17	4×9	16×6	4	40	15	37†	19882
Bacanora	21	3.5×9	19×5&19	3.5–3	40	18	44†	16623
	22	3.5×9	19×5&20	3.5–3	40	18	45†	16623
nizandensis								
Nizanda, Oax.	15	3×7	16×4&15	3&2	37	14	34*	22567
	11	4×7	16×4&15	4&3	18	13	33*	22567
ocahui								
Guasabas, Son.	15	4	15	4	40		35*	16637
Sierra Baviso, Son.	13	1.5×7	15×4.5	1.5–2	27	13	30†	s.n.
	13	2×7	16×4.5	2&3	37	9	31†	s.n.
o. var. *longifolia*								
Sierra de Matapé, Son.	17	2×8	16×5&16	2–1	40	14	36†	11610
	20	2×8	17×5&17	2–1	44	13	38†	11610
pedunculifera								
Colomas, Sin.		4	12–15×5&6				*	1713
Palmito, Sin.	21	3	17	3	30	16	41†	18182
	21	3	18	3	38		42†	18182
Sierra Manantlan, Jal.	27	3×9	22×5–6	2–3	45	22	52†	23507
	27	4×10	22×5–6	3–4	55	23	52†	23507
Arcelia, Mich.	20	1×6	19×5&18	1	38	19	39†	23371
	21	2×8	17×5	1–2	37	13	40†	23371
Tacuichamona, Sin.	22	6×12	15×6&15	6–5	22	14	44†	5692
	21	6×12	16×6&16	6–5	27	15	42†	5692
vilmoriniana								
Sierra Tecurahui, Son.	15	3×8	17×4&16	3–3.5	38	15	35†	2016
	19	4×8	17×4.5	4	40	16	39†	2016
Sierra Charuco, Son.	20	4×8	16×4&15	4	35	16	40†	10232
	20	4×8	15×5&15	4	32	16	39†	10232
Hunt. Bot. Gard., Cal.	20	3–4×10	16×5	3–4	32	16	40†	19677
	20	4×10	17×4	4	34	16	41†	19677
San Ramón, Dur. ††	22	5	16×4.5				40–42*	135
yuccaefolia								
Mórtola Garden	20	8	15–16	7–8			38–40*	Berg.

* Measurements from dried flowers relaxed by boiling.

† Measurements from fresh or pickled flowers.

†† Type of *eduardi*.

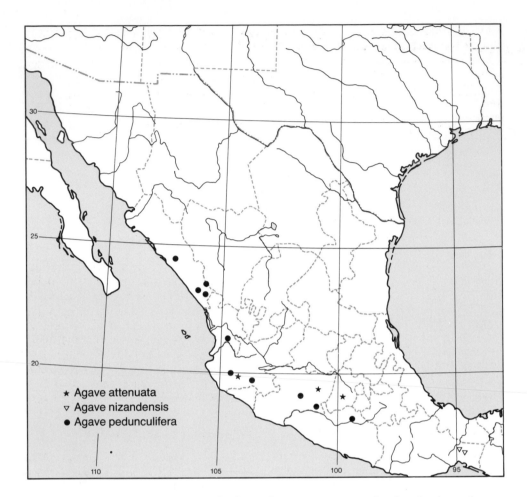

Fig. 4.4. Map of known distributions of *Agave attenuata, A. nizandensis,* and *A. pedunculifera.*

Affinities

Agave attenuata with its tubular flowers appears more suitably aligned with the Amolae, than with *A. bracteosa* and *A. ellemeetiana* in Berger's mixed Section Anoplagave. It is closely related to *A. bakeri* and especially to *A. pedunculifera,* which altogether form a broad-leaved group within the Amolae. It is easily distinguished by its long stems eventually becoming naked except for the crowning leaf rosettes. Berger separated *A. cernua* as having more concave acuminate leaves, but it appears to this writer as only a vigorous form of *A. attenuata. A. attenuata* var. *serrulata* Teracc. with serrulate-denticulate leaf margins, as a homologous variant, appears to have more taxonomic use. *Agave pruinosa* appears to be a serrulate-leaved form of *A. attenuata* and is probably equivalent to Terracciono's variety.

Distribution

The natural range of *Agave attenuata* was long unknown, but is here indicated in Fig. 4.4, from the state of Mexico west to central Jalisco on the volcanic mountains that transverse the broad Mexican plateau. It favors the high rocky outcrops in pine forests between 1,900 and 2,500 m, where it forms small widely scattered colonies. It has been observed in few localities and can be considered as relatively rare; see also the Exsiccatae.

A. attenuata is now a wide-spread ornamental, responding particularly well to garden culture in frostless climates. It is a soft voluptuous harmless plant luxuriating in patio corners. It also does well in greenhouses and will tolerate more shade than many agaves. It does not do well in the Desert Botanical Garden where summer sun is intense and shade temperatures commonly reach 115°F (46°C), and it has been frozen by temperatures of 24–25°F (−4° to −5°C).

Agave bakeri
(Fig. 4.5; Table 4.2)

Agave bakeri Hook. fil., Bot. Mag. 1903: tab. 7890, 1903.
  A. gilbertii Berger, Monatsschrift, f. K. 1904: 126, 1904.

Single, short-stemmed, non-suckering, monocarpic, green, multi-leaved rosettes; leaves 90–100 x 10–12 cm, lanceolate, succulent, recurving, broadest in the middle, narrowed and thickened near the base, concave to plane above, convex beneath, glaucous green, coriaceous, with thin brown margin, no teeth, and slender spine 0.5–2 cm long; raceme cylindric, to 3 m tall, densely flowered from near base, with linear lanceolate bracts longer than lowest flowers; flowers greenish, 50–60 mm long; ovary slender, fusiform, ca. 20 mm long, 6-sulcate above the slender neck; tube 11–12 mm long, 6-furrowed; tepals 20 mm long, reflexed-rolled, greenish outside, whitish within, the inner broader with dark green keel, obtuse; filaments whitish, 40–50 mm long, inserted in orifice of tube at 11 mm above the tube base; anthers 14 mm long, yellow (measurements from dried relaxed flowers; capsules and seeds not seen).

  Type: From Kew Gardens in Kew Herb., Jan. 22, 1903, entry no. 255 1889, Fig. 4.5.

Stated by author to come to Kew from ''the late Mr. Peacock's noble collection of Cactuses, Aloes, and Agaves in 1889, with no indication of its native country or collector. It flowered in the Mexican division of the Temperate House in January of 1902.''

I have seen no further account of this distinct species since Berger's monograph in 1915. He relied on the Kew materials for his account. The species apparently passed out of horticulture with the maturation of Peacock's plant, and it can be inferred that it failed to sucker or to seed.

Except for the deeper flower tube, it is closely akin to other species in the group Amolae. Although the leaves are very broad, they are otherwise quite in keeping with the distinctive unarmed leaves in this section. Whoever refinds this singular plant in nature will make a most interesting discovery.

Agave chrysoglossa
(Figs. 4.1. 4.2, 4.6, 4.7, 4.8, 4.14, 4.17; Tables 4.1, 4.2)

Agave chrysoglossa I.M. Jtn., Proc. Calif. Acad. Sci. (4) 12: 998, 1924.

Single, short-stemmed, openly spreading, few-leaved, green rosettes, 1–1.3 x 2–2.4 m; leaves 70–120 x 4–7 cm wider at the base, straight or slightly curved, flat above, convex below, deflexed at maturity, light green, smooth, linear-lanceolate, toothless, with a thin brown fragile margin 1 mm wide; spine 2–4 cm long, acicular, brown, aging grayish, with a short fine groove at base above; spike mostly 2–4 m tall, densely flowered in upper ¾; bracts dry, chartaceous, the lower ones exceeding the flowers; pedicels bifurcate, 10–15 mm long, their bracteoles small, white-papery, curling; flowers yellow,

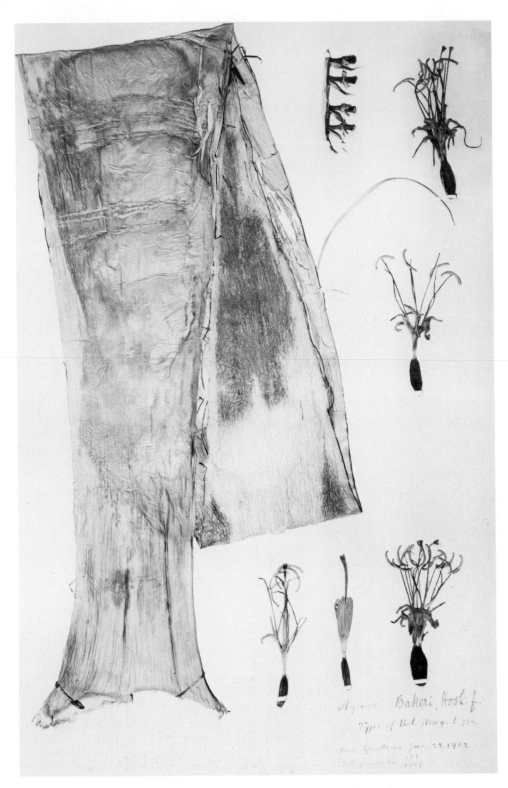

Fig. 4.5. Type specimen of *Agave bakeri* in the Kew Herbarium.

Fig. 4.6. *Agave chrysoglossa* flowering spike.

Fig. 4.7. *Agave chrysoglossa* in natural habitat at San Carlos Bay, Sonora.

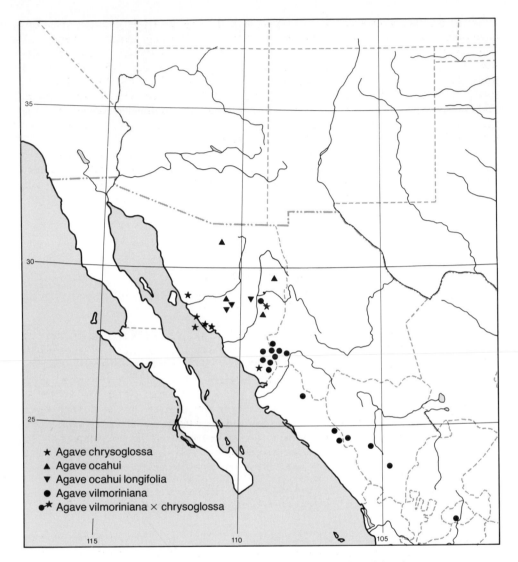

Fig. 4.8. Distribution map of the northwestern species of Amolae.

35–45 mm long, geminate; ovary 16–20 mm long including neck 3–5 mm long, slender, scarcely grooved; tube shallow, 4–4.5 mm deep, 9 mm wide; tepals ca. equal, ovate, plane, outcurving at anthesis, clasping filaments in post-anthesis, thin, the outer 14–16 x 6 mm, keeled, the inner wider (8 mm) and unkeeled; filaments 40–45 mm long, inserted on the rim of tube at base of tepals; anthers 14–18 mm long; capsules mostly oblong, 2 x 1 cm, short-apiculate, parallel-veined and slightly cross-ridged; seeds lunate, mostly 4–4.3 x 2.5–2.8 mm, 1 to several wrinkles on faces, marginal wing low, firm, the hilar notch sharply angled (Gentry, 16611).

Type: *I. M. Johnston 3123, CAS,* which he found flowering April 17, 1921, on the rocky, granitic slopes of San Pedro Nolasco Island, Gulf of California.

A. chrysoglossa is closely related to *A. vilmoriniana,* but the straight, narrow, plane leaves and non-bulbiferous inflorescence are usually sufficient to distinguish *A. chrysoglossa.* The population along the Río Yaqui, latitude of Sahuaripa, appears to consist of intergrades of the two species. *A. chrysoglossa* is the coastal or lowland relative of this alliance in Sonora, more xerophytic and at home upon the hot, frequently bare, rock surfaces. (See Fig. 4.8.) It flowers from mid-March to mid-May during their spring dry season when air temperatures during the day usually rise to 100–110°F (38°–44°C). I found the inland population about Bacanora in May to have prolific sets of maturing black seeds. Estimates based on partial seed and capsule counts indicated as many as 500,000 to 750,000 seeds per plant. It appears to be a free-seeding, cross-pollinated species, with the nectar attracting animals during the thirsty season. Numerous insects, hummingbirds, and other kinds of birds were observed about the flowers.

This *Agave,* also known as "amole," is as sapogenous as *A. vilmoriniana* and has the same local use for washing clothes (Table 4.1). Small plants were transplanted from Bacanora to my Murrieta garden, where they have done well but did not flower until 20 or more years. Plants from the Guaymas monadnocks and the coastal Sierra Seri may sucker profusely.

Agave nizandensis
(Fig. 4.4; Table 4.2)

Agave nizandensis Cutak, Cact. & Succ. J. 23: 143, 1951.

Small, unarmed, acaulescent surculose, monocarpic, few-leaved, succulent rosette with open habit; leaves 20–30 x 1.5–2.5 cm, linear-lanceolate, green with pale mid-stripe, plane above, convex below, patulous, about straight, sparsely fibrous, rather brittle or pliant, the margins finely serrulate; spine small 4–8 mm long conic, non-pungent; raceme 1–2 m tall, sparsely flowered in upper ¼ of shaft, peduncular bracts 6–10 cm long ascending; flowers pale yellow, 35–40 mm long, in twos or fours on dichotomous bracteolate pedicels 6–10 mm long; ovary 12–15 mm long, cylindric, with short unconstricted neck; tube 3–4 mm long, short-funnelform, rimmed by tepals and filaments; tepals 15–16 mm long, narrowly elliptic to lanceolate, ascending, involuting, the outer more apiculate and narrower (4 mm wide) than the inner; filaments 35–40 mm long, inserted with tepals on orifice of tube; anthers 13–14 mm long, yellow; capsules.

Type: *Cutak 19, MO.* Nizanda, Oaxaca, km 233 along Tehuantepec Nat. RR., 21 Feb. 1947; flowered at Mo. Bot. Gard. July 1951.

The taxonomic position of this distinct species is anomalous. It has no close relatives and does not fit well into any section or group. The small umbellate form of inflorescence indicates it belongs in the subgenus **Agave,** but the rosette and leaf are more Littaeaoid in character.

Cutak (loc. cit.) also reported it from Santa Domingo Petapa and San Miguel Chimalapa, neighboring localities on the Isthmus of Tehuantepec. His article carries good photographs.

Agave ocahui
(Figs. 4.1, 4.2, 4.8, 4.9, 4.10; Tables 4.1, 4.2)

Agave ocahui Gentry var. *ocahui,* U.S. Dept. Agri. Hb. 399: 72, 1972.

Short-stemmed, single, dense, green, yuccoid rosettes, 30–50 cm tall, 50–100 cm broad; leaves numerous, 25–50 cm long, 1.5–2.5 cm broad, widest at base, linear-lanceolate, plane above, green, mostly stiff, erect to ascending, some older declined or falcate, the margin straight and lined with a narrow, reddish brown, firm border detachable on dried leaves, surface smooth, minutely, densely punctate in fine lines; teeth none;

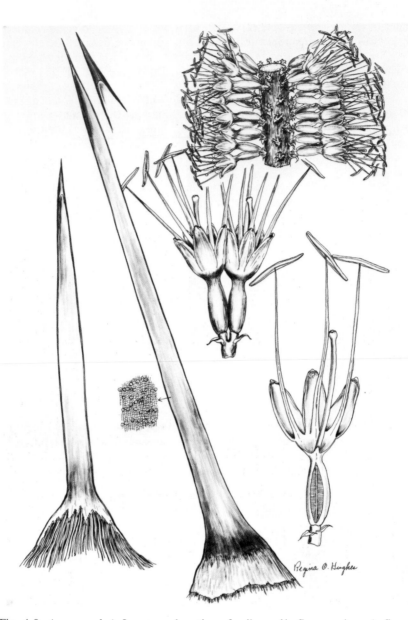

Fig. 4.9. *Agave ocahui.* Leaves and section of spike, x ⅓; flower pair, x 1; flower section slightly enlarged; cuticular pattern, x 20. Drawn from *Gentry 16637.*

spine weak, rather brittle, 1–2 cm long, pruinose gray over brown; spike ca. 3 m tall, slender, densely flowered from 1 to 1.5 m above the base, with copious, narrow, chartaceous bracts; flowers 30–38 mm long, yellow, on short bifurcate pedicels; ovary 15–20 mm long with constricted neck; tube 2–4 mm deep, openly spreading; tepals subequal, 14–16 mm long, 4.5 mm wide, oblong, the abruptly acute tips outcurving, clasping filaments in post-anthesis; filaments 35–40 mm long, pale yellow, inserted at base of tepals; anthers 10–15 mm long, yellow, centric; capsules small, 12–15 x 7–8 mm, ovoid, apiculate, yellowish tan to reddish, tardily dehiscent; seeds small 2.5 x 1.5–1.7 mm, rather thick, crescentic.

Fig. 4.10. *Agave ocahui* on tuffaceous rocks on a sun slope of Sierra Baviso, near Magdalena, Sonora.

Type: *Gentry & Arguelles 16637.* Deep volcanic canyon above (E of) Guasabas, Sonora, Mexico, May 21, 1957; alt. ca. 4,000 feet. Holotype US, isotypes DES, MEXU.

Known only in northeastern Sonora, where it occurs on cliffs and other outcrops of volcanic rocks between 1,500 and 4,500 feet elevations. Also noted as "scattered on limestone crest of Murrieta Mountain," southwest of Sahuaripa. Such sites have dry poor soils inhibiting growth.

The smooth narrow leaves with fine, brown, unarmed margins, the prolifically flowering spike with its small slender yellow flowers with shallow tube and filament-clasping tepals relate this species with the other Sonoran amoles, *A. vilmoriniana* and *A. chrysoglossa.* The leaves and fruits of *A. ocahui* were found to contain 0.5–1.4 percent sapogenins, mainly smilogenin, which also relates it to the amoles (Table 4.1). From the other species it is easily distinguished by its smaller rosettes with more numerous but smaller leaves and smaller flowers with small stamens on long filaments. Also, in contrast to the other Sonoran amoles, which have fine weak fibers in the leaves, the leaf fibers of *A. ocahui* are coarser, abundant, strong, and employed locally for cordage. The wild plants seed abundantly and all the plants observed in the several populations visited were nonsurculose.

Neither the chemistry nor the fibers of *A. ocahui* were assayed sufficiently to assess the prospect of this plant as a potential resource. Considering the fiber-sapogenin industry of the cultivated but frost-sensitive fiber agaves, *A. ocahui,* because of its apparent cold-hardiness, its superior fibers, and good genin content, merits investigation for cultivation in the United States. The length of generation is apparently 20 years or more, but may be shortened by selective breeding and cultivation. Its cultural requirements have not been tested in detail. Garden plants in Arizona and southern California have responded well without special care, and flowering specimens have enabled me to improve the original flower description. The small plants in my Murrieta garden have grown very slowly since first planted in 1952. It is suitable and handsome as an ornamental for our southwestern gardens. The flowering season is indicated in Fig. 4.1.

"Ocahui," "ojahui," and "amoliyo" are local names for this agave. The first two terms are of Indian origin and signify fiber or cordage. A related term "majahui" carries the same meaning and is employed by country people in Sinaloa and southeastward, but it is also applied to such bast fibers as those from the Malvaceae and Sterculiaceae, *Ceiba,* etc.

The longer-leaved specimens first noted in the field, and interpreted as due to site or shade, have maintained their long-leaved form along side the short-leaved form in my Murrieta garden. I am therefore describing the long-leaved form as a variety.

Agave ocahui var. *longifolia* var. nov.*
(Figs. 4.2, 4.8, 4.11; Tables 4.1, 4.2)

Single plants with large round stems; mature leaves 60–80 (–90) cm long, 2–3 cm wide near base, linear-lanceolate, straightly ascending or recurving, sometimes falcate; inflorescence as in var. *ocahui.* Like the species but differs in developing larger rosettes with larger longer leaves (Fig. 4.11).

Type: Grown in author's garden near Murrieta, California; flowered July 1977; from small plant collected on Sierra de Matape, Sonora, 16 Feb. 1952; elev. ca. 3,500 feet, *Lysiloma-Quercus-Nolina* woodland. *Gentry 11610*, US, isotype DES.

This larger more robust form of the species occurs widely scattered in the mountainous region between Matape and Sahuaripa of eastern central Sonora (see Fig. 4.11 and Exsiccatae). It does not appear to be geographically separated from the short-leaved

---

*Ab *A. ocahui* var. *ocahui* caulis majioribus et foliis longioribus differt.

forms of the species and therefore does not merit subspecies status. The flowers and habit are those of the species. Because of the longer leaves and longer fibers, it would have more promise for development as a cultivated fiber plant.

Agave pedunculifera
(Figs. 4.1, 4.2, 4.4, 4.12, 4.13; Table 4.2)

Agave pedunculifera Trel., U.S. Nat. Herb. Contr. 23: 134, 1920.

Single, caulescent, multiannual, soft-leaved rosettes; leaves mostly 50–70 x 15–18 cm, ovate-acuminate, or 80–90 x 11–15 cm and lanceolate, soft-succulent, symmetrically ascending-horizontal, plane to concave, thickened, narrowed and convex at base, pale green to glaucous white, the margin narrowly lined brown or white, closely denticulate, the denticles 0.5–2 mm long; spine weak, acicular, ca. 1 cm long; spikes racemose, 2–3 m long, erect or recurving (on cliffs), flowered from near leaf crown; bracts chartaceous, very narrow, the lower ones longer than the flowers; flowers geminate or in fours on dichotomous pedicels 2–3 cm long, yellow, slender, 37–52 mm long; ovary 20–27 mm long, slender, cylindric, with constricted ungrooved neck; tube shallow funnel-form, 2–6 mm deep, slightly furrowed; tepals equal, oblong to elliptic, thin, ascending to recurving, obtuse, over-lapping at base, the inner wider (7–8 mm), with broad low keel; filaments

Fig. 4.11. *Agave ocahui* var. *longifolia* in the author's garden near Murrieta, California, July 1977. The type collection was prepared from this plant.

Fig. 4.12. *Agave pedunculifera* on volcanic rock near Arcelia, Guerrero; narrow leaf form rosette and weathered spike.

30–55 mm long, inserted with tepals on orifice of tube; anthers 14–22 mm long, yellow, centric; capsules 15–20 mm long, apiculate; seeds not seen.

Type: *Rose 1713, US.* Near Colomas, Sinaloa, 16 July 1897 (2 sheets, 1, cap).

Agave pedunculifera is closely related to *A. attenuata,* as expressed in both leaf and flower. However, the single, nearly stemless plant habit is a basic and enduring characteristic throughout the populations observed. The early suckering of plants in the University of Michigan greenhouse is the only observed instance of this habit and may have been stimulated by their transplanting. *A. pedunculifera* shows considerable variability in leaf form and size and the depth of the flower tubes. The long leaves and very short tubes (2 mm) in the Arcelia population in the southern limits of range along the Río Balsas contrast strongly with the small plants with relatively deep tubes (6 mm) in its northern limits on the cliffs of Sierra Tacuichamona in central Sinaloa. These two extremes, however, are linked together by intermediate variants in Jalisco.

Distribution and Habitat

Figure 4.4 outlines the natural distribution of *A. pedunculifera* along the mountain slopes of Pacific drainage from middle Sinaloa to the Río Balsas country of Michoacán and Guerrero. Although it ascends to over 1,800 m on warm rock faces, it is primarily a species of low and middle elevations between 600 and 1,500 m. The cliff faces that it frequents are in the hot tropical deciduous forest or in the more equable oak forest zones, sometimes with the lower pines as well. The rainfall plotted in Fig. 4.1 indicates the summer rainfall season contrasting with the prolonged spring dry season, which must

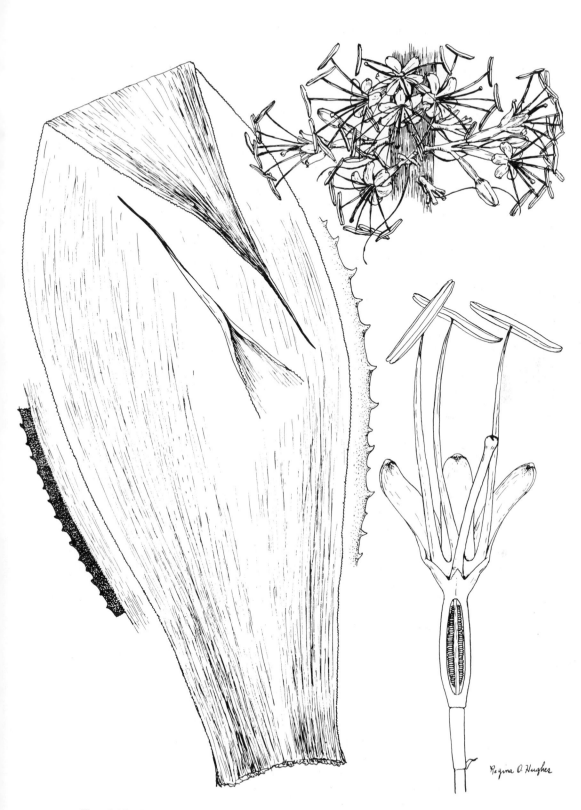

Fig. 4.13. *Agave pedunculifera*. Leaf and marginal detail of two leaves enlarged; section of inflorescence reduced; long section of flower, x 2.

impose intense stress on rock-dwelling plants. The habitat is frost free and *A. pedunculifera* has proven sensitive to the light frosts in my Murrieta garden.

A. pedunculifera is one of the most beautiful of agaves, composed by its natural symmetry and attitude of leaf with soft gray coloring and marginal outlining. When first introduced into culture at the University of Michigan Botanical Garden in 1940, it was considered by many to be the handsomest plant in the garden. Visitors frequently offered generous sums for one plant. However, in 1974 I found one plant remaining, which had flowered several months previouly. It was a gross monstrosity deformed by many years of over-watering and reduced light in the greenhouse; scarcely recognizable. In frost-protected sunny sites with restrained watering, it makes a fine garden ornamental enduring well beyond 20 years before flowering and dying. It is a winter bloomer and our cool winters apparently inhibit flowering. The single plant in my Murrieta, California, garden endured for 30 years before flowering.

Agave vilmoriniana
(Figs. 4.1, 4.2, 4.8, 4.14, 4.15, 4.16; Tables 4.1, 4.2)

Agave vilmoriniana Berger in Fedde, Repertorium 12: 503, 1913.
> *A. eduardii* Trel., Contr. U.S. Nat. Herb. 23: 134, 1920.
> *A. houghii* Hort. ex Trel., Contr. U.S. Nat. Herb. 23: 134, 1920.
> *A. mayoensis* Gentry, Rio Mayo Plants, Carn. Inst. Wash. Publ. 527: 94, 1942.

Short-caulescent, light green to yellowish green, single rosettes, 1 m tall x 2 m broad with toothless, arching, deeply guttered, pliant leaves, 90–180 x 7–10 cm, broadest at base, linear-lanceolate, long acuminate, heavily thickened toward the base, concave to conduplicate above; margin unarmed with a fine brown continuous border about 1 mm wide, scaly in age; epidermis smooth; spine acicular, brown to grayish brown, 1–2 cm long; spike 3–5 m tall, densely flowered from 1 to 2 m above base, bulbiferous or non-bulbiferous; lower bracts 10–20 cm long, scarious, acicular, brownish toward the brittle tip; upper bracts 1–2 cm long; bracteoles of the pedicels deltoid, 2–3 mm long; pedicels 8–20 mm long, 1–2–bifurcate; flowers yellow, 35–40 mm long with a shallow open tube; ovary 15–20 mm long including neck (3–4 mm); tube 4 mm long x 8–9 mm wide; tepals 14–17 x 4–5 mm nearly plane, ascending to spreading at anthesis, clasping the filaments in drying, equal; filaments 30–40 mm long, inserted at upper edge of tube; anthers 16 mm long.

> Lectotype: In barranca, Jalisco, July 9, 1899, *Rose & Hough 4833* (l, cap) in US.

Berger cited no type, his description being based on living plant or plants at Paris. In an undated letter to J. N. Rose (fid. copy in U.S.), Berger wrote that the plant was collected by ''Mr. Diguet for the Jardin des Plantes at Paris and which I named *Agave vilmoriniana* in 1913 in honour of M. Maurice de Vilmorin, an excellent man, in whose garden at Verreves I found this Agave for the first time.—The fruits in your photo resemble greatly those of *A. ellemeetiana* etc. of the same group 'Anoplagave.' It was not in cultivation at La Mortola and of course if I were still there I should be delighted to receive your plants.'' Rose had found the plant on the rim of the barranca near Guadalajara, Jalisco, July 9, 1899, *Rose & Hough 4833,* Hough having risked his life in obtaining the specimen. As Diguet is known to have collected in the vicinity of Guadalajara, it is probable that this barranca is the source of his collection. There are two spines and a pencil sketch by Berger in the Nat. Herb.

A cliff-dweller, the typical habitat of this handsome plant is the volcanic brecciated cliffs of the barrancas from southern Sonora south through Sinaloa and Durango to Jalisco and Aguascalientes, at altitudes between 2,000 and 5,500 feet (600 and 1,700 m). In the larger deeper canyons it forms extensive vertical colonies, which when viewed from a

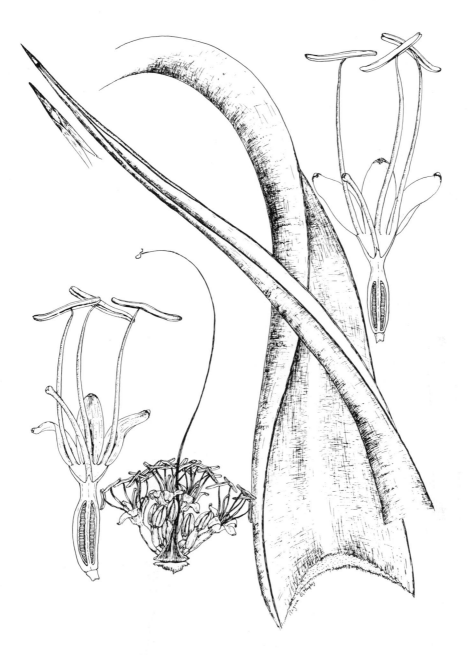

Fig. 4.14. *Agave vilmoriniana* drawn from *Gentry 10232*. Upper and lower thirds of leaf (with tip enlarged); flower cluster from lower part of spike, ca. x ⅓; flower section of *A. chrysoglossa* slightly enlarged.

distance resemble giant spiders on a wall. Individuals may be found upon gentler adjacent slopes in open light. The cliff situations are virtually frostless, but the open heights above 4,000 feet (1,250 m) have sharp frosts. In southern California the plants have demonstrated tolerances to over-night freezes as low as 24°F ($\approx$4.5°C). They withstand protracted drought easily and will take in water during the chilling winter rains in California. Many of the cliff plants, for lack of soil, remain stunted through their lives, their maturation inhibited, and the inflorescence is feeble when at last it does appear.

Fig. 4.15. *Agave vilmoriniana* along arroyo Gochico in Río Mayo country. The dried flowers hang bleached and beard-like on fruiting spikes.

This species is distinguished by its unarmed, large, gracefully arching, deeply guttered leaves. In Sonora and elsewhere it is known as "amole," and, like other plants of the name, is employed in washing clothes and other fabrics. For this use, the old dried leaf bases on dead flowered plants are cut off about 6 or 8 inches (15 or 20 cm) above the base. The fibers in the lower butt are beaten and loosened by pounding with a rock, forming an effective brush with a built-in soap or detergent, which foams and cleans in water as one brushes. These brushes have been noted for sale in local markets of the Mexican west coast. The cleaning agent is principally the sapogenin, smilagenin, which by analysis was found to form 3–4.5 percent of the dry leaf, among the highest known for any agave leaf (see Table 4.1). Younger green leaves contain much less.

As an ornamental it lends itself well to design plantings and it propagates readily from bulbils. The clone in the Huntington Botanic Garden in San Marino, California, forms mature flowering plants in seven or eight years from bulbils. However, young plants collected in Sierra Charuco, *Gentry 10232,* and planted at Murrieta in 1951 did not flower until 1966; 15 years. The spike emerged soon after April 1 and reached about its full height in early May. The more rapid maturation in the Huntington Gardens is apparently due to better culturing there and, perhaps, also to genetic early maturation factors. Capsules and seeds are also produced; hence, a given population may consist of both apomictic and sexual generations. This was noted in the Arroyo Gochico population. No rhizomatous offsets have ever been observed.

Fig. 4.16. Bulbils growing out of flower axils of *Agave vilmoriniana;* a seedless inflorescence.

Fig. 4.17. Mature capsules on fruiting spike of *Agave* ➤ *chrysoglossa* near Bacanora, Sonora, a freely seeding species.

Agave yuccaefolia
(Fig. 4.18; Table 4.2)

Agave yuccaefolia DC., in Redoute, Les Liliacaes, pl. 328, 329, 1812.

Small to medium-sized, short-stemmed or stemless, suckering, open, monocarpic rosettes with rather few leaves; leaves 50–65 x 3–3.5 cm, linear, recurving with maturity, concave above, convex below, mostly green with pale medium stripe, sometimes reddish or purple spotted, soft, pliant, scarcely succulent, weakly and finely fibrous, the margin finely serrulate with unequal denticles; spine 3–8 mm long, conic to subulate, brown; raceme 2–3 m tall, slender, arching, with triangular-lanceolate, long-acuminate, erect bracts; flowers short-pedicellate, mostly geminate, 40 mm long, greenish yellow with red-tinged stamens, malodorous; ovary 16–18 mm long, roundly 3-angled with a short neck; tube ca. 8 mm long, narrowly cylindric; tepals 15–16 mm long, linear, obtuse, spreading to reflexed, the inner narrower; filaments 40–45 mm long, reddish, inserted in

Fig. 4.18. *Agave yuccaefolia* in glass house No. 5, Royal Botanical Gardens at Kew, June 1969.

orifice of tube, 7–8 mm above tube base; anthers 15 mm long, yellow; capsule 20 mm long, 17 mm broad, obovate, broadly 3-angled, thinly woody, light gray glaucous, short-beaked; seeds nearly half-round, 5 mm long, black, shining.

The type can be taken as the illustrations, pl. 328, 329, in Redoute's work cited above.

Hooker described and illustrated this species in Curtis Botanical Magazine in 1860, tab. 5213. As Berger (1915) remarked (p. 44), that plant was abnormal in its long length of stem, numerous leaves, and long inflorescence. This attenuate habit was probably caused by greenhouse conditions in cool English climate, which retarded flowering maturation while it fostered continued growth. Fig. 4.18 shows a more normal appearing plant several years old in glasshouse No. 5, Royal Botanic Gardens, Kew, in 1969. The above description is largely drawn from that of Berger in his herbarium specimens (US) prepared in the Mórtola Gardens in western Italy, Jan. 1900.

The species was introduced to European gardens in the early 1800s and first bloomed in the Malmaison garden near Paris in 1810–11. It flowered in Kew in 1829 and repeatedly at La Mórtola garden after 1887. The suckering habit provides offsets for continuing propagation and the species is still common in European gardens. Besides Palermo and Kew, I found it at Munchen and Berlin botanical gardens.

Its specific American origin is unknown. There are no original documented field collections. Hooker (loc. cit.) stated the species "was received by us (Kew) from Rio del Monte district, Mexico." but he probably meant Real del Monte, site of a large mine near Pachuca, Hidalgo, and which is quoted as a source of several other Agavaceae introductions at the turn of the 19th century. Several years ago I went to Real del Monte in search of such plants but saw only a barren hillslope surrounding the mining settlement, long since bereft of everything but the most fleeting non-perennial vegetation.

This plant is not closely related to the others included in this section, but is even less suited for inclusion in other groups.

Amolae Exsiccatae

Agave attenuata

HORTICULTURE. *Kimnach 7,* DES. Cult. Huntington Botanical Gardens, San Marino, California, 2 Dec. 1976 (f).

JALISCO. *Kimnach & Boutin 3019,* DES. Sierra Minantlan above Valley of Durasno, elev. 7,200 ft. Growing with *A. pedunculifera.* (l, f, cult. at Hunt. Bot. Gard.).

MEXICO. *Gentry 23372,* DES. Ixtapantonga, Valle del Bravo, ca. 1959, "large population on cliffs," Cult. Dudley Gold's garden in Cuernavaca, fl'ed Feb. 1974 (l, f, cap).

MICHOACAN. *Gentry & Arguelles 10430.* 6–8 miles W of Mil Cumbres along road from Morelia to Guadalajara, 8 May 1951; elev. 7,000–7,500 ft. "old plants on naked caudices 4–8 dm long" (live pl.).

Agave bakeri

HORTICULTURE. Entry no. 255 1889 K. Type. Kew Gardens, Jan. 22, 1903. From "the late Mr. Peacock' noble collection of Cactuses, Aloes, and Agaves in 1889. It flowered in the Mexican division of the Temperate House in January of 1902."

Agave chrysoglossa

SONORA. *Felger 20484,* ARIZ. Sierra Seri, 17 May 1972; elev. 1,500 ft.

Felger & Stronk 3113, ARIZ. Ca. 2 miles NW of Bahía San Carlos along road to Bahía Algodones. 9 Feb. 1960; riparian canyon, desert shrub and thorn scrub.

Gentry 11349, DES, MEXU, MICH, US. Isla San Pedro Nolasco, 16 Dec. 1951; steep granitic mountain slope with scattered trees and giant cacti.

Gentry 16611, DES, MEXU, US. 26 miles NW of Sahuaripa, 14 May 1957; alt. ca. 1,000 feet, volcanic cliffs, (l, cap, s).

Gentry 16623, DES, MEXU, US. 14 miles S of Bacanora along road to Tonichi, 16 May 1957; alt. ca. 2,500 ft., volcanic rocks and cliffs (l, f).

Gentry & Arguelles 19882, DES, MEXU, US. Cerro 4–5 miles N of Bahia San Carlos, 29 March 1963; arid palm canyon in volcanic rocks (l, f).

Johnston 3123, CAS. Type. San Pedro Nolasco Island, 17 April 1921.

Johnston 4338, CAS, US. San Pedro Bay, 7 July 1921.

Agave nizandensis

OAXACA. Cutak 19, MO. Type. Nizanda, km. 233, Tehuantepec Nat. RR., 21 Feb. 1947; flowered at Mo. Bot. Guard. July 1951.

Gentry & Weber 22567, DES. Cult. Des. Bot. Gard., Dec. 1967, from seed col. by T. MacDougal at Type Locality, 1958. (l, f).

Agave ocahui

SONORA. *Felger 3301,* ARIZ, US. Southern rim of Canyon Cruz de Peñasco, 7 miles E of Río Bavispe along road to Bacadehuachi, 5 June 1960; alt. ca. 3,650 ft. (l, f, cap).

Gentry & Arguelles 16637, type, DES, MEXU, US. Mountain pass above (E of) Guasabas, 21 May, 1957; volcanic cliffs, alt. ca. 4,000 ft. (l, f).

Gentry & Arguelles 16607, 6–8 miles NE of Matape, Sierra Batuc (pl, s).

Gentry & Arguelles 19886, DES, MEXU, US. Sierra Baviso, 16 miles Se of Magdalena near Canyon de Palmas, 31 March, 1963 (l, cap).

Gentry s.n., DES. Cult. young plant in Des. Bot. Gard., flowering 15 May– 15 June 1973 (l, f, cap).

Gentry & Kaiser D.B.G. 62-7269, DES. Cult. Des. Bot. Gard., originally col. by Kaiser on Sierra Baviso along road to Cucurpe, flowered May 1975 (l, f).

Agave ocahui longifolia

SONORA. Gentry 11610, DES, **US.** Type. Grown in Gentry's garden near Murrieta, Calif., (fl'ed July 1977), from small plant collected on Sierra Matape, 16 Feb. 1952; elev. ca. 3,500 ft., *Lysiloma-Quercus-Nolina* woodland (l, f, photo).

Gentry & Arguelles 16603, DES. Ahuaje Pescado, 8 miles E of Matape, 13 May, 1957; elev. ca. 2,800 ft., limestone rocks. Leaf from plant cult. in Gentry garden.

Gentry 16612, DES. Ca. 4 miles W of Sahuaripa, 15 May 1957; elev. 1,500–1,800 feet, on volcanic rocks. Plant cult. in Gentry garden.

Gentry & Arguelles 16626, DES, MEXU, US. Sierra Murrieta near ranche Toribusi, 15 miles W of Bacanora, 17 May 1957; elev. ca. 4,500 ft., limestone, oak woodland.

Agave pedunculifera

GUERRERO. *Gentry 22547,* DES, MEXU, US. 5–6 miles W of Arcelia, 14 Dec. 1967; elev. ca. 1,500 ft., sun side of volcanic cliffs (l, cap, photo).

Gentry 23371, DES, MEXU, US. Same loc., 23 Feb. 1974 (l, f).

Rzedowski & McVaugh 289, ENCB, MICH. Cruz de Ocote, Mun. Chihihualco, ca. 43 km W de Chilpancingo, 31 Jan. 1965; alt. 2,000 m, Calizas karsticas (l, f).

MICHOACAN. *Nelson 6923,* US. La Salada, 15–22 Mar. 1903 (l, cap).

JALISCO. *Gentry & Gentry 23503,* DES, INIF, MICH, US. 6 miles by road SW of Chiquilistlan, 30 Jan. 1975; elev. 5,700–6,000 ft., on limestone outcrop in oak forest (l, pl, photo).

Gentry & Gentry 23507, DES, MEXU, MICH, US. 6 miles W of Chante along road to Sierra Manantlan, 3 Feb. 1975; elev. ca. 4,000 ft., cliff of volcanic rock (l, f).

McVaugh 25983, MICH. 5–6 km (airline) S of Chiquilistlan; elev. 1,800– 1,950 m, oak forest with *Pinus lumholtzii* (l, infr).

SINALOA. *Bye 7563,* DES, GH. W of El Alazan ca. km 206 on Mazatlán-Durango road, 19 July 1977; elev. 1,840 m, pine-oak forest, "large colony clinging to vertical W-facing wall."

Gentry & Gilly 10523, DES, MEXU, MICH, US. 3–4 miles E of El Batel along Mazatlán-Durango hwy, 28 May 1951; elev. 6,000–6,500 ft., mixed oak and pine forest, cliffs.

Gentry & Arguelles 18182, DES, MEXU, US. Palmito and vicinity along Durango-Mazatlán hwy, 14– 15 Nov. 1959; elev. 2,000–2,200 m, cliffs (l, f).

Gentry 5692, DES. Cliffs of Cerro Tiburón, Sierra Tacuichamona, 19 Feb. 1940 (f from live plant cult. in Gentry garden, April 1977).

Rose 1713 Type, **US.** Near Colomas, 16 July, 1897 (2 sheets, l, cap).

JALISCO. *Gentry & Gilly 10893,* DES, US. Barranca de Colimilla, ca. 4 miles NE of Guadalajara, 19 July 1951; alt. ca. 4,000 ft., rocky volcanic slope.

Pringle 3740, MEXU, US. Cliffs of Río Grande de Santiago near Guadalajara, 20 May 1891 (l, cap & an extraneous f).

Rose & Hough 4833, lectotype, **US.** Barranca near Guadalajara, 9 July 1899 (l & old cap). Hunt. Bot. Gard. has topotypic or clonal material of this col., e.g. "No. *10844,* HNT, received 6/18/42 from USDA, Washington, D.C. Rose succulent collection."

SINALOA. *Breedlove 19079,* CAS. Mouth of Canyon de Tarahumaras along road from Mocorito to Surutato, Mun. Sinaloa y Vela, 3 March 1971; on rocks along stream in forested slopes of Sierra Surutato (l, f in bud).

SONORA. *Barclay & Arguelles 2016,* DES, US. Rancho San Antonio, Sierra Tecurahui, 20 May 1966 (l, f, photo).

Gentry 3673, ARIZ, *CAS,* DES, MO. (Type of *A. mayoensis).* Arroyo Gochico, Río Mayo, canyon cliffs in barranca forest, 5 April 1938 (l, f).

Gentry 16531, DES. Arroyo Gochico, 5 miles above San Bernardo, 1 May 1957 (bulbils, photo, l for anal.).

Agave vilmoriniana

CHIHUAHUA. *Gentry 10232,* DES, MEXU, MICH, US. High rocky rims around Arroyo Honda, Sierra Charuco, 6– 7 March 1951, transpl. flowered in Murrieta May 1966 (l, f).

Hartman 551, GH. Apajeachi, 15 May 1892–93 (f). "wa-we-ke, Tarah."

Palmer 135, US. San Ramón, April 12–May 18, 1906 (l, f, cap, type of *A. eduardi).*

Agave yuccaefolia

HORTICULTURE. *Berger s.n.,* US. Cult. La Mórtola, Jan. 1900, 1901 (l, f).

Gentry 22605, DES. Cult. Orto Botanico, Palermo, Sicily, 12 April 1969 (l, photo).

Gentry 22736, DES. Cult. Royal Botanic Garden, glass house no. 5, Kew, England, 23– 27 June 1969 (photo).

Type, non vidi.

5.

Group Choritepalae

Single multiannuals or polycarpic perennials with axillary or rhizomatous branching, forming large clusters of rosettes; freely seeding. Leaves succulent, viscous-fleshy, armed or unarmed, with or without terminal spines. Inflorescence in spicate racemes, relatively short, densely flowered; flowers small, tubeless, the tepals essentially distinct, spreading, white to greenish yellow; filaments long to very long exserted, thread-like, some spiraling in age; anthers small. Capsules elliptic to ovate, tardily dehiscent, freely seeding. Calciphytes. Northeastern Mexico to isthmus of Tehuantepec.

Typical species, *Agave ellemeetiana* Jacobi, Hamb. Blumenz. 21: 457, 1865.

The Choritepalae is a small distinctive group distinguished by the separate tepals arising from a discoid receptacle, rather than from a tube. This distinctive flower structure together with the unarmed leaves without terminal spines could justify removal from *Agave* to a separate genus. However, *A. guiengola*, although it has the flower characteristics, does have terminal spines and usually well-developed marginal serrations. It therefore bridges the group to *Agave*.

Berger included *A. ellemeetiana* and *A. bracteosa* in his Section Anoplagave, along with the less related *A. attenuata*, *A. cernua*, and *A. vilmoriniana*, which have unarmed leaves but lack the flower structure of the Choritepalae.

Agave bracteosa occupies a small area of the northern Sierra Madre Occidental, where it is nearly confined to cliffs. *A. guiengola* is known only from two hilly localities near the Isthmus of Tehuantepec. The native habitat of *A. ellemeetiana* is still unknown but reported from Mexico. Figure 5.1 depicts rainfall and flowering seasons of *A. bracteosa* and *A. guiengola*. Figure 5.2, by flower sections and ideographs, illustrates some striking features of flower structure.

Key to Species of the Choritepalae

1. Leaves without teeth or terminal spines; filaments very long, to 50–60 mm   2
1. Leaves with teeth and terminal spines; filaments 30–40 mm long
guiengola, p. 97
2. Rosettes polycarpic with epigeous offsets; leaves narrow (3–4 cm), widest at base; flowers white to pale yellowish   *bracteosa*, p. 91
2. Rosettes monocarpic with hypogeous offsets; leaves broad (15–20 cm), widest in middle; flowers greenish yellow   *ellemeetiana*, p. 94

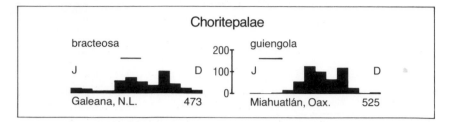

Fig. 5.1. Rainfall (silhouettes) and flowering seasons (bars) of two species of Choritepalae. Relevant meteorological stations with average annual rainfall in millimeters. Data from Atlas Climatológico de México (1939). Flowering periods based on herbarium specimens and field observations.

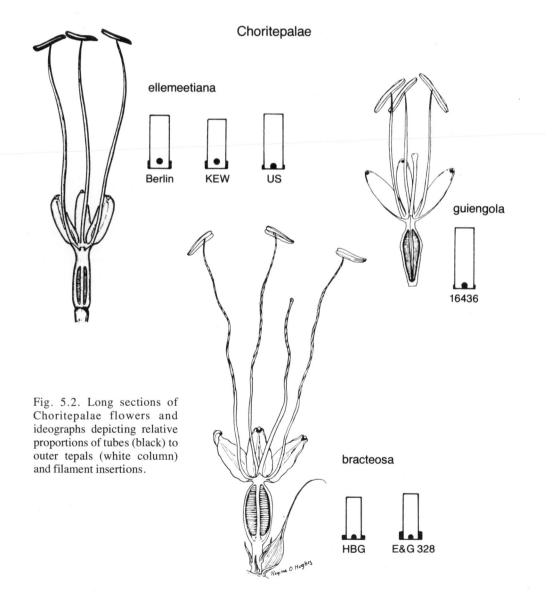

Fig. 5.2. Long sections of Choritepalae flowers and ideographs depicting relative proportions of tubes (black) to outer tepals (white column) and filament insertions.

Agave bracteosa
(Figs. 5.1, 5.2, 5.3, 5.4, 5.5; Table 5.1)

Agave bracteosa S. Wats. ex Engelm., Gard. Chron. n. ser. 18: 776, 1882.

Small to medium-sized, open, graceful rosettes with relatively few leaves, forming cespitose mounds by epigeous axillary budding, flowering repeatedly from upper leaf axils. Leaves 50–70 x 3–5 cm, long lanceolate, widest near base, yellow-green, arching and recurving, succulent, with weak fibers, asperous, plane above, convex below in the basal 1/3, the margin minutely serrulate; spine absent, the tip drying early, yellowish, friable, wind-scuffed; spike 1.2–1.7 m tall, ascending to erect, densely flowered in upper third of shaft, the peduncle with erect triangular-acuminate bracts; flowers 22–26 mm long, white to pale yellow, geminate, persistent; bractlet 50–70 mm long, chartaceous, base boat-shaped, long-caudate; ovary fusiform, 12–14 mm long, virtually neckless; tube reduced to a short receptacle; tepals 11 mm long, ovate, spreading, hyaline, distinct; the outer over-lapping inner, lanceolate, acute, the inner broadly ovate, obtuse, both flocculose at tips; filaments, long-exserted, 50–60 mm long, white, persisting, elongating in post-anthesis, inserted on receptacle; anthers sagittate, 7–8 mm long, yellow; pistil eventually exceeding stamens; capsules 2 x 1 cm, obtuse, thin-walled, 4 mm stipitate; seeds 3–3.5 x 2–2.3 mm, half-round.

Neotype: Tab. 8581, Curtis Botanical Magazine, 1914, here reproduced as Fig. 5.3. Described from a plant in the Cambridge Botanical Garden collected by Edward Palmer near Monterrey, Nuevo León, 1879–80.

Agave bracteosa, with its unarmed curling leaves and white flowering spikes, is so distinctive that it has never been confused with other species, not even with its sectional relatives. With *A. ellemeetiana* it could well be regarded as a separate genus, charac-

Table 5.1. Flower Measurements in the Choritepalae (in mm)

TAXON & LOCALITY	OVARY	TUBE	TEPAL	FILAMENT INSERTION & LENGTH		ANTHER	TOTAL LENGTH	COLL. NO.
bracteosa								
Hunt. Bot. Gard., Cal.	13	0×5	11×5 & 11	0–1	50	7	24†	A. '59
	13	0×5	11×5 & 11	0–1	54	7	24†	A. '59
	10	0×5	11×6 & 11	0–1	32	8	24†	J. '77
	11	0×5	12×6 & 12	0–1	43	8	24†	J. '77
Paso San Lazaro, Coah.	16	0–2×6	12×4 & 12	1–2	45	8	28†	328
	16	0–2×5	11×4	1–2	45	8	28†	328
ellemeetiana								
Berlin	16	2	13	2	50	8	31*	1876
	20	2	15	2	50	8	37*	1876
Kew Bot. Gard.	13	2	13×5	2	54	8–9	27*	1877
US	14	1	15 & 14	0	56		28*	Berg.
	14	1	14 & 13	0	60		28*	Berg.
guiengola								
Tehuantepec, Oax.	17	0	17–18	0	35		33*	16436
	18	1	16–17	0			35*	16436

* Measurements from dried flowers relaxed by boiling.

† Measurements from fresh or pickled flowers.

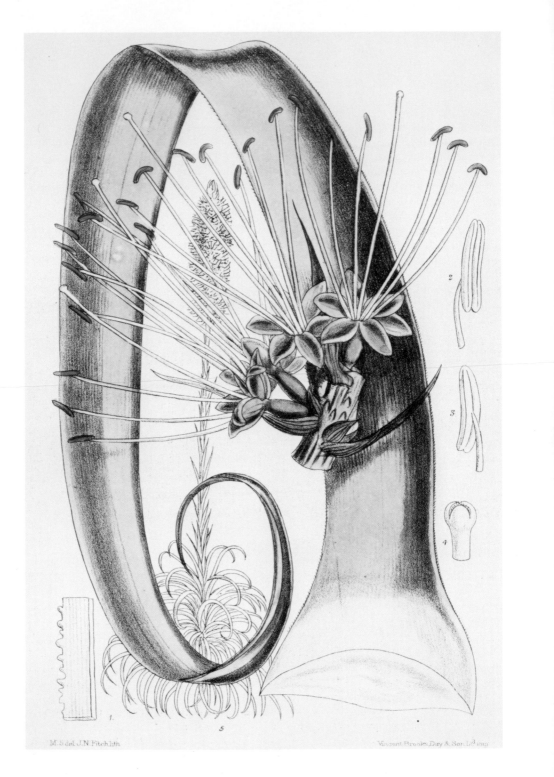

Fig. 5.3. *Agave bracteosa* as illustrated in Curtis Botanical Magazine Tab. 8581, 1914.

Fig. 5.4. *Agave bracteosa* in native habitat, 12–16 miles (19–25 km) NE of Saltillo, Coahuila, Mexico along road to Monterrey, June 1963. Photo by Arthur S. Barclay.

terized by tubeless perianth, elongate filaments, and unarmed leaves, were it not for *A. guiengola* with similar flowers, but which, with its armed leaves, bridges the section to other agaves. *A. bracteosa* is endemic to the northern end of the Sierra Madre Oriental, where it is scattered on the limestone cliffs and rocky slopes between 3,000 and 5,500 feet (900 and 1,700 m); see map, Fig. 5.5. It is a true xerophyte of the Coahuilan Desert growing healthily with only 400–500 mm of annual precipitation on the average, falling irregularly in the winter and late summer months.

The growth of *A. bracteosa* is slow. Single plants may flower when attaining only ca. 30 leaves, but, as the rosettes continue to grow and flower, the old ones develop many leaves. As the rosettes mature, new branches form in the axils of the lower leaves, adding clustered rosettes until large mounds are formed (Fig. 5.4). When several of the bright white, densely flowered spikes are produced in one clump, the effect is strikingly ornamental.

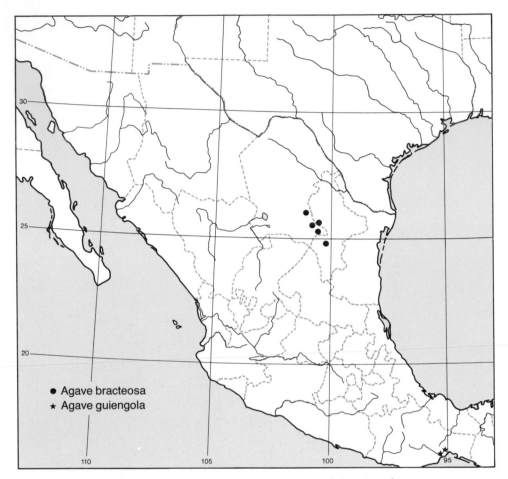

Fig. 5.5. Distribution of *Agave bracteosa* and *A. guiengola*.

Agave ellemeetiana
(Figs. 5.2, 5.6, 5.7; Table 5.1)

Agave ellemeetiana Jacobi, Hamb. Gart. Blumenz. 21: 457, 1865.

Nearly stemless, monocarpic, surculose, open, unarmed rosettes 35–50 cm tall, 70–100 cm broad, with rather few leaves; leaves 50–70 x 12–20 cm, unarmed, light bright green, thickly soft succulent, smooth, ovate to oblong, acuminate, somewhat recurved, reclining with maturity, concave to plane above, plane below beyond the thick base, widest in the middle, the short-acuminate tip slightly calloused, the edge with a thin, friable, smooth margin sometimes reddish and finely serrulate toward apex; raceme 3–4.5 m tall, erect, thickly flowered from near the base; peduncular bracts ovate, caudate-acuminate, 8–10 cm long; flowers 28–40 mm long, pale greenish yellow, campanulate, mostly in fours on a pair of dichotomous pedicels 15–20 mm long; ovary 13–20 mm long, flask-shaped with conspicuously long neck; tube very short, 1–2 mm long, receptacular; tepals 13–15 mm long, 5 mm wide, lanceolate, concave, somewhat hooded at tip, the

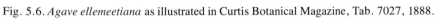

Fig. 5.6. *Agave ellemeetiana* as illustrated in Curtis Botanical Magazine, Tab. 7027, 1888.

Fig. 5.7. *Agave ellemeetiana* in glass house No. 5, Royal Botanical Gardens, Kew, June 1969.

inner somewhat wider; filaments 50–60 mm long, erect, inserted on rim of tube with tepals; anthers 10–12 mm long, yellow; capsules 13–15 x 10 mm, trigonous, woody, light brown, rounded at base, beaked; seeds black, shining, 3 mm long, numerous.

Neotype: In Kew Herb. "Hort. Kew. March 10, 1877, sheet 1, 2, 3."

Tab. 7027, Bot. Mag. 1888, "We have had it at Kew for many years, and flowered it several times. Our drawing was made from a plant at Kew in May, 1888. Fortunately the present plant produces seed copiously in cultivation so that there is no danger of it being lost.'' Jacobi named the plant after his friend Jonge van Ellemeet, a plant fancier in Zeeland, Holland, but he was not credited with its introduction.

The species is reported to have been introduced from Mexico to Europe about 1864, and it became familiar in many European gardens. However, details of its introduction and specific source are lacking. Trelease (Contr. U.S. Nat. Herb. 23: 134, 1920) suggests it may have come from the vicinity of Jalapa, Veracruz. Obviously, it is another rare agave endemic, or its native habitat would have been discovered by modern botanists during the present century.

Berger's illustration (fig. 31, *Die Agaven,* 1915) shows a mature clone in a close thicket growing on an open slope of the well-known La Mórtola Gardens on the Mediterranean coast near Ventimiglia in western Italy. As this was an excellent site for culturing *Agave* growth, we can from this photograph envisage how mature plants may appear or may have appeared in their native habitat. I did not find this clone in La Mórtola during my visit there in 1955. There is no record of *Agave ellemeetiana* in North American gardens, but it is illustrated in Breitung's recent account (1968, Figs. 93, 94). It was still maintained in culture at Kew in 1969 (Fig. 5.7).

Agave guiengola
(Figs. 5.1, 5.2, 5.5, 5.8, 5.9; Table 5.1)

Agave guiengola Gentry, Brittonia 12: 98, 1960.

Acaulescent, mostly single, few-leaved, light gray or white-glaucous; open rosette with ca. 30 leaves at maturity; leaves ovate to ovate-lanceolate, short-acuminate, openly ascending, nearly plane above but briefly and narrowly channeled apically, the margins variously serrate with flattened, blunt, 1-2-cuspidate, dark brown, fine or coarse teeth; epidermis finely and densely papillate; spine acicular, dark brown, rounded above and below, not decurrent or decurrent for about its own length in a corneous margin; inflorescence spicate, 1.6-2 m long, the stalk flowering from near the base, erect; upper bracts and bracteoles similar, 1-2 cm long, 3 mm wide at the deltoid base, scarious, recurved, long attenuate, with a dark midvein; flowers inconspicuous, 33-35 mm long; tepals pale yellow or yellowish white, elliptic, openly ascending, straight; anthers yellow, excentrically attached, versatile; capsules (immature) 22-24 mm long, oblong, thin-walled.

Type: *Gentry 16436,* 25-27 km NW of Tehuantepec, Oaxaca, Mexico, 22 Mar. 1957, on Guiengola limetone at 125 m Alt. Holotype US, isotypes DES, MEXU.

Agave guiengola is a strikingly ornamental species, distinct because of its broad, thick, white glaucous leaves serrated by numerous broad, low, blunt, dark-colored teeth. The light shade in which some individuals grow suggests that it will tolerate some shade without the etiolation that deforms other *Agave* leaves in similar situations. The virtually tubeless flower with its consequently distinct tepals, and the insertion of the filaments at the base of the tepals, show relation to *Agave bracteosa* and *A. ellemeetiana.* Its broad,white, ovate leaves with their conspicuous coarse teeth, and its monocarpic rather than polycarpic habit, set off *Agave guiengola* from either of the species mentioned. The population observed at the type locality showed individuals varying in the form and size of both leaves and teeth. On one variant the lower two-thirds of the leaf margins were without teeth. Altogether, however, the plants appeared to constitute a well-integrated species, indicating normal sexual reproduction via seed dissemination rather than through bulbils or offsets. Some of the plants had rhizomatous offsets, and several were established at the Huntington Botanical Gardens in San Marino, California.

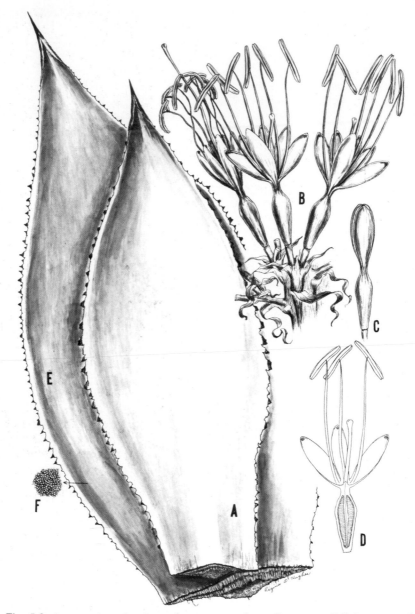

Fig. 5.8. *Agave guiengola,* drawn from type specimen. Leaves ca. 2.5 times natural size; flowers natural sizes; epidermal detail much enlarged.

The specific name is taken from the limestone formation to which this *Agave* appears to be endemic. Thomas Macdougall, who has long been a student of the Oaxacan flora, stated, on seeing the plant at the collection site, that it was known only from the Guiengola limestone between elevations of 100 and 1,000 meters. A note on the collector's label indicates it is used locally for mucilage. The mucilaginous character of the thick sap contained in the turgid leaves is immediately apparent when a leaf is split open, and the dressed leaves adhered tightly to the news sheets during curing. The flowering period in Tehuantepec appears to be during February and March.

Fig. 5.9. *Agave guiengola* in native habitat on guiengola limestone, Isthmus of Tehuantepec, Oaxaca, May 1957.

Choritepalae Exsiccatae

Agave bracteosa

COAHUILA. *Engard & Getz 328,* DES. San Lazaro Pass on hwy. 57 S of Monclova, 11 July 1974; alt. ca. 4,000 ft., limestone (l, f).

Gentry et al. 20020, DES, MEXU, US. 12–16 miles NE of Saltillo along road to Monterrey, 10 June 1963; alt. 4,000–5,000 ft (l, f).

Wind & Muller 161, US. Rocky slopes of El Puerto de San Lazaro, 17 June 1936 (l, f).

HORTICULTURE. *Gentry 23681,* DES. Cult. Hunt. Bot. Gard., San Marino, Calif., July 1977 (l, infl).

NUEVO LEON. *Endlich 900,* MO. Cañon de Sta. Catarina, near Monterrey, 22 Sep. 1905, "amole de castillo" (l, cap).

Gentry & Arguelles 11517, DES, MEXU, MICH, US. Canyon Huasteca ca. 8 miles W of Cd. Monterrey, 30 Jan. 1952; limestone cliffs in deep narrow canyon (l, cap).

Pringle 2523, MO, US. Dry calcareous walls near Monterrey, June–July 1889 (l, f).

Agave ellemeetiana

HORTICULTURE. *Berger s. n.,* US. Cult. La Mórtola, 31 Nov. 1910 (f).

KEW s. n., K, neotype. "Hort. Kew March 10th 1877." teste N. E. Brown (3 sheets, f, l).

Gentry s. n. DES. Cult. Inst. Orto Botanico, N. 265, 10–12 April, 1969 (photo).

Gentry 22737, DES. Glass house No. 5, Kew, 23–27 June 1969 (photo).

Agave guiengola

OAXACA. Gentry 16436, DES, MEXU, **US.** Type. 16 miles NW of Tehuantepec along Panamerican hwy, 22 March 1957; foothills with short-tree forest, limestone, alt. ca. 500 feet (l, f).

Gentry 12241, DES, MEXU, MICH, US. Same loc., 30 Sep. 1952 (l, photo).

6.

Group Filiferae

Small single or cespitose, short-stemmed, multiannual plants with monocarpic rosettes. Leaves narrow, filiferous, white bud-printed, generally pliant, frequently recurved, unarmed, the terminal spines small, weak, scarcely wounding; inflorescence spicate, erect, densely bracteate with long-caudate, chartaceous bracts; flowers small, green, yellow, red, or purplish, usually geminate on short dichotomous pedicels, rarely somewhat clustered; flower tubes short to medium, exceeded in length by tepals; filaments inserted with tepals on rim of tube, rarely below; stamens small, varicolored; capsules variable, small slender, long dehiscing, seeding abundantly; seeds small, relatively thick, crescentic to deltoid. Mainly Occidental Sierra Madrean.

Typical species: *Agave geminiflora* (Tagl.) Ker-Gawler, Roy. Inst. Gr. Brit. J. Sci. Arts 2: 86–90, 817.

The Filiferae is one of two groups characterized by unarmed, filiferous, white bud-printed leaves. They are separable from the Parviflorae by their larger leaves and larger flowers with relatively short tubes and the filaments inserted on the rim or orifice of the tube. The species as here defined are not always distinct. Morphological separation is difficult because of intraspecific variation in leaves and flowers. The flower tube is inconstant, for instance. One is forced to use growth habits, as simple or cespitose, for some specific separations. *A. schidigera* is the most extensive complex both geographically and in variation but is maintained separately from its close relative, *A. filifera,* on habit and minor leaf differences.

Except for the more eastward *A. filifera,* the taxa are nearly limited to the Sierra Madre Occidental Region of northwest Mexico. However, they extend from the Sonoran Desert in the north to the Coliman tropical forest in the south. They range from sea level *(A. felgeri and A. colimana)* to elevations of 2,000–2,500 m *(A. multifilifera, A. schidigera)*, all of them being limited to cliffs and rocky outcrops. This diversity of habitat must have had a strong impact on their evolutionary development, but their constitutional diversity is not reflected in their morphology: e. g., the general similarity in the temperate high montane *A. multifilifera* and the tropical coastal *A. colimana*. The climatological extremes are not manifest in the morphological criteria here used to separate them. The habitat differences are barely suggested in the rain silhouettes of Fig. 6.1. Note also the variability in the flowering seasons depicted in the same figure, for which I have no causal explanation. The factors controlling flowering in agave species are quite unknown; all puzzles are awaiting an inquiring studious mind.

The Filiferae are long-lived plants and do not flower frequently. This together with their generally inaccessible sites seldom allows their collection. Flowering specimens of definite populations are still inadequate to solve some of the taxonomic problems remaining.

Key to Species of the Filiferae

1. Leaves very narrow, 4–6 mm wide in mid-leaf   2
1. Leaves broader, 1–4 cm wide in mid-leaf   4
2. Leaves 45–60 cm long; flower tube 7–12 mm long   3
2. Leaves 25–30 cm long; flower tube 3–4 mm deep   *felgeri*, p. 107
3. Stem large with symmetrical rosette and numerous leaves; not surculose; flower shaft thick, (over 40 cm at rosette top level)   *geminiflora*, p. 112
3. Stem small with asymmetrical rosette and few leaves; surculose; flower shaft slender, (less than 30 cm at rosette top level)   *ornithobroma*, p. 117
4. Leaves elongate, 40–90 cm long   5
4. Leaves shorter, generally 20–40 cm long   6
5. Flower tube deep and slender, 10–17 mm long. 5–9 mm broad. Coastal to the Jaliscan Plateau Region   *colimana*, p. 102
5. Flower tube short and broad, 5–6 x 10 mm. Northern Sierra Madre Occidental   *multifilifera*, p. 112
6. Leaves narrow, 1 cm or less wide, frequently falcate; flower tube 3–4 mm deep   *felgeri*, p. 107
6. Leaves wider, mostly 2–4 cm wide, usually not falcate; flower tube 5–10 mm deep   7
7. Plants cespitose; leaves short, 20–30 cm long, thickened below middle, not flexible, with fine marginal threads; flower tube short, 5–6 mm deep. Central Mexican Plateau   *filifera*, p. 110
7. Plants single; leaves longer, 25–40 cm, not noticeably thickened below middle, thin, flat, flexible, frequently with coarse marginal threads; flower tube 8–10 mm deep. Sierra Madre occidental   *schidigera*, p. 119

Agave colimana
(Figs. 6.1, 6.2, 6.3, 6.4, 6.5, 6.6; Table 6.1)

Agave colimana Gentry, Cact. Succ. J. Am. 40: 212, 1968.
  Agave angustissiuma var. *ortgiesiana* Trel., U.S. Nat. Herb. Contr. 23: 141, 1920.
  Agave schidigera ortgiesiana Baker, Gard. Chron. n. ser. 7: 303, 1877?

Short-caulescent, monocarpic, multi-leaved, single, nonsurculose rosettes, 4–6 x 10–12 dm; leaves straight, linear, thin and flat above, 1–2.5 x 40–70 cm, slightly narrowed above the base and widest through the middle, smooth, green, with a narrow brown margin filiferating with fine, long brown threads; spine short, weak, 5–8 mm long, flattened or somewhat hollowed above, broadly rounded below, grayish brown to dark brown, decurrent into the leaf margin; inflorescence spicate; shaft 2–3 m tall, slender, straight flowering from ca. 1 m above the base; bracts dark brown, or purplish, linear-acicular, the lower exceeding the flowers in length, the upper somewhat shorter than the flowers; fresh flowers 40–50 mm long, not crowded, on forked pedicels 10–15

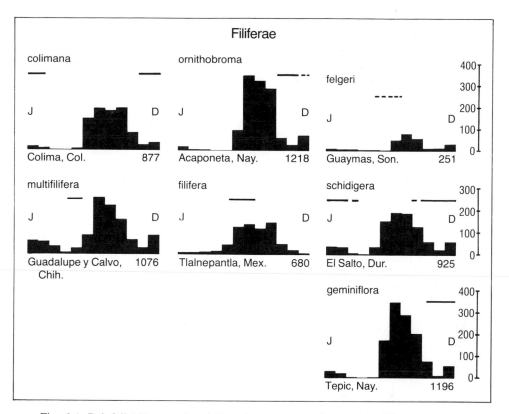

Fig. 6.1. Rainfall (silhouettes) and flowering perimeters (bars) of the Filiferae. Relevant meteorological stations with average annual rainfall in millimeters. Data from Atlas Climatológico de México (1939). Flowering seasons are based on herbarium specimens and field observations, supplemented by plants in cultivation.

mm long, the latter bearing 3 or 4 alternate, short, brownish, deltoid bracteoles ca. 2 mm long; ovary 14–20 mm long, greenish yellow, round; neck 4–7 mm long, slightly constricted; tube 9–17 mm long, narrow, slightly 6-furrowed, the insertion of the stamens not prominent and 1–2 mm below tepal sinuses; tepals pale yellow or lavender, recurved at anthesis, 14–19 x 4–6 mm, the outer tepals 1 mm longer than the inner, thin, obtuse and with thickened papillate tip; filaments 30–40 mm long, pale reddish, inserted 8–11 mm above base of tube; anthers yellow, 17–21 mm long; stigma not clavate, roundly trigonous with papillae decurrent on angles below tip; capsules 20–25 mm long, oblong, conspicuously constricted at base, beaked, the forked pedicel stout, 10–15 mm long; seeds small, hemispherical 2–3 x 3.5–4 mm, rugulose.

Type: *Gentry 18325* collected on sea cliffs at Manzanillo, Colima, Mexico, December 5, 1959, US. Duplicates, DES, MEXU.

Typical *Agave colimana* is distinctive with its elongate leaves and deep narrow flower tube. However, there are reductions in the length of tube, aligning it with *A. schidigera*, and its relatively large elongate leaves appear much like those of the montane *A. multifilifera*. Without both leaf and flower, specimens may be hard to separate and geographic location becomes important.

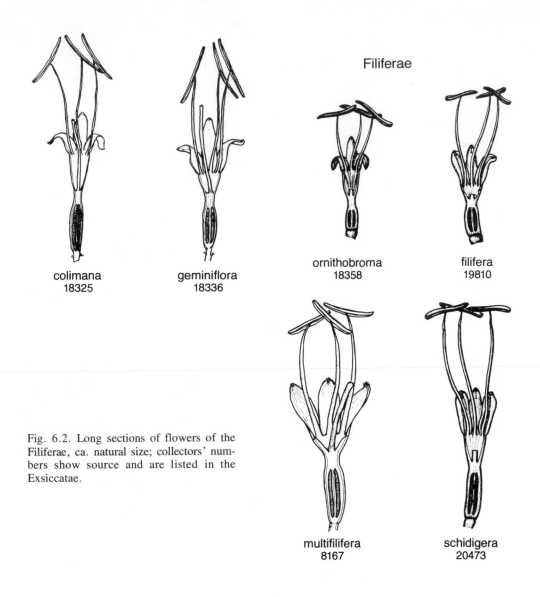

Filiferae

colimana
18325

geminiflora
18336

ornithobroma
18358

filifera
19810

Fig. 6.2. Long sections of flowers of the
Filiferae, ca. natural size; collectors' num-
bers show source and are listed in the
Exsiccatae.

multifilifera
8167

schidigera
20473

A. colimana is primarily a littoral species, scattered along the rocky sites by the sea in
Jalisco and Colima, as near Chamela and at Manzanillo, where sea spray at high tides may
fall upon the plants. It is also found inland on rocky heights (still coastal) in the tropical
deciduous forests. A large colony thrives on the gypsiferous cliffs 13–16 miles (20–26
km) southwest of Cd. Colima. Ecologically it is well separated from other members of the
group, although it may well be introgressed with coastal outlying populations of *A.
schidigera*, as some specimens appear to indicate. It makes a fine garden ornamental. I
found no local use of the plant, but the now vanished Indians may well have used it for
food and fiber.

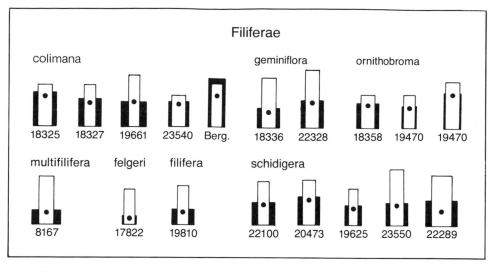

Fig. 6.3. Ideographs of Filiferae flowers; compare with Fig. 6.2. They depict relative size of tubes (black) to outer tepals (white columns) and filament insertions.

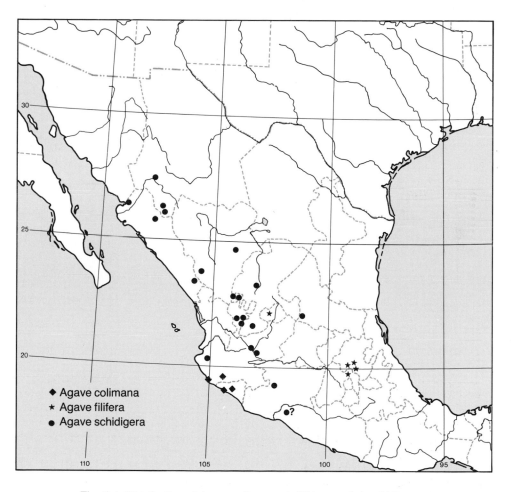

Fig. 6.4. Distribution of *Agave colimana*, *A. filifera*, and *A. schidigera*.

Table 6.1. Flower Measurements in the Filiferae (in mm)

Taxon & Locality	Ovary	Tube	Tepal	Filament Insertion & Length		Anther	Total Length	Coll. No.
colimana								
Manzanillo, Col.	20	12×7	15×6 & 14	11	38	17	47†	18325
	20	13×8	16×6 & 15	11	40	17	47†	18325
	18	10	16×4 & 15	8	30		42†	18327
Colima, Col.	18	10	15×4	8	30		40†	18327
	13	9×9	19×4.5	9–10	20	20	42†	19661
	14	9×9	18×4	9–10	40	21	40†	19661
Chamela, Jal.	14	9×5	11×5	8–9	22	13	34†	23540
	16	10×5	12×5	9–10	32	13	37†	23540
La Mórtola	20	18×6	15×4	12	28	18	52*	Berg.
	20	17×6	16×3	12	37	19	53*	Berg.
felgeri								
Bacachaca, Son.	16	3×5	12×4 & 11	3	20	13	30*	17822
	15	3×5	13×4 & 12	3	20	13	30*	17822
filifera								
Hunt. Bot. Gard.,	14	5.5×8	14×4	5.5	32	12	34†	19810
Cal.	13	6×8	14×4	6	32	12	33†	19810
geminiflora								
	17	6×8	18×4.5	6	38		41†	18336
Ocotillo, Nay.	16	7×8	18	7	40		42†	18336
	20	10×8	21×5 & 20	9–10	38	22	50†	22328
	20	11×8	21×5 & 20	10–11	46	22	52†	22328
ornithobroma								
	12	10	12×4.5	10	20		35†	18358
	12	9	12	9	19		34†	18358
Esquinapa, Sin.	12	7×6	10×4.5	7	23	12	29†	19470
	13	8×6	13×4	8	20	16	35†	19470
	17	13×6	17×5	13	19	20	47†	19470
multifilifera								
Sierra Charuco,	21	5×10	17×5	4–5	25	17	43†	10247
Son.	19	5×10	16×5	4–5	32	16	40†	10247
	21	5×10	17×5	4–5	37	17	42†	10247
schidigera								
Fresnillo, Zac.	12	8×8	15×4	7–8	38	14	38†	22100
	14	7×8	16×4	6–7	32	13	35†	22100
Palmito, Dur.	20	10×8	16×5	10	50	16	45†	20473
	20	10×7	16×4	9	46	17	46†	20473
Cañón del	12	6×6	13×3.5	5–6	20	13	30†	19625
Marquez, Jal.	13	8×6	13×3.5	7–8	20	13	33†	19625
Sierra Cuale, Jal.	12	8×8	20×5–6	8	38	21	41†	23550
	13	9×8	19×5	8	40	21	41†	23550
Cerros del Fuerte,	20	10×12	18×6	5–6	38	18	45†	22289
Sin.	20	9×11	18×7	5–6	37	18	44†	22289

* Measurements from dried flowers relaxed by boiling.

† Measurements from fresh or pickled flowers.

Fig. 6.5. *Agave colimana*, A–C, drawn from the type, *Gentry 18325*. Leaves and inflorescence section x ⅓, flower section slightly enlarged. *Agave geminiflora* D, flower section slightly enlarged.

Agave felgeri
(Figs. 6.1, 6.3, 6.7, 6.8; Table 6.1)

Agave felgeri Gentry, U.S. Dept. Agri. Hbk. 399: 60–62, 1972.

Small, green to yellow-green, surculose rosettes with rather few leaves forming rather closely cespitose clones; leaves 25–35 x 0.7–1.5 cm, linear to narrowly lanceolate, straight or falcate, plane above, convex below, widest at the base, faintly bud-printed and frequently with a pale median stripe, epidermis rugose or scabrous above, the margin with a narrow brown border weakly filiferous, smooth; spine small, weak, 8–15 mm long, gray; spikes 1.5–2.5 m tall, flowering in upper ¼ of shaft; bracts chartaceous, the lower

Fig. 6.6. *Agave colimana* on rocky beach above the sea near Chamela, Jalisco, February 14, 1975.

Fig. 6.7. A broad-leaved form of *Agave felgeri* at San Carlos Bay, Sonora. The fruiting spikes were carried a few meters to the beach for photographing.

ones 2–3 cm long, very narrow; pedicels strong, single or bifurcate, 2–5 mm long; old
dry flowers (relaxed) 25–30 mm long; ovary 12–14 mm long; tube 2–4 mm deep; tepals
10–12 mm long, about equal, linear; filaments 20–25 mm long inserted in orifice of tube;
capsules oblong or obovoid, narrowed toward the base, apiculate, 15–20 x 9–12 mm;
seeds irregular, 4–5 x 3 mm, thick, angulate, wrinkled, the hilar notch narrow and deep.

  Type: *Gentry 11343* Bahia San Carlos, District Guaymas, Sonora, Dec. 13, 1951 (l & cap) with coastal
shrub on volcanic agglomerate (Fig. 6.7), deposited in US, isotypes DES, MEXU, MICH.

  Agave felgeri strongly resembles *A. schottii* of the Parviflorae in habit, size, and leaf
form. However, because of the open shallow flower tube and long tepals, it is aligned with
the Filiferae. The narrow-leaved form on the Rancho Bachaca, 12–13 miles (about 20 km)
east of Navojoa, is similar though shorter-leaved, to *A. geminiflora*. Otherwise, *A. felgeri*

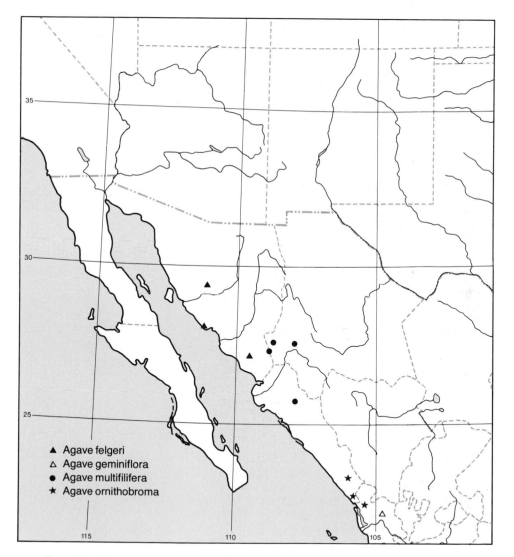

Fig. 6.8. Distribution of *Agave felgeri*, *A. geminiflora*, *A. multifilifera*, and *A.
ornithobroma*.

is distinctive among the Filiferae. Its closest geographic relative is the broad-flowered form of *A. schidigera* on the Cerros del Fuerte in northern coastal Sinaloa; a more robust nonsuckering colony.

The type-source colony of *A. felgeri* at San Carlos Bay was destroyed during the 1960s in clearing for recreational development at that tourist-attracting bay. It is known only from coastal Sonora but occurs many miles inland on the coastal plain (Fig. 6.8); see also the Exsiccatae. It occupies the driest habitat of all the Filiferae and is a rather rare plant.

At Rancho Bachaca (Mayo, meaning bad water) the people called this plant "mescalito" and said they did not use it. I found it there in one isolated colony upon one of the bald volcanic domes that characterize the place. Analyses of the rosettes yielded about 1 percent sapogenins, that from Rancho Bachaca consisting wholly of chlorogenin, while that from the Bahía San Carlos material contained 67 percent chlorogenin and 33 percent tigogenin.

Agave filifera
(Figs. 6.1, 6.2, 6.3, 6.4, 6.9; Table 6.1)

Agave filifera Salm., Hortus Dyckensis 1834: 309 and Lemaire, Ill. Hort. 7: pl. 243, 1860.

Agave filamentosa Salm, Bonplandia 7: 94, 1859.
Agave filamentosa var. *filamentosa* Baker, Gard. Chrom., n. ser. 7: 303, 1877.

Mature plants in large clumps of small green dense rosettes with many thick filiferous leaves. Leaves straight, lanceolate, 15–30 x 2–4 cm, broadest in middle, thickened and convex above and below from base to mid-blade, acuminate, white bud-printed, smooth, the margin finely filiferous; spine 1–2 cm long, grayish, flat above, rounded below; inflorescence a tapered spike 2–2.5 m tall, densely flowered through upper half of shaft, densely set with long-caudate recurving bracts; flowers 30–35 mm long, reddish, mostly geminate, ascending-outcurving on thick short pedicels; ovary 13–15 mm long, fusiform, angulate, with furrowed neck and tube; tube 5–6 mm deep, 8 mm wide, funnelform, the apex bulging with filament insertions; tepals equal, 14 x 4 mm, recurving, lanceolate, apiculate, the inner a little wider, with prominent keel; filaments 30–35 mm long, slender, reddish, with a high knee, inserted on rim of tube; anthers 7–12 mm long, reddish, regular, centric; capsules not seen, similar to *A. schidigera,* which see.

Neotype: Planche 243, Illustration Horticole VII, 1860.

Agave filifera is well known in European horticulture, but it is a clonal thing, stemming from one or two introductions there in the last century. I have seen no wild populations but know it from the Huntington Botanical Gardens and from the private garden of Dudley Gold in Cuernavaca, who collected it on Sierra Guadalupe north of Mexico City. It is closely related to *A. schidigera,* from which it is separable by its cespitose habit, its shorter thicker leaves, and smaller flowers with a shorter tube. It has also been collected in the state of Hidalgo; see Exsiccatae. Sanchez Mejorada reports (1978, p. 119) that it is found "sporadically at the rim of the Barranca (de Metztitlán) in the basaltic cliffs." Figure 6.9 is a reproduction of Lemaire's illustration, here designated as neotype. It shows a somewhat deeper tube and shorter tepals than the Huntington Botanical Gardens clone.

The variety *filamentosa* is reported to have leaves twice as long, without increase in width, and to bear suckers with denticulate leaves.

The question has been raised if *A. schidigera* Lem. is specifically distinct from *A. filifera* Lem. Scheidweiler (in Koch, Wochens. 1861) reduced the former to *A. filifera*

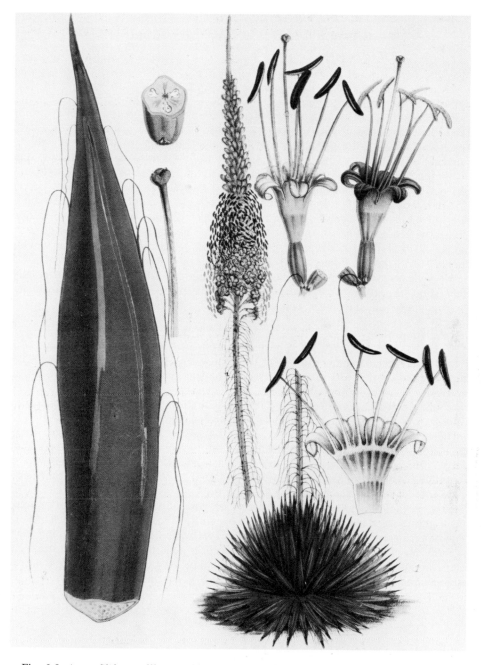

Fig. 6.9. *Agave filifera* as illustrated by Lemaire in Illustration Horticole VII, pl. 243, 1860.

var. *ornata*. Lemaire argued against this reduction (Illus. Hort. 12: n. ser., 49, 1865), saying his species, *schidigera*, differs from *A. filifera* in having a definite stem, and that a plant of Tonel's was just flowering and showed the flower is distinct from *A. filifera*. However all that was, I myself am doubtful if the two are specifically distinct, but we need more evidence of *A. filifera* from its natural habitat and more study.

Agave geminiflora
(Figs. 6.1, 6.2, 6.3, 6.8, 6.10, 6.11; Table 6.1)

Agave geminiflora (Tagl.) Ker-Gawler, Roy. Inst. Gr. Brit. J. Sci. & Arts 2: 86–90, 1817.
   Littaea geminiflora Tagliabue, Bib. Ital. 1: 100, 1816.
   Agave angustissima Engelm., St. Louis Acad. Trans. 3: 306, 1875.

Single, short-stemmed, monocarpic plants forming dense, green, rosettes 7–10 dm tall and somewhat broader, with many arching, narrow, flexible, unarmed leaves. Leaves green, linear, 45–60 x 0.6–0.8 cm, pliant, eventually arching, smooth, roundly convex above and below, abruptly acute, the margin finely filiferous or rarely nonfiliferous; spine 5–7 mm long, short-subulate, grayish; spike 4–6 m tall, stout at base (9–12 cm diam.), long tapering, flowering in upper 2/3–3/4; bracts narrow, ascending-reflexed; flowers 40–52 mm long, greenish below, flushed above with red or purple, mostly geminate on slender dichotomous pedicels 5–8 mm long; ovary 16–20 mm long, slender, fusiform-angulate, with grooved neck; tube 6–11 mm deep, narrowly funnelform, 6-grooved; tepals 18–21 x 4–5 mm, linear-lanceolate, recurving, red-flushed, the inner slightly shorter, distinctly keeled; filaments 35–46 mm long, colorless or reddish, inserted on rim of tube; anthers 20–22 mm long, excentric; capsules 18–20 x 9–10 mm, trigonous, oblong, short-pedicellate, short-beaked, tough, persistent; seeds 3–4 x 2–3 mm, lunate, thick, irregularly veined on faces, with expressed marginal wing.

Lectotype: Illustration of *"Littaea geminiflora"* Tagliabue in Biblioteca Italiana, 1816; see Fig. 6.10, here.

I reviewed the taxonomic history of this plant several years ago (Gentry, 1968). It is distinct from other members of the Filiferae by the combination of the following characters: relatively large simple stem, bearing innumerable, very narrow, pliant, smooth leaves, and large spike with relatively remote long flowers. Its closest relative is the novelty, *A. ornithobroma,* described and distinguished below.

The native habitat of this agave was long unknown. Ker-Gawler presumed it was from South America. Its discovery in Nayarit in 1951 has been described previously (Gentry, 1968). It is still known only in a small area of oak woodland, 2–10 miles (3–16 km) north of Ocotillo at elevations between 1,000 and 1,400 m. This area has an equable climate with around 1,000 mm of annual rainfall from summer to winter followed by a dry spring. *A. geminiflora* grows along the rocky arroyos and scattered with grass on rocky ground among the deciduous attractive oaks as well. It makes a long-lived handsome rosette ornamental for patios and gardens. In the 1950s a few plants lent special interest to the central plaza in Tepic, where they were placed as pedestal plants in large urns. It is rarely seen in the wild and in horticulture but should be very suitable for landscaping. As in other gardens, my plants flowered without leaving offsets and are gone.

Agave multifilifera
(Figs. 6.1, 6.2, 6.3, 6.8, 6.12, 6.13, 6.14; Table 6.1)

Agave multifilifera Gentry, U.S. Dept. Agri. Hbk. 399: 46–50, 1972.

Single, monocarpic, nonsurculose, green rosettes, about 1 m tall; 1.5 m broad, with a short trunk at maturity and 200 or more leaves; leaves 50–80 x 1.2–3.5 cm, linear-lanceolate, broadest at base, erectly ascending to declined, long-filiferous, plane above, slightly convex below, firm but pliant, smooth, light green, spine 1–1.5 cm long, subulate, flattened above, castaneous to aging gray; spike to 5 m tall, 18–21 cm thick with flowers, densely flowered from level of upright upper leaves; lower flowering bracts about

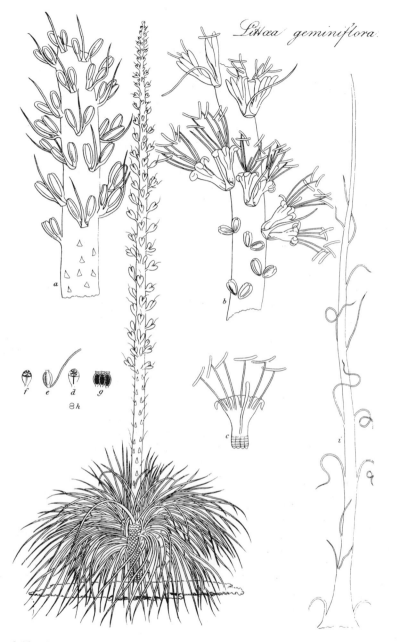

Fig. 6.10. *Agave geminiflora* as it appeared in the Biblioteca Italiana in 1816, here taken as type of species.

twice as long as flowers, grayish, subulate, the upper bracts much shorter; buds green with lavender luster; flowers pale waxy green with pink tinge on tepals, 40–43 mm long; ovary 20–21 mm long with constricted, faintly grooved neck 5 mm long; tube 5 mm long, 10 mm broad at apex, short-funnelform, the nectariferous innerliner about filling the tube; tepals 16–17 mm long, subequal, narrowly elliptic, some erect, some recurving at anthesis, apiculate, the papillate area broad and decurrent on the inner tepals, the outer tepals

Fig. 6.11. *Agave geminiflora* in the oak woodland near Ocotillo, Nayarit, in 1959.

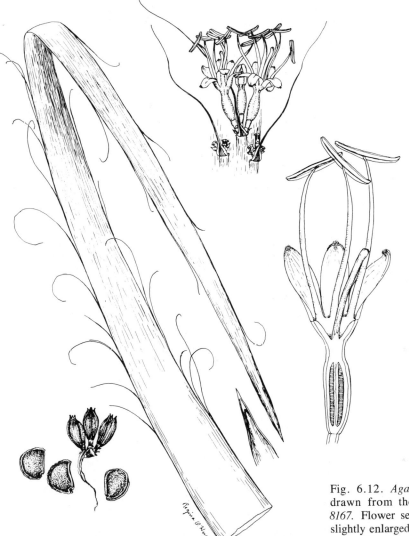

Fig. 6.12. *Agave multifilifera* drawn from the type, *Gentry 8167*. Flower section and spine slightly enlarged, seeds, x 2, the rest, x ½.

narrower, 5–5.5 mm wide, and also with a low keel; filaments 35–40 mm long, slender, red, elliptic in cross section, anthers yellow, 16–17 mm long; pistil slender, whitish; capsule 20–25 x 10–12 mm, slender ovate, apiculate; seeds small, 3.5–4.5 x 2.5–3 mm, the curved margin with a wrinkled or erose winglike edge, hilum notch abrupt.

Type: *Gentry 8167*, north rim of Arroyo Hondo, Sierra Charuco, Chihuahua, live plants collected April 1948, flowered at Murrieta, Calif., June 10–July 18, 1966, deposited in U.S. Natl. Herbarium No. 2558493 and No. 2558494. Isotypes, DES, MEXU.

Agave multifilifera with good soil makes one of the most robust plants in the Filiferae, developing definite stems and long, copiously filiferous leaves (Fig. 6.12). It is rivaled in size only by *A. colimana,* which is separable by its deep narrow flower tube (Figs. 6.5, 6.6). Depauperate plants on rocky sites may be taken for narrow-leaved forms of *A.*

schidigera. The distributions of the two species overlap in the northern Sierra Madre (see map, Fig. 6.4). Such cases are difficult to determine, as with so many other agaves.

Agave multifilifera ranges along the Sonora-Chihuahua border from central Chihuahua to northern Durango and northern Sinaloa at elevations between 1,400 and 2,200 m above sea level. Its habitat is the cliffs and rocky sites of the pine and oak forests, where snows are a common winter occurrence. The rainfall profile given for it in Fig. 6.1, as the Guadalupe y Calvo station, is probably wetter than average for its range.

The name "chahuiqui" was applied to the plant on Sierra Guicorichi and is obviously of Indian origin, but I have no notes on local uses. The normal lifespan of this species appears to be 18 to 25 years, as the young plants, assumed to be 3 to 5 years old when collected in 1948, did not flower until 18 and 21 years later. In 1966 two plants at the Huntington Bontanical Gardens also flowered, but starting in May, about 1 month earlier than my plant at Murrieta (Fig. 6.13). These were probably from my 1951 collection. It forms a handsome plant in the garden, the copious white threads among the leaves being curiously attractive, and because of the larger size would be more suitable than other Filiferae for landscaping grounds. It should be more cold hardy than other members of the Filiferae, as it grows in sites of the higher Chihuahua mountains where snow falls in winter. It showed no frost damage during the 15 years in Murrieta where winter lows reached 24°F (−4.5°C).

Fig. 6.13. *Agave multifilifera* in natural habitat on brecciated volcanic rock on Sierra Charuco with single plant of *A. vilmoriniana* at left center of field.

Fig. 6.14. An 18-year-old plant of *Agave multifilifera* in the Gentry garden.

Agave ornithobroma
(Figs. 6.1, 6.2, 6.3, 6.8, 6.15, 6.16; Table 6.1)

Agave ornithobroma Gentry, sp. nov.

Small, short-stemmed, single to cespitose, asymmetrical rosettes with few leaves, suckering sparingly at maturity; leaves 60–75 x 0.5–0.8 cm, narrowly linear, straight-ascending to curving, frequently to one side, or falcate, convex above from base to midleaf, convex from base to apex below, smooth, light green to reddish, short-acuminate, the margin reddish to white, filiferous; spine weak, 6–10 mm long, subulate, fraying with grounding; spike slender, 2.5–4 cm diam. near base, 2.5–3 m tall, narrowly bracteate, laxly flowered above the middle; pedicels dichotomous, 5–8 mm long, thickening with fruit; flowers geminate, 30–48 mm long, green with reddish flush, slender; ovary small, 12–17 mm long, fusiform-angulate; tube 9–13 x 5–6 mm, narrowly funnelform, triquetrous, finely grooved; tepals about equal, 10–17 x 4–5 mm, recurving, linear-lanceolate, obtuse, rather thin, the inner with low rounded keel; filaments 20–25 mm long, red to purplish, inserted with tepals on rim of tube; anthers 12–20 mm long, green to yellow, slender, somewhat excentric; pistil reddish; capsules and seeds not seen.

Type: *Gentry 18358*, 15–16 miles SE of Escuinapa, Sinaloa, along hwy. to Acaponeta; elev. 200–400 ft, on volcanic rock in tropical savanna (l, f), holotype *US*, isotypes DES.

Planta graminiforma, asymmetrica, primo singula, postremo surculosa, pauci-foliata. *Folia* viridia, linearia, albo-filifera, 60–75 cm longa, 5–8 mm lata, ascendentia vel recurvata plerumque secunda, basim versus convexa utrinque; *Spina terminali* parva, grisea, 6–10 mm longa, debili, evanescente. *Scapus* inflorescentia inclusiva 2.5–3 m altus virgatus. *Flores* 30–48 mm longi, geminati, gracile, virides vel purpurei. *Ovarium* parvum fusiforme, 12–17 mm longum. *Perianthii tubo* 9–13 mm longo, 5–6 mm lato, triquetro, 6-sulcato; segmentis ca. aequalibus 10–17 mm longis, recurvatis, lineari-lanceolatis, obtusis; *staminum* filimenta 20–25 mm longa, rubra vel purpurea, ad apicem tubi inserta; *antheris* viridibus vel luteis, 12–20 mm longis, gracilibus. *Capsula* et semina non vidi.

This grass-like agave is closely related to *A. geminiflora,* from which it is separable by its cespitose habit, small few-leaved rosettes, and slender spikes with small flowers. It grows in the lowland hot tropical savanna of southern Sinaloa and Nayarit, but limited to the dome-like outcroppings of massive volcanic rocks that are also characteristic of the region.

Notes made at the type locality state that plants offset only with maturity, the younger plants being single rosettes, the leaves commonly bowered over to one side, not radiately deployed as in *A. geminiflora.* The flowers are variously colored, green to reddish or purplish. Small parrots were observed on the spikes and had eaten many of the flowers and buds. It was necessary to search for undamaged spikes to obtain specimens. From this observation comes the specific name; from Greek, ornithos-bird and broma-food.

Fig. 6.15. *Agave ornithobroma* at the type locality ca. 16 miles (25 km) southeast of Escuinapa, Sinaloa, November 1961.

Fig. 6.16. Section of flowering spike of *Agave ornithobroma*.

Agave schidigera

(Figs. 6.1, 6.2, 6.3, 6.4, 6.17, 6.18; Table 6.1)

Agave schidigera Lem., Illus. Hort. 8: Plantes Recommendees, 1861, and Pl. 330, 1862.
  Agave vestita S. Wats., Amer. Acad. Proc. 25: 163, 1890.
  Agave disceptata Drum., Curt. Bot. Mag., Tab. 8451, 1912.
  Agave wrightii Drum, Curt. Bot. Mag., Tab. 8271, 1909.

Single, symmetrical, short-stemmed, monocarpic, freely seeding, variable plants with green, filiferous, bud-printed leaves. Leaves mostly 30–40 (–50) x 1.5–3 (–4) cm, linear, widest at or below the middle, green to grayish green or yellowish green, rarely reddish, smooth, straight, sometimes falcate, relatively thin, pliant, plane above, convex below toward base, acuminate, the margin brown to white, coarsely white filiferous; spine 5–16 (–20) mm long, brown to gray in age, flat above, round below, short-decurrent; spikes 2–3.5 m tall, slender, with narrow peduncular bracts, laxly flowered in upper half of shaft; flowers geminate on short dichotomous pedicels, 30–45 mm long, green to yellow or purplish; ovary 12–20 mm long, fusiform, angulate; tube 7–10 mm deep, narrowly funnelform, grooved; tepals equal, 13–20 mm long, linear, acute, recurving, greenish to yellow or purple-flushed, thin, over-lapping at base; filaments 35–50 mm long, lender, inserted on rim of tube, pale-colored to purple; anthers 13–21 mm long, very slender, frequently excentric; yellow to purplish; capsules variable in size and shape,

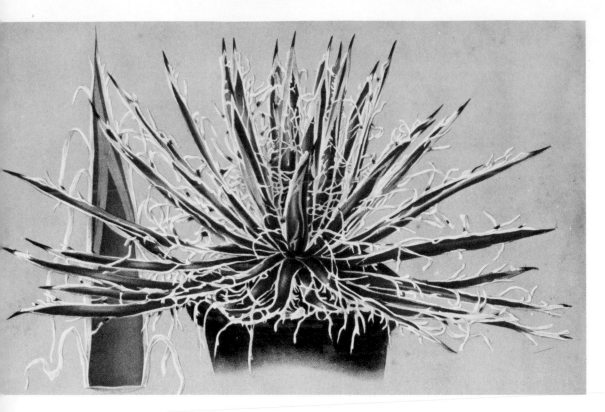

Fig. 6.17. *Agave schidigera* Lemaire, as illustrated in Illustration Horticole, pl. 330, 1860, which is here taken as lectotype.

ovoid 12 x 6–7 mm or oblong 18–20 x 6–9 mm, short-stipitate, usually beaked; seeds thick, deltoid 2 x 1.5 mm or lunate 3.5 x 2 mm, sharply cornered rather than winged, rugose or irregularly veined on faces.

Lectotype: Pl. 330, Illustration Horticole, 1860; here reproduced as Fig. 6.17.

A. schidigera appears closely related to *A. filifera,* but is separable by its non-surculose habit, longer, thinner, more pliable leaves, coarse rather than finely filiferous margins. The leaves are quite variable, but the variants are so randomized in various populations and so intergrading within localities that it does not appear practicable, at this stage, to delineate varieties. The isolated population on the lowland Cerros del Fuerte in northern Sinaloa is particularly distinct by its relatively broad bulging tubes with the filaments attached well below the orifice of the tubes.

The taxa placed in synonymy above do not appear sufficiently represented by specimens or origins for me to recognize. *A. wrightii* Drummond with its longer leaves, so beautifully illustrated in Tab. 8271, Curt. Bot. Mag. 1909, resembles *A. multifilifera* but differs from the latter in its shorter leaf, longer flower tube, and long filaments. The type at Kew consists of 2 incomplete leaves and a couple of withered flowers.

The source of *A. schidigera* Lemaire, the author gave as Verschaffelt's garden in Ghent, Belgium. It is known that Ghiesbreght collected for Verschaffelt in Mexico and visited Jalisco and Volcán Colima just previous to 1860. It seems reasonable, therefore, to accept Lemaire's name and his figure as type of an *Agave* species inhabiting the southern part of the Sierra Madre Occidental. This species is again well illustrated in Curt. Bot. Mag. Tab. 5641, 1867. Berger's figure 12 (Agaven, p. 75), which he calls *A. schidigera,* I place rather as *A. colimana,* after reviewing his specimen in US; see also ideograph, Fig. 6.3. This group is difficult to interpret and needs a thorough study.

Fig. 6.18. *Agave schidigera* in natural habitat on volcanic rock on Sierra Morones, Zacatecas.

Filiferae Exsiccatae

Agave colimana

COLIMA. Gentry 18325, Type, DES, MEXU, **US.** Manzanillo, 5 Dec. 1959; on cliffs facing the sea (l, f).

Gentry 18327, DES, MEXU, US. 18–20 miles SW of Cd. Colima, 5 Dec. 1959; elev. ca. 1,000 ft., gypsiferous rocky slopes and cliffs (l, f).

Jones 452, US. Manzanillo, June 1982.

Palmer 1070, MO, US. Manzanillo, Dec. 1890 (l, f, cap).

McVaugh & Koelz 1064, MICH. Ca. 11 miles SW of Colima on road to Manzanillo, 25 Nov. 1959; on gypsum & slate in deciduous woodland with *Juliana, Bursera, Caphalocerous,* etc., Elev. 400–500 m. (l, f).

McVaugh 26205, MICH. 19–21 km. S-SW of Colima, 9 Feb. 1975; deciduous woodland with *Cassia, Comocladia, Rhacoma,* etc., elev. 300–400 m (l, cap).

HORTICULTURE. *Gentry 19661,* DES. Cult. Desert Botanical Garden, Phoenix, Ariz. 26 March 1974 (l, f). Orig. from SW of Colima, 1961.

Rose 1132, 4128, US. Wash, Bot. Gard. 1897, 1898 (l, f, without data).

JALISCO. Gentry 10942, DES, MEXU, MICH,US. 8–10 miles SW of Autlan.

Gentry & Gentry 23540, DES, MEXU, MICH,US. Rancho Paraiso, S of Chamela, 14 Feb. 1975; elev. 20–40 feet, rocks above the sea (l, f, cap).

Agave felgeri

SONORA. *Felger 650,* ARIZ. Bahía San Carlos, 7 Nov. 1954.

Felger et al. 11456, 12121, ARIZ. Ensenada Grande (Bahía San Pedro), ca. ¼ mile inland at N end of bay, Nov. 1964 & 12 Jan. 1965 (l, cap).

Gentry 11343, Type, DES, MEXU, MICH, **US.** Bahía San Carlos, 13 Dec. 1951; coastal desert shrub on tuffaceous agglomerate (l, cap).

Gentry 11366, DES, MEXU, US. Bachaca, ca. 15–18 miles E of Navojoa, 19 Dec. 1951; thorn forest, open rock ledges. Same loc., later date, *Gentry 17822* (l, old f).

Gentry & Arquelles 11620. Rancho Grande, ca. 15 miles NW of Hermosillo, 19 Feb. 1952; dense clumps scattered along arroyo (cap.)

Kimnach 1965, HNT. San Carlos Bay, 3 Nov. 1977 (l, cap).

Agave filifera

AGUASCALIENTES. *Rzedowski & McVaugh 852*, MO. 2 km E of La Congoja, camino a San Jose de Gracia, 18 Oct. 1975; alt. 2,500 m. (l, cap).

HIDALGO. *Miranda 413*, MEXU. El Chico, 9 Junio, 1940.

Purpus s.n., US. Pachuca, on rocks, 1905. Cult. at La Mórtola, 1913.

Rose & Hough 4449, US. Mts. near Pachuca, 1 June 1899.

Rose & Hough 11482, US. Sierra de Pachuca, 23 Sep. 1906.

HORTICULTURE. *Gentry 19810, 23682*, DES. Huntington Botanical Gardens, July 2, 1962 & 14 July 1977 (l, f).

MEXICO. *Gentry & Gold 23384*, DES. Cult. transplant from E side of Sierra Guadalupe; elev. ca. 8,000 ft. ''Multileaved green suckering rosettes with bud-printed leaves.''

Agave geminiflora

NAYARIT. *Gentry & Gilly 10502, 10504*, DES, MEXU, MICH, US. 25 miles SE of Tepic along hwy. to Guadalajara, near Ocotillo, 24 May 1951; elev. 3,500–4,000 ft, oak woodland (l, cap, pl).

Gentry 18336, DES, MEXU, US. Ca. 2 miles N of Ocotillo, along old mule trail to Tepic, 7 Dec. 1959; elev. 3,500–4,000 ft., arroyos in oak woodland (l, f).

Gentry 22328, DES, MEXU, US. 2½ miles N of Ocotillo along old road to Tepic, 9 Oct. 1967; along arroyo stream and lateral arroyos in oak woodland (l, f, cap).

Gregg s.n., US. Near Ocotillo, direction of Tepic, in western Mexico (fid. Engelm., loc. cit.), May 1849 (l frags.). Type of *Agave angustissima*.

Agave multifilifera

CHIHUAHUA. *Gentry 8167*, DES, *US*, MICH. N. rim of Arroyo Honda, Sierra Charuco, Chihuahua-Sonora border, April 1948. Transplant flowered Gentry garden, Murrieta, June 10–July 18, 1966. Type (l, f, cap).

Gentry 10247, DES, MEXU, MICH, US. Same loc., 6–7 March 1951.

Gentry & Bye 23362, DES, MEXU, US. 1¾ miles SW of Quirare along road to La Bufa, Batopilas Barranca, 10 Oct. 1973.

SINALOA. *Gentry & Arguelles 18378*, DES, MEXU, US. Sierra Surotato, 13 Dec. 1959; alt. 6,000 ft., pine and oak forest.

Agave ornithobroma

NAYARIT. *Rose 3421*, US. Between Pedro Paulo (Pablo?) and San Blasito, 4 Aug. 1897.

SINALOA. Gentry 18358, Type, DES, MEXU, US. 15–16 miles SE of Escuinapa along hwy. to Acaponeta, 9 Dec. 1959 (l, f).

Gentry 19470, DES. Same loc., 5 Nov. 1961 (f, photo).

Kimnach & Lyons 721, HNT. 15 mi. E of Rosario on road to Matatan, 19 Jan. 1966 (fl'ed. Hunt. Bot. Gard.)

Rose 1625, US. Between Rosario and Colomas, 12 July 1897 (l, cap).

Agave schidigera

AGUASCALIENTES. *Rzedowski y McVaugh 852*, ENCB, MICH. 2 km al E de la Congoja, sobre el camino a San Jose de Gracia, 18 Oct. 1973; alt. 2,500 m, bosque de *Quercus* abierto (l, cap, wide form of leaf).

CHIHUAHUA. *Gentry s.n.*, DES. Ca. 10 miles SW of Nabogame, 11 Oct. 1959; alt. 5,000–6,000 ft., barranca (cap).

Gentry et al. 17990, DES, MEXU, US. Guadalupe y Calvo, 12 Oct. 1959; alt. 7,000–7,500 ft., rocky sun slope and cliff.

Hartman 556, GH, K, US. San Jose, 1892–93. ''Produces the best native beer of any century plant in Tarahumare land.''

DURANGO. *Gentry 8371*, DES, US. Canyon Cantero, Sierra Gamon, 21 Sep. 1948; elev. 7,000–8,000 ft., rocky promontory in oak woodland. Live pl. col., flowered Murrieta, Calif. spring 1958.

Gentry et al. 20473, DES, MEXU, US. 32 km. E of Palmito along Durango-Mazatlán road, 25 Sep. 1963; elev. 8,000–8,500 feet, sun side of volcanic cliffs (l, f).

Quero 47, MEXU. Km. 1147 along Durango-Mazatlán Hwy, 16 July 1967 (f).

GUERRERO. *Nelson 6989*, US. Growing on cliffs near La Junta, 4 April 1903. Identity uncertain.

HORTICULTURE. *Berger s.n.*, US. Cult. Mórtola Bot. Gard. 3 VIII 1909. (l, f).

Kellick 423–93, K. Cactus House, Hort. Kew, 3 Oct. 1911. Type *Agave disceptata*, Drum., Bot. Mag. Tab. 8451.

———— K, Cult. in Hort. Kew, 12 Nov. 1908. Type of *Agave wrightii* Drum. (l, f).

JALISCO. *Boutin & Brandt 2611*, HNT. Along road to Talpa de Allende past turnoff to Mascota, 27 Nov. 1968; elev. 4,500 ft. (seed col., fl'ed. Hunt. Bot. Gard. Oct. 1977).

Gentry & Gentry 23457, DES, MEXU, MICH, US. 20 miles SW of Valparaiso along road to Huejuquilla, 11 Jan. 1975; elev. ca. 6,700 feet, oak woodland.

Gentry & Gentry 23471, DES. MEXU, MICH, US. Rancho Cruz de las Flores, 6 miles SW of Villa

Guerrero, 17 Jan. 1975; elev. ca. 5,300 feet, rim of volcanic barranca with oaks (l, cap).

Gentry & Gentry 23479, DES, MEXU, MICH, US. 8–10 miles W of Bolaños on S end of Sierra de los Huicholes, 22 Jan. 1975; elev. ca. 7,500–8,000 feet; pine-oak forest on chalky volcanics.

Gentry & Gentry 23550, DES, MEXU, MICH, US. Sierra Cuale, ca. 5 miles NE of Tuito, 16 Feb. 1975; elev. ca. 3,000 feet, pine-oak forest with volcanic rocks (l, f, cap).

Pringle 2432, GH, MEXU, MO, US. Dry porphyritic ledges near Guadalajara, 8 Nov. 1889 (l, f, cap; type of *A. vestita* Wats.).

Rose & Painter 7463, US. Near Guadalajara, 30 Sep. 1903 (2 sheets, l, f, cap; f on one resembles *A. colimana*; slender, long tube, 8–12 mm).

MICHOACAN. *Gentry 10448,* DES, MEXU, MICH, US. Canyon del Marquez between Uruapan and Apatzingan, 12 May 1951; elev. ca. 500 m, hot canyon in volcanic rock with semi-arid short-tree forest.

Gentry et al. 19625, DES, US. Same loc., 10 Dec. 1961 (l, f, cap).

SAN LUIS POTOSI. *Gentry et al. 20464,* DES, MEXU, US. 12 miles SW of San Luis Potosí along road to Guadalajara, 23 Sep. 1963; elev. ca. 7,300 feet, volcanic arid slope.

Gentry 11520, DES, US. 14 miles SW of San Luis Potosí, 26 Jan. 1952 (photo. pl).

SINALOA. *Breedlove & Thorne 18533,* CAS, DES. Bufa de Surutato, 3 miles SE of Los Ornos, Mun. of Badiraguato, 4 Oct. 1970; elev. 7,200 feet (l, cap).

Gentry & Gilly 10522, DES, MEXU, MICH, US. 3 to 4 miles E of El Batel along hwy from Mazatlán to Durango, 28 May, 1951; elev. 6,000–6,500 feet, cliffs and talus.

Gentry 12561, DES, MICH, US. Ca. 8 miles N of Rio Fuerte in Cerros del Fuerte, 15–20 miles N of Los Mochis, 31 March 1953 (l, f).

Gentry 22289, DES, US. Same loc., 9 Dec. 1966. Transplant flowered Des. Bot. Gard. spring, 1975 (l, f, cap).

Kimnach & Lyons 654, HNT. Same loc., May 1978 (a good specimen with l & f).

ZACATECAS. *Gentry 18286,* DES, US. Sierra Morones, 30 Nov. 1959.

Gentry & Gentry 23463, DES, MEXU, MICH, US. 7.5 miles W of San Juan Capistrano, Zacatecas-Durango border, along road to Jesús María, 12 Jan. 1975. Elev. ca. 4,900 feet, thorn forest on volcanic tuff.

Gentry & Gentry 23464, DES, MEXU, MICH, US. 48 miles W of Huejuquilla along road to Sierra de los Huicholes, 12 Jan. 1975; elev. ca. 7,850 feet, pine-oak forest.

Gentry & Gentry 23496, DES, MEXU, MICH, US. Sierra Morones, 13 miles W of Jalpa, 24 Jan. 1975; elev. ca. 8,500 feet, open oak forest (l, photo).

7.

Group Marginatae

Plants small to medium with short stems or subcaulescent, single, surculose, or cespitose and forming large dense mounds of rosettes. Leaves stiff, mostly thick fleshy, armed with small to large teeth, rarely toothless, usually with continuous corneous margins from decurrent terminal spines to near base. Inflorescence generally spicate with flowers in twos or threes, rarely racemose with flowers clustered on numerous short lateral branches, the axis set with quick drying narrow bracts. Flowers small, variously colored; ovaries small, generally fusiform, round in cross section; tubes very short open funnelform; tepals much exceeding tube in length, ca. equal, narrow, linear to elliptic, acute or lightly cucullate at apex, involuting and clasping filaments at or after anthesis; filaments slender, inserted on or near rim of tube; anthers small, mostly centric; pistil exceeding stamens in post-anthesis. Capsules rather large, mostly oblong, beaked, freely seeding. From southern New Mexico and Texas south through Mexico to Guatemala.

Typical species: *Agave lophantha* Schiede, Linnaea 4: 581, 1829 and Schlechtendal, Linnaea 18: 413, 1844.

Taxonomic Comment

The Marginatae are characterized by the continuous horny margins on the leaves and by the small flowers with short open tubes with proportionally long tepals that clasp the filaments. This flower structure is quite constant throughout the group, (Figs. 7.1, 7.2). In fact, the flowers are so uniform as to be of little or only secondary use in the delineation of species. The leaves, however, occur in great variety of form, shape, size, color, and armature. The evolution of flower has been very conservative or inactive, while the leaf has greatly diversified and provides combinations of characters for taxonomic readout. The lack of corroborative flower characters leaves the taxonomist in an uncomfortable, if not untenable, position. However, one can find some supportive criteria for leaf character in growth habits and inflorescence structure. Calling upon these criteria, as well as upon the gross morphology of both leaf and flower, I have laid out the following relation for the Marginatae, justifying by rationale case by case.

There are some poorly defined specific boundaries in large variable intergrading complexes, such as that of *ghiesbreghtii-kerchovei* and *obscura-horrida*. Such complexes need closer study, bringing in more criteria in leaf anatomy, chemistry, cytology, and so on. The problems offer challenging studies and provoke evolutionary philosophy as well. For instance, why have flowers remained uniform, while leaves have diversified in the Marginatae?

My grouping of Marginatae (21 spp.) includes Berger's sections Pericamptagave (28 spp.) and Brachysolenagave (6 spp.). He separated the latter section as having more ample flower tubes, for which I see little evidence, but states they are otherwise like the Pericamptagave. He included *Agave utahensis,* but it has a quite different flower structure and I relegate it to a group of its own, the Urceolatae. Jacobi (1864: 499) called this group Maginatae Dentatae, while Baker (1888: 167) included many of these taxa in the Group Marginatae, an appropriate name I am following. Trelease (1920: 138–39) added five species, two of which I maintain below along with three new species. Many older names fall out as synonyms, or as *nomina dubiae* for lack of specimens or adequate descriptions to place them e.g. *Agave gilbeyi* Haage and Schmidt and *A. grandidentata* Jacobi.

Environment

The climate of the Marginatae is essentially tropical, generally being arid to semiarid (Fig. 7.3). However, a few taxa grow in moist mountainous areas, as southern forms of *A. lophantha, A. horrida,* and *A. huehueteca* of the *A. ghiesbreghtii* complex, but their particular habitats are virtually limited to the drier porous sites of lavas and other rock outcrops. *A. lechuguilla* is a true xerophyte, occupying some of the driest areas of the Chihuahuan Desert. Unlike the Amolae and many other agaves, there is no general restriction of the Marginatae to cliff sites. They are frequent on open slopes and plains. Perhaps the development of well-armed leaves has had some survival value, but toxic substances in the leaves is probably even more important for survival. *A. lechuguilla* is well known as poisonous to cattle (see p. 157). Hungry herds of goats over-run and over-graze the country side around Tehuacán, Puebla, but they do not feed on the foliage of the many Marginatae that abound in this region. It is unlikely that the spines and teeth alone would repel these ever-hungry destroyers of plant life. Inflorescences, however, are frequently eaten by goats and other cattle, when they can reach them. Considering the many ungulates that existed through the Tertiary Period (Scott, 1937), the browsers may have been a strong factor in promoting cloning so common in the Marginatae. Seeding forms that relied on seed dispersal for perpetuation would tend to be eliminated, leaving the field for the vegetative reproducers.

Chemistry

A scattering of samples in Table 7.2 indicates sapogenin content in the Marginatae. Sapogenins appear to be virtually lacking in several species, but present in others. Smilagenin is the most common compound. It is dominant in relatively high percentage in *Agave lechuguilla*. Combined with the fiber value of this species, the smilagenin would appear to have a latent commercial value and is discussed below under that species heading.

Marginatae

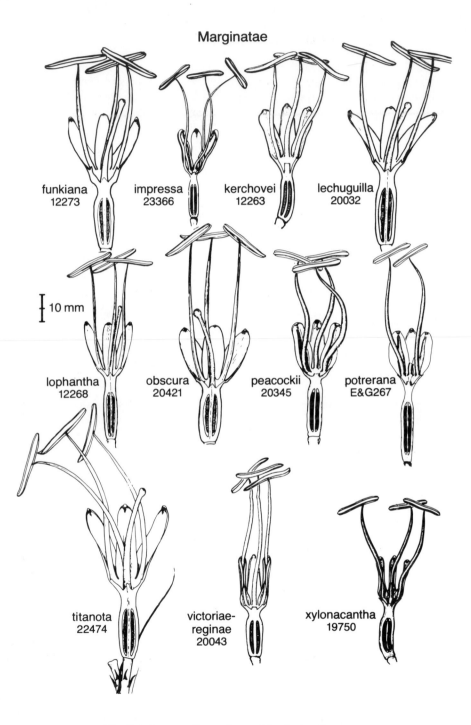

funkiana
12273

impressa
23366

kerchovei
12263

lechuguilla
20032

10 mm

lophantha
12268

obscura
20421

peacockii
20345

potrerana
E&G267

titanota
22474

victoriae-
reginae
20043

xylonacantha
19750

Fig. 7.1. Long sections of some Marginatae flowers.

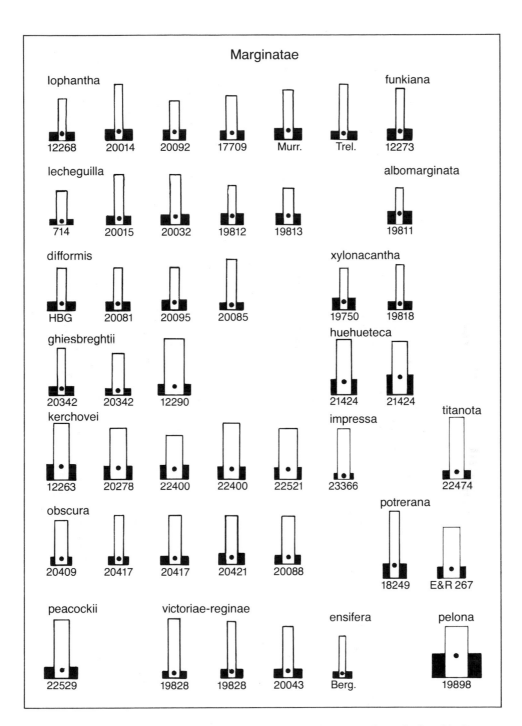

Fig. 7.2. Marginatae flower ideographs depicting relative proportions of tubes (black) to outer tepals (white column) and filament insertion in tube.

Key to Species of the Marginatae

1. Inflorescence racemose or subumbellate, the flowers in clusters on short lateral peduncles   2
1. Inflorescence not as above, spicate   3
2. Plants larger; leaves 60–140 x 9–14 cm; teeth very large (8–20 mm long), rarely lacking   *peacockii*, p. 165
2. Plants smaller; leaves 35–55 cm long, linear deltoid and 5–6 cm wide or lanceolate and 5–10 cm wide; larger teeth mostly 5–8 mm long
                              glomeruliflora, p. 142
3. Leaves with conspicuous white budprinting   4
3. Leaves without white bud-printing   5
4. Leaves 40–60 cm long with regular soft whitish teeth on whitish margins
                                       impressa, p. 146
4. Leaves 15–25 cm long, without teeth below apex   *victoriae-reginae*, p. 183
5. Plants large, leaves generally over 35 cm long, mostly much longer   6
5. Plants small, leaves generally less than 35 cm long, frequently deltoid   19
6. Leaves narrow, 10–20 times longer than wide, mostly 2–4 cm wide in mid-blade; spikes more laxly flowered
                                                                  7
6. Leaves broader, 4–10 times longer than wide, mostly 5–15 cm wide in mid-blade; spikes densely flowered, except *A. angustiarum*   12
7. Leaf margins straight, frequently detachable; teeth generally directed downward; leaves sometimes check-marked on back; leaves narrow, generally 2.5–4 cm wide in mid-blade   8
7. Leaf margins undulate or sinuous, firm; teeth variously curved, frequently upcurved or straight; leaves not check-marked on back; leaves wider, generally 4–6 cm wide in mid-blade   9
8. Leaves 25–60 cm long, yellowish green to pale green, margin gray to brownish   *lechuguilla*, p. 154
8. Leaves 100–125 cm long, grayish green; margin white
                                     albomarginata, p. 129
9. Teeth coarse, gray, irregular in size and spacing or the leaves toothless
                                        difformis, p. 135
9. Teeth slender, brown, more regular in size   10
10. Teeth irregular in size, spacing, and curvature   *ensifera*, p. 139
10. Teeth regular in size, spacing, and curvature   11
11. Margin sinuous to crenate, the teeth on low broad teats, occasionally 2 teeth per teat   *lophantha*, p. 157
11. Margin straight   *funkiana*, p. 139
12. Leaves toothless entirely or toothless on upper or lower portions of leaves 17
12. Leaves regularly with teeth through most or all their lengths   13
13. Leaves white glaucous, broadly lanceolate, only 2–4 times longer than wide; spine short and broad-based
                                        titanota, p. 176
13. Leaves green to glaucous green, deltoid to lanceolate, mostly 5–10 times longer than wide; spines more slender
                                                                14
14. Leaves with a wide corneous margin, markedly sinuous with large coarse teeth, frequently with 2 or more together on prominent marginal teats
                                   xylonacantha, p. 187
14. Leaves not as above, the margins about straight; teeth more regularly spaced   15
15. Leaves more linear, 40–100 cm long; teeth remote, mostly 3–6 cm or more apart; spine with open or narrow groove   *kerchovei*, p. 149
15. Leaves more deltoid or oblong, 30–50 cm long; teeth mostly proximal, 1–3 cm apart; spine grooved or hollowed to flat above   16
16. Leaves mostly deltoid, straight, rigid, frequently mottled olivaceous in color, rarely green, with few straight teeth or without teeth; spine grooved above
                                     triangularis, p. 181
16. Leaves broadly lanceolate, green, widest in middle; teeth more numerous, variously curved or flexed; spine grooved or ungrooved above
                                     ghiesbreghtii, p. 141

17. Leaves always without teeth, the margin smooth, white; spine white, 5–8 cm long, acicular; flowers red
pelona, p. 169

17. Leaves usually with teeth either on upper or lower portions of borders, rarely entirely toothless; margins and spines brown to gray; spines 3–5 cm long; flowers red or yellow   18

18. Flowers pale yellow; base of spine below deeply intruding into leaf flesh. Southern Mexico
angustiarum, p. 134

18. Flowers red or yellow; base of spine not intruding leaf flesh. Northern Mexico   *potrerana,* p. 172

19. Plants of 2 forms: juvenile with leaves 2–4 cm long, ovate to orbicular, and mature with leaves 25–35 cm long, deltoid, rigid (known only in hort.)
pumila, p. 174

19. Plants not as above   20

20. Leaves generally deltoid with teeth frequently lacking on upper margins; small forms of   *potrerana,* p. 172

20. Leaves broadly lanceolate to oblong, generally well armed throughout   21

21. Leaves broadly lanceolate, narrowed below the middle, 7–10 cm wide in mid-blade, small forms of
ghiesbreghtii, p. 141

21. Leaves more oblong, generally less than 7 cm wide in mid-blade   22

22. Plants single, leaves generally about 80 in a rosette; spikes relatively small, laxly flowered, 2–3 m tall; pedicels 4–10 mm long. Morelos and Puebla
horrida, p. 144

22. Plants single or cespitose; leaves more numerous, 100 or more in a rosette; spikes large, more densely flowered, 3–5 m tall; pedicels 1–2 mm long. San Luis Potosí, N. Puebla and adjacent Veracruz   *obscura,* p. 161

Agave albomarginata
(Figs. 7.2, 7.4; Table 7.1)

Agave albomarginata Gentry, sp. nov.

Freely suckering, subcaulescent, open, few-leaved, grayish green rosettes. Leaves long lance-linear, 100–125 cm long, 4 cm wide near base, 2.5 cm wide in middle, straightly ascending, nearly flat above, convex below, somewhat keeled toward base, grayish green, with a thin, white, corneous, somewhat friable margin, the distal third toothless; teeth in mid-blade white like the margin, 2–4 mm long, remote, 3–5 cm apart, thin, recurved, those toward base blunt, 1–2 cm apart; spine 1.5 cm long, subulate, gray with dark tip, roundly grooved above, thinly decurrent; inflorescence spicate, 4–6 m tall, slender, laxly flowered; flowers 35–40 mm long, greenish yellow, in two or threes on short, thick, bracteolate pedicels; ovary 18–22 mm long, fusiform, with thick grooved neck; tube 4–5 mm deep, 9 mm wide, openly spreading; tepals equal, 13–14 x 3 mm, pale yellow, strictly erect, clasping filaments; filaments to ca. 30 mm long, reddish, inserted on rim of tube; anthers 16–17 mm long, centric; capsule and seed unknown.

Type: *Gentry 19811, US,* DES. Huntington Botanical Gardens, San Marino, California, July 3, 1962; cultivated in several parts of the succulent garden.

Planta rhizomati-cespitosa, brevicaulis, 1.2–1.5 cm alta, gracilis; folia angusta pauca lineari-lanceolata, 1–1.25 m longa, 4 cm lata prope basim, 2–3 cm lata in medio, radialia, supra fere plana infra convexa grisei-viridia; margine albocorneo, aliquantum friabili parte tertia distali, inormi; dentibus in medio foliorum 2–4 mm longis, remotis, tenuibus obtusis; *spina terminali* 1.5 cm longa, grisea, subulata. *Inflorescentia* scapo incluso 4–6 m alta, spicata laxiflora; floribus viridi-luteis 35–40 mm longis, geminatis vel ternatis. *Ovarium* 18–22 mm longum, fusiforme. *Perianthii tubo* 4–5 mm longo, 9 mm lato; *segmentis* equalibus 13–14 mm longis, 3 mm latis, strictis, involutis filamenta amplectentibus. *Filamenta* ca. 30 mm longa, rubella, ad basim segmentorum inserta; *antheris* 16–17 mm longis. *Capsula* et semina non vidi.

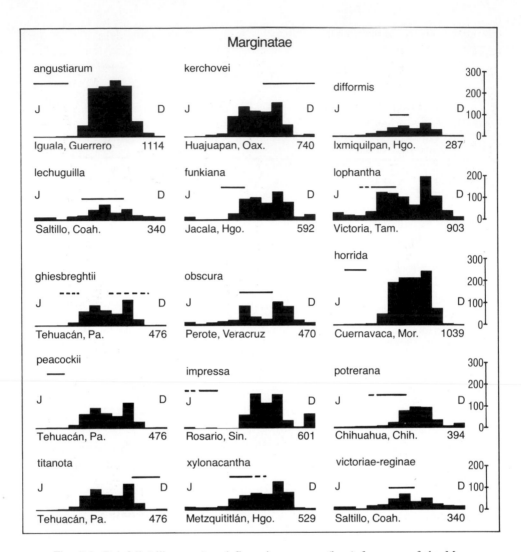

Fig. 7.3. Rainfall (silhouettes) and flowering seasons (bars) for some of the Marginatae. Relevant meteorological stations with average annual rainfall in millimeters. Data from Atlas Climatológico de México (1939). Flowering periods are based on herbarium specimens and field observations.

This plant is closely related to *Agave lechuguilla* and may be more cogently treated as a variety or subspecies. However, it is more extreme in morphologic characters than any other varieties I have noted in that polymorphic species: viz, the elongate leaf with conspicuous white friable margin, and the long toothless terminal third of the leaf. As with most other Marginatae, there is little notable distinction in the flowers. It is known only as a horticultural novelty cultivated in the Huntington Botanical Gardens. A more thorough taxonomic study of the *A. lechuguilla* complex should be undertaken by one of our promising young agave-oriented students.

Fig. 7.4. *Agave albomarginata,* an isotype in DES. From same plant as the holotype in US.

Table 7.1. Flower Measurements in the Marginatae (in mm)

Taxon & Locality	Ovary	Tube	Tepal	Filament Insertion & Length		Anther	Total Length	Coll. No.
albomarginata								
Hunt. Bot. Gard., Cal.	18	5×9	14×3 & 14	4	22	17	38†	19811
	18	5×9	13×3	4	22	16	35†	19811
	21	5	13	4	30	16	38†	19811
difformis								
Hunt. Bot. Gard., Cal.	18	3×10	16×4	3	24	17	37†	19826
	18	3×10	15×3	3	29	16	36†	19826
Sierra Alvarez, S.L.P.	21	4×9	15×4	3.5	32	16	41†	20095
	20	3.5×9	16×4	3	34	16	39†	20095
	19	2×8	18×4	2	44	18	39†	20095
Río Tula, Hgo.	16	3×8	15×4	3	46	15	35†	20081
ensifera								
Grimaldi, Italy	25	3×7	14×2	2–3	43	10	41*	Berg.
	25	2.5×7	14×2	2–2½	44	10	41*	Berg.
funkiana								
Jacala, Hgo.	20	3×10	19×5	3	35	20	42†	12273
	20	4×10	18×4	4	29	20	41†	12273
	21	4×9	18×3	4	34	20	43†	12273
Gentry Gard., Cal.	23	3.5×8	17×5	2–3	31	18	43†	s.n.
	23	3.5×8	17×3	2–3	38	18	42†	s.n.
ghiesbreghtii								
Tehuacán, Pue.	21	3×9	17×3	3	37	17	40†	20342
	14	2.5×9	15×5	2.5	32	14	32†	20342
	19	4×12	21×8	4	40	22	43†	12290
	16	3	19	4	32	21	40†	12290
glomeruliflora								
Chisos Mts., Tex.	15	8	22×7	7–8	45		45	840
Boquillas, Coah.	31	9×15	20×6 & 19	7	50	20	61†	11459
	35	9×15	20×4 & 18	6–7	55	21	64†	11459
horrida								
N of Cuernavaca, Mov.	18	5×9	16×4.5	4–5	36	16	39†	23368
	17	3×8	16×4	3	42	14	36†	23368
	18	4×8	15×6	3–4	38	14	36†	23368
	19	4×9	16×5	3–4	40	16	39†	23368
huehueteca								
E of Huehuetenango, Guat.	19	5×9	21×6	4–5	50	20	45†	21424
	17	7×9	19×6	7	45	19	44†	21424
	19	4×10	19×4 & 18	4	45	18	42†	23629
	18	7×10	18×4 & 17	6–6½	50	18	42†	23629
impressa								
E of Escuinapa, Sin.	17	1.5×7	17×5 & 16	1–2	38	16	35†	23366
	19	2×7	18×4.5	1–2	38	16	37†	23366
	18	2×7	17×4	1–2	40	16	36†	23366
kerchovei								
Huajuapán, Oax.	18	5×10	21×6	5	20	23	46†	12263
	19	6×10	20×6	6	37	22	45†	12263
	17	5×10	16×6.5	4.5	41	18	38†	22400
	21	4.5×9	20×6.5	4.5	46	20	45†	22400
Mitla, Oax.	20	4×10	17×7	4	42	18	40†	22521
lechuguilla								
Big Bend Nat. P., Tex.	11	2×9	11×4	1	25	11	24†	714
	12	2×8	12×4	1	24	10	25†	715

Table 7.1. cont.

Taxon & Locality	Ovary	Tube	Tepal	Filament Insertion & Length		Anther	Total Length	Coll. No.
lechuguilla (cont.)								
Sierra de la Silla, N.L.	23	3×9	17×4	2.5	35	19	44†	20015
	27	3×9	22×2	2.5	40		54†	20015
Saltillo, Coah.	23	4×10	17×6	3	34	18	42†	20032
	23	3.5×10	18×5	3	38	18	44†	20032
Hunt. Bot. Gard., Cal.	20	4×8	14×3	3	28	16	38†	19812
lophantha								
Cd. Mante, Tamp.	22	3×9	15×2.5	2–3	31	17	40†	12268
	21	2×8	14×3	1–2	33		38†	12268
Monterrey, N. L.	22	4×10	20×3	3	45	21	47†	20014
Río Verde, S.L.P.	18	3×8	14×4	2.5	42	15	34†	20092
Hunt. Bot. Gard., Cal.	20	3×	16	2	25	16	37†	17709
Gentry Gard., Cal.	22	3×10	18×4	2–3	37	18	43†	s.n.
Naulinco, Ver.	18	3×9	20×3	2–3	43	13	38*	Trel.
obscura								
Zacatepec Dist., Pue.	17	2.5×7	15×4.5	2–2.5	30	14	34†	20409
	17	3×7.5	16×5	3	45	14	35†	20409
Limón, Ver.	16	3×7	17×2.5	3	36	18	36†	20417
	16	3×10	16×5	2–3	40	17	35†	20417
Zacatepec Dist., Pue.	16	4×9	17×7	4	45	16	37†	20421
	17	4×9	17×5	3–4	50	17	37†	20421
Sierra Alvarez, S.L.P.	21	3×9	17×5	3	38	15	41†	20088
peacockii								
Tehuacán, Pue.	25	4×13	21×6 & 19	2–3	40	21	49†	22529
	25	4×13	20×6 & 18	2–4	44	22	48†	22529
pelona								
Cerro Quituni, Son.	21	8–9×16	18×8	8–9	38	16	47†	19898
potrerana								
	24	6×8	20×4	5	50	20	50*	18249
	28	5	24 & 23	5–6	50	21	55*	18249
Sierra Campana, Chih.	31	4×10	18×6	4	48	16	53†	267
	32	3×10	19×6	3	50	16	54†	267
	24	5×11	17×7	5	48	13	48†	269
	25	5.5	16.5×6	5	47	14	48†	269
titanota								
Tambor, Oax.	23	3×9	22×5	2–3	48	20	47†	22474
	22	4×10	22×5	3–4	55	22	48†	22474
victoriae-reginae								
Saltillo, Coah.	18	2×10	19×6	2	50	20	40†	20043
	24	3×10	18×5	3	45	21	46†	20044
Hunt. Bot. Gard., Cal.	21	3×8	20×3	3	47	18	44†	19828
xylonacantha								
Cd. Mante, Tamp.	27	5×10	16×5	4	42	18	47†	11516
Hunt. Bot. Gard., Cal.	20	4×9	15×	2–3	40	18	40†	19750
Gentry Gard., Cal.	22	4×10	20×3	3–4	38	22	46†	10117

* Measurements from dried flowers relaxed by boiling.

† Measurements from fresh or pickled flowers.

Agave angustiarum
(Figs. 7.3, 7.5, 7.6; Table 7.2)

Agave angustiarum Trel., U.S. Nat. Herb. Contr. 23: 139, 1920.

Single, subcaulescent, open, few-leaved rosettes. Leaves 50–80 x 6–7 cm, linear to lanceolate, long-acuminate, firm, straight, thick, green or glaucous, plane to concave above, convexly thickened below, the corneous margins continuous and characteristically without teeth below apex for ¼ to ⅓ the leaf length; teeth mostly 4–7 mm long, 1–3 cm apart, straight or down-slanted but commonly upcurved or flexed, flattened, the largest somewhat scattered, brown to gray; spine 3–4.5 cm long, acicular, well-rounded, narrowly grooved above, with a conspicuous median protrusion below, long decurrent to the upper teeth; spikes 2–4 m tall, slender; "flowers glaucous greenish white, 35–40 mm long" (Trelease); ovary 15 mm long, with narrow neck; tube 4–5 mm deep, 6–7 mm wide; tepals 16 x 3 mm, clasping filaments; filaments 40–45 mm long; "capsules 12–15 mm broad, 25 mm long; seeds 2–3 mm wide, 3–5 mm long" (Trelease).

Type: *Trelease 17 & 77,* MO. Iguala Canyon, 5 Mar. 1905, between Naranjo and Los Amates, Guerrero (4 sheets, l, f, cap, s).

Agave angustiarum is distinguished by its long narrow leaves, generally toothless along the long-tapering apex, and the protruding base of the spine. The teeth, although having a downward angle, frequently are curved upward. It is among the minority of the Marginatae in being a simple multiannual, a habit that appears to have some correlation with its cliff-dwelling predilection. Pruinose glaucous and non-glaucous individuals are

Fig. 7.5. *Agave angustiarum* in the type locality on limestone cliff in Iguala Canyon, Guerrero, August 1952.

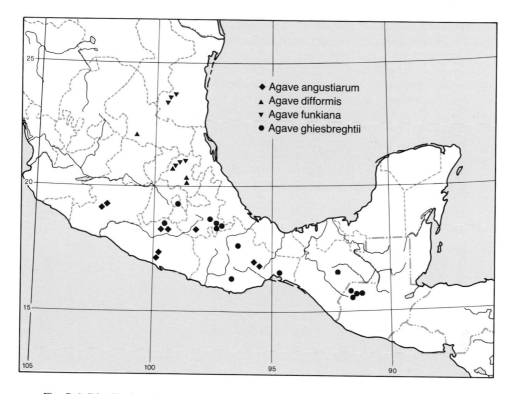

Fig. 7.6. Distribution of *Agave angustiarum, A. difformis, A. funkiana, A. ghiesbreghtii.*

found in the same local populations. The species may be confused with some forms of *A. kerchovei*.

The main geographic incidence of *A. angustriarum* appears to center in eastern Guerrero, but it occurs sporadically on rock abutments and cliffs in the tropical deciduous forest to the east and as far south as southern Oaxaca (Fig. 7.6). The elevation range is low to median, or from 600 to 1,500 m. I have observed it as far west as the Canyon del Marques in Michoacán, on the walls of that deep canyon, by the spanning spectacular bridge south of Uruapán. The fine fiber of the leaves is used locally for rope and twine.

Agave difformis
(Figs. 7.2, 7.3, 7.6, 7.7, 7.8; Tables 7.1, 7.2)

Agave difformis Berger, Die Agaven 1915: 95.
  A. haynaldii Tod. ? Hort. Panorm. 1: 88, 1876.

Rather vigorous, subcaulescent, freely suckering, open, variable rosettes 7–10 dm tall, 10–15 dm broad. Leaves 50–80 x 4–6 cm, polymorphic, green to yellow-green, stiffly ascending, thickly convex below, concave above, straight or falcate or sinuous, the margin predominantly light gray, firm or detachable, straight or undulate; teeth variable, generally 5–10 mm long and 2–3 cm apart, dark brown to gray, rarely double, sometimes with smaller interstitial teeth, or reduced or entirely lacking; spine stout, 1.5–3 cm long, conic-subulate, with a short open groove above, dark brown to gray; spike slender 3.5–5 m tall, flowering in upper ⅓ to ½ of the waxy glaucous shaft, the thin, narrow, chartaceous bracts deflected, those subtending the flower mostly longer than flowers; flower 30–40 mm long, color various, light green to yellow and pink; ovary 15–21 mm long waxy green, with short neck; tube 2.5–3.5 mm, spreading; tepals equal, 15–18 x 3–4 mm, pale yellow or tinged pink, mostly erect and clasping filaments at anthesis, sharply hooded at apex; filament pale to pinkish 35–45 mm long inserted on rim of tube; anthers 15–18

Table 7.2. Sapogenin Content in the Marginatae
(given in percentages on dry weight basis)

COLL. NO.	SOURCE LOCALITY	MONTH COLL.	PLANT PART	SAPOGENINS*				
				TOTAL	HEC.	SMIL.	TIG.	YUCC.
angustiarum								
5853	W of Iguala, Gro.	Aug.	leaf	0.2				
11984	Cañón Iguala, Gro.	Aug.	leaf	0.0				
12247	Camaron, Oax.	Sep.	leaf	0.0				
difformis								
	Hunt. Bot. Gard., Cal.		leaf	0.7		?		
funkiana								
12270	Jacala, Hgo.	Oct.	leaf	0.6		85		15
12273	Jacala, Hgo.	Oct.	leaf	0.6				
ghiesbreghtii								
12014	Taxco, Gro.	Aug.	leaf	0.0				
12107	Miahuatlan, Oax.	Sep.	leaf	0.9				
12195	Comitan, Chis.	Sep.	leaf	0.2				
12234	Tehuantepec, Oax.	Sep.	leaf	0.0				
12257	S. Mateo Cajones, Oax.	Sep.	leaf	0.0				
horrida								
12037–39	Cuernavaca, Mor.	Aug.	leaf	0.0				
impressa								
11487	Escuinapa, Sin.	Jan.	leaf	0.0				
kerchovei								
12263	Huajuapan, Oax.	Oct.	infl.	0.0				
lechuguilla								
11504	San Luis Potosí, S.L.P.	Jan.	leaf	0.23		90		
11521	Cañón Huasteca, N.L.	Jan.	leaf	0.63		100		
	Belle Grade, Flor.	Aug.	leaf	1.2–1.7		100		
	Belle Grade, Flor.	Aug.	bud	1.6		100		
	Belle Grade, Flor.	Aug.	bagas.	3.4–3.6		100		
lophantha								
	Hunt. Bot. Gard., Cal.	Oct.	leaf	0.0–1.3		?		
12268	Mante. Tamp.	Oct.	leaf	0.4	√		√	
victoriae-reginae								
	Kingston, Jamaica	Oct.	leaf	0.0				
	Hunt. Bot. Gard., Cal.	Oct.	leaf	0.0				
11525	Saltillo, Coah.	Jan.	leaf	0.0				
xylonacantha								
10117	Hunt. Bot. Gard., Cal.	Jan.	leaf	0.6		?		

SOURCE: Wall et al. (1954–61).

*Hec. = hecogenin; Smil. = smilogenin; Tig. = tigogenin; Yucc. = yuccogenin.

Fig. 7.7. *agave difformis* in native habitat near the Río Tula between Ixmiquilpan and Zimapán, Hidalgo.

mm long, yellow or bronze-colored, excentrally affixed; capsules 22–26 x 12–15 mm, oblong, strongly walled, transversely rugose, sessile, short-beaked, dark brown; seeds 5 x 3 mm, rather long crescentic, somewhat wavy on faces, the circumferential wing low, hilar notch small.

Type: *Berger 10-IX-1914*, K, *US*. La Mórtola, Cult. (l, infl).

This robust agave is characterized by polymorphic long ensiform leaves. Berger's name presumably refers to the presence or absence of teeth, which characterizes the clone he knew. Had he known the fuller variations to be found in wild populations, he might better have called it *polyformis*. His type specimens at Kew and the U.S. National Herbaria consist of two or three leaves with parts of an inflorescence, the leaves showing both toothless and armed margins, as I have observed in a single clone in the Huntington Botanical Gardens. *A. haynaldii* Tod. may belong to this complex, but without something more tangible than Todaro's handsome but idealized illustration, I hesitate to recognize its priority over Berger's later name, which is represented by herbarium specimens. One leaf collection *(Gentry 20081)* from the Río Tula has double teeth suggesting *A. xylonacantha*, but it lacks the sinuous margin and fine blade of that species, the margin being brittle and separable from the fleshy leaf, like *A. lechuguilla*. Berger stated (loc. cit.) that *A. difformis* had long been cultivated at La Mórtola and other European gardens. I found a large series of it in pots at Palermo in 1969, one of the forms being that of the sinuous leaf. When given room in the ground in warm climates it builds large clonal thickets.

Fig. 7.8. A sinuous leaf form of *Agave difformis* in the same locality, June 1962.

Native region

Agave difformis is native to the coarse lime-rocky soils on the desert side of the Sierra Madre Oriental in San Luis Potosí and Hidalgo, between elevations of 5,000 and 6,000 feet, (1,560 and 1,875 m); rainfall and flowering season are indicated in Fig. 7.3. The few wild collections available outline its native distribution (Fig. 7.6).

The population by Río Tula and highway 85 in Hidalgo is particularly rich with leaf variability (Figs. 7.7, 7.8). However, I did not observe the toothless form there. Such populations in the state of Hidalgo must have been visited by earlier European collectors, and their collections may have provided materials for some of the earlier names, but for which no herbarium specimens were prepared. Some of the forms in the Río Verde region of San Luis Potosí resemble those of *A. lechuguilla,* as though genes of the latter from neighboring populations had penetrated the generative cells of *A. difformis.* By the Río Tula there is also a form with sinuous leaves, suggesting a gathering of snakes, with their heads stuck in a common feeding jug with spiny tail stuck out in a vegetable discipline, as though to say, ''Don't bother us, we're busy!'' (Fig. 7.8).

In the National University Botanical Garden in Mexico City in 1965 they had a highly diversified collection growing, collected by Gomez-Pompa from near Ixmiquilpán. He gave its local name as ''xixi'' and stated that the macerated fiber was employed as soap.

Agave ensifera
(Fig. 7.2; Table 7.1)

Agave ensifera Jacobi, Nachtrage zu dem Versuch einer systematischen Ordnung der Agaveen 1: 138, 1868.
Agave lophantha var. *latifolia* Berger, Agaven 1915: 92.
Agave heteracantha Baker, Gard. Chron. (1): 369, fig. 59, 1877.

Cespitose, densely leaved rosettes; leaves ensiform, 50–60 x 4–5 cm, linear-lanceolate, leathery-fleshed, near the base 3.8–4 cm wide, 1.5–1.7 mm thick, strongly convex below and above, upward becoming concave to the apex, smooth and dark green with a clear light stripe 5–7 mm wide up the middle; margins with a narrow gray border 0.5–1 mm wide, closely set with light gray, mostly antrorsely curved teeth 4–6 mm long, 1–2 cm apart, interspersed with smaller teeth, altogether 30 to 40 teeth per side; spine short, 1–1.5 cm long, brown to gray, the basal groove above short and opening broadly with the decurrent border; spike 2–2.5 m tall, with deflected subulate bracts 6–10 cm long, the flower subtending bractlets shorter than the flowers; flowers 35–42 mm long, mostly in pairs, on pedicels 2–3 mm long, light green, the tepals light yellowish; ovary 20–24 mm long with constricted neck ca. 3 mm long, tube short, open, 2–3 mm long; tepals subequal, 14–17 mm long, linear, involute around filaments at anthesis, the outer overlapping the inner base; filaments slender, 40–45 mm long, inserted on rim of tube, anthers 16 mm long, yellow; pistil whitish, equaling anthers (description partly from Berger); capsules and seeds not known.

Lectotype: Berger, Chateau Grimaldi di Mórtola, Italy, 19, VI, 1909, *US* 1023791, 1023763.

The origin of this taxon is unknown. Berger stated it was common in culture along the Mediterranean Riviera. Jacobi (loc. cit.) stated that he found the plant growing in the Jardin des Plantes in Paris. I have recognized no living representatives. Berger's *Agave lophantha* var. *latifolia* represented by the type, US 1023795, cultivated at La Mórtola (Berger, June 1912), does not appear separable from *A. ensifera*. Both collections consist of leaf margins and well-pressed flowers. It appears related to the *lechuguilla-difformis* groups. A leaf margin on the lectotype is partly detached. It is recognized specifically here by its long narrow leaves with numerous teeth of two sizes regularly spaced along the narrow margins, and the short mostly weak spines. Perhaps it will some day be found and recognized more clearly among the wild populations in San Luis Potosí, Hidalgo, or farther south.

Agave funkiana
(Figs. 7.1, 7.2, 7.3, 7.6, 7.9; Tables 7.1, 7.2)

Agave funkiana Koch & Bouché, Wochens. Gart. Pflanz. Ver. Beford. Gart. Konig. Preuss. 3:47, 1860.
?A. haynaldii Tod.?, Hort. Bot. Panorm. 1: 88, 1876.

Small to medium, freely suckering, yellowish green to dark green, acaulescent, open, radial, self-fertile rosettes, 60–90 cm tall, 120–80 cm broad. Leaves linear, mostly 60–80 x 3.5–5.5 cm, firm, straight, or somewhat falcate, patulous, concave above, frequently with pale median stripe, convexly thickened below, the base broadly clasping, the corneous margin thin, brown to gray, nearly straight, firm, with regular slender teeth 3–5 mm long, 1–2.5 cm apart, mostly directed downward, a few small irregular interstitial teeth; spine 1–3 cm long, conic-subulate, with narrow to open groove above, brown to white; spike slender, 3.5–4.5 m tall, glaucous gray, laxly flowered in upper half of shaft, the small bracts above apparently caducous; flowers 40–45 mm long, pale glaucous

Fig. 7.9. *Agave funkiana* in the Desert Botanic Garden, 1977.

green, on dichotomous pedicels ca. 1 cm long; ovary 20–24 mm long, oblong-fusiform with constricted grooved neck; tube 3.5–4 mm deep, conspicuously grooved and knobby angled; tepals 18–19 mm long, at first spreading to ascending, appressed and clasping filaments after anthesis, linear, mucronate with a small hood; filaments 30–35 mm long, inserted on rim of tube, red or pink; anthers 20 mm long, yellow with pink flush toward tips; pistil red; capsules 2.5–3 x 1.5 cm, strong-walled, transversely wavy-ridged, light pruinose gray, abruptly apiculate; seeds 5–6 x 3.5–4.5 mm, thick, shape very variable, wavy-lined on faces, hilar notch large.

Neotype: *Gentry 12273*, ca. 43 miles S of Jacala, Hidalgo, along rt. 85, in open woods on limestone, alt. ca. 1,600 m, Oct. 12, 1852, live plants, grown at Murrieta, California, flowered May 1962.

Agave funkiana, obviously related to *A. lophantha,* is distinguished by its larger size, the regular form of its linear, slightly concave leaves with nearly straight fine margin, and numerous regular fine teeth. It is not well known in the field and the few individuals observed show little variation. However, some potted well-developed specimens in the botanical garden in Palermo, with broader leaves and coarser teeth, appeared to belong to this species. One of them was listed as *A. haynaldii* Tod. and may indeed have been a direct descendent of that taxon. In its natural range, both the conical bud and mature leaves are cut for the fine strong fiber. Trelease (1920) stated it was called ''ixtle de Jaumave.'' It has one of the best quality fibers in the short-leaved class, much more spinable than the coarse stiff fibers of *A. lechuguilla.*

A. funkiana readily established itself in the foreign gardens of western Europe and southwestern U.S. (Fig. 7.9). It tolerates the hot dry summers and winter lows of 20°F (−6.7°C) in the Desert Botanical Garden and has built up densely stacked rosettes in the 40 years since it was first planted. John Weber informed me that all the large clumps in the garden are from one introduction from Mrs. G. D. Webster's garden. These clumps flower and seed every year. No doubt they are good soil builders; single plants, however, are more suitable and attractive in the garden but require annual pruning of their many offsets.

Agave ghiesbreghtii
(Figs. 7.2, 7.3, 7.6, 7.10; Tables 7.1, 7.2)

Agave ghiesbreghtii Lem. ex Jacobi, Versuch System Ord. Agaveen, Hamb. Gart. & Blumenz. 20: 545, 1864.

Agave roezliana Baker, Gard. Chron. n. ser. 528, 1877.
Agave purpusorum Berger, Agaven 111, Fig. 25, 1915.
Agave huehueteca Standl. & Steyerm., Field Mus. Bot. 23: 4, 1943.

Short, open, few-leaved, copiously suckering, green, strongly armed rosettes. Leaves of mature plants generally broadly lanceolate, ovate or deltoid, 30–40 x 7–10 cm, narrowed above base and widest in middle, or more rarely broadly linear 35–38 x 5.5–6 cm, thick, rigid, straight or upcurving, dark green to light green, convex below, plane to

Fig. 7.10. *Agave ghiesbreghtii (A. huehueteca)* in characteristic clonal growth habit on a south-facing slope of Sierra Cuchumatanes, NE of Huehuetenango, Guatemala, January 1976.

slightly hollowed or guttered above, the acuminate apex with reduced teeth; corneous margin relatively narrow, brown; teeth moderate, the larger 5–8 (–10) mm long, 1–3 cm apart, frequently straight, sometimes curved upward or downward, brown to grayish; spine 2–4 cm long, subulate, brown to gray, shortly shallowly grooved above, rounded below; spike 3–4 m tall, densely flowered; flowers 40–50 mm long, greenish brown to purplish, paler within; ovary 16–20 mm long, cylindric, with neck at length constricting; tube 3–5 (–10?) mm long, broadly funnelform, grooved; tepals subequal, 15–21 x 8–5 mm, at first outcurving, later involute and filament clasping, broadly linear, apex rounded, thickly apiculate, the inner narrower and with broad low keel; filaments 35–45 mm long, inserted on rim of tube at base of tepals; anthers 15–22 mm long, brownish, centric; capsules and seed unknown.

Neotype: *Berger s.n., US.* Cult. Mórtola 16 July 1906 and 1909 (l, f.).

The perimeters of this taxon are not clear to me. Berger described the tube as 10 mm long and apparently used this character to split off his small section Brachysolenagave, which otherwise, he wrote, is like his Pericamptagave or the Marginatae. The suckering clones of *A. ghiesbreghtii* rarely flower and I failed to find an adequate series of flowering specimens from the richly diverse Marginatae about Tehuacán and elsewhere. The names I have put into synonymy above appear to be only some of the many forms or varieties that are common to a diverse species complex. I am unable to separate the robust Guatemalan *A. huehueteca* from the rest of *ghiesbreghtii* populations. The more moist Guatemalan climate appears to account for the larger size.

Agave ghiesbreghtii (fid. Berger, erroneously spelled *ghiesbrechtii*) is closely related to *A. kerchovei,* from which it is generally distinguished by its shorter broader leaves with more proximal smaller teeth on narrower darker corneous margins (compare Fig. 7.10 with Fig. 7.19). Some of the small-leaved forms may not be separable at sight from A. horrida and *A. obscura,* but the latter are usually separable by their larger, more closely spaced teeth which continue nearly to the base of the spine and by the flat ungrooved base of the spine.

A. ghiesbreghtii is frequently cultivated as a fence plant. In time, and when kept trimmed, it forms dense impenetrable obstacles up to 1 m tall and 2–3 m broad, as at Tlacotepec, Puebla, and Miahuatlán, Oaxaca. The leaves of such forms are usually linear or deltoid, stiff, and strongly armed (see *Gentry 12290* and *12107).*

Agave glomeruliflora
(Figs. 7.11, 7.12; Table 7.1)

Agave glomeruliflora (Engelm.) Berger, Agaven 95, 1915.
> *Agave heteracantha* var. *glomeruliflora* Engelm., Gard. Chron. n. ser. 19, Fig. 6, 1883, Coll. Works 325, date ?
> *Agave lechuguilla* f. *glomeruliflora* (Engelm.) Trel., U. S. Contr. Nat. Herb. 23: 136, 1920.
> *Agave chisosensis* C. H. Mull., Am. Midl. Nat. 21: 763, 1939.

Single or surculose or cespitose, small to medium, few-leaved, light green rosettes, with tall stout, racemose, freely seeding inflorescences. Leaves variable, linear 40–55 x 5–6 cm, or lanceolate to deltoid 36–50 x 7–10 cm, rigid, thick, concavo-convex, light green, the margin corneous, separating easily with drying, straight or mammillate; teeth 5–10 (–15) mm long, 1–3 cm apart, generally declined and flexed or curved, brown to grayish; spine stout 2.5–4 cm long, conical to subulate, broadly grooved above, rounded to sharply keeled below and protruding into leaf flesh, grayish with tip persisting brown; inflorescence racemose, 4–6 m tall, stout, the peduncle long, strongly bracteate with persistent deltoid bracts; flowers ca. 45 mm long, yellow, short-pedicellate, clustered or umbellate on stout numerous branches 4–10 or more cm long; ovary ca. 15 mm long; tube

Fig. 7.11. Lectotype of *Agave glomeruliflora* as it appeared in the Collected Works of G. Engelmann, 1911. "AGAVE HETERACANTHA (*Zucc.*), FORMA GLOMERULIFLORA.—The investigations of Dr. V. Havard of the medical service of the U.S. Army have brought out some interesting morphological facts in regard to this species, which may teach us a lesson about the variability of characters taken from the inflorescence of Agaves. The species (or the form which is known as *A. Poselgeri*) is abundant in the mountain regions of West Texas and along the Rio Grande between El Paso and Presidio, and has there usually narrow leaves ¾ to 1½ inch wide, and a foot or less long, and an inflorescence of geminate flowers, but occasionally some more vigorous specimens are found with much larger leaves over 2 inches wide; others bear the flowers in clusters instead of pairs, three to six, and even ten, in number, on stout, flattened peduncles, ½ to ¾ inch long, which seem to form an approach to the paniculate character. The figure...represents a cluster of ten capsules, mostly denuded of the remnants of the flowers, and the diagram shows their arrangement." The agglomerated capsules on short lateral peduncle are one of the key characteristics of this taxon.

7–8 mm long, funnelform; tepals 20–22 x 6–7 mm; filaments 40–50 mm long, inserted at bases of tepals; capsules 2.7–3.7 x 1.3–1.5 cm, thick-walled, oblong, short-stipitate or sessile, beaked; seeds 5–6 x 3.5–4.5 mm, crescentic to deltoid, glossy black, with wavy marginal wing.

Lectotype: Englemann's illustration in Gardener's Chronicle, 1883; or Neotype: **Sperry et al. 840,** US. Basin, Chicos Mountains, Brewster Co., Texas, 13 July 1937.

This taxon is obviously of hybrid origin and has been found in numerous situations reflecting morphological gradations between *A. lechuguilla* and several other species, such as *A. gracilipes, A. neomexicana,* and *A. havardiana* of the subgenus **Agave.** The genetics of such intergrades are unknown and I can but deal with the situations only by observation of the morphological phenotypes. We now have several other collections at hand besides the original on which Muller based his name. Engelmann and Berger left no specimens. I am continuing the earliest name to stand for all those cases of the racemose inflorescence seconded by the loose corneous margins of leaves intermediate in width between the wide leaves of the above cited species and *A. lechuguilla.* Many of these intermediates also reflect the down-slanted teeth on the straight margins of *A. lechuguilla.* Some of the small scattered populations appear to represent seed-viable perpetuating populations, as though we have a recent or on-going case of new species formation by hybridization. They present unusual opportunity for evolutionary studies. On the whole, I find Berger's name useful as recognition of this young dynamic taxon.

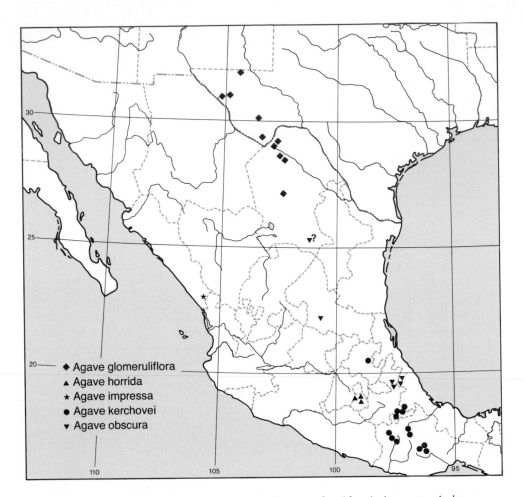

Fig. 7.12. Distribution of *Agave glomeruliflora, A. horrida, A. impressa, A. kerchovei,* and *A. obscura.*

Agave glomeruliflora inhabits the grassland slopes of the Big Bend mountains of Texas and the Sierra del Carmen of northern Coahuila. These grasslands include the oak-juniper communities at elevations around 4,000–5,000 feet (1,250–1,560 m) and the mesquite-*Larrea* grasslands as low as about 2,000 feet (620 m). The colonies of *A. glomeruliflora* fringe the upper northern limits of *A. lechuguilla* (Fig. 7.12).

Agave horrida

(Figs. 7.3, 7.12, 7.13; Tables 7.1, 7.2)

Agave horrida Lem. ex Jacobi, Versuch System Ord. Agaveen, Hamb. Gart. & Blumenz. 20: 546, 1864.

Small, single, compact, spreading, formidably armed, light green rosettes with 80 to 100 leaves at maturity. Leaves generally 18–35 x 4–7.5 cm, ovate to elliptic-lanceolate, short-acuminate, slightly narrowed above base, patulous, rigid thick fleshy, plane to hollowed above, convex below, yellowish green to green; margins thickly corneous, straight to sinuous between teeth; teeth large, generally 10–15 mm long, 5–10 mm apart, broadly flattened at base, straight to variously flexed, even hooked, light gray, continuing

Fig. 7.13. *Agave horrida* on the lava field northeast of Cuernavaca, Morelos, with a representative series of leaves.

to near base of spine, rarely much smaller; spine 2.5–4 cm long, half-round subdeltoid, flattened rather than grooved above, very pungent; spikes 2–2.5 m tall, slender; peduncle 1–1.5 m long, with deltoid-attenuate, narrow appressed bracts; flowers varicolored, geminate or single, 35–40 mm long, on slender pedicels 4–8 mm long; bractlets shorter than flowers; bracteoles 4–5 mm long; ovary 17–20 mm long, fusiform, with constricted, smooth or slightly grooved neck; tube 3–5 mm deep, 8–9 mm broad, short-funnelform; tepals 15–16 x 4–6 mm, equal, linear, obtuse with small hood, the inner broader, with prominent keel; filaments 36–42 mm long, slender regular, flattened, mostly reddish, inserted near rim of tube; anthers 14–16 mm long, centric, regular, bronze to red or purple; capsules 2–2.5 x 1.2–1.4 cm, ovoid, the valves thin, beaked, or strong jointed pedicels 10–12 mm long; seed not seen.

Neotype: *Pringle 8206, US.* Pedregal above Cuernavaca, Morelos, 2 Feb. 1899; elev. 7,500–8,000 feet, on S slope of lava beds (l, f).

Agave horrida is hardly separable from *A. obscura* and some small-leaved forms of *A. ghiesbreghtii.* Indeed, they are all closely related and whether *A. horrida* deserves a specific or subspecific status is a moot point. At best, it identifies the extensive populations on the volcanic rocks and mountains of Morelos. I have separated it on the differences outlined in the key to species; from *A. ghiesbreghtii* by the larger more numerous teeth continuing nearly to the base of the spine, and from *A. obscura* by the longer flower pedicels, somewhat fewer leaves in the rosette and the smaller more laxly flowered spikes. I have followed Berger's lead in identifying the Lemaire-Jacobi name of *horrida* with *Pringle 8206* collection of the Morelos lava field population. However, neither Jacobi nor Berger recognized *A. obscura* Schiede, it apparently having gone out of culture with the demise of Salm-Dyck's garden. The nonsuckering habits of both *A. obscura* and *A. horrida* do not lend these plants to long continuous horticulture unless special effort is made to renew them with seed propagation. Berger lists a number of horticultural synonyms and varieties, which are taxonomically meaningless and pointless to name here.

Lemaire apparently fancied *Agave horrida* as a repulsive plant, but I see it as an attractive, although formidable, bizarre succulent ornamental suitable for both garden and pot culture (Fig. 7.13).

Agave impressa
(Figs. 7.1, 7.2, 7.3, 7.12, 7.14, 7.15, 7.16; Tables 7.1, 7.2)

Agave impressa Gentry, sp. nov.

Small to medium, single, spreading, subcaulescent, yellow-green, openly spreading rosettes with racemose spike. Leaves 40–60 x 5–9 cm, linear to lanceolate, rigidly spreading, plane to hollowed above, convex below, thickly fleshy with viscid adhesive sap and little fiber, pale yellowish green with conspicuous white bud-printing on upper surface; margin flakey corneous, continuous, 2–3 mm wide, light to dark gray, straight to sinuous between teeth; teeth regular, mostly 3–5 mm long, 1–1.5 cm apart, straight or slightly curved, flattened, blunt, gray like margin; spine stout, 3–5 cm long, subulate, sharp to blunt at tip, flat and broad at base above, rarely channeled, rounded below and sometimes protruding from base into leaf flesh; inflorescence racemose, 2–3 m tall, erect, flowering from near base, the base closely bracteate; flower green in bud, yellow with anthers, 35–40 mm long, in twos and threes on slender dichotomous pedicels 2–2.5 cm long; ovary slender, 17–20 mm long, 4 mm in diameter, fusiform, slightly angled, with smooth unconstricted neck; tube short, spreading, 1.5–2 mm deep, 7 mm wide at apex; tepals equal, thin, 17–18 x 4–5 mm, ascending, partially recurved, linear-elliptic, apex

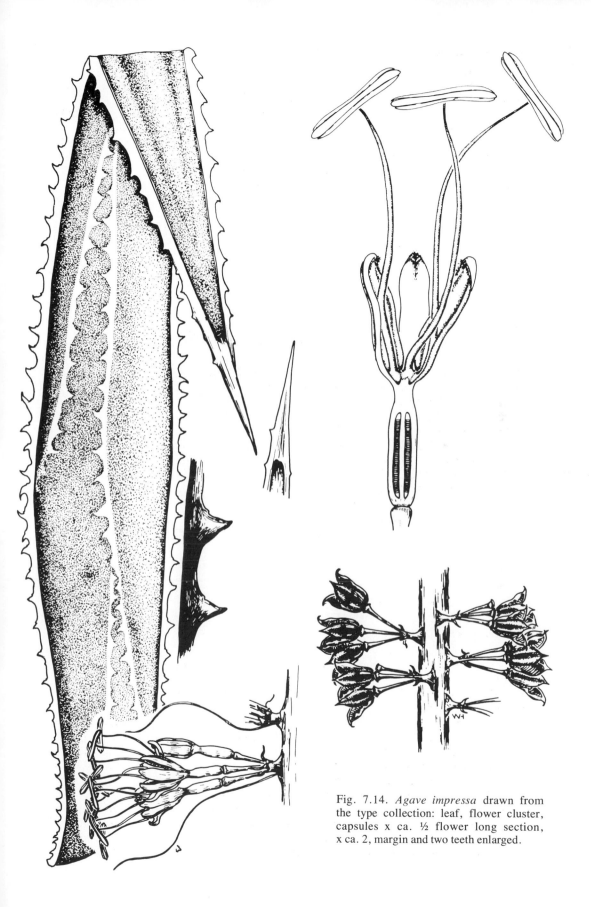

Fig. 7.14. *Agave impressa* drawn from the type collection: leaf, flower cluster, capsules x ca. ½ flower long section, x ca. 2, margin and two teeth enlarged.

Fig. 7.15. *Agave impressa* with young flowers in natural habitat on a volcanic ridge east of Escuinapa, Sinaloa, February 1974.

rounded, apiculate, the outer flat, the inner wider, involute, with broad low keel; filaments 35–40 mm long, very slender, flattened, white, inserted 1–2 mm above base; anthers 15–16 mm long, regular, centric, yellow; capsules 15–18 x 8–10 mm, sharply beaked; seeds unknown.

Type: *Gentry 23366, US,* isotypes ARIZ, DES, MEXU. 13 miles E of Escuinapa, Sinaloa, 16 Feb. 1974; elev. 500–1,000 feet, on volcanic rocks with *Plumeria, Bombax,* Bromeliads (l, f).

Planta non-surculosa, caule brevi, infra rosulam terminalem foliorum vivorum. *Rosula* 50–60 cm alta, 1 m lata. *Folia* lineari-lanceolata, 40–60 cm longa, 5–9 cm lata, viridia, utrinque diagonaliter albo-maculata, margine albo-corneo, continuo, dentato; *spina terminali* 3–5 cm longa, crassa, subulata, pungenti vel obtusa, plerumque non-canaliculata. *Inflorescentia* scapo incluso 2–3 m alta, racemosa, erecta, densiflora cylindrica; *floribus* 35–40 mm longis geminatis vel ternatis viridibus postea pallescentibus luteolis. *Ovarium* 17–20 mm longum, 4 mm latum, fusiforme. *Tubo* expanso vadoso 1.5–2 mm altitudine; *segmentis* equalibus 17–20 mm longis, 4–5 mm latis, proparte recurvatis, lineari-ellipticis involutis. *Filamenta* 35–40 mm longa, gracilia, ad tubi orificio inserta; *antherae* luteae 15–16 mm longae. *Capsulae* oblongae vel ovatae, 15–18 mm longae 8–10 mm latae; semina non vidi.

Agave impressa (so-named from Latin *impressas,* to print or impress, in reference to the bud-printed leaves) is a distinctive plant with no close relatives. The symmetrical rosettes with spreading, white-marked, light green, gracefully proportioned leaves with blunt teeth and spines, are quite unlike any other plant in the genus, (Figs. 7.14, 7.15, 7.16). It may be as closely related phylogenetically to *A. pedunculifera* of the Amolae

Fig. 7.16. Detail of leaves of *Agave impressa*.

Group, as to members of the Marginatae, as the long pedicellate flowers and their structure indicate. However, as a taxonomic convenience, I leave it with the Marginatae, until such time as studies in cytology, chemistry, etc., can teach us better.

With field glasses plants can be seen on the high cliffs from the highway about 12 miles (19 km) east of Escuinapa. One plant was observed with secondary shoots 4–5 dm long from the lower axils of an old rosette. This may have been due to an earlier injury. It appears to be a plant of the hot lowlands. Most of the small plants taken to gardens in Miami, Florida, and the Huntington Botanical Gardens did not grow. It presents difficulties in culture and is doubtless sensitive to frost.

I was given the name "masparillo" for this agave, which the informant in Tepic stated was abundant in the Sierra de los Huicholes toward Bolaños, Jalisco. The viscid leaves are used as a paste for curing wounds in horses. However, I failed to find the plant on the Sierra Madre west of Bolaños. Analysis of the leaves showed no sapogenin (Table 7.2).

Agave kerchovei
(Figs. 7.1, 7.2 7.3, 7.12, 7.17, 7.18, 7.19, 7.20; Tables 7.1, 7.2)

Agave kerchovei Lem., Illus. Hort. 11:64, 1864.
> *Agave convallis* Trel., U.S. Nat. Mus. Contr. 23: 138, 1920.
> *Agave dissimulans* Trel., ibid.
> *Agave inopinabilis* Trel., ibid.
> *Agave expatriata* Rose, Mo. Bot. Gard. Rep. 11: 82, 1900.
> *Agave noli-tangere* Berger, Agaven p. 103, 1915.

Medium size, single or cespitose, short-stemmed, light green, well-armed, openly spreading rosettes with 80–100 or more leaves at maturity, commonly branching from

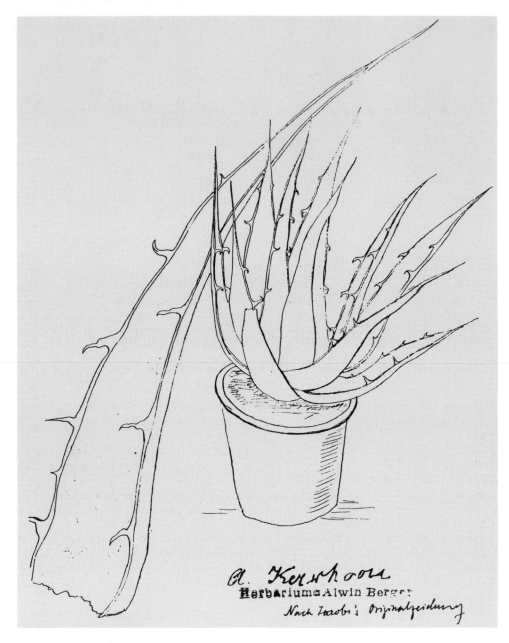

Fig. 7.17. Lectotype of *Agave kerchovei;* copy of a sketch made by Jacobi in U.S. National Herbarium.

lower leaf axils. Leaves 40–100 (–125) x 5–12 cm, generally lanceolate, the long-acuminate apex usually toothless, straight to slightly curved, thick at base, rigid, plane to hollowed above, convex below, light yellowish green to green, rarely pruinose, the margin generally heavy, continuous, straight, rarely toothless; teeth generally large, variable, the larger 8–15 mm long (from margin), remote, 2–5 (–7) cm apart, smaller teeth irregularly occurring, straight to variously curved, broadly flattened, gray; spine stout, 3–6 cm long, brown to gray, deeply channeled above, narrowly or broadly, nearly to tip,

Fig. 7.18. *Agave kerchovei* in natural habitat on the volcanic hills east of Mitla, Oaxaca, October 1967.

broadly rounded below; inflorescence spicate, 2.5–5 m tall, thickly flowered in upper ½–⅔ of shaft, the peduncle set with thin narrow bracts, quickly drying; flowers greenish to purplish, 38–46 mm long; ovary 18–21 mm long, fusiform, constricted in neck; tube openly spreading, 4–6 mm deep, 9–10 mm wide at apex, lightly grooved; tepals sub-equal, 15–20 x 6–7.5 mm, green to reddish, linear, acute, at first erect to spreading, clasping filaments after anthesis, the inner with low broad keel; filaments 40–50 mm long, paler than tepals, inserted on rim of tube; anthers 18–20 mm long, centric, yellow to reddish bronze; capsules long oblong, 2.5–3.5 x 1–1.2 cm, strongly walled, persistent, beaked or simply rounded; seed 3.5–4 x 2.5–3 mm, thick, lachrymiform to crescentic, smoothly surfaced, smoothly winged, glassy black.

Lectotype: Jacobi's sketch as traced by Berger in US, here reproduced as Fig. 7.17.

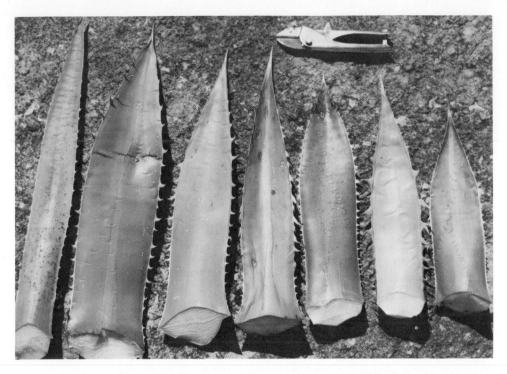

Fig. 7.19. Leaf series of *Agave kerchovei* collected near Mitla, Oaxaca.

Relationship

   Agave kerchovei is among the more robust of the Marginatae group. The typical form is easily distinguished by the long lanceolate leaves prominently armed with large variable teeth, remotely spaced, usually, but not always, lacking along the long acuminate apex. The stem is thick, to 3–5 dm tall. The dark-colored flowers have rather wide tepals, which are late in clasping the filaments. It is apparently closely related to *A. ghiesbreghtii,* the two cohabit over much of southern Puebla and Oaxaca, and intermediate forms are hard or impossible to distinguish. Along the Puebla-Oaxaca highway, 13 to 14 miles (about 20 km) south of Huajuapán, a large vigorous form occurs with or without teeth *(Gentry, Barclay & Arguelles 20278).* Considerable variation exists in nearly all populations observed. This includes habit, whether single or cespitose, as well as numerous differences in size, shape, and armature of leaves (Figs. 7.19, 7.20).

   A. kerchovei occurs frequently with *A. marmorata,* a distinctive species in the subgenus **Agave** (Fig. 7.29). Both have the same blooming period. Morphological combinations of these two species were observed west of Tehuacán. These are represented by, a *kerchovei* form with wide floppy leaves, and *marmorata* forms with the straight margins and large hooked teeth of *kerchovei.* One plant had a racemose inflorescence (an intermediate type between the spicate **Littaea** and the umbellate **Agave**) with short lateral peduncles over the broad *marmorata* type of leaves. Such examples suggest intersubgeneric crosses and a mode of origin for the distinctive *A. peacockii* in the same locality, which has a racemose inflorescence (Fig. 7.30). In any case, these intergrading forms leave the monographer wondering what really constitutes close relationship in this nefarious genus.

Fig. 7.20. A rosette of *Agave kerchovei* showing axillary budding, near Mitla, Oaxaca, October 1967.

Range and habitat

A. kerchovei ranges from central Hidalgo to southern Oaxaca at elevations generally between 4,500 and 6,000 feet (1,400 and 1,875 m). Although a highland plant, its varied habitats are semi-arid, being in the rain shadow of the Sierra Madre Oriental. Annual rainfall ranges from around 450 to 800 mm, with an extended dry period from October to June (Fig. 7.1). It grows on valley alluvium, coarse-soil detrital slopes, and steep rocky slopes. The plant communities are generally a semi-open short-tree woodland with a rich assortment of succulent Cactaceae, Agavaccac, Bromeliaceae, and sarcophytic trees and shrubs.

The region has long been inhabited and many of the valleys have a denuded aspected of over-use by man and his animals. *Agave kerchovei* is among the minority of plants that persist on eroded areas and there are striking examples of their rosettes perches on remnant pinnacles of soil, as west of Tehuacán. The flowers are eaten by people, but the leaves are apparently inedible to livestock. I observed no general use of the species. At Mitla and Huajuapán the plant was called "rabo de león" (lion slobber). Near Tehuacán it was called "cacaya" and "maguey pichomel."

Agave lechuguilla
(Figs. 7.1, 7.2, 7.3, 7.21, 7.22; Tables 7.1, 7.2)

Agave lechuguilla Torr., U.S. & Mex. Bound. Bot. 1859: 213.
>   *A. poselgeri* Salm., Bonplandia 7: 92, 1859.
>   *A. multilineata* Baker, Handb. Amaryllideae 1888: 168.
>   *A. heteracantha* Hort.

Small, widely suckering, rather open, few-leaved rosettes, mostly 30–50 x 40–60 cm, with yellow or reddish flowers, freely seeding; leaves generally 25–50 x 2.5–4 cm, linear lanceolate, light green to yellow green, mostly ascending to erect, sometimes falcately spreading, concave above, deeply convex below, sometimes green check-marked, thick, stiff, the margin straight continuous, light brown to gray, easily separable from dry leaf; teeth typically deflected, regular 2–5 mm long, brown or mostly light gray and weak-friable, mostly 1.5–3 cm apart, 8 to 20 on a side; spine strong, conical to subulate, 1.5–4 cm long, grayish, the short groove above at base open or closed; spike 2.5–3.5 m tall, the shaft generally glaucous, the flowers short-pedicellate in twos or threes, rarely on longer (2–15 cm) paniculate several-many-flowered ascending laterals; flowers 30–45 mm long, yellow or frequently tinged with red or purple; ovary 15–22 mm long, fusiform, roundly angled, constricted at neck; tube 2.5–4 mm long, shallow, open; tepals subequal, linear, 13–20 mm long, acending, involute around filaments, the outer widely over-lapping the inner at base, thickly hooded; filaments 25–40 mm long, spreading; anthers 15–20 mm long; capsules oblong to pyriform, 18–25 x 11–18 mm, abruptly very short pedicellate or sessile, rounded and short-beaked at apex, glaucous; seeds 4.5–6 x 3.5–4.5 mm with small hilar notch and low fluted wing around curved side.

Type: *Charles Wright 682*, Western Texas at El Paso, May–Oct. 1849 US 125459, (l, cap).

The abundant *Agave lechuguilla* is usually easily recognizable by its widely suckering habit and narrow leaves with down-slanted teeth on straight margins. The margins are easily detached on dried specimens. Vertical dark green dash marks may be present on the under or abaxial leaf surface, but they are frequently absent and should not be taken as species markers. In localities where *A. lechuguilla* grows together with *A. lophantha* there are intermediate forms, which are assignable as hybrids. Such situations were encountered on the mountain passes through the Sierra Madre Oriental, as along the road from Saltillo to Monterrey and from Galeana to Linares.

Fortunately, *A. lechuguilla* was described with a type specimen, collected early in the botanical exploration of the Southwest, and the species is unusually well founded for an *Agave*. Wright's 1849 specimen leaves no doubt regarding the application of Torrey's name. The name was misspelled, but under Article 73 of the International Rules of Nomenclature, such orthographic errors may be corrected. European botanists have generally not recognized the name of *Agave lechuguilla*, applying instead other names without representative specimens and of questionable interpretations or brief descriptions, such as *A. poselgeri*, and *A. heteracantha*. Berger applied the name of *A. lophantha* to the *lechuguilla* complex, because he did not know the real *lophantha* that Schiede collected and named. This will be clarified under my following account of *Agave lophantha*. Earlier botanists, up to and including Berger, paid little attention to name priority, usually not bothering to give the year of author citations.

Berger's six varieties under *A. lophantha* (1915: 92–94), and Jacobsen and Rowley's (1973: 4) recent transfer of them under *A. univittata*, would probably be assignable to varieties of *A. lechuguilla*, whenever herbarium specimens are prepared of them.

Distribution and habitat

Agave lechuguilla has one of the most extensive ranges of the agaves (Fig. 7.22), and the number of individual rosettes probably exceed those of all other native agaves. It is

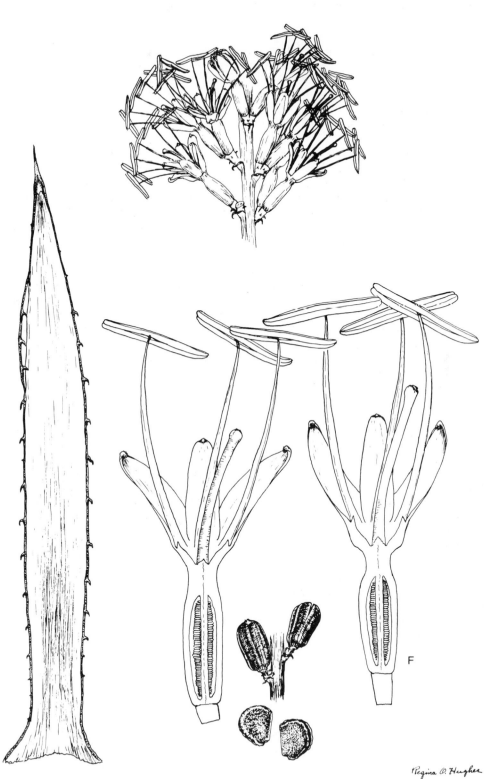

Fig. 7.21. *Agave lechuguilla*. Leaf, flower cluster, and capsules, ca. x ½; flower section x 1.2 and seed enlarged. F, *Agave funkiana* flower section for comparison.

abundant from southeastern New Mexico and southwestern Texas southward through the Chihuahuan Desert to the states of Mexico and Hidalgo. It is a characteristic succulent component of the varied plant communities of the Chihuahuan Desert between 3,000 and 7,500 feet (950 and 2,300 m) elevations, and has been collected at near 1,500 feet elevation in Uvalde County, Texas. It is partial to limestone soils, either on primary sedimentaries or on the derived "caliche," which is a leached-out and subsoil redeposit type of lime over much of the arid west on mountain slopes and bajadas. *A. lechuguilla* is frequently lacking on volcanic areas. The annual rainfall of the lechuguilla area generally ranges between 300 and 500 mm, but there are local exceptions and some differences in rainfall regimes (Fig. 7.3).

Uses

Agave lechuguilla is the principal source of the hard fibers known as "istle" or "ixtle," used for rope, twine, and other artifacts. In the export trade it is also known as "Tampico fibre" and is still used for manufacturing brushes in the United States and Europe. The industry is maintained by the hand labor of country villagers and rancheros who live in the lechuguilla areas. Both the mature leaves and the immature leaves in the

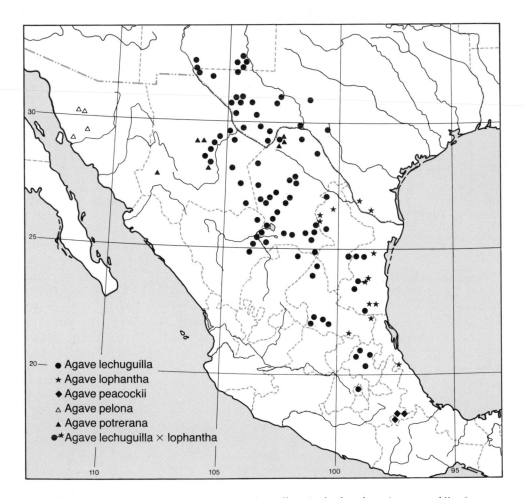

Fig. 7.22. Distribution of *Agave lechuguilla, A. lophantha, A. peacockii, A. potrerana.*

conical bud are employed. The bud fiber is finer, less strong, but easier to work with than the harder less twillable leaf fiber. Some twist the fiber into rope and twine for their own use, while others sell their harvested fiber to the town merchants, who in turn pass it on with profit to merchants of the large cities along the gulf ports of Mexico.

Martinez (1936:261) reported that dense stands of lechuguilla may number up to 30,000 plants (rosettes) per hectare, but that an average would be about 21,000 per hectare. Such stands are not easy to walk through, as the sharp spines stab the moving legs and ankles. Most of the stands are on rocky broken terrain. The fiber is extracted with simple hand tools and carried out by hand or on burros to the villages. Attempts to grow the plants in plantations have never been sustained, it being cheaper and easier to work the wild stands, which grow on land generally unsuitable to tillage; a ready resource for the poor peasant. Martinez calculated that some 86,000 hectares of lechuguilla growing within easy reach of seaports had an estimated value of 36,120,000 pesos (ca. 10 million US dollars) in 1936. Today's value in our cheap dollars would exceed this figure several fold. If we were to include the several million hectares occupied by lechuguilla throughout the range of the plants, there exists a possible resource of several billion dollars in gross value of labor, material, and capital—proceeding from Martinez' basic estimate. Unless new uses for the plant are found, we could not use it all.

Agave lechuguilla has been considered as a source for steroid drug manufacture, as its leaves contain significant amounts of the steroidal precursor, smilogenin (Table 7.2). But lechuguilla fiber is not extracted in decorticating mills like sisal and henequin, which are plantation grown. Just the transport of lechuguilla from its rough inaccessible terrain is cost prohibitive. There are other agave species better suited to cultivation that produce more sapogenins in less time than lechuguilla. It is probable that lechuguilla will remain a country hand industry for some time. As a writer in my study, I visualize lechuguilla fiber gathering as a pleasant occupation, sitting out in peaceful nature under a mesquite tree skinning out agave fiber, but those who are habitually forced to it must regard it as a drudgery.

Agave lechuguilla is poisonous to cattle. Cows, goats, and sheep have died from feeding on the leaves. Generally, livestock avoid the plants, but when there is little else to eat, they will take lechuguilla. The toxic principle has been identified as a saponin, which Mathews (1938) calls a "hepato-nephro-toxin." It is activated by a photodynamic agent, not identified, that intensifies the effect of the saponin. Bright sunlight on cattle contributes to the toxicity of lechuguilla. Stockmen would like to eliminate lechuguilla and replace it with a palatable xerophytic forage. But until such substitution can be made, lechuguilla should be regarded as a soil saving asset and a protective agent of the range, penalizing those stockmen who, through force of circumstance or lack of foresight, decimate their resource by over-use. The toxicity of lechuguilla is a built-in protective factor, which, I have no doubt, helped lechuguilla survive the thousands of years it had to contend with a long procession of hungry ungulates and other animals, known to us today as fossil camels, horses, antelopes, ground sloths, and pachyderms.

Agave lophantha
(Figs. 7.1, 7.2, 7.3, 7.22, 7.23, 7.24; Tables 7.1, 7.2)

Agave lophantha Schiede, Linnea 4: 582, 1829. Kunth, Ennum. Plantarum 5: 838, 1850.
  Agave univittata Haw., Phil. Mag. 10: 415, 1831.
  Agave heteracantha Zucc., Act. Acad. Caes. Leop. Carol. 16(2): 675, 1833?
  Agave vittata Regel, Gartenflora 7: 312, 1858.
  Agave mezortillo Hort.

Small, radiate, single or surculose rosettes, 30–60 x 50–100 cm, the old sometimes with visible stems; leaves numerous, generally 30–70 x 3–5 cm, patulous, light green to yellow-green, with or without pale mid-stripe, linear to lanceolate, rather thin, pliant,

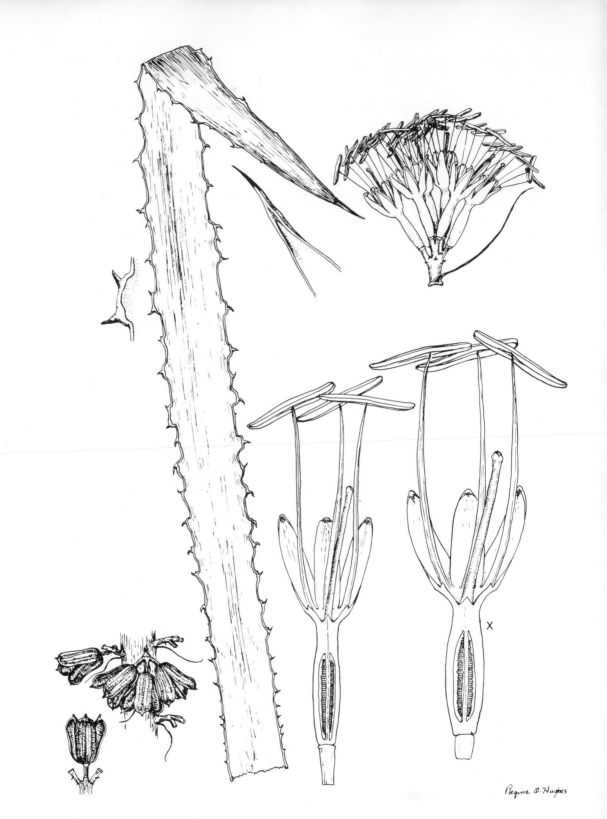

Fig. 7.23. *Agave lophantha* drawn from *Gentry 12268*. Leaf, flower cluster, and capsules, ca. x ½; flower section, x 2. Enlarged bicuspid tooth on teat, which distinguishes *A. lophantha* from *A. lechuguilla*. X, *A. xylonacantha* flower section for comparison.

somewhat thick toward base and rounded below, plane to concave above; margins corneous, undulate to crenate, the teeth single or occasionally double on broad low teats,
straight or mildly curved, slender, mostly 4–8 mm long and 1–2 cm apart; spine small,
1–2 cm long, subulate, ferrugineous to gray, flattened above at base; spike 3–4.5 m tall,
slender, with flowers in upper half of shaft, single or paired on single or dichotomously
branched pedicels 5–10 mm long, or in clusters of 3–7 on short lateral peduncles; flowers
light gray glaucous green to yellow, 35–47 mm long; ovary fusiform, 18–22 mm long
with neck short or long (5–7 mm) and constricted; tube short, open, 2–4 mm long, 8 10
mm wide; tepals subequal, erect to ascending, 14–20 mm long, persisting erect around
filaments; filaments 30–45 mm long, greenish or lavender, spreading, inserted at level of
inner tepals on rim of tube; anthers 15–20 mm long, pale yellow; capsules oblong, 18–24
x 10–12 mm or orbicular 15–20 x 12–18 mm, sessile or on short (2–3 mm) slender stipe;
seeds crescentic, 5–6 x 3–4 mm, the faces with wavy ridges, the edges with raised
margins.

Nomenclature and affinities

Schiede, when he named this plant, gave no description, writing only that it is one of
two spicate agaves growing on the Malpais de Naulingo, Veracruz, Mexico. Kunth added
little more: "Caulis simplex, 2-pedalus, floribus flavescentibus dense coopertus, Schiede.
Affinis A. univittatae C. Bouche." Jacobi in 1864 used Schiede's name, but his description applies to A. lechuguilla Torr. The same applies to Berger's use of the name
in 1915. Trelease (1920) appears to be the first to write a plausible description of
A. lophantha Schiede, which, however, is so brief (4 lines) and general it could be
applied to two or three other species as well. It is remarkable that Trelease gave
no flower description, because I found three sheets of dried flower specimens in
Missouri Botanical Garden labeled "Trelease 2/28/05 Malpais de Naulingo (type loc.)-
6." I found no leaves at MO and the species does not appear well founded until the
current account with my leaf collection (Gentry 20410) and photographs (Fig. 7.24).
Haworth's description of A. univittata appears applicable to the species under discussion
and has been taken up in modern accounts (Breitung, 1968; Jacobsen & Rowley, 1973).
Neither name nor species has been founded with specimens until the present time. The
rule of priority in nomenclature appears to mandate the use of Schiede's earlier name.
Agave lophantha has several close relatives. From A. lechuguilla it is distinguished
by its flatter leavers with sinuous to undulate firmer border, the teats usually with at least
one double set of teeth, frequently more (Fig. 7.23). A. xylonacantha is a coarser plant
with relatively few leaves, the margins highly teated and with large variously flexed teeth.
Some wide-leved forms of A. lophantha with margins crenate with large teats look very
much like A. xylonacantha, and indeed may be inseparable. The teeth of A. lophantha
are consistently more slender and more closely set than in A. lechuguilla and A.
xylonacantha, and resemble more the fine teeth of the straight-margined A. funkiana.
These four taxa comprise a closely related complex. All of them have leaf forms with pale
broad center stripes, but it is not constantly present in any of them. Some populations of
A. lophantha show mostly simple habits, while others sucker or grow in clumps, all
developing apparent stems.

Distribution and habitat

The natural distribution of Agave lophantha is now known to extend from southeastern Texas southward in Mexico along the east coast of Mexico to central Veracruz
(Fig. 7.22). It is frequent on limestone, as on cliffs and rocky outcrops, where tropical
forests are not too dense to prevent adequate sunlight, at elevations from 100 to 5,000 feet

(30 to 1,500 m). The rainfall indicated (Fig. 7.1) in the middle of its range at Ciudad Victoria is considerably above what is received north of that station, and the rocky sites it frequents have little moisture holding capacities. It is inferrable, however, that light but frequent rains are sufficient for optimum growth.

The identity of the name *lophantha* did not become certain until it was recollected at the type locality by Trelease in 1905 and Gentry in 1963. Fortunately Schiede, when he named the plant, stated that it came from the ''malpais de Naulinco'' in Veracruz. When I relocated the topotypic plants in 1963, I had to look for the old mule trail from Xalapa to Naulinco (also spelled Naulingo), which more or less parallels the modern auto road at lower elevations along the forested slopes of the Huatusco Region.

Near a sugar mill village, La Concepción, the trail enters the lava bed far below the town of Naulinco. The lava makes a large open expanse of black rough rock with scattered trees and clumps of shrubbery along the edges. The old mule trail enters the lava field through a cleft in the lava. Along this rough trail I soon found the agaves shown in

Fig. 7.24. *Agave lophantha* in its type locality, growing on the lava field below Naulingo, Veracruz, September 1963.

Fig. 7.24. I have little doubt that it was about here that Schiede collected his live specimens and the locality to which he alluded in his brief note of origin.

A. lophantha has responded well in our southwestern gardens. Some of the tropical forms may be frost sensitive, but in the Desert Botanical Garden we have collections from San Luis Potosí and elsewhere that have survived frost down to 20°F (−7°C) without damage. The leaves are bright green color.

Agave obscura
(Figs. 7.1, 7.2, 7.3, 7.12, 7.25, 7.26, 7.27; Table 7.1)

Agave obscura Schiede, Linnaea 5:464, 1830. Schlecht., Linnaea 18: 413, 1844.

Small to medium, single or cespitose, compact, light green to dark green, freely seeding rosettes with numerous heavily armed leaves. Leaves 25–40 x 5–8 cm, variable, broadly linear to ovate, short-acuminate, rigid, straight to upcurved, plane to somewhat hollowed above, thickly convex below, smooth, pale green to green; margin heavy, continuous, straight to undulate between teeth, gray; teeth variable, straight to curved or flexuous, frequently downslanted and curved, small (3–5 mm long) or large (10–15 mm long), less than 1 cm apart or 2–3 cm apart, flattened, all gray; spines mostly 3–5 cm long, conical to subulate, flat above, keeled below, broadly decurrent, gray; inflorescence a long tapering spike, 3–5 m tall, densely flowered in upper ⅔ in a spiraling sequence; bracts broad at base, long-caudate, deflexed, longer than flowers; flowers 34–41 mm long, geminate on short pedicels, dark purple or red or yellow; ovary 16–21 mm long, with short thick neck; tube short, spreading, 3–4 mm long; tepals linear, ca. equal, 15–17 x 4–7 mm, rounded at apex to an apiculate hood, erect to slightly recurved, not filament clasping at anthesis; filaments 40–50 mm long, slender, inserted on rim of tube; anthers 14–18 mm long, centric, varicolored; capsules 2.5–3 x 1–1.4 cm, oblong, rounded at base and apex, without beak; seed crescentic, 3–4 x 1.5–2.5 mm, dull black, the marginal wing low, narrow.

Neotype: *Gentry, Barclay & Arguelles 20417*, on malpais 4–5 miles N of El Limón (RR station) Veracruz, 4 Sep. 1963; alt. ca. 7,800 feet, in US, dupl. DES, MEXU.

The first mention of this name, *A. obscura*, by Schiede (loc. cit.) hardly qualifies as a diagnosis: "In the Malpays de la Jolla (I) found a straight spicate-flowered agave in bloom, of which I have spoken earlier. The flowering shaft is scarely as tall as a man. Since all the flowers have a dark reddish color, they show why I have called it *Agave obscura*." Schlechtendal (1844:413) provides a better description, which with Schiede's location assures us of the population for which his name stands. My description is drawn from several collections (see Exsiccatae) on the malpais or lava fields above La Jolla in highland Veracruz and the adjacent volcanic highlands of Puebla.

Subsequent authors of horticultural agaves in Europe made no accounts of *A. obscura*, from Salm-Dyck through Jacobi to the modern work of Berger. Apparently Schiede's introductions failed to offset, and there was no progeny to distribute to other gardens. However, it is one of the earliest names in the Marginatae and is an abundant and interesting plant in northern Puebla, adjacent Veracruz, and in eastern San Luis Potosí.

Trelease traveled through *A. obscura* area in 1903 but left no herbarium specimens. However, he recognized the species in his account of the agaves of Mexico (Standley, 1920), stating that it was "common on the lava beds about Limón." I have found no historical specimens of *A. obscura* and am left no choice but to stake the name down after

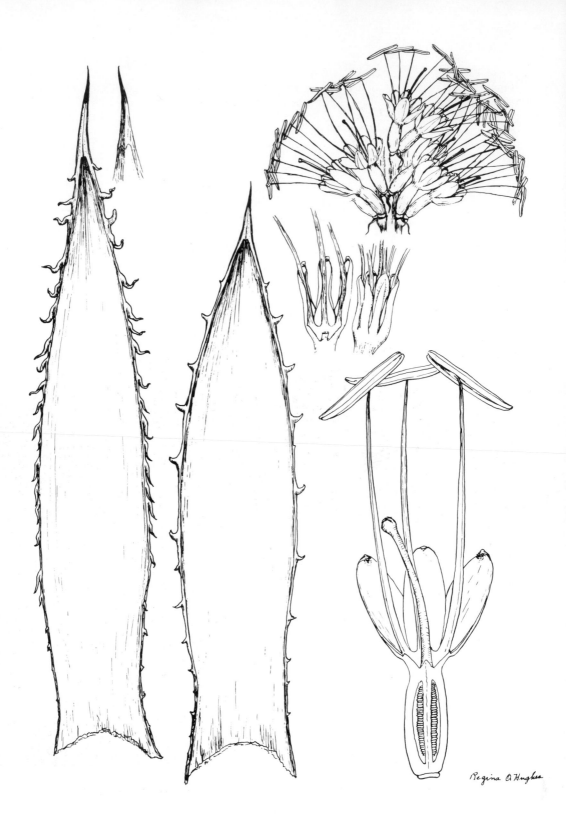

Fig. 7.25. *Agave obscura*. Leaves and flower cluster, x ½; flower section, x 2. Detail of tepals clasping filaments in post-anthesis.

Fig. 7.26. *Agave obscura* on the lava rocks north of El Limón, Veracruz, November 1963.

Fig. 7.27. A pair of mature rosettes of *Agave obscura,* both with retrorse slanting teeth; lava field of El Limón.

these 150 years with a collection of my own. It is a fine little agave, the variability of which could add interest to any garden.

Arturo Gómez Pompa, of the Instituto Nacional de Investigaciones sobre Recursors Bióticos in Jalapa, has called by attention to the possibility that Schiede's name, *Agave obscura,* may not really have been meant to apply to *A. obscura* as I have elected to use it. Pompa had a student look for ''my'' *obscura* on the lavas near La Jolla, but the student did not find it, the *Agave* there being what is here called *Agave polyacantha* and its variety *jalapensis.* Compare recent account (Chazaro Basañez, 1981). Just what did Schiede mean by the ''malpays de La Jolla''? which is a small village between Jalapa and Las Vigas. In Schiede's time there was no railroad and such railroad settlements as Las Vigas and El Limón, only a few miles west of La Jolla, did not exist then. There is a spicate *Agave* abundant on the lavas about El Limón and this is the species I have elected to bear Schiede's early name ''*obscura.*'' The name application appears rational and to do least violence to other established names.

Agave peacockii
(Figs. 7.1, 7.2, 7.3, 7.22, 7.28, 7.29, 7.30, 7.31; Table 7.1)

Agave peacockii Croucher, Gard. Chron. 1873; 1400, fig. 283. Hooker in Curtis Bot. Mag. Tab. 7757, 1901.

> *Agave roezliana* Baker var. *peacockii* Trel., U.S. Nat. Herb. Contr. 23: 137, 1920.
> *Agave macroacantha* Zucc. var. *latifolia* Trel., Mo. Bot. Gard. Rep. 18: 251, pls. 27, 28, 1907.

Single, medium to large, multiannual, subcaulescent, openly spreading rosettes with large well-armed leaves and racemose freely seeding inflorescence. Mature leaves 60–145 x 10–16 cm, linear to lanceolate, usually widest near middle, rigid, straightly ascending to horizontal, thick at base, toughly fibrous, plane to concave above, convex below, green, the margin usually corneous, or discontinuously so, rarely lacking, undulate to crenate; teeth highly variable, the larger 10–15 (–25) mm long, frequently with small lateral spurs, mostly 1–3 (–5) cm apart, thickly flattened, variously curved and shaped, dark brown to gray; spines 2.5–9 cm long, subulate to acicular, dark brown to gray, deeply, narrowly grooved above, sharply keeled below, the base sometimes protruding into the leaf (in one case forming a medium corneous keel 7 cm long with 2 or 3 median prickles); inflorescence a racemose panicle, 3–5 m tall, the flowers borne in clusters of 5 to 20 on stout lateral peduncles 4–10 cm long; bracts 3–7.5 cm long, narrowly deltoid; flowers green to yellow, the tepals tinged with purple or "blood-red" spotted, 48–55 mm long; ovary 24–28 mm long, 3-angulate-cylindric, with short thick neck; tube 4 mm deep, broad, shallow, thick-walled with base of filaments; tepals 18–21 x 5–6 mm, erect to incurved, thick, somewhat involute but not clasping filaments, the outer longer, rounded on back, the tip somewhat corneous-glandular, the inner thickly keeled; filaments slender, 45–50 mm long, inserted with tepals on rim of tube; anthers 21–22 mm long, yellow; capsules 2.5–3 x 1.2–1.4 cm, oblong, short-stipitate, beaked; seed not seen.

> Type: Tab. 7757, Curtis Bot. Mag. 1901.

"Fortunately there can be no doubt as to the *Agave* here figured being that to which the name *Peacockii* is given, for it is the type specimen purchased by the Royal Gardens at the sale in 1889 of the rich collection of Succulents formed by the late Mr. Peacock, of Hammersmith. It flowered in the Palm House of the Royal Gardens in December, 1899, having thrown up a scape which, with the inflorescence, was fourteen and a half feet high. It is a native of the province of Tehuacán, in central Mexico, whence it was imported by Mr. Roezl.—J.D.H."

Affinities and character

Agave peacockii is a remarkably distinct plant with no close relatives, unless it is a hybrid offspring of *A. kerchovei* with an unrecognized member of the subgenus **Agave,** as suggested by the racemose panicle; possibly *A. marmorata*, which grows with *A. kerchovei* in the environs. However, the flower structure is characteristic of the Marginatae (Fig. 7.1). The rosettes are the largest found in the Marginatae, one individual being 1.5 m tall and nearly twice as wide with its large spreading leaves *(Gentry et al. 20348)*. This same large plant showed a peculiar structure in the leaf apex. The lower base of the heavy spine protruded deep (5–7 cm) into the flesh of the leaf, ending in 2 or 3 denticles in the mid-line of the lower leaf surface.

The teeth of *A. peacockii* are generally very large, an extreme being 2.5 cm long, and are frequently characterized by small lateral spurs on the base of the main cusp. These teeth characters together with the racemose type of inflorescence are sufficient to identify the species, regardless of other variations in leaf form, size, and margins.

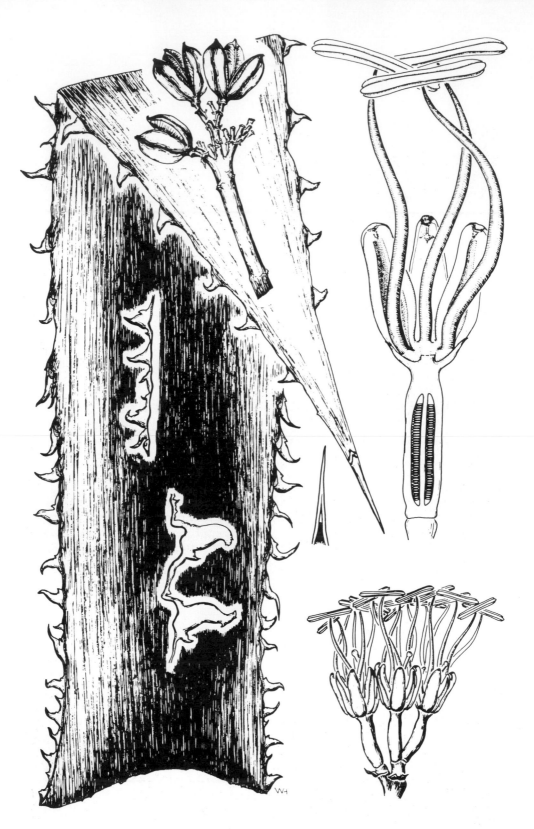

Fig. 7.28. *Agave peacockii* drawn from Gentry collections near Tehuacán; leaf, flower cluster, capsules, x ca. ½, flower long section, x ca. 2, detail of teeth inset.

Fig. 7.29. *Agave peacockii* on the rolling mesa southwest of Tehuacán, Puebla, with *A. kerchovei* behind on left, and *A. marmorata* close on the right.

Fig. 7.30. *Agave peacockii,* central section of a raceme with dehisced capsules.

Fig. 7.31. Individual rosette of *Agave peacockii* with large formidable teeth.

Distribution

Agave peacockii is known only from a small area around Tehuacán, Puebla; from about 4 miles west to 15 miles southwest of Tehuacán. There I have observed perhaps 40 to 50 plants in all scattered widely on the dry calcareous stony hills with thin shrubbery and tree cactus (Fig. 7.29). Elevations range from near 5,000 feet to 6,000 feet (1,500 to 1,900 m), the climate is frostless, the sun bright, and annual rainfall is about 500 mm per annum (Fig. 7.3).

Native name and use

The local countrymen call this agave "tlalometl" or "capulixtl," the former a Nahuatl term, the latter apparently deriving from Mayan. They stated that capulixtl has the finest and strongest fiber of any of the numerous agaves growing in the region and that it was employed for making rope, nets, and other woven products. This may partly account for its scarcity, but I also observed several old flowering shoots that had been bitten off by cattle. It is a rare endemic in need of protection.

Agave pelona
(Figs. 7.2, 7.22, 7.32, 7.33, 7.34; Table 7.1)

Agave pelona Gentry, U.S. Dept. Agri. Hb. 399: 76, 1972.

Subcaulescent, single, dark green to purplish, shiny, many-leaved, toothless, compact rosettes 40–60 cm tall, 60–80 cm broad; nonsurculose; freely seeding. Leaves 35–50 x 3–5 cm, linear-lanceolate, long-acuminate, sometimes slightly narrowed toward the base, plane above, rounded below, thick, stiff, erect to ascending, shiny dark green, turning reddish to purplish with drought or maturity; epidermis smooth, waxy, minutely circle-punctate; margin toothless, with a smooth, white, firm border becoming brittle and detachable on dried specimens; spine strong, sharp, 4–7 cm long, white to reddish, sharply angled below, grooved or plane above, decurrent as a white border down the leaf; inflorescence racemose, 2–3 m tall, flowering through upper half of shaft; flowers 45–50 mm long, dark red, campanulate, on bifurcate pedicels 30–50 mm long; ovary including neck about 20 mm long, cylindric, slender, light green, faintly striate; tube 8–9 mm long, 15–17 mm wide, openly funnelform, lined with a nectariferous disk; tepals 18 mm long, 8–10 mm wide, linear-ovate, plane and unkeeled, the inner widest, 8–9 nerved, recurved at the tip; tip apiculate, white-pappillate; filaments 35–40 mm long, red, inserted on rim of tube; anthers red, 15–17 mm long, centrally affixed; pistil small; capsules 25–30 mm long, 12–15 mm broad, oblong, light pruinose, thin-walled, rounded and short-apiculate; seeds irregular in size and shape, mostly 4 x 4 to 5 x 5 mm, thick, rugose, sooty black, numerous.

Type: *Gentry & Arguelles 19898*, US. Cerro Quituni, about 26 miles south of Caborca, Sonora, April 7, 1963, on arid limestone mountain; alt. 1,500–3,000 feet; dups. DES, MEXU.

This beautiful and distinct, endemic xerophyte has no close relative. The rosettes cling firmly to the rough limestone rocks and cliffs. The rigid, dark-green leaves, often tinged brightly with red and purple, are strikingly outlined by the white, smooth margins and the long, strong, white spine. The long-pedicellate, dark-red campanulate flowers with spreading recurved tepals are distinct in the genus. The funnelform tube with its thickened, nectariferous inner lining with 6-pointed apex of filaments is like that in the section *Filiferae*, but the leaf is more like agaves of the section Marginatae. One tooth on one leaf was observed on one plant. The nearest relative appears to be *A. potrerana*, as evidenced by the red corolla and by the strong tendency of *A. potrerana* toward reduction of teeth on its long-acuminate leaves. *A. pelona* is a disjunctive member of the Marginatae, separated by several hundred km, and its distinctive morphology appears due in large measure to long isolation from other relatives.

Flowering appears to be normal during April, but in dry years flowering is sparse, or not at all. In the dry spring of 1963, all of the several inflorescences found in Sierra del

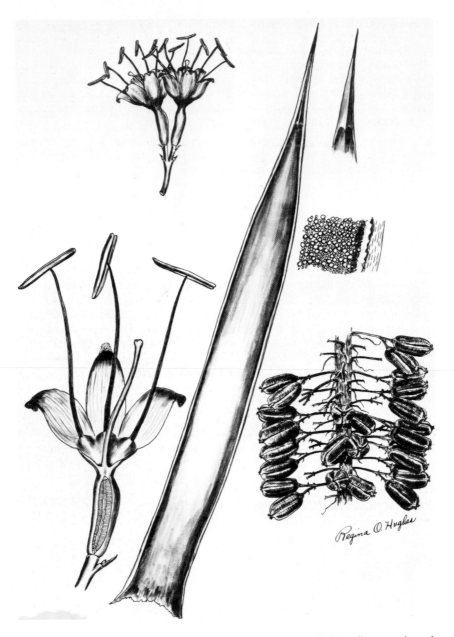

Fig. 7.32. *Agave pelona*. Leaf with spine offset of ventral side, flower pair and capsules, x ½; flower section slightly enlarged with apex of outer tepal raised; section of leaf cuticle and margin greatly enlarged. Drawn from type collection.

Viejo had the buds completely gutted by an insect, assumed to be a Lepidopteran larva. The flowers on the one flowering plant found on Cerro Quituni the same season were being eaten by a rodent. Several of the buds were collected and placed in a small water jar. These opened the following day and it is from these that my flower measurements and description and Regina Hughes' drawings are taken.

I was first guided to this agave by a ''mescalero'' encountered along the streets of Caborca, who called this plant ''mescal pelón'' in reference to the bald or toothless leaf margins. He stated that it was a sweet kind and suitable for making the ''moonshine'' called ''mescal,'' except its ''cabezas'' (heads) were too small for the labor involved in cutting and trimming. For this reason, he preferred the larger ''mescal lechuguilla'' in the same region, *A. zebra*. The mescaleros of Mexico have frequently been very helpful in my investigations of agaves. Presumably, *A. pelona* was formerly eaten and its strong fibers were used by the local Amerindians, such as the Papagos and Seris who still inhabit adjacent areas, but there are no reports of uses other than those I have already given. Analyses of the leaves of steroids showed 0.06 percent sapogenin from Sierra del Viejo and 0.3 percent smilagenin from Sierrita de Lopez.

Fig. 7.33. *Agave pelona* on Sierra del Viejo; old dry plant with seeded spike.

Fig. 7.34. The simple rosette habit of *Agave pelona*. Note the long white spines. Sierra del Viejo, 1963.

Agave potrerana
(Figs. 7.1, 7.2, 7.3, 7.22, 7.35; Table 7.1)

Agave potrerana Trel., U.S. Nat. Herb. Contr. 23: 138, 1920.

Single, light green, thick-stemmed, regularly spreading rosettes, 0.7–1 m tall, 1.5–2 m broad. Leaves numerous, mostly 40–80 x 6–7 cm, lanceolate, straight, rigid, roundly guttered above, convex below, widest below the middle, glaucous to light green, the corneous margin continuous, straight, brown toward the base, gray above, firm; teeth generally small, 2–4 mm long, commonly 2–3 cm apart, lacking or reduced to serrations below the mid-blade, mostly straight; spine 2.5–4 cm long, acicular, sharply angled below, flat to broadly canaliculate above, light brown to gray; inflorescence spicate, stout, 4–7 m tall with narrow scarious bracts, densely flowered through upper ⅔ of length, straight or arching; flowers pink to red or yellow, 46–58 mm long, in twos to fours on dichotomous pedicels 4–15 cm long; bractlets long attenuate, shorter than flowers; ovary slender, cylindric, 25–32 mm long, with smooth constricted neck; tube 3–6 mm deep, thick-walled; tepals 17–24 x 5–7 mm, erect, linear to ovate, obtuse, clasping filaments after anthesis, the outer ca. 1 mm longer and closely over-lapping inner, the inner broader, with a wide keel; filaments ca. 50 mm long, yellow or red, inserted at base of tepals; anthers 14–21 mm long, yellow, centric or excentric; capsules 2.5–4 x 1–1.5 cm, oblong, rarely pyriform, glaucous, strong-walled, rounded at base, shortly beaked; seeds 4–5 x 3–3.4 mm, crescentic, with low broad circumferential margin, hilar notch shallow.

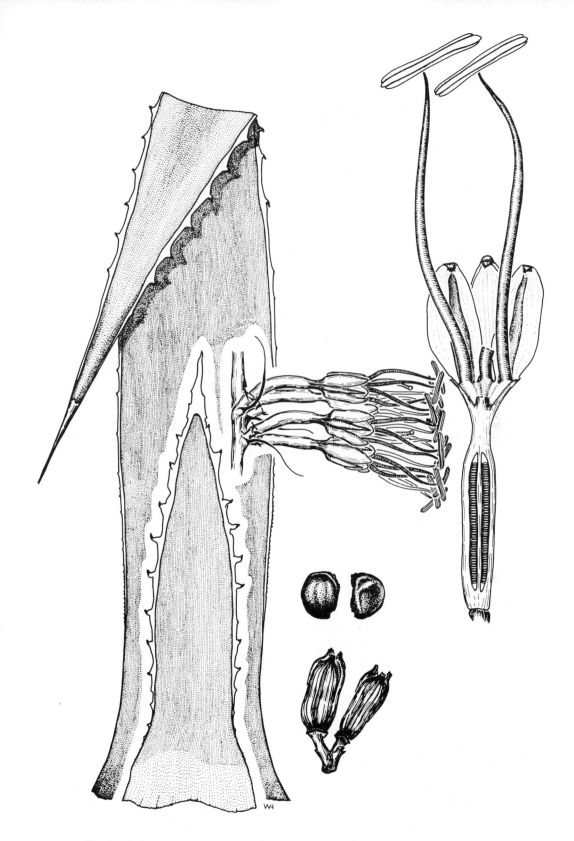

Fig. 7.35. *Agave potrerana*, drawn from several collections from the Sierra Campana, Chihuahua. Leaves, flower cluster, capsules, x ⅔; flower section, x 2; seed, x 2.

Type: *Pringle 802, MO,* B, UC, US. Potrero Peak (Sierras de Santa Eulalia, SE of Cd. Chihuahua), Chihuahua, 10 Sep. 1886; alt. ca. 7,400 feet (f, l, cap). "We went by train to Mapula, thence walked some three hours to the highest peak of the Santa Eulalia Mountains, on the summit of the mountains collecting...on the ledges of the peak 802 *Agave kerchovei* Lem." (Davis 1936: 34).

The habitat of *Agave potrerana* is the oak-pine-grassland communities of the volcanic ranges between 5,000 and 8,000 feet (1,500 and 2,500 m) elevations of northern Coahuila and Chihuahua. Recent collectors report it as abundant in the high elevations of the Sierras del Carmen in northern Coahuila. See Fig. 7.22 and the Exsiccatae.

Agave potrerana is a distinct species without a close relative, distinguished among the Marginatae by its single habit, tall husky spikes with large red flowers, and long acuminate leaves with teeth much reduced or lacking along the lower half of the leaf margins. On Sierra Campana of central Chihuahua in 1959 there was a large scattered population. Some of the individual plants had leaves up to 1 m long and spikes to 7 m, freely seeding. No suckering plants were observed. The leaves have an abundance of strong fiber, longer and more pliable than its commercial relative, *A. lechuguilla,* of the same general region. Anyone interested in the fiber and sapogenin commercial potentials of *A. potrerana* should consider test plantings of selections of seed in the United States. It grows in a habitat with hard winter freezes, far beyond what commercial fiber species can tolerate. A local informant of Sierra Campana stated it was called "lechuguilla," the fiber was used locally, and it was common all over the area. However, when visiting the area again in 1974, I found the plants much scarcer, the range there being heavily stocked with cattle. Heavy cropping of the inflorescence by cattle can soon decimate a non-surculose agave population. A very long-leaved specimen (to over 1 m) developed in Huntington Botanical Gardens during the 1970s. This demonstrated further its potential as a fiber plant under professional cultivation, suggesting that natural selections could still be made from the wild Mexican populations. See *Kimnach & Brandt 900* under Exsiccatae.

Agave pumila
(Figs. 7.36, 7.37)

Agave pumila De Smet ex Baker, handbook Amaryllideae, 172, 1888.

Plants dimorphic, persisting in juvenile form for 8–12 years, as small surculose rosettes 5–8 cm wide. Leaves ovate-orbicular, broader than long, 2–4 x 3–4 cm, deeply concave above, rounded below, the base broadly clasping, thickly succulent, grayish green, check-striped below, the margin thin, white, friable with several weak small teeth; spine small, conical, flexuous. Mature form a short, thick-stemmed, nonsuckering, open rosette, 40–50 cm tall, 60–70 cm wide. Leaves 30–38 x 4–4.5 cm, deltoid-lanceolate, rigid, patulous, thickened at base, concavo-convex, grayish green, without check marks below; margin narrowly white corneous, detaching, with small weak teeth 1–2 mm long, 1–1.5 cm apart; spine 1.5 cm long, slender conical, slightly grooved above, decurrent along leaf edges and the keel decurrent in median. Inflorescence unknown.

Neotype: *HBG 16230, US.* Cult. Hunt. Bot.Gard., 1963–1979, dups. DES, HBG. (l, photos by Trelease & Gentry; letter Kimnach.)

Agave pumila has long been known only in juvenile form as a pot cultivate (Fig. 7.36). During the past two decades, however, plant fanciers in southern California, e.g., Hummel, Glass, Kimnach, set young plants out of pots into roomier garden soil. So treated, the plants have greatly changed form, as appears in Fig. 7.37. This is a dimorphism I have not personally seen, nor seen equaled in other agaves, but the veracity of observation is confirmed by each of the above cited gentlemen.

The narrow detachable margin with weak teeth and the check-marked striping on the under side of young leaves indicate relation to *Agave lechuguilla*. The plant may be a *lechuguilla victoriae-reginae* hybrid. More certain relationship could appear with its eventual flowering. The origin of this plant is unknown. Baker reported the Kew plant was obtained from the Dutch plant trader De Smet in 1879, who probably received it from Mexico. If hybrid, the plant may never again reproduce in nature. If I were to search for it, I would look first in the *A. lechuguilla* and *A. victoriae-reginae* hybrid populations east of Saltillo, Coahuila.

"Yes, we have about 2 dozen or so plants of *A. pumila* in the ground and they are all growing out of the dwarf, juvenile form. I agree with your hypothesis that it is . . . a hybrid of *lechuguilla,* possibly from the north edge of Laguna de la Viesca where we collected the compact or dwarfish form of *Agave victoriae-reginae* (together with *lechuguilla*)" (Glass, letter, Jan. 23, 1979).

Fig. 7.36. *Agave pumila* in juvenile form. From Trelease, Pop. Sci. Monthly 1911, Fig. 14. Photo from Mo. Bot. Gard. 1978.

Fig. 7.37. *Agave pumila* in more mature form in Hunt. Bot. Gard., 1978, after 15 years of growth in garden location.

Agave titanota
(Figs. 7.1, 7.2, 7.3, 7.38, 7.39, 7.40, 7.41; Table 7.1)

Agave titanota Gentry, sp. nov.

Medium-sized, single or sparingly surculose, openly spreading, glaucous white, sub-caulescent rosettes with freely seeding spikes. Leaves broad, linear-ovate, short-acuminate, 35–55 x 12–14 cm, rigid, thick toward base, plane or concave above, convex below, the apex involute above, keeled below, alabaster white, finely granular; margin corneous, widest toward apex (3–5 mm), continuous to base or nearly so, undulate to crenate, the teeth variable, the larger 8–12 (–20) mm long, variably spaced, the leaf apex sometimes toothless for 8–12 cm below spine; spine broadly conical, 3–4 cm long, with deep inrolled groove above, keeled and protruding below, dark brown to gray; spike ca. 3 m tall, erect, the peduncle with deltoid, long attenuate ascending bracts, flowering in upper half of shaft; flowers on bifurcate pedicels 1–2 cm long, 45–50 mm long, yellow or with lavender flushes on tepals and anthers; ovary pale greenish, 22–25 mm long, slender cylindric, with constricted neck; tube 2–4 mm deep, abruptly spreading; tepals 21–24 x 5

Regina O. Hughes

Fig. 7.38. *Agave titanota*. Leaf and leaf margins of three plants, flower cluster, capsules, x ½; flower section, x 1.6, 1 capsule, x 1. Drawn from the type collection.

Fig. 7.39. *Agave titanota* in native habitat at Rancho Tambor in northern Oaxaca, November 1967.

mm, ascending, linear, acute, thickly succulent, not clasping filaments at anthesis, the outer 1 mm longer, the inner with prominent keel protruding with apex; filaments 45–55 mm long, flattened, finely tapered toward apex, inserted near rim of tube; anthers 20–23 mm long, regular, centric; capsules (old, weathered), oblong, 2.5 x 1 cm, acute at apex, thin-walled; seed not seen.

Type: *Gentry & Tejeda 22474, US,* isotypes DES, MEXU. Rancho Tambor, ca. 17 miles W of San Antonio, Dist. Teotitlán, Oaxaca, 25 Nov. 1967; elev. 3,200–4,000 feet, cliffs and ledges in limestone canyon. ''Rabo de león ceniza.''

Planta singula vel surculosa, caule brevi infra rosulam terminalum foliorum vivorum albo-glaucorum. *Folia* 35–55 cm longa, 12–14 cm lata, lineari-ovata brevi-acuminata rigida, basim versus crassa apicesuperne involuta infra carinata; *marginis dentibus* prominentibus plerumque 6–12 mm longis, cinerascentibus, curvatura et interspatio variabili; *spina terminali* 3–4 cm longa, lati-conica, superne canaliculata

Fig. 7.40. The glaucous white mature rosette of *Agave titanota* at the same locality.

infra carinata decurrente. *Spica* erecta densiflora cylindrica scapo incluso ca. 3 m alto; *floribus* 45–50 mm longis in pedicellis bifurcatis 1–2 cm longis, luteis vel purpureis suffis. *Ovarium* 22–25 mm longum cylindricum. *Tubo* 3–4 mm longo expanso; *segmentis* 21–24 mm longis, 5 mm latis, subequalibus, linearibus acutis crassis, adscendentibus. *Filamenta* 45–55 mm longa, gracilia tubi orificio inserta; *antheris* 20–23 mm longis luteis vel purpureis. *Capsulae* 2.5 cm longae 1 cm latae, oblongae, fragilae. *Semina* non vidi.

The short spreading tube, long narrow tepals, and high insertion of the filaments relate *Agave titanota* with the Marginatae. It is distinctive with its broad glaucous white leaves, whence the Greek name signifying alabaster white.

It is known only from the Rancho Tambor, where I noted a large population in the limestone canyon there. Tambor is a watering place for goats until the shallow water hole

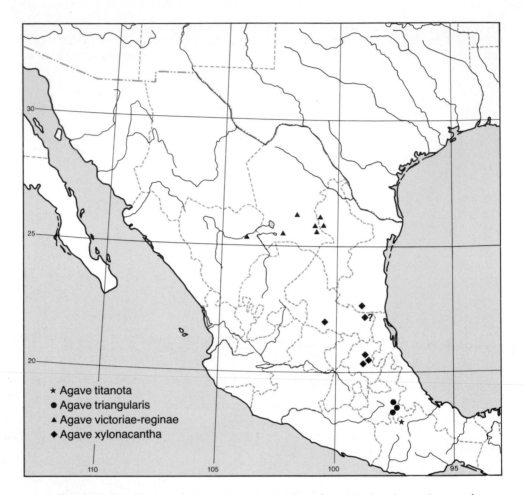

Fig. 7.41. Distribution of *Agave titanota, A. triangularis, A. victoriae-reginae,* and
A. xylonacantha.

goes dry after the summer rains. The surrounding country is arid, uninhabited, and
covered with an extensive virgin thorn forest, which on some rocky ridges features the
distinctive *Fouquieria purpusii* and many indigenous cacti. Further exploration of this
area would likely find other populations of *A. titanota.* It was called ''rabo de león
ceniza'' by our local constable guide.

Agave triangularis
(Figs. 7.41, 7.42, 7.43)

Agave triangularis Jacobi, Zweiter Nachtrag, zu dem Versuch einer systematischen
Ordnung der Agaveen 149, 1869.

Agave rigidissima Jacobi, ibid, 150, 1869.
Agave hanburyi Baker, Kew Bull. 1892: 3, 1892.

Slow-growing, short-stemmed, widely surculose, seldom flowering, rigid rosettes,
forming open clones; mature leaves deltoid-lanceolate, 30–60 x 5–7 (mid-leaf) cm, rigid,
straight, long-acuminate, thick at base, concavo-convex, olivaceous or light yellowish

green finely flecked with brownish red, finely asperous; margin corneous continuous, 1–2 mm wide, grayish, straight, with or without teeth, the teeth small, few (2–3 mm long), and remote (3–5 cm apart), or large, 5–9 mm long, 1–2 cm apart, gray, straight or curved; spine 2.5–4 cm long, conical to subulate, usually straight, grayish, grooved above, somewhat keeled below; inflorescence unknown.

Neotype: *Gentry 23399*, DES, MEXU, *US*. 6 miles SW of Tehuacán, Puebla, long road to Huajuapán, 11 March 1974; elev. 5,750 feet, on sedimentaries.

This taxon is recognizable by its thick, rigid, deltoid, olivaceous leaves. Toothless forms are common. Fig. 7.42 shows a characteristic open clone near Tehuacán, Puebla.

Fig. 7.42. *Agave triangularis* as a dispersed clone on the limey mesa southwest of Tehuacán, Puebla, March 1974.

Fig. 7.43 shows the variable leaves from as many differrent clones. Jacobi (loc. cit.) reports it was introduced to Europe in 1868 by the botanist Besserer, who collected the plant from Cerro Colorado near Tehuacán. Trelease (1920) recognized *A. triangularis* over *A. rigidissima* Jacobi published under the same date. The unarmed form he recognized as *Agave triangularis* var. *subintegra* Trel., U.S. Nat. Herb. Contr. 23: 138, 1920 (*A. kerchovei inermis* Baker, Gard. Chron. n. ser. 7: 527, 1877).

Agave triangularis is common on the arid calcareous mesa west of Tehuacán in southern Puebla. Figure 7.42 shows this heavily grazed area with stony ground and tree forms of *Yucca periculosa* and *Lysiloma* sp. Here also are numerous tree cactus and several other agaves. This table land is 5,500 to 6,000 feet (1,700 to 1,900 m) in elevation with 475–500 mm of rain, annual average (Fig. 7.3 as for *A. ghiesbreghtii* and *A. peacockii*). The winters are dry and frostless. *A. triangularis* is uncommon in American gardens, but once established it should endure indefinitely.

Fig. 7.43. *Agave triangularis* leaf series from as many clones; southwest of Tehuacán.

Agave victoriae-reginae
(Figs. 7.1, 7.2, 7.3, 7.41, 7.44, 7.45; Tables 7.1, 7.2)

Agave victoriae-reginae T. Moore, Gard. Chron., n. ser. 4: 484, 1875.
   Agave consideranti Carr., Rev. Hort. 1875: 429.
   Agave fernandi-regis Berger, Agaven 90, 1915.
   Agave nickelsii R. Gosselin, Rev. Hort. 1895: 579.

Small, compact, single or surculose or cespitose, acaulescent to short-stemmed in hort., very variable plants. Leaves short, green with conspicuous white markings, generally closely imbricate, 15–20 (–25) x 4–6 cm, linear ovate, rounded at apex, rigid, thick, plane to concave above, rounded to sharply keeled below; margin white corneous, usually toothless, 2–5 mm wide, continuous to base; terminal spines 1–3, 1.5–3 cm long, trigonous-conical, subulate, very broad at base, with broad open groove above, roundly keeled below, black; inflorescence spicate, 3–5 m tall, erect, densely flowered in upper half of shaft, the peduncle with deltoid long attenuate chartaceous bracts; flowers in pairs or triads on short, forking, stout pedicels, 40–46 mm long, varicolored, the tepals and stamens frequently tinged with red or purple; ovary 18–24 mm long, thickly fusiform, with short neck; tube shallow, spreading, 3 x 8–10 mm; tepals ca. equal, 18–20 x 5–6

Fig. 7.44. *Agave victoriae-reginae* on right growing with *A. scabra* on limestone ridge ca. 15 miles (24 km) east of Saltillo, Coahuila, July 1963.

mm, linear, rounded apiculate, spreading, then clasping filaments in post-anthesis and erect, the inner strongly keeled; filaments 45–50 mm long, inserted on rim of tube; anthers 18–21 mm long, yellow or bronze-colored, centric or excentric; capsules ovoid to oblong, 17–20 x 10–13 mm, rounded at base, apiculate; seed 3–5 x 2.5–3.5 mm, hemispherical to lacrimiform, veined on faces, the marginal wing low.

Neotype: *Gentry, Barclay & Arguelles 20043*, DES, MEXU, US. 12–16 miles NE of Saltillo, Coahuila, along road to Monterrey, June 10–July 5, 1963; elev. 4,000–5,000 feet, shrub and succulent desert on limestone. The general distribution is mapped, Fig. 7.41.

Typical *Agave victoriae-reginae* is easily identified by its toothless, small, green, thick, rigid leaves with white markings on both faces. The corneous leaf margins and flowers with shallow tubes and filament-clasping tepals clearly mark its inclusion in the Marginatae. It is, however, very variable, especially in the small forms scattered about the Chihuahuan Desert, many of which have appeared in horticulture during the last two decades. Breitung has named several forms, all finely illustrated in the Cactus & Succulent Journal (U.S.) 32: 35–38, 1960. They are relisted here with Breitung's figure numbers and his types deposited in the California Academy of Sciences, which I have not reviewed. They all appear to be submature specimens, based on leaf morphology without flower descriptions.

Agave victoriae-reginae f. *dentata* Breitung

Leaves short, broad, the white margin serrate with small retrorse teeth, Fig. 54. Cult. by Monmonier, Los Angeles, California, from seed collected in Mexico.

Type: No. 18161, CAS.

Agave victoriae-reginae f. *latifolia* Breitung

Leaves short, broad (4–6 cm), ovate, dark green, with conspicuous white markings and short black terminal spines, Fig. 58. Cult. by Monmonier, Los Angeles, California, from seed collected in Mexico.

Type: No. 18164, CAS.

Agave victoriae-reginae f. *longifolia* Breitung

Leaves long, acuminate, 20–30 cm long, thick, rigid, dark green, Fig. 60. Cult. by Monmonier, Los Angeles, California, from seed collected in Mexico.

Type: No. 18165, CAS.

Agave victoriae-reginae f. *longispina* Breitung

Leaves acuminate with elongate terminal spines, 2.5–3.5 cm long, black, sinuous, Fig. 57. Cult. by Monmonier in Los Angeles, California, from seed collected in Mexico.

Type: No. 18163, CAS.

Agave victoriae-reginae f. *ornata* Breitung

Leaves small, rather narrow, finely variegated with white markings in addition to regular bud printing.

No type designated but his Fig. 59 is taken as lectotype.

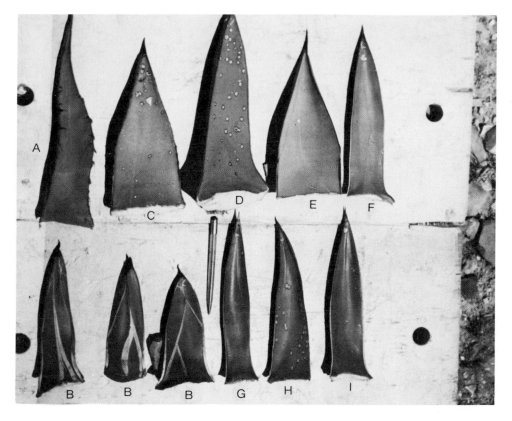

Fig. 7.45. A series of leaves intergrading from *Agave scabra* (above) to *A. victoriae-reginae* (below).

Agave victoriae-reginae f. *viridis* Breitung

Leaves linear elongate, white markings pale or lacking, Fig. 56.

Type: No. 18162, CAS.

Agave victoriae-reginae f. *nickelsii*
(Gosselin) Trel., U.S. Natl. Herb. Contr. 23: 140, 1920.

Differs from the typical species in having fewer leaves in a more open rosette.

Neotype: Fig. 53, p. 26, in Breitung loc. cit.

Agave victoriae-reginae is also known to vary radically due to apparent introgression with other species in the subgenus **Littaea** and the subgenus **Agave** as well. Gentry described putative hybrids between *A. victoriae-reginae* and *A. asperrima* (= *A. scabra*) of the subgenus **Agave** (Journ. Hered. 58: 32–36, 1967). Figures here 7.44 and 7.45 show the two dissimilar parents and the clinal leaf variants of individual plants that link the two distinctive leaf forms of the species together. Some of the intermediate forms were found to bear seed, while others showed sterile fruiting. This population observed ca. 15 miles (24 km) northeast of Saltillo, Coahuila, appeared to form a ''hybrid swarm,'' with some individuals rather monstrous. Others have reported introgression of *A. victoriae-reginae* with the promiscuous *A. lechuguilla* (word of Charles Glass). The variable forms of *A. victoriae-reginae* provide a novel array in garden collections, in or out of pots, they are quite cold-hardy, and recently have grown in popularity as ornamentals.

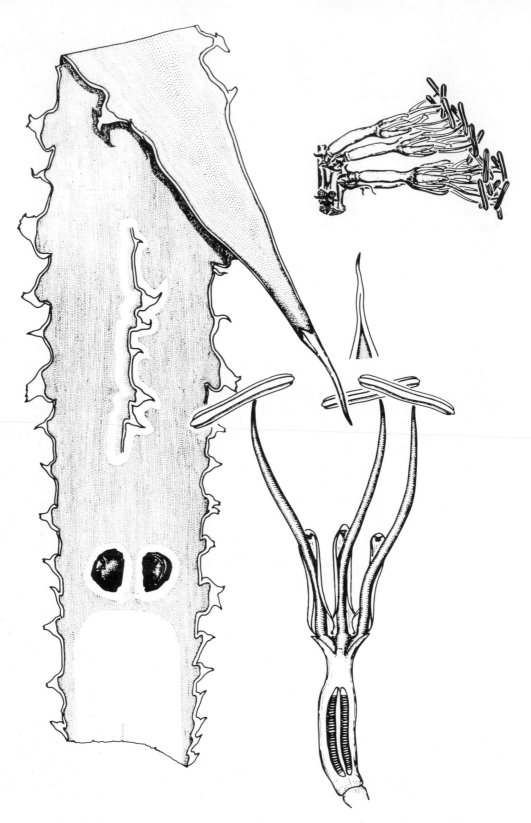

Fig. 7.46. *Agave xylonacantha*. Leaf and flower cluster, x ca. $^3/_5$, long flower section, x 1.7, seed enlarged.

Agave xylonacantha
(Figs. 7.1, 7.2, 7.3, 7.41, 7.46, 7.47; Tables 7.1, 7.2)

Agave xylonacantha Salm., Bonplandia 7: 92, 1859.
   Agave carchariodonta Pampanini, Nuevo Giornale Bot. It. 1907: 591.
   Agave kochii Jacobi, Hamb. Gart. & Blumenz. 22: 117, 1866.
   ?Agave noli-tangere Berger, Agaven 103, 1915.
   ?Agave splendens Jacobi, Abh. Schl. Ges. Vaterl. Cult. 1870: 147.
   ?Agave vittata Regel, Gartenflora 7: 313, 1858?

Single or cespitose, short-stemmed, openly spreading rosettes with long tapering, freely seeding spikes. Leaves 45–90 x 5–10 cm, ensiform-lanceolate, broadest through the middle, long-acuminate, rather rigid, plane to concave above, explanate in shade, rounded below, green to yellowish green, sometimes glaucous, with or without pale center stripe; margin continously corneous, straight between remote prominent teats, but looping over the teats; teeth broadly flattened, thickly capping the broad teats, frequently

Fig. 7.47. *Agave xylonacantha* in a dispersed extensive population on a limestone slope in the Barranca de Metztitlán, Hidalgo, March 1974.

3–5-cuspid, commonly 2–5 cm apart, 8–15 mm long, light gray; spine 2.5–5 cm long, trigonous-subulate, stout, flat to slightly grooved above, keeled below, light gray; inflorescence spicate, 3–6 m tall, erect, long tapering, flowering in upper ⅔ to ½ of shaft, the peduncular bracts very narrow, chartaceous; flowers in pairs or triads, 40–50 mm long, greenish to pale yellow, filaments sometimes red or purple; ovary 20–27 mm long, fusiform, 4-angulate; tube 3–5 mm long, 8–10 mm broad; tepals ca. equal, 15–20 x 4–5 mm, narrowing and clasping filaments at anthesis, erect; filaments 38–42 mm long, slender; anthers 17–22 mm long, excentric; capsules and seeds not seen.

> Neotype: Tab. 5660, Curtis Bot. Mag., 1867. Fide Hooker, cult. at Kew from plant sent by Mr. Repper in 1846 from Real del Monte, Hidalgo, Mexico.

Salm-Dyck (loc. cit.) states that he received the species from the "Pariser botanischen Garten," where it was grown from seed. There appear no data to connect Salm-Dyck's plant with that at Kew; thus, the type designation must be "neo."

Agave xylonacantha is related to *A. lophantha*, its highly convoluted leaf margins with large flattened, multicuspid teeth being like an exaggeration of a character started by *A. lophantha* (see detail in Fig. 7.23). Indeed, some shade forms in the San Luis Potosí lowlands are difficult to designate as one or the other. Fig. 7.23 shows the sectioned flowers of the two taxa, featuring their basic similarity, but the flowers are not specifically definitive considering the basic similarity of most Marginatae flowers.

Figure 7.41 indicates the distribution of *A. xylonacantha*. Generally, it occupies the drier slopes and valleys of limestones on the desert side of the Sierra Madre Oriental, at elevations above 3,000 feet (900 m), in the Mexican states of Nuevo León, Tamaulipas, San Luis Potosí, and Hidalgo. It is abundant on the limetone slopes in the Barranca de Meztitlán, not far from the great mines of Real del Monte (Fig. 7.47).

Marginatae Exsiccatae

Agave albomarginata

HORTICULTURE. Gentry 19811, DES, US Type. Cult., Huntington Botanical Gardens, San Marino, California, 3 July 1962 (l, f, photo).

Agave angustiarum

GUERRERO. *Fox & Gentry 5853,* DES, US. Milestone gorge 24 miles W of Iguala, 9 Aug. 1952; elev. ca. 4,000 ft. (l, photo).

Gentry 11984, DES, US. Canyon Iguala, E of Naranjo, 14 Aug. 1952 (l, photo).

Ogden 5116, DES, US. 15 km N of Chilpancingo, 21 Oct. 1951.

Sharp 441426, MO. Bluff W of Chilpancingo, 21 Oct. 1944; elev. ca. 6,000 ft. (f, br).

Trelease 17 & 77, MO. Type. Iguala Canyon, 5 Mar. 1905 (4 sheets, l, f, fr, br).

MEXICO. *Hinton 7561,* UC, US. Villa Neda, Dist. Temascaltepec, 27 Mar. 1953 (l, f).

MICHOACAN. *Nelson 6931,* US. In canyon, 20 miles S of Uruapan, 22 Mar. 1903.

OAXACA. *Gentry 12247,* DES, MEXU, MICH, US. Km 660 along Oaxaca-Tehuantepec hwy, 30 Sep. 1952; elev. ca. 3,000 ft.

Gentry 21333, DES, US. In km 654 along hwy to Tehuantepec, near San Luis Viejo, 29 Sep. 1965; elev. 3,000–4,000 ft., cliffs.

PUEBLA. *Gentry 12120,* DES, MEXU, MICH, US. Ca. 7 miles NW of Tehuitzengo along hwy to Oaxaca, 6 Sep. 1952; rocks in short-tree forest.

Agave difformis

HIDALGO. *Gentry 23377,* DES, MEXU, US. Km 55 along rt. 105, 15–20 km N of Atotonilco, 6 Mar. 1974; elev. ca. 5,275 ft. (l, photo).

Gentry 23382, DES, MEXU, US. 3 km NE of Metztitlán along rt. 105, Barranca de Metztitlán, 6 Mar. 1974; elev. ca. 4,700 ft.

Gentry et al. 20081, 20081B, DES, MEXU, US. By Río Tula between Ixmiquilpán and Zimapán, 25 June 1963; elev. ca. 5,000–5,500 ft. (l, f, photo).

Gomez Pompa 800. Cult. Jardin Botanico Nacional at UNAM, Mexico, from vicinity of Ixmiquilpán, May 1965 (photo).

HORTICULTURE. Berger, s.n. Type, K, US. Cult. La Mórtola, 11 Aug., 1909 (l, f).

Gentry 19826, DES, US. Cult. Huntington Botanical Gardens, Block 20 (l, f, photo. Matches the type material very well; same clone with and without teeth).

Gentry 10170, ibid., 22 Jan. 1951.

SAN LUIS POTOSI. *Gentry et al. 20095,* DES, MEXU, US. 35 miles E of San Luis Potosí, along road to Rio Verde, 16–18 July 1963; elev, 5,500–6,800 ft. (l, f, cap, photo).

Agave ensifera

HORTICULTURE. Berger s.n. Neotype, US. Chateau Grimaldi, bi Mórtola, Italy. 19 June, 1909 (l, f, 5 sheets, leaf margins only). US 1023791, 1023763.

Berger s.n., US. Cult. Mórtola, 25 Mar. 1911, Jan. 1913 (6 sheets; leaves ca. 1 m long).

Berger s.n., US. Cult. Mórtola, 17 July 1909 (l, f, which Berger had identified as *A. lophantha,* probably belongs here also).

Agave funkiana

HIDALGO. Gentry 12273, Neotype, DES, US. 43 miles W of Jacala along hwy. to Mexico, Sierra Madre Oriental, 12 Oct. 1952, transplant flowered in Gentry garden, May 1962 (l, f, photo).

Moore 1751, BH. Jacala, Minas Viejas, km 255, 28 Oct. 1946; elev. ca. 1,800 m.

Moore 3525, BH, US. Dist. Jacala, Puerto de la Zorra near km 284 on hwy. E of Jacala, 3 Aug. 1947; alt. 5,000 ft. (l, f).

Ogden et al. 51144, DES, US. At km 251 on Laredo-Mexico hwy, just S of La Placita, 27 Feb. 1951; rocky hillsides in *Quercus* forest.

TAMAULIPAS. *Dewey s.n.* MO., UC. Hacienda Soldanos, Jaumave, 1 June 1903.

Dewey 586, WIS. ibid.

Nelson 4442, US. Between Victoria and Jaumave valley, 31 May 1898; elev. 800–2,500 ft. (l, f) Doubtfully referred here.

Ogden et al. 51111. 17 miles from Cd. Victoria on road over Sierra Madre Oriental to Jaumave, 17 Feb. 1951 (cap).

CHIAPAS. *Gentry 12195,* DES, MEXU, US. 15 miles S of Comitan along highway, 20 Sep. 1952; elev. ca. 2,000 ft., mixed woodland with open grass slopes with limey rocks.

Agave ghiesbreghtii

HORTICULTURE. Berger s.n., Neotype, US. Cult. Mórtola, 16 July 1906 and 1909 (l, f).

GUATEMALA. *Gentry 21424,* DES, US. 7 miles E of Huehuetenango along road to Aguacatán, 14 Oct. 1965; elev. ca. 6,500 ft., limey rock slope (l, f, photo, topotypic).

Gentry & Gentry 23639, DES, US. ibid, 1 Jan. 1976; elev. 7,000–7,500 ft. (l, f, cap, photo).

Gentry & Gutierrez 21026, DES, US. Mountain grade S of Cunen, Depto. Quiche, 6 May 1965; elev. ca. 6,500–7,000 ft., limestone with pine and oak (l, photo).

Harmon & Fuentes 4869, MO. 3 km N of Chiantla, Depto. Huehuetenango, 12 Nov. 1970 (l, f).

Standley 82039, F. Along Aguacatán road E of Huehuetenango at km 13–14. (Type of *A. huehueteca* Stand. & Steyerm.)

Steyermark 50940, US. Cumbre Papal, between Cuilco and Ixmoqui, 19 Aug. 1942; elev, 1,400–3,000 m (infruc., doubtfully referred here. Has much larger bracts than other collections).

GUERRERO. *Gentry & Fox 12014,* DES, MEXU, MICH, US. Taxco, 17 Aug. 1952.

MEXICO. *Gentry 23370,* DES, MEXU, US. 10 miles N of Zacualpán, 22 Feb. 1974; elev. ca. 6,500 ft.

OAXACA. *Gentry 12234,* DES, MEXU, MICH, US. 36 miles W of Sanatepec and ca. 26 miles E of Tehuantepec along hwy, 29 Sep. 1952; cliff near estuary.

Gentry 12107, DES, MEXU, MICH, US. Miahuatlán. 2 Sep. 1952.

Gentry & Halberg 12257, DES, US. San Mateo Cajones, Dist. Villa Alata, 26 Sep. 1952; elev. ca. 3,200 ft.

PUEBLA. *Gentry 12290,* DES, MEXU, US. 15–20 miles N of Tehuacán along hwy to Puebla, 16 Oct. 1952 (l, f).

Gentry 20406, DES, MEXU, US. Tlacotepec, between Tehuacán and Puebla, 2 Sep. 1963.

Trelease s.n., MO. Tehuacán and vicinity, 13–14 Aug. 1903. As *A. triangularis* Jacobi.

Trelease s.n., MO. Tehuacán, 1903–05. (5 sheets of leaves as *A. roezliana.)*

Agave glomeruliflora

COAHUILA. *Gentry & Engard 23117,* DES, MEXU, US. 28 miles S of La Cuesta, along Muzquiz-Boquillas road, 11 Oct. 1972; grassland on limestone (l, photo).

Gentry & Engard 23127, DES, MEXU, US. Los Cojos Minas, SW slope of Sierra del Carmen, 12 Oct. 1972; elev. 6,000–6,500 feet, oak woodland, limestone (l, cap, photo).

Gentry & Engard 23131, DES, MEXU, US. Same loc. & date; should be regarded as *A. havardiana X lechuguilla;* with broad leaves.

Gentry & Engard 23134, DES, MEXU, US. 30 miles S of La Cuesta, along Muzquiz-Boquillas road, 12 Oct. 1972; grassland on limestone with *A. lechuguilla.*

Henrickson 11392, DES. 22 air miles E of Boquillas, 1.2 miles NW of La Mula on road to El Jardin, 27 July, 1973; elev. 4,500 ft., limestone alluvium (l, cap, photo).

Pinkava et al. 13603, ASU, DES. Slopes between Canyon de Hacienda y Canyon de Agua, Sierra de la Madera, NW of Cuatro Cienegas, 22 June 1976 (l, cap, photo).

TEXAS. *Gentry & Correll 20626,* DES, TEX, US. The Basin near cottages, Chisos Mts., Brewster Co., 17 June 1964.

Muller 5007, ARIZ. 6 miles NE of Nickel Creek Camp on Boger Canyon, E slope of Guadalupe Mts., Culberson Co., 16 May 1942.

Sperry et al. 840, Neotype, **US.** Basin, Chisos Mountains, Brewster Co., 13 July 1937 (l, f, cap, photo, type of *A. chisosensis*).

Stanton s.n. in 1898 col. near Kent a robust form and at Sierra Blanca, a depauperate form, both of which are probably assignable to *A. glomeruliflora.*

Warnock 20885, UC. Gilliland Flat, Glass Mountains, Brewster Co., 4 July 1940 (f).

Agave horrida

HORTICULTURE. *Berger s.n.,* US. Cult. Mórtola Bot. Gard., 16 July 1906 and 1909 (l, f). Also 3 sheets labeled as from *Pringle 8206.*

MORELOS. *Gentry 23368,* DES, MEXU, US. Pedregal above and NE of Cuernavaca, 20 Feb. 1974; elev. 7,500–8,000 ft., on S slope of lava bed (l, f, photo).

Gentry & Fox 12037, 12038, DES, MEXU, MICH, US. Ca. 11 miles NE of Cuernavaca along new hwy to Mexico, 20 Aug. 1952; elev. ca. 7,000 feet, lava (l, cap, photo)

Pringle 8206, Neotype, UC, **US.** Pedregal above Cuernavaca, 2 Feb. 1899 (l margin, f). Also *6854,* US, same loc. a year earlier.

Agave impressa

SINALOA. Gentry 23366, type, DES, MEXU, **US.** 13 miles E of Escuinapa, near hwy, 16 Feb. 1974; elev. 500–1,000 ft., on volcanic rocks with *Plumeria, Bombax,* Bromeliads, etc. (l, f).

Gentry et al. 11487, 19467, 22561, DES, MEXU, MICH, US. Same loc. in 1952, 1961, 1967, respectively. (l, cap, photo)

Agave kerchovei

HIDALGO. *Gentry & Sanchez-M. 23383,* DES. Cult. garden of Sanchez Mejorada in Velasco; originally from Peña del Aigre, near Calera, Barranca de Metztitlán, 8 March 1974.

OAXACA. *Delgadillo 206, US.* 4 km NE de Mitla, 3 Feb. 1966 (l, f).

Ernst 2438, US. On San Lorenzo road near Mitla, 3 Feb. 1966 (l, f, fr).

Gentry 12263, DES, MEXU, MICH, US. 20–25 miles SE of Huajuapán along hwy, 5 Oct. 1952 (l, f).

Gentry et al. 20278, DES, MEXU, US. Ca. 14 miles SE of Huajuapán along rt. 190, 10 Aug. 1963; alt. 5,600–6,000 ft., brushy rocky slope (l, f).

Gentry 22521, 22355, DES, MEXU, US. 2 miles E of Mitla, 7 Dec. 1967; alt. 6,000 ft (l, f, photo)

Griffiths s.n., US. Tomellín Canyon, 1 Oct. 1909.

Kirby 2727, US. Barranca de Rio Grande near Mitla, 11 Feb. 1966; alt. 1,700 m (l, fr).

Ogden & Gilly 51195, 51107, DES, US. Along hwy between Cd. Oaxaca and Acatlán, 9.5 miles N of Tamazulapán, 18 Apr. 1951 (l, photo).

Trelease 4, 10, MO. El Parian, Tomellín Canyon, 2 Dec. 1905 (l, f, 2 sheets). Type of *A. convallis* Trel.

Trelease 81–82, MO. Mexia (toward Tomellín), 2 Nov. 1905 (l, cap). Type of *A. dissimulans* Trel.

PUEBLA. *Endlich 1927,* MO. Tehuacán, April 1907.

Endlich 1910, MO. Tehuacán, April 1907.

Gentry et al. 20223, 20343, DES, MEXU, US. 4–10 miles SW of Tehuacán along road to Zapotitlán, Aug. 1963 (l, cap). "cacaya."

Gentry et al. 20351, US. Near Acatepec along road to Huajuapán, 19–27 Aug. 1963 (l, with large blunt confluent teeth).

Gentry 22418, DES, MEXU, US. 6 miles SW of Tehuacán, 12 Nov. 1967.

Ogden et al. 5175, DES, US. 4 miles NE of Tehuacán road fork along side road from Córdoba hwy, 4 Feb. 1951; "maguey pichomel."

Rose & Rose 11266, US. Near Tehuacán, Sep. 1906.

Smith et al. 3679, US. On and around Petlanco Hill; alt. 1,000–1,500 m, 12 July 1961; "cola de león."

Trelease 1856, MO. Tehuacán, 2/5/05 (l, 2 sheets).

Trelease 51, 69, 33, 34, 27, MO. Tehuacán, 5–6 Feb. 1905.

Trelease s.n., MO. Tehuacán, 8 Dec. 1903. As type of *Agave inopinabilis* Trelease.

Agave lechuguilla

CHIHUAHUA. *Chiang et al. 8845,* TEX. Sierra de la Pampas, W of Hacienda Berrendo, 25 Aug. 1972; elev. ca. 1,600–1,850 m, 27°20'N, 104°43'W (l?, f).

Gentry & Engard 23272, DES, US. Ca. 15 miles NE of Aldama along road to Ojinaga; 16 May 1973; limestone, elev. ca., 4,125 ft.

Pringle 157, US. Santa Eulalia Mts., May–Sep. 1885 (l, f, cap).

Rose & Hough 4219, US. Santa Eulalia Mts., 11 May 1899 (l, cap).

Shreve 8090, ARIZ, US. 6 miles W of Cuchillo Parado, 31 July 1937.

COAHUILA. *Chiang et al. 7573A*, TEX. Puerto de la Bufa, pass through limestone hills, 9 June 1972; alt. 1,250 m. 27°46'N, 102°50'W (f).

Chiang et al. 7510, TEX. 6 km SW of Rancho San Miguel near mouth of Canyon de los Burros, NE part of Serranias del Burro, 6 June 1972 (f). Elev. 600 m. 29°11'N, 101°31'40"W.

Chiang et al. 7942, TEX. 4 km SE of Tanque Escondido, 18 June 1972; alt. 1,875 m, 24°44'N, 101°05'W, (l, f).

Gentry et al. 20032, DES, MEXU, US. 13 miles SW of Saltillo along rt. 54, 12 June 1963; alt. 6,500 ft., highland desert shrub.

Gregg 699, MO. Cienaga Grande, E of Parras, 18 May 1847.

Henrickson 6002, DES. 28.4 miles NW of San Pedro along Chihuahua hwy. 30, in Puerto de Ventanillas, 1st pass NW of Laguna Mayran, 24 Aug. 1971; elev. 3,750 ft., in rocky limestone. 26°02'N, 102°43'W. (l, f).

Henrickson 11393, DES. Ca. 22 (air) miles E of Boquillas near Puerta de Boquillas, 1.2 miles NW of La Mula on road to Jardin; elev. 4,500 ft., limestone alluvium, 27 July 1973. 103°05'W, 29°02'N. (l, cap).

Henrickson 12571, DES. Ca. 22 (air) miles NE of Torreón, 6.6 miles SW of turnoff to Las Delicias on Cuatro Cienagas–San Pedro hwy. in northern Puerto de Ventanillas, 19 Aug. 1973; limestone slopes. 26°45'N, 102°07'W. (l, f, cap).

Johnston & Muller 658, TEX. Sierra del Pino; dry E ridge ca. 4 miles NE of camp at La Noria, 23 Aug. 1940 (l, f).

Nelson 6148, US. Monclova, 14 May 1902 (l, f).

Palmer 131, US. Saltillo and vicinity, 1880 (l, cap).

Palmer 227, UC, US. Saltillo, 1898 (l, f, infl). Also s.n. Nov. 1902, mountain sides and canyons.

Pinkava et al. 3758, DES, ASU. paso de la Becerra, ca. 10 miles S, SW of Cuatro Cienagas, 13 Aug. 1967 (l, cap).

Pinkava et al. 5081, ASU, DES. NW Tip of Sierra San Marcos, 8 June 1968 (l, f).

Pinkava et al. 5669, ASU, DES. 3.3 miles NW of Los Fresnos, 14 June 1968.

Pinkava et al. 9747, ASU, DES. Along rt. 40 at Higueras, 1.9 miles W of Coah.–Nuevo León state line, 2 Aug. 1972 (l, f).

Wislizenus 306, MO. Angostura, S of Saltillo, 21 May 1847.

Wislizenus 316, MO. Saltillo.

DURANGO. *Gentry 8289*, DES, US. 5–6 miles SW of Cuencame, 17 Sep. 1948; elev. ca. 1,700 m (l, cap).

Gentry & Gilly 10597, DES, MEXU, MICH, US. 6 miles SW of Pedricena along hwy. from Torreón to Durango, 12 June 1951; elev. ca. 4,900 ft., calcareous soil (l, cap).

Gentry et al. 20467, DES, MEXU, US. 9–10 miles NW of Cuencame along rt. 40, 24 Sep. 1963; caliche soil (l, cap, photo).

HIDALGO. *Gonzales Q. 2558*, CAS, WIS. 3.5 km N of Zimapán, 22 July 1965; alt. 3,000 m.

Moore 3525, UC. Puerto de la Zorra, near km 284 on hwy. NE of Jacala, 3 Aug. 1947; alt. 5,000 ft (l, f).

Ogden 5146, DES, US. Along hwy. between Pachuca and Real del Monte, 18 Jan. 1951.

MEXICO. *Ogden 5140*, DES. UC. Hill W of Mexico–Laredo hwy. just N of Fed. Dist. line, 17 Jan. 1951 (with broad leaves).

NEW MEXICO. *Dole Jr. 76*, UC. Carlsbad Nat. Park, E of Park Hdqts., 14 June 1937; alt. 4,200 ft (l, f).

Gentry & Engard 23052, DES, US. 30 miles SW of hwy. 285 along rt. 137, N end of Guadalupe Mts., 25 Sep. 1972; elev. ca. 4,750 ft. (l, f).

NEW MEXICO. *Parry et al. 143a*, US. Doña Ana (on Emory Comm.).

Wright 1907, UC, US. New Mexico, 1851–52 (l, f).

NUEVO LEON. *Barkeley & Barkeley 17M314*, TEX. 2 miles W of Casa Blanca, 8 June 1947 (l, cap).

Gentry et al. 20015, DES, MEXU, US. Sierra de la Silla, near Monterrey, near upper cable lift station, 10 June 1963 (l, f).

Gentry 11521, Canyon Huasteca, 8–10 miles W of Monterrey, 31 Jan. 1952; limestone cliff (live pl., photo).

Taylor 320, CAS., ARIZ. Rancho Resendez, Lampazos, 23 June 1927.

SAN LUIS POTOSI *Chiang et al. 8234B*, TEX. 12 km N of El Cubo, 43 km N of Charcas on road to Catorce, 3 July 1972; alt. 1,700 m (l, f). 23°26'N, 101°00'W.

Gentry et al. 11504, DES, MEXU, MICH, US. 11 miles SW of Cd. San Luis Potosí, 26–28 Jan. 1952 (l, photo).

Gentry et al. 20093, DES, US. Ca. 55 miles SE of San Luis Potosí along road to Rio Verde, 30 June, 1963.

Palmer 72, US. Alvarez in 1902.

Rzedowski 5415, ENCB, US. San Jose, SW of Guadalcazár, 2 Nov. 1954; alt. 1,650 m, "hojas de verde muy intenso."

TAMAULIPAS. *Dewey 586*, UC. Hacienda de los Soldaños, Jaumave, 1 June 1903.

Ogden et al. 5185, DES, MEXU, MICH, US. 28 miles W of Cd. Victoria on road to Jaumave, 15 Feb. 1951 *(lechuguilla X lophantha ?)*.

TEXAS. *Barclay 714*, DES. Near Tornilla Creek, Big Bend Nat. Park, 1 May 1960, on dry exposed rocky hills (l, f).

Blakeley s.n., DES. Near El Paso and cult. in Des. Bot. Gard. (l, f, cap).

Cutler 726, MO. 2 miles E of Castelon, Brewster Co., 4 Mar. 1937.

Ferris & Duncan 2419, CAS. Scenic Drive, Franklin Mts., El Paso, 21 July 1921 (l, f).

Cutler 1860, ARIZ., CAS, MO, UC. Dog Canyon, Santiago Mts., Brewster Co., 27 May 1938 (l, f).

Gentry & Barclay 18450, DES, US. 1½ mile W of Laguna along road 334, Uvalde Co., 24 Mar. 1960.

Gentry & Correll 20621, DES, TEX, US. 5½ miles NE of McCamey, Upton Co., 14 June 1964.

Gentry & Correll 20635, DES, TEX, US. 12 mile N of Allamore, turn-off of rt. 80, Hudspeth Co., 19 June 1964 (photo).

Gentry & Engard 23062, DES, US. Hueco Hills, 5 miles E of Hueco along rt. 180, 27 Sep. 1972; elev. ca. 5,350 ft. (l, photo).

Gentry & Ogden 9876, DES, MICH, US. Rio Pecos and hwy. 90, Oct. 1950.

Gentry & Ogden 9897a, DES, MICH, US. 13 miles N of Van Horn, Culberson Co., Oct. 1950 (l, cap).

Hinckley s.n., ARIZ. Near Sonora, Sutton Co., 24 Sep. 1939.

Jones 25925, CAS. Hot Springs, 24 April 1930 (f, cap).

Mueller 7960, UC. Chisos Mts., 19 June 1931 (l, f).

Rose 1119, US. Near El Paso, Tex.–Mex., 1 June 1897,

Semple 395, MO, US. 2 miles N of main road along Grapevine Spring Road, Brewster Co., 23 Aug. 1970 (l, f, cap).

TEXAS. *Sperry 1584*, UC. Persimmon Gap, Brewster Co., 5 May 1939 (l, f).

Tharp 43-522, 43-524, CA, UC. Pecos Co., June–July 1943.

Vasey 580 & s.n., CAS, US. El Paso, Mar. 1881 (l, cap).

Warnock Lot 4, US. 4 miles S of Marathon, Brewster Co., 24 Sep. 1950.

Warnock Lot 38, US. End of Quitman Mts., ca. 5 miles of Sierra Blanca, Hudspeth Co., 23 Sep. 1950.

Warnock Lot 48, US. Green Gulch of Chisos Mts., Brewster Co., 24 Nov. 1950.

Wright 682, Type, MO, **US.** Western Texas to El Paso, May–Oct., 1849 (US 125459, l, cap).

Wright 1907, MO, UC. Mts. near El Paso, Tex., and N. Mex., 1951–52.

York 269, CAS. Ft. Stockton, 14 June 1907 (l, f).

ZACATECAS. *Kirkwood 8*, MO. Cedros, 1908.

Lloyd 126, MO, UC, US. Hacienda de Cedros, 1907–08, "footslopes and hills, especially slopes facing S2" (l, f).

Agave lophantha

NUEVO LEON. *Gentry et al. 20014*, DES, MEXU, US. Ca. 40 miles NE of Cd. Monterrey along road to Roma, 9 June 1963 (l, f).

Mueller 8, MEXU. Monterrey.

Pringle 2517, MEXU. Mountains near Monterrey, 18 June–1 Aug. 1889 (l, f, cap).

SAN LUIS POTOSI. *Gentry et al. 20092*, DES, MEXU, US. Ca. 38 miles E of Rio Verde along road to Cd. Valles, 29 June 1963 (l, f).

Ogden et al. 51133, DES, MEXU, US. Ca. 10 km E of Cd. Valles along road to Tampico, 21 Feb. 1951.

TAMAULIPAS. *Crutchfield & Johnston 5240*, MEXU. 13 miles W of Gonzales along road to Mante, 11 Mar. 1960; elev. 300 ft. (l, cap; atypical with heavy teeth like *A. xylonacantha*).

Gentry 12268, DES, MEXU, MICH, US. Ca. 22 miles E of Mante, 11 Oct. 1952 (l, f).

Kimnach & Lyons 1442A, HNT. 13 miles along road from Cd. Victoria to Jaumave; alt. 4,400 ft., limetone oak zone.

Nelson 6616, US. San Fernando to Jimenez, 26–27 Feb. 1902.

Ogden et al. 51111, DES, MEXU, US. 17 miles from Cd. Victoria along road to Jaumave, 17 Feb. 1951 (l, cap, photo).

Ogden et al. 5199, DES, MEXU, US. 17.8 miles from Jaumave along road to Cd. Victoria, 16 Feb. 1951 (l, cap).

TEXAS. *Correll & Rollins 20941*, DES. 8 miles E of Rio Grande City along rt. 83, Starr Co., 19 April 1959; in flat brushland (l, f).

Gentry & Barclay 20009, DES, MEXU, US. 5 miles S of Zapata, along rt. 83, Zapata Co., 5 June 1963. Scattered colonies on sand hills; observed also 4 miles SE of Zapata on sandy plain (l, cap).

VERACRUZ. *Gentry 20410*, DES, MEXU, US. Topotype. By Rancho Concepción on lava beds below and SE of Naulinco, 3 Sep. 1963; alt. ca. 3,000 ft. (l, photos).

Trelease 6, 8, DES, MO. Topotype. Malpais de Naulinco, 28 Feb. 1905 (f).

Agave obscura

PUEBLA. *Gentry et al. 20407*, DES, MEXU, US. 1½–2 miles off (SE) Puebla-Jalapa hwy. along road to Guadalupe Victoria, 2 Sep. 1963; elev. 7,500–8,000 ft., outwash of volcanic gravels (l, cap, pl).

Gentry et al. 20409, DES, MEXU, US. Ca. 9 miles NE of Zacatepec along road to Jalapa, 2 Sep. 1963; elev. ca. 7,500 ft., open limestone ridge (l, f).

Gentry et al. 20421, DES, MEXU, US. Cerro 5 miles NW of Zacatepec, 5 Sep. 1963; elev. ca. 8,000 ft., volcanic hill slope (l, f, photo).

SAN LUIS POTOSI. *Gentry et al. 20088*, DES, MEXU, US. 21–22 miles E of San Luis Potosí along road to Río Verde, 16–18 July, 1963; elev. 500–6,800 ft., limestone mountain with pine and oak (l, f, cap).

Orcutt 1798 & s.n., US. Alvarez (SE of San Luis Potosí) Dec. 1924.

VERACRUZ. *Gentry, Barclay & Arguelles 20417*, Neotype, *US*, DES, MEXU. On malpais 4–5 miles N of El Limón (RR Station), 4 Sep. 1963; alt. ca. 7,800 ft. (l, f, pl, photo).

Agave peacockii

PUEBLA. *Gentry et al. 20345, 20360*, DES, MEXU, US. 4–10 miles SW of Tehuacán along road to Zapotitlán, Aug. 1963; elev. 5,600–6,000 ft., arid Thorn forest over limestone hills (l, photo).

Gentry et al. 20348, 20351, DES, MEXU, US. Near Acatepec along road to Huajuapán, 19–27 Aug. 1963; elev. 5,500–6,000 ft., arid thorn forest over limestone hills (l, cap, photo).

Gentry 22529, DES, US. (MEXU ?). 5½ miles SW of Tehuacán along road to Huajuapán, 9 Dec. 1967; elev. ca. 5,700 ft., ft., arid calcareous hills.

Tejeda & Gentry s.n., DES. Ca. 5 miles SW of Tehuacán, Feb. 1968 (f pickled, col. by Narciso Tejeda from plant marked by Gentry).

Smith et al. 4062, US. Near Cerro Colorado beyond Santa Cruz, Tehuacán area, July 1961; alt. ca. 1,000–1,800 m (l, infruct. with short branches of 5–8 cap).

Agave pelona

SONORA. *Barclay & Arguelles 2023*, DES, MEXU, US. Cerro Aquituni, S of Caborca, 30 May 1966.

Felger et al. 18121, 20530, ARIZ. Sierra Seri, 2 Feb. 1969 & 18 Dec. 1972; elev. ca. 500 m; stoloniferous. Seri name "assoot."

Felger & Strickland 20475, ARIZ. Sierra Seri; mostly on N and E-facing slopes; plants profusely stoloniferous.

Gentry & Arguelles 19898, Type. US, MEXU, DES. Cerro Aquituni Ca. 26 miles S of Caborca, 7 April 1963; arid limestone mt., alt. 1,500–3,000 feet. (l, f, photo).

Gentry 10202, DES, MEXU, MICH, US. Sierra del Viejo, 25 Feb. 1951; limestone cliffs ca. 2,500 ft. elev.

Gentry & Arguelles 19896, DES, MEXU, US. Cerro del Viejo, ca. 25 miles S of Caborca, 5 April 1963; arid limetone mt., alt. 3,500–4,000 ft. (l, infl, pls, photo).

Gentry 11615, DES, MEXU, MICH, US. Sierrita de Lopez, about 35 miles NW of Hermosillo, 19 Feb. 1952; N side of rocky cliff.

Agave potrerana

CHIHUAHUA. *Engard & Getz 267, 269*, DES. W end of Canyon de la Santa Clara, Sierra Campana, 16 miles W of Mexico hwy. 45 at Restaurante Parrita km. 9, 6 July 1974 (l, f, photo).

Engard & Haughey 959, DES. Canyon de Santa Clara, Sierra Campana, 5 miles W of Mexico hwy. 45, 8 Aug. 1976; elev. ca. 5,600 ft. (l, f, cap).

Engard & Gentry 71, DES. Arroyo Sauces, Sierra Campana, 4 miles W of hwy. 45, 16 May 1976; elev. ca. 5,700 ft. (cap).

Gentry 18249, 18250, DES, MEXU, US. Las Tapias, Sierra Campana, N of Cd. Chihuahua, 25 Nov. 1959; elev. 5,000–5,500 ft., oak grassland (l, f, cap, photo).

Gentry & Bye 23361, DES, US. 1.7 miles SW of Quirare along road to Lu Bufa, Batopilas Barranca, 10 Oct. 1973; elev. ca. 6,200 ft., on volcanic rock (l, photo).

CHIHUAHUA. *Kimnach & Brandt 900*, DES. Cult. Hunt. Bot. Gard., small plant col. along road from Creel to La Bufa, 6.4 miles NE of Batopilas Bridge at La Bufa, 10 Mar. 1967; elev. 6,000 ft. (Flwred. Hunt. Bot. Gard. June–July 1978). Apparently belongs here, but the leaves are 1.3 m long, lack teeth on upper half of margins, and the flower tube unusually deep (10–11 mm).

Parfitt & Reeves 2303, ASU, DES. Topotype. Canyon on N-facing slope of Pico Chihuahua Viejo Santa Eulalia Mts. 20 July 1977; elev. 6,200 ft. (l, f. cap)

Pringle 802, Type, B, MEXU, **MO,** UC. Potrero Peak, Sierras de Santa Eulalia, SE of Cd. Chihuahua, 10 Sep. 1886; alt. ca. 7,400 ft. (l, f, cap).

COAHUILA. *Johnston & Muller 711*, A, US. Sierra del Pino, Aug. 1940 (l, cap).

Nelson 6731, US. General Cepeda, 20 April 1902.

Riskind 1812, DES. Mun. de Ocampo, Sierra Maderas del Carmen, in upper portion of Carboneras Canyon, 27 May 1951. 28°51′N, 102°34′W. Western edge of high scarp, assoc. with *Pinus, Quercus, Pseudotsuga* and *Muhlenbergia* grass (l, f).

Stewart 1502, TEX. Canyon de la Madera, W side of Sierra de los Guajes, 4 km E of Rancho Buena Vista, 7 Sep. 1941 (l, cap).

Wauer s.n., DES. Above Los Cojos Mine, Fronteria Mts., (Sierras del Carmen), 5 Aug. 1970, S of Boquillas; alt. 7,300 ft. (f, cap).

ZACATECAS. *Rose 2404*, US. Near San Juan Capistrano, 18 Aug. 1897 (2 ls.).

Agave pumila

HORTICULTURE. Hunt. Bot. Gard. Acces. No. 16230, DES, **US.** Neotype. Cult. Hunt. Bot. Gard. 1963–1979. (l, photo. by Trelease, Gentry, letter Kimnach).

Agave titanota

OAXACA. Gentry & Tejeda 22474, Type, DES, MEXU, **US.** Rancho Tambor, ca. 17 miles W of San Antonio, Dist. Teotitlán, 25 Nov. 1967; elev. 3,200–4,000 ft., cliffs and ledges in limestone canyon (l of several, f of several pl., cap, photo).

Agave triangularis

HORTICULTURE. Mórtola Gardens, Vientimiglia, Italy. Sep. 1955 (photo).

Gentry 22746, Royal Botanical Garden, Kew, England, 23–27 June 1969, cult. Glass House No. 5 (photo, toothless).

PUEBLA. *Endlich 1927,* MO. Tehuacán.

Gentry 23399, Neotype, DES, MEXU, **US.** 6 miles SW of Tehuacán along road to Guajuapan, 11 Mar. 1974; elev. 5,750 ft. (l, photo).

Gentry et al. 20224, 203422A, DES, MEXU, US. 4 miles SW of Tehuacán along road to Zapotitlan, Aug. 1963; elev. 5,600–6,000 ft., on limestone hills (l, series, photo).

Griffiths s.n., MO. Tehuacán, 9 April 1909.

Trelease s.n., MO. Tehuacán, 2 June 1905 & 8 Nov. 1903.

Agave victoriae-reginae

COAHUILA. *Gentry & Arguelles 11525, 11530, 11532,* DES, MEXU, MICH, US. 16 & 17 miles E of Saltillo, 31 Jan. & 1–3 Feb. 1952; elev. 4,000–5,000 ft., rocky limestone slope with low dispersed desert shrub (cap, l, photo).

Gentry, Barclay & Arguelles 20043, Neotype, & *20344,* DES, MEXU, **US.** 12–16 miles NE of Saltillo along road to Monterrey, 10 June–5 July 1963; alt. 4,000–5,000 ft., shrub and succulent desert (l, f, photo).

DURANGO. *Henrickson 12443,* CSLA, DES. In narrow region 7.1 miles W of hwy. 40, on road to Presa Francisco Zarco along Rio Nazas, ca. 24 air miles SW of Torreon, 15 Aug. 1973; elev. 4,300 ft., limestone.

HORTICULTURE. *Breitung,* types of forms: *18161, dentata; 18164, latifolia; 18165, longifolia; 18163, longispina; 18162, viridis* all in CAS. See also Cact. & Succ. J. (U.S.) 32: 35–38, 1960.

Gentry 19828, DES, US. Cult. Hunt. Bot. Gard. 6 July 1962 (l, f, photo).

NUEVO LEÓN. *Endlich 898,* B, MO. Canyon de Sta. Catarina near Monterrey, 22 Feb. 1905. ''noa.''

Marroquin s.n., MEXU. Canyon de la Huasteca, 3 June 1967 (l, f).

Palmer s.n., MO. Monterrey, 1880.

Parry s.n. MO. Monterrey, 1878.

Powell 26, TEX. Canyon Huasteca, 14 July 1946; limestone (l, f).

Agave xylonacantha

HIDALGO. *Gentry 23378,* DES, MEXU, US. 3 km S of Metztitlán, Barranca de Metztitlán, 6 Mar. 1974; elev. ca. 4,200 ft. (l, photo).

Gentry 23381, DES, MEXU, US. 3 km NE of Rio Metztitlán on rt. 105, 6 Mar. 1974; elev. ca. 4,700 ft., sun slope sedimentaries.

Moore 2498, BH, GH, UC. Barranca walls near Los Venados, Dist. Metztitlán, 24 Mar, 1947; elev. 1,300 m (l, f).

Rzedowski 9046, ENCB, US. Venados, 25 km al S de Metztitlán, cerca el Puente de Venados, 10 Junio 1957; alt. 1,400 m.

Rzedowksi 19518, CAS, ENCB. Same loc. 4 April 1965; alt. 1,450 m (f).

HORTICULTURE. *Berger s.n.,* US. Cult. La Mórtola, 30 Sep. 1904, as type of *A. noli-tangere* Berger (f, ls. with typical margins of *xylonacantha,* but wider than average.)

Berger s.n., US. Cult. La Mórtola, 16 Aug. 1909. Also marked as type of *A. noli-tangere.* (f, ls of 2 different plants; 1 with gray corneous border and upcurved teeth, the other with blunt rounded teeth.)

Gentry 10117, 17709, 19818, 18750, DES. US. Cult. Hunt. Bot. Gard., col. from 1951 to 1962 (l, f).

SAN LUIS POTOSI. *Ogden & Gilly 51133,* DES, MEXU, US. Ca. 10 km E of Cd. Valles along highway from Tampico, 21 Feb. 1951; rocky bluffs. Doubtfully referred here.

Purpus 1, MO. Guasconia (= Guascamá, Mun. Villa Juarez), 8 April 1911.

TAMAULIPAS. *Gentry & Arguelles 11516,* DES, MEXU, US. 8–10 miles S of Cd. Mante, 29 Jan. 1952; limestone cliffs of canyon in tropical forest. (Live plant flowered in Gentry garden, May 1964, l, f, photos.)

Johnston & Crutchfield 5240, TEX. 13 miles W of Gonzales on road to Mante, 11 Mar. 1960; elev. ca. 300 ft. (l, cap).

8.

Group Parviflorae

Small, freely suckering, compact rosettes with filiferous, small, bud-printed, narrow, firm but not rigid leaves, with small weak spines. Inflorescence racemose or spicate, rather lax with small chartaceous, deciduous bracts; flowers in twos, threes, fours, rarely single, 10–40 mm long; tube generally cylindric, well-developed; tepals short to greatly reduced, erect to spreading; stamens short-exserted, inserted mostly deep in the tube, with small anthers; capsules small, strong-walled, persistent. Populations mostly outbreeders with bat, bird, and insect pollinators; some not protandrous. Northern Sierra Madre Occidental and adjacent Arizona.

Typical species: *Agave parviflora* Torr., U.S. & Mex. Bot. 214, 1859.

Taxonomic comment

The Parviflorae are notable for their small highly modified flowers (Fig. 8.1). Although their filiferous leaves appear to relate them to the Filiferae, where Berger placed them (his section Xysmagave), their specialized flowers indicate the relationship is not close. The modification of these sexual organs show they are the most specialized group in the subgenus Littaea. In this evolution they are comparable to the specialized flowers of the Ditepalae and Salmianae of the subgenus *Agave,* although the flower evolution is very different in the two subgenera. In addition to the small flower of the Parviflorae, the elongation of the tube in comparison with the relatively short segments is the most extreme in the genus. However, the dimorphism between the outer and inner tepals is less than in several groups of the subgenus *Agave*.

The subspecific taxa of the Parviflorae appear to indicate that speciation is active and in continuance. As in many other phanerogamic groups, the evolution of the flowers appears linked with pollinators. The flowers of *Agave parviflora* and *A. toumeyana* are obviously pollinated by carpenter bees and bumblebees (Schaffer and Schaffer, 1977), while those of *A. schottii* are visited by both bats and bees (Howell, 1972). The reflexed or recurved attitudes of the flowers during anthesis appears to assure that the insects will contact and carry pollen as they brush against anthers and stigmas, as they seek the nectar bait. This design and activity has been well reported by Schaffer and observed by others.

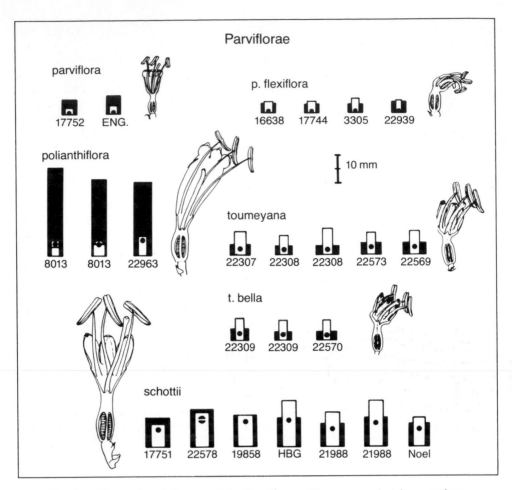

Fig. 8.1. Ideographs of flowers of the Parviflorae with correspondent long sections of flowers.

There is little doubt that this biotic relationship has had much to do with the evolution of agave flowers. It is certainly well marked in the Parviflorae. The habit of flowering from late spring to mid-summer may reflect a general correlation with their temperate climate (Fig. 8.1).

Distribution and environment

The distribution of the Parviflorae indicates a Madro-Tertiary origin in Axelrod's meaning of that paleobotanical term (1950). From southern Sonora and Chihuahua they range northward through the mountains to central Arizona. An unidentified taxon, located recently by Myron Kimnach in coastal Sinaloa, is a notable tropical exception. Figure 8.3 outlines the taxa distributions. The question mark for the Sinaloa plant signifies question of identity, not the presence of Parviflorae there. *Agave schottii* is the most abundant, some of the populations in the grama grassland of southeastern Arizona being extensive. *A. toumeyana* exists in considerable numbers on Fish Creek Hill on the Pinal Mountains and again near Sunflower. The remainder of the Parviflorae taxa are very scattered and rare, the small populations being numbered in the dozens. They should be protected to avoid their possible extinctions by acquisitive man and other more innocent predators.

Considering the high elevation montane occupancies, the general habitat of the Parviflorae is temperate, rather than subtropical. *A. schottii* does, however, also grow below 3,000 feet (900 m) elevation. The principal plant communities in which they are found include the oak-juniper-grama grasslands, the Arizona and Sierra Madre chaparrals, and the oak-pine forests at 5,000–7,000 feet (1,500–2,200 m) elevations. Rainfall is predominantly of summer-fall incidence in Mexico, but more binary in Arizona (Fig. 8.2). A protracted dry spring is common to all. All are cold tolerant in our gardens.

Uses and chemistry

The Parviflorae were used for soap and food by the Indians, as noted under the following accounts of species. Table 8.2 reports the results of the United States Department of Agriculture analyses for sapogenin content. *Agave schottii*, locally used as a crude soap or cleaning agent, contains significant amounts of sapogenin: tigogenin and gitogenin. *Agave toumeyana* was found relatively high in content of hecogenin, several analyses showing around 1.5 percent on a dry weight basis. As this is a manageable precursor for cortisone and the sex hormones, the USDA made a small experimental planting at Sacaton, Arizona, in 1950. Several hundred small offsets were set out and irrigated to encourage rapid growth. However, there was little growth and the planting was soon abandoned at the request of the Pima Indians, who owned the land. Current use of these small agaves is as ornamentals; *Agave parviflora* and *A. polianthiflora* make attractive house plants, if not over-watered.

The more abundant species, *A. schottii* and *A. toumeyana,* are sapogenous and relatively unpalatable. This together with prolific suckering habits may have constituted survival factors protecting them from man and animals. The palatable *A. parviflora* and *A. polianthiflora,* on the other hand, are rare and weak in populations.

Key to Species and Subspecies of the Parviflorae

1. Flowers small, 12–25 mm long; filaments inserted in or near base of tube 2
1. Flowers larger, 30–42 mm long; filaments inserted above base of tube 3
2. Flowers 12–17 mm long, tepals much shorter than or about equaling tube; leaves small, 6–18 x 0.8–1.2 cm
    parviflora, p. 200
    A. Flowers not reflexed, tube twice the length of tepals; leaves not dimorphic
    ssp. *parviflora*, p. 200
    B. Flowers reflexed; tube about equaling tepals; leaves dimorphic, short form and long form   ssp. *flexiflora*, p. 201
2. Flowers 18–25 mm long, tepals twice the length of tube   *toumeyana*, p. 209

A. Leaves larger, 20–30 x 1.5–2 cm, only 40–70 per mature rosette; margins without denticles
    ssp. *toumeyana*
B. Leaves smaller, 6–15 x 0.6–0.8 cm, 100–200 per rosette; margins with denticles below mid-leaf
    ssp. *bella,* p. 211
3. Flowers pink, tube 22–30 mm long; leaves linear lanceolate, 7–18 cm long
    polianthiflora, p. 203
3. Flowers yellow, tube 8–14 mm long; leaves linear, 20–50 cm long, 0.8–1 cm wide   *schotti,* p. 205
    Leaves 1.7–2.5 cm wide
    schottii var. *treleasii,* p. 207

Table 8.1. Flower Measurements in the Parviflorae (in mm)

Taxon & Locality	Ovary	Tube	Tepal	Filament Insertion & Length		Anther	Total Length	Coll. No.
parviflora								
Sierra Pajarito, Ariz.	7	5.5	2	1	10	5	14†	17752
	6	5×4	2.5×3	1	9	6	14†	17752
Patagonia, Ariz.	7	6×4.5	3.5×2.5	1	14	8	17†	Eng.
	7	7×4.5	3×2.5	1	13	8	17†	Eng.
p. ssp. *flexiflora*								
Guasabas, Son.	7	4×5	5×2.5	0	14	7	15†	22939
	7	3.5	3.5	0	11	6	14*	16638
Bacadehuachi, Son.	9	3–4	4–3.5	0	12		17*	16643
	7	3–4	4	0	13	6	14*	16643
Matape, Son.	7.5	2.5	4–5	0	15	5	14†	17744
	7	3	4	0	13	5	13†	17744
Nacori Chico, Son.	6	3×5	5×2.5	0	12	6.5	13†	3305
polianthiflora								
Sierra Charuco, Son.	10	30×6	3×3	4	34	13	40*	8013
	9	24×5	4×4&3	4	37	11	37*	8013
Majalca, Chih.	11	22×5	7×3.5	5–6	25	9.5	40†	22963
	10	25×7	5×3.5	6–7	27	10	41†	22963
schottii								
Sierra Pajarito, Son.	12	10	7	6	18		29†	17751
	14	9	9	6	17		31†	17751
Hunt. Bot. Gard., Cal.	14	11×8	11×6	9	20	11	36†	19858
	11	10×8	16×5	9.5	22	15	37†	s.n.
Benson, Ariz.	12	14×8	13×4.5	11&10	20	14	39†	22578
	10	11×8	13×5	9&7	21	13	34†	22578
Rcho. Primavera, Son.	13	8×8	12×5	5–6	16	16	32†	21988
	13	11×8	16×5	8	22	17	40†	21988
Sierrita Picu, Son.	14	8×7	10×4	6	13	10	31†	Noel
toumeyana								
Queen Creek, Ariz.	13	4×8	8×4&7	2	15	9	25†	22307
	13	3.5	7×4&6	2	12	9	23†	22307
Fish Creek Hill, Ariz.	10	3×6	9×3	2	17	9.5	22†	22308
	16	3×8	9×4	2	16	10	28†	22308
Sunflower, Ariz.	11	4×8	7×3.5	3	13	9	21†	22573
	12	4×7.5	8×3.5	3	15	9	23†	22573
Camp Creek, Ariz.	12	4×7	7×3.5	3	13	9	22†	22569
	14	4×10	8×3.5	3	13	9	25†	22569
Winkelman, Ariz.	11	4×6	8×3	3	14	9	23†	14609
t. ssp. *bella*								
Parker Creek, Ariz.	9	4×6	7.5×3	3	13	9	21†	22309
	10	3×6	6×3	2	11	7	18†	22309
New River Mt., Ariz.	12	3×7	6×2.5	2	12	7	20†	22570
	8	3×6	6×2.5	2	11	7	18†	22571

* Measurements from dried flowers relaxed by boiling.

† Measurements from fresh or pickled flowers.

SOURCE: Wall et al. (1954–61).

Table 8.2. Sapogenin Content in the *Parviflorae*
(given in percentages on dry weight basis)

COLL. No.	SOURCE LOCALITY	MONTH COLL.	PLANT PART	SAPOGENINS*					
				TOTAL	TIG.	HEC.	CHL.	MAN.	GIT.
parviflora									
10022	S. of Ruby, Ariz.	Dec.	pls.†	0					
polianthiflora									
10229	Sierra Charuco, Son.	Mar.	leaf	0					
11334	Sierra Saguaribo, Son.	Dec.	pls.	0.2	55	45			
schottii									
10014	Peloncillo Mts., N.M.	Dec.	leaf	0.05	100				
10210	El Datil, Son.	Feb.	pls.	0.35			100		
10020	Catalina Mts., Ariz.	Dec.	pls.	0.75					100
toumeyana									
9982	Superior, Ariz.	Nov.	pls.	0.3		?			
10033	Winkelman, Ariz.	Dec.	pls.	0.13		100			
1027	Queen Creek, Ariz.	June	leaf	0.9		65		30	5
11623	Fish Creek Can., Ariz.	Mar.	pls.	1.5		100			
11625	Fish Creek Can., Ariz.	Mar.	pls.	1.37		50		25	25
3437	Magma Copper Mines, Ariz.	Feb.	leaf	1.5					
3492	Sacaton, Ariz.	Mar.	pls.	1.8		?		?	?
3555	Kingman (?), Ariz.	Oct.	leaf	1.5		?		?	?

SOURCE: Wall et al. (1954–61).

* Tig. = tigogenin; Hec. = hecogenin; Chl. = chlorogenin; Man. = manogenin; Git. = gitogenin.

† Whole plants.

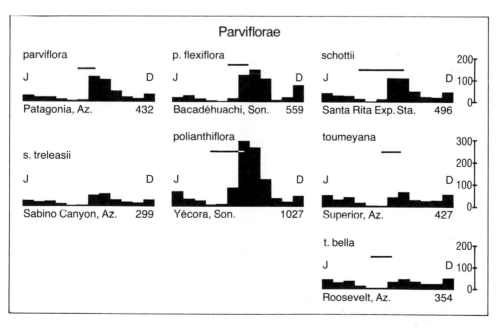

Fig. 8.2. Rainfall (silhouettes) and flowering seasons (bars) of the Parviflorae. Relevant meteorological stations with average annual rainfall in millimeters. Data from U.S. Weather Bureau and Hastings et al., 1969. Flowering periods based on herbarium specimens (see Exsiccatae) and observations.

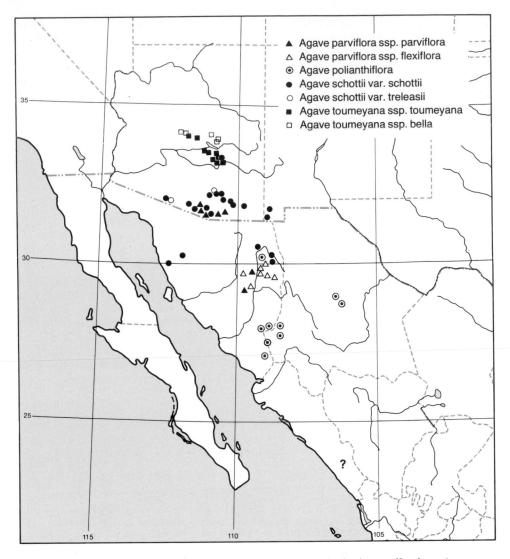

Fig. 8.3. Distribution of the Parviflorae; based on herbarium collections (see Exsiccatae).

Agave parviflora

(Figs. 8.1, 8.2, 8.3, 8.4, 8.5; Tables 8.1, 8.2)

Agave parviflora Torr. subsp. *parviflora*

   Agave parviflora Torr., U.S. & Mex. Bound. Bot. 214, 1859.

Very small, single or cespitose rosettes 10–15 cm high and 15–20 cm broad; leaves 6–10 x 0.8–1 cm, oblong-linear, widest at or above the middle, plane above, convex below, green, white bud-printed above and below, the margin conspicuously white-filiferous and minutely toothed near the base; spine weak-subulate, 5–8 mm long, brown to grayish white; spike 10–18 dm tall, laxly-flowered through upper half of shaft which is frequently reddish; bracts with lowest flowers 1–3 cm long, much smaller above, scarious, long-subulate from a broad deltoid base, ephemeral; flowers pale yellow, in 2s, 3s, or 4s, 13–15 mm long; ovary proper 4–5 mm long; neck 2 mm long; tube 5 mm long, urceolate; tepals 2–3 mm long, the outer slightly longer than the inner, the latter much broader than long, 1.5 x 3–4 mm and abruptly narrowed below in the tepal sinus, all erect to incurved; filaments erect 10–12 mm long, inserted at base of tube; anthers 5–6 mm long; capsules orbicular to oblong, 6–10 mm in diameter, sessile to short pedicellate, shortly beaked; seeds half-round, black, wedge-shaped, 3 mm long on the thin flat edge, 2.5 mm in diameter, much thickened on the curved side.

Type: *Schott* in 1855, U.S. "Pajarito Mountains, Sonora," which was located by Trelease (1911) as near international boundary marker No. 129, between Sonora and Arizona, 10–15 miles W of Nogales (l, infl). Locally, these mountains are referred to as the "Peña Blanca Mountains" and on some road maps are shown as the "Atasco Mountains."

Agave parviflora is well named for its small flowers, the smallest in the genus. The tepals are no more than broad lobes around the apex of the small cylindrical tube (Fig. 8.4). The white bud-printed filiferous leaves are very similar to those of its relative *A. polianthiflora,* and the species are certainly separable only by their distinctive flowers, as given in the key above. *A. parviflora* produces rhizomatous offsets sparingly during its slow growth to maturation. The rosettes remain green for 1 to 2 years after flowering, but I have not observed offsetting in the post-flowering period. The flowers mature in late spring to summer and are pollinated by wild bumblebees and carpenter bees. The seeds are gradually released from the drying capsules during fall and winter. These diminutive leaf succulents make attractive potted plants on window sills and patios, but water must be rationed to maintain their natural compact habit.

Plants in native rocky habitat on Sierra Pajarito, December 1950, are shown in Fig. 8.5.

Agave parviflora ssp. *flexiflora*
(Figs. 8.1, 8.2, 8.3, 8.4; Table 8.1)

Agave parviflora ssp. *flexiflora* Gentry, U.S. Dept. Agr. Handb: 56, 1972.

Small, single or cespitose rosettes with filiferous, dimorphic, white bud-printed, green leaves. Leaves linear or lanceolate, 6–10 x 1 cm or 15–18 x 1.2 cm, plane above, convex below, frequently widest above the middle, short-acuminate, the margin minutely toothed near the base; spine small, weak, whitish, 5–8 mm long; spikes slender, 15–25 dm tall, laxly flowered, with small, triangular, scarious, caducous bracts; flowers 1–3 at a node, mostly geminate, short-pedicellate, yellow, sacate, 14–17 mm long, the corolla with anthers and pistil flexed downward at anthesis; ovary 6–8 mm long, fusiform; tube 3–4 mm long; tepals 3.5–5 mm long, broadly lanceolate, obtuse; the inner sharply keeled and with broad hyaline margins, slightly shorter than the outer, filaments 14–16 mm long, inserted in base of tube; anthers 5–6 mm, long, yellow; capsules orbicular, 8–10 x 8–10 mm, nonstipitate, apiculate, on short stout pedicels; seed as for species, 3 x 2 mm, hemispherical or irregular.

Type: *Gentry 16638, US,* dups. DES, MEX. Ca. 15 miles E of Guasabas along road to Huachinera, Sonora, Mexico, May 20, 1957, on volcanic rocks in oak woodland, elev. ca. 4,000 feet (l, f). The habitat is the short, sparse, open, grama grasslands with gravelly clay soils derived from volcanics between 2,000 and 4,500 feet elevations in eastern Sonora (Fig. 8.3).

A. parviflora flexiflora differs from *A. parviflora parviflora* in the downward flexed flowers, in the longer tepals, and the tendency to develop longer leaves. The latter character is conspicuous in the collections from west of Guasabas, but it is seldom apparent in the collections from west of the Bavispe and Moctezuma rivers, the leaves of which closely resemble those of typical *A. parviflora.* The downward flexing of the

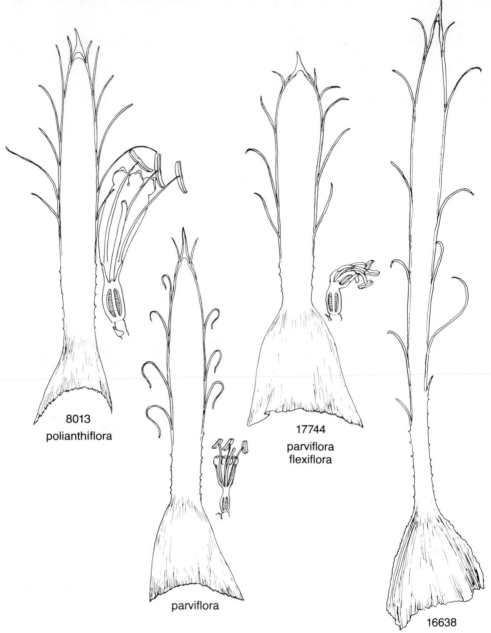

Fig. 8.4. The similar leaves of *Agave polianthiflora*, *A. parviflora*, and *A. parviflora flexiflora*, with respective distinguishing flower sections. *A. parviflora flexiflora* has two leaf forms.

flower develops with anthesis, becoming marked with the downward flexure of the growing pistil, a position that assures contact of stigma and anthers with the visiting bees, bearing pollen from other flowers. The bees enter the flowers from below, poking their heads up between pistil and anthers to drink nectar from the tilted flower tube. Bumblebees were observed working the flowers assiduously east of Guasabas at the time of specimen collections in May and June.

Local names for these diminutive agaves were reported as "tauta," "tautilla" and "sóbali" or "sóbari." They were said to be among the sweetest and most edible of the wild "mescales," when properly cooked, but rarely used because of their small size. All forms of the species are attractive as pot ornamentals. A specimen plant in the Huntington Botanical Gardens that had flowered bore also basal flowering shoots and bulbils in July.

Agave polianthiflora
(Figs. 8.1, 8.2, 8.3, 8.4, 8.6, 8.7; Tables 8.1, 8.2)

Agave polianthiflora Gentry, U.S. Dept. Agr. Handb. 399; 51, 1972.

Small, single or cespitose rosettes 10–20 cm high and 20–30 cm broad; leaves 10–20 x 1–1.3 cm, linear lanceolate, widest in the middle, plane above, convex below, green, white bud-printed above and below, the margin conspicuously white-filiferous and minutely toothed near the base; spine 7–10 mm long, grayish, weak; spike 12–20 dm tall; the stalk red, bearing flowers above the middle; bracts 1–2 cm long, deltoid-lanceolate, subulate, chartaceous; flowers 37–42 mm long, short-pedicellate, usually in pairs, or 1 or

Fig. 8.5. *Agave parviflora* near the type locality on Sierra del Pajarito, Dec. 10, 1950.

Fig. 8.6. *Agave polianthiflora* in flower in the Gentry garden near Murrieta, California. June 1957.

Fig. 8.7. *Agave polianthiflora* drawn from the type collection; spike and leaves, x 1.3, spine, flower cluster, and flower section enlarged.

3, at a node, pruinose pink, long-tubular, subtended by scarious, triangular bractlets 3–6 mm long; ovary 9–12 mm long, red, ovoid; tube 22–32 mm long, 6–8 mm wide at the top, very narrow and curved below; tepals ovate to deltoid, subequal, 4–7 mm long, filaments 27–37 mm long, inserted 4–6 mm above base of tube; stamens short, reddish before anthesis, yellow afterward, shortly exserted; pistils exserted before anthesis, slender; capsules orbicular to ovoid, 10–15 mm long, apiculate, pruinose, rugose; seeds black, lucid, thick, 3.5–4 x 2.5–3 mm.

Type: *Gentry 8013, US,* dups. DES, MEXU. High rocky rims around Arroyo Hondo, Sierra Charuco, Chihuahua, April 17–25, 1948 and cultivated at Murrieta, California, flowering there in June, 1957.

It occurs infrequently in small clones on rocky outcrops in the pine and oak forests between 4,000 and 6,500 feet (1,250 and 2,000 m) elevation from the Río Bavispe south on both sides of the Sonora-Chihuahua border and eastward to latitude of Cd. Chihuahua (Fig. 8.3 and Exsiccatae).

The long tubular red or pink flower with its short tepal lobes distinguish this small agave from all others. The red pigment is made dusky with a pruinose over-lay. The flowering shaft is also reddish. The flowers resemble those of the genus *Polianthes*, whence the specific name. Were it not for the agavoid leaves and the exserted filaments, this plant could be placed in *Polianthes*. It is also unlike other *Agave* in not having a protandrous flower, as the pistil in this plant is exserted before anther dehiscence.

The leaves on the smaller rosettes of *A. polianthiflora* are scarcely separable from those of *A. parviflora* (Fig. 8.4). In moist or more fertile situations and if watered frequently in cultivation, the leaves become much larger than in its natural habitat. The Felger collection from Yecora has unusually broad coarse leaves. In Río Mayo plants (Gentry, 1942), it was reported doubtfully as *A. hartmanii* S. Wats. Watson described that taxon from a growing plant, in the now extinct Cambridge Botanical Garden, collected originally by Hartman and Lloyd from an unspecified locality in the northern Sierra Madre Occidental of Sonora or Chihuahua on the Lumholtz expedition in the 1890s. Since no herbarium specimen was prepared and the garden plant disappeared, it is not possible to orient Watson's species. Watson's brief description could apply to several taxa, all growing in the same region. These taxa can be identified certainly only by their flowers. Wiggin's citation of *A. hartmanii* (Shreve and Wiggins, 1964), based on a collection near Matape, is referable to *A. parviflora flexiflora*. It seems appropriate, therefore, to drop Watson's name as a *nomen confusum*.

The Warihio Indian name for this plant is "taiehcholi." They reported it as a good source of sweet food when pit-baked, and the flowering stalks were formerly employed as arrow shafts. The Río Mayo Mexicans refer to the small plants as "mescalitos." It makes a distinctive pot ornamental, but should be watered very sparingly and given full sunlight, or the leaves become etiolated, even floppy, and the rosette loses its diminutive and attractive compactness. The filiferous leaves are green with white diagonal brushlike markings.

Agave schottii
(Figs. 8.1, 8.2, 8.3, 8.8; Tables 8.1, 8.2)

Agave schottii Engelm. var. *schottii*, Acad. Sci. St. Louis Trans. 3: 307, 1875.
  Agave geminiflora sonorae Torr., U.S. & Mex. Bound. Bot. 214, 1859.
  Agave schottii var. *serrulata* Mulford, Mo. Bot. Gard. Rep. 7:73, 1896.
  Agave mulfordiana, Trel., U.S. Nat. Herb. Contr. 23:140, 1920.

Densely cespitose, small, yellowish green to green rosettes. Leaves narrowly linear, 0.7–1.2 x 25–40 (–50) cm, widest at the base, straight, incurved, or falcate, pliant, flat or somewhat convex above, deeply convex below, smooth above and below, margins with a narrow brown border and sparse brittle threads; spine 8–12 mm long, grayish, fine, rather weak and brittle; spikes 1.8–2.5 m tall, slender, frequently crooked, flowering in upper ⅓ to ¼ of shaft; bracts straw-colored, filiform, acicular, 2–4 cm long; flowers yellow, 30–40 mm long, 1, 2, or 3 on stout pedicels 3–5 mm long with setaceous bracteoles 10–15 mm long or lower spike but shorter above; ovary greenish yellow, 10–14 mm long including neck 4–6 mm long; tube deeply funnelform, 9–14 mm long, very narrow below; tepals 10–16 mm long, yellow, unequally spreading at anthesis, the outer without a keel; filaments 15–22 mm long, inserted high in tube at 6–9 mm above base of tube; anthers light yellow, 10–15 mm long; capsule 10–20 mm long, rounded to apiculate; seeds 3–3.5 mm along the straighter thinner edge, with hilar notch at one corner, variably thickened.

Type: *Arthur Schott* in 1855, *US*. Sierra del Pajarito in southern Arizona (l, f).

Fig. 8.8. *Agave schottii* on the south end of Peloncillo Mountains in southwestern New Mexico.

Taxonomic comment

The flower of *Agave schottii* has a slender deep tube, which, together with its narrow ovarian neck, give the flower a characteristically long tubular appearance. The ratio of tube length to tepal length, however, is variable, the tube equaling or exceeding the tepals in length (Fig. 8.1). The densely cespitose plants, 1 m or more broad in natural habitat, with congested narrow linear leaves, form the typical aspect of this plant. The flowers are apparently pollinated by both bats and bees, and there appear to be structural features of the flower, such as the ascending nectariferous tube and short filaments, to accommodate this symbiotic relationship. In gardens, *A. schottii* is easily confused with narrow-leaved forms of *A. felgeri,* but they are readily separated by the short flower tube of the latter.

Distribution and use

A. schottii has a relatively extensive distribution across southern Arizona and adjacent Sonora to southwestern New Mexico and northwestern Chihuahua (Fig. 8.3). The low elevation disjunct populations near El Datil and Sierrita Picu in northwestern Sonora are ecologically exceptional, as the main habitat is on elevations above 3,000 feet (900 m) with grama grassland and oak woodland biomes. Here they are so abundant on some cattle ranges that cattlemen regard them with disfavor, as they are not edible to livestock. However, they contribute to soil building and hold soil on the steep rocky slopes (Fig. 8.8).

Among Spanish-speaking people these small nondescript agaves are known as "amole" or "amoliyo" and are used locally for washing clothes. They are not eaten because of the small size of the "heads" (stems) or because of bitter constituents. The plants were found to contain up to 1 percent sapogenins: chlorogenin, manogenin, and tigogenin (Table 8.2). Van Etten et al. (1967) reported the seed to contain 21.4 percent oil and 29.4 percent protein with 75.2 percent of the nitrogen as amino acids. The plants have little if any value as ornamentals.

Agave schottii var. *treleasei*
(Figs. 8.2, 8.3, 8.9)

Agave schottii var. *treleasei* (Toumey) Kearney & Peebles, Jour. Wash. Acad. Sci. 29: 474, 1939.

Agave treleasei Toumey, Mo. Bot. Gard. Rep. 12: 75, 1901.

This variety differs from the rest of the species with its larger, deep green, thicker, wider leaves (15–25 mm). Toumey reported type "in herb. J. W. Toumey" and living cotypes in gardens at the University of Arizona, Missouri Botanical Garden, and Dept. of Agriculture, Washington, D. C. I have seen none of these in the herbaria of the cited institutions. A drawing and two photos published with his description portray the plant, which is certainly closely allied to *A. schottii*. His type came from Castle Rock on the southern slope of the Catalina Mountains, Pima County, Arizona, at ca. 6,500 feet (2,000 m) elevation. A later collection from Organ Pipe National Monument, Pima Co., Ariz., was regarded by Kearney & Peebles as this variety. Further notes are given in the Exsiccatae. Good herbarium specimens are needed to better establish this taxon. Plants in the Boyce Thompson Southwestern Arboretum closely resemble the type clone (Fig. 8.9).

Fig. 8.9. *Agave schottii* var. *treleasei* as found in Boyce Thompson Southwest Arboretum, Oct. 1978. Probably from the original clone cult. at Univ. Ariz., col. by Toumey.

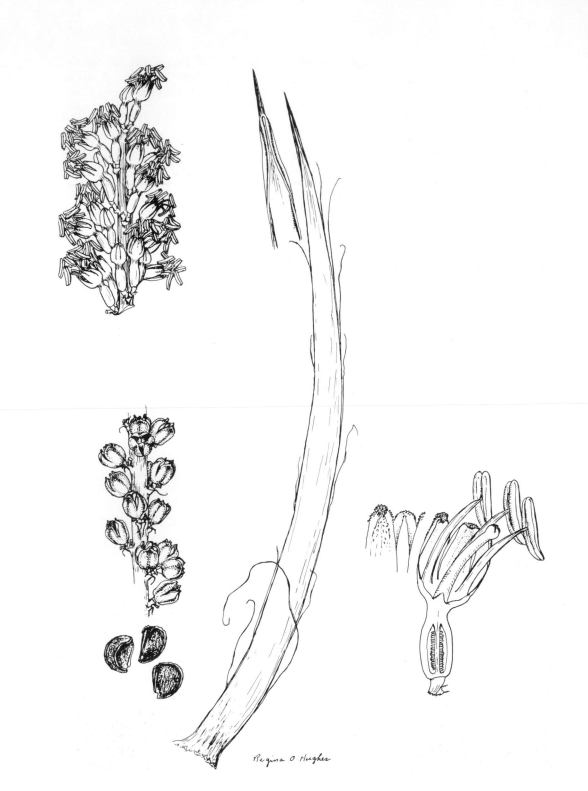

Fig. 8.10. *Agave toumeyana* ssp. *toumeyana*. Leaf, sections of flowering spike and fruiting spike natural size; spine, seeds, and flower sections, x 3.

Agave toumeyana
(Figs. 8.1, 8.2, 8.3, 8.10, 8.11, 8.12; Tables 8.1, 8.2)

Agave toumeyana Trel., ssp. *toumeyana,* Contr. U.S. Nat. Herb. 23: 140, 1920.

Small, densely cespitose, light green or yellowish rosettes with rather few (40–70) leaves. Leaves generally 20–30 x 1.5–2 cm, linear lanceolate, straight or falcate or upcurving, plane above, thickly convex toward base, rather rigid and of unequal length, bud-printed above and below, smooth, the edges with a fine brown margin and white threads, sometimes serrulate at base, spine 1–2 cm long, subulate, brown to grayish, with a short narrow groove above; spike racemose, 1.5–2.5 m tall, densely or laxly and mostly geminately flowered in upper third of shaft; pedicels short, bifurcate; flowers 18–25 mm long, the perianth saccate, green with whitish tepals and filaments, flexed downward; ovary 10–15 mm long with slender bent neck 3–5 mm long; tube 2–4 mm long, angled, broadly spreading, 7–8 mm broad; tepals subequal, appressed to filaments, 7–9 mm long, lanceolate, thin, the outer closely over-lapping the inner; filaments shortly exserted,

Fig. 8.11. *Agave toumeyana* ssp. *toumeyana* in natural habitat on limestone above Superior, Ariz., with saguaro, ocotillo, jojoba, and *Agave chrysantha.* November 1950.

Fig. 8.12. An old clone of *A. toumeyana* in same locality as Fig. 8.11, showing thick clustering habit.

13–17 mm long, inserted 2–3 mm above base of tube; anthers 9–10 mm long, dun-colored; capsules short oblong, 12 x 15 to 8 x 10 mm, sessile, short-beaked, the walls thin, persistent; seeds thick, 2 x 3 mm.

Type: *Toumey 442,* US. Pinal Mountains, Arizona, July 31, 1892 (three split immature rosettes only 8–10 cm long and spikes with capsules).

Endemic in central Arizona in and about the Pinal Mountains from 2,000 to 4,500 feet (625 to 1,400 m) on limestone and volcanic rocks with highland desert vegetation to the chaparral and lower pines; on open rocky ledges (Fig. 8.3).

Agave toumeyana suggests a large edition of *A. parviflora,* having a similar habit, but the former is more variable in leaf form and leaf attitude. The leaves are always more acuminate. The flowers of the former differ consistently in form, size, and proportions from *A. parviflora. Agave toumeyana* is also larger, and the broad shallow tube has more elongate tepals, and the filament insertion is higher in the tube. Within *A. toumeyana* there is little variation in size of flowers and tube-tepal proportions and the overall characters are consistent. A small colony of *A. t. toumeyana* along Camp Creek in the New River Mountains was observed in late November of 1974 with numerous bulbils. They initiated in the axils of the pedicels and were present on both damaged and undamaged spikes. This form of reproduction is not common in the species. Like *A. parviflora* it also has a subspecific satellite, which is distinguished by the rosette characters (Figs. 8.13, 8.14).

Young plants of *A. t. toumeyana* are very attractive. An isolated rosette develops in single symmetry, but in a few years it ruins its own form with prolific close suckering. Rosette growth is slow. Flowering is sparse, irregular, and not showy, but seed is eventually produced abundantly in wild stands. With seed, suckers, and bulbils it appears that *toumeyana* is more devoted to reproduction than to growth.

Analyses of plants by the U. S. Department of Agriculture chemists found a consistent content of sapogenins, 0.5–1.5 percent (Table 8.2). A trial planting set out at Sacaton, Arizona, a former U.S.D.A. experiment station, had to be abandoned after two years because of Pima Indian needs for their land. The growth response there from transplanted offsets was disappointing even with irrigation. The main value of the species is undoubtedly its soil building and holding capacities on the rocky slopes it inhabits.

Agave toumeyana ssp. *bella*
(Figs. 8.1, 8.2, 8.3, 8.13, 8.14; Table 8.1)

Agave toumeyana Trel., ssp. *bella* (Breitung) Gentry stat. nov.
   Agave toumeyana var. *bella* Breitung, Cact. & Succ. Journ. 32: 81, 1960.

Small, compact, filiferous, light green, surculose rosettes, differing from *A. t. toumeyana* in having 100 or more leaves at maturity vs. 40–70, in the smaller, more equal, linear leaves 9–20 cm long vs. 20–40 cm, and in having the brown leaf margin replaced by denticles on the lower half of the leaf. The spikes of *A. t. bella* are smaller than *A. t. toumeyana,* but the flower measurements (Table 8.1), organ proportions (Fig. 8.1), and general form (Fig. 8.4) show the flowers of the species.

Type: *Breitung & Gibbons 18153,* CAS. Indian Coral mesa near Parker Creek, Sierra Ancha, Gila County, Arizona.

Distribution

The *bella* subspecies has a confined distribution in central Arizona between 4,000 and 5,000 feet (1,250 and 1,560 m) elevations (Fig. 8.3). Its limited habitat is within the Arizona chaparral with juniper on open stony slopes and benches. It is known from the southwest slope of the Sierra Ancha in the Salt River drainage and the New River Mountains to the west of the Verde River. It has recently been located in the Bradshaw Mountains south of Prescott (letter of F. W. Reichenbacher, 1980).

Taxonomic comment

A. toumeyana bella shows considerable distinction from the typical species and because of this and its natural nearly separate distribution, I rank it as a subspecies, not to be confused with the forms and varieties common to many agave populations and species. No morphological intermediates have been observed. The short leaves discussed by Weber (1965) are, after all, just small leaves of subspecies *toumeyana,* which are highly variable.

The short leaf form and flexed flowers occurring in *Agave tourmeyana* ssp. *bella* form an interesting parallel in these characters as they occur also in *A. parviflora flexiflora.* The former, however, has a larger flower with wider tube, longer tepals, and the filaments are inserted higher in the tube. These are the same critical characters that

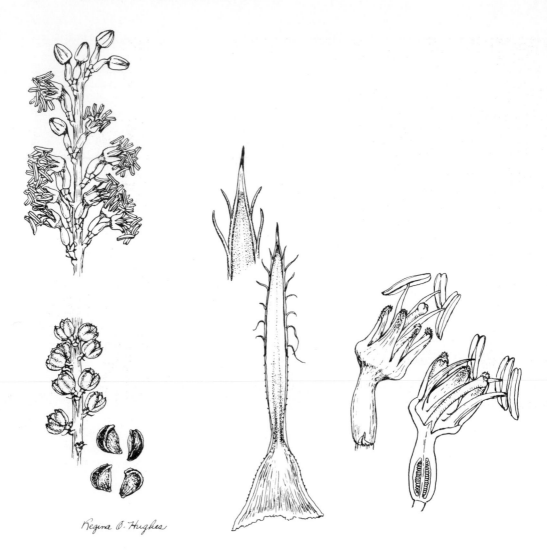

Fig. 8.13. *Agave toumeyana* ssp. *bella*. Leaf, inflorescence, and infructescence, x 1; leaf tip, flower sections, seeds, x 3.

separate the species *A. toumeyana* from *A. parviflora* and prevent merging of the two satellites under either species. The forms with longer leaves and flexed flowers of the respective subspecies, hence, are regarded as isolated developments in different gene pools, or homologous variations. It is probably also biogenetic because carpenter or bumble bees were observed working the flexed flowers of both subspecific taxa. The respective flower postures can be compared in Fig. 8.1.

Agave toumeyana bella forms fairy rings (Fig. 8.14). This together with the large number of leaves in mature rosettes indicates a long sexual generation age; clones probably exceed 100 years old. It is well-named ''bella'' for beauty and forms a most attractive ornamental in rock gardens or pots.

Fig. 8.14. *Agave toumeyana* ssp. *bella* at the type locality near Parker Creek, Sierra Ancha, Arizona, June 1967.

Parviflorae Exsiccatae

Agave parviflora parviflora

ARIZONA, *Engard & Haughey 1063a,* DES. Las Guijas Mts., R 10E, T 21S, S 5, Pima Co., 19 Aug. 1977; elev. ca. 4,000 ft. on rocky slopes in desert grassland (l, cap).

Gass s.n., ARIZ. 2 miles W of Castle Rock on road to Ruby, Oct. 1973.

Gentry 10022, DES, MEXU, MICH, US. Ca. 11 miles SE of Ruby, Pajarito Mountain, 10 Dec. 1950; elev. 4,500 ft., oak woodland (l, cap, topotypic).

Gentry 17752, DES. Ibid, 4 July 1959 (f, 6 pl).

Gentry & Engard 22999, DES. Patagonia near Sonoita Creek; flowered in D.B.G. 1972 (l, f, cap).

Gentry & Kaiser 19900, 16 miles NE of Nogales along Arroyo Sonoita, 10 April 1963; elev. ca. 4,000 ft., oak woodland.

Lehto & McGill 20490, ASU, DES. Between Peña Blanca and Ruby, 2 miles W of stream crossing from Peña Blanca Lake, 9 Sep. 1976 (l, cap).

Marshall & Blakely H-2152, DES. Florita Canyon, Santa Cruz Co., 18 Aug. 1950 (l, f).

McGill & Kirk s.n., ASU, DES. 4 miles SW of Peña Blanca Lake, Sta. Cruz Co., 24 Nov. 1978 (l, cap).

Peebles 11444, ARIZ, US. 10 miles SE of Ruby, Sta. Cruz Co., 4 May 1935; elev. 4,600 ft. (l, f).

Pinkava et al. K11111A, ASU, DES. Thumb Rock Picnic Area, Peña Blanca Lake, 6 Sep. 1975; elev. ca. 4,100 ft; grassland.

Schott s.n. Type, **US.** ''Pajarito Mountains, Sonora'' in 1855 (l, infl).

SONORA. *Felger 3492,* ARIZ, DES. 18.3 miles W of Moctezuma along road to Mazochahui, 18 July 1960; elev. ca. 3,000 ft., lower edge of oaks.

Gentry 16655, DES. Ures to Moctezuma, 22 May 1957; elev. ca. 3,500 ft., oak-*Lysiloma* woodland (pl).

Van Devender s.n., ARIZ. Rancho La Brisca, ca. 4 miles N of Agua Fria on trib. of Río Saracachi, 7 June 1978; elev. ca. 1,000 m (l, f).

A. parviflora flexiflora

SONORA. *Felger & Marshall 3305,* ARIZ, DES. 18.3 miles NW of Nacori Chico, 6 June 1960; elev. ca. 3,450 ft., rocky low hills, grassland (l, f).

Gentry & Arguelles 16638, Type, DES, MEXU, US. Ca. 15 miles E of Guasabas along road to Huachinera, 20 May 1957; elev. ca. 4,000 ft., on volcanic cliffs in oak woodland (l, f).

Gentry 16601, 17744, DES, MEXU, US. 3–4 miles NE of Matape, May 1957 (l, bud), July 1959 (l, f); elev. 2,000–2,500 ft., "savanilla" grassland.

Gentry & Arguelles 16643, DES. MEXU, US. 6–8 miles NW of Bacadehuachi, 21 May 1957; elev. ca. 3,500 ft., rocky slope with *Lysiloma* (l, f).

Gentry & Arguelles 22938, DES. MEXU, US. 17 miles E of Moctezuma along road to Guasabas, 3 June 1971; elev. ca. 3,600 ft., with oak.

Gentry & Arguelles 22939, DES, MEXU, US. 9–10 miles E of Río Bavispe along road to Huachinera, 4 June 1971; elev. ca. 3,750 ft., volcanic rocky short grassland with oaks (l, f).

Nabhan et al. X493, ARIZ. 2 miles S of Aribabi, Río Bavispe, 21 Aug. 1976.

Wiggins & Rollins 413, ARIZ, CAS. Western slope of Sierra Batuc, 5 miles N of Matape, 8 Sep. 1941 (l, cap).

Agave polianthiflora

CHIHUAHUA. Gentry 8013, Type, DES, MEXU, US. Around Arroyo Hondo, Sierra Charuco (Son.?), 17–25 April 1948, cult at Murrieta, Calif., fl'ed June 1957 (l, f, photo).

Gentry 2850, DES. San Jose de Pinal, 22 Sep. 1936; rocky pine slope (l, cap).

Gentry & Arguelles 22963, 14 miles W of route 45 along road to Majalca, 14 June 1971; elev. ca. 5,700 ft., rocky volcanic canyon, rare (l, f).

Knobloch 5816, MSC. Mojarachic (Batopilas Dist.), 22 Aug. 1939 (l, cap).

Le Šueur 572, MO. Majalca, 24 June 1936 (l, f).

Nabhan 959, ARIZ, DES. Rancho Quemado, Sierra Charuco, 15 Sept. 1978; elev. 4,500–5,000 ft., S-face slope in pine-oak assoc. (l cap).

Pennington 37, TEX. Yepachic, 12 July 1970 (l, f).

Pringle 1995, MO, US. Dry porphyritic hills near Cd. Chihuahua, 6 Sep. 1888 (l, one f, cap. The flower in US definitely identifies this early collection).

SONORA. *Gay 3649,* HNT. 8 miles E of Santa Rosa, Mun. de Yecora, June 1974; alt. 5,400 ft. (Fl'ed. Hunt. Bot. Gard. June 1978).

Gay 3651, HNT. 7 miles W of Yecora, June 1974; alt. 5,700 ft. "flowers red." (Fl'ed Hunt. Bot. Gard. July 1977).

Gentry 10229, DES, MEXU, MICH, US. Arroyo Hondo, Sierra Charuco, 6–7 Mar. 1951.

Gentry 11384, Sierra Saguaribo, 26 Dec. 1951 (pl only).

Pennington 123, TEX. Maicoba, June 1968; elev. 5,000 ft. (l, f).

White 2820, ARIZ, MICH. Canyon de la Escalera (loop of the Bavispe River), 23 June 1940 (l, f).

Agave schottii

ARIZONA. *Bailey s.n.,* US. Sabino Canyon, Sta. Catalina Mts., 1913 (l, cap).

Brass 1420, A. Tanque Verde Canyon, 18 miles E of Tucson, Mar. 1940.

Gentry 10020, DES, MEXU, MICH, US. Along road to Mt. Lemmon, S slope of Catalina Mts., Dec. 1950; elev. ca. 4,000 ft. (l, cap).

Gentry 22578, DES, US. Near roadside rest area, 8–12 miles S of Benson on road to Huachuca, 16 June 1968; elev. ca. 4,500 ft. (l, f, cap).

Goldman 2525, US. Coyote Mt., 2 Sep. 1915; elev. ca. 4,000 ft.

Griffiths 4786, US. Sta. Rita Forest Reserve, 24–26 June 1903.

Earle & McGill s.n., DES. Peña Blanca Lake, Sta. Cruz Co., 8 May 1964 (l, f).

Harris C16280, US. The Basin, Sta. Catalina Mts., 9 July 1916 (l, infl).

Lehto & McGill 20504, ASU, DES. Grassy bajada of Whetstone Mt., 8.8 miles N of hwy. 82–90, along hwy. 90, 11 Sep. 1976 (l, bud).

Lehto & McGill 20331, ASU, DES. Road to Bear Canyon Camp Ground, 1.5 miles above Coronado Nat. For. Bound., Catalina Mt., 9 Sep. 1976.

Lehr 1602, DES. Molina Basin Campground, Sta. Catalina Mt., Pima Co., 13 July 1975; elev. ca. 4,000 ft., open rocky slopes with oak, juniper, manzanita.

McKelvey 1642, A. Pajarito Mts., along road to Ruby from Nogales, 26 Mar. 1930 (l, f, pl).

McKelvey 1589, 1608, A. Near Colossal Cave in Rincon Mts., March 1930 (l, cap, bul).

Lemmon & wife s.n., US, 298, A. Santa Catalina Mts., May 1881 (l, f).

Mearns 2686, US. Pajarito Mts., Warsaw Mill, Pima Co., 3 Dec. 1893 (l, cap).

Morton 97, US. Vicinity of Ruby, Sta. Cruz Co., alt. ca. 1,375 m.

Neally 148, Sabino Canyon, Sta. Catalina Mt. in 1891; elev. 5,000 ft.

Peebles & Harrison 530, US. Baboquivari Canyon (Baboquivari Mt.), 29 Oct. 1925 (l, infl).

Peebles et al. 7982, US. Rincon Mts., 2 Aug. 1931 (l, f, cap).

Pringle s.n., US. Rincon Mts., 19 June 1884 (l, f).

Schott in 1885, US. Type. Sierra del Pajarito in southern Ariz. (l, f).

Toumey s.n., A. Santa Catalina Mts., 29 July 1894 (l, f).

Wetmore 689, US. Dragoon Mts., 17 July 1919 (l, infl).

NEW MEXICO. *Gentry & Ogden 10014,* DES, MEXU, MICH, US. Southern end of Peloncillo Mts., ca. 2 miles N of US-Mex. boundary, 3 Dec. 1950 (l, photo).

Hershey 3383, A. Canyon Guadalupe, Hidalgo Co., 6 Oct. 1944 (l, cap).

Mearns 575, A. US. Guadalupe Canyon, 24 July 1892.

SONORA. *Felger 3276,* ARIZ. Ca. 16 miles SE of Magdalena along road to Cucurpe by Sierra Baviso, 22 May 1960 (l, f).

Gentry & Weber 21988, DES, MEXU, US. Rcho. Primavera, W of Sierra Jojoba, 24 July 1966 (l, f).

MacDougal & Shreve 14, ARIZ. Picu Pass, 18–20 miles NE of Libertad.

MacDougal & Shreve s.n., ARIZ. Port Libertad.

White 628, MICH, US. Between Sta. Rosa Canyon and Bavispe, 20 July 1938 (l, f).

White 4466, MICH, US. W. foothills of Sierra de la Caballera, ca. 10 miles E of Col. Morelos, 22 Sep. 1941; alt. 3,400 ft. (l, f).

Wiggins 6097, CAS, US. Low hills 18 miles NE of Libertad, 26 Oct. 1932.

Agave schottii treleasei

ARIZONA. *Toumey s.n.* Type "in herb. Toumey." Castle Rock, SW slope of Santa Catalina Mt., Pima Co., Dec. 1896 (pl).

Forrts s.n., US. Organ Pipe Cactus Nat. Mon., Pima Co., 1 May 1948 (l, although large, the leaves do not appear typical of the Toumey variety).

Agave toumeyana

ARIZONA. *Darrow s.n.,* ARIZ. On S side of Superstition Mt., Don's Trail, Pinal Co., 10 April 1938.

Darrow s.n. ARIZ. 2 miles E of Coolidge Dam, Pinal Co., 2 June 1943 (f).

Eastwood 17357, MO, CAS. Top of Fish Creek Hill, 21 May 1929.

Gentry 22573, DES, US. 1 mile N of Sunflower, Maricopa Co., 12 June 1962; elev. 3,500–3,700 ft. (l, f).

Gentry 22307, DES, US. Queen Creek Canyon, Pinal Mts. above Superior, 19 June 1967 (l, f).

Gentry & Ogden, DES, US, MICH. Ibid. 24 Nov. 1950 (l, cap).

Gentry & Weber 22308, DES, US. Top of Fish Creek Hill on Apache Trail, Maricopa Co., 20 June 1967 (l, f), elev. 2,000–2,600 ft.

Gentry & Weber 22569, 23685, DES, US. Camp Creek, Maricopa Co., 10 June 1968 & June 1978; open chaparral slope, elev. ca. 3,500 ft. (l, f).

McKelvey 795, A. Top of Fish Creek Hill (l, f).

McKelvey 1110, A. Ibid. 21 May 1929.

McKelvey 949, A. Near top of divide on road from Globe to Winkelman, Pinal Range, 10 June 1929 (infl).

McKelvey 1103, A. Road to Horse Mesa Dam off Apache Trail, 21 May 1929 (infl).

Peebles 10773, US. Fish Creek, Apache Trail, Maricopa Co., 16 Mar. 1935.

Peebles 12937, US. Devil's Canyon, Pinal Mts. above Superior, 18 Feb. 1936.

Peebles 13308, US. Ibid, 18 May 1936 (l, f).

Peebles & Parker 14609, US. 15 miles N of Winkelman, Gila Co., 18 April 1940; alt. 4,000 ft (l, f).

Toumey 442, Type, **MO.** Pinal Mountains, 31 July 1892 (rosettes, infruc. Small depauperate plants, but otherwise typical).

Agave toumeyana bella

ARIZONA. Breitung & Gibbon 18153, CAS. Type. Indian Coral mesa near Parker Creek, Sierra Ancha, Gila Co.

Gentry & Weber 22570, DES, US. 13 miles W of Magazine Spring on side road to Bloody Basin, New River Mts., 11 June 1968 (l, f); elev. ca. 4,300 ft.

Gentry & Weber 22571, DES, US. 14.5 miles W of Magazine Spring on side road to Bloody Basin, New River Mts.; elev. ca. 5,000 ft. (l, f).

Gentry & Weber 22309, DES, US. 2.1 miles S of Parker Creek F. S. Camp, Gila Co., 20 June 1967; elev. ca. 4,500 ft. (l, f, cap, topotypic).

Little s.n., US. Ibid, 21 June 1938.

McGill s.n., DES. Red Rover Mine, ca. 4.5 miles NE of Jctn. with Seven Springs Road along F.S. road 254, Aug. 1977; elev. ca. 4,500 ft. (l, f, cap).

McKelvey 773, 945, A. Road to Pleasant Valley, Sierra Ancha, 16 June 1929 (l, infl. First botanical collections, annotated by Trelease as *Agave mckelveyana* Trel., but was never published).

Weber s.n., DES. Rim of Boulder Basin, New River Mts., Maricopa Co., 10 May 1972; decomposed granite (l, f).

9.

Group Polycephalae

Plants generally perennial with axillary branching, forming with age large clusters of many rosettes, or some simple multiannuals; rosettes small to medium-sized. Leaves soft-fleshy, thick, tender, ovate to lanceolate, with or without narrow margins, serrulate to dentate; spines small to moderate, acicular. Inflorescence spicate, lax or congested, relatively short; flowers 1–several, usually geminate, in axils of persistent bractlets, fleshy, yellow to brown or red or purplish; ovary and tube trigonous; tepals ca. equal to or longer than tubes, spreading to reflexed, dimorphic, the inner usually broader, with thick broad keel; stamens long-exserted, with small anthers; pistil in post-anthesis exceeding stamens; capsules trigonous.

Typical species: *Agave celsii,* Hook., Curtis Bot. Mag. III, 12: pl. 4934, 1856.
In part: Sect. Subcarinatae Jacobi 1864: 501; Group Aloideae Baker 1888: 165; Sect. Anacamptagave Berger 1915: 38.

Taxonomic comment

This group is distinguished from leaves to flowers by its soft succulent structures, by a growth habit of forming dense mounds or mats of tender spongy rosettes (Fig. 9.6), and by the 3-angled form of ovary and tube conspicuously grooved. The teeth are usually very small. The color of leaves and flowers and the form and attitude of leaves are variable. The simple habit of *Agave polyacantha* is atypical, and some forms of it appear to be cespitose. It nevertheless appears to belong with this group. A helpful guide to relationships rests with the ideographs in Fig. 9.1 and with Table 9.1.

The Polycephalae, like the Rigidae in subgenus *Agave,* has been fouled by too many names. Berger (1915) in his Section Anacamptagave recognizes 27 specific names. Fourteen of these were provided by Jacobi (1865–70), while the rest were published by various European botanists beginning with Haworth in 1821. Trelease (1920) recognized only six species, but listed 41 binomials as synonyms or of dubious identity. Like Trelease, I am unable to place taxonomically many of these apparent forms and varieties, which are without type and with few exceptions are unrepresented by specimens. The following account of five species and two varieties summarizes my limited acquaintance with this

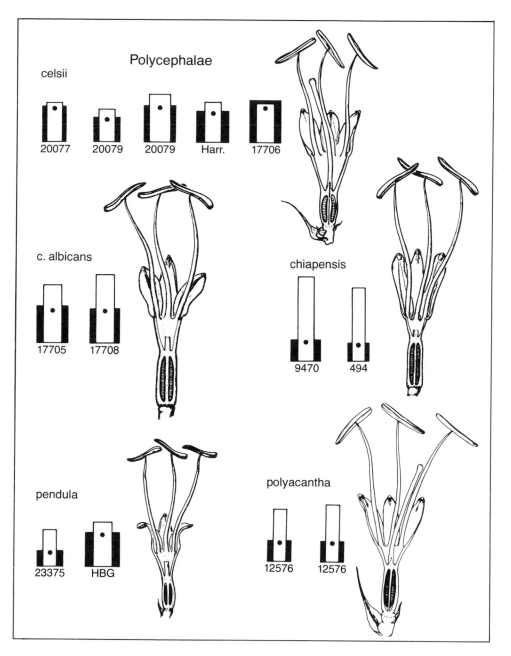

Fig. 9.1. Ideographs of flowers of the Polycephalae with correspondent long sections of flowers.

group. *A. bakeri* and *A. yuccaefolia,* included in Berger's Anacamptagave, I have placed in my Group Amolae. Because of inadequate recognition criteria, some of my synonymy allocations are tentative and should not be taken as definitive. What this group needs is a competent botanist resident with the wild populations who can carry forth detailed studies in field and laboratory.

Table 9.1. Flower Measurements in the Polycephalae (in mm)

TAXON & LOCALITY	OVARY	TUBE	TEPAL	FILAMENT INSERTION & LENGTH		ANTHER	TOTAL LENGTH	COLL. NO.
celsii								
S. Cucharas, Tamp.	13	13×10	13×6	12–13	30	21	40†	20077
	15	14×9	15×6	13–14	45	21	45†	20077
Jacala, Hgo.	15	10×10	13×6	9–10	45	17	37†	20079
	18	14×12	18×8	13–14	55	22	49†	20079
S. Cucharas, Tamp.	15	12×12	15×6	11–12	40	21	43†	Harr.
	12	17×13	15×7 & 13	12–13	43	18	46†	17706
Hunt. Bot. Gard., Cal.	17	18×12	12×5 & 10	12–14	44	15	45†	17707
	17	16×12	12–15×5–6	12			46†	17707
c. var. *albicans*								
Hunt. Bot. Gard., Cal.	19	14×12	21×7 & 18	13	47	20	54†	17705
	20	14×12	23×7 & 20	13	47		56†	17705
	20	13×12	27×7–8	12–13	60	21	60†	17708
	19	12×11	26×7 & 25	11–12	58	22	58†	17708
chiapensis								
S. Cristobal Casas, Chis.	22	8×11	33×6	7–8	70	30	64*	9470
	24	8×12	32×6	7–8	70	30	65*	9470
Mun. Zinacatan, Chis.	18	7×8	28×3	7–8	55	22	53*	494
La Mórtola	30	11×11	30×9	10–11	80	32	70	Berg.
La Mórtola (US)	17	11	22	11	60		50	22/VI
pendula								
SE of Japala, Ver.	15	5×10	14×6	5–4	33	14	34†	23375
	14	6×10	13×5	6–5	32	13	34†	23375
Hunt. Bot. Gard., Cal.	12	13×13	18×7 & 17	12	45	18	42†	s.n.
	13	13×13	15×14	10	44	18	40†	s.n.
polyacantha								
Hunt. Bot. Gard., Cal.	20	8×9	21×5 & 20	7	50	20	48†	12576
Murrieta, Cal.	20	8×9	22×5 & 21	7–8	70	21	51†	12576
warrelliana								
La Mórtola	40	14×14	35×10	13–14	85	32	90	Berg.
La Mórtola (US)	36	12	32 & 30	12	75		80*	21/VI
	35	15	35	15			82*	21/VI

* Measurements from dried flowers relaxed by boiling.

† Measurements from fresh or pickled flowers.

Habitat

The Polycephalae are wholly tropical in distribution and are the most hydrophytic group of the agaves. The relatively heavy rainfall of their area is indicated in Fig. 9.2, taken from stations near or within the respective species area. This region comprises the moist Caribbean slopes of the Sierra Madre Oriental and the forested lands of Chiapas from 200- or 300- to 1,500-m elevations. The climate is frostless and average annual rainfall is between 30 (750 mm) and 100 (2,500 mm) inches, with most of it falling during summer and fall. It is one of the major coffee zones of Mexico. The plant communities range from the tropical deciduous forests through open tropical savannas to the high cloud forests. Even more than elsewhere, the agaves are limited to the more open sites on rocks and cliffs; consequently, the populations are fragmentary and disjunctive. However, I have observed *A. polyacantha* perched like an epiphyte on tree trunks, and the species appears more tolerant to shade and to greenhouse culture than most other agaves. Distributions of the taxa are indicated in Fig. 9.3.

Key to Species of the Polycephalae

1. Leaves ovate or oblong to lanceolate, soft fleshy; flowers thick fleshy, tube and ovary trigonous, 6-furrowed 2
1. Leaves lanceolate, firmly fleshed; flowers slender, not trigonous, not markedly furrowed on tube and ovary, smooth *polyacantha* p. 228
2. Bractlets subtending the flowers large with broad bases; flowers large, 60–90 mm long, the tepals 3–4 x length of the tubes 3
2. Bractlets small, narrow, inconspicuous; flowers smaller, 35–60 mm long; tepals 1–2 x the tube length 4
3. Teeth larger, 3–10 mm long, not

united on a continuous red margin *chiapensis*, p. 224
3. Teeth closely serrate, 2–3 mm long, united on a continuous reddish margin *warelliana*, p. 230
4. Leaves lanceolate, long-acuminate, length 7–10 x width; spikes slender, laxly flowered; flowers colorless, small, 35–42 mm long *pendula*, p. 226
4. Leaves ovate to oblong or spatulate, short-acuminate, length 4–6 x width; spikes densely flowered; flowers yellow to purple, larger, 40–60 mm long *celsii*, p. 220

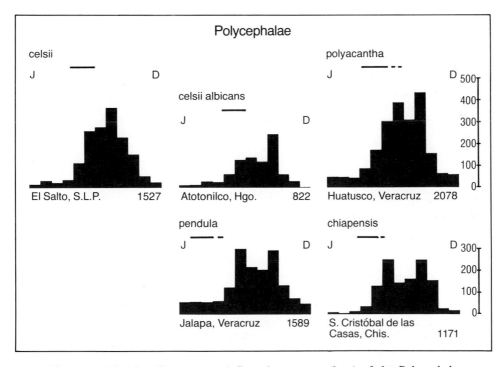

Fig. 9.2. Rainfall (silhouettes) and flowering seasons (bars) of the Polycephalae. Relevant meteorological stations with average annual rainfall in millimeters. Data from Atlas Climatológico de México, 1939. Flowering periods based on herbarium specimens (see Exsiccatae) and observations.

Agave celsii

(Figs. 9.1, 9.2, 9.3, 9.4, 9.5, 9.6; Table 9.1)

Agave celsii Hook var. *celsii*, Curtis Bot. Mag. III, 12; pl. 4934, 1856.

  Agave botterii Baker, Curtis Bot. Mag. III, 32: pl. 6248, 1876.
  Agave bouchei Jacobi, Hamb. Gart. Blumenzeit. 21: 217; 1865.
  Agave haseloffii Jacobi, Ibid, 22: 220, 1866.
  Agave micracantha Salm, Bonplandia 7: 93, 1855.
  Agave mitis H. Monac. ex Salm, Ibid.

Plants perennial by axillary branching and flowering, forming large long-lived dense clumps of soft succulent rosettes, freely seeding. Leaves 30–60 (–70) x 7–13 cm, ovate or oblong or spatulate, short-acuminate, ascending to out-curving, guttered or concave above, convex below, thick soft fleshy, green to light gray glaucous, the margin straight to undulate with small closely spaced teeth 1–3 mm long, frequently bicuspid, sometimes with ciliated crests, whitish to reddish brown; spines weak, acicular, 1–2 cm long, brownish, decurrent along apex of leaf for 1–6 or more cm; inflorescence spicate, 1.5–2.5 m tall, densely bracteate and flowered, becoming lax in fruit; bracts chartaceous, deltoid long-caudate, persistent, 1.5–7 cm long; bractlets reduced, 1 long, 2 very short (3–4 mm); flowers protandrous, fleshy, 40–60 mm long, green outside, yellow to reddish or lavender to purplish within, including pistil and filaments; ovary 13–20 mm long, 3-angulate cylindric, 6-grooved, neckless, truncate at base and apex; tube funnelform, 10–17 mm deep, 10–12 mm broad at apex, thick-walled, 3-angulate, 6-furrowed; tepals linear-lanceolate, thick-fleshy, dimorphic, 12–18 x 5–8 mm, ascending to recurving,

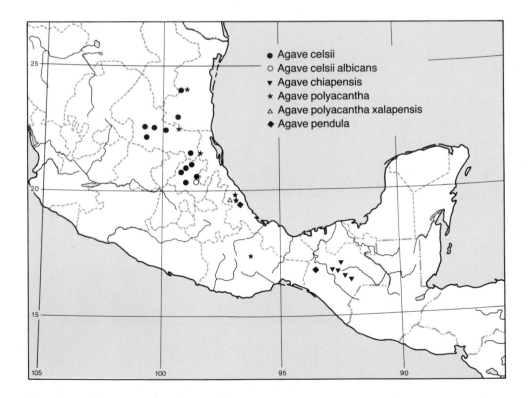

Fig. 9.3. Distribution of the Polycephalae, based on herbarium collections (see Exsiccatae).

Fig. 9.4. Illustration of the type plant of *Agave celsii* in Curtis Botanical Magazine, Tab. 4934, 1856.

short-acuminate, cucullate, glandular floccose at tip, the inner with broad fleshy keel; filaments 40–60 mm long, inserted in apex of tube, variously colored; anthers 17–22 mm long, mostly excentric, yellow to varicolored; capsules at first fleshy, drying rough dark brown, 18–25 x 9–12 mm, angular ovoid, beaked, thick-walled; seeds 4 x 3 mm to 3 x 2 mm, hemispherical, with well developed complete marginal wing.

Type: Hort. Kew, June 1856, *K.* (leaf, 6 fls., sketch of cross section of ovary). Apparently from the plant figured in Tab. 4934, Curtis Bot. Mag. 1856, regarding which Hooker wrote. "This fine *Agave* was received many years ago from the garden of M. Cels at Paris, as an unknown species, and equally unknown as to its native country—probably Mexico.—It flowered in May and June of the present year, for the first time with us."

Fig. 9.5. *Agave celsii* in native cloud forest habitat on Sierra de las Cucharas, above Gomez Farias, Tamaulipas, June 1963.

Taxonomic comment

Agave celsii is distinguished by its small compact rosettes, broad tender denticulate leaves, and dense clavate flowering spikes. The bicuspid teeth appear to be a good specific indicator, but some forms of the species do not have them. The tube is more ample than in other members of the group (Fig. 8.1) and sometimes exceeds the tepals in length.

In latter June of 1963 I reviewed this species in its natural habitat and found it to occupy the limestone outcrops of the "Huasteco" region from Nuevo León and eastern San Luis Potosí to about Jacala in Hidalgo. Such peaks and outcrops have small isolated populations in what was or still is extensive forestland. At Rancho Cielo on Sierra Cucharas they grew in moist cloud forest. Several small populations were noted on cliffs northeast and southwest of Jacala along route 85 toward Zimpapán.

The inflorescence appeared to be more variable than the leaves, as the spikes vary in being dense or lax, the latter especially lax in fruiting stage. The flowers varied in size and color. They are green without while varying from yellow to reddish or purple

Fig. 9.6. *Agave celsii* in Huntington Botanical Gardens in 1951. Reported by William Hertrich, Superintendent, to be 40 years old. Its dense mound habit contrasts with the open habit on Sierra Cucharas.

within. Since such variables occur within the small populations from plant to plant, they should not be interpreted as specific differences, as Mr. Harrison of Rancho Cielo did. All of these populations appeared to be in the latter stage of flowering, there being many fruiting spikes and few flowering spikes in June.

There are more synonyms of this species than I have listed above. Trelease (1920) listed seven other names as positive or probable synonyms of *A. celsii,* and under *A. micracantha* Salm, which I regard as *A. celsii,* he lists six others. Among his synonyms is *A. albicans,* which I find useful to retain as a glaucous whitish-leaved variety supported by flower characters.

Figure 9.5 shows *Agave celsii* in native habitat on Sierra Cuchara. They have a more spreading open habit, probably reflecting the moist cloud forest conditions, than does the 40-year-old clump in the Huntington Botanical Gardens (Fig. 9.6).

Agave *celsii* var. *albicans*
(Figs. 9.1, 9.2, 9.3, 9.7; Table 9.1)

Agave celsii Hook. var. *albicans* (Jacobi) Gentry, stat. nov.

Agave albicans Jacobi, Hamb. Gart. Blumenzeit, 21: 256, 1865.
Agave mitis Salm var. *albicans* Terrac., Primo Contributo Mon. Agave 1885: 25, 1885.

Variety *albicans* differs from the typical species in its pale glaucous leaves and especially in the larger tepals, which are 20–27 mm long, vs. 12–18 mm long in var. *celsii*. It has been known only in garden culture since the middle of the last century and large clumps of 60 or more years old were seen in Huntington Botanical Gardens in 1978.

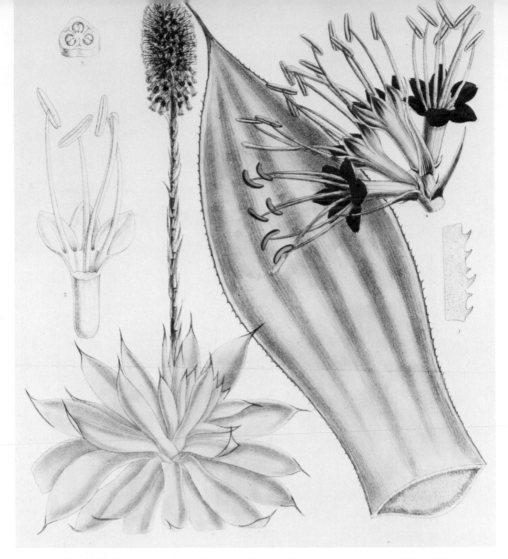

Fig. 9.7. Neotypic *Agave celsii* var. *albicans,* Curtis Bot. Mag. Tab. 7207, 1891.

A small colony scattered along the eastern rim of the Barranca de Metztitlán (*Gentry 23380* in 1974) appears assignable to this variety. None of the plants in that rocky habitat, however, approached in size the large old clumps in the fertile soil of the Huntington Botanical Garden.

Neotype: Hort. Kew Gard., 22 May 1891. Curtis Bot. Mag. Tab. 7207.

Agave chiapensis
(Figs. 9.1, 9.2, 9.3, 9.8; Table 9.1)

Agave chiapensis Jacobi, Hamb. Gart. Blumenzeit, 22: 213, 1866.

Medium-sized, robust, cespitose, short-stemmed, light shiny green, openly spreading rosettes. Leaves variable, mostly 30–50 x 7–16 cm, ovate, narrowed near base, short-acuminate, thickly smooth, light gray green, plane to slightly hollowed above and upcurving, rounded below, the margin slightly undulate to crenate, the larger teeth deltoid, upcurved, 3–4 mm long, closely spaced, or more remote and subulate, 5–10 mm long,

Fig. 9.8. Neotypic plant of *Agave chiapensis* in native habitat near San Cristóbal de las Casas, Chiapas, September 1952.

dark brown to graying; spine strong, subulate, 2–3.5 cm long, openly grooved above, rounded to broadly keeled below, straight to sinuous; spike ca. 2 m tall, long pedunculate, flowering in terminal ⅓ to ¼, closely bracteate, the flowers obscured in large tufts of broad-based bractlets, the longer equaling flowers; flowers 60–70 mm long, yellow or green flushed with reddish or purple, trigonous, fleshy; ovary 20–30 mm long, truncate at base and apex, roundly 3-angled, grooved to base; tube funnelform, 8–12 mm deep, thick-walled, 3-angulate, grooved; tepals linear elongate, 30–32 x 7–8 mm, unequal, fleshy, cucullate and glandular at tip, copiously papillate on and below hood, the inner with prominent keel and involute; filaments 70–80 mm long, very broad at base, inserted in orifice of tube; anthers ca. 30 mm long; capsules ca. 3 x 1.5 cm, oblong to obovoid, rounded truncate at base, acute at apex, thick-walled; seed mostly 4 x 5 mm, deeply hemispherical, with regular complete marginal wing.

Neotype: *Gentry 12178, US,* dups. DES, MEXU, MICH (l, cap, s). 1 mile S of San Cristóbal de las Casas, 18 Sept. 1952; Limestone cliff (l, cap, photo).

The leaves of the neotype collection closely match those shown by Berger (1915: Fig. 6). The capsules are old and weathered, with the dried rind of the valves shrunken, scurfy, moldy, and after 25 years still emit a moldy odor, an effect of the moist climate on the Sierra of San Cristóbal de las Casas. However, the large bractlets persist. Flowering collections of Breedlove and Laughlin (see Exsiccatae) also show the large broad bractlets subtending the flowers. These bractlets and the elongate tepals distinguish this taxon from the rest of the Polycephalae. It appears closely related to *A. warelliana,* which is separable by its closely serrate teeth on a red margin. Source of the neotypic collection is shown in Fig. 9.8.

Agave pendula
(Figs 9.1, 9.2, 9.3, 9.9, 9.10; Table 9.1)

Agave pendula Schnitts., Zeitschrift Gartenbau-Vereins, Darmstadt 6: 7, 1857.
> *Agave sartorii* Koch, Wochenschr. Ver. Beford. Gartenb. Konig. Preus. Staat. 3: 37, 1860.
> *Agave aloina* Koch, Ibid.
> *Agave coespitosa* Tod., Hort. Bot. Pan., Palermo. Tab. 8, 1876.

Plants perennial by axillary branching with open, short-stemmed, few-leaved spreading rosettes and axillary, slender, freely seeding, laxly flowered spikes. Leaves 20–30 per rosette, slender lanceolate, 50–75 x 5–11 cm, ascending to somewhat out-curving, plane to concave above, rounded below, soft fleshy, green to yellow-green, frequently with a pale yellow center stripe, the margins unlined, denticulate with brown denticles ca. 1 mm long; spines small, 5–8 mm long, brown, nondecurrent; spikes nodding, slender 1.3–1.8 m long, laxly flowered in upper half to third of shaft, the peduncle with narrow ascending bracts; flowers 30–45 mm long, single or geminate, greenish or tinged with lavender, whitish inside, in axils of subulate chartaceous bractlets; ovary 10–15 mm long, roundly 3-angled, grooved above, cylindric with unconstricted short neck; tube 6–13 mm long, funnelform, sulcate, thick-walled; tepals 14–16 x 5–7 mm, ca. equal, broadly linear, with small acute hood, recurving, outer plane, over-lapping inner, the inner with broad flat keel; filaments 30–50 mm long, whitish to lavender, inserted within orifice of tube; anthers 12–18 mm long, green to yellowish; capsules orbicular to oblong, 20 x 10–12 mm, thin-walled, rounded at base, obtuse at apex; seed 4 x 2.5–3 mm, hemispherical, sharply angled, shiny black, marginal wing pronounced.

Neotype: Hort. Kew, March 1877, teste N. E. Brown, *K.,* Curtis Bot. Mag. Tab. 6292, 1877 (l, infl).

Figure 9.9 illustrates *Agave pendula;* a reproduction of the fine lithograph of C. Visconti, drawn by A. Ficarrotta, for Todaro's *Agave coespitosa.* The flowers are whitish

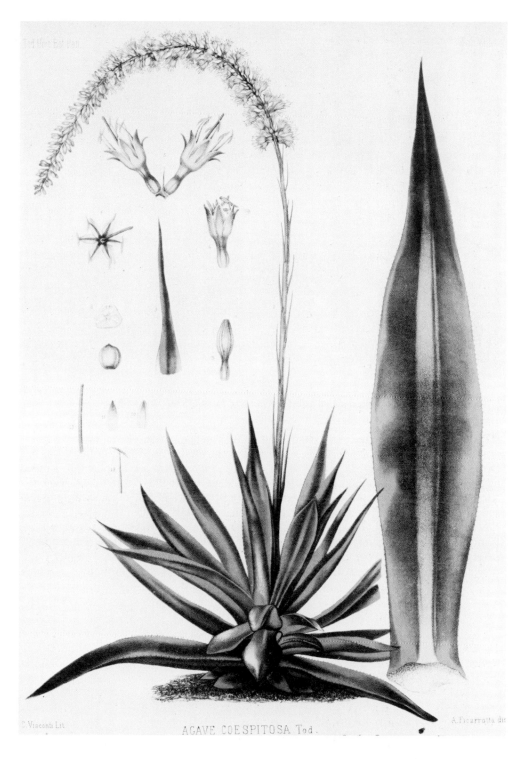

AGAVE COESPITOSA Tod.

Fig. 9.9. *Agave pendula* Schnitts., as reproduced in Hort. Bot. Pan., Tab. VIII, 1876–78, as *Agave coespitosa* Tod.

Fig. 9.10. *Agave pendula* on a cliff southeast of Jalapa, Veracruz, February 1974.

with lavender anthers in the early stage of anthesis. The neck is here given more constriction than I have seen, including the probable descendants of this plant, collected in the Orto Botánico in Palermo in 1969 *(Gentry 22606)*. Figure 9.10 shows plants in natural habitat on a cliff in Veracruz. Baker (Curtis Bot. Mag. Tab. 6292) wrote that *A. pendula* "was first introduced to the Berlin botanic garden by Dr. Rohrbach about 1850, and has since been received from the district of Orizaba, in Mexico. Our first notice of its flowering is by Dr. Schnittspahn in 1857 in the Zeitschrift des Gartenbauvereines zu Darmstadt." Mature clumps flower frequently if not every year, the one at Palermo had seven spikes when photographed in April. It is, therefore, an uncommonly prolific flowerer among ornamental agaves. It has not yet gained much place in American gardens.

Agave polyacantha
(Figs. 9.1, 9.2, 9.3, 9.11; Table 9.1)

Agave polyacantha Haw., Rev. P. Succ. 1821: 35, 1821.
    Agave chloracantha Salm, Bonplandia 7: 93, 1859.
    Agave densiflora Hook., Curtis Bot. Mag. 1857, Tab. 5006.
    Agave uncinata Jacobi, Hamb. Gart. Blumenzeit. 21: 165, 1865.
    Agave englemannii Trel., Mo. Bot. Gard. Rep. 3: 167, 1892.
    ?*Agave flaccifolia* Berger, Agaven 42, 1915.
    ?*Agave muilmannii* Jacobi, Nachtrage II; in Abh. Schles. Ges. Naturw. 160, 1870.

Plants medium-sized, single or cespitose (surculose?), green, with openly spreading rosettes, the spikes freely seeding, rarely bulbiferous. Leaves 35–65 x 7–10, lanceolate-acuminate to oblong short-acuminate, straightly ascending to upcurving, usually plane,

Fig. 9.11. *Agave polyacantha* on lava field along·road from Jalapa to Naulinco, Veracruz, September 1963.

narrowed above base, broadest in mid-blade, green or yellow-green, transiently glaucous, flesh firm, finely fibrous; margin generally straight, noncorneous except for thin spinal decurrency in apex; teeth closely spaced, 2–3 mm long, deltoid, reddish to dark brown; spine small, 0.5–2.5 cm long, acicular, rounded above and below, sometimes flatly grooved above, dark brown; spikes 2–3 tall, laxly or densely flowered in upper half to third, with long-caudate deltoid bracts and reduced bractlets 10–50 mm long; flowers 46–51 mm long, mostly geminate on short stout pedicels 2–3 mm long, tepals and stamens reddish over green ovaries; ovary 17–20 x 7 mm, cylindric, unconstricted in short neck; tube 7–9 x 9 mm, funnelform, scarcely grooved, smooth, glandular-thickened with and between filament strands; tepals unequal 19–23 x 5 mm, ascending to outcurved, lanceolate, the outer rounded on back, inner usually narrower, with large lanceolate keel, broad at base, long tapering to tip; filaments 50–70 mm long, reddish, inserted on rim of tube; anthers 19–21 mm long, nearly centric; capsules and seeds (see *A. p.* var. *xalapensis*).

Neotype: Hort. Kew, June 8, 1878, *K* (L, infl., as *Agave micrantha*). See also Tab. 5006, Curtis Bot. Mag. 1857 (as *Agave densiflora*).

I recognize this species by its more slender flowers, rounder ungrooved ovary, and well-developed teeth, 3–6 mm long, on more elongate lanceolate leaves. However, there are described forms that are too difficult to separate from the variable complex of *A. celsii*. A better knowledge of the branching habits, whether wholly axillary or surculose, would be of assistance in separating some of the taxa. It would be helpful if collectors noted this feature. The large tufted floral bractlets and long tepals of *A. chiapensis* and *A. warelliana* separate them from the *polyacantha-celsii* complex. Berger's horticultural specimens in U.S. and Kew, made up from plants growing at Mórtola and other nearby gardens, have been helpful in assigning synonyms, according to my broader concept of species. Uncertain cases are indicated by question marks.

Agave polyacantha is scattered along the drier rocky slopes of the moist tropical Huasteco Region of the Sierra Madre Oriental from central Tamaulipas to central Veracruz (Fig. 9.11). I have noted it particularly in the semi-open pine and hardwood mixed forest between 1,000 and 1,500 m. Epiphytic examples were observed roosting on the broad limbs or in crotches of trees. Their tolerance of light shade and frequent watering make the members of the Polycephalae suitable for greenhouse culture and they have been widely cultivated in European gardens for the past 150 years. The strongly toothed variety, annotated below, is particularly attractive.

Agave polyacantha var. xalapensis
(Figs. 9.3, 9.12)

Agave polyacantha Haw., var. *xalapensis* (Roezl ex Jacobi) Gentry, stat. nov.
  Agave xalapensis Roezl ex Jacobi, Hamb. Gart. Blumenzeit. 21: 60, 1865.

Like var. *polyacantha,* but differing in its generally shorter more ovate leaves with reddish brown larger teeth, 3–6 mm long, 5–12 mm apart, deltoid, frequently curved up or down, with occasional small interstitial teeth; spine 2–3.5 cm long, reddish brown, straight, long-grooved above, rounded below; spikes 2–3 m tall, stout, capsules 1.8–2.4 x 1 cm, obovoid, constricted at base, obtuse at tip, dark brown; seed 3.5–4 x 2–3 mm, lunate, dull black, hilar notch on rounded end, marginal wing low.

> Neotype: *Gentry & Dorantes 23376, US,* dups. DES, INIF. 4–6 km SE of Las Vigas along hwy, 140 to Jalapa, Veracruz, 1 March 1974 (l, cap). "Lava with pines; perennial plants branching several times from leaf axils." Fig. 9.12.

Agave warelliana
(Fig. 9.13; Table 9.1)

Agave warelliana Baker, Gard. Chron. II, 1877: 264, fig. 53, 1877.

Rosettes relatively robust, acaulescent or subcaulescent, sparsely surculose and axillary branching, densely leaved, ca. 1 m tall, 1.7 m broad. Leaves 70–75 x 13–14 cm, light pale green, or shiny glaucous, erect to spreading, plano-convex, thickly fleshy, lanceolate-spatulate, acuminate, slightly constricted above base, 6–7 cm thick at base, the margin brown, finely serrulate with denticles 1 mm long at 2 mm intervals; spine 18–20 x 3–4 mm, straight, brown, broadly grooved above, long-decurrent; spike ca. 5 m tall, the peduncle ca. 2 m long, densely bracteate; bracts 35 cm long, graduating smaller to apex, deltoid, long acuminate, appressed; flowers 90–95 mm long, subtended by small, pale, lanceolate bractlets 35 mm long; ovary smooth, light green, 40 mm long, trigonous, narrowed at base and apex; tube 14–15 mm long and as wide, 6-grooved, light green; segments linear-lanceolate, 35 mm long, yellow within, violet brown spotted on back, the inner 10–11 mm broad, slightly concave, with broad thick keel; filaments 85 mm long, colored like filaments; pistil 140 mm long, of same color; capsule 4 cm long, somewhat

Fig. 9.12. Neotypic plant of *Agave polyacantha* var. *xalapensis* on lava with pines near Las Vigas, Veracruz, March 1974.

obovoid, 3-grooved, narrowed at base, beaked; seed 5–6 x 4–5 mm, shiny black. (Desc. drawn from Berger, 1913, 1915.)

Neotype: Sheet 1, *Berger 21, VI, 1912, K.* Mórtola Bot. Gard., dupl. US. Berger prepared text for Tab. 8501, Curtis Bot. Mag., 1913.

"*Agave warelliana* was first described by Mr. Baker from the famous collection of Mr. Wilson Saunders. It is still an uncommon but very attractive plant in gardens. During the summer of 1912 it flowered at La Mórtola in the garden of Lady Hanbury, and also in the garden of Professor G. Roster at Ottonella on the Island of Elba. From the plant which flowered at La Mórtola was derived the material from which our figure has been prepared.'' Here reproduced as Fig. 9.13.

Agave warelliana appears to be the largest of the Polycephalae, the 90–95 mm long flowers are especially notable. It seems close to *A. chiapensis* but differs in the larger flowers with smaller bractlets and the finely serrulate brownish leaf margins. I have seen no living specimens. A plant so labeled in Kew Greenhouse No. 5 obviously did not agree with Baker's and Berger's descriptions. Berger's specimens at K and US were reviewed and are listed in the Exsiccatae. The native habitat of *A. warelliana* is still unknown, but can be expected in or near the presently known range of the Polycephalae.

Fig. 9.13. *Agave warelliana* Baker as drawn from Tab. 8501, Curtis Bot. Mag., 1913. The plant is still unknown in native habitat.

Polycephalae Exsiccatae

Agave celsii

HIDALGO. *Gentry et al. 20079*, DES, MEXU, US. 22–23 miles NE of Jacala along rt. 85, 20 June 1963; alt. 4,000–4,500 ft., limestone cliff (l, f).

Gentry 12269, DES, MEXU, US. Ca. 15–17 miles SE of Jacala along rt. 85, 12 Oct. 1952; elev. 1,500–1,600 m, cliffs (l, cap).

Moore 1733, BH, GH. Barranca de San Vicente near km. 238, on hwy. between Zimapan & Jacala, 28 Oct. 1946; alt. 1,800–2,000 m, limestone ledges and streamside thickets (l, cap).

Moore 3351, 3352, BH. El Capulin near km. 134 on hwy. between Actopan and Ixmiquilpan, 15 July 1947; alt. 1,800 m, slopes of low limestone hill (l, f).

Moore & Wood 3991, BH. Barranca de San Vicente near km. 238 on hwy. between Zimapan & Jacala, 13 July 1948; alt. 1,800–2,000 m, limestone ledges (l, f).

Moore & Wood 4389, BH. Barranca de Toliman above mines on road from Zimapan to Mina Loma del Toro and Balcones, 8 Aug. 1948; alt. 5,000 ft., cliffs on rocky slopes.

HORTICULTURE. *Berger s.n.*, US. Bot. Gart. Dahlem, 1913.

Berger s.n., US. Mort. Bot. Gard., 1905, 1909 (l, f, cap, 5 sheets).

Berger s.n., US. Cult. Mórtola Bot. Gard., 1909, 1911, 1904 (l, f, 6 sheets—small editions of *A. celsii*, one sheet with relatively large teeth, but some bicuspids as is typical).

Berger s.n., K. Mórtola Bot. Gard., 19 May 1896 (as *Agave bouchei*).

Berger s.n., K. Mort. Bot. Gard., 1 July 1914 (as *A. macracantha*).

Gentry 10142, 17706, 17707, DES, US. Cult. Hunt. Bot. Gard., Jan. 1951, June 1959 (l, f, cap, photo).

Hort. K Type. Cult. Kew Bot. Gard., June 1856 (l, f).

Hort. K. Cult. Kew Bot. Gard., 8 June 1878 (l, infl. as *A. micracantha*).

SAN LUIS POTOSI. *Gentry et al, 20461*, DES, MEXU, US. Sierra 25–30 miles E of Cd. San Luis Potosí along road to Rio Verde, 22 Sep. 1963; alt. ca. 5,600 ft., rocky limestone outcrops (l, cap).

Kimnach 287, HNT. Same loc., July 1962.

Palmer 235, US. Near Los Canos, 15–20 Oct. 1902 (l, seed).

Parry & Palmer s.n., in 1875, MO. Sierra Alvarez at 8,000 ft. (l, cap).

Pringle 3739, K, MO, UC, US. Limestone ledges, Tomasopa Canyon, 24 June, 1891 (l, cap).

Purpus s.n., MO, Minas de San Rafael, Cerritos, 30 Nov. 1913 also US via Berger Bot. Gard. Darmstadt, 1913.

Rzedowski 7105, ENCB, US. 5 km NW de Soledad de Zaragosa, Mun. Xilitla, 27 Jan, 1956; alt. 2,000 m.

TAMAULIPAS. *Dressler 1828*, MO, Rancho del Cielo above Gomez Farias, 8 July 1957 (l, cap).

Gentry 20077, DES, MEXU, US. Rancho Cielo, Sierra de las Cucharas, 8 miles SW of Gomez Farias, 19 June 1963; elev. 3,500–4,000 ft., limestone outcrops over forest (l, f).

Gilbert 46, TEX. Same loc. 4 July 1965; elev, 4,400 ft. (l, f).

Meyer & Rogers 2554, MO. N. L.–Tamps. border, NE of Dulces Nombres, alt. 1,850 m (l, f).

Agave celsii albicans

HIDALGO. *Gentry 23380*, DES, INIF, US. 8 km N of Mesquititlan along rt. 105, 5 March 1974; elev. ca. 6,100 ft., volcanic rock.

Kimnach 359A, HNT, also collected in this vicinity, 1977.

HORTICULTURE. *Gentry 17705, 17708*, DES, US. Hunt. Bot. Gard., Block 20–175 and block 20–183, 13 June 1959 (l, f).

Gentry 22739, DES. Photo of plant in Kew Glass house No. 5.

Folder No. 109, K. Cult. Kew Gard., 22 May, 1891. Bot. Mag. Tab. 7207. Neotype.

Agave chiapensis

CHIAPAS. *Alava s.n.*, HNT. Original col. at Yola. Cult. flowered May 1973. *Breedlove 9470*, CAS, US. Cerro San Cristobal, S. C. de la Casas, 28 Mar. 1965; elev, 7,100 ft. (l, f).

Breedlove 15133, CAS. Hy. 190 near Rancho Nuevo, 8 miles SE of San Cristobal Casas, 20 Aug. 1966, elev. 9,000 ft. (cap).

Breedlove 16119, CAS. Limestone cliff near Tenejapa Center, Mun. Tenejapa, 24 April 1968; elev. 6,700 ft. (l, f).

Gentry 12178, Neotype. DES, MEXU, MICH, **US.** 1 mile S of San Cristobal de las Casas, 18 Sep. 1952; limestone cliff (l, cap, photo).

Laughlin 494, CAS, MICH. Kampana ch'en along rt. 190, 3 miles W of Paraje Navenchauk, Mun. Zinacantan, 22 Mar. 1966; elev. 6,000 ft., steep slope with *Quercus & Pinus* (l, f). Teeth very long (to 10 mm) on teats; questionably assigned here.

Also *Laughlin 638*, CAS. With flowers in April.

Matuda 37648, MEXU. Km 1189 entre San Cristobal y Teopisca.

Miranda 2674, MEXU. Rancho de Chacajoconte, arriba Teopisca, 19 Abril 1943. "Lugares rocosos en encinares humedas" (l, f).

Agave pendula

HORTICULTURE. *Gentry 10106*, DES, US. Cult. Hunt. Bot. Gard., 9–15 June 1951.

Gentry 17711, DES, US. Cult. Hunt. Bot. Gard., block 20–161, June 1959 (l, f, photo).

Gentry 22606, DES, US. Cult. Ist. Orto Bot., Palermo, Sicily, 12 April 1969; topotype of *A. coespitosa* Tod., type clone ?.

Gentry 22921, DES. Charles Mieg Gard., Phoenix, 24 May 1971 (l, f, col. 10 years ago ca. 20 miles S of Manzanillo, Colima?). Hort. B. "Flor 1876, 13 Juni 1876 h. Berol" (f, photo, as *A. coespitosa* Tod.).

Hort, K. "March 1877. Bot. Mag. t. 6292. teste N. E. Brown" (2 sheets, l, infl). Also here in Kew sheets from Wash. Bot. Gard., April 1897; Edinburg Bot. Gard. June 1897.

CHIAPAS. *Breedlove 34328*, CAS. Choreadora near Derna, Rio de la Venta, Mun. Ocozocuatla, 20 Mar. 1973 (l, f).

VERACRUZ. *Gentry & Dorantes 23375*, DES, MEXU, US. Cerro by rt. 140, 10–20 miles SE of Jalapa, 28 Feb. 1974; elev. 1,400 ft., on volcanic breccia below limestone (l, f, cap).

Purpus s.n., US. Zacuapán, July 1926; on steep rocks. (Very small l, immature?, openly fruiting spike).

Agave polyacantha

HORTICULTURE. Hort. K. Cult. Mórtola Gard., 14 May 1908 (l, f).

Hort. K. June 8, 1878 (l, infl, as *Agave micracantha*).

Berger numerous sheets in US from European Gardens, some as forms or hybrids or as *densiflora*. He used Jacobi as species author.

Berger s.n., US. Cult. Mórtola, 5 VII 1910. (l, f, is a fine-toothed form of *A. polyacantha*, as *A. terraccianoi* Pax).

Berger, s.n., US. Cult. Italy, 1909, 1910 (l, f, as *A. uncinata* Jacobi).

Berger s.n., US. Cult. Mórtola Gard., June 1907 (l,

f, as type of *A. flaccifolia*. Under this name also from Mórtola in 1907, 1911).

Berger s.n., US. Cult. Mórtola Gard. in 1909 (l, with thin brown margin and close-set denticles as *A. muilmannii*).

Hort. Mórtola Gard., 1 July 1914 (l, infl, as *A. flaccifolia*, 2 ls, infl, as *A. chloracantha* Salm).

Berger s.n., B. Cult. Mórtola Gard., 21 VII 1913 (f only, as *A. flaccifolia*).

OAXACA. *Gentry & Halberg 12251*, DES, US. 4 km SE of Zacatepec Mixe, 23 Sep. 1952; elev. ca. 5,800 ft.; limestone hill slope.

SAN LUIS POTOSÍ. *Ogden & Gilly 51136*. DES, MEXU, MICH, US. Between Mora Quemada & Reten, 17 miles W of Xilitla, along road to Queretaro Boundary, 22 Feb. 1951; crevices & ledges of rocky bluffs (l, cap).

Ogden & Gilly 51141, DES, MEXU, MICH, US. 38 miles W of Antigua Morelos along hwy. to Cd. Mais & San Luis Potosí, 24 Feb. 1951 (l, cap).

TAMAULIPAS. *Meyer 2792*, DES, NA. Along road from Adelaida to Dulces Nombres; elev. 1,500 m, shaded limestone cliffs (photo).

VERACRUZ. *Gentry 12576*, DES, US. 14 miles W of Jalapa along road to Puebla, 17 April 1953. Transplants grown in Gentry garden & Hunt. Bot. Gard., flowered April, 1972 (l, f).

Gentry et al. 20412, DES, MEXU, US. Malpais below Jilotepec along road to Naulinco, 3 Sep. 1963; alt. ca. 4,500 ft., lava with pines.

Agave polyacantha xalapensis

VERACRUZ. *Fox 5799*, DES, MEXU, US. 4 miles E of Las Vigas along road to Jalapa, 5 July 1952; elev. ca. 8,000 ft., crumbled lava with pine and oak (l, cap).

Gentry & Dorantes 23376, DES, MEXU, US. 4–6 km SE of Las Vigas along rt. 140, 1 Mar. 1974; elev. 7,000–7,500 ft., lava with pines (l, cap).

Ogden & Gilly 51160, DES, MEXU, MICH, US. Pedregal de las Vigas along hwy. between Las Vigas & La Joya, just below Toxlacuaya, 20 Mar. 1951; lava and pines.

Agave warelliana

HORTICULTURE. Berger s.n., K, US. Neotype. Cult. Mórtola Bot. Gard., 21 VI, 1912 (l, f).

10.

Group Striatae

Plants perennial, single or cespitose forming large clumps by axillary branching, generating by seeds. Leaves numerous, striate, narrow, linear, hard, thick but firmly fleshed, the margins scabrous serrulate with thin yellow borders, grayish green to yellow green, reddish or purplish, the tip pungent but relatively weak; inflorescence spicate, 1.5–3 m tall, long-pedunculate, bracteate, flowering in upper ½ to ¼ of shaft; flowers mostly geminate, campanulate to cylindric, variously colored, persisting dry on capsules; tubes well developed, frequently longer than the tepals; filaments elongate, inserted in mid-tubes, frequently at 2 levels; capsules 3-angulate, the valves with indented rib on backs; seeds small thick. Arid regions of central Mexico; Coahuilan Desert south to Oaxaca.

Typical species: *Agave striata* Zucc., Nov. Act. Leopold-Carol. 16(2): 678, 1833.
Group Striatae, Baker, Handbook Amaryllidac 184, 1888.
Sect. Schoenoagave & Sect. Chonanthagave, Berger, Agaven 78, 83, 1915.

The Striatae is a distinctive group of perennial habit with hard firm serrulate leaves and small flowers with deep tubes and small tepals. The short trigonous ovaries become conspicuously 3-angulate in capsule. They show relationship with the Polycephalae in flower structure, especially the ovary, which in some cases is not completely inferior. Both groups have short trigonous ovaries forming deeply sutured valves in capsule maturity. The Polycephalae evolved mainly with the moist tropical climate of the Sierra Madre Oriental, while the Striatae became xerophytes of the arid canyons and plains in the rain shadow west of the mountains (compare Figs. 10.3 and 9.2).

Distributions are outlined in Fig. 10.4. *Agave dasylirioides* in several features is the most generalized species and I have selected it (MS) as an ancestral type of *Agave;* this is discussed below under that species heading.

Chemistry

Sapogenins in leaves of the Striatae were found to be virtually absent. Of the eight samples collected 1950–52, only one *(A. stricta)* showed a minute amount by micro screening test. I consider this a sectional characteristic further distinguishing this natural group of taxa.

Key to Species and Subspecies of the Striatae

1. Rosettes single; leaves 1.5–4 cm wide, pliant; spikes arching; flowers campanulate, tepals greenish to yellow *dasylirioides*, p. 237
1. Rosettes cespitose or suckering; leaves 0.8–1.5 cm wide, hard, frequently rigid; spikes erect; flowers usually cylindric, tepals multicolored, greenish to yellow or purple   2
2. Segments of the perianth about as long as the tube; leaves shorter, 25–35 cm long. Tehuacan Valley *stricta*, p. 248

2. Segments of the perianth ca. ¼ as long as the tube or less; leaves frequently longer, 40–80 cm long. Hidalgo northward   *striata*, 3
A. Rosettes with numerous leaves; leaves 1 cm or less wide
          striata ssp. *striata*, p. 242
B. Rosettes with relatively few leaves; leaves more than 1 cm wide, thickly keeled, frequently falcate
          striata ssp. *falcata*, p. 245

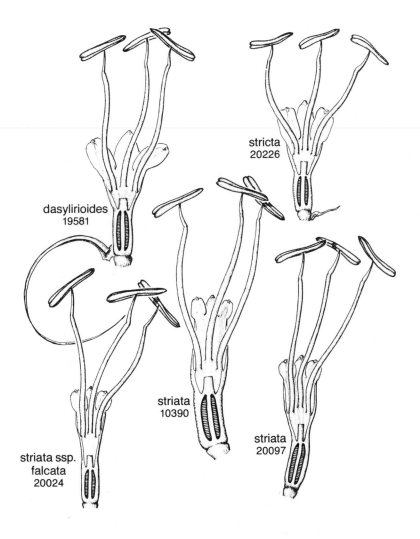

Fig. 10.1. Long flower sections of species and subspecies of Striatae.

Agave dasylirioides
(Figs. 10.1, 10.2, 10.3, 10.4, 10.5, 10.6; Table 10.1)

Agave dasylirioides Jacobi et Bouché Hamb. Gart. Blum Zeitg. 21: 344, 1865.
  Agave dealbata Lem. ex Jacobi loc. cit. p. 346, 1865.
  Agave intrepida Greenm. Proc. Amer. Acad. Arts 34: 567, 1899.

Perennial, generally single, symmetrical rosettes 30–50 cm tall, 60–100 cm broad with 70–100 leaves; leaves glaucous green, mostly 40–60 x 2–3 cm, straightly spreading but pliant, linear-lanceolate, smoothly striate above and below, relatively thin, plane above, scarcely succulent, margin 1 mm wide, pale yellowish white, minutely serrulate; spine 5–15 mm long, acicular, rounded below, flat near base above, reddish brown; spike 1.5–2 m long, arching, the peduncle with conspicuous light-colored, persistent bracts 10–15 cm long; flowers in apical ½–⅓ of shaft, greenish yellow with pink filaments and

Table 10.1. Flower Measurements in the Striatae (in mm)

Taxon & Locality	Ovary	Tube	Tepal	Filament Insertion & Length		Anther	Total Length	Coll. No.
dasylirioides								
Berlin Bot. Gard.	13	13	13	6–7	41	12	39*	
	13	13	13	7 & 5	46	13	37*	
Tepoztlán, Mor.	9	7×10	9×7	5–4	40	13	25†	22498
	10	9×9	9×6	6–5	33	12	28†	22498
	11	11×12	10×7	6 & 5	40	15	31†	19581
	12	11×12	10×7	6 & 5	37	14	33†	19581
striata								
Querétaro, Quer.	14	21×8	5×3–4	9 & 7	45	10	40†	10390
	14	22×8	5×3–4	9 & 7	45	10	40†	10390
Ixmiquilpan, Hgo.	13	15	5×3	8	25	16	34*	8951
	14	16	5×3	10 & 8	28	16	34*	8951
Matehuala, Hgo.	14	17×9	5×4	8–7	38	16.5	35†	20097
	12	12×7	5×4	7–6	42	14	28†	20097
Saltillo-Monterey, Coah.	11	14×8	4×4 & 4	8–7	28	12	30†	20017
	10	15×8	5×4 & 5	9–8	17	12	30†	20017
ssp. *falcata*								
Saltillo, Coah.	9	16×7	6×4 & 5	9–8	30	15	30†	20024
	10	15×8	6×4 & 5	8–7	38	17	31†	20024
Pedriceña, Dur.	14	16×11	5×5	7	42	15	35†	10601
	15	15×11	6×5	7–6	57	15	36†	10601
Las Barrancas, Coah.	12	18×10	4.5×4	9–10	40	15	35†	23106
	12	19×10	5×4.5	11	38	15	36†	23106
Sierra Maderas, Coah.	15	19×10	7×5.5	12	52	17	41†	23256
	15	18×10	7×5	11	47	18	40†	23256
stricta								
Tehuacán, Oax.	8	10×10	10×6 & 9	7–6	28	12	28†	20226
	10	8×10	11×6 & 10	7–6	30	13	28†	20226
	8	9×9	10×5.5	7	33	12.5	27†	20226
Hunt. Bot. Gard., Cal.	11	9×9	9×5	6	26	12	29†	19846
	11	8×9	9×5	5	30	12	28†	19846

* Measurements from dried flowers relaxed by boiling.

† Measurements from fresh or pickled flowers.

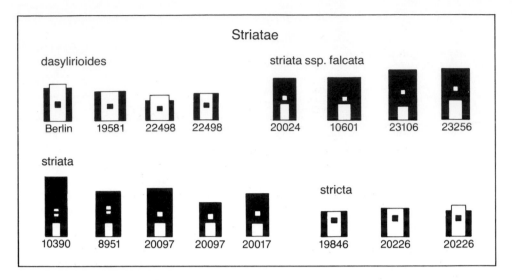

Fig. 10.2. Ideographs of flowers on the taxa in Striatae; based on measurements in Table 10.10.

yellow anthers, persistent, drying chartaceous on capsules; ovary neckless intruding the tube, 9–12 mm long, linear-tapered, 3-angulate, truncate at base; tube funnelform, 8–12 mm long, 8–12 mm broad, shallowly grooved below tepal sinuses; tepals 9–11 mm long, 6–7 mm wide, equal, spreading, ovate to oblong, mucronate, the outer plane, the inner with a broad keel and convergent costae within; filaments 35–50 mm long, nearly round, inserted in mid-tube; anthers 12–15 mm long, yellow, persistent, centrally affixed; capsules ovoid, ca. 2 x 1 cm, thin-walled, beakless; seeds 4 x 3 mm.

Lectotype: In Berlin Herbarium (Herbarium Berol) "*Agave dasylirioides* v. Jacobi h. Berol, Jan. 1863." Two sheets of specimens of leaves and flowers and the additional note, "*Agave dasylirioides* Jacoby et Bouché Prope Quessaltenango Guatemala ad montem ignivomum a Waxsiwiczio leifan."

Two flowers from the lectotype specimens were relaxed and measurements made (Table 10.1). These flowers are more slender than those shown by Hooker in the Bot. Mag. 23 (3rd ser.), pl. 5716, 1868 and as found in the closely related *A. intrepida* Greenm. from Tepoztlan, Morelos, but the size and proportions are otherwise very similar; compare also the ideographs (Fig. 10.2). Hooker in his account of *A. dasylirioides* (loc. cit.) stated, "My authority for the specific name *dasylirioides* is General Jacobi, who says that the same species is cultivated in the Vienna Botanic Garden. Koch, however, considers it the same with *A. dealbata* Lemaire, a plant I have no means of comparing it with." Jacobi (Hamb. Gart. Blum. Zeitg. 22: 269, 1866) rebutted Koch's opinion, separating *A. dasylirioides* by its longer, more numerous, arching leaves from *A. dealbata*. I found no herbarium specimens of *A. dealbata* in the herbaria at Berlin, Munich, Palermo, Kew, and the British Museum. However, in the Orto Botanico at Palermo they had a specimen, growing outside in the open mild climate of Sicily, which had the shorter leaves like the native Morelos plants. Hooker's illustration and doubtless Jacobi's specimen were grown in humid glass houses, and were abnormally etiolated. I give small credence to Jacobi's argument for a distinct species and see but one species under the three names listed above. If it proves desirable to recognize a short-leaved variety, then Baker's combination *Agave dasylirioides* var. *dealbata* can be re-employed.

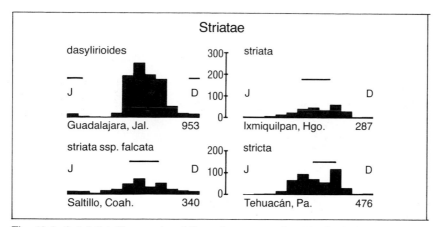

Fig. 10.3. Rainfall (silhouettes) and flowering seasons (bars) in the Striatae. Relevant meteorological stations with average annual rainfall in millimeters. Data from Atlas Climatológico de México (1939). Flowering periods are based on herbarium specimens and field observations.

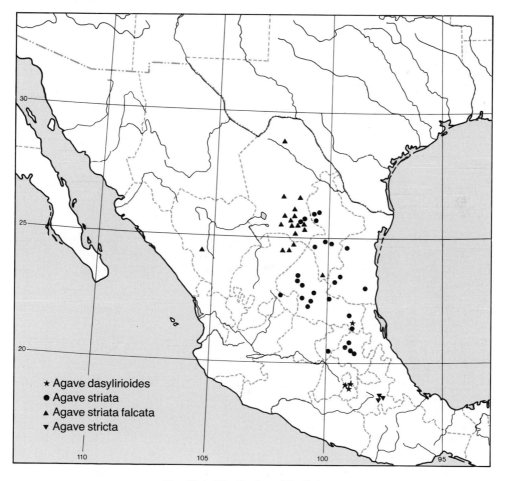

Fig. 10.4. Distribution of the Striatae.

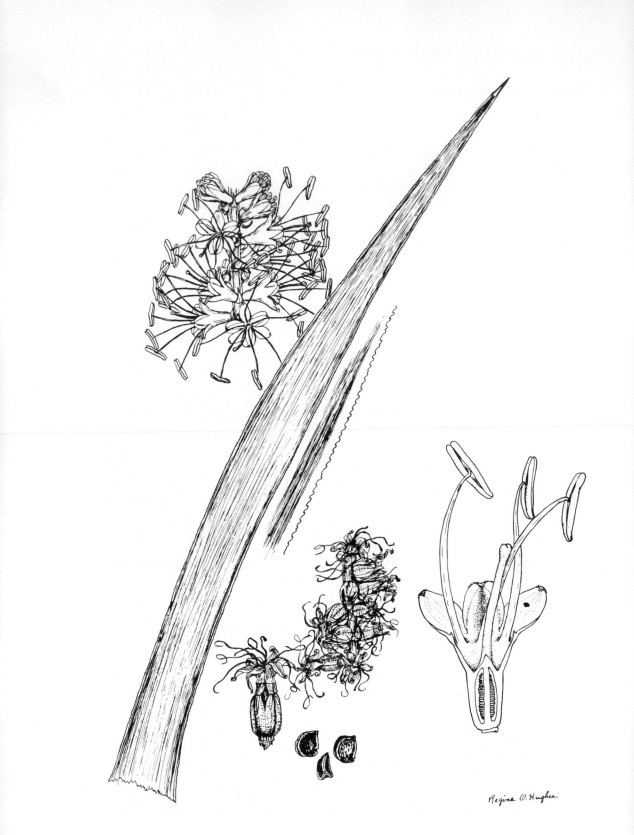

Fig. 10.5. *Agave dasylirioides* with leaf, section of flowering spike and fruiting spike reduced; leaf margin, flower long section, 1 capsule, and seeds enlarged.

Phylogenetic relationship

Agave dasylirioides occupies a paramount position phylogenetically in the genus *Agave*. It appears closest to what I conceive as the most generalized or ancestral form. The following characteristics align it with early or less specialized members in the Liliales and Amaryllidales: (1) Perennial habit. (2) Non-dimorphic perianth segments; the outer and inner tepals are similar in *dasylirioides*. (3) Tube and tepals are about equally proportioned, which is to say there has been little modification of either. (4) Ovary is incompletely inferior and protrudes into the tube; perhaps a survival character relating it to the Liliales. (5) Leaves are serrulate and scarcely succulent, much like the general structure common to many members of the perennial Liliaceae and Agavaceae. (6) The inflorescence is a relatively simple spike with flowers geminate on a relatively short axis. (7) Reproduction is entirely by seed, which I view as a conservative type of reproduction,

Fig. 10.6. *Agave dasylirioides* on the volcanic cliffs above the valley of Tepoztlán, Morelos. *Agave horrida* in silhouette.

persisting with cliff-dwellers, in contrast to the suckering and bulbiferous devices developed in other agaves, theoretically to compensate for predation by animals both large
and small.

Distribution and habitat

In spite of the Berlin specimens and Jacobi's report, it is doubtful that *Agave
dasylirioides* occurs in Guatemala or was ever collected there. Standley and Steyermark in
their Flora of Guatemala (24, pt. 3: 111, 1952) state: "Possibly it may occur on the dry
rocky mountain sides above Quezaltenango, which we have not explored carefully, but it
is somewhat strange that it has not been introduced into cultivation in Quezaltenango if
growing wild in that region." In 1966, with car and spotting telescope, I made a short
search for the plant upon the lava slopes of the old volcanoes about Quezaltenango without
success. However, in the US National Herbarium there is a specimen from San Salvador
(see Exsiccatae) prepared from a cultivated plant. But it could as likely have been taken
there from Mexico.

Hooker's report of the Kew material of *A. dasylirioides* (loc. cit.) gives a Mexican
origin. "Mr. Sowerby informs me that it was raised from a packet of seeds sent from
Mexico about twenty-five years ago (ca. 1843), amongst which seeds were also those of
Fourcroya longaeva." Although Jacobi attributed the collection of his type specimen to
Waxsiwiczio in Guatemala, it is more likely that Waxsiwiczio collected it on the same trip
through Mexico, and then through some labeling error it was attributed to Guatemala. It is
very unlikely the precise origin of Jacobi's specimen will ever be known. Specimens I
have seen are listed in the Exsiccatae and plotted on the distribution map in Fig. 10.4.

Agave dasylirioides is ordinarily a cliff dweller on the mountainous slopes of central
Mexico. It is scattered on the brecciated tuffaceous cliffs around Tepoztlán, Morelos, and
was also observed on rocky volcanic slopes just above and north of Cuernavaca, associated with a mixed pine and hardwood forest. This is a well-watered region with
around 900 mm (36 inches) of rain during the summer and fall (Fig. 10.3). At elevations
between 1,500 and 2,200 m (5,000–7,200 feet) this is a cool tropical habitat constant in its
annual rhythm of temperature, sunlight, and water. This is well below the Tropic of
Cancer in a region that is old geologically, and, except for local orogeny, has been for a
long time a constant kind of mesophytic habitat. The endemic mesophytes have been
relatively free of the climatic stress as imposed elsewhere by drought and freezes. It is in
such a region, fairly central to the *Agave* area, that we would expect a conservative kind of
species, little changed by geologic time. *Agave dasylirioides* appears to be such a persisting prototype, still clinging precariously on the mossy cliffs by Tepoztlan. Altogether, it is
a handsome remarkable plant, deserving special place on every southern campus of
botanical learning. It would foster learning and provoke discussion.

Agave striata
(Figs. 10.1, 10.2, 10.3, 10.4, 10.7, 10.8; Table 10.1)

Agave striata Zucc., ssp. *striata,* Nov. Act. Acad. Leopold-Carol. 16 (2): 678, 1833.
> *Agave recurva* Zucc., Abh. Akad. Wiss. Muenchen 4: 22, 1845.
> ?*Agave echinoides* Jacobi, Nachtrage I in Abh. Schles. Ges. Vaterl. Cult. Abth. Naturwiss. 1868: 163,
> 1868.?

Plants perennial with compact, short-stemmed, many-leaved rosettes 50–100 cm tall,
50–120 cm broad, often forming large dense clusters by axillary branching, 2–3 m broad,
pale green to red or purplish. Leaves mostly 25–60 x 0.5–1 cm wide in the middle, linear,
striate, thick, rather rigid, straight to arching, convex above, smooth or scabrous along the

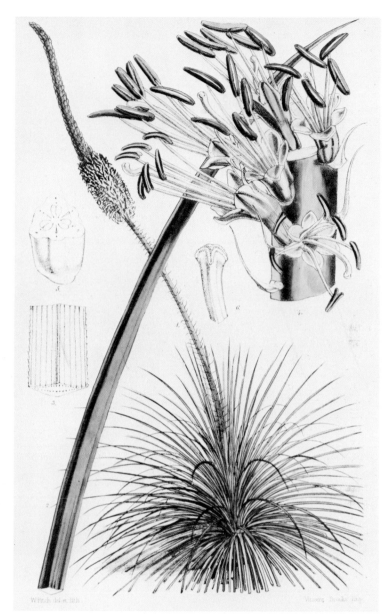

Fig. 10.7. *Agave striata* as illustrated in Curtis Botanical Magazine Tab. 4950, 1856.

keels above and below, brownish at apex below spine, the margin cartilaginous, 1 mm or less wide, pale yellow, scabrous or minutely serrulate; spine 1–5 cm long, subulate, rounded below and above, very pungent, reddish brown to dark gray; spikes 1.5–2.5 m tall, erect, rather laxly flowered above a long peduncle, the peduncle bracts 5–10 cm long, floral bracts shorter than flowers, deciduous; flowers mostly geminate, 30–40 mm long, tubular, greenish yellow or red to purple with bronze or brownish anthers; ovary 12–15 mm long, neckless, intruding the tube, cylindric-triangulate, grooved; tube 14–20 mm long, 8–11 mm broad, shallowly grooved from tepal sinuses; tepals 5–7 mm long, 3–5 mm wide, ovate-oblong, about equal, spreading with inflexed tips, the inner a little wider and with acuminate keel; filaments 30–50 mm long, with a high knee, inserted in

Fig. 10.8. *Agave striata* on the dry calcareous plains north of Matahuala, Zacatecas, but in the state of Nuevo León.

mid-tube, usually on two levels, oval in cross section; anthers 12–16 mm long, centrally affixed, bronze or brownish, yellow with dehiscence; capsules 13–16 x 8–10 mm, trigonous, dark brown, truncate at base, apiculate; seed 3 x 3.5 mm, crescentic, thick, especially on curved side.

Neotype: ''Hort. Kew, Dec. 15, 1880,'' *K*. Annotated possibly by Baker as, ''*Agave striata* Zucc.! Bot. Mag. t. 4950.''

However, that illustration shows the tepals and tube in proportion more suitable to *A. stricta,* making the specific reference doubtful, but it may be that the artist did not show the flowers in true perspective. No sectional drawing was given. The leaves appear like *A.*

striata and the flowers of the above cited herbarium specimen does show the short tepals and long tube of *A. striata*. Selection of this historic specimen as representative of *A. striata* provides a basis for identifying the widespread northern complex and separating it from the short-tubed relative in the Tehuacán Valley, assigned to *A. stricta*. *A. echinoides* Jacobi is a short-leaved form, the flowers of which are unknown, and without which the name remains of doubtful affiliation. *A. falcata* Engelm. has a more definite appearance and is here recognized as a subspecies, described below.

Agave striata is represented by extensive populations varying in growth habit, leaf forms, and to a lesser extent in flower structure; see the two flower sections in Fig. 10.1. Some local populations consisting of single plants may represent recently established colonies, not old enough to exhibit the large clumps developed elsewhere. The species inhabits a large area of northeastern Mexico on both sides of the Sierra Madre Oriental (Fig. 10.4). It is limited, however, to the drier valleys and plains with annual average rainfall below 500 mm (20 inches) (Fig. 10.3).

The names "espadin, espadillo, guapilla, soyate, and sotolito" have been applied to this species. The fiber is used locally, but it is inferior to that of *A. lechuguilla*. It also has a desultory local use as a hedge or fence plant.

Agave striata falcata
(Figs. 10.1, 10.2, 10.3, 10.4, 10.9, 10.10; Table 10.1)

Agave striata Zucc., ssp. *falcata* (Engelm.) Gentry, comb. & status nov.
>*Agave falcata* Engelm., Trans. Acad. Sci. St. Louis 3:304–305, 1875.
>*Agave paucifolia* Todaro, Hort, Bot. Panorm. 1: 77, pl. 19, 1877.
>*Agave californica* Jacobi, Nachtrage I in Abh. Schles. Ges. Vaterl. Cult. Abth. Naturwiss. 1868: 151, 1868 = *Yucca whipplei* Torr.
>*Agave californica* Baker, Gard. Chron. II: 56, 1877 = *Yucca whipplei*
>*Agave striata* var. *californica* Terr., Primo Contributo ad una monografia delle Agave. Napoli, 1885 = *Yucca whipplei.*

Single or cespitose, nearly acaulescent, pale green glaucous green to reddish rosettes with relatively few leaves. Leaves 30–60 x 0.8–1.8 cm, straight to falcate, rigid, smooth, striate above and below, rounded above, angled below, firmly fleshed, coarsely fibrous, the very narrow margin serrulate; spine 2–4 cm long, acicular, rounded below, flat above, dull gray; spikes erect, 1.5–3 m tall, with narrow peduncular bracts 5–8 cm long, flowering in upper ¼ of shaft; floral bracts shorter than the flowers; flowers geminate on short pedicels, 28–40 mm long, green or yellow or reddish purple; ovary 10–15 mm long, intruding tube, 5-angulate, truncate at base; tube 14–20 mm long, 7–9 mm wide, somewhat curved, lightly sulcate from tepal sinuses; tepals 5–7 mm long, 4–5 mm wide, erect, firm, the outer deltoid-acute, inner ovate to oblong, rounded, broadly keeled; filaments 30–60 mm long with high knee, inserted near mid-tube; anthers 15–17 mm long, slate-colored to yellow; capsules 12–15 x 12–13 mm, ovoid, trigonous, truncate at base, brown, transversely ridged, with persisting perianths; seeds 4–5 x 2.5–3 mm, crescentic, cuneate or flat, the circumferential wing wavy.

Lectotype: *Wislizenus 312, MO*. Saltillo, Coahuila, 23 May, 1847 (l, f, 2 sheets). Engelmann refers to specimens collected by Wislizenus and Gregg at Buena Vista near Saltillo.

Agave falcata is too closely related to *A. striata* for me to maintain it as a species. The flowers of the two are very similar and the subspecific variation in size and detail do not help in their separation. The broader heavier leaves of the Coahuilan *falcata* become narrower and less stiff southward, grading into the narrower *striata* leaf form in San Luis Potosí. The *falcata* rosettes mature with fewer leaves than typical *striata* and this is true

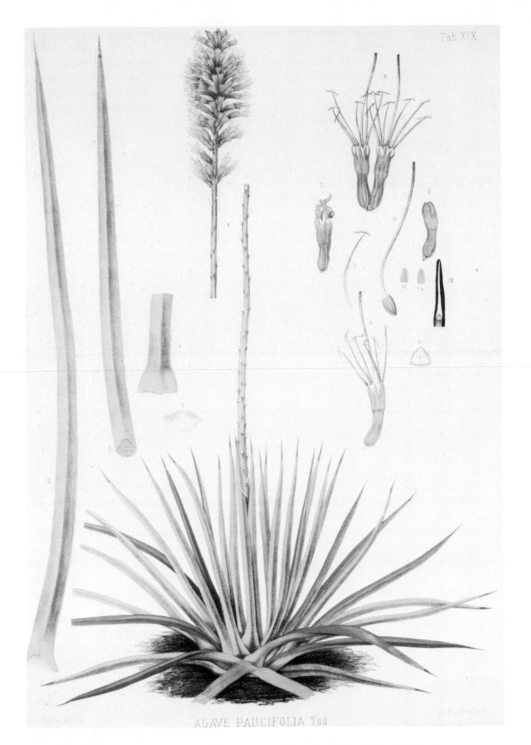

Fig. 10.9. *Agave striata* ssp. *falcata,* an excellent depiction in Hort. Bot. Panorm. Tab. XIX, 1876–78, as *Agave paucifolia* Todaro.

Fig. 10.10. *Agave striata* ssp. *falcata* in a disjunct colony on the Pacific slope south of Cd. Durango, Valle de Río Mesquital. Note the few-leaved rosettes with stiff up-curving leaves.

of the larger clumping plants as well as the single plants of *falcata*. However, the core of the northern desert populations, with their broader stiffer leaves and more xerophytic nature, appear worthy of subspecific recognition. *A. striata* develops longer stems and larger clumps built up by axillary branching. Separation of herbarium specimens is frequently difficult and nominal.

Agave striata falcata occurs on sandy, coarse, rocky soils on bajadas, slopes, and plains between 1,000- and 2,000-m elevations in shrub and succulent deserts, from Nuevo León and Durango to southeastern Tamaulipas and San Luis Potosí (Fig. 10.4). It appears to be fond of both primary and secondary limestones, thriving in areas with only 200 to 300 mm of irregular annual rainfall (Fig. 10.3).

The common local name is "espadin" and the local uses are about the same as those noted under the species above. The tough coarse leaves appear to be inedible to cattle.

Fig. 10.11. *Agave stricta* 3–4 miles southwest of Tehuacán, Puebla. This is a large old colony dominating the vegetation at this site.

Agave stricta

(Figs. 10.1, 10.2, 10.3, 10.4, 10.11; Table 10.1)

Agave stricta Salm-Dyck, Bonplandia 7: 94, 1859.

?*Agave echinoides* Jacobi, Nachtrage in Abh. Schles. Ges. Vaterl. Cult. Abth. Naturwiss. 1868: 163, 1868?

Agave striata stricta Baker, Gard. Chron. n. ser. 8: 556, 1877.

Plants perennial, often densely cespitose with elongate branching stems decumbent in age, 1–2 m long, with deep crowns of innumerable leaves. Leaves 25–50 x 0.8–1 cm, linear, long-lanceolate, upcurved to straight, widest near base, striate, green, rigid, sometimes keeled above and below, rhombic in cross section, the margin thin, pale yellow, cartilaginous, scabrous-serrulate; spine acicular, 1–2 cm long, gray, decurrent with margin, flat above, angularly keeled below, bordered at base with brownish leaf tip; spike 1.5–2.5 m tall, straight or crooked, with bracteated peduncle longer than flowering section; flowers geminate, short-pedicellate, 25–30 mm long, red to purplish, ascending to outcurved; ovary 8–11 mm long, neckless, intruding base of tube, sharply 6-angulate; tube 8–10 mm long, 9–10 mm wide at apex, funnelform, angulate, grooved; tepals equal, 8–10 mm long, 5–6 mm wide, erect to spreading with incurved tips, deltoid to oblong, apiculate, outer and inner similar, the inner broadly keeled; filaments 28–30 mm long, reddish, inserted at 2 levels in mid-tube; anthers 12–13 mm long, bronze to brownish red, opening yellow, centric; capsules ovoid, trigonous, 12–14 x 9–10 mm, truncate at base, spiculate, black; seed 3.5–4 x 2.5–3 mm, hemispherical, thickly discoid.

Neotype: *Gentry, Barclay & Arguelles 20226,* US. Dupl. in DES, MEXU. Ca. 4 miles southwest of Tehuacán along road to Zapotitlan, Puebla, Mexico, Aug. 1953; elev. 5,600–6,000 ft. (1,700–1,850 m), arid thron forest over limestone hills (l, f).

[248]

Agave stricta is separable from *A. striata* by its short flower tube, equaled or exceeded by the tepals in length, and from *A. dasylirioides* by its short stiff narrow leaves on elongate stems. The flower proportions are similar to the latter, but growth habit and leaves are quite distinct and the species are not close. There is a closer affinity between *A. striata* and *A. stricta* and the populations between Tehuacán and Pachuca should be studied and sampled, if such can be found; compare distributions in Fig. 10.4.

The habitat of *A. stricta* is semiarid around Tehuacán, where it appears limited to the open calcareous slopes with a sparse cover of xerophytic shrubs, cactus, and other agaves (Fig. 10.3). *A. stricta* is one of the agaves there called "rabo de león." The flower buds are cooked and eaten with eggs, "como cacayitas." Livestock do not eat the foliage. In fact, man and his animals, by their obvious over-use of the vegetation, inadvertently may have made more room for increase of the *stricta* populations scattered in the area. Figure 10.11 shows their colonial dominance in this area.

Striatae Exsiccatae

Agave dasylirioides

HORTICULTURE. Anonymous s.n., B, Lectotype, Berlin Botanic Garden, Jan. 1863 (2 sheets, l, f). "Agave dasylirioides v. Jacobi h. Berol ...Prope Quessaltenango Guatemala ad montem ignivomum a Waxsiwiczio leifan." (?!)

Calderon 1635, US. Cult. in San Salvador, El Salvador, 1923.

Gentry 22604, DES. Cult. Orto Botanico, Palermo, Sicily, 12 April 1969 (l, photo).

In MO only drawings, photos, and specimens from other European botanic gardens; 1 sheet McDowell from "Mexico," 2/23/1896 (l, f).

Regents Park Botanic Garden, London, K, 1/1868.

MORELOS. *Gentry and Fox 12033,* DES, MEXU, US. Mountain ca. 1 mile W of Tepoztlán, 19 Aug. 1952.

Gentry et al. 19581, DES, MEXU, US. Volcanic monadnock by Tepoztlan, 1 Dec. 1961: alt. ca. 6,000 ft., volcanic cliffs (l, f).

Gentry & Tejeda 22498, DES, MEXU, US. Tuffaceous cliffs by Tepoztlan, 1 Dec. 1967; alt. 550–6,500 ft., shade site of cliffs (l, f, photos, cap).

Pringle 6868, MO, US. On mossy cliffs, Parque Station, NE of Cuernavaca, 2 June 1898; elev. 7,000 ft. (l, f, type of *Agave intrepida* Greenm.).

Pringle 8095, 8685, MO, US. Cliffs, Sierra de Tepoztlán, 1899 and 1902.

SAN LUIS POTOSÍ. *Rzedowski 7128a,* ENCB, US. 5 km al NW de Soledad de Zaragosa, mun. Xilitla, 24 Jan. 1956; alt. 2,000 m, ladera caliza-bosque de pino y encino.

Agave striata ssp. *striata*

COAHUILA. *Gentry et al. 20017,* DES, MEXU, US. 12–16 miles NE of Saltillo along road to Monterrey, 10 June 1963; alt. 4,000–5,000 ft., shrub and succulent desert (l, f).

Valdez y Miranda A-108, MEXU. Between Ramos Arispe & Paredón, 25 Nov. 1955.

HIDALGO. *Endlich 1002,* B, MO. Near Tizapán on mountain, 22 Sep. 1905; alt. 1,500–1,800 m (f).

Gonzales Q. 2005, CAS. Ca. 3 km al Oeste de Villagran, Mun. Ixmiquilpan, 28 Jan. 1965 (l, f, cap).

Moore 3351, 3352, BH, UC. El Capulin, near km. 134 on hwy, between Actopan & Ixmiquilpan, Dist. Actopan, 15 July 1947; alt. 1,800 m, low limestone hill (l, f).

Purpus 49, MO, US. Sierra de la Mesa. Cult. at Mórtola Gard.

Rose et al. 8951, MO. Ixmiquilpan, July 1905 (l, f, cap).

Troll 411, M. Actopan–Ixmiquilpan, 4 March 1954; alt. 2,200 m.

HORTICULTURE. Baker s.n.?, K Neotype. Cult. Kew Bot. Gard. From Real del Monte. (A long-leaved form with short tepals. All are hort. specimens, none from the field.)

NUEVO LEON. *Gentry et al. 11519,* DES, MEXU, MICH, US. Canyon Huasteca, 8–10 miles W of Monterrey, 30 Jan. 1952; shade side of cliffs in deep narrow limestone canyon (l, cap).

Gentry et al. 20097, DES, MEXU, US. 37 miles N of Matehuala along hwy. 57, 1 July 1963; elev. ca. 5,300 ft., Coahuilan Desert shrub.

Kimnach & Lyons 1400, HNT. 16.6 miles along road to Rayones from Hi 85, 22 May 1971; alt. 2,200 ft., steep shady cliff.

Meyer & Rogers 2767, MO, NA. Dulces Nombres, N. L.-Tams. border, 13 July 1948; alt. 1,690 m, limestone ledges (l, f).

Mueller 2070, MO. Rancho Las Adjuntas, Mun. Villa Santiago, 27 June 1935 (l, f).

Nelson 6086, US. Cerro de la Silla near Monterrey, 20 March 1902 (l, cap).

Pringle 1953, MO. Limestone ledges of Sierra Madre above Monterrey, 27 June 1888 (l, f).

Ramirez L. 1219, MEXU. 20 km SE of Galeana, 13 Aug. 1936.

Taylor 159, MO. Hacienda Pablillo, Galeana, 13 Aug. 1936 (l, f).

QUERETARO. *Gentry 10390,* DES. Along rt. 45, near km 183, limestone hill slope, 22 April 1951 (live pl. col., flowered Gentry Garden summer, 1974).

SAN LUIS POTOSI. *Gentry 11508,* DES, MEXU, MICH, US. 50 miles NE of Cd. San Luis Potosí along hwy. 80, 26–28 Jan. 1952; limestone, highland desert with low shrub cover (l, cap).

Ogden et al. 51125, 51127, DES, MEXU, US. 13 miles NW of Cd. del Mais along hwy. to San Luis Potosí, 19 Feb. 1951; slopes with *Larrea* and cacti.

Rzedowski 6034, ENCB. 3 km N de Guadalcazar, 24 Junio 1955; alt. 1,700 m (l, cap).

Rzedowski 7179, ENCB, US. 5 km al N. de Soledad de Zaragosa, Mun. Xilitla, 28 Jan. 1956; alt. 2,200 m.

Rzedowski 8729, ENCB, US. Sierra de Alvarez, km 34 del la carretera San Luis Potosí a Rio Verde, 24 March 1957; alt. 2,150 m; short-leaved form.

Rzedowski 8875, ENCB. Cerca de Corazones, Mun. Villa Hidalgo, 2 June 1957; alt. 1,950 m (l, f).

Troll 474, M. Sierra de Abra, oste Ciudad Valles, 5-5-54; alt. 200 m.

Whiting 590, CAS. Charcas, July–Aug., 1934 (l, f).

TAMAULIPAS. *Bartlett 10533,* CAS. Vicinity of San Jose, 20 July 1930 (l, f).

Dressler 1920, MO. Rock slopes of Cerro de las Yuccas, NNW of Aldama, 20 July 1957 (l, f, cap).

Nelson 4457, US. Jaumave Valley, 1 June 1898; alt. 2,000 ft (l, f).

Ogden et al. 5191, DES, MEXU, US. Along road to Tula, about 5 km from Jaumave, 15 Feb. 1951.

ZACATECAS. *Gentry 23394,* DES. 55 miles NW of San Luis Potosí on rt. 49 to Zacatecas, 17 March 1974; elev. ca. 7,000 ft; limestone (cap).

Agave striata ssp. *falcata*

COAHUILA. *Endlich 879,* B. Im Suden der Sierra de la Paila, 7 June 1905; alt. 1,300 m (l, buds, cap).

Gentry et al. 20024, DES, MEXU, US. 40 & 118 km N of Saltillo along rt. 57, 11 June 1963 (l, f).

Gentry & Engard 23106, DES, MEXU, US. Las Barrancas, W of Saltillo, 8 Oct. 1972; elev. ca. 5,400 ft., desert shrub on limestone (l, cap).

Gentry & Engard 23256, DES, MEXU, US. Canyon de la Hacienda, Sierra de los Maderos, 10 May 1973; elev. ca. 5,500 ft., forested canyon on limestone (l, f).

Johnston 8370, A. West base of Picacho del Fuste, NE of Tanque Vaioneta, 23–25 Aug. 1941; clumps on limestone (cap).

Nelson 3891, US. Sierra Encarnación, 28 July 1896 (l, f).

Palmer 716, UC, MO, US. Chojo Grande, 27 miles SE of Saltillo, 16 July 1905 (l, f).

Palmer 1314, US. Parras, 8–28 June 1880.

Pimentel s.n., US. Carretera Saltillo to Monclova.

Pringle 7, 465, US. Dry hills, Jimulco, 13 May 1885 (l, f, cap).

Trelease 3/23/05, ARIZ, MO. Carneros Pass.

Wind & Mueller 69, MO, US. Ca. 20 miles N of Hipolito en el Desierto de Paila, 15 June 1936 (l, f).

Wislizenus 294, MO. Rancho Nuevo near Parras, 18 May 1847.

Wislizenus 312, Type, **MO.** Saltillo, May 23, 1847 (l, f, 2 sheets).

DURANGO. *Gentry 22091A,* DES. Ca. 34 miles S of Cd. Durango along road from Nombre de Dios to Mesquital, 3 Nov. 1966; alt. ca. 5,000 ft., desert chaparral (photo).

Gentry & Gilly 10601, DES, MEXU, MICH, US. 6 miles SE of Pedriceña along hwy. from Torreón to Durango, 12 June 1951; elev. ca. 4,900 ft., desert "chaparrillo."

Thompson s.n., MO. 16 km from Valardina toward Copper Queen Mine, 23 Aug. 1910 (l, cap, long leaves).

NUEVO LEON. *Endlich 630,* B, Mo. In der umgebung von Dr. Arroyo, 22 April, 1905.

ZACATECAS. *Dewey 4, 5, 6,* Mo. Mazafil, 1908.

Henrickson 6229, DES. .4 miles S of Zacatecas-Coahuila border along hwy. 54, 30 Aug. 1971; elev. 6,100 ft., desert shrub on limestone shale (l, f).

Lloyd 78, MO, US. Hacienda Cedros, 1908. Sierra Ramirez, June 1908.

Agave stricta

HORTICULTURE. *Gentry 19846,* DES, US. Cult. Hunt. Bot. Gard., San Marino, California (l, f).

PUEBLA. *Endlich 1930,* MO. Tehuacan, April 1907.

Gentry et al. 20226, Neotype, DES, MEXU, **US.** 4–10 miles SW of Tehuacán along road to Zapotitlan, Aug. 1963; alt. 5,600–6,000 ft., arid thorn forest over limestone hills (l, f).

Gentry 22528, DES, MEXU, US. 5½ miles SW of Tehuacán along road to Zapotitlán, 9 Dec. 1967; elev. ca. 5,500 ft., arid calcareous hills (l, cap).

Gentry 23402, DES. 3 miles SW of Tehuacán, 11 March 1974 (photo).

Miranda 4431, MEXU. Camino to Zapotitlán, SW of Tehuacán, 23 May, 1948.

Smith s.n., DES. Flank of Cerro Zapotitlán, W of Tehuacán, 28 Aug. 1962; succulent desert on limestone mesa (photo).

Trelease 76, MO. Tehuacán 2/5/05. SW of Hotel.

11.

Group Urceolatae

Plants small, generally cespitose, forming tight clusters of numerous acaulescent rosettes, sometimes simple with short stems, freely seeding. Leaves succulent, linear lanceolate, with long acicular to subulate spines and mostly small weakly attached teeth; inflorescence spicate to racemose or subumbellate with lateral peduncles; flowers protandrous, small, yellow, in clusters of two to several on unequal branching pedicels, urceolate to tubular; ovary small, with constricted neck; tube short, spreading, much exceeded in length by the broadly over-lapping tepals, which are frequently appressed to the filaments; filaments inserted within tube; anthers small, shortly exserted; capsules persistent, roundly ovoid to oblong; seeds small, thick, hemispherical to deltoid. Northwestern Arizona and adjacent Utah, Nevada, and California.

Typical species: *Agave utahensis* Engelm. in King, Rep. Geol. Expl. 40th Par. 5: 497, 1871.

The Urceolatae are a distinct marginal group, both geographically and morphologically, distinguished by their small size and small flowers with a marked tendency for the urceolate (urn-shaped) flower form, the corolla being widest near the base and the apex of the over-lapping tepals appressed or connivent about the filaments. The principal species, *A. utahensis,* varies in leaf armature and growth habit. One of the variants, *A. utahensis kaibabensis* is the most distinct and appears to be diverging phylogenetically to a species; I am elevating it to a subspecies in this treatment. The other variants are less distinct and tend to intergrade, especially in overlapping areas. These have come to be regarded as varieties and I am following such segregates here, as outlined in the following key to the Urceolatae. The morphological boundaries are sometimes fuzzy.

The rare *Agave arizonica* is very distinct from *A. utahensis* in the form of its corolla and detail of leaf margin. The small size of *arizonica,* its narrow umbelliform inflorescence and flower structure suggest some affinity with the Deserticolae, as, for instance, with *A. mckelveyana,* its geographic neighbor. However, the flower form and especially the leaf morphology appear to align it with the Littaeoid Urceolatae; a group far removed phylogenetically from the more generalized *A. dasylirioides* in the Striatae.

Berger (1915) placed *A. utahensis* in his Section Pericamptagave (= Marginatae), but it is out of sectional character there. I doubt if he ever saw plants or flowers of *A. utahensis,* and *A. arizonica* was unknown in his time.

Long flower sections and correspondent ideographs of the Urceolatae are shown in Fig. 11.1. Distributions are charted in Fig. 11.3. These, as well as the flowering seasons depicted with the rainfall silhouettes in Fig. 11.2, are based primarily on herbarium collections, which are listed in the Exsiccatae.

Key to Species and Subspecies of the Urceolatae

1. Leaves with a horny margin extending nearly to the base; flowers not markedly urceolate, the tepals erect to ascending, subumbellate on short lateral peduncles; filaments inserted in mid-tube   *arizonica,* p. 254
1. Leaves without a horny margin except for spine decurrency, not reaching to mid-leaf; flowers urceolate, the tepals connivent, spicate to racemose, the clusters when pedunculate not umbellate; filaments inserted near bottom of tube   2
2. Plants larger, rosettes 40–60 cm tall, usually single, rarely with few suckers; inflorescence stout, 3.5–5 m tall, the peduncle 4–6 cm thick
    utahensis ssp. *kaibabensis,* p. 259
2. Plants smaller, the rosettes 15–30 cm tall, profusely surculose; inflorescence slender, 2–3 m tall, the peduncle 2–3 cm thick   3
3. Spines 2–3 (–4) cm long, the green length of leaf 5 to 10 times the length of the spine; leaf margins usually straight, with relatively small teeth, the larger in mid-blade 2.5–5 mm long. Utah-Arizona
    utahensis ssp. *utahensis,* p. 257
3. Spines 4–15 cm long, the green length of leaf only 1 to 4 times the spine length; leaf margins usually undulate with teated teeth, the teeth in mid-blade 4–12 (–15) mm long. Nevada-California   4
4. Terminal spine 3–8 cm long, slender, brown to whitish; teeth relatively small
    utahensis var. *nevadensis,* p. 261
4. Terminal spine 10–20 cm long, large, thick, ivory white; teeth larger, coarse, crooked
    utahensis var. *eborispina* p. 262

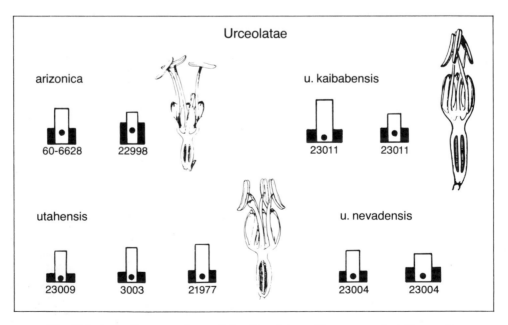

Fig. 11.1. Long flower sections of the Urceolatae with correspondent ideographs, showing proportions of tube (black) and tepals (white column) with insertion of filaments in tube (dots).

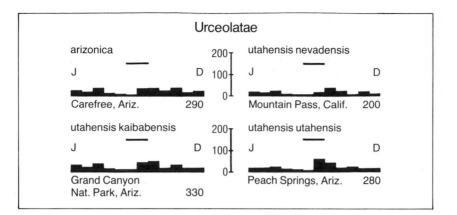

Fig. 11.2. Rainfall (silhouettes) and flowering season (bars) in the Urceolatae. Relevant meteorological stations with average annual rainfall in millimeters. Data from U.S. Dept. Commerce, Environmental Data Service.

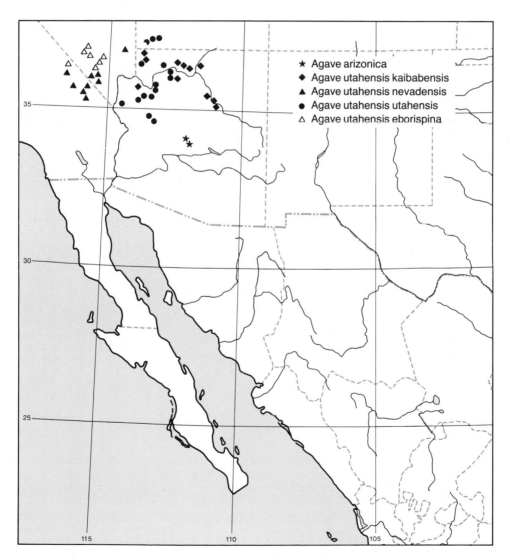

Fig. 11.3. Distribution map of the Urceolatae.

Agave arizonica
(Figs. 11.1., 11.2, 11.3, 11.4, 11.5, 11.6, 11.7; Table 11.1)

Agave arizonica Gentry & Weber, Cact. & Succ. J. (US) 42: 223, 1970.

Small, acaulescent, single or cespitose rosettes ca. 30 cm high, 40 cm broad, with many leaves in a depressed globose form; mature leaves mostly 17–24 x 2–4 cm, broadest in middle, linear-lanceolate, acuminate, patulous, rigid fleshy, dark green, fibrous, smooth above and below, shallowly concave above, thickly convex below, the margin 1–2 mm wide, reddish brown to light gray, continuous nearly to leaf base; teeth variable, the larger 2–5 mm long, mostly down-flexed, rarely horizontal, proximal or to 1.5–2 cm apart, dark-colored to lighter; terminal spine 1–2.5 cm long, subulate, shallowly grooved above, thickly convex below, dark brown aging gray, decurrent with margin; inflorescence 3–4 m tall, slender racemose-paniculate, very narrow with 35–50 short, small lateral branches in upper ¼–⅓ of shaft; peduncular bracts long-deltoid, 8–12 cm long; flowers small, durable, 25–32 mm long, pale yellow, cylindric-urceolate, short-

Table 11.1. Flower Measurements in the Urceolatae (in mm)

Taxon & Locality	Ovary	Tube	Tepal	Filament Insertion & Length		Anther	Total Length	Coll. No.
arizonica								
New River Mts., Ariz.	13	4×8	8×4.5&7	3	24	9	25†	59-6510
	13	4–5×8	11×4&10	3–4	26	11	27†	60-6628
	14	4–5×7	8×3.5&7	3–4	18	10	26†	61-6732
	13	5×7	8×4&7	3–4	18	10	26†	68-9200
	15	6×7	9×3	3–4	20	12	31†	22998
utahensis								
St. George, Utah	15	2×7	9×4	1			26*	Palmer
Beaver Dam Mts., Utah	14	2.5×7	9×3.5	1	14	9	24†	23009
	13	2.5×7	9×3	1–2	12	9	24†	23009
	17	3×9	12×4	1–2	15	12	30†	4232
Tuweep, Ariz.	19	3×8	10×3&9	1–2	17	10	30†	3003
	15	3×8	9×4&9	2–3	16	10	27†	3004
	14	3×9	9–10×5	1	17	9	25†	21977
Peach Springs, Ariz.	17	3.5×9	9×4.5	2	15	9	28†	21977
	18	3.5×9	11×4	2	18	11	31†	21977
Burro Creek, Ariz.	16	3×12	10×5	1–2	20	11	27†	2237
u. ssp. *kaibabensis*								
Cameron, Ariz.	17	3–4×10	12×5	1–2	20	11	31†	23011
	20	4×8	8×4	2–3	18	9	32†	23011
Little Colorado	18	3	10	1–2	18		32*	13337
u. var. *nevadensis*								
Ivanpah Mts., Cal.	16	2.5×7	9×4	1–2	19	8	28†	23004
	14	3×8	9×4	1–2	18	9	25†	23004
	15	3.5×8	8×4	1–2	20	8	25†	23004
Goodsprings, Nev.	12	4×9	8×4.5	2–3	18	8	23†	23007
	15	4×10	8×5	1	15	8	25†	23007

* Measurements from dried flowers relaxed by boiling.

† Measurements from fresh or pickled flowers.

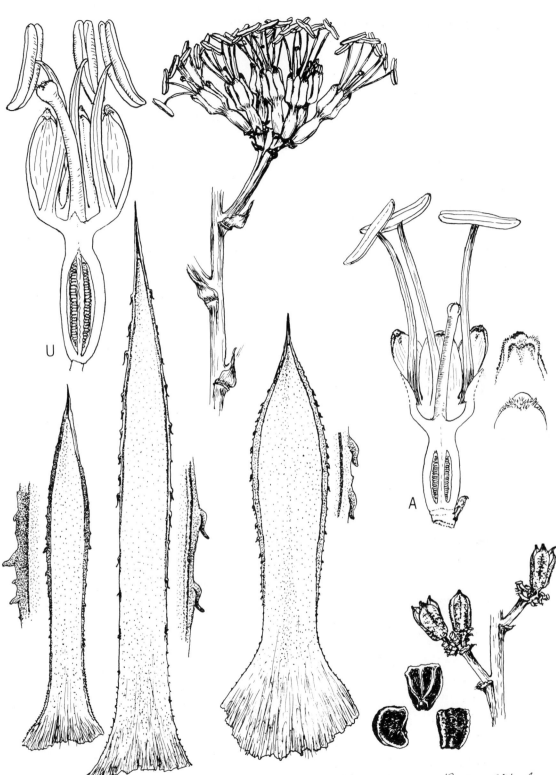

Fig. 11.4. *Agave arizonica.* A, three leaves, x 3.5, with variant detail of margins enlarged; long flower section, x ½, with tepal tips enlarged; clusters of flowers and capsules ca. ³/₅ nat. size; seeds, x 3 with face and rim or edge views. *A. utahensis,* U, flower section, x 2.4.

Regina O. Hughes

Fig. 11.5. Inflorescence of *Agave arizonica* with some flowers in early anthesis; note the strictly erect tepals in this early stage.

Fig. 11.7. *Agave arizonica* in natural habitat on New River Mountains in central Arizona.

Fig. 11.6. Inflorescence of *Agave arizonica* with very elongate peduncle.

pedicellate, in close-set clusters of 10–20; ovary 12–15 mm long, slender, slightly angled, constricted in neck; tube 3.5 x 8 mm to 5 x 9 mm, shallow, broadly spreading; tepals 8–11 mm long, unequal, closely overlapping, erect to ascending, broadly ovate with blunt tips, rather thin, rounded on back and involuting after anthesis, the outer with rugose hood, sometimes red-darkened, the inner with low keel; filaments 25–30 mm long, inserted just above mid-tube; anthers 9–11 mm long, yellow to dull-colored; capsules 15–20 x 8–9 mm, elliptic to ovate, strongly beaked, with strong walls; seeds 3 x 2 mm, almost as thick as wide.

Type: *Weber s.n., US.* Near summit of the New River Mountains, near Maricopa-Yavapai county line; specimen prepared from transplant grown in Desert Botanical Garden, Phoenix, Ariz., accession No. 60-6628, where it flowered June 1–7, 1968. Isotypes DES, ASU.

Although *Agave arizonica* is variable in leaf characters, the few specimens available indicate a distinct specific taxon, distinguished from *A. utahensis* by its marginate leaves, subumbellate flower clusters, and the more cylindric flower form with deeper tube and longer exserted stamens. *A. arizonica* suckers sparingly and does not build the dense rosette clumps prevalent in most forms of *A. utahensis*. The subumbellate panicles suggest it is a product of past introgression with subgenus *Agave* genes, but this is a loose speculation without evidence.

Distribution and Habitat

Since the discovery of *A. arizonica* in the 1960s, a few additional plants have been located. The most significant is a site east of Payson below the Tonto Rim, a range extension of about 70 miles (110 km) to the northeast. The plant community is similar to that of the New River Mountains, being an association of Arizona chaparral and juniper grassland communities over volcanic pediments between 3,000 and 4,500 feet (900–1,400 m). See Fig. 11.3 and Exsiccatae. Figure 11.2 indicates the amount and pattern of rainfall, as recorded at Carefree, the nearest or relevant meteorological station available.

A. arizonica has grown well in the Desert Botanical Garden and several plants have flowered during the last decade. It has withstood our low winter temperatures down to 19°–20°F (ca. −6° C) without damage. I suggest that several hundred more plants should be propagated and transplanted back to the original habitat, where this very rare species is presently at an endangered population level. It makes a fine little garden ornamental.

Agave utahensis
(Figs. 11.1, 11.2, 11.3, 11.8, 11.9, 11.10; Table 11.1)

Agave utahensis Engelm., ssp. *utahensis,* Trans. Acad. Sci. St. Louis 3: 308, 1875.
Engelm. in King, Geol. Expl. 40th Par. 5: 497, 1871.
Agave newberryi Engelm., Trans. Acad. Sci. St. Louis 3: 309, 1875.
Agave haynaldii Tod., var. *utahensis* Terrac., Primo Contr. Mon. Agave, Napoli 1885: 28, 1885.
Agave scaphoidea Greenm. & Roush, Ann. Mo. Bot. Gard. 16: 391, 1929.
Agave utahensis Engelm. var. *discreta* M. E. Jones, Contr. West. Bot. 17: 19, 1930 (in part).

Small, cespitose, light grayish to yellow green, rather compact rosettes 18–30 cm tall, 25–40 cm broad with 70–80 leaves. Leaves mostly 15–30 x 1.5–3 cm, linear-lanceolate, stiff, straight or falcate or upcurving, plane to concave above, convex below; teeth brown-ringed around bases, blunt, thick, light gray, the larger mostly 2–4 mm long, 1–2.5 cm apart, detachable; spine 20–40 mm long, acicular, decurrent for 1–3 cm, openly grooved above, light gray; inflorescence spicate, racemose, or paniculate, 2–4 m tall, lax or congested, the flowers in clusters of 2 to 8 on sessile pedicels or on

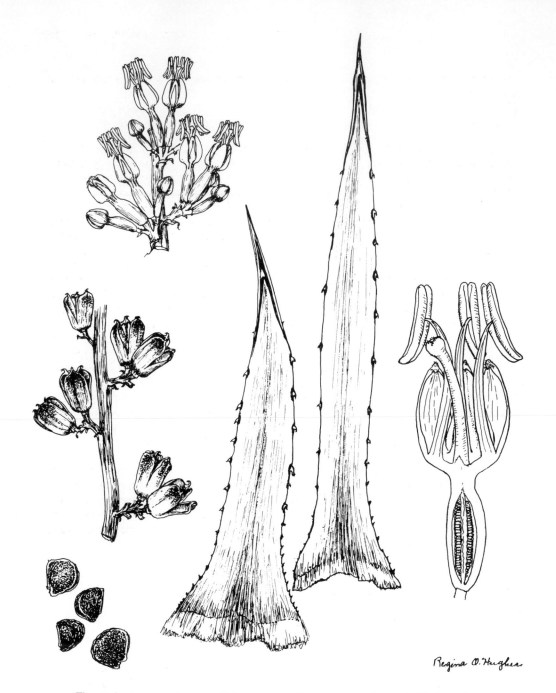

Fig. 11.8. *Agave utahensis* with leaves and inflorescence sections x 2.13; long flower section and seeds x 2.

short lateral branches 5 cm or more long; flowers 25–31 mm long, urceolate with yellow connivent segments; ovary 15–20 mm long with long (4–6 mm) constricted neck; tube very short, widely spreading, 2.5–4 mm long, 9–12 mm broad, bulging at tepal bases; tepals 9–12 mm long, nearly equal, the outer thin, deeply concave and over-lapping the inner, the inner strongly keeled, all persisting erect, conniving around fila-ments; filaments inserted near bottom of tube, 18–20 mm long; anthers small 9–12 mm long; capsules ovoid to oblong, 1 x 1 cm to 2.5 x 1.2 cm, short-beaked; seeds thick, crescentic, variable in size, 2 x 3 mm to 4 x 4 mm, hilar notch usually conspicuous, dull or shiny black.

Type: *Palmer in 1870, MO,* US. About St. George, Utah (l, f. The leaves a broad-leaved form, 10 x 2 cm, deltoid lanceolate, broadest at base with 6 teeth per side in brown eyelets; spine 2–2.5 cm long.)

Agave utahensis is a variable complex with some extreme developments in teeth, spine, and inflorescence. Earlier botanists, having very few and incomplete specimens, were led to interpret some of the variants as species, as the above synonymy attests. The simple spicate inflorescence when compared with the paniculate type is confusing, for instance, unless the two are seen together in natural habitat. With study of the natural populations, such variation becomes a species character, bringing the whole into perspective. The brown ring around the tooth bases is like that characterizing the *A. cerulata* complex of Baja California, but this expresses no close affinity of the two complexes.

Agave utahensis is partial to limestone and may form extensive populations, as on the broad open stony slopes between Ash Fork and Kingman in northwestern Arizona. In May and June the slender yellow-flowered shoots are conspicuous along the Santa Fe railway. The large colony at Peach Springs (Figs. 11.9, 11.10) has been the source of several collections and represents in variable form the variety *discreta* of Jones.

A common Indian name for this plant was "yant." McKelvey (collection notes) adds that the flowering stalk, used for prodding cattle, was called a "Yant-stick" and the Paiute Indian name for the plant is "Oose." Like other Agaves in the Southwest in historic times it was commonly known as "mescal." Castetter et al. (1938: 65) state, "The Walapai made large ropes by twisting four to eight rovings of fiber obtained from the dried tips of mescal (*A. utahensis*)." The numerous mescal pits and quids found in the area chronicle its common use as food (1938: 37 with map). Harrington reported (1933) quids of this agave in the Gypsum Cave of southern Nevada, showing it as a food of the prehistoric people there. It is doubtful if any such use, including the ceremonials connected with its preparation, are continued today, as the reservation peoples have ready access to modern groceries. Modern man employs it to enhance his garden, and the several distinct leaf forms are intriguing to the agave fancier. I have found no notes on the faunal associates of this plant, but they must be there. It has been making little rocks out of big ones much longer than man has, and as a pioneer at soil making should be conserved and encouraged.

Agave utahensis ssp. *kaibabensis*
(Figs. 11.1, 11.2, 11.3, 11.11; Table 11.1)

Agave utahensis ssp. *kaibabensis* (McKelvey) Gentry, stat. nov.
  Agave kaibabensis McKelvey, J. Arn. Arbor. 30; 230, 1949.
  Agave utahensis var. *kaibabensis* Breitung, J. Cact. Succ. Soc. Amer. 32: 21, 1960.

Plants robust, usually single, with bright green, numerous leaves on a short trunk at maturity, the inflorescence stout, narrowly paniculate with many small slender lateral peduncles. Leaves mostly 30–50 x 3–5 cm, lanceolate, long-acuminate, light green, the younger frequently pruinose glaucous, firm-fleshed, rigid, patulous; teeth 3–5 mm long, 1–4 cm apart, blunt, deltoid, little curved, grayish to white; spine stout, subulate, 3–4 cm long, flat to hollowed above, roundly angled below, gray to white, long decurrent to upper teeth; inflorescence 3.5–5 m tall, paniculate with numerous slender lateral branches 7–10 cm long; peduncle ca. as long as the panicle, stout, 4–6 cm diam.; flowers 4–12 per panicle, as for the species, but larger.

Type: *McKelvey 4381, A.* North side of Grand Canyon (near the rim), Kaibab Plateau, Coconino Co., Arizona, 15 May 1934 (l, cap; leaves of these specimens measured 25–40 cm length. McKelvey gave 0.4 m as leaf length in her type, but that included the clasping butt. Cult. plants in the Desert Botanical Garden develop larger leaves, up to 54 cm, from axil to tip of spine.)

Fig. 11.9. *Agave utahensis* near Peach Springs in typical clumped habit among basaltic rocks. June 1966.

Fig. 11.11. *Agave utahensis kaibabensis* in the Virgin River Canyon, extreme northwestern Arizona. Photo from Geirsh.

Fig. 11.10. *Agave utahensis*, showing detail of leaves and closely bunched rosettes. June 1966.

Fig. 11.12. *Agave utahensis nevadensis* in the type locality on the Ivanpah Mountains of southeastern California. June 1972.

Agave utahensis var. *nevadensis*
(Figs. 11.1, 11.2, 11.3, 11.12; Table 11.1)

Agave utahensis var. *nevadensis* Engelm. in Greenm. & Roush, Ann. Mo. Bot. Gard. 16: 390, 1929.

> *Agave nevadensis* (Engelm.) Hester, J. Cact. Succ. Soc. Amer. 15: 133, 1943.

Variety *nevadensis* is like var. *utahensis* but differs with smaller rosettes, 15–25 cm tall; leaves frequently bluish gray glaucous, generally with larger teeth; spines elongated especially in proportion to the fleshy part of the leaf. Inflorescence as for the species with both simpler spicate forms and paniculate forms. The main core of *nevadensis* exhibits a morphological difference from the general populations in the eastern areas of the species, but there are transitional forms between the two in southwestern Utah and perhaps elsewhere.

Type: *S. B. & W. F. Parish 41, CAS, MO.* Ivanpah Mts., Mohave Desert, California, May 1882 (l, f, s).

Agave utahensis nevadensis is a montane endemic in the southern Mohave Desert, frequenting limestone outcrops between 4,000 and 6,000 feet (1,240 and 1,850 m) elevations; southern Nevada and southeastern California. This is an arid environment with freezing temperatures in winter, down to 0°F (−18°C). Rainfall is indicated in Fig. 11.2, and general distribution is mapped in Fig. 11.3. Specific localities of collections and the names of mountains are listed in the Exsiccatae.

Agave utahensis var. *eborispina*
(Fig. 11.3)

Agave utahensis var. *eborispina* (Hester) Breitung, Cact. & Succ. J. (US) 32: 20–23, 1960.

> *Agave eborispina* Hester, Cact. & Succ. J. (US) 15: 131–33, 1943.

Variety *eborispina*, when well developed is very distinctive, the large, elongate, ivory white spines making it easily recognizable (Fig. 51, Breitung 1968). Recent plant explorations in southern Nevada by S. A. Cochrane and others have established this taxon in a small area on limestone mountains as the northwesternmost segregate of the genus *Agave* (Fig. 11.3). See also Exsiccatae.

Urceolatae Exsiccatae

Agave arizonica

ARIZONA. *Gentry & Weber 22998*, DES, US. New River Mts., W of Seven Springs Road along road to Bloody Basin, Maricopa Co., May 1972; rocky arroyo margin with dispersed chaparral (l, f, cap).

Weber s.n., ASU, DES, **US.** Type. Live plant col. near summit of New River Mts., Maricopa-Yavapai Co. line, flowered Des. Bot. Gard., Access. No. 60-6628, Phoenix, Arizona, June 1–7, 1968 (l, f).

Weber s.n., DES. Cult. Des. Bot. Gard., Access. Nos. 59-6510, 61-6732, 61-6738, New River Mts., near Maricopa-Yavapai Co. line, 1960s (l, f, cap).

Agave utahensis eborispina

CALIFORNIA. *Cochrane et al. 1006*, UNLV. A. 4 miles S of Stewart Valley summit on Hwy. 178, W side of Nopah Range, Inyo Co., 23 April 1978; elev. 1,128 m, limestone.

NEVADA. *Ackerman s.n.*, UNLV. Peek-a-Boo Canyon (Sheep Range Mts.), Clark Co., 24 June 1973; elev. 5,900 ft., rocky crevices (f).

Cochrane 604, UNLV. Northwest of Vabm Curry, W. Mercury Ridge, Nye Co., 8 June 1977; elev. 1,372 m, limestone.

Cochrane 820A, UNLV. Road to Lee Canyon on Hwy 52, Charleston Mts., 15 Feb. 1978; elev. 1,219 m, Larrea vegetation. Clark Co.

Cochrane & Bostick 538, UNLV. W slope at top of Peak 5485, SW of Aysees Peak, W. Buried Hills, Lincoln Co., 13 May, 1978; elev. 1,672 m (l, infl.).

Hester s.n., CAS, MO, US. Peek-a-Boo Peak, Sheep Range Mts., ca. 35 miles NW of Las Vegas, 22 July 1942 (l, f, cap, photo). Type of *A. eborispina* Hester. [A fine series of specimens; in US are 11 sheets.]

Agave utahensis kaibabensis

ARIZONA. *Brown 1029*, ASU, DES. 3.2 miles SE of canyon mouth. Elbow Canyon, Mohave Co.,

"Arizona Strip," 31 May 1979; elev. 4,400 ft., with pine, oak, *Opuntia* & scattered shrubs. Rare in locality.

Coville 1692, US. Bright Angel Trail, Grand Canyon Nat. Park, 22–23 Feb. 1903.

Dudley s.n., CAS. Bright Angel Trail above Indian Gardens, Grand Canyon, June 1904 (l, f).

Eastwood 3598, 5773, CAS. Grand View Trail and Hermit Trail, Grand Canyon Nat. Park, 16 June 1916 (l, bud).

Eastwood & Howell 1103, CAS. Trail to Roaring Springs, Grand Canyon Nat. Park.

Ferris & Duncan 2275, CAS. Grand View Point, Grand Canyon Nat. Park, 16 June 1921 (l, cap).

Gentry 23011, DES, US. 9 miles W of Cameron on S rim of little Colorado River, 14 June 1972 (l, f, cap). Elev. ca. 4,900 ft.

Goldman 2064, US. Grand Canyon, 13 May, 1913.

Grater 251, UC. Rim of Quartermaster Canyon, W end of Grand Canyon, Mohave Co., 27 June 1939; elev. 5,500 ft. (l, cap).

Hitchcock 80, US. Grand Canyon, 30 June 1913 (l, f).

MacDougal 194, US. Grand Canyon.

Maguire 12242, UC. E side of Grand Canyon, 27 June 1935 (l, f).

McKelvey 4381, 4381A, A. Type. On N side of Grand Canyon (near rim), Kaibab Plateau, Coconino Co., 15 May 1934 (l, cap).

Peebles 13337, ARIZ, CAS, US. Gorge of the Little Colorado River, Coconino Co., 8 June 1937; elev. 5,400 ft. (l, 1 to 5 cm wide).

Toumey 446, ARIZ. E side of Grand Canyon, 27 June 1935.

Ward s.n., US. Red Canyon Trail, Grand Canyon Nat. Park, 10 June 1901; elev. 4,200 ft. (l, f).

Woodbury 1032, DES. Lime Kiln Canyon, Virgin River Canyon, Mohave Co., 31 July 1978 (l, cap).

Wooton s.n., US. Grand Canyon, 7 July 1892.

Agave utahensis nevadensis

CALIFORNIA. *Alexander s.n.,* UC. Clark Mts., San Bernardino Co., 1939.

Gentry 23004, DES, US. N slope of Ivanpah Mts. above Mountain Pass on rt. 15, San Bernardino Co., 12 June 1972; elev. ca. 5,000 ft., limestone with juniper sagebrush community (l, f).

Hester s.n., UC. Ivanpah Mts., mt. pass S side of hwy 91, July 1943 (l, cap).

Jones s.n., UC. Clark Mts., San Bernardino Co., 14 Sept. 1932 (cap).

McElroy s.n., DES. Clark Mts., San Bernardino Co., 25 Mar. 1962. Cult. Des. Bot. Gard. 62-7051, flowered May 1973 (l, f).

Munz 12,869, UC. Clark Mts., San Bernardino Co., 14 Sep. 1952; elev. 5,000 ft. (cap).

Parish 414, CAS, **MO.** Type, from plant brought from Ivanpah Mts., San Bernardino Co., July 1882 (l, f, s).

Wolf 6814, CAS, UC. Kingston Mts., 1¼ miles NW of Beck Spring on Tecopa Road, San Bernardino Co., 14 May 1935; elev. 3,500 ft. (l, bud).

Wolf 6859, CAS, UC. NE base of pass between Kingston & Francis Spring, Kingston Mts., San Bernardino Co., 15 May 1935; elev. 3,500 ft. (l, f).

Wolf 7065, 7609, CAS. Clark Mt., San Bernardino Co., 2 4/10 miles W of Coliseum Mine, 28 May 1935; elev. 4,600 ft. (l, bud, cap). Oct. 7.

NEVADA. *Clokey 7881,* CAS, UC. Wilson's Ranch, Charleston Mts., 19 June 1938; elev. 1,200 m (f, cap).

Clokey 7882, UC. S of Indian Springs, Clark Co., 13 June 1938; elev. 1,200 m (f).

Gentry 23007, DES, US. 6 miles NW of Goodsprings, Bird Spring Range, 12 June 1972; elev. 5,100–5,200 ft., rocky limestone slope (l, f).

Langenheim 3726, UC. Tributary to Porter Wash, near Monte Cristo Mine, Goodsprings Quadrangle, Clark Co., 24 June 1954; elev. ca. 3,900 ft. (l, f).

Maguire 18047, UC. Calcite Mine, 10 miles SW of Wilsons Ranch, Charleston Mts., 4 May 1939 (l, f).

McKelvey 4141, A. Yellow Pine Mine, 45 miles from Las Vegas, 3 May 1934; elev. 4,800 ft. (l, infl in bud).

Munz 16760, CAS. Yant Pit Canyon, S end of Virgin Mts., Clark Co., 4 June 1941; elev. 4,600 ft. (l, f).

Purpus 6135, UC. Sheep Mts., May–Oct. 1898; elev. 4,000–5,000 ft.

Train 1923, UC. Hackberry Spring, Mormon Mt., 30 miles N of Moapa, Clark Co., 7 June 1938; elev. 5,000 ft. (l, f).

Agave utahensis utahensis

ARIZONA. *Barclay & Mason 3003, 3004, 3005,* ARIZ, DES, US. North rim of Grand Canyon, vicinity of Tuweep, Grand Canyon Nat. Mon., Mohave Co., 8 June 1968; elev. ca. 4,000 ft. (l, f).

Blakley B-1763, DES. 1 mile N of Valentine, Mohave Co., 8 June 1953; elev. 3,850 ft. (l, f, cap).

Clover 5212, ARIZ, MICH. Havasupai Canyon, Coconino Co., 21 July 1940.

Eastwood 18388, CAS. Aquarius Mts., 14 May 1931 (l, f).

Eastwood 18446, CAS. Peach Spring, 17 May 1928 (f).

Ferris & Duncan 2233, CAS. Rocky hills at Peach Spring, Mohave Co., 15 June 1921.

Gentry 21977, DES, US. ½ mile W of Peach Spring, 26 June 1966; open rocky sun slope (l, f, cap, photo).

Gentry 21978, DES, US. Valentine along rte. 66, 26 June 1966; steep granitic, sunny mountain slope (l, f).

Gentry & Ogden 9967, DES, MICH, US. 32 miles E of Kingman along hwy. 66, 16 Nov. 1950; among granite boulders on W slope (l, cap).

Goldman 2939, US. Colorado River, mouth of Diamond Creek, 7 Sep. 1917; alt. 2,000 ft. (l, cap).

Howell 26421, CAS. Havasu Canyon, Grand Canyon, Coconino Co., 23 May 1950.

Jones 25167, CAS, POM. Oatman, 14 June 1930 (l, f, cap, type of var. *discreta* Jones at Pomona includes cap of *A. mckelveyana* Gentry).

Lemmon s.n., UC. Peach Springs, June 1884 (f).

Mason & Phillips 2891, CAS, US. Pass N of Cottonwood Wash, Mohave Co., 19 June 1969 (l, f).

McKelvey 1514, 1650, A. Near Burro Creek, Aquarius Mts., March 1930; "dense clumps" (l, f, cap).

McKelvey 2237, A. *ibid.,* 14 May 1931.

McKelvey 2740, A. Diamond Creek, 5 May 1932 (f preserved).

McKelvey 4091, A. New Water Point, 29 April 1934 (l, cap).

McKelvey 1655, 1657, 2276, A. Peach Spring, 31 March 1930 (l, cap).

Rusby 835, M, UC. Peach Spring, 6 June 1883 (l, f).

NEVADA. *Niles & Leary 1901.* UNLV. Paradise Gap, Gold Butte area.

UTAH. *Gentry 23008,* DES, US. 21 miles W of St. George along rt. 91, Beaver Dam Mts., 12 June 1972; elev. ca. 4,700 ft., rocky limestone slope with juniper.

Gentry 23009, DES. US. 17 miles W of St. George along rt. 91, Beaver Dam Mts., 13 June 1972; elev. ca. 4,200 ft., limestone with juniper (l, f).

McKelvey 4232, A. Beaver Dam Mts., near St. George, 8 May 1934 (l, f, cap).

Palmer in 1870, MO, US. Type. About St. George (l, f).

PART III

Systematic Account of Genus and Species

Subgenus **Agave**

Subgenus **Agave**

(Subgenus **Euagave** of previous authors.)

The subgenus **Agave** is distinguished by the flowers being borne on horizontal (more or less) secondary peduncles in flat-topped or rounded clusters. The 123 taxa I have been able to identify with mainland North America are organized into 12 groups or sections. This classification brings many of the related species together, helps distinguish relationships, and assists in identifying the highly variable species.

The following keys leave much to be desired. It is difficult to find simple contrasting characters, especially to identify plants without flowers or old inflorescences. I have used size categories with the images of the general wild populations in mind. Measurements or sizes given are approximate and generally prevailing, but there are exceptions, such as depauperate plants in nature or pots, and excessive growth with rich culture in botanical gardens. If one size group does not work, try another; several key pathways may need to be explored. All is relative.

Synoptical Key to Groups of Subgenus **Agave**

1. Leaves ensiform, linear, patulous, 10 to 20 times longer than wide; tepals drying reflexed on tube; capsules broadly ovate   *Rigidae,* p. 551
1. Leaves generally not ensiform, lanceolate to ovate, much less than 10 times longer than wide; capsules more oblong   2
2. Plants large to very large, rosettes 1.5–2.5 m tall; leaves mostly 1.2–2.5 m long, 12–40 cm wide; flowers large, 70–110 mm long   3
2. Plants smaller, rosettes generally less than 1.6 m tall, leaves less than 1.5 m long   4
3. Axis of inflorescence with large fleshy bracts; panicles pyramidal in outline with widest branches below; ovary generally longer than tube and tepals; leaves massive, generally green. Central Mexican highlands
                 Salmianae, p. 594
3. Axis of inflorescence with smaller chartaceous bracts; panicles linear to deeply oval in outline with widest branches in middle; ovary shorter than corolla; leaves usually less massive, light gray glaucous
                 Americanae, p. 270
(With green leaves and spherical bracteolate flower clusters, see also *A. atrovirens* of the Hiemiflorae).
4. Plants medium to large, rosettes 1–1.6 m tall; leaves 1–1.5 m long   5

4. Plants small to medium; rosettes 0.4–1 m tall; leaves 20–100 cm long   10
5. Leaves without teeth or teeth reduced to prickles or irregularly occurring. Horticultural   *Sisalanae*, p. 619
5. Leaves with teeth   6
6. Plants generally glaucous gray to light green; stems short, freely surculose   7
6. Plants green, not suckering from the base   8
7. Peduncle with large fleshy bracts   *Salmianae*, p. 594
7. Peduncle with small chartaceous bracts   *Americanae*, p. 270
8. Plants budding from leaf axils, frequently developing long branching stems; leaves generally short, less than 70 cm long. Baja California   *Umbelliferae*, p. 635
8. Plants single, unbranched multiannuals (except *A. capensis*); stems short; leaves longer, 70–150 cm   9
9. Flowers with thin-walled broad tubes (15–22 mm wide) and short tepals but little exceeding tube in length; teeth uniform in size and spacing. Baja California   *Campaniflorae*, p. 309
9. Flowers with narrow tubes and long tepals 2 to 4 times longer than tubes; teeth irregular in size and spacing, frequently with small interstitial teeth. Sierra Madre Occidental   *Crenatae*, p. 323
10. Plants generally surculose; spring to summer flowering   11

10. Plants generally not surculose; winter to early spring flowering   13
11. Flowers with deep tubes; tepals short, dimorphic, the outer conspicuous, larger; filaments inserted on two levels. Sierra Madre Occidental   *Ditepalae*, p. 416
11. Flowers with deep or shallow tubes, the tepals much longer than the tubes, subequal; filaments generally inserted on one level   12
12. Panicles stout, with 15 to 35 large umbels flushed with red in the bud; flower tubes deep, equaling to half as long as tepals. SW U.S. and NW Mexico.   *Parryanae*, p. 520
12. Panicles slender, open, with 10 to 18 small, greenish yellow umbels; flower tubes very shallow, ⅓ to ⅙ as long as the tepals. Sonoran Desert   *Deserticolae*, p. 354
13. Leaves usually asperous, marble-colored, with short (1–2.5 cm) end spines; flowers in small flat-topped umbels with few minute caducous bracts; flowers 30–50 mm long; spring flowering   *Marmoratae*, p. 507
13. Leaves various, not markedly asperous, green to glaucous, with longer end spines (3–6 cm); flowers mostly congested in spherical clusters, often conspicuously bracteate; flowers 50–90 mm long, winter to spring flowering. S. Mexico and Central America   *Hiemiflorae*, p. 465

Sectional List of Species

Americanae (16 taxa)
americana
franzosini
lurida
oroensis
scabra
scaposa

Campaniflorae (3 taxa)
aurea
capensis
promontori

Crenatae (9 taxa)
bovicornuta
caldodonta
cupreata
hookeri
inaequidens
jaiboli
maximiliana

Deserticolae (17 taxa)
avellanidens
cerulata
deserti
gigantensis
margaritae
mckelveyana
moranii
sobria
subsimplex
vizcainoensis

Ditepalae (12 taxa)
applanata
chrysantha
colorata
durangensis
flexispina
fortiflora
murpheyi
palmeri
shrevei
wocomahi

Hiemiflorae (13 taxa)
atrovirens
congesta
hiemiflora
hurteri
lagunae
pachycentra
parvidentata
potatorum
pygmae
seemanniana
thomasae
wercklei

Marmoratac (4 taxa)
gypsophila
marmorata
nayaritensis
zebra

Parryanae (9 taxa)
gracilipes
guadalajarana
havardiana
neomexicana
parrasana
parryi

Rigidae (25 taxa)
aktites
angustifolia
breedlovei
cantala
datylio
decipiens
fourcroydes
karwinskii
macroacantha
panamana
rhodacantha
stringens
tequilana

Salmianae (8 taxa)
macroculmis
mapisaga
ragusae
salmiana
tecta

Sisalanae (6 taxa)
desmettiana
kewensis
neglecta
sisalana
weberi

Umbelliflorae (3 taxa)
sebastiana
shawii

12.

Group Americanae

Handbook of the Amarillidae 1888: 175.

Medium to large, suckering, non-bulbiferous, rosettes; leaves lanceolate to linear, light glaucous gray to pale green with well-developed teeth on variable margins and strong, mostly subulate spines; inflorescence with slender shaft with appressed or reflexed, chartaceous, triangular bracts and open panicle, elliptical to oval in outline, the lateral branches in upper half of shaft horizontal to ascending; umbels decompound, not crowded, small in the wild, larger and spreading in cultivars; flowers long-pedicellate, yellow, rather slender; tepals generally longer than the tube, not wilting until after anthesis; stamens with long filaments and anthers, the filaments inserted near middle of the furrowed tube, generally on one level; pistil exceeding the stamens in post-anthesis.

    Typical species: *Agave americana* L.

Baker was the first to use the term Americanae for a miscellaneous non-systematic group including two taxa of the subgenus **Littaea.** Of the 26 species he lists under this heading only one, *Agave americana,* is recognized today as a sectional relative. Berger in 1915 continued use of the group name for his series Americanae. Of the 17 binomials he lists in this group, only three taxa are retained in my account. The rest I have allocated to other sections, or have abandoned as unrecognizable horticultural names. Trelease, the last serious student of *Agave,* did not define this section of the genus but did propose several new names that are dealt with in turn below.

    Agave americana, a phylogenetically advanced member of the genus, epitomizes the section. It and its sectional congenators are generally recognizable by their long lanceolate grayish leaves, by their strong surculose habit, tall open decompound panicles on straight small-bracted shafts, and slender yellow flowers with long tepals and long stamens. The ovary is apt to be relatively short and tapering to the base. The flower proportions are similar to the Crenatae, but the color and chemistry of that section are very different. Phylogenetically, there appears to be a connection with Section Ensiformae through *A. lurida,* which I am including with the Americanae.

    Agave americana and *A. scabra* form two principal branches of the Americanae. The former, because of its better edibility and fiber, has been aided and greatly extended geographically by man, while *A. scabra* shows more indigenous adaptability in its more

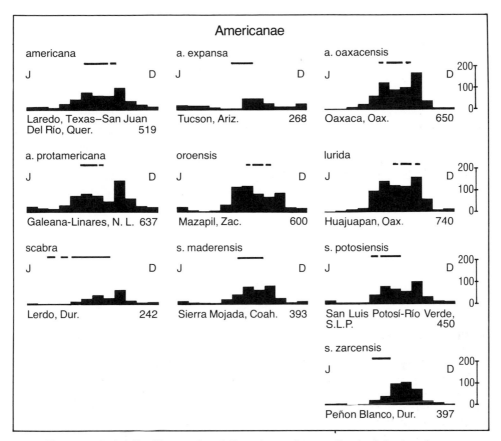

Fig. 12.1. Rainfall (silhouettes) and flowering perimeters (bars) of the Americanae; relevant meteorological stations with annual rainfall in millimeters. Data from Atlas Climatológico de México, 1939, and U.S. Weather Bureau. Flowering periods based on specimens and field observations; uncertainty expressed by dotted lines.

extensive natural distribution over the Coahuilan Desert. Both species have forms that appear to be native to southern Texas. The other species, *lurida* and *oroensis*, are less closely related and relatively rare. The separation of taxa in the Americanae has been difficult, and, as with other sections of the genus, the characters in all organs are necessary for a well-balanced system. A rather full representation of flowers is catalogued in Table 12.1 and illustrated in Figs. 12.3, 12.4. The taxa are genetically transgressive with gene flow usually fluid, as indicated by the chromosome studies reported below.

The Americanae appear to have had their genesis in the arid and semi-arid subtropical climates of northeastern Mexico. Rainfall generally ranges from 250 to 800 mm with a well-defined dry spring season (Fig. 12.1). Temperatures are high during late spring and summer. Frosts are common during winter and are severe in the higher elevations in the north. Most of the forms respond well to garden culture through the southern United States. As far as I can judge, the extension of the group to the Jaliscan Plateau Region and into Oaxaca was most likely by the Amerindians, *A. americana expansa* and *A. americana oaxacensis* being respective examples. The horticultural varieties of *A. americana* are widely scattered around the world. They have retained their xerophytic nature and will tolerate protracted drought. The agave habit of daytime closure of stomates provides a high degree of water conservancy in leaf tissue, as ably demonstrated by Ehrler's study (1967).

Chromosomes of the Americanae

The chromosomes of the Americanae are of special interest for the simple reason that a number of them have been studied and their karyotypes published (Granick 1944; Cave, 1964). Most of them are reproduced here (Fig. 12.2). The series of *Agave americana* is from Granick, while *A. scabra* was reported by Cave as *A. asperrima*. Her source of the latter was the living clone in Huntington Botanical Gardens, block 20—clone No. 192, the same source as *Gentry 19807,* which is verifiable as *A. scabra* Salm ssp. *scabra* in this treatise.

While few of the varieties of *A. americana* reported by Granick are certainly identifiable to variety, her series established polyploidy as extensive in this complex species. Her series shows diploids, tetraploids, and hexaploids. She observed that tetraploids and hexaploids tend to have thicker, coarser, more succulent leaves than diploids. She also reports some observations of H. H. Bartlett, who collected *Agave* in the northern Sierra Madre Oriental. "The closely related 'magueys' (americanae Baker) of each collective species are merely a swarm of interbreeding and constantly segregating Mendelian combinations." The partial or complete mixing of chromosome sets in the generative cells between taxa is apparently the basis for much of the variability in this and other groups of *Agave*. Some of the *americana* varieties appear to be sterile clones, others appear fertile. The evolution of forms was no doubt greatly accelerated with the man-agave symbiosis, as many genetic combinations could survive with cultivation that would otherwise have died.

Key to Species of the Americanae

1. Leaf surface smooth to slightly rough; teeth straight to variously curved, not conspicuously deflected; panicles generally with straight shafts and more branches, 15 to 40; flower tube generally ½ to ⅓ as long as tepals   2

1. Leaf surface mostly scabrous; teeth below the mid-blade generally deflected; panicles very open, sometimes crooked, with only 8 to 15 branches (Fig. 5); flower tube deep, little exceeded by tepals   *scabra*, p. 296

2. Plants medium to very large, 1–2 m tall; leaves frequently light gray glaucous and reflexed, lanceolate, variable in size, armor, coloring, the margin frequently undulate; flower buds green to light yellow   3

2. Plants medium-sized, to 1 m tall; leaves light green, linear, the margin nearly straight, varying little; flowers pink in bud   *oroensis*, p. 294

3. Flowering peduncle elongate with panicle congested, ca. ¼ as long as peduncle; leaves broadly lanceolate with closely spaced teeth, confluent or 1–2 cm apart   *scaposa*, p. 303

3. Peduncles not so elongate, the panicle ½–⅓ as long as penduncle; leaves with teeth 2–5 cm apart, if only 1–5 cm apart, then leaves linear   4

4. Plants very large, to 2 m tall; leaves spatulate, much broader above the middle, narrowed toward base, white glaucous with green markings at and below mid-blade. Cultivar.
franzosini, p. 290

4. Plants not as above; leaves sometimes variegated with yellow to whitish markings   5

5. Plants sparingly surculose; leaves ensiform with uniform, regularly closely spaced teeth (1–2 cm apart), the larger along the mid-blade 5–7 mm long. Wild.   *lurida*, p. 292

5. Plants surculose, often prolifically so; leaves lanceolate, broadest in mid-blade or above; teeth not uniform, irregularly spaced (2–5 cm apart), varying in size, 4–10 mm long, wide and cult.   *americana*, p. 278

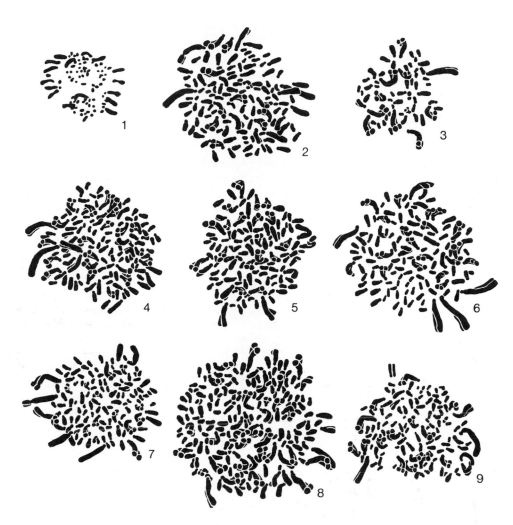

Fig. 12.2 Chromosomes in the Americanae.

1, *Agave scabra, n* = 87, with one large chromosome extra, microspore mitosis (Cave, 1964, p. 165). 2–10, somatic chromosomes from root tips (Granick, 1944, p. 288).

2, *A. americana* var. *expansa?, 4n* = 120, from Sonora.

3, *A. americana* var., *2n* = 60, from Tucson, Arizona.

4, *A. americana* var., *4n* = 120, origin unknown.

5, *A. americana* var., *4n* = 119, origin unknown.

6, *A. americana* var. *marginata, 4n* = 118, origin unknown.

7, *A. americana* var. *marginata, 4n* = 120, San Luis Potosí.

8, *A. americana* var., *6n* = 180, Webb County, Texas.

9, *A. americana* var. *expansa, 4n* = 119, probably Mexico.

Table 12.1. Flower Measurements in the Americanae (in mm)
(See p. 41.)

Taxon & Locality	Ovary	Tube	Tepal	Filament Insertion & Length		Anther	Total Length	Coll. No.

americana

Taxon & Locality	Ovary	Tube	Tepal	Filament Insertion & Length		Anther	Total Length	Coll. No.
Laredo, Tex.	32	9 × 16	31 × 9 & 28	5	72	36	72†	20005
	33	7 × 16	32 × 8 & 29	5–6	90	37	73†	20005
	34	8 × 16	31 × 8 & 29	5–6	85	36	72†	20005
	30	14 × 18	25 × 8 & 24	8–9	50	27	70†	20022
Saltillo, Coah.	36	13 × 18	26 × 6 & 24	8–9	58	29	75†	20022
	31	14 × 18	26 × 8 & 24	8–9	55	28	71†	20022
	32	11 × 20	33 × 8 & 30	7–8	83	35	76†	20154
	34	10 × 20	30 × 8 & 28	8	85	34	74†	20154
Río Verde, SLP	26	12 × 18	32 × 5 & 31	7–8	78	34	70†	20090
	27	11 × 18	33 × 5–31	6–7	78	34	72†	20090
San Juan del Río, Quer.	37	15 × 17	30 × 7–28	9–10	58	34	81†	20084
	35	13 × 17	27 × 7 & 25	9–10	70	32	77†	20084
Jacala, Hgo.	39	12 × 15	24 × 6 & 22	7–9	58	32	76†	20080

a. var. *expansa*

Taxon & Locality	Ovary	Tube	Tepal	Filament Insertion & Length		Anther	Total Length	Coll. No.
Amado, Ariz.	39	14 × 19	31 × 9 & 29	8–9	68	28	85†	21983
	38	14 × 12	28 × 7 & 26	8–10	46	31	81†	21983

a. var. *marginata*

Taxon & Locality	Ovary	Tube	Tepal	Filament Insertion & Length		Anther	Total Length	Coll. No.
Hunt. Bot. Gard., Cal	37	14 × 14	25 × 5	9	65	33	77†	19864
	33	15 × 14	26 × 5	9	53	33	74†	19864

a. var. *oaxacensis*

Taxon & Locality	Ovary	Tube	Tepal	Filament Insertion & Length		Anther	Total Length	Coll. No.
Valle de Oaxaca, Oax.	47	14 × 18	38 × 9 & 36	8–9	63	37	99†	
	52	15 × 19	35 × 6	8–9	85	38	102†	
	53	14 × 21	36 × 6	8–9	85	38	104†	

a. ssp. *protamericana*

Taxon & Locality	Ovary	Tube	Tepal	Filament Insertion & Length		Anther	Total Length	Coll. No.
Galeana-Linares, N. L.	46	15 × 15	28 × 5 & 25	10	68	29	87†	20156
	38	18 × 13	30 × 5 & 28	11–12	62	33	86†	20156
	41	19 × 14	29 × 5 & 27	12–13	70	33	90†	20156
	38	20 × 15	29 × 5 & 26	10–11	65	36	86†	20157
	42	20 × 16	30 & 27	10–11	73	35	92†	20157
S of Galeana, N. L.	40	17 × 14	20 × 6 & 17	9–10	57	26	77†	20160
	41	15 × 14	20 × 6 & 17	9–10	72	25	76†	20160

franzosini

Taxon & Locality	Ovary	Tube	Tepal	Filament Insertion & Length		Anther	Total Length	Coll. No.
Hunt. Bot. Gard., Cal.	36	18 × 17	32 × 6 & 30	9–10	65	38	87†	19866
	39	20 × 17	30 × 6 & 28	10	70	38	89†	19866

lurida

Taxon & Locality	Ovary	Tube	Tepal	Filament Insertion & Length		Anther	Total Length	Coll. No.
Huajuapán, Oax.	30	9 × 13	18 × 4	7–8	40	21	58†	20279
	31	10 × 13	19 × 4	8–7	45	22	60†	20279
	31	9 × 13	20 × 4	8–7	45	22	59†	20279
La Mórtola	50	18 × 16	17 × 5	7–8	55	27	80*	Berg.
	38	16 × 14	20 × 5	7–8	56		78*	Berg.

oroensis

Taxon & Locality	Ovary	Tube	Tepal	Filament Insertion & Length		Anther	Total Length	Coll. No.
Concepción de Oro, S.L.P.	34	18 × 12	20 × 5 & 18	11–10	53	30	72†	23592
	37	17 × 13	20 × 4.5 & 18	11 & 10	55	29	75†	23592
	37	16 × 13	21 × 4.5 & 19	11 & 10	53	30	74†	23592

Table 12.1 cont.

Taxon & Locality	Ovary	Tube	Tepal	Filament Insertion & Length		Anther	Total Length	Coll. No.
scabra								
	33	14 × 15	22 × 4.5	8–9	60	26	69†	23268
	36	15 × 15	20 × 4	8–9	60	26	70†	23268
Sierra Sarnosa,	30	13 × 15	19 × 4.5	10	50	27	62†	23268
Dur. (Neotype)	31	12 × 15	19 × 4	9	50	26	62†	23268
	39	16 × 16	19 × 5	9–10	58	23	74†	23268
	42	16 × 16	20 × 5	10	60	25	79†	23268
Laredo, Tex.	31	15 × 16	24 × 6	10–11	55	31	70†	20004
Zapata, Tex.	37	16 × 14	21 × 5 & 18	8–9	68	29	74†	20006
	36	13 × 15	26 × 5 & 25	8–10	65	29	76†	20008
Rio Grande City, Tex.	30	12 × 16	20 × 5 & 18	9–10	55	26	63†	20012
Monterrey-Saltillo, Coah.	36	17 × 16	19 × 5 & 17	11–12	58	28	72†	20016
Saltillo, Coah.	28	16 × 13	18 × 5 & 16	9–11	58	24	61†	20018
Cuatro Cienegas, Coah.	35	20 × 16	21 × 6	12–14	63	30	76†	23259
	36	18 × 16	21 × 5	11–12	48	29	73†	23259
Cuencame, Dur.	41	18 × 14	22 × 5	12	58	31	81*	11549
Pedriceña, Dur.	32	16 ×	18	9–11	50	24	67*	10595
Hunt. Bot. Gard., Cal.	40	17 × 14	20 × 5 & 18	11–13		28	77†	19807
s. ssp. *maderensis*								
	36	12 × 15	18 × 5	9–7	53	23	66*	E.309
Sierra Madera,	36	11 × 12	17 × 5	8 & 7	56	23	65*	E.309
Coah.	35	16 × 17	15 × 4 & 14	10–11	48	17	66	E.
	38	16 × 17	16 × 4 & 14	10–11	46	15	70	E.
s. ssp. *potosiensis*								
S of Matehuala, SLP	32	11 × 15	27 × 8 & 26	7–8	58	26	70†	20075
Huizache, SLP	30	15 × 14	20 × 5 & 18	7–9	63	28	64†	20162
Sto. Domingo, SLP	44	16 × 17	22 × 6 & 21	9–11	60	30	82†	11510
	34	18 × 16	18 × 5 & 16	11–12	66	25	70†	20176
S. S. Manuel, SLP	46	17 × 16	27 × 6 & 25	9–10	74	33	90†	20176
	53	15 × 16	24 × 5 & 22	7–8	68	28	82†	20176
Río Verde, SLP	42	22 × 16	23 × 7 & 20	11–12	73	32	89†	20091
s. ssp. *zarcensis*								
	50	20 × 18	21 × 5 & 20	16 & 13	62	34	91†	8632
Zarca Mesa, Dur.	37	16 × 17	19 × 4 & 18	12 & 10	55	27	72†	8632
	50	18 × 18	19 × 4 & 18	13 & 11	58	29	76†	8632
	35	17 × 15	17 × 5.5	10–12	56	26	68†	DBG

* Measurements from dried flowers relaxed by boiling.

† Measurements from fresh or pickled flowers.

Americanae

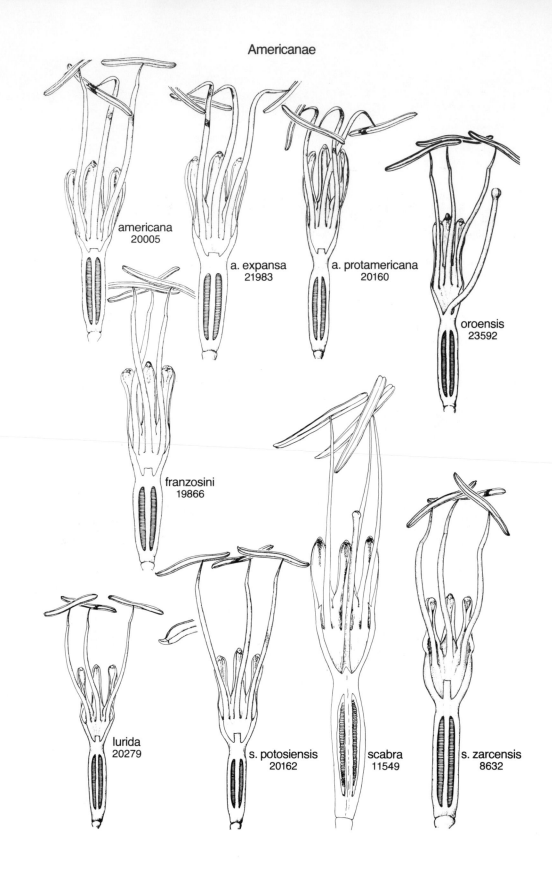

americana
20005

a. expansa
21983

a. protamericana
20160

oroensis
23592

franzosini
19866

lurida
20279

s. potosiensis
20162

scabra
11549

s. zarcensis
8632

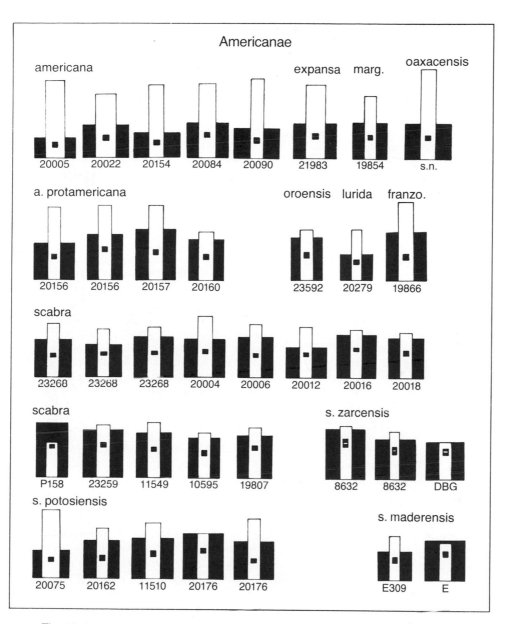

Fig. 12.4. Ideographs of Americanae flowers, showing proportions of tube (black) to outer tepals (white) and locus of filament insertion (black squares) in tube. From measurements in Table 12.1.

◄ Fig. 12.3 *(opposite page)* Long sections of flowers of the Americanae, drawn from pickled flower collections.

Agave americana

(Figs. 12.1, 12.2, 12.3, 12.4, 12.6, 12.7, 12.8, 12.9; Table 12.1)

Agave americana L., var. *americana*. Sp. Pl. 1753: 461.

    Agave complicata Trel. ex Ochoterana, Mem. Rev. Soc. Ci. Alsate Mex. 33: 100, 1913.
    Agave gracilispina Engelm. ex Trel. in Bailey Stand. Cycl. Hort. 1: 234, 1914.
    Agave melliflua Trel., ibid.
    Agave zonata Trel., ibid.
    Agave felina Trel., Contr. U. S. Nat. Herb. 23: 128, 1920.
    Agave rasconensis Trel., ibid. p. 122.
    Agave subzonata Trel., ibid. p. 129.

Plants medium to large, freely suckering and sometimes seeding, short-stemmed, the rosettes 1–2 m tall, 2–3.7 m broad; leaves mostly 10–20 x 1.5–2.5 dm, lanceolate, narrowed above thickened base, plane or guttered or reflexed, usually acuminate, light gray glaucous to light green, sometimes variegated; cuticle smooth to slightly asperous; margin undulate to crenate; teeth variable, the larger 5–10 mm long, the slender cusps straight to flexuous or curved, from broad low bases, 2–6 cm apart, brown to pruinose gray; spine mostly 3–5 cm long, conic to subulate, shallowly grooved above for ca. half its length, shiny brown to pruinose gray; inflorescence 5–9 m tall, the shaft slender, straight, with scarious, rather small triangular bracts, the panicles generally long oval in outline, rather open, with 15–35 spreading decompound umbellate branches in upper ½–⅓ of shaft; flowers 70–100 mm long, long pedicellate, slender, opening yellow over greenish ovary; ovaries mostly 30–45 mm long with grooved neck, tapering to narrower base; tube 8–20 mm deep, 16–20 mm broad, funnelform, thick-walled, deeply grooved; tepals unequal, the outer larger, 25–35 mm long, thicker, linear-lanceolate, the apex cucullate, rugose, sometimes red-tipped, conduplicately narrowing at anthesis, the inner 2–3 mm shorter, conduplicate, with narrow high keel, prominently costate within; filaments frequently very long, 60–90 mm long, somewhat flattened, long tapering, inserted near mid-tube, 5–10 mm above base; anthers 30–36 mm long, yellow, centric to excentric; pistil stout, with trilobate stigma, exceeding anthers in length in post-anthesis; capsules oblong, 4–5 cm long, short-stipitate, short-beaked; seeds lunate to lacrimiform, 7–8 x 5–6 mm, shiny black.

    Type: Sheet No. 443.1, Herb. Linn. Soc. London (3-flowered partial inflorescence; photo only seen). Linnaeus listed it as being from "America calidiore."

The flowers of *A. americana*, both in typical form and in the numerous variants cultivated in northeastern Mexico, are characterized by a short, tapering ovary, shorter than the perianth; large, thick-fleshy, elongate outer tepals; smaller inner tepals with thick keels, thin involute margins, and paired prominent inner costae; and long, strong filaments divergent above the tepals.

    The habitat of *Agave americana* is largely an artifical one, since its varieties are cultivated in several distinct climates around the world. They respond well in the Mediterranean type of climate, provided they are given a little water in summer. The regional habitat in northeastern Mexico has a regime of summer-fall rainfall and is relatively dry in winter-spring (Fig. 12.1). A common variety is spontaneous along the highways near Laredo, Texas. They are glaucous gray plants up to 2 m tall with reflexed leaves, abundantly suckering from the root crown and forming small to large clones. The older central rosettes are closely surrounded by offsets of varying sizes. It does not occur away from the roads in competition with native desert plants, but, tolerated by road men and fed somewhat by run-off from the pavement, it is building itself a population in a narrow artificial habitat in southern Texas. Generally the varieties are drought resistant, have some frost resistance in southern temperate climates, and exhibit broad tolerance to different soil types. The distributions of *Agave americana* segregates in Mexico are indicated in Fig. 12.8.

Fig. 12.5. *Agave scabra zarcensis* (left) and *A. americana protamericana* flowering at Murrieta, California. The shaft of *scabra* is frequently crooked and has fewer umbels than *americana*.

A putative progenitor of the cultivars, ssp. *protamericana,* grows naturally on the more open slopes of the Sierra Madre Oriental in tropical deciduous forest and thorn forest zones between elevations of 500–1,400 m. It encroaches along the eastern edges of the Chihuahuan or Coahuilan Desert, but does not enter the desert proper. *Agave scabra,* a true desert xerophyte, also extends into this marginal zone and morphological intermediates have been observed here, seeming to express interspecific hybrids. Figure 12.1 indicates the rainfall patterns and flowering seasons for the Americanae.

Fig. 12.6. *Agave americana* near Laredo, Texas, planted along an electric power line. Photo, June 1963.

Fig. 12.7. A single nearly mature rosette of *Agave americana,* showing the structurally weak leaves which characteristically fall downward.

Agave americana is the "type" of section and the "type" of the whole genus, although it is typical of relatively few species, except in a broad or basic way. The foregoing description is made to represent what I consider to be a polymorphic species. It broadens species limits much beyond the horticultural variety described by Linnaeus and subsequent students. As the following account will show, with close study, the relationship of A. americana to other species is close and involved, particularly with A. scabra. Both wild and cultivated forms appear to represent introgressions between the two species. Some of the Mexican cultivars appear like hybrids between A. americana and A. salmiana, both of which are widely cultivated. A. quiotifera appears to be such a case. However, the two sections are fairly distinct in morphological characters.

The polymorphic nature of Agave americana is shown in the subspecific segregates described below and outlined in the following synoptical key. A. americana ssp. protamericana is a wild complex, varying in size, leaf form, and armature. From this complex in the Sierra Madre Oriental of northeastern Mexico, I believe that several to many of the cultivated forms have been selected by man during the several thousand years of his patrimony of the magueys. The flowers of ssp. protamericana (compare Fig. 12.4 of ideography) show a similar basic structure to that of var. americana, with exceptions attributable to genes of A. scabra. The cultivated variety americana itself appears to have had its origin as a selection from the wild complex of ssp. protamericana. This work does not account for the numerous cultivars or varieties to be found in Agave americana. However, several of the common horticultural varieties are recapitulated here. I should repeat that measurements reported are generally taken of mature specimens, whether of leaf, inflorescence, or flower, and the exceptions of extreme forms are rarely given.

Key to Varieties and Subspecies of *Agave americana*

1. Spines relatively broad and short, 2–4 cm long; leaves straight, guttered, sometimes valleculate, the margins straight or crenate; rosettes commonly on trunks 3–5 dm long   2
1. Spines subulate, 3–6 cm long; leaves frequently reflexed and otherwise not as above   3
2. Rosettes with stems to 6 dm long; leaves glaucous gray, the margin crenate with teeth along the mid-blade on sharply angled mammae; flowers 80–85 mm long   var. *expansa*, p. 283
2. Rosettes on shorter stems; leaves glaucous white, the margins nearly straight with close-set non-mammillate teeth; flowers 100 mm long
    var. *oaxacensis*, p. 285
3. Leaves relatively short, 80–135 cm long, 4 to 6 times longer than wide, plane or guttered, straight to curving, green to glaucous gray; panicles shorter with 15–20 branches; wild
    ssp. *protamericana*, p. 287

3. Leaves longer, 100–200 cm, 6 to 10 times longer than wide, frequently reflexed, glaucous gray or green variegated; panicles longer with 25–35 branches; cult.   4
4. Leaves glaucous gray to light green, narrowed above base, some reflexed above the mid-blade
    var. *americana*, p. 278
4. Leaves gray or green, variegated yellow or whitish   5
5a. Leaf with whitish or yellow margins
    var. *marginata* Trel. in Bailey, Stand, Cycl. Hort. 1: 235, 1914.
5b. Leaf with yellow center stripe
    var. *medio-picta* Trel., ibid.
5c. Leaf lined yellow or whitish
    var. *striata* Trel., ibid.
5d. Leaf dark green and yellow, twisted
    var. *variegata* Trel., ibid.
5e. Leaf larger, all green or green with yellow margins, guttered and straight or curved   var. *picta*

Agave americana var. *picta*

Agave americana var. *picta* (Salm) Terrac., Primo Contributo ad una Monografia delle
  Agave. Naples, 1885.
  Agave picta Salm-Dyck, Bonplandia 7: 88, 1859.
  Agave ingens Berger, Die Agaven 1915: 154.

These varieties, segregated largely on the basis of the yellow or whitish striations, are
inconstant in their color patterns. What should we call a variant that has a yellow center
stripe *(medio-picta)* and is also lined with yellow stripes along the margins (like *striata*)?
These plants appear as inconstant changeable forms from one generation to the next and
are hardly worth varietal designations. However, the names are convenient for agave
fanciers, census takers, or others for distinguishing and discussing their collections and
so on. I am therefore repeating their segregation here as outlined by Trelease and also
indicate my opinion that *A. picta* is only a variety of *A. americana*.

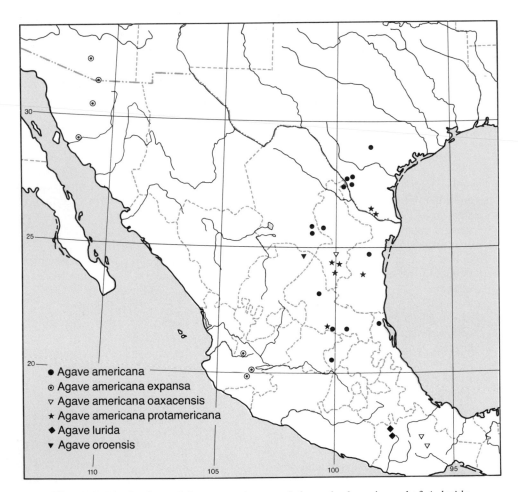

Fig. 12.8. Distributions of *Agave americana* varieties and subspecies and of *A. lurida*
and *A. oroensis*. Based on specimens and photographs.

Agave americana var. *expansa*
(Figs. 12.1, 12.2, 12.3, 12.4, 12.8, 12.9, 12.10, 12.11; Table 12.1)

Agave americana L., var. *expansa* (Jacobi) Gentry, U.S. Dept. Agr., Agricultural Handbook 399: 80, 1972.

> *Agave expansa* Jacobi, Nachtrage I in Abh. Schles. Ges. Vaterl. Cult., Abth. Naturwiss. 1868: 151. 1868.
> *Agave abrupta* Trel., Contr. U.S. Nat. Herb. 23: 132, 1920.

Large, light-gray rosettes, 1.5–2 m tall, 2 m broad, with a short trunk in age, suckering abundantly from early age; leaves 12–15 dm long, 18–24 cm wide, lanceolate, rather abruptly acuminate, narrowed toward the base, deeply rounded below, the upper ⅔ of leaf more plane, firm, thick, erect or ascending, frequently cross-zoned, sometimes valleculate, the margin nearly straight, with several sharply angled, low teats in the mid-blade; the larger teeth 5–8 mm high, 1–4 cm apart, shortly cuspidate from low broad bases, brown becoming gray; spine short, conical, 2–3 cm long, narrowly grooved above, brown to gray; panicle large, 7–9 m tall, deeply oval in outline, with 20–30 long, horizontal laterals; bracts of the peduncle 10–15 cm long, triangular, chartaceous, reflexed; flowers thick and fleshy, 70–85 mm long; ovary green, short, tapering from unconstricted neck; tube green, 13–14 mm deep, 12–20 mm broad, scarcely bulging, but deeply grooved and thick-walled; tepals pale yellow, erect, thick, 26–32 mm long, 7–9 mm broad, linear, involute, cucullate, the outer 2–3 mm longer, the inner with thick keel, 2 strong costae within, and thin involute margin; filaments about 65–70 mm long, flattened on adaxial side and bowed toward pistil, inserted 8–9 mm above tube base; anthers yellow, 28–32 mm long, excentric; capsules and seeds not seen.

This variety is known only as a cultivar, introduced in western Europe, where it was described by Jacobi as growing at "St. Germain en Laye bei Paris," France. What is

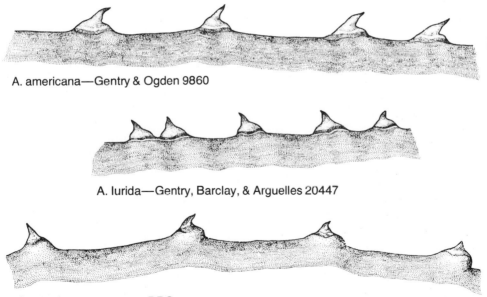

A. americana—Gentry & Ogden 9860

A. lurida—Gentry, Barclay, & Arguelles 20447

A. americana expansa—DBG

Fig. 12.9. Some mid-leaf margins in the Americanae, reduced ca. ¼. Note the sharp angle on lower side of teat of the variety *expansa;* a ready recognition character.

taken for this taxon has been observed from southwestern United States to Jalisco, where it probably originated. Wiggins (42) treats *A. expansa* as a synonym of *A. americana*. Indeed, the large, glaucous gray, prolifically suckering rosettes and tall, diffuse panicle resemble the type of the species and genus in its Mediterranean form. Trelease (43) regarded the Arizona form as a separate species, distinguishable by its short, heavy, conic spine and unreflexed leaves vs. the long subulate spine and reflexed leaves of *A. americana*. Until 1963 only a few, incomplete herbarium specimens of *A. americana* were available in American herbaria and only two leaf specimens and one poor, withered flower collection of *A. expansa*. The recent flower collection (Gentry & Weber 21983) confirms the close relationship of *A. expansa* to *A. americana*.

In Jalisco this plant has been observed planted about country houses and fields. The photo in Fig. 12.10, taken south of Zacoalco along the road to Cd. Guzman, shows the valleculate form of leaf. The plant also appears to be spontaneous in the disturbed countryside, but I have not seen seeding capsules there or in the American southwest. *Agave abrupta* Trel., described from La Barca, Jalisco, and photographed in 1961 at La Primavera, 20–25 miles (32–40 km) west of Guadalajara (Fig. 12.11) appears to be another form of *A. americana expansa*.

Fig. 12.10. *Agave americana expansa* growing near Zacoalca, Jalisco, December 1966. The man is Juan Arguelles, who assisted the author for many years.

Fig. 12.11. *Agave abrupta* Trelease, a cultivated variety of *A. americana* photographed in Primavera, west of Guadalajara, Jalisco, 1961. By leaf conformation it appears closely related to *expansa*.

Agave americana var. *oaxacensis*
(Figs. 12.1, 12.4, 12.8, 12.13; Table 12.1)

Agave americana L., var. *oaxacensis* Gentry var. nov.

Large, single or sparingly surculose, white-glaucous rosettes with broad short stems, 1.5–2 m tall; leaves 120–200 x 18–24 cm, linear, rather rigidly outstretched, little narrowed above base, somewhat guttered, thickly fleshy and convex toward base, sometimes cross-zoned; margins repand or undulate with rather small uneven teeth, 3–5 mm long, 1½–3 cm apart, straight or flexuous, dark brown; spine 3 cm long, long conical, openly grooved above for ½ its length, decurrent as a thin corneous margin for several cm; panicle to 10 m tall with 30 or more wide-branching decompound umbels; flowers greenish yellow, succulent fleshy, 95–105 mm long; ovary thick, cylindric, slightly 6-angled, scarcely constricted in short neck; tube 14–17 mm deep, 18–20 mm wide, funnelform, deeply grooved, thick-walled; tepals unequal, thick, linear, erect, the outer 35–38 mm long, broadly rounded on back, involutely over-lapping inner at base, the

Fig. 12.12. *Agave americana* ssp. *protamericana* showing introgression with *A. scabra* (plant on left) in a valley south of Galeana.

inner ca. 2 mm shorter, with thick narrow keel, prominently 2-costate within; filaments 80–90 mm long, stout, inserted near middle of tube on one level; anthers 37–40 mm long, excentric, yellow; capsules very large, 6.5–8 x 3–3.5 cm, ovoid to obovoid, woody, dark brown, short-stipitate, short-beaked. Seeds not seen. (Measurements of flowers and capsules from type only.)

Type: *Gentry & Arguelles 12260,* Tlacolula, Oaxaca, 5 Oct. 1952, offset collected, which flowered at Murrieta, Calif., June–July, 1971; US, isotypes MEXU, DES. (Fig. 12.4, 12.13.)

Planta caule singula vel caespitosula, grandis. *Folia* albida, 120–200 cm longa, 18–24 cm lata, lineari-lanceolata reacta semiconduplicata, *margine* repando vel undulato, *dentibus* plerumque parvis 3–5 mm longis, 1.5–3 cm separatis; *spine terminali* subulata ca. 3 cm longa supra brevi-canaliculate, decurrente. *Inflorescentia* panaliculata ad 10 m alta ramis lateralibus 30 vel plus; *floribus* luteo-viridibus, 95–105 mm longis; *tubo* 14–17 mm longo, profunde sulcato, carssi-succulento; *segmentis* inaequalibus exterioribus 35–38 mm longis. *Filamenta* 80–90 mm longis, tubi medio inserta; *antheris* 37–40 mm longis, luteis. *Capsulae* grandes ligneae, 6.5–8 cm longae, 3–3.5 cm latae, oblongae brevi-stipitatae, brevi-rostratae. *Semina* non vidi.

Fig. 12.13. *Agave americana* var. *oaxacensis* in the Oaxaca Valley, September 1952.

Agave americana oaxacensis is distinguished from other varieties by its very large, non-reflexed, spreading, white glaucous leaves, large inflorescences with large flowers, and large thick-walled capsules. It has been observed only in cultivation, especially in the Oaxaca Valley, but similar plants were observed also as cottage plantings near Huajuapán *(Gentry 21300)* and as far north as the Río Verde in San Luis Potosí *(Gentry, Barclay & Arguelles 20090)*. In Oaxaca, it was formerly grown for fiber and mescal but has generally been abandoned as a mescal plant, because it requires 18–20 years to mature. It has largely been replaced for that use by a bluish glaucous sword-leaved form of the *A. angustifolia* alliance, looking very much like *A. tequilana,* which matures in 7 to 8 years. Informants in Oaxaca gave me two names for *oaxacensis,* "maguey cenisa" and "maguey de pulque." It may still be used as a source of pulque.

Agave americana ssp. *protamericana*
(Figs. 12.1, 12.3, 12.4, 12.5, 12.8, 12.12, 12.14, 12.15; Table 12.1)

Agave americana L., ssp. *protamericana* Gentry ssp. nov.*

Rosettes acaulescent or short-stemmed, usually surculose, free seeding, openly spreading; leaves broadly lanceolate, 80–135 x 17–22 cm, rather rigid, shortly narrow above thick fleshy base, convex below, becoming plane to guttered above, light glaucous

---

*proto-americana = earliest form of *americana.*

Fig. 12.14. *Agave americana protamericana* wild in canyon of the Sierra Madre Oriental, east of Galeana, summer 1963.

gray to pale green, sometimes cross-zoned, the margin crenate to repand; teeth generally regular in size and spacing, the larger 5–10 mm long with slender flattened cusps from low broad bases, straight to curved, dark brown to gray; spine 3–6 cm long, broad at base, subulate, dark brown to grayish brown, shortly decurrent, openly grooved above; panicles 6–8 m tall with nearly straight small-bracted shaft and 15–20 wide-branching laterals in upper ½ to ⅓ of shaft; flowers 75–90 mm long, on slender 2-bracteolate pedicels; ovary 38–45 mm long, cylindric or fusiform; tube deeply funnelform, 15–20 mm long, thick-walled, deeply grooved, tepals unequal, narrowing by involution during anthesis, the outer 20–30 mm long, linear to lanceolate, the apex slightly hooded, rugose papillate, sometimes dark-colored; filaments 60–70 mm long, inserted in mid-tube; anthers 25–35 mm long, excentric; capsules (Ogden et al. 51104) small, 3.5–4 x 1.6–2 cm, oblong, thin-walled, light brown, stipitate, short-beaked; seeds 7–7.5 x 5–6 mm, lacrimiform, black, hilar notch subapical, marginal wing low, even.

Type: *Gentry & Barclay 20156*, 20–22 miles east of Galeana along road to Linares, Nuevo León, 14 July 1963; alt. ca. 3,000 feet, tropical short-tree forest (2 sheets, l, f), US, isotypes DES, MEXU.

Planta brevicaulis surculosa. *Folia* effusa lati-lanceolata 80–135 cm longa, 17–22 cm lata, glauco-viridia aliquando zonatim crassisucculenta, *margine* crenato vel repando, *dentibus* 5–10 mm longis atrobadis vel griseis; *spina terminali* 3–6 cm longa subulata brunnea vel grisea supra canaliculata brevi-decurrente. *Inflorescentia* 6–8 m alta, lati-paniculata ramis lateralibus 15–20; *floribus* luteo-viridibus, 75–90 mm longis; *tubo* 15–20 mm longo sulcato nectarifero; *segmentis* inaequalibus, exterioribus 20–30 mm longis, apice aliquando rubro. *Filamenta* 60–70 mm longa tubi medio inserta; *antheris* 25–35 mm longis luteis.

Subspecies *protamericana* differs from the varieties of *americana* in its greater variability in form and color of the leaves and their armature, in the proportions of the flower tubes to tepals, the tube being relatively deeper. The leaves of *protamericana* are generally shorter and the inflorescence generally has fewer branches.

Agave americana protamericana is a wild complex scattered along the Sierra Madre Oriental, notable as much for its variability as for its *americana* character. On the coastal side of the mountains, where it receives more rainfall, the rosettes reach large sizes, as in the canyon leading from Galeana to Linares, Nuevo León, and again inland from Cd. Victoria in Tamaulipas. The natural vegetation here is mainly tropical deciduous forest, or quite comparable as a plant community to the short-tree forest of the Mexican Pacific coast (Gentry, 1942). Along the eastern drier slopes of the Sierra Madre Oriental, there is a noticeable reduction in length of leaf. In this region it cohabits with *A. scabra* morphological intermediates between the two species are displayed in both leaf and flower characters (Fig. 12.12). The species have the same flowering season (Fig. 12.1). This apparent introgression has been mentioned above and elsewhere (Gentry, 1967). The

Fig. 12.15. *Agave americana protamericana* flowering along the road Galeana–Linares in July 1963. Flowers for the type specimen were collected from this plant.

population ca. 56 miles (90 km) S of Galeana along the road to Dr. Arroyo may be described as a ''hybrid swarm'' and similar variable populations occur in other localities.

There is a race of small agaves in the lower Rio Grande Valley, as 4–5 miles SE of Rio Grande City *(Correll 14887A, Gentry & Barclay 20010)*, which is probably *A. americana protamericana,* but no inflorescence is available for certain identification.

Some notes penned in the type locality area, July 14, 1963, give revealing details about the habitat and the agaves there. Agaves noted in the big canyon include *A. americana, A. lophantha, A. bracteosa,* and upward from the Cuesta de la Reyna to Galeana *A. falcata* and *A. scabra* become abundant. The canyon, large and narrowing with high spectacular limestone cliffs, has more luxuriant vegetation on the lower slopes. *Dioon edule* is abundant along the cliffs. The *Agave americana* alliance is abundant on the rocky limestone slopes from Puerta de la Reyna to the foothill plains; 12 to 20 miles (20 to 32 km) E of the Galeana crossroad. It varies in size, color, glaucous gray to glaucous green, in armature, and inflorescence; lateral branches counted as 10, 20, 24 per inflorescence. Some flowers were quite slender. Although suckering, it is a free-seeding population showing many variants; many very handsome. There appears to be some intergrading with *A. scabra* in the upper part of the canyon.

Agave franzosini

(Figs. 12.3, 12.4, 12.16; Table 12.1)

Agave franzosini Baker, Kew Bull. Misc. Inf. 1892: 3, 1892.*

Rosettes very large, widely spreading, freely suckering, 2–2.7 m tall (to 3 m tall, 4.5 m wide, fid. Berger*); leaves lanceolate, narrowed above base, 1.8–2.2 m long, 22–35 cm wide, light glaucous gray or bluish glaucous variously marked with green below mid-blade, spreading, recurved, or sharply reflexed, hollowed above, thickened and concave below toward base, the cuticle somewhat asperous; margin straight to repand with remote, dark brown teeth, the larger along the mid-blade 8–10 mm long, on fleshy prominences; spine 3–6 cm long, dark brown, with short open groove above, decurrent along inrolled apex; inflorescence ca. 8–11.4 m tall (Berger) with a strong shaft, the panicle deep, broadly cylindric, to 2.9 m broad, with broad spreading decompound umbels of large yellow flowers; ovary 3.5–4.5 cm long, 10–13 mm thick, with slightly narrower neck, light bright green; tube 18–22 mm long; tepals 30–32 mm long, soon withering, the outer linear, the inner keeled on the back, grooved within (= 2-costate within); filaments inserted in the mid-tube, stout, yellow, 65–80 mm long; anthers 38–40 mm long, yellow; pistil to 12 cm long, stout, 3-lobed with clavate stigma; capsule elongate-claviform, woody. 55–70 mm long; seeds black, shiny, to 12 x 8–9 mm.

Neotype: *Gentry 10163* (l) & *19866* (f), *US,* DES. Hunt. Bot. Gard., San Marino, Calif., Jan. 1951 & Aug. 1962; cult. in block 20 and elsewhere in succulent garden.

This plant is one of the most handsome of the large agaves. The nearly white leaves variegated with dark green perform variously curved attitudes to form large graceful rosettes. Berger wrote that it suckers and seeds freely and was widely dispersed in European gardens. It is rare in the United States, but the Huntington Botanic Gardens in San Marino, California, has had several fine examples for many years. There is some

---

*Description drawn partly from Berger, *Die Agaven,* 1915. No type has been designated, having been described from European gardens, named after Italian representative Francesco Franzosini, who had a fine garden by Lake Maggiore, Italy.

Fig. 12.16. *Agave franzosini* in Huntington Botanical Gardens, San Marino, California, with six-foot Myron Kimnach for comparison.

difference in the Huntington plants and Berger's description, but this could well be due to seedling variation. The plant is distinctive and not easily confused despite minor variations of size and armature. It is obviously related to *A. americana*. If of hybrid origin, it would be an excellent example of sudden species saltation, which De Vries for one would have been eager to extol. Its native country is unknown. I have not found it in Mexico or Central America and it is not included in Trelease's account of Caribbean *Agave*.

Agave lurida

(Figs. 12.1, 12.3, 12.4, 12.8, 12.17, 12.18; Table 12.1)

Agave lurida Aiton, Hort. Kew. 1: 472, 1789.

  Agave vernae Berger, Die Agaven 1915: 245.
  ?*Agave vera-cruz* Mill., Gard, Dict. ed. 8; Agave No. 7, 1768.
  ?*Agave verae-crucis* Haw., Syn. Succ. 1812: 72.

Plants single, nonsurculose (or rarely so) radially symmetric, short stemmed, 1.2–
1.7 m tall and twice as broad; leaves linear-lanceolate, 110–150 x 12–18 cm, stiffly
ascending to outcurving, concave to guttered and thinning beyond the slightly narrowed
base, dull green to glaucous gray, the margin nearly straight; teeth very regular, the larger
4–6 mm long, mostly 1–2 cm apart, smaller and closer together toward leaf base, the low
bases black, on low protuberances, the cusps mostly deltoid-flattened, straight or curved,
brown to grayish; spine 3–4.5 cm long, conic-subulate, 6–8 mm wide at base, grayish
brown, shallowly grooved above, decurrent for several cm; panicle 6–7 m tall, the shaft

Fig. 12.17. *Agave lurida,* spontaneous in semi-open thorn forest north of Huajuapán,
Oaxaca, August 1963.

Fig. 12.18. *Agave lurida*. The long linear leaves with close-set teeth are characteristic.

with small chartaceous bracts, with 20 or more ascending, diffusely spreading, decompound, open, umbellate branches in upper ½ to ⅓ of shaft; flowers greenish yellow, 58–65 mm long, on slender, minutely bracteolated pedicels; ovary 28–34 x 7–8 mm, fusiform, grooved in constricted neck; tube 9–11 mm long, funnelform, grooved; tepals erect, ca. equal, 18–24 mm long, incurved at tip and hooded, wilting after anthesis, the inner with narrow thin margins; filaments 45–60 mm long, inserted toward top of tube; anthers 20–22 mm long, bright yellow; capsules 5.5–6 x 2.5 cm, stipitate.

Neotype: *Dn. Masters, June 1883, K* Ricasoli's garden, Casa Blanca, Porte Ercole, Mt. Argentario, Tuscany, (sheets I, II, III, IV, l, f, cap.).

Drummond (Kew Bull. Misc. Info. 1910: 344–349) wrote a historical account of this plant, tracing it back through European gardens to Kerr-Gawler and the younger Aiton in the early 19th century, as the plant described by the elder Aiton in 1789. *A. lurida* was confused with *A. vera-cruz* of Miller (1768), which Drummond says was closely related, but distinguished from *A. lurida* by amber-colored flowers and a *revolute leaf apex*, a phenomenon I have never observed in *Agave*. I rather think that *A. vera-cruz* was some form of *A. americana*, a complex species apparently not understood by Drummond. The excellent specimens sent to Kew by Ricasoli and still preserved at Kew afford a splendid opportunity to establish the old name of *A. lurida*. I have, accordingly, designated it as the neotype.

In the summer of 1963 an excellent match of the Kew neotype was found growing spontaneously in the tropical highlands of the Mexican state of Oaxaca (Gentry, Barclay & Arguelles 20279, 20447) (Figs. 12.17 and 12.18). Thus, after two centuries of its introduction in Europe, the area of origin becomes known! Several scattered individuals were found growing 20–24 miles (32–38 km) N of Huajuapán along the road of Tehuacán, on volcanic rocky soils in semi-arid tropical forest at an elevation of about 6,000 feet (1,850 m). The 5 or 6 individuals seen were consistently alike with dull green to slightly glaucous, stiff, narrowly lanceolate leaves, the proximal teeth set on low broad bases. No suckers and no young plants could be found. No intermediates between it and the polymorphic widespread *A. angustifolia sensu lato,* which abounds in the region, could be found. Both species were blooming in August. This is the only area in which I have seen *A. lurida* and I count it as a rare species. I very much regretted having cut down the only inflorescence for flowering specimens. Although the leaves are somewhat ensiform, the nature of the small flowers, drying with tepals unreflexed, do not lend themselves well for inclusion in the Section Rigidae. The short, broad, diffusive panicle and leaf form appear to conform to the Section Americanae, but I see no close specific relative of it. Berger's *A. vernae* (1915, Fig. 73) shows a deeper but similar diffusive panicle and appears to me as only a white glaucous form of *A. lurida*.

Agave oroensis

(Figs. 12.1, 12.3, 12.4, 12.8, 12.19; Table 12.1)

Agave oroensis Gentry, sp. nov.

Plants acaulescent, single or suckering, in openly spreading low rosettes; leaves 80–100 x 8–10 cm, linear-lanceolate, straight to recurving, narrow and thickly convex below toward base, guttering upward, long-acuminate, green, slightly asperous, the margin straight to repand; teeth along the mid-blade 3–6 mm long, mostly 2–3 cm apart, smaller and more closely spaced toward leaf base, grayish, mostly straight; spine 2.5–3 cm long, acicular, narrowly grooved above for half its length, grayish, finely decurrent to uppermost teeth; inflorescence 5–6 m tall, with slender, small-bracted peduncle; panicle on upper half of shaft, of 12–16 loosely flowered spreading umbels, extending 50 cm or more; pedicels slender, to 1 cm long, with ligulate bracteoles; flowers pink in bud, opening yellow, very slender, with connivent or adherent tepals, 70–75 mm long; ovary 34–37 mm long, fusiform, the neck grooved, greenish; tube 16–18 mm deep, urceolate or constricted at apex, grooved, walls thickened with filament insertions, thin above; tepals 20–21 x 4.5–5 mm, unequal, linear, scarcely involute, connivent or adherent, rather thin, the outer bluntly apiculate with reddish rugose cap, the inner with low keel, non-costate within; filaments yellow, slender, flattened, long-tapering, inserted on slightly different levels (1 mm), 11 and 10 mm above base of tube; anthers 29–30 mm long, yellow, slender, excentric, curved or sinuous; pistil much thicker than filaments, soon overreaching anthers.

Type: *Gentry & Engard 23592, US*. Estación Margarita, near Concepción del Oro, Zacatecas, 1 August 1975; border cultivate on silty valley land derived from limestone; elev. ca. 6,000 feet [1,850 m], isotypes MEXU, DES.

Planta brevicaulis singula vel caespitosula. *Folia* 80–100 cm longa, 8–10 cm lata, lineari-lanceolata basim versus crassissima asperulosa viridia; *marginis dentibus* 3–6 mm longis, plerumque 2–3 cm separatim cinerei; *spina terminali* 2.5–3 cm longa aciculari, supra anguste canaliculata decurrente. *Inflorescentia* paniculata scapo incluso 5–6 m alta ramis 12–16 remotis diffusis; *floribus* gemiatis rubris postea luteo-viridibus gracilibus; *ovario* 34–37 mm longo fusiformi. *Tubo* 16–18 mm longo urceolata; *segmentis* inaequalibus linearibus conniventibus, exterioribus 20–21 mm longis, 4.5–5 mm latis, apice galeato rubro. *Filamenta* 50–55 mm longa supra medio tubi inserta; *antheris* ca. 30 mm longis, luteis. *Pistillum* postremo quam filamentis longius. *Capsular* et *semina* non vidi.

Agave oroensis, a local cultivate in the mining region about the Concepción del Oro mine in northern Zacatecas, is unique in the Americanae. It is well characterized by the thick-fleshy, narrow, green leaves, and especially by the broad, open, pink-budded panicles, the flowers of which are slender with a tube constricted at the orifice. The erect tepals have a strong tendency to adhere to one another, but this habit has been observed in other unrelated species, as with *A. aktites* of the Section Ensiformae. It is a weakness

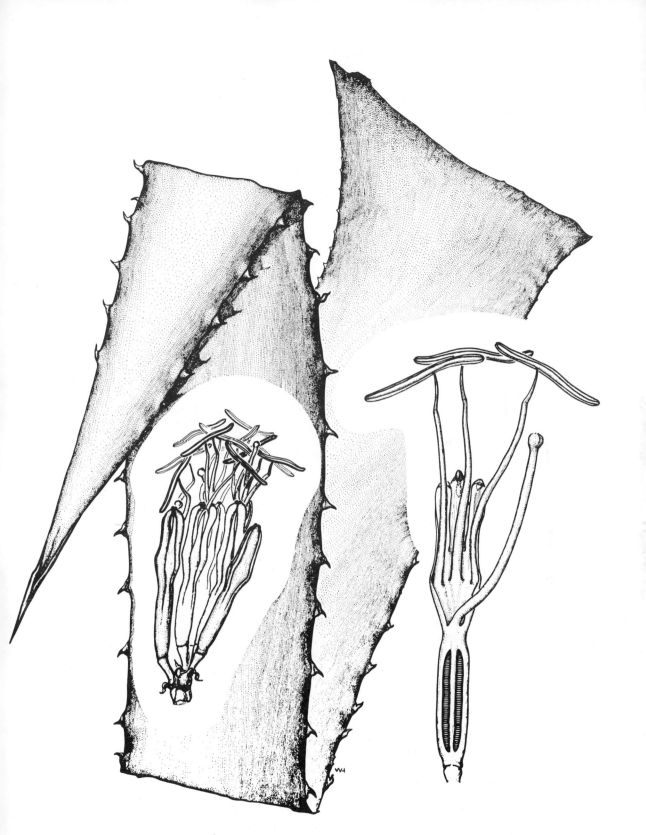

Fig. 12.19. *Agave oroensis*. Drawn from the type; leaf and flower cluster reduced ca. 4/7, long flower section ca. natural size.

perhaps related to insufficient water and turgor pressure at the extreme upper end of these long succulent structures.

The species was also observed a few miles northward of Estación Margarita along highway 54 on the brushy desert plain. This site showed signs of having been an abandoned domicile many years past, and, although the plants appeared spontaneous with the shrubbery, they were most likely originally planted there. Not observed elsewhere.

Agave scabra

(Figs. 12.1, 12.2, 12.3, 12.4, 12.12, 12.20, 12.21, 12.22, 12.23; Table 12.1)

Agave scabra Salm-Dyck ssp. *scabra*
  Agave scabra Salm-Dyck, Bonplandia 7: 89, 1859.
  Agave asperrima Jacobi, Hamb. Gart. Zeit. 20: 561, 1864.
  Agave caeciliana Berger, Die Agaven, 147, 1915.

Plants acaulescent, freely suckering and seeding, the rosettes rather open with 30–40 leaves at maturity, 7–10 dm tall, nearly twice as broad; leaves (mature) generally 60–110 x 12–16 cm, rigid, lanceolate, very broad at base, constricted just above base, very thickly convex below, flat above, then deeply guttering through mid-blade, long-acuminate, the surface scabrous, light green to glaucous gray, heavily armed, sometimes with corneous margin along upper half; teeth below the mid-blade generally deflected, the larger 8–15 mm long, the cusps from broadly rounded bases, brown to pruinose gray; spine 3.5–6 cm long, subulate to acicular, very narrowly grooved above, scabrous at base, long decurrent on involute leaf margin; panicles mostly 4–6 m tall, openly spreading with 8–12 branches in upper third of shaft, with small compact umbels, the peduncle with chartaceous bracts usually reflexed; flowers 60–80 mm long, yellow above greenish ovary; ovary 30–40 mm long, cylindric or fusiform, slender, with constricted furrowed neck; tube 13–20 mm deep, cylindric, thick-fleshy, grooved; tepals 18–25 x 4–6 mm, unequal, erect, lanceolate-linear, fleshy, the outer 2 mm longer, the inner broadening at base, inlapping the outer, with narrow prominent keel, both involute, cucullate-papillate at apex; filaments regular, equally inserted just above mid-tube, 50–65 mm long, slender, tapering, oval in x-section; anthers 24–30 mm long, centric, regular; pistil over-reaching stamens in post-anthesis, broadly clavate toward apex, with trilobate stigma; capsules 4–5 x 1.7–2 cm, oblong, short-stipitate, beaked, rather thin-walled; seeds 5 x 6–7 mm, lunate, the margin with wavy wing.

   Neotype: *Gentry & Engard 23268*, E. bajada of Sierra Sarnosa near Dinamita, Durango, May 15, 1973; *Larrea* shrub desert, elev. ca. 4,000 feet, US, dups. MEXU, DES. It can also be considered topotypic, as it is near the old Hacienda San Sebastian, where Wislizenus originally obtained seeds or offsets (Gentry, 1975).

   Agave scabra, next to *A. lechuguilla*, is the most widely spread and abundant *Agave* in the Chihuahuan Desert of northern Mexico. It is unusual among wild agaves in that it inhabits the broad valleys and plains as well as the stony slopes of the mountains. It grows in the driest areas of the desert, generally between 1,200- and 1,900-m elevations, and seems fond of limestone and caliche subsoils. It is commonly found in the *Larrea-Flourensia* community, and with many other common desert shrubs, as *Acacia, Mimosa, Fouquieria, Prosopis, Opuntia,* and *Yucca*. Figures 12.23 and 12.1 indicate its distribution and low rainfall requirements.

   Agave scabra is clearly related to *A. americana* as evidenced by the narrow-bracted inflorescence, general conformation of the flowers (see sections and ideographs Figs. 12.3, 12.4), the karyotypes (Fig. 12.2), and the apparent introgressions of the two species complexes. *A. scabra* is generally distinguished from *A. americana* by its more scabrous

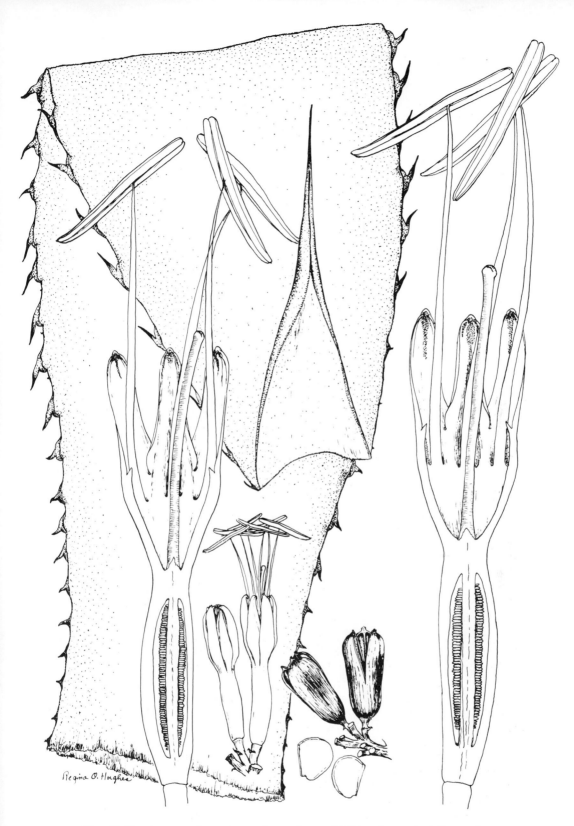

Fig. 12.20. *Agave scabra*. Drawn from *Gentry 20016* and a second flower from *Gentry 11549;* leaf reduced, flower cluster and capsules x ½, long flower section x 1.5, seeds x 2.

Fig. 12.21. *Agave scabra,* a long-leaved form growing spontaneously near Zapata in the lower Rio Grande Valley, Texas.

Fig. 12.22. *Agave scabra,* a short-leaved form on the open plain near Saltillo, Coahuila.

and shorter leaves, panicles with fewer branches, deeper flower tubes, and smaller tepals. Berger placed *A. scabra* in Section Applanatae, *A. asperrima* and *A. caeciliana* in Section Salmianae, but obviously was unfamiliar with the inflorescences of the latter two variants. No definitive taxonomic work has yet been done with *A. scabra*.

Agave scabra is a variable complex, usually recognizable by the characters given above in key and comments, but in interspecific hybrid localities identities become blurred. Introgressive variants have been noted between *A. scabra* x *victoriae-reginae* (Gentry, 1967) and *A. scabra* x *americana,* which has been discussed above (p. 289, Fig. 12.12). I have noted no mixing between *A. scabra* and *A. salmiana,* which meet in the southern Chihuahuan Desert. Some of the more distinct intra-specific varieties that appear to represent geographic populations are separated and described below. Since they are all wild forms, not cultivates, they are designated as subspecies. The respective populations are readily recognized in the field, but isolated specimens in botanic gardens and herbaria may be difficult to identify, especially when they are marginal variants. In a rough way this taxonomic layout should represent the *scabra* genesis in the overall systematic picture.

Key to Subspecies of *Agave scabra*

1. Mature leaves longer, 65–110 cm, scabrous, long acuminate, glaucous green to glaucous white; teeth usually larger   2
1. Mature leaves shorter, 55–65 cm long, asperous to smooth, short-acuminate, or if long-acuminate then smooth, green; teeth frequently smaller   3
2. Rosettes pale green to glaucous gray with rigid ascending leaves, the back of leaf about straight on the abaxial axis. Northern Coahuilan Desert
                      s. *scabra,* p. 296
2. Rosettes white glaucous, the leaves less rigid, frequently recurving. Southern Coahuilan Desert
                   s. *potosiensis,* p. 300
3. Leaves asperous, glaucous gray, short ovate-acuminate, larger teeth 5–7 mm long. Eastern Durango; shrub grassland   s. *zarcensis,* p. 302
3. Leaves smooth, green, triangular-acuminate, larger teeth 8–12 mm long. Montane, Sierra de la Madera, Coahuila   s. *maderensis,* p. 300

Fig. 12.23. Distribution of *Agave scabra* and its ➤ subspecies. Based on herbarium specimens and sightings (see Exsiccatae).

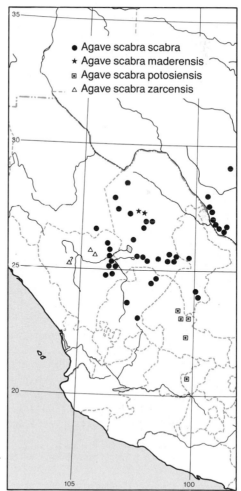

Agave scabra ssp. *maderensis*
(Figs. 12.1, 12.4, 12.23, 12.24; Table 12.1)

Agave scabra Salm, ssp. *maderensis* Gentry ssp. nov.

Rosettes single, with short thick stems, low spreading, ca. 7.5 dm tall, 14–15 dm broad, with 30 to 40 leaves at maturity; leaves 50–60 x 7–12 cm, rigid, triangular-linear-lanceolate; very thick and broad at base, guttered above, green to yellow-green, the cuticle relatively smooth; teeth slender, the larger 5–8 mm long, mostly 2 to 3 cm apart, brown to grayish; spine slender, 3–4 cm long, brown, decurrent as a corneous border along leaf apex; panicle 5–6 m tall with up to 12 or more spreading, decompound, large umbels of yellow flowers; flowers 65–70 mm long; ovary 30–40 mm long; slender; tube 11–16 x 12–15 mm; tepals linear, 15–20 mm long; filaments 46–57 mm long, inserted above mid-tube; anthers ca. 23 mm long; capsules narrowly oblong 4.5–6 x 1.7 cm, narrowly short-stipitate, apex rounded, scarcely beaked; seeds 4.5 x 6.5 mm, lunate to lacrymiform, shining black.

Type: *Gentry & Engard 23251,* May 10, 1973, Canyon de la Hacienda, Sierra de la Madera, NW of Cuatro Cienegas, Coahuila; limestone canyon with pine, oak, madroño, and palm; 6,000–6,500 feet (1,850–2,000 km) elev. Leaves and photos (Fig. 12.24) were collected in 1973; Engard returned the following year to collect flowers and fruits, *Engard & Getz 309.*

Planta caulibus crassis forsan non-surculosa paucifolia. *Folia* 50–60 cm longa, 7–12 cm lata, viridia rigida ad basim crassissima latissima triangulo-lanceolata semi-conduplicata; *marginis dentibus* validis grandioribus 5–8 mm longis plerumque 2–3 cm distantibus, gracilibus brunneis vel albis; *spina terminali* 3–4 cm longa, gracili brunnea decurrente. *Inflorescentia* paniculata 5–6 m alta ramis lateralibus umbelliformibus 12 vel plus; *floribus* 65–70 mm longis luteo-viridibus; *tubo* 11–16 mm longo, 12–15 mm lato; *segmentis* 15–20 mm longis linearibus. *Filamenta* 46–57 mm longa, supra medio tubi inserta; *antheris* ca. 23 mm longis luteis. *Capsulae* oblongae 4.5–6 cm longae, 1.5–1.7 cm latae, brevistipitatae; seminibus 4.5 x 6.5 mm lacrimiformibus nigris.

No suckers were observed on the several plants seen. If this habit is consistent, it would be another distinguishing character and, when combined with the short thick stem and smooth green leaves, this plant would be even more distinct as a local endemic of the Coahuilan Desert mountains.

Agave scabra ssp. *potosiensis*
(Figs. 12.1, 12.3, 12.4, 12.23, 12.25; Table 12.1)

Agave scabra Salm, ssp. *potosiensis* Gentry ssp. nov.

Rosettes sparingly surculose, acaulescent, openly spreading to urceolate in outline, mostly 8–11 dm tall, ca. twice as broad; leaves 65–110 x 14–20 cm, broadly lanceolate, narrowed and thickened toward base, somewhat infolded above and outcurving with sigmoid apex, glaucous gray to nearly white, frequently cross-zoned, asperous to nearly smooth; teeth variable (as for the species), sometimes on mammaeform margins; spine 4–6 cm long, thickly subulate, brown to grayish; panicle 4–6 m tall with 10–18 branches of small yellowish umbels; ovary 32–50 mm long, slender, cylindric-fusiform; tube large 15–22 mm deep, 14–17 mm broad, furrowed; tepals 18–27 x 5–7 mm, unequal, lanceolate, involute; filaments 60–74 mm long, inserted in mid-tube; anthers 25–33 mm long, sometimes apiculate at apex; capsule & seeds not seen.

Type: *Gentry, Barclay & Arguelles 20162,* 16–17 miles E of Huisache Junction along route 80, San Luis Potosí, 16 July, 1963; silt desert plain, elev. ca. 3,700 feet; holotype US, dupls. MEXU, DES.

Planta rosulatim plerumque urceolata, parce-surculosa. *Folia* 65–110 cm longa, 14–20 cm lata, late lanceolata arcuata, grisea vel albida aliquando zonata; *dentibus* variabilibus, ut in subspecie sypica; *spina*

Fig. 12.24. *Agave scabra* ssp. *maderensis* in natural habitat in Canyon de la Hacienda, Sierra de la Madera, Coahuila.

Fig. 12.25. *Agave scabra* ssp. *potosiensis* on the plain east of Huisache, San Luis Potosí, July 1963.

terminali 4–6 cm longa, crassi-subulata, brunnea ad grisea. *Inflorescentia* 4–6 m alta ramis lateralibus umbelliformibus 10–18; *floribus* 65–90 mm longis, *tubo* quam ovario multo latiore, 15–22 mm longo, 14–17 mm diametro; *segmentis* 18–27 mm longis, 56–7 mm latis, inaequalibus lanceolatis involutis. *Filamenta* 60–74 mm longis, medio tubi inserta; *antheris* 25–33 mm longis luteis, aliquando apice apiculatis. *Capsula* et semina non vidi.

Subspecies *potosiensis* comprises the southern populations of *Agave scabra*, occupying the plains and hills of the southern Coahuilan Desert. It is a semi-open arborescent plant community with small tree forms of *Acacia, Prosopis, Yucca filifera*, and numerous shrubs, such as *Larrea tridentata* and *Opuntia* spp. Taxonomically, it is not altogether satisfactory, as some individuals are not clearly separable from the main matrix of the species. In the field *potosiensis* is distinguished by the more open, spreading, graceful rosettes (Fig. 12.25), with broad, less rigid, light glaucous leaves, some nearly white, and the slender ovaries under the large tubes. Some individuals show *americana* characters, as No. 20075, the ideograph of which (Fig. 12.4) is like those of *americana*, 2005. However, the leaves of 20075 are distinctly *scabra* in character. Conversely No. 20091 has the ideograph of *scabra*, but the leaves more like *A. americana protamericana*. I presume such mixing of characters is due to gene flow between the two species. It is optional with which they should be identified. Such cases cause general confusion and embarrassment to the identifier.

Agave scabra ssp. *zarcensis*
(Figs. 12.1, 12.3, 12.4, 12.5, 12.23, 12.26; Table 12.1)

Agave scabra Salm. ssp. *zarcensis* Gentry ssp. nov.

Rosettes short-stemmed, openly spreading, surculose forming large clones, freely seeding; leaves 55–60 x 15–20 cm, linear-ovate, thick, rigid, hollowed above, rounded below, grayish green, rather short-acuminate; teeth moderate, the larger 5–7 mm long, 1–2 cm apart, mostly reflexed, the margin straight to repand; spine 3–4 cm long, conical, brown, to grayish, narrowly grooved above through lower half; panicles 4–6 m tall, broad, open, with 8 to 14 sigmoid lateral peduncles in upper third of shaft; flowers yellow, 68–92 mm long; ovary 35–50 mm long, cylindric, 3-angled and 6-grooved in constricted neck; tube 16–20 mm deep, 15–18 mm broad, thick-walled, deeply furrowed, thickly 12-ridged within; tepals unequal, linear-lanceolate, the outer thick, rugose on back, involute, the hooded tip brown, white papillate, the inner conduplicate, with thick keel and thin involute margins; filaments 55–65 mm long, slender, flattened on inner face, inserted above mid-tube mostly on 2 levels; anthers 26–34 mm long, yellow, centric to excentric; capsules large, woody, 5–6 x 2–2.5 cm, rarely to 8 cm long, oblong to obovoid, stipitate, short-beaked, pedicels stout, 1–2 cm long; seeds 6.5–7 x 5 mm, deeply lunate, the marginal wing low and even.

Type: *Gentry & Arguelles 22084, US.* Along route 45 (15–20 miles S of La Zarca) on sun slope above Río Nazas, Durango, 31 Oct. 1966; desert edge of grassland, rocky slope, alt. ca. 4,000 feet (l, cap), dups. MEXU, DES.

Planta caule breva caespitosa diffusa. *Folia* 55–60 cm longa, 15–20 cm lata, lineari-ovata, rectifolia usque ad apicem involutum, crasse rigentia brevi-acuminata, pallida; *dentibus* plerumque amplitudine moderatis 5–7 mm longis, 1–2 cm separatis in margine recto vel undulato; *spina terminali* 3–4 cm longa, crassi-subulata, brunnea vel grisea, supra anguste canaliculata. *Inflorescentia* 4–6 m alta axe frequenter distorto 8–14 ramis lateralibus umbellatis; *floribus* 68–92 mm longis luteis. *Ovarium* 35–50 mm longum apice constricto incluso. *Perianthii tubo* 16–20 mm longo, crasso sulcato; *segmentis* inaequalibus exterioribus 17–20 mm longis, 4–5.5 mm latis, involutis, apice obscure galeato. *Filamenta* 55–65 mm longa, gracilia, in duobus planis supra medio tubi inserta; *antheris* luteis, 26–34 mm longis. *Capsulae* lignescentes, 5–6 x 2–2.5 cm, oblongae vel obovatae stipitatae. *Semina* 6.5–7 x 5 mm semilunata, nigra.

This taxon appears to be a highland ecotype, observed only on and about the Zarca Mesa country (whence the name) north of the Río Nazas. It grows in the brushy transition plant community, called "chaparillo" (Gentry, 1957), between desert proper and the grama grasslands, and also on rocky slopes within the grassland up to about 5,500 feet (1,700 m) elevation. It is better established on the rocky sunny slopes of the Nazas affluents with such associates as *Acacia, Opuntia,* and *Fouquieria.* It flowers in June and July during the start of the summer rain season (Fig. 12.1).

Among the members of the *Agave scabra* complex, *zarcensis* is distinguished by its mature short broad leaves with moderate teeth, the large yellow flowers, the 2-level insertion of the filaments, and the large woody capsules. The 2-level filament insertion, suggests affinity with *A. havardiana* of the Parryanae, in which this character sporadically appears. However, a 2-level insertion is also apparent in the neotype of *A. scabra.* The tepals and leaf characters also align *zarcensis* with *A. scabra.* Apparently, the 2-level filament insertion, so prominent in the Ditepalae, is another case of homologous variation in *Agave.*

In cultivation it has resisted 10°F (−6°C) of freezing without damage. It is slow to grow and mature, like other *A. scabra,* the young transplant at Murrieta requiring 18 years to flower (see Fig. 12.26).

Fig. 12.26. *Agave scabra* ssp. *zarcensis* flowering in the Gentry garden near Murrieta, California, summer 1966. Young plant collected originally from Mesa de la Zarca, Durango, October 1948.

Agave scaposa
(Fig. 12.27)

Agave scaposa Gentry sp. nov.

Large single, short-stemmed, light green to yellowish green, frequently glaucous, broad rosettes, 1.5–1.7 cm tall, with 60–70 outcurving heavy succulent leaves. Leaves 100–115 x 20–25 cm, broadly lanceolate, nearly plane above to concavo-convex, outcurving to spreading, coriaceous, slightly narrowed above broad thick base; margins straight to crenate with close-set, numerous, dark brown teeth, sometimes mammillate, confluent or 1–2 cm apart, the cusps 3–8 mm long from broad flattened bases, smaller interstitial teeth few at random; spine 2.5–6 cm long, dark brown, subulate, conic at base, with deep narrow groove above, decurrent to ¼–½ of leaf; panicle ca. 2 m long on long scape ca. 5–7 m tall with narrow chartaceous, reflexed bracts below panicle; umbellate branches 25–40 in upper ¼ of shaft, bractlets small, scarcely persistent; pedicels 0.5–1 cm long, slender; capsules 5–5.5 x 2–2.5 cm, oblong, stipitate, shortly beaked, strongwalled; seeds lachrymiform, 7–9 x 5–6 mm, dull black. Flowers unknown.

Type: *Gentry 22472, US.* Sierra de Mahuisapán, 10 miles by car and 1 day by mule west of San Antonio, northern Oaxaca, 24 Nov. 1967; alt. 6,500–7,000 feet, limestone with oak-palm grassland. Dup. in DES.

Planta grandis, plerumque caule singulo, 1.5–1.7 m alta, 2–2.5 m lata. *Folia* 60–70, lati-lanceolata, 100–115 cm longa, 20–25 cm lata, pallidi-viridia, concavo-convexa, radulos vel recurvata, coriacea; *margine* recta vel undulata, *dentibus* fere confluentis vel 1–2 cm distantibus, brunneis, 4–5 mm longis, latioribus quam altis, denticulibus interjectis; *spina terminali* 2.5–6 cm longa, valida subulata longi-

[303]

Fig. 12.27. *Agave scaposa* on top of Sierra Mahuisapan in northernmost Oaxaca, November 1967. Inflorescences of two individual plants, both showing the characteristically long peduncles with short compact panicles.

decurrenti, brunnea, supra anguste canaliculata. *Inforescentia* panaliculata 7–9 m alta scapo inclusio; panicula ca. 2 m longa ramis 25–40 congestis; *bracteis* siccis triangulis reflectis. *Capsulae* 5–5.5 cm longae, 2–2.5 cm latae, oblongae, stipitatae, breviapiculatae. *Semina* lachrymiformia 7–9 x 5–6 mm. *Flores* incogniti.

This large, generally light green glaucous maguey, was observed rarely in northern Oaxaca. *Rose & Hough 4706* from southern puebla with its two photos appears assignable here. The tall narrow panicles with dry small scapose bracts relate it with the Americanae. The seeds in the well-developed capsules in the type collection, together with the scattered single large rosettes observed in the locality, indicate a fertile out-breeding species. Another plant *(Gentry 20446)* noted with suckers, but without inflorescence, is tentatively assigned here (see Exsiccatae). Many large variable plants were also observed and photographed along the northeastern footslope of Sierra Mahuisapán, where we started climbing the high slope. The leaves differed in being light glaucous to pale green and with larger more widely spaced teeth. No inflorescences were observed in this lower elevation population and it may not belong with this proposed species; such plants are perhaps referrable to one or two of the numerous unnamed varieties of *A. americana*.

Americanae Exsiccatae

Agave americana

BAJA CALIFORNIA SUR. *Gentry 11231*, US, DES, MICH. Rancho Burrera at W base of Sierra Laguna, 1–4 Oct. 1951; short-tree forest over granitics, elev. ca. 2,000 feet.

CHIHUAHUA. *Rose & Hough 4222*, US, Chihuahua City–Santa Eulalia Mts., 11 May 1899.

COAHUILA. Gentry, Barclay, Arguelles 20022, US, MEXU, DES. Rancho 37 miles N of Saltillo along rt. 57, 11 June 1963; alt. ca. 3,300 feet. (l, f).

Gentry et. al. 20154, US, MEXU, DES. 50 km S of Saltillo along rt. 57, 13 July 1963; cult. along roadway (l, f).

Palmer 200, US. Saltillo and vicinity. 1898. Cult. for pulque.

Palmer 316, US. Saltillo and vicinity, 10–20 Nov. 1902.

Trelease 143, MO. Cult. S of Monterrey, 20 Mar. 1900.

DURANGO. *Trelease 142*, MO. Pueblito, 11 April 1900 (l, photo).

Trlease 145, 149, MO. Pueblito near Durango, April 1901. [l, f, photo. Probably the type collections of *A. compluviata* Trel. ex Ochoterana, which is definitely of the *A. americana* alliance.]

HIDALGO. *Gentry et. al. 20080*, US, MEXU, DES. Puerto Colorado, 6–7 miles NE of Jacala, 20 June 1963; cult. for pulque. (l, f).

R. Hernandez M. 638, MEXU. Cerca de Actopa, 17 Oct. 1969.

Ramirez L. 1219, MEXU. Actopan, March–June, 1936 (f).

JALISCO. *Rose 3559*, US. Near Huejuquilla, 24 Aug. 1897.

Rose & Painter 7637, US. Near Chapala, 5 Oct. 1903.

NUEVO LEON. *Anonymous 219*, MEXU. 20 km SE of Galeana, Canyon de San Francisco (s, l).

Muller 6, TEX. Sierra Madre Mts., Monterrey, 19 July 1933 (l, f).

Trelease s.n., MO. Monterrey, 5 Dec. 1902 (lf. cuttings with spine, series of photos. Bulbiferous, light gray glaucous.).

OAXACA. *Gentry et. al. 20275*, US, MEXU, DES. 3–5 miles S of Nochistlan along highway, 10 Aug. 1963; spontaneous along open fields.

Rose & Hough 4647, US. Near Cd. Oaxaca, 16–21 June 1899.

QUERETARO. *Gentry et. al. 20084*, US, MEXU, DES. 3 miles N of San Juan del Río along rt. 57, 27 June 1963; cult. as fence, elev. ca. 6,000 feet (l, f).

SAN LUIS POTOSI. *Dewey 580*, UC, US. Cerritos, 25 May 1903. Cult. for juice and pulque.

Eschauzier 75a, MO. Rascon, July, 1905 (f) (Trelease note "Same locality as my No. 75 x 103/05 and original of 150/03.") (Fls. show it definitely = *A. americana* L.)

Gentry et. al. 20089, DES. 4 miles W of Río Verde along road from San Luis Potosí, 28 June 1963; cult. as fence. (photo)

Gentry et. al. 20455, US, MEXU, DES. Tamasopa, Rio Verde District, 21 Sep. 1933; cult. as ornamental.

Trelease s.n., MO. 16 Mar. 1905. Rascon, "glaucous Euagave." (Rather slender light leaf with regular moderate deltoid teeth, little teated.) Type of *A. rasconensis* Trel.

Whiting 776, CAS. Charcas, July–August, 1934 (f).

TAMAULIPAS. *Bartlett 10314*, US. (as *A. melliflua* Trel.) Vicinity of San José, 1930; alt. 2,680 ft. (f).

Dressler 2137, GH, MO. Las Yucas, guayabal scrub near ranch house, ca. 40 km. NNW of Aldama, 2 Aug. 1957 (l, f).

Kastelic s.n., MO. Nuevo Laredo, 28 July 1908.

Kastelic s.n., MO. Reynosa, June 1908, cult.

Kastelic s.n., MO. Nuevo Laredo, 28 July 1908.

Nelson 6647, MO, US. Soto la Marina, 2 Mar. 1902.

Rose & Russell 24337, US. N of San Fernando, 2 Nov. 1927.

Stanford et. al. 2043, US. 30 km SW of Cd. Victoria, 3 km N of Huisachal; 22 June 1949; limestone (l, f).

TEXAS. *Gentry & Barclay 20005*, US, MEXU, DES. 3 miles S of Laredo along rt. 83, Webb Co., 5 June 1963 (l, f).

Gentry & Ogden 9860, US, DES, MEXU. Ca. 5 miles E of Laredo along highway 50, 18 Oct. 1950.

McKelvey 1793, A. San Antonio, 11 April 1931.

Reed 1206, DES. Near Laredo, 27 July 1951; cult. along highway (l, f).

VERACRUZ. *Gomez-P y Nevling 1261*, MEXU. Mata de Caña, Emilio Carranza, alt. 20 m, 21 Junio 1970 (l, f).

Rose & Hough 4287, US. Las Vegas, 17–22 May 1899.

Agave americana expansa

ARIZONA. *Gentry & Ogden 9940*, US, DES. Tumacacori, 6 Nov. 1950; mesquite grassland. Spontaneous.

Gentry & Weber 21983, US, DES. Amado, S of Tucson along road to Nogales, 23 July 1966 (f, photo). Cult.

CALIFORNIA. *Gentry s.n.*, DES. Los Angeles, spring, 1951 (photo only).

Gentry 11328, DES. Fallbrook along highway 395 (Glover's avocado farm), 19 Nov. 1951. (photo).

JALISCO. *Gentry 22303*, DES. S of Zacoalco along road to Cd. Guzman, Dec., 1966 (photos).

Gentry s.n., DES. Primavera, 20–25 miles W of Guadalajara, Nov., 1961; cult. in fence. *Agave abrupta* of Trelease. (photo).

Rose & Hough 4776, US. Near Tequila, 5–6 July 1899.

Trelease s.n., MO. La Barca. 21 Mar. 1903, cult. (2 sheets, cuttings of lvs.). Type of *Agave abrupta* Trel.

Agave americana oaxacensis

OAXACA. Gentry & Arguelles 12260, Type **US,** DES. Tlacolula, 5 Oct. 1952; transplant flowered Murrieta, Calif., July 1971. (l, f, cap., photos). Cult.

Gentry 21300, DES. 3 miles N of Huajuapán road junction to Tehuacán, 24 Sep. 1965; cult., alt. ca. 5,000 feet (cap., photos).

Gentry et. al. 20264, DES. 10 miles E of Cd. Oaxaca, 6 Aug. 1963; alt. 5,000 feet; cult. (photo).

Gentry 12080, US, DES. 6 to 8 miles NE of Cd. Oaxaca along road to Ixtlan, 28 Aug.–1 Sep. 1952; elev. 5,000–6,000 feet (l, photo).

SAN LUIS POTOSI. *Gentry et. al. 20090*, US, DES. 4 miles W of Río Verde along road from San Luis Potosí, 28 June 1963; cult. as fence. (l, photos).

Agave americana protamericana

NUEVO LEON. Gentry & Barclay 20156, US, DES, MEXU. Type:—20–22 miles E of Galeana along road to Linares, 14 July 1963; elev. ca. 3,000 ft.; short-tree forest (l, f, photos).

Gentry & Barclay 20157, US, MEXU, DES. 16–17 miles E of Galeana along road to Linares, 14 July 1963; elev. ca. 3,500 feet; tropical short-tree forest (l, f, photo).

Gentry & Barclay 20160, US, MEXU, DES. 56 miles S of Galeana along road to Dr. Arroyo, 15 July, 1963; elev. ca. 5,400 feet, (l, f, photos).

SAN LUIS POTOSI. *Gentry et al. 20091*, US, MEXU, DES. 15–16 miles E of Rio Verde along road to Cd. Valles, 29 June 1963; elev. 3,500–4,000 feet, thorn forest (l, f).

Rzedowski 3162, ENCB, US. Sierra de San Miguelito, 21 June 1954. (doubtfully assigned, 1 margin only).

Rzedowski 6033, ENCB, US. 12 km NE de Guadalcazar, cerca terrero, 23 June 1955, alt. 1,900 m. (doubtfully assigned, 1 margin only).

TAMAULIPAS. *Harriman 11058*, OSH. In desert scrub on route 101, 16 miles N of Jimenez. Not cultivated but plantations of it are present within a few miles, 4 April 1975 (l, f).

Ogden et al. 51104, US, MEXU, DES. 20.3 miles from Jaumave along road to Cd. Victoria, 16 Feb. 1951 (l, cap).

TEXAS. *Correll 14887A*, DES. About 5 miles SE of Rio Grande City, Starr Co., 5 Oct. 1952.

Correll & Johnston 25630, TEX. Ca. 3 miles E of Rio Grande City, Starr Co., 28 June 1962 (1, cap).

Gentry & Barclay 20010, US, DES. 4 miles SE of Rio Grande City along rt. 83, Starr Co., 6 June 1963.

Agave franzosini

HORTICULTURE. *Berger s.n.*, US. In 1910. From La Mórtola (l, f).

Gentry 10061, 10163, 19866, US, DES. Huntington Botanic Gardens, San Marino. Jan., 1951 (l, photos, f), Aug., 1962 (f).

Agave lurida

HORTICULTURE. Berger's La Mórtola material, US, 1911, 1912, 1911, show a larger fl. with deeper tube and fils. inserted just below mid-tube! A sheet of fls. from Kew, 6374, appear more like Gentry's.

Dn. Masters s.n., K. Neotype:—Ricasoli's garden, Casa Blanca, Porte Ercole, Mt. Argentario, Tuscany. (l, f, sheets I, II, III, IV).

Ex. Herb. Hort. Reg. Kew 6375, US. Cult. Kew (1 seg. & fs. = *A. lurida*, 2 sheets).

OAXACA. *Gentry et. al. 20279*, US, MEXU, DES. 16 miles N of Huajuapán along road to Tehuacán, 11 Aug. 1963; elev. ca. 6,000 feet, volcanic rocky slope (l, f).

Gentry et. al. 20447, US, MEXU, DES. 21 miles NE of Huajuapán along road to Tehuacán, 12 Sep. 1963; open volcanic slope.

Agave oroensis

ZACATECAS. Gentry & Engard 23592, US, MEXU, DES. Type:—Estación Margarita near Concepción del Oro, 1 Aug. 1975; elev. ca. 6,000 feet, silty soil derived from limestone; cult. (l, f).

Agave scabra

CHIHUAHUA. *Gentry & Engard 23143*, US, MEXU, DES. Ca. 30 miles SE of Jimenez along rt. 49, 17 Oct. 1972.

White 128, TEX. Juarez, 5 Nov. 1911.

COAHUILA. *Barkley & Barkley 17M313*, TEX. Near Casa Blanca, 8 June 1947.

Berger s.n., US. Bot. Gard. Darmstadt, 1904, from San Pedro.

Gentry 11533, US, MEXU, DES, MICH. Ca. 15 miles E of Saltillo, 1–3 Feb. 1952: elev. ca. 4,000 feet, open limestone slope.

Gentry 20044, US, DES. 12–16 miles NE of Saltillo and 1½ miles E off road to Monterrey, 10 June– 5 July 1963; elev. 4,000–5,000 feet, limestone hill. Hybridizing with *A. victorae-reginae*.

Gentry et. al. 20016, 20018, US, MEXU, DES. 12–16 miles NE of Saltillo along road to Monterrey, 10 June 1963 (l, f).

Gentry & Engard 23259, US, MEXU, DES. 18 miles S of Cuatro Ciengas along road to San Pedro, 11 May 1973; elev. ca. 2750 feet; *Larrea* silt flat (l, f).

Gentry & Engard 23261, US, DES. Parras, 12 May 1973; elev. ca. 5,000 ft.

Gentry & Koch 19807, US, DES. Huntington Botanic Gardens, San Marino, Calif. Block 20–194 (l, f).

Gomez-Pompa 201, MEXU. Between Saltillo and Torreón, 4 April 1960 (l, f).

Gregg s.n., GH. Near Saltillo, 1 Jan.

Henrickson 6154, DES, TEX. 35.1 miles NE of San Pedro along rt. 30 in Puerto de Ventanillas; elev. 3,500 feet; limestone slopes with *Larrea, Fouquieria, Opuntia*, sandy clay. (l, f). 29 Aug. 1971.

Henrickson 7918, TEX, DES. 3 miles W of Est. Juncal, 36 air miles W of Cuatro Cienegas, 21 Sep. 1972; elev. 3,850 feet; *Larrea, Opuntia, Fouquieria*, etc. (l, f).

Johnston & Muller 725, A, TEX. Sierra del Pino, southern canyon below oak and pine belts, 26 Aug. 1940 (l, f). May belong with *A. zarcoensis*?

Palmer 315, GH. Vicinity of Saltillo, 10–20 Nov. 1902.

Palmer s.n., MO. Saltillo, April, 1880.

Palmer 1308, 1309, US. Saltillo, 1880, July (l, f).

Pinkava et. al. 5223, ASU. 1.8 miles SSW of Poso de la Vecera, Cuatro Cienegas Basin, 9 June 1968 (l, f).

Pinkava et. al. 5671, ASU. 5.3 miles NW of Los Fresnos, Cuatro Cienegas Basin, 14 June 1968 (l, bud).

Pringle 891, MO. Mesas near Jimulco, 9 April 1886, (f, l)

Pringle 158, 468, US. Jimulco, 17 May 1885 (l, f).

DURANGO. Gentry & Engard 23268, US, MEXU, DES. Neotype:—E. bajada of Sierra del Sarnosa near Dinamita, 15 May 1973; elev. ca. 4,000 feet; *Larrea* shrub succulent desert (l, f).

Gentry & Gilly 10595, US, MEXU, DES. 6 miles SW of Pedriceña along road to Durango, 12 June 1951; elev. ca. 4,900 ft.; Coahuilan desert shrub. (l, f).

Gentry 11549, DES. Near Cuencamé along highway to Durango, 4 Feb. 1952 (f).

Gentry et. al. 20468, US, DES, MEXU. 9–10 miles N of Cuencamé along rt. 40, 24 Sep. 1963; desert shrub on caliche soil.

Medrano & Quero 1380, MEXU. 20 km N de Cuencamé, carretera á Torreón, 21 Sep. 1966; alt. 1,300 m.

Palmer 228, US. Tobar, 28–31 May 1906 (f, cap).

Pringle 158, B. In large canyon in mountains near Jimulco. 17 May 1885. (Sheet in Berlin annotated by W. T. as *Agave oblita* Trel.)

NUEVO LEON. *Trelease s.n.*, MO. S. of Monterrey, 20 Mar. 1900.

TEXAS. *Gentry & Barclay 20004*, US, MEXU, DES. Ca. 7 miles N of Laredo along rt. 81, Webb Co., 4 June 1963 (l, f).

Gentry & Barclay 20006, US, MEXU, DES. 2 miles N of Zapata, Zapata Co., 5 June 1963 (l, f).

Gentry & Barclay 20008, US, MEXU, DES. 5 miles S of Zapata, Zapata Co., 5 June 1963 (l, f).

Gentry & Barclay 20012, US, DES. 5 miles SE of Saus, Starr Co., 7 June 1963; open mesquite slope (l, f).

Gentry & Barclay 18414, DES. 4 miles N of rt. 83 on Farm Road 649, Starr Co., 16 Mar. 1960; sandy open chaparral desert (live pls., photos).

Reed 1205, DES. 21 miles S of Zapata, 25 July 1952; wild stand (l, f).

Reed 1348, DES. Near Laredo, June or July 1951 (l, f).

Reed 1349, DES. Near Zapata, 8 June 1952 (l, f).

Reed 1350. DES. Catarina, 8 June 1952, "native" (l, f).

ZACATECAS. *Gentry & Engard 23609*, US, INIF, DES. 12 miles N of road junction 49 & 45, along rt. 45 to Durango, 5 Aug. 1975; elev. ca. 7,350 feet; rocky limestone hill slope of S exposure (l, f).

Kirkwood 9, GH, MO. Cedros, May, 1908, (f, l), "flats, foot slopes and high ridges, forming dense patches, propagating by stolons."

Lloyd 127, MO, US. Cedros, Spring, (l, f), "Footslopes, chiefly Cedros."

Nelson 6511, US. Zacatecas, 19 Dec. 1902.

Agave scabra maderensis

COAHUILA. Gentry & Engard 23251, US, MEXU, DES. Type:—Canyon della Hacienda, Sierra de la Madera, NW of Cuatro Cienegas, 10 May 1973; elev. 6,000–6,500 feet, limestone canyon with pine, oak, madroño, and palm (l, cap).

Engard & Getz 309, DES. Ibid., 10 July 1974.

Pinkava et al. P13625, ASU, DES. Canyon del Agua, Sierra de la Madera, NW of Cuatro Cienegas, 23 June 1976; elev. 5,000–6,000 feet, oak chaparral (l, f).

Agave scabra potosiensis

QUERETARO. *Gentry et al. 20452,* US. MEXU, DES. 2–3 miles N of Cd. Queretaro along rt. 57, 17 Oct. 1963; with *Opuntia* and shrubs on rocky volcanic slope.

SAN LUIS POTOSI. Gentry et. al. 20162, US, MEXU, DES. Type:—16–17 miles E of Huisache junction along rt. 80, 16 July 1963; elev. ca. 3,700 feet, silt desert plain. (l, f, photo)

Gentry 11510, US, DES. Near Santo Domingo, rt. 80, 26–28 Jan. 1952; pass between limestone hills, transplant flowered at Murrieta, 12–30 July 1966.

Gentry et al., 20075, US, MEXU, DES. 30 miles S of Matehuala along rt. 57, 17 July 1963; elev. 4,000–4,500 feet, silty plain with *Yucca* and *Larrea* (l, f).

Gentry et al. 20176, US, MEXU, DES. CA. 35 miles E of San Luis Potosí along road to Rio Verde, 16–18 July 1963; elev. 5,000–6,800 feet, limestone with pine and oak forests (l, f).

Agave scabra zarcensis

DURANGO. Gentry & Arguelles 22084, US, MEXU, DES. Type. Along rt. 45 on sun slope above Río Nazas, 31 Oct. 1966; elev. ca. 4,000 feet, desert edge of grassland (l, cap).

Gentry 8632, US, DES, MEXU. Zarca Mesa, 5 Oct. 1948; limestone, grama grassland (live pls.). Transplant flowered Murrieta, Calif. June 10–July 10, 1966.

Des. Bot. Gard. 5642, DES. 123 miles S of Parral along rt. 45 = ca. 15 miles S of La Zarca (live pl.) flowered Desert Botanical Garden, Phoenix, May 1972.

Agave scaposa

OAXACA. *Gentry 20446,* DES, US. 17 miles NE of Huajuapán along road to Techuacán, 12 sep. 1963; roadside limestone (40 teeth in up. ⅔ lf.).

Gentry 22472, Type, DES, **US.** Sierra de Mahuisapan, 10 miles by car and 1 day by mule W. of San Antonio, 24 Nov. 1967; alt. 6,500–7,000 ft., oak-palm grassland, limestone (l, cap., photos, ca. 50 teeth over 3 mm long in upper ⅔ of l).

PUEBLA. *Rose & Hough 4706,* US. Between Tepeaca and Santa Rosa, 27 June 1899 (apex with 1 margin only, with fewer teeth, ca. 20, spine heavier, but photos show the long scape and narrow panicles). Inflo. suggests Americanae Group.

13.

Group Campaniflorae

Trelease, Missouri Bot. Gard. Rep. 22:44. 1912.

Large to small, perennial or multiannual plants with rather open, short-stemmed rosettes; leaves green, rather soft succulent, long-lanceolate, frequently sigmoid toward apex, with regular, close-set, moderate teeth on noncorneous margins. Panicles large, diffuse, in upper half of shaft subtended by small scarious bracts; flowers campanulate with deep, ample tubes, deepset filaments, red to purple on outside, yellow within; capsules short-oblong to ovoid, rather thin walled. Southern half of peninsular Baja California.

Typical species: *Agave aurea* Brandegee

The Campaniflorae occupy one of the most isolated areas of the continental Agavaceae (Fig. 13.3). The pre-Cretaceous granitic pediment on which they grow has long been isolated from the contemporary formation of the Sierra Madrean mainland. This long isolation seems to be reflected by the Campaniflorae in their consistent conformation to the distinct campanulate flowers with broad tubes and thin-margined tepals, all with a certain succulent texture difficult to describe. The regular, moderate, close-set teeth on the soft succulent leaves are also consistently maintained through the group. The flower proportions appear unusually variable (Fig. 13.2 and Table 13.1). They do not appear

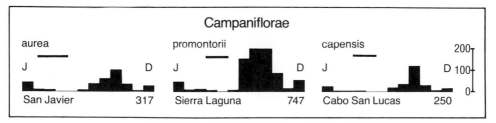

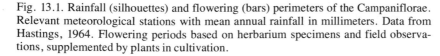

Fig. 13.1. Rainfall (silhouettes) and flowering (bars) perimeters of the Campaniflorae. Relevant meteorological stations with mean annual rainfall in millimeters. Data from Hastings, 1964. Flowering periods based on herbarium specimens and field observations, supplemented by plants in cultivation.

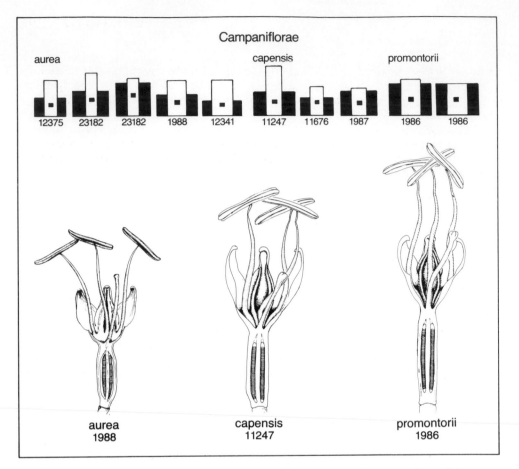

Fig. 13.2. Floral ideographs of the Campaniflorae, showing relative proportions of the tube (black) to outer tepal (white column), and level of insertion of filament (black square). Measurements are listed in Table 13.1.

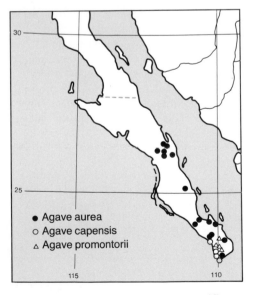

Fig. 13.3 Distribution of the Campaniflorae, based on herbarium specimens, pp. 321–22.

Key to Species of the Campaniflorae

1. Rosettes large, non-surculose multi-annuals; leaves 60–150 cm long; panicles broad; filaments without apical gland.   2
1. Rosettes smaller, perennials by axillary budding; leaves mostly 35–60 cm long; panicles narrow; filaments with an apical gland at anther fixation. Cape District   *capensis*, p. 316
2. Rosettes generally 0.7–1 m tall; leaves smaller, more pliant, 7–12 cm broad. Sierra de la Giganta and Cape District   *aurea*, p. 311
2. Rosettes 1.5–2.3 m tall; leaves larger, thicker, relatively rigid, 11–17 cm broad. Sierra Laguna, Cape District   *promontorii*, p. 319

useful for species separation. The connective gland with the filament-anther fixation in *A. capensis* has not been observed elsewhere (Fig. 13.6). The habit of axillary budding with resultant large clusters of many rosettes, found in *A. capensis,* is regarded as a relict character, widely scattered in the genus.

The rainfall silhouettes in Figure 13.1 show that the Campaniflorae occupy moister habitats than other agave groups of the Peninsula. Rain falls predominantly in summer-fall from storms of tropical origin. The large growth-form of *Agave promontorii* appears to reflect the higher rainfall.

Agave aurea
(Figs. 13.1, 13.2, 13.3, 13.4, 13.5; Tables 13.1, 13.2, 13.3)

Agave aurea Brandegee. Proc. Calif. Acad. Sci. ser. 2. 2:207. 1889.

Single short-stemmed, rather open, green, graceful rosettes 10–12 dm tall, 15–20 dm broad, with widely arching leaves; leaves 63 x 7 cm to 110 x 12 cm, average 86.3 x 8.6 cm, linear to long-lanceolate, pliant, guttered, rounded below, thickly fleshy toward base, green to somewhat glaucous, the margin straight to undulate; teeth moderate, regular, mostly 4–7 mm long, 1–2 cm apart, straight or moderately curved cusps from low angular bases, dark brown to light brown; spine subulate, 25–35 mm long, dark brown or grayish red, with a short narrow groove above, shortly decurrent or decurrent as a dark corneous

Table 13.1. Flower Measurements in the Campaniflorae (in mm)
All localities are in Baja California del Sur.

TAXON & LOCALITY	OVARY	TUBE	TEPAL	FILAMENT INSERTION & LENGTH		ANTHER	TOTAL LENGTH	COLL. NO.
aurea								
	29	8 × 14	16 × 6	6	35	19	54†	12375
	20	9 × 14	16 × 6	5	38	20	45†	12375
Comondú	34	11 × 16	19 × 5	8–7	45	23	63†	23182
	36	15 × 15	16 × 5	10–9	46	20	66†	23182
	39	14 × 15	17 × 5	9–8	45	21	70†	23182
	30	9 × 18	17 × 9	7 & 5	41	21	55†	1988
Todos Santos	30	10 × 18	15 × 9	7 & 5	40	20	54†	1988
	20	7 × 8	17 × 10	6–5	45	19	43†	12341
	21	6 × 18	15 × 10	5–4	43	19	43†	12341
Cape Region	32	7 × 8	14 × 7	4–5	40	19	52*	Bdge.
capensis								
	34	12 × 20	20 × 7	6	38	28	65†	11247
	31	10 × 20	23 × 7	7	43	27	63†	11247
Cabo San Lucas	27	8 × 13	13 × 6	6–7	26	15	48†	19676
	24	8 × 14	13 × 6	5–6	27	16	44†	19676
	25	11 × 16	12 × 7	5–6	32	19	48†	1987
	28	11 × 16	13 × 7	5–6	40	20	52†	1987
promontorii								
	42	14 × 20	16 × 9	8 & 6	54	26	72†	1986
Sierra Laguna	36	14 × 20	14 × 9	8 & 6	50	24	64†	1986
	41	14 × 20	15 × 9	8 & 6	50	24	70†	1986

* Measurements from dried flowers.

† Measurements from fresh or pickled flowers.

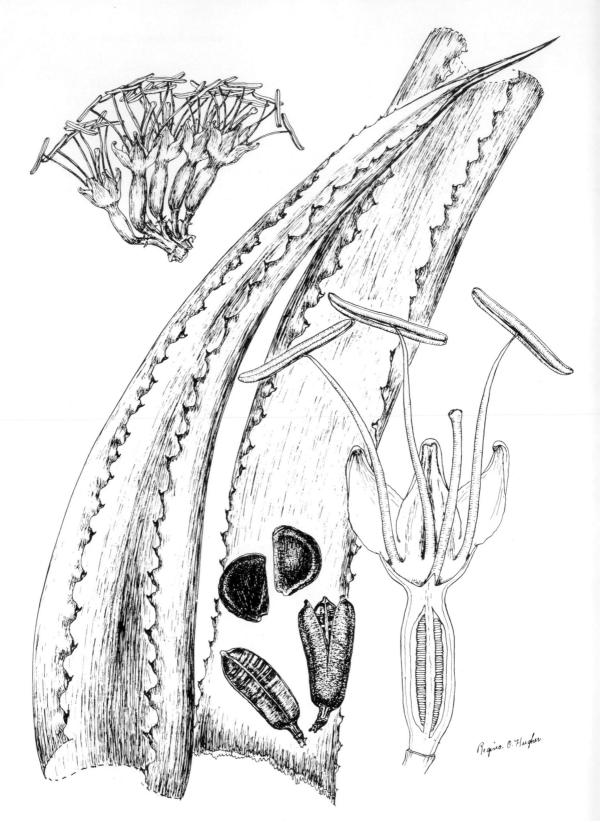

Fig. 13.4. *Agave aurea*. Leaf from *Gentry 11295*, flower cluster from *Barclay & Arguelles 1988*, capsules x ½, flower section x ¼, seeds x 2.

margin through 8 to 10 teeth bases; panicles 2.5–5 m tall, broad, the peduncle reddish and with remote, quickly drying lanceolate bracts; umbels broad, congested, 15 to 25 in upper half of shaft, red to purplish in bud and fruit; flowers opening yellow to orange-yellow, 43–70 mm long, campanulate; ovary reddish, 25–35 mm long, slender, angular-cylindric, with constricted furrowed neck 6–10 mm long; tube 8–14 mm deep, 14–18 mm broad, evenly grooved from tepal sinuses; tepals 16–19 mm long, lanceolate, acute, incurving, the outer larger with tufts of white papillae below distinct hood, the inner with long keel and yellow, thin margin involuting at anthesis; filaments stout, 35–45 mm long, those of outer tepals inserted 5–10 mm above base of tube, slightly higher than those of inner tepals; anthers 19–23 mm long; capsules oblong, 45 x 17 mm to 35 x 15 mm, non-stipitate, rounded apiculate, reddish, drying light brown; seeds 7–8 x 5–5.5 mm, irregularly lunate, dull black.

Type: *Brandegee s.n.*, Purísima, Baja California, 13 Feb. 1889, UC. Consists of marginal sections of a large leaf, a small leaf, 5 flowers, pieces of 2 capsules. The plant does not grow in the sandy valley at La Purísima; doubtless collected on the volcanic mesas near that settlement.

Characteristics and Distribution

Agave aurea is the most abundant species in the group Campaniflorae. It is easily recognized by the long narrow lanceolate green leaves arching out to form an open, spreading rosette, by the broad, rather diffuse, reddish panicles, and by the bright yellow flowers from reddish buds and ovaries. The teeth are moderate and regularly spaced (Figs. 13.4, 13.5). It forms scattered and sometimes massive populations on the extensive lava fields of the western slopes of the Sierra de la Giganta, mostly between elevations of 1,000 and 3,500 ft. (ca. 300 and 1,070 m) above sea level. It is also widely scattered on the granitic lower slopes of the Cape District at lower elevations. It is lacking or rare on the sandy Magdalena plain.

In the latitude of Todos Santos, an *A. aurea* population shows forms similar to both *A. promontorii* and *A. capensis*. Here there are forms like *aurea* in the size of their rosette and in the color and form of their leaves, but with the softer leaves and more proximal teeth of *promontorii*. Here also are small cespitose-rosette forms with softer narrower leaves of low fiber content like those of *capensis* at Cape San Lucas. Large inflorescences with well-developed fruits and ovules are among these forms, indicating interfertility. It appears that *A. aurea* may have invaded the islandic Cape District after it was connected to the peninsula in Pleistocene times and is now receiving genes from the insular *promontorii* and *capensis* populations.

Table 13.2. Fiber Ratings of *Agave aurea* and *A. promontorii*,
as Compared to Henequen, *A. fourcroydes*

CHARACTERISTICS	Agave aurea	Agave promontorii
Total weight (4 lvs.)	± 2000 grams	± 2000 grams
Average weight/leaf	500 grams	500 grams
Total fiber yield	26 grams	65 grams
Yield gross	1.21%	3.25%
Milling characteristics	good	good
Texture of fiber	excellent	fine
Spinning factor	excellent	fine
Break test	20% weaker	20% weaker
Length of recovered fiber	60 cm = short	80 cm = good
Bleaching color	good	good
Pulp adherence	minimum = good	minimum = good

SOURCE: Grady Venable, Tampico Fiber Co.

Fig. 13.5 *Agave aurea* on the lava beds north of Comondú, April 1973 (Bruce Gentry photo).

Fiber of *A. aurea* and *A. promontorii*

The fiber of *Agave aurea* was at one time harvested in the vicinity of Comondú, where this agave grew in abundance upon the lava beds. The *aurea* fiber is fine and of excellent spinning quality, but yield of fiber per leaf is low, and compared with henequen or sisal could not be competitive for twine and rope. In 1952 leaf samples of *A. aurea* and *A. promontorii* were sent to the Mexican Fibre Company in Tampico, Tamaulipas, for testing. Grady Venable, in charge of that operation, made the following report. See also Table 13.20.

1. *Agave promontorii*. The fiber is excellent and, it is believed, would be acceptable to the market. The yield of fiber per leaf, however, is too low for commercial operation, but it could probably be made to produce more fiber under proper planting conditions. If the chemical properties are more valuable than the henequen now grown on El Carrizal, Tamaulipas, it might be cultivated commercially.

2. *Agave aurea*. This fiber, while of excellent texture, cannot be produced commercially unless it is found that the yield can be increased. It cannot be commercially produced unless its chemical content should prove exceptionally valuable.

Chemistry

Tables 13.3, 13.4, and 13.5, which give the sapogenin contents of the Campaniflorae, enable one to judge further any commercial possibilities. The selection of individual plants with abundant fiber together with high sapogenin content for breeding could materially raise the commercial potential.

Table 13.3. Sapogenin Content in *Agave aurea*
(given in percentages on dry weight basis)

COLL. No.	BAJA CALIFORNIA SOURCE LOCALITY	MONTH COLL.	PLANT PART	SAPOGENINS*				
				TOTAL	HEC.	GIT.	MAN.	TIG.
11198	La Paz	Sep.	leaf	0.7	100			
11253	San José del Cabo	Oct.	leaf	0.5	70			20
11253	San José del Cabo	May	leaf	0.4	100			
11255	Las Cuevas	Oct.	leaf	0.0				
11283	NW of La Paz	Oct.	leaf	0.0				
11295	Comondú	Oct.	leaf	0.1	100			
11295	Comondú	May	leaf	0.0	100			
11295	Comondú	Oct.	leaf (dead)	0.3				
11297	Comondú	Oct.	leaf	0.55	90		10	
11297	Comondú	May	leaf	0.34	87		13	
11297	Comondú	Dec.	leaf	1.4	x	x	x	x
11823	Todos Santos	May	leaf	0.0				
11823	Todos Santos	May	fruit	1.6	45		40	15
11886	Comondú	May	fruit	0.5	5	30		65
12383	San Javier	Dec.	leaf	1.2	100			
12338	La Paz	Nov.	leaf	1.2	x	x	x	x
12339	La Paz	Nov.	leaf	1.1	100			
12342	Todos Santos	Nov.	leaf	0.0				
12341	Todos Santos	Nov.	leaf	0.7	x		x	
12420	Comondú	Dec.	leaf	0.1	x		x	

SOURCE: Wall (1954) and Wall et al. (1954a, 1954b, 1955, 1957).
* Hec. = hecogenin; Git. = gitogenin; Man. = manogenin; Tig. = tigogenin.

Agave capensis

(Figs. 13.1, 13.2, 13.3, 13.6, 13.7; Tables 13.1, 13.4)

Agave capensis Gentry, Calif. Acad. Sci. Occ. Pap. 130: 72. 1978.
  Agave brandegeei in herb. & hort.

Plants perennial by axillary budding, eventuating in large clusters with short-stemmed, small, open rosettes, 6–8 dm tall, 8–12 dm broad; leaves mostly 30–60 x 4–7 cm, narrowly lanceolate, straight to arching, commonly sigmoid toward apex, concave above, convex below, light glaucous green, soft, brittle, succulent, with undulate non-corneous margin; teeth regular, 4–5 mm long, mostly 1–2 cm apart, mildly curved, with short mammillate bases, reddish brown to grayish; spine 1.5–3 cm long, subulate, dark brown, short decurrent for 1–2 cm; panicles mostly 2.5–3.5 m tall with 15–24 lateral branches in upper ⅔ to ½ of shaft, the laterals up to 30 cm long, ascending; umbels small; bracteoles very small, 2–3 mm long, narrow-triangulate; flowers in bud reddish brown or purplish, opening yellow inside, 50–65 mm long; ovary green, 25–35 mm long, thick, 3-angled, furrowed in unconstricted neck; tube 8–14 mm deep, 15–20 mm broad, 6-bulged, 6-furrowed from tepal sinuses; tepals equal, 13–23 x 6–7 mm, lanceolate, incurved and rather thin; filaments 30–43 mm long, yellow, inserted 5–7 mm above base of tube, the apex swollen with colorless glands; anthers 15–27 mm long, excentric, yellow or bronze; capsules 25–35 x 15–17 mm, ovoid, scarcely stipitate, apiculate; thin walled; seeds not seen.

> Type: *Gentry & Fox 11247,* Cabo San Lucas & vicinity, Baja California. Young plant collected 5 Oct. 1951; flowered in Murrieta, California, July 1964; deposited in US.

Agave capensis appears to be an islandic endemic of the Cape District. It grows with the sparse open shrubbery on the arid, neutral, well-aerated soils on granitic slopes from near sea level to 1,000 ft. (ca. 300 m) or more above sea level. Scattered single rosettes northwest of Punta Frailes appeared similar to those of *A. capensis,* but their immaturity at time of sighting leaves their identity doubtful. This is also true of a collection (*Gentry 11301*) in the shady canyon above La Purísima, which probably represents a narrow-leaved shade form of *A. aurea.*

The flower structure of *Agave capensis* is very similar to that of its sectional relatives, *A. aurea* and *A. promontorii,* but it is distinguished by its small narrow leaves and clustered growth habit. The photographs in Fig. 13.7, taken in the same year at San José del Cabo, show both single and branching rosettes and are both very similar to the clone in Huntington Botanical Gardens. A specimen transplanted from San José del Cabo in 1951

Table 13.4. Sapogenin Content in *Agave capensis*

COLL. No.	BAJA CALIFORNIA SOURCE LOCALITY	MONTH COLL.	PLANT PART	SAPOGENINS*				
				% TOTAL	% HEC.	% GIT.	% MAN.	% TIG.
11247	Cabo San Lucas	Oct.	leaf	0.4	35			65
11247	Cabo San Lucas	Oct.	stem	0.0				
11247	Cabo San Lucas	May	leaf	1.0	100			
11247	Cabo San Lucas	Nov.	leaf	0.0				
11250	Cabo San Lucas	Oct.	leaf	0.8	80			20
11250	Cabo San Lucas	May	leaf	0.55	20			80
11250	Cabo San Lucas	Nov.	leaf	3.6	x		x	x
11847	Cabo San Lucas	May	fruit	0.3	20		30	50

SOURCE: Wall (1954) and Wall et al. (1954a, 1954b, 1955, 1957).
* Hec. = hecogenin; Git. = gitogenin; Man. = manogenin; Tig. = tigogenin.

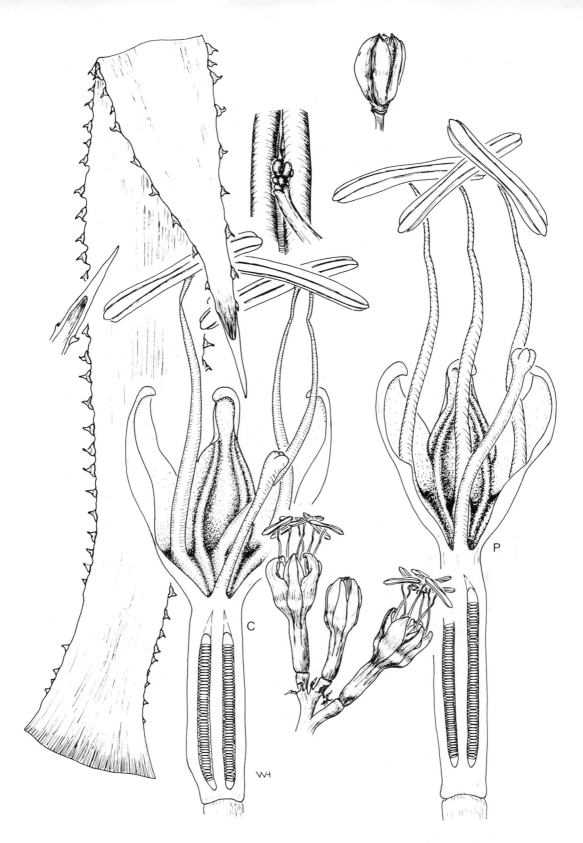

Fig. 13.6. *Agave capensis* drawn from type; C, leaf, flower cluster, capsule x ³/₅, flower section x 1.5, apex of filament enlarged to show gland; *A. promontorii* flower section x 1.5. P.

Fig. 13.7. *Agave capensis* at Cabo San Lucas; upper, a cluster of rosettes showing axillary branches, and lower, a single rosette.

flowered at Murrieta in July 1964 and is selected as the type. Subsequent to the July flowering, some of the axillary rosettes also flowered, but none of the flowers at Murrieta developed fertile fruits.

The connective gland or swelling found on the stamen where the anther is affixed to the filament (Fig. 13.6) is a structure peculiar to the Campaniflorae. While well developed in *A. capensis,* it was also noted in minor form on a putative hybrid, x *A. aurea.* These two species appear to be close, even genetically compatible, as discussed under *A. aurea,* but the origin of the filament gland appears to lie with *A. capensis.*

Agave promontorii
(Figs. 13.1, 13.2, 13.3, 13.8; Tables 13.1, 13.2, 13.5)

Agave promontorii Trel., Missouri Bot. Gard. Rep. 22:50. 1912.

Large, single, green, open rosettes, 1–2 m or more tall, 2–2.5 m broad, with thick stems; leaves 10–15 dm long, 11–17 cm wide, lanceolate fleshy succulent, stiff, usually concave above, thick at base, straight to arching, green to soft glaucous green, the margins about straight and closely set with regular straight to curved teeth, mostly 4–8 mm long, 5–10 mm apart, reddish brown; spine 3–5 cm long, conic-subulate, dark brown, short-decurrent, narrowly sulcate above; inflorescence 5–9 m tall, massive, the peduncle reddish and with conspicuous deltoid bracts; panicles large, broad, with 25–30 diffuse umbels on recurving laterals in upper half of shaft, the buds and ovaries red to purplish; flowers short-pedicellate, 60–75 mm long, campanulate; ovary 36–42 mm long, 3-angled-cylindric, scarcely narrowed at base, 6-furrowed and narrowed in neck; tube 14–15 mm long, 20 mm broad, cup-shaped, 6-bulged, with deep furrows between bulges; tepals 14–16 x 9 mm, triangular-lanceolate, equal, thin, reddish purple on outside, yellow within, the outer plane and overlapping inner at base, the inner with yellow margins and low, broad, 4-lined keel, with inner costae pronounced and wide apart; filaments 50–55 mm long, inserted unequally at 8 and 6 mm, oval in cross section, yellow; anthers 24–26 mm long, yellow; capsules shortly pyriform-oblong, 15–20 x 30–35 mm, rather stipitate, beaked; seeds narrow, 4–5 x 6–9 mm. (Fl. description from *Barclay and Arguelles 1986*; cap. and seed description from Trelease 1912.)

Type: *Nelson & Goldman 7437,* Sierra de la Laguna, Baja California, 21 Jan. 1906, US.

Table 13.5. Sapogenin Content in *Agave promontorii*

COLL. NO.	SOURCE LOCALITY	MONTH COLL.	PLANT PART	SAPOGENINS*				
				% TOTAL	% HEC.	% GIT.	% MAN.	% TIG.
	Baja California:							
11218	Sierra Laguna	Oct.	leaf	0.3	55			55
11218	Sierra Laguna	Oct.	stem	0.0				
11229	La Burrera	Oct.	leaf	0.75	90		10	
11229	La Burrera	May	leaf	0.0				
11218	Sierra Laguna	Nov.	leaf	3.4	x		x	x
12346	Sierra Laguna	Nov.	leaf (dead)	0.4			x	x
12347	Sierra Laguna	Nov.	infl.	0.2	x			x
12353	Sierra Laguna	Nov.	leaf	0.8	x		x	
12349	Sierra Laguna	Nov.	leaf	0.7	x			x
12351	Sierra Laguna	Nov.	leaf (dead)	0.9			x	x
12352	Sierra Laguna	Nov.	leaf	0.9	x		x	
12350	Sierra Laguna	Nov.	leaf	0.5	100			
12356	Sierra Laguna	Nov.	leaf	0.5	x		x	x
12344	Burrera	Nov.	leaf	0.4	100			
12345	Burrera	Nov.	leaf	0.8	x	x	x	x
	California:							
3233	Balboa Park, San Diego	Nov.	leaf	0.0				

SOURCE: Wall (1954) and Wall et al. (1954a, 1954b, 1955, 1957).
* Hec. = hecogenin; Git. = gitogenin; Man. = manogenin; Tig. = tigogenin.

Fig. 13.8. *Agave promontorii* in Balboa Park, San Diego, California, with large but sterile infructescences.

Agave promontorii is a large handsome plant, becoming massive with thick stems and thick juicy leaves in fertile situations (Fig. 13.8). The arching attitude of the leaves combined with their soft glaucous green and regular close-set teeth make it an attractive ornamental. It is, however, rare in gardens because it does not offset and must be renewed with seedlings. The plants growing in Balboa Park, San Diego, and my plants near Murrieta, California, did not set seed. Night freezes at 24°F (−4°C) injured the leaves causing apical dieback.

The species is known only from the granitic mountains of the Cape District. It attains best development in the higher elevations from 3,000 to 6,000 ft. (ca. 900–1,800 m) above sea level, coincident with the better moisture conditions about the Sierra de la Laguna. It flourishes, for instance, on the open, precipitous, rocky, southwest slopes above the Rancho Burrera. Although it was observed in the shade of trees, it does not thrive in the crowded chaparral thickets common on the Sierra Laguna. Strays from the montane populations are common along the rocky arroyos issuing from the canyons on the west, as at Rancho La Burrera. Due to its limited and remote range, *A. promontorii* is another rare *Agave* scarcely known to man.

Except for variations in color and teeth, *Agave promontorii* appears well-knit morphologically. The flowers show it to be closely related to the other members of the Campaniflorae. However, its large size in both rosette and leaf distinguish it clearly from both *A. aurea* and the still smaller cespitose *A. capensis*. It was noted earlier that some of the *aurea* population around Todos Santos bears resemblance to *promontorii*, but no individuals of the latter were observed in the vicinity. The apparent introgression there was probably only between *A. aurea* and *A. capensis*, whose populations meet in the low coastal elevations there and near Cabo San Lucas.

Campaniflorae Exsiccatae

Agave aurea

BAJA CALIFORNIA DEL SUR. *Barclay & Arguelles 1988*, DES, MEXU, US. 5 miles N of Todos Santos along road to La Paz, 20 April 1966 (l, f).

Brandegee s.n., UC, DS. Type. Purísima, 13 Feb. 1889 (margin of large l, small, l, 5 f, pieces of 2 cap. Has typically close-set, regular teeth on regular teats forming undulate margin, as growing on lava mesas about Comondú and E of Purísima.).

Brandegee s.n., MO. Cape Region mountains, 20 Sep. 1899 (l, f, mixed with f of *A. sobria;* type of *A. brandegeei* Trel.).

Carter 5132, UC. Mesa de San Geronimo, N from Rancho Viejo (on road from Loreto to San Javier), 8 May 1966; alt. ca. 1,110 m.

Carter 5484, US. Arroyo de Puerta Vieja on road from Loreto to Comondú, Sierra de la Giganta, 4 July 1970; alt. ca. 1,400–1,500 ft.

Carter 5779, UC. Mesa de Humi, a mesa on crest of Sierra de la Giganta opposite N end of Isla San José, 20 March 1973; alt. ca. 750 m.

Gentry 11299, DES. Polymorphic population on mesa NW of Comondú (photo).

Gentry 12341, DES, MEXU, US. 5 miles N of Todos Santos, Cape Dist., 22 Nov. 1952 (l, f).

Gentry 12375, DES, MEXU, US. 3 miles N of Comondú, 30 Nov. 1952 (f).

Gentry 10321, DES, MEXU, US. Comondú, 2 April 1951 (l, f, cap).

Gentry 11253, DES, MEXU, US. Ca. 10 miles W of San José del Cabo, 5 Oct. 1951 (l, photo).

Gentry 11198, DES, US. Ca. 12–15 miles E of La Paz, Cape Dist., 29 Sept. 1951 (l, photo).

Gentry 4272, ARIZ, DES, DS, US. Cerro de la Giganta, 1 March 1939 (f). Volcanic slopes and crags up to 4,000 ft.

Gentry 12383, DES, MEXU, US. Ca. 3 miles N of Mission San Xavier, Sierra Giganta, 2 Dec. 1952.

Gentry & Cech 11255, DES, MEXU, US. Las Cuevas, Cape District, 7 Oct. 1951 (l, photo).

Gentry & Cech 11283, DES, MEXU, US. 35 miles NW of La Paz, 11 Oct. 1951 (l, photo).

Gentry & Cech 11295, DES, MEXU, US. 3 miles NW of Comondú, 18 Oct. 1951 (l, photo).

Gentry & Cech 11301, DES, MEXU, US. 10 miles W of Canipolé along road to Purísima (l, photo).

Gentry & Gentry 23182, DES, MEXU, US. 3 miles N of Comondú on lava fields, 10 April, 1973 (l, f, photo).

Harbison s.n., SD. Mesa 2 miles N of Comondú; elev. 400 m, 7 Oct. 1967.

Harbison s.n., SD. 6 miles N of Todos Santos.

Moran 7144, CAS, DS, MEXU, SD. Cape District, 4 km N of La Huerta, 400 m, 25 Jan. 1959 (l, f).

Purpus s.n., MO, UC. San José del Cabo, Jan.– March 1901 (l, f).

Wiggins 14470, CAS, DS, MEXU. 100 miles W of Los Planes, alt. 490 m, south-facing slopes, 21 Dec. 1958 (l, f).

Wiggins 14531, CAS, DS, MEXU, UC. 6.4 miles N of Todos Santos along road to La Paz, 25 Dec. 1958 (l, f).

Agave capensis

BAJA CALIFORNIA. *Barclay & Arguelles 1987,* MEXU, US. Vicinity of Cabo San Lucas, 19 April 1966.

Brandegee s.n., UC. Cabo San Lucas, 18 March 1892 (l, cap).

Gentry 10080, DES, MEXU, US. Huntington Botanical Gardens, San Marino, Calif., Jan. 9–15 (l, cap, photo).

Gentry 19676, DES, MEXU, US. Huntington Botanical Gardens, 17 April 1962 (l, f).

Gentry & Fox 11247, 11250, DES, MEXU, **US.** Type. Cabo San Lucas & vicinity, 5 Oct. 1951 (l, f, photo).

Gentry & Fox 11823, DES, MEXU, US. 3 miles N of Todos Santos, Cape Dist., 4 May 1952.

Rose 18326, GH. ? Magdalena Island, March 1911.

Agave promontorii

BAJA CALIFORNIA. *Barclay & Arguelles 1986,* DES, MEXU, US. Western summit of Sierra Laguna, 15 April 1966 (l, f).

Brandegee s.n., UC. Sierra de la Laguna, 24 April 1892 (l, f).

Brandegee s.n., UC. San Jose del Cabo, cultivated in San Diego, 1903.

Gentry 10164, DES, MEXU, US. Huntington Botanical Gardens, San Marino, Calif., 9–15 Jan. 1951.

Gentry 11218, DES, MEXU, US. Rancho Laguna and vicinity, Sierra Laguna, Cape District, western summit, 3 Oct. 1951.

Gentry 11229, DES, MEXU, US. Rancho Burrera at west base of Sierra Laguna, 1–4 Oct. 1951.

Gentry 19671, DES. Sierra Laguna, Cape District, 1952 (photo).

Moran 7451, CAS, SD. La Aguja, elev. 1,900 m, Cape District, 18 May 1959 (l, f, cap).

Nelson & Goldman 7437, US. Type. From San Bernardo to El Sauz, Sierra La Laguna, 21 Jan. 1906, alt. 2,400–5,000 ft. (l, f, photo).

14.

Group Crenatae

Die Agaven 1915: 194, 1915.

Plants free-seeding outbreeders, rarely surculose, with short-stemmed, single, medium to large rosettes; leaves thickly succulent, green to yellow-green, pruinose when young, usually clearly bud-printed, lanceolate to lance-spatulate, the margins deeply crenate and undulate, the large teeth prominently teated, frequently also with smaller interstitial teeth; spine broadly grooved above, decurrent, ligulate below from the base; inflorescence generally tall, narrowly paniculate, with diffuse umbels of flowers tinged red or purple in the bud, opening yellow over green ovaries; tepals 2–4 times longer than the tubes, erect-ascending, incurving, conduplicating in anthesis; anthers large, frequently excentric, yellow; filaments slender, varicolored, inserted in mid-tube or above; capsule short ovoid or short oblong, stipitate, short-beaked or beakless; seeds large, deeply lunate.

Typical species: *Agave inaequidens* Koch

Berger made the Crenatae a subgroup of his large section, "Reihe Scolymoides," which included some members of my section Hiemiflorae and several others unknown to me. Among these is his *A. calodonta* of unknown origin, which matured in 1897 without leaving seed or other offspring, and so far as known has not been seen since. However, as it appears distinct from other *Agave,* was well described, figured, and is represented by type specimen (Fig. 14.8), I am including it in this revision. Other taxa, *A. conjuncta* and *A. crenata,* appear as synonyms of earlier names, while *A. inaequidens* Koch and *A. hookeri* Jacobi, from unspecified origins, are now recognized as field populations; I am transferring them from Berger's Americanae group to this amended Crenatae. *A. longisepala* Todaro, described with the filaments inserted in the base of the tube, is excluded from this section. Perhaps, as Berger suggests, it belongs in the Americanae. The name, *A. mescal* Koch, has long been associated with this group, but is predated by *A. inaequidens,* was not diagnosed by Koch, and should not be used; see Gentry (1972, p. 86).

The members of the Crenatae are characterized by the deeply crenate-mammillate margins of the leaves (whence the name Crenatae), with richly assorted teeth and deep narrow panicles. The flowers are homogeneous in structure, the rather shallow tubes being universally well exceeded by long tepals that conduplicate and narrow during anthesis (see ideographs, Fig. 14.3). They are of minor use in separating the species, which are rather narrowly defined in this treatment, mainly on leaf characters. Generally the plants are bright green to yellowish green in color, the younger leaves with a glaucous or pruinose bloom accented with bud-printing patterns. The epidermis is generally smooth to the touch, but the microscope shows the usual agavoid translucent cuticular overlay. The leaves generally have graceful attitudes, and the bright castaneous to coppery colored teeth of some species add to their exotic attractiveness. Altogether, they are among the most beautiful agaves, but because of their large size, some have limited use in the ornamental gardens.

The Crenatae are free-seeding multiannual mesophytes of the Sierra Madre Occidental and the central trans-Mexican sierras. They are montane plants growing between elevations of 800 and 2,500 m, as in the oak woodlands and more open or rocky slopes of the higher pine forest zones. Precipitation ranges generally from 25 to 40 inches (300–500 mm) annually, and frosts in these latitudes, even in the higher elevations, are light and fleeting (Fig. 14.1). The more northern *Agave bovicornuta* is resistant to light frosts in southwestern U.S., but some of the more southern species, as *A. maximiliana,* are frequently damaged by frosts of a few degrees. The Crenatae agaves grow on both the acid soils derived from volcanics and on the alkalinic soils of limestone origin. As simple seed-plants, the wild populations usually consist of widely dispersed individuals. Some of

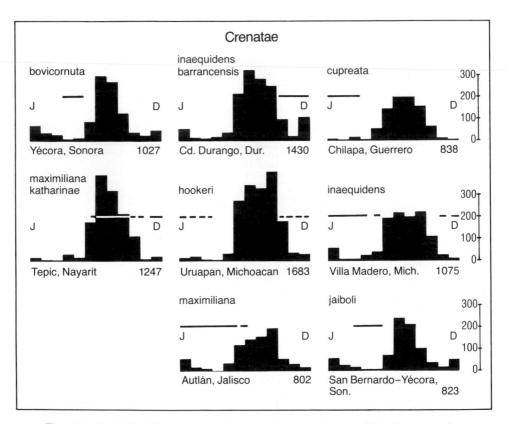

Fig. 14.1. Rainfall (silhouettes) and flowering (bars) perimeters of the Crenatae; relevant meteorological stations with annual rainfall in millimeters. Data from Atlas Climatológico de México, 1939, Sec. Agr. y Fom., Mexico, D. F. Boletin Meteorológico No. 2, 1968, Com. Fed. Elect., México. Flowering periods are based on herbarium specimens and field observations; uncertainty expressed by dotted lines.

the thicker colonies are represented in the habitat photos under the following account of species.

In the southern Crenatae range, the country name for agave is generally "maguey" and modified locally, as "maguey silvestre" (wild forms), or "maguey de pulque." Through the northern Sierra Madre the plants are known as "lechuguilla" or "lechuguilla verde" to distinguish them from the pale gray broad-leaved agaves, as *A. shrevei*. The chemistry of the Crenatae seems to be complex and little known. Some species were used as sweet food, *A. jaiboli*, or for drink as aguamiel and pulque, *A. hookeri*. Some were not eaten, as *A. bovicornuta*, which was employed to stupefy fish in fresh water streams (Bye

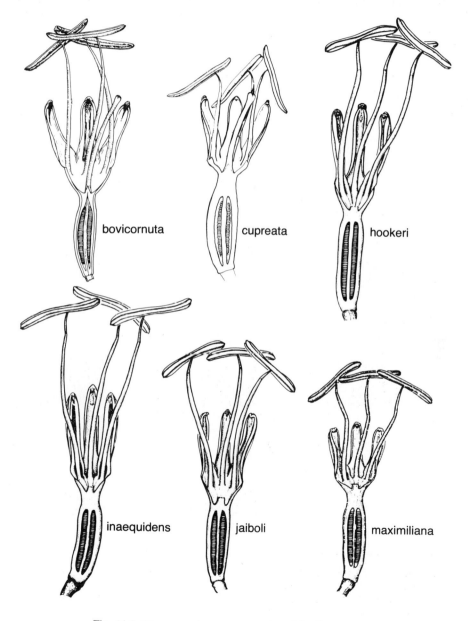

Fig. 14.2. Flower sections representative of the Crenatae.

et al., 1975). This plant and several other Crenatae species have caustic juice in the leaves and when handled cause itching irritation on the skin; the local name "maguey bruto" of Michoacán seems to refer to this undesirable quality. The group contains little sapogenin (Table 14.1). Notes on specific uses are given in the following account of species.

Table 14.1 shows the results of analyses to determine sapogenin content in 6 species of this section. The group is rather uniform in its general lack of these compounds. Only one sample of *Agave maximiliana* showed an appreciable content. This was reported for a leaf sample, but I suspect a clerical error and it may represent rather the inflorescence sample sent at the same time. The group is not a good prospect for finding precursors for cortisone and the sex hormones.

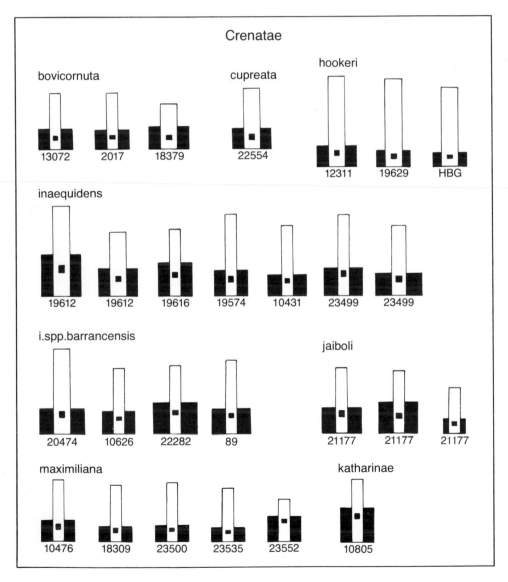

Fig. 14.3. Ideographs of Crenatae flowers, showing proportions of tube (black) to outer tepals (white) and locus of filament insertion (black squares) in tube. From measurements in Table 14.2.

Table 14.1. Sapogenin Content in the *Crenatae*

COLL. No.	SOURCE LOCALITY	MONTH COLL.	PLANT PART	SAPOGENINS*		
				TOTAL%	GIT.	SMIL.
bovicornuta						
10236	Sierra Charuco, Chih.	Mar.	leaf	0		
10265	Tepopa, Sierra Saguaribo	Mar.	whole pl.	0		
10252	Sierra Charuco, Chih.	Mar.	infl.	0		
11061	Sierra de Alamos, Son.	Aug.	leaf	0		
cupreata						
5106	Chilpancingo, Gro.	Jan.	leaf	0		
5106	Chilpancingo, Gro.	Jan.	infl.		0.06	
hookeri						
12311	Cheran, Mich.	Oct.	leaf	0		
inaequidens						
10431	Quiroga-Patzcuaro, Mich.	May	leaf	0		
51151	Patzcuaro, Mich.	Apr.	leaf		0.17	
jaiboli						
10256	San Bernardo, Son.	Mar.	leaf	0		
11374	Arroyo Guajaray, Son.	Dec.	leaf	0		
maximiliana						
10476	Plan de Barrancas, Jal.	May	leaf	0		
10520	El Batel, Sin.	May	leaf	0		
10751	Tepic-Jalcocotan, Nay.	July	leaf	0		
10950	Autlan, Jal.	July	leaf	0		
51172	Cocula-Tecolotlan, Jal.	Apr.	leaf			0.86
51172	Cocula-Tecolotlan, Jal.	Apr.	infl.	0		

SOURCE: Wall et al. (1954a, 1954b).

* Git. = gitogenin; Smil. = smilogenin.

Key to Species of the Crenatae

1. Rosettes large, leaves generally 1–2 m long (rarely shorter in *A. inaequidens*) ... 2
1. Rosettes medium, leaves generally less than 1 m long ... 4
2. Leaves 4–7 times longer than wide, the apex with teeth to within 5–10 cm of spine; spine 3.5–5 cm long
inaequidens, p. 340
2. Leaves 6–10 times longer than wide, the apex typically toothless for 12–20 cm below base of spine; spine 4–6 cm long ... 3
3. Leaves generally glaucous gray, massive, 20–25 cm wide; tepals very long, 4–5 times longer than tube. Cultivated
hookeri, p. 338
3. Leaves bright green to yellow green, sometimes slightly glaucous, narrow, 10–17 cm wide; tepals 2–3 times longer than tube. Wild
inaequidens ssp. *barrancensis,* p. 342
4. Leaves linear lanceolate, 8–10 times longer than wide   *jaiboli,* p. 344
4. Leaves broadly lanceolate, 2–5 times longer than wide ... 5
5. Leaves ovate, 2–3.5 times longer than wide; teeth and spine copper colored. Balsas River Basin   *cupreata,* p. 335
5. Leaves broadly lanceolate to spatulate, 4–5 times longer than wide; teeth not copper colored. ... 6
6. Leaves bright green to yellow green, only the young ones pruinose; spine 3–5 cm long; filaments inserted in mid-tube   *bovicornuta,* p. 328
6. Leaves paler, glaucous gray green to bluish green; spine 2–4 cm long; filaments inserted above mid-tube
maximiliana, p. 346

Agave bovicornuta

(Figs. 14.1, 14.2, 14.3, 14.4, 14.5, 14.6, 14.7; Tables 14.1, 14.2)

Agave bovicornuta Gentry, Carnegie Inst. Wash. Pub. 527, p. 92. 1942.

Medium to large, single, nonsuckering, light green rosettes, 8–10 dm tall, 15–20 dm wide; leaves 60–80 x 14–17 cm, lanceolate to spatulate, widest at or above the middle, much narrowed toward the base, yellowish green to green, the younger leaves frequently satiny glaucous, smooth, bud-printed, the margins crenate with prominent teats under the larger teeth; teeth dimorphic, the larger mostly 8–12 mm long, flexuous and slender above a broad base, mostly 2–4 cm apart, the smaller mostly 2–5 mm long, 1-several between the larger, all castaneous or dark brown to grayish brown in age;

Table 14.2. Flower Measurements in the Crenatae (in mm)

Taxon & Locality	Ovary	Tube	Tepal	Filament Insertion & Length		Anther	Total Length	Coll. No.
bovicornuta								
Curohui, Son.	22	7×12	21×4	4–5	42	20	50*	3672
	24	8		19 & 18 3–4	44		50*	3672
Sierra Tecurahui, Son.	35	7×13	20×5 & 19	4–5	45	24	62†	2017
	35	7×13	21×5	4–5	45	24	63†	2017
	34	6×12	19×5 & 18	4–5	45	23	58†	2017
Surutato, Sin.	22	9×14	15×6–7	4	30	18	46†	18379
	23	8×14	16×6–7	4	34	20	45†	18379
cupreata								
Tzitzio, Mich.	32	7×15	21×6 & 20	4–5	35	24	57†	22554
	31	7×15	20×5 & 19	3–5	37	23	58†	22554
	34	6×14	21×6 & 20	4–3	35	24	59†	22554
hookeri								
Cheran, Mich.	34	7×14	32×5 & 30	5	50	34	73†	12311
	35	8×14	31×3.5	6	60	34	74†	12311
	35	6×14	32×4 & 29	4	53	33	76†	12311
	34	6×12	31×6 & 28	5	60	32	70†	19629
	41	5×12	30×5 & 29	4	50	32	77†	19629
	33	5×12	28×5 & 27	4	50	26	67†	H.B.G.
inaequidens								
Temascaltepec, Mex.	41	15×14	34×7 & 31	11 & 8	68	34	90†	19612
	37	10×14	23×7 & 21	7 & 6	54	26	70†	19612
Mil Cumbres, Mich.	37	11×14	25×4 & 23	7 & 6	50	31	73†	19615
	40	12×13	24×4 & 22	8 & 7	54	28	76†	19616
	38	14×14	26×4 & 24	9 & 8	55	29	77†	19616
Coajomulco, Mor.	29	9×12	30×5 & 28	6–5	58	30	69†	19574
	41	10×12	28×4 & 26	8–7	52	30	81†	19574
Patzcuaro, Mich.	26	5×13	25×5 & 23	4	53	26	57†	10431
	36	8×14	26×4 & 24	5–6	45	29	70†	10431
	32	7×14	24×3 & 22	5	48	27	62†	10431
Sierra Tapalpa, Jal.	40	10×15	29×4.5 & 27	8–9	57	34	78†	23499
	41	9×18	32×5 & 30	7–8	58	35	83†	23499
	28	8×17	25×4–5	6	58	29	60†	23499

panicles 5–7 m tall, narrow and deep with 20–30 short laterals in upper half of shaft; bracts below the laterals 20–30 cm long, triangular-acuminate, soon drying reflexed, those above smaller; flowers small in compact umbels, greenish yellow, 55–65 mm long, on pedicels 5–10 mm long; ovary pale green, 30–35 mm long, including neck 4–6 mm long; tube 6–8 mm deep, 12–14 mm wide; tepals yellow, 18–21 mm long, 4–6 mm wide, linear-lanceolate, ascending-spreading, conduplicate, involute, broadly overlapping sinuses, the inner wider and 2-costate within; filaments yellow, 40–45 mm long, inserted 4–5 mm above base of tube; anthers 20–23 mm long, yellow; capsule stipitate, 40–50 mm long, 15–20 mm in diameter, oblong, the valves thin; seeds 7 x 15 mm, finely punctate, the curved side with a flange or wing, the hilar notch shallow.

Type: Curohui, Sierra Saguaribo, Rio Mayo, Apr. 4, 1938, *Gentry 3672*. Holotype DS, isotypes ARIZ, DES.

Table 14.2. cont.

Taxon & Locality	Ovary	Tube	Tepal	Filament Insertion & Length		Anther	Total Length	Coll. No.
i. ssp. *barrancensis*								
Palmito, Dur.	51	9×16	30×6&27	8&6	50	33	89†	20474
	50	9×18	29×6&26	8&6	57	33	89†	20474
	51	9×15	30×6&28	6–7	50	33	90†	20474
	24	8×13	23×5&21	6	45	28	54†	10626
Revolcaderos, Dur.	27	8×12	23×3&21	5–6	50	28	57†	10626
	38	11×16	24×3&23	7–8	52	30	73†	22282
	36	9×16	26×3&25	7–8	53	28	71†	22282
	32	11×16	23×3&21	7–8	55	28	65†	22282
S of El Salto, Dur.	25	8×14	25×4&24	7	48	25	57†	89
	28	8×15	26×4	6–7	48	25	60†	89
jaiboli								
Sierra Ventana, Son.	19	4×8	15×4&14	3–4	35	13	36*	21177
	28	4.5×8	16×4&15	3–4	35	13	47*	21177
Murrieta transpl., Cal.	27	10×15	23×4&21	6–8	45	24	61†	21177
	29	9×15	23×4&21	6–8	46	23	61†	21177
	28	11×15	22×3.5	6–7	45	23	61†	21177
maximiliana								
Plan de Barrancas, Jal.	30	8×12	22×6&21	6	48	24	59†	10476
	32	7×12	21×5&20	6	45	23	59†	10476
Autlan, Jal.	29	5×12	20×6–7	5	36	24	56†	18309
	33	6×13	20×4&19	4–5	40	23	58†	18309
Sierra Tapalpa, Jal.	38	5×14	20×4&19	4	40	24	68†	23500
	33	7×13	21×3&20	4–5	35	22	59†	23500
	29	6×13	21×3&20	4–5	37	21	58†	23500
Sierra Minatitlan, Jal.	30	5×12	17×4	3–4	34	18	52†	23535
	30	5×14	21×4&20	3–4	45	23	56†	23535
	31	6×14	19×4	4–5	45	23	57†	23535
Sierra Cuale, Jal.	28	9×12	15×4	7–8	30	22	52†	23552
m. var. *katharinae*								
Sierra San Juan, Nay.	34	12×12	23×4.5	10–9	60	25	69*	10805
	29	12×12	22×4.5	9–10	40	26	62*	10805

* Measurements from dried flowers relaxed by boiling.

† Measurements from fresh or pickled flowers.

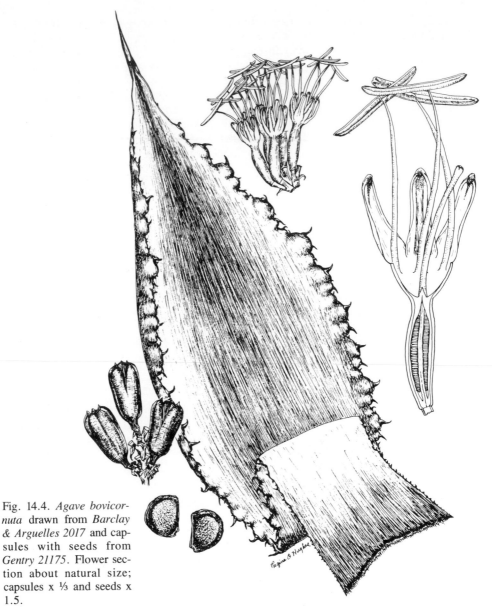

Fig. 14.4. *Agave bovicornuta* drawn from *Barclay & Arguelles 2017* and capsules with seeds from *Gentry 21175*. Flower section about natural size; capsules x ⅓ and seeds x 1.5.

Generally it occurs at elevations between 3,000 and 6,000 feet (930 and 1,850 m) on the rocky open slopes in oak woodland and the pine-oak forest zones. Although common and widespread in the northern Sierra Madre Occidental, the populations are usually local with limited numbers of individuals, and one may travel for many mountain miles without seeing any (Fig. 14.7).

Agave bovicornuta is distinguished from other Crenatae by the light green to yellowish color of the leaves with narrowly hafted bases, its relatively small flowers, and the low insertion of the filaments in the mid-tube. Its nearest relative appears to be among the Durango and Nayarit populations of *A. maximiliana*. The latter, however, are more glaucous light colored rosettes with more heteromorphic teeth, and large flowers. *A. bovicornuta* may be confused with *A. wocomahi* of the Ditepalae. However, the latter has darker green leaves with generally longer teeth flexed downward from the mid-

Fig. 14.5. *Agave bovicornuta*. Depauperate plants in the cracks of tuffaceous volcanic rock on Sierra de la Ventana, Sonora.

blade below, and a different flower structure; compare respective flower sections and ideographs (Figs. 14.2, 14.3 and Figs. 16.4, 16.5).

Local names for *A. bovicornuta* have been reported as (Spanish) "lechuguilla verde"; (Warihio) "sapari" (Gentry, 1963); (Tarahumara) "sapuli" (Bye et al., 1975). Bye reports that the flowers are eaten after being washed to remove the bitterness and are preferred to those of other species for making tortillas. He also reports that the stems are pit-baked in Tarahumara country and distilled to make "pisto" or mescal. This is probably due to the growing scarcity of other less astringent sweeter species, as earlier investigations reported that the Río Mayo peoples avoid the use of *A. bovicornuta* stems for food and beverage (Gentry, 1942, 1963). It is still employed for poisoning fish (Bye et al., 1975; Bennett and Zingg, 1935; Pennington, 1958). The juice is caustic and produces a temporary dermatitis on tender skin, with a burning sensation, inflammation, and white welts or blisters. Some people are more susceptible than others. Washing with soap only increases the burning. It is better to wipe dry or wash with clear water only. *A. bovicornuta* was among many samples of Section Crenatae analyzed for sapogenins; like the others, it contained little or no sapogenin. The constituents toxic to fish and skin have not been identified chemically. On the Sierra Surotato, Sinaloa, I was given the name "noriba" for *A. bovicornuta*.

A. bovicornuta forms a handsome plant in the succulent garden, especially the more glaucous individuals, where a satiny sheen appears on the new leaves along with crenulated bud-printing. Because they do not sucker or form bulbils and apparently require

Fig. 14.6. *Agave bovicornuta,* showing detail and a leaf unfolding from the white terminal bud.

cross-pollination, garden stocks are exhausted in one generation, 12–18 years, unless new field collections are made. The fertile black seeds germinate readily and can easily be cultured in flats or pots during early stages. Some wild stands in the native habitat should be preserved as perennial seed sources for future plantings.

Agave calodonta
(Fig. 14.8)

Agave calodonta Berger, Hort. Mortol. 1912: 364, 1912.
  Agave scolymus Berger, Gartenwelt 2: 603, 1898.

Rosette stemless, single, multi-leaved, ca. 15–16 dm in diameter, hemispherical, with thick conic terminal bud. Leaves outstretched, the older prostrate, fleshy, convex on both sides at the base, upwards markedly thin, light green, with light gray bloom, very clearly bud-printed on both sides, 80 cm and more long, spatulate, short-

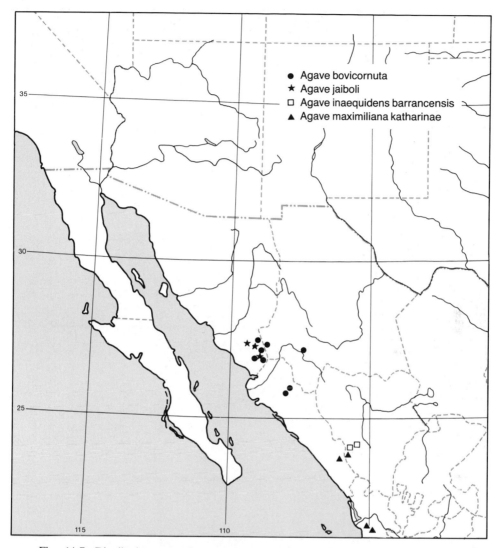

Fig. 14.7. Distribution map of *Agave bovicornuta, A. jaiboli, A. inaequidens* var. *barrancensis,* and *A. maximiliana* var. *katharinae.*

acuminate, in upper third ca. 20–21 cm broad, narrowed toward base, 7½–8 cm wide at the narrowest near the base, upper side shallowly hollowed, somewhat keeled on the back under the apex; spine 3–4 cm long, decurrent to the upper 3–4 teeth, broadly, deeply furrowed above, keeled on back and with deltoid protrusion; margin in the mid-leaf sinuous, heavily toothed; teeth irregular, the middle and upper largest on broad fleshy prominences, with broad horny bases and deltoid cusps hooked forward or backward, 10–13 mm long, 2½–3½ cm apart, in the hollows much smaller teeth; teeth in lower half of leaves much smaller, straight or reflexed, all teeth light brown, inflorescence tall with strong shaft and long pyramidal panicle, at the base with numerous deltoid reflexed bracts, the lowest leaf-like but with pale clear broad horny margin with small teeth and strong spine; flowers yellow, without stamens 8½ cm long (dried);

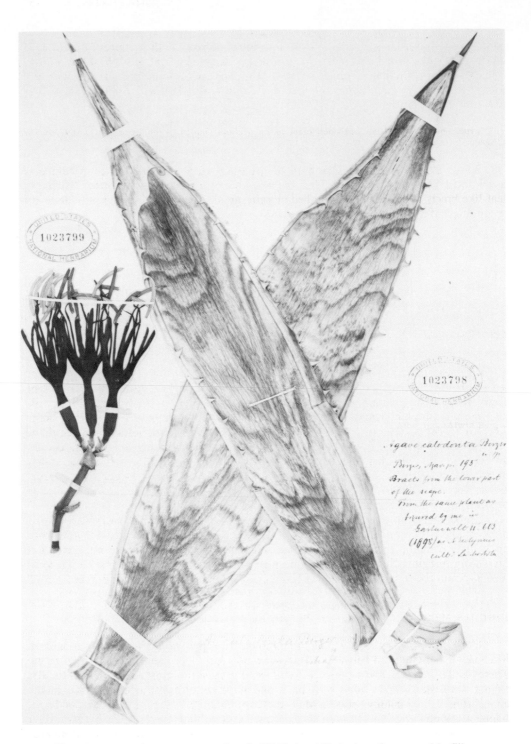

Fig. 14.8. *Agave calodonta*. Type sheet in US National Herbarium; flowers and leaflike bracts, Berger "Gartenwelt 2: 603, 1898, as A. scolymus cult. La Mortola."

pedicels to 10 cm long; ovaries 35–40 mm long, narrow, both sides (ends?) strongly constricted; tube broadly funnelform, ca. 10 mm long; segments ca. 35–40 mm long, linear, acute yellow-green, green on apex, the inner with broad keel; filaments 6 cm long, inserted above mid-tube; anthers 28–32 mm long; pistil with thick blunt, trique-trous stigma. (Translated from German.)

Type: *Berger s.n.* from cultivated plant in La Mortola Garden, Ventimiglia, Italy, 1897, in *US* (2 sheets, lower bracts, f), Fig. 14.8.

Berger's description shows the plant to be closely related to other Crenatae members, as he noted. How and from where it came to La Mortola Gardens is unknown. The large leaf-like bracts with wide horny toothed margins are distinctive and should provide ready recognition, if it is ever rediscovered.

Agave cupreata
(Figs. 14.1, 14.2, 14.3, 14.9, 14.10, 14.11, 14.21; Tables 14.1, 14.2)

Agave cupreata Trel. & Berger, Agaven 197, 1915.

Rosettes single, caulescent, medium-sized, bright shiny green, openly spreading; leaves broadly lanceolate or ovate, 40–80 x 18–20 cm, strongly narrowed at base, bright green, thick-fleshy, plane to slightly concave above, the margin deeply crenate mammil-late; teeth dimorphic, strongly flattened, straight to curved, cupreus to gray, the larger 10–15 mm long on prominent teats, 3–6 cm apart, the smaller in hollows of the margin of varying sizes; spine 3–5 cm long, slender, sinuous, light brown to grayish, openly grooved above, with sharp borders decurrent to upper teeth; panicle 4–7 m tall, rather broad in outline, with 14 to 25 lateral peduncles in upper half of shaft; buds rufous; flowers in loose diffusive umbels, orange-yellow, 55–60 mm long, on dark-bracteolate pedicels; ovary 30–35 mm long, olive green, fusiform, with constricted double 3-grooved neck; tube 6–7 mm deep, 14–15 mm broad, broadly funnelform, knobby, grooved, thick-walled; tepals subequal, erect, linear-lanceolate, acute, the outer 20–21 mm long, apex rusty-colored, broader, thicker than inner, the inner with narrow keel and thin involuting margins; filaments 35–40 mm long, thickened toward base, inserted in mid-tube 3–5 mm above tube base; anthers 23–24 mm long, yellow, excentric, curved.

Type: *Langlasse 867*, Sierra Madre, versant oriental, 15–1,700 m, Michoacán and Guerrero, 15 Feb. 1899 (l, f) B, MEXU, US. The sheet in the Berlin Herbarium was annotated as *Agave cupreata* Trel.

Agave cupreata is widely scattered on mountain slopes about the Río Balsas basin in Michoacán and Guerrero between 4,000 and 6,000 feet (1,220 and 1,850 m) elevations. There is a large population on the south and east facing slopes west of Chilpancingo. The habitat is a frostless zone below the Tropic of Cancer with 30–35 inches (73–86 mm) of annual rain, falling mainly between May and November. There is a long dry season during winter and spring, during which this agave flowers and fruits. Langlasse noted that it grew on "terrain granitique" and was called "magues de mescal. Fournit un alcool nomme mescal."

Agave cupreata is a beautiful plant distinguished by the broad, highly teated, shiny green leaves, highly bud-printed, with the teeth bright copper-colored in early stages (Figs. 14.9, 14.10).

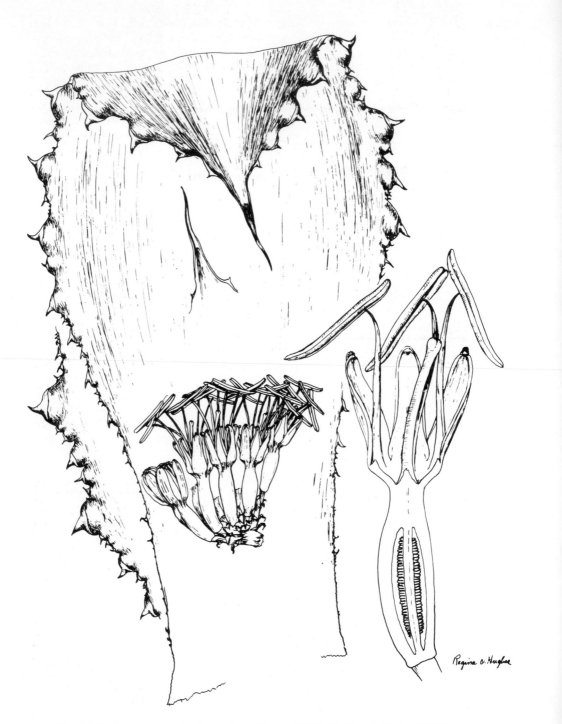

Fig. 14.9. *Agave cupreata,* drawn from *Gentry 22553* (leaf) and *Gentry 22554,* collected near Tzitzio, Michoacán, 15 Dec. 1967.

Fig. 14.10. *Agave cupreata,* a flowering plant growing with oaks and other shrubbery near Tzitzio, Michoacán, 15 Dec. 1967.

Fig. 14.11. *Agave cupreata,* detail of teeth and bud-printing.

Agave hookeri
(Figs. 14.1, 14.2, 14.3, 14.12, 14.13, 14.14, 14. 21; Tables 14.1, 14.2)

Agave hookeri Jacobi, Hamburger Garten-Blumenzeitung 22: 168, 1866.

Rosettes large, single, glaucous to green or yellow-green, up to 2 m tall, with short thick stem, variable; leaves 120–175 x 20–25 cm, lanceolate, gradually narrowed toward base and apex, thick-fleshy, generally concave above and arching in age, the margin undulate to crenate, especially through the mid-blade, nearly straight below with small teeth; teeth largest through the mid-blade, 8–12 mm long, 2–5 cm apart, straight or curved, broadly based on fleshy prominences, dark brown to grayish brown, a few smaller interstitial teeth, much reduced and closely spaced toward base; spine 3.5–6 cm long, subulate, shortly, openly grooved above, the edges decurrent sharply as a smooth corneous leaf border for 15–20 cm below spine, roundly keeled below and with linguiform protrusion into leaf; panicle 7–8 m tall with reflexed narrow triangular bracts and with 20–40 large compact umbels in upper half of shaft; flowers 63–80 mm long, slender, red to pink in bud and on tepals, opening yellow, long pedicellate, with caducous bractlets; ovary 34–41 mm long, cylindric, smooth, angulate, with long constricted grooved neck; tube 5–8 mm deep, 13–14 mm broad, grooved; tepals unequal, erect, elongate, linear, conduplicating, the outer 28–32 mm long, rounded on back, cucullate, the inner 2–3 mm shorter, with prominent keel; filaments 50–60 mm long, slender, flattened, inserted above

Fig. 14.12. *a. Agave hookeri.* "Type specimen from the plant from which Jacobi made his description !! Dec. 15, 1880. N. E. Brown. Hort. Kew." *b.* Base of leaf of type specimen of *A. hookeri.*

Fig. 14.13. *Agave hookeri*, flowering plant at Cheran, Michoacán, Dec. 11, 1961.

Fig. 14.14. Hedge row of *Agave hookeri* southwest of Jiquilpán, Jalisco.

mid-tube; anthers 26–34 mm long, yellow, excentric; capsule (22555) 5–5.5 x 2.5 cm, oblong, stipitate, rounded at apex, thick-walled; seeds broadly lunate 8–9 x 6–7 mm, rugose, lucid black, the hilar notch broad, the marginal wing broad but little raised.

Type: N. E. *Brown s.n.*, Hort. Kew, 15 Dec. 1880, K (1, 3 sheets). "Type specimen from the plant which Jacobi made his description."

Among the Crenatae, *Agave hookeri* is recognizable by its large size, glaucous leaves with a strong tongue-like projection from the spine base, short tube and the very long tepals. It has been confused with *A. fenzliana* Jacobi, but that large agave has a conspicuously bracteate diffusive umbel, suggesting the Hiemiflorae, and does not belong here. The native origin of *A. hookeri* was unknown until I reviewed the material in the Kew Herbarium and recognized it as representative of plants I had collected in Michoacán.

Agave hookeri is mostly encountered as a cultivate in central Michoacán at 6,500 to 7,500 feet (2,000 to 2,300 m) elevations, especially in the pine country. Occasional spontaneous individuals occur, some perhaps as escapes from cultivated sources. They are frequently used as fence plants, which are also tapped for their sweet juice and to ferment into pulque. Such rows are conspicuous lining fields about Cheran along the road to Uruapán (Fig. 14.14). One sometimes finds plants of *A. inaequidens* mixed in these rows and they may be hard to distinguish, as both species show variation in leaf color, shape and size of leaf, and armature. East of Morelia a woman stated that young plants are brought in from wild stands to plant along fields. Such plantings are, therefore, seedling variable populations. A strongly mutilated stem, with some old leaves still attached, was proliferating numerous small rosettes when observed at Cheran in 1961. Such plantlets could be the source of some agave rows, which have the uniform appearance of vegetative propagation or clones.

Agave inaequidens

(Figs. 14.1, 14.2, 14.3, 14.15, 14.16, 14.21; Tables 14.1, 14.2)

Agave inaequidens Koch ssp. *inaequidens*
    Agave inaequidens Koch, Wochenschr. Ver. Beford Gartenb. 3: 28, 1860; Jacobi, Hamburger Garten-Blumen. 20: 554, 1864.
    Agave megalacantha Hemsl. Diag. Plantae Mex. 55, 1880.
    Agave mescal Koch, ibid. 8: 94, 1865.
    Agave crenata Jacobi, Hamburger Garten-Blumenz. 22: 176, 1866.

Plants medium to large, single, short-stemmed, openly spreading; leaves variable, mostly 75–150 x 11–21 cm, broadly or narrowly lanceolate to oblanceolate, ascending to outcurving, concave above, thick-fleshy, especially toward the rounded base, light green to yellow-green, rarely faintly glaucous, the margin undulate to repand and crenate; teeth dimorphic, the larger on broad prominences, commonly 8–10 mm high, the flattened bases longer than height of teeth, 2.5–4 cm apart, straight or variously curved, castaneous to dark brown, the smaller interstitial teeth few; spine stout, 2.5–5.5 cm long, broadly, deeply channeled above, dark brown, smooth, protruding into flesh of leaf below, sharply decurrent to uppermost teeth; panicles 5–8 m tall, in deep narrow outline, with 30–50 compact umbels in upper half of shaft; buds and tepals reddish purple; flowers 60–90 mm long, opening yellow; ovary 30–40 mm long, trigonous cylindric, with short furrowed neck; tube 5–12 (–15) mm deep, deeply grooved, bulging with filament insertions, thick-walled; tepals unequal, 25–30 (–34) mm long, narrow, linear, erect, conduplicate, involute, strongly cucullate and papillate within apex; filaments 50–60 mm long, stout, ovate in cross section, inserted above mid-tube; anthers large, 26–34 mm long, centric to excentric; capsules 4–4.5 x 2 cm, oblong, stipitate, rounded to apiculate at apex, beak-

Fig. 14.16. *Agave inaequidens*, a large rosette at same locality.

◄Fig. 14.15. *Agave inaequidens* in native habitat; plant with characteristically tall narrow panicle about 20 miles (32 km) northeast of Temascaltepec along road to Toluca, Mexico, 7 Dec. 1961.

less, brown, strong-walled; seeds 6–7.5 x 4.5–5.5 mm, hemispherical, shiny black, finely punctate, the marginal wing curvaceously half-rised or erect, hilar notch broad.

Neotype: *Gentry, Barclay & Arguelles 19612, US*. 20 miles (32 km) NE of Temascaltepec along road to Toluca, State of Mexico, 7 Dec. 1961; alt. ca. 7,500 feet (2,300 m), open pine slope. Dups. DES, MEXU.

The proper taxon for the wild complex has been confused with *A. mescal* by Trelease and others. Koch really did not describe *A. mescal*, but used the name in discussing a young potted plant and several other vague taxa, representing young growing plants in European gardens. However, Koch did describe *A. inaequidens* five years earlier (1860), and its characteristics are certainly applicable to our Crenatae; the earliest description I have found to do so. Accordingly, I have selected *A. inaequidens* as typical of the group and nominated a representative specimen as neotype of species.

Agave inaequidens is closely related to *A. hookeri* of the same region, the obvious distinguishing characters being the bright yellowish green color of the former and the glaucous light gray of the latter. *A. hookeri* is also more consistently massive and its short flower tube contrasts more strongly with the long narrow tepals. There appears to be some mixing of the species, not only in plantings, but also in genetic crosses. Both are free-seeding outbreeders with little or no natural vegetative reproduction. *A. hookeri* is primarily a selected edible cultivate, while *A. inaequidens* is a caustic, less edible wildling of seed-variable disposition. The local name ''maguey bruto'' has been applied to these dermatitic plants.

Agave inaequidens is the predominant agave in the trans-Mexican mountains from Morelos through Michoacán to Colima and Jalisco. The primary habitat is the pine forested slopes between 6,000 and 8,000 feet (1,850 and 2,480 m), also characterized by oaks and other hardwoods. Annual rainfall averages about 1,000 mm, or over 40 inches, the mountain slopes are generally well timbered and agaves are restricted to the more open rocky slope (Fig. 14.15). During the winter of 1961 I traced the flowering populations of this robust complex across the trans-Mexican sierras from Morelos and the Distrito Federal to western Michoacán.

Agave inaequidens ssp. *barrancensis*
(Figs. 14.1, 14.3, 14.7, 14.17, 14.18; Table 14.2)

Agave inaequidens ssp. *barrancensis* Gentry, ssp. nov.

Rosettes short-stemmed, single, open, large, 1.5–2 m tall, 3–3.5 m broad; mature leaves mostly 1–1.7 m x 10–16 cm, narrowly lanceolate, widest in middle, long-acuminate, the apex without teeth for 15–20 cm below spine, at base thick but little narrowed, ascending to outcurving, hollowed above, convex below, green somewhat glaucous, margin nearly straight to undulate; larger teeth through the mid-blade 5–10 mm long, 1–3 or 4 cm apart, with smaller occasional interstitial teeth, all light brown to dark brown, straight to curved; spine 4–6 mm long, subulate, dark brown, openly grooved above, ligulately protruded at base below, long decurrent; panicle 6–8 m tall with 20 to 30 broadly spreading umbels in upper half of shaft, the peduncular bracts narrow, reflexed, remote, early drying; flowers in diffuse umbels, yellow, sometimes tinged reddish, over pale green ovaries, 60–90 mm long; ovary 30–50 mm long, trigonous angled, fusiform, furrowed in short neck; tube 8–11 mm deep, funnelform, deeply grooved, thick-walled; tepals unequal, 23–30 mm long, 6 mm wide narrowing to 3 mm with anthesis, involuting, conduplicating, rather thin, linear-lanceolate, ascending-incurved, cucullate, closely overlapping at base, the inner with prominent keel; filaments 48–57 mm long, slender, somewhat flattened, inserted above mid-tube 6–8 mm above tube base; anthers 25–33 mm long, yellow, excentric, about straight; pistil slender, with small stigma head; capsules broadly ovoid 4–5 x 2.2–2.5 cm, stipitate, nearly beakless, strong-walled; seeds 7–9 x 5–6 mm, lachrymiform, smoothly black, hilar notch shallow, irregular.

Types: *Gentry & Arguelles 22282*, Revolcaderos, Durango, along Mazatlan-Durango road, 8 Dec. 1966; elev. ca. 6,500 feet (2,000 m), on cliffs and rocky sun slopes, pine-oak forest zone (l, f); in *US* (holotype), MEXU, DES (isotypes).

Planta plerumque non-surculosa, 1.5–2 cm alta, 3–3.5 m lata. *Folia* lineari-lanceolata, 11–18 dm longa, 1–1.7 dm lata, viridia, recti-adscendentia, laevia; *dentibus marginalibus* brunneiscastaneis, inaequalibus dimorphis, grandioribus 5–10 mm longis, 1–4 cm distantibus, minoribus paucis 1–2 mm longis; *spina terminali* 4–5 cm longa brunnea subulata, supra canaliculata, longidecurrente. *Inflorescentia* paniculata 6–9 m alta, angusta, ramii 20–35 lateralibus; floribus 60–90 mm longis, luteis vel rubipurpureis; ovarium 30–50 mm longum trigoni-angulatum fusiforme viride; *perianthii* tubo 8–11 mm longo, 12–18 mm lato, infundibuliformi sulcato; *segmentis* inaequalibus 23–30 mm longis, 6 mm latis, per anthesin 3 mm latis involutis, linearis ad apicem galeatum. *Filamenta* 48–57 mm longa supra medium tubi inserta; *antheris* 25–33 mm longis luteis. *Capsulae* ovoideae vel latioblongae, 4–5 x 2.2–2.5 cm stipitatae. *Semina* 7–9 x 5–6 mm lachrymiformia nigra.

This agave is known only from along the mountainous slopes of the deep barrancas of the Sierra Madre Occidental along the Durango-Mazatlán road. The elevations it occupies are between 1,800 and 2,400 m with oak and pine being the most familiar zone indicators,

Fig. 14.17. *Agave inaequidens* ssp. *barrancensis*. A colony of plants on canyon slope near Revolcaderos, Durango, 8 Dec. 1966.

Fig. 14.18. *Agave inaequidens barrancensis* at same locality with leaves wider than average.

but there is also a rich variety of tropical deciduous trees and shrubs in the canyons, including *Tilia* and the singular *Coriaria thymifolia* of the Coriariaceae. Annual rainfall, according to the 9–11 year records available (1968), ranges from 1,200 to 1,450 mm, with frosts lacking or light. With the high rainfall, *barrancensis* makes a large and vigorous growth (Fig. 14.18). However, it also reached large proportions in my Murrieta garden under much drier conditions.

The ssp. *barrancensis* looks like a large edition of *A. jaiboli,* but the closer relation is with *A. inaequidens. Barrancensis* is distinguished by its narrow leaves with nearly straight margins, long acicular spines, and broader shorter panicles than the species proper. Some of the more extreme variants have linear leaves up to 2 m long and may drop reflexed with their own weight.

Agave jaiboli

(Figs. 14.1, 14.2, 14.3, 14.7, 14.19, 14.20; Tables 14.1, 14.2)

Agave jaiboli Gentry, Agave Family in Sonora. U.S.D.A. Agr. Handb. 399: 89, 1972.

Medium-size, single, nonsuckering, green to yellowish green, usually open rosettes, 6–10 dm tall, 14–20 dm broad; leaves 60–100 x 8–12 cm, linear to lanceolate, widest at or above the middle, gradually narrowed below, usually straightly ascending to spreading, long-acuminate, sometimes incurved, plane to conduplicate, the margins noncorneous or narrowly corneous with the decurrent spine for less than its length; the larger teeth mostly 2–3 cm apart, on small regular teats, 5–8 mm long, flexed downward or upward, reddish brown, the smaller interstitial teeth 1 to several, 1–4 mm long, terminal spine 3–4 cm long, subulate, terete, reddish brown, shiny, smooth, openly or narrowly grooved in lower ½ to ⅔ of length; panicle 6–8 m tall with narrow scarious bracts and 12–15 lateral peduncles in upper ⅓ or ½ of shaft, ascending, rather sigmoid, with small diffuse clusters of yellow flowers flushed ferruginous above green ovaries, on long (1–2 cm) bifurcate pedicels with discoid apices in fruits; flowers ca. 60 mm long; ovary angular-fusiform, 25–30 mm long including 3–4 mm neck, shortly tapered at base; tube 9–11 mm deep, 15 mm broad, funnelform, bulging at filament insertions, deeply grooved; tepals unequal, rather thin, conduplicating and involuting with anthesis, the outer longer 22–23 x 4–5 mm, ascending, incurved, linear-lanceolate, acute, slightly valicullate, the inner with broad keel; filaments inserted above mid-tube, ca. 45 mm long, pink, somewhat flattened; anthers 23–24 mm long, nearly centric; pistil stout, over-reaching anthers in post-anthesis; capsules 4–5 x 1.8–2 cm, long-stipitate, shortly beaked, brown; seeds 7 x 5 mm, shiny black, rugose, very finely punctate, the straight edge as well as the rounded ones with a prominent wing.

Type: *Gentry 21177,* Sierra de la Ventana, Río Mayo, Sonora, May 27, 1965, near the Sonora-Chihuahua border in oak-*Nolina* grassland at ca. 1,000 m elevation, in US (2 sheets, 1, old f, cap, photo). A transplant from this locality under the same number flowered in Murrieta, Calif., April–May, 1974, and from this the present description is drawn, the pickled flowers showing much larger than the old dried originals.

Agave jaiboli is known from the outer Pacific slopes of the northern Sierra Madre Occidental in southeastern Sonora and adjacent Chihuahua at elevations between 300 and 1,000 m. Its small habitat is embraced by the upper slopes of the Short-tree Forest (see Gentry, 1942) and open grassy slopes of oak woodland, a nearly frostless zone with ca. 500 mm of annual rainfall. The open, rocky, grassy slopes of the Guajaráy country and the adjacent Sierra de la Ventana appear to be the heartland of this agave. It is, however, quite scarce. Only one to a dozen individuals or so have been observed in one locality, all as single, non-suckering rosettes (Figs. 14.19, 14.20).

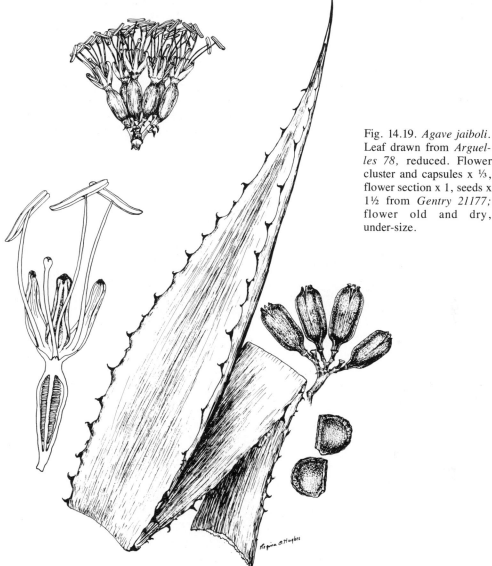

Fig. 14.19. *Agave jaiboli.* Leaf drawn from *Arguelles 78,* reduced. Flower cluster and capsules x ⅓, flower section x 1, seeds x 1½ from *Gentry 21177;* flower old and dry, under-size.

The flower structure and dimorphic teeth on the soft, fleshy, crenate leaves place *A. jaiboli* in Section Crenatae. The narrow, almost ensiform leaves distinguish this species from its relatives. From its Sonoran relative, *A. bovicornuta,* it is further distinguished by the unnarrowed haft of the leaves so characteristic of the latter. The natives have no difficulty in separating "jaiboli" or "temeshi," Warihio Indian names, from the unpalatable *A. bovicornuta.* Jaiboli is much esteemed for its sweet edible qualities.

Both the young flowering shoots and the headlike stem of jaiboli are cooked and eaten by the peoples of the Río Mayo country; the succulent flowering shoots are usually boiled, the heads pit-baked over hot stones and coals. The roving eye of the man with a ready machete is quick to discover a new shoot, and it takes but a moment to cut and lay it upon the shoulder to carry home. The species appears to become ever rarer, as it reproduces only by seeds, and my search for flowering specimens has been frustrated during

Fig. 14.20. *Agave jaiboli*. Mature rosette in natural habitat on east slope of the Sierra de la Ventana, Sonora.

the many years I have known the plant. The non-Indian people, ''gente de razón,'' living on Sierra de la Ventana frequently decapitate the flowering shoots, so that the rosette will remain green and mature until such time that it can be conveniently collected for making mescal. This is a common practice in many parts of Mexico.

The young plants are easily cultured and have grown well in my Murrieta collection, showing no ill effects of the 5 to 8 degrees of frost occurring there. However, rabbits and gophers are fond of the plants, and completely destroyed three of the plants introduced there. The cottontail rabbit, *Sylvilagus,* ate the leaves of half-grown plants down to the base. The food qualities of this plant appear to merit detailed investigation as a potential resource. It is becoming very rare.

Agave maximiliana
(Figs. 14.1, 14.2, 14.3, 14.21, 14.22, 14.23, 14.24; Tables 14.1, 14.2)

Agave maximiliana Baker, Gard. Chron. 1877: 201, 1877.
   Agave katharinae Berger, Agaven, 197, 1915.
   Agave conjuncta Berger, ibid., 158, 1915??

Rosettes single, nonsuckering, acaulescent to short-stemmed, medium-sized, freely seeding, balled or open, mostly pale glaucous green; leaves generally 40–80 x 10–20 cm, usually broadly lanceolate or oblanceolate, curved or straight or slightly recurved, soft fleshy, mostly pale glaucous pruinose over yellow-green to green, or bluish glaucous, the margin variously repand to crenate, mammillate; teeth heteromorphic, the larger in mid-blade 6–10 mm long, 1.5–3 cm apart, compressed, the cusps slender, variously flexed, from elongate low bases that are sometimes confluent, the interstitial

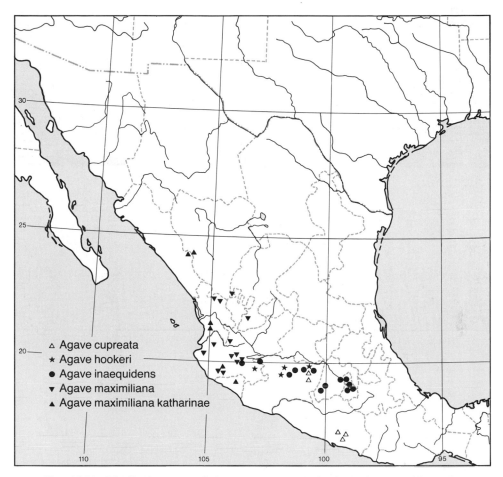

Fig. 14.21. Distribution map of *Agave cupreata, A. hookeri, A. inaequidens, A. maximiliana,* and *A. maximiliana* var. *katharinae.*

teeth numerous and variable; spine 2.5–4 cm long, straight, slenderly conic, smooth, brown or castaneous to gray, openly grooved above, rounded below and shortly protruding at base; panicles 5–8 m tall, forming a deep narrow outline with 15–25 or even 30 rather rounded small umbels in upper half of shaft; flowers 52–65 mm long, slender, greenish yellow, frequently rufous flushed; ovary 28–35 mm long, angular cylindric to fusiform, with short or long furrowed neck; tube 5–9 (–12) mm deep, 12–14 mm broad, openly funnelform, grooved; tepals sub-equal, ascending to incurved, 15–22 mm long, narrow, linear, conduplicate, roundly cucullate, the inner with high or low keel, involute; filaments 28–35 mm long, sometimes pink, inserted above mid-tube; anthers 20–24 mm long, yellow, regular, mostly centric; capsules (10476) 3.5–5 x 1.7–2 cm, short-oblong, stipitate, apex rounded (23525 = 6 x 2 cm, oblong); seeds 5.5–6 x 4.5–5 mm, the testa wavy, finely punctate, marginal wing abruptly raised.

Type: "Hort. Kew, Feb. 22 & April 5, 1881," K.

The type leaf, except for the atypical narrowness, conforms with spine and teeth characteristics to the Jaliscan populations herewith allocated. I attribute part of the leaf narrowness of the type to etiolation, as it was grown as a potted greenhouse plant, a

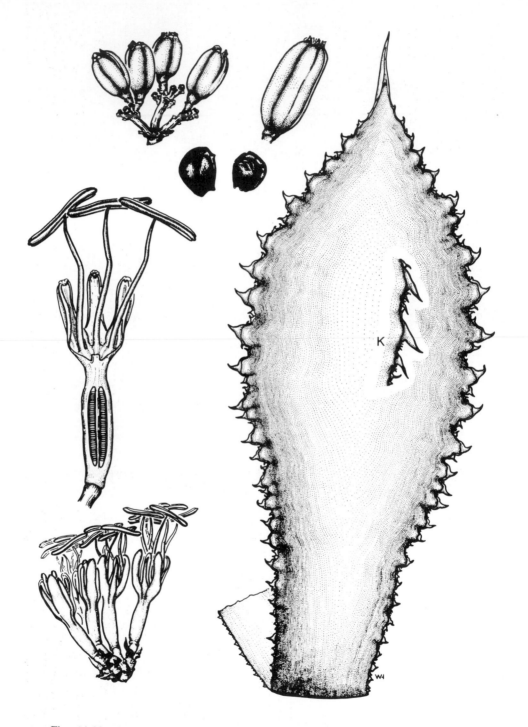

Fig. 14.22. *Agave maximiliana*. Leaf reduced from *Gentry 10950;* flowers from
Gentry 23500; capsules and seed from two other numbers. K, leaf margin inset of
var. *katharinae*.

Fig. 14.23. *Agave maximiliana*. A colony on open rocky slope in natural habitat with oaks near Chiquilistlán, Jalisco, 30 Jan. 1975.

Fig. 14.24. A single gray high-teated plant of *A. maximiliana* in the same locality.

practice still followed at Kew. Baker did not know or did not give the origin, but stated, ''Described from specimens seen at Kew and Reigate, the former sent by Mr. Justice Corderoy'' (Baker, 1888).

Agave maximiliana is commonly scattered on the rocky mountain slopes of the Jaliscan Plateau region, including Nayarit and southern Durango. It occupies elevations between 3,000 and 6,500 feet (930 and 2,000 m), rarely higher, commonly in oak and pine forest zones, on rather dry calcareous or igneous derived rocky soils. Most of the 30 to 40 inches (750 to 1,000 mm) of annual rainfall occurs during summer-fall; there is a marked dry season from January to May or June. Frosts are almost lacking, and *A. maximiliana* is sensitive to the light freezes of southern California and Arizona.

Berger described *A. conjuncta* and *A. katharinae* without flowers or fruits and their specific placement appeared uncertain. However, the type of *A. conjuncta* in US shows that he added flowers after his text was written; ''15 Sept. 1914. Flower from same plant'' (as leaf on separate sheet), ''cult. La Mortola 1913.'' The type represents a common form of *A. maximiliana* with regularly spaced teeth on undulate margin with very few interstitial teeth. The leaf of *A. katharinae* strongly resembles the large forms found in Nayarit and western Durango, which are striking in leaf margin forms with rich displays of assorted interstitial teeth (Fig. 14.25). These forms are generally larger plants with greener color, and at least *Gentry & Gilly 10805,* the only flowering specimen available, has a deeper flower tube (Fig. 14.3). It is here recognized as a variety in the variable complex of *A. maximiliana*.

Agave maximiliana var. *katharinae*
(Figs. 14.1, 14.3, 14.7, 14.21, 14.25; Table 14.2)

Agave maximiliana var. *katharinae* (Berger) Gentry, stat. nov.
  Agave katharinae Berger, Agaven 197, 1915.

Differs from the species with larger rosettes, greener leaves with more undulate-repand margins with numerous variable interstitial teeth, and deeper flower tube (Fig. 14.25).

Type: **Berger s.n.**, cult. La Mórtola, Ventimiglia, Italy, 1914, **US.** Consists of leaf margin, a peduncular bract, and leaf cross section.

Crenatae Exsiccatae

Agave bovicornuta

CHIHUAHUA. *Bye et al. 5806,* GH. Between Cusarare & Nopalero, Mun. Guazapares y Chinipas, ca. 3,500 feet, 8 Nov. 1973; associated with oaks.

Gentry 10236, 10252, US, MEXU, MICH, DES. High rocky rims around Arroyo Hondo, Sierra Charuco; pine-oak forest 5,000–5,500 feet elev., 6–7 March 1951.

Kimnach & Brandt 996, HNT. Ca. 3 km S of Parajes along Rio Cuiteco & Chihuahua-Mochis RR., 21 Mar. 1967; elev. 4,300 ft.

SINALOA. *Breedlove 19297,* CAS, DES. Canyon de Tarahumaras below La Jolla, Sierra Surutato; elev. 4,500 feet, slope with *Quercus, Ipomoea, Bursera, Lysiloma.* 9 March 1971 (l, f).

Gentry & Arguelles 18378, 18379, DES. Sierra Surutato; elev. ca. 6,000 feet; rocky crags in pine-oak forest. 13 Dec. 1959 (pls, f).

SONORA. *Barclay & Arguelles 2017,* US, MEXU, DES. Vicinity of Rancho San Antonio along road to Milpillas, Sierra Tecurahui, May 20, 1966. ''Abundant on slopes in Pine-oak woods'' (l, f).

Felger 376, ARIZ. La Tinaja, N side of Sierra de Alamos, 20 Feb. 1954; elev. ca. 3,000 ft. Short-tree forest & oaks (l, f).

Gentry 3672, CAS, ARIZ, DES. Type, Curohui, Sierra Curohui; rocky slopes in oak forest, elev. ca. 3,500 feet. 4 April 1938 (l, f).

Gentry et al. 21175. Sierra de la Ventana, 27 May 1965. Elev. ca. 3,500–4,500 feet; oak-*Nolina* grassland (photo, cap).

Gentry 11061. Barranca del Salto, Sierra de Alamos, 12 Aug. 1951 (l for anal.).

Kimnach 1261, HNT. San Pedro Canyon of Arroyo Cuchujaqui, 15 May 1969 (l, f, cult in Hunt. Bot. Gard.).

Fig. 14.25. *Agave maximiliana* var. *katharinae*. Detail of the living leaves and pale conal bud, near Cerro San Juan, Nayarit.

Nabhan et al. 956, ARIZ. DES. Pine pass above Los Algodones (Sierra Charuco), Río Mayo, 14 Sept. 1978; pine and oaks (l, cap).

Agave cupreata

GUERRERO. Langlasse 867, B, MEXU, US. Type. Sierra Madre, versant oriental, Michoacan & Guerrero, alt. 15–1,700 m, 15 Feb. 1899 (l, f).

Matuda 37248, MEXU. Cerca Rincon de la Via, 15 Oct. 1960, 1,700 m.

Moore & Valiente 6192, MEXU. Dry hills above Petaquillas, 28 March 1952 (f).

Ogden 5106, MEXU, US, DES, MICH. 15 km S of Chilpancingo in dry soil, 1,700 m elev., 3 Jan. 1951 (bud, l).

Ramirez L. s.n., MEXU. Barranca de Huacapa, Sep. 1932.

Verity s.n., HNT. 7 miles S of Chilpancingo, 31 Aug. 1964.

MICHOACAN. *Gentry 22553,* US, MEXU, DES. 2 miles S of Tzitzio, 15 Dec. 1967; rocky oak slopes, elev. ca. 4,250 feet. Abundant, all flowers in bud.

Gentry 22554, US, MEXU, DES. 6 miles N of Tzitzio, 15 Dec. 1967; rocky wash & ledges in open grassland, elev. ca. 6,000 feet (l, f).

Agave hookeri

JALISCO. *Gentry s.n.,* DES. SW of Jiquilpan along road from Colima, Nov. 1961; cult. in fence rows & tapped for pulque (photo).

MICHOACAN. *Gentry 10427,* US, DES. 6–10 miles W of Mil Cumbres along road to Morelia, 8 May 1951; mixed pine and deciduous forest; elev. ca. 7,500 feet.

Gentry 12311, DES. Cheran, along highway to Uruapan, 21 Oct. 1952; elev. ca. 7,000 feet. Cult. as fence plant (f, photo, pls).

Gentry et al. 19617, DES. Between Mil Cumbres & Morelia along route 15, 9 Dec. 1961; elev. 7,000–7,500 feet.

Gentry et al. 19629, DES. Cheran along road to Uruapán, 11 Dec. 1961; elev. ca. 7,000 feet. Cult. as fence plant (l, f, photo).

Gentry 22555, DES. 2–4 miles E of Uruapán road fork along route 15, 16 Dec. 1967 (cap).

Agave inaequidens

JALISCO. *Gentry 23499,* US, INIF, DES, MICH. Mt. slope above Amacueca along road to Tapalpa, 29 Jan. 1975; elev. ca. 6,500 feet, oak and second growth shrub (l, f).

MEXICO. *Bourgeau 1020, 1399,* M. Pedregal, Valle de Mexico, 1965–66.

M Garcia 83, MEXU. Xoxhitepec, D. F., 12 Aug. 1962.

Gentry 19612, US, DES. Neotype. 20 miles NE of Temascaltepec along road to Toluca, 7 Dec. 1961; open pine slope, elev. ca. 7,500 feet (l, f).

J. Guerrero s.n., MEXU. Cerro Gordo, cerca San Juan Teotihuacán, 8 April, 1962, alt. 2,500 m (l, f).

Moore & Cetto 5490, MEXU. Between Amanalco & Santa Maria Pipilotepec on road to Toluca, Valle del Bravo, 2 Nov. 1949 (f).

Rose & Hough 4513, MEXU. Pedregal near San Angel, 9 June, 1899.

Rose & Painter 8042, 8045, 8053, 8055, MEXU. Near Santa Fe, Valle de Mexico, 23 Aug. 1903 (l margins).

Hinton 7512, MICH. Chorrera, Dist. Temascaltepec. 13 March, 1935.

MICHOACAN. Gentry 10431, US, DES, MEXU, MICH. Neotype. Between Quiroga & Patzcuaro along highway, 9 May 1951. Transplant fl'ed. Murrieta, Calif. summer 1964 (l, f, cap).

Gentry 19615, 19616, 19617, US, DES. Between Mil Cumbres & Morelia along route 15, 9 Dec. 1961; elev. 7,000–7,500 feet, steep rocky slope (l, f).

Ogden & Gilly 51151, US, MICH, MEXU, DES. Along road to Patzcuaro, ca. 1 mile S of Tzintzuntzan. 9 March 1951 (l, f).

Ogden & Gilly 51179, US, MEXU, MICH, DES. About 11 miles W of Jiquilpán along road to Guadalajara, 3 April 1951.

MORELOS. *Gentry 19574,* US, MEXU, DES. Near Coajomulco along toll highway to Cuernavaca, 29 Nov. 1961; alt. ca. 8,500 feet, lava with pine and oak.

Gomez P. 340bis, MEXU. Bajada a Cuernavaca on lava, 11 March 1961.

Kimnach 363, HNT. Lava flow above Cuernavaca along toll highway to Cd. Mexico, 5 Aug. 1962 (f, l, cult. Hunt. Bot. Gard.).

Pringle 6677, GH. Pedregál, Serrania de Ajusco, 17 Aug. 1897, alt. 8,500 feet (l, f).

Rose & Hough 4384, GH, US. Near Cuernavaca, 27–30 May, 1899.

Agave inaequidens barrancensis

DURANGO. Gentry & Arguelles 22282, US, MEXU, DES. Type. Revolcaderos along Mazatlan-Durango road, 8 Dec. 1966; elev. ca. 6,500 feet, on cliffs and rocky sun slopes, pine-oak forest zone (l, f).

Gentry & Barclay 20474, US, MEXU, DES. 8–9 miles E of El Palmito along road from Durango-Mazatlán, 24 Sep. 1963; elev. ca. 7,000 feet, rocky volcanic slopes & cliffs (l, f).

Gentry & Gilly 10618, US, MEXU, DES. 15–17 miles NE of Palmito along Mazatlan-Durango highway, 16 June 1951; elev. 7,000–7,500 feet, pine-oak forest zone (l, pl).

Gentry & Gilly 10626, DES, US. Ca. 45 miles SW of El Salto along road to Mazatlan. 16 June 1951, Transplant fl'ed. in Gentry garden near Murrieta, Cal., Oct.–Nov. 1975 (l, f).

Mieg, McCleary & Diaz 89, DES. 55 miles from Durango along road to Mazatlan, June 1956, elev. ca. 8,000 feet.

Agave jaiboli

SONORA. *Arguelles 78,* DES. San Bernardo & vicinity, Río Mayo. May, 1959.

Gentry 21177, US, MEXU, DES. Type. Sierra de la Ventana, 27 May 1965, alt. ca. 3,500 feet, oak-*Nolina* grassland (l, cap, old f, pls).

Gentry 10256, DES, US, MEXU, MICH. Near San Bernardo, 8 March 1951; volcanic cliffs in short-tree forest.

Gentry 11374, US, DES. 2 miles NW of Conejos along Arroyo Guajaray, N of San Bernardo, 22–25 Dec. 1951; Savanilla on broken volcanic terrain with *Lysiloma, Quercus, Acacia,* etc. 2,500–3,500 ft.

Agave maximiliana

HORTICULTURE. "Hort. Kew, Feb. 22 & April 5, 1881." K. Type.

JALISCO. *Boutin & Kamnach 3068,* HNT. Along road to Sierra de Parnaso y La Cumbre, 2.1 mi. from road to Mascota, Feb. 1970 (l, f).

Gentry & Gilly 10476, 10476A, US, MEXU, MICH, DES. Plan de Barrancas, 19 May 1951; subtropical mixed woodland, elev. 2,500–3,500 feet (l, f).

Gentry 10920, US, DES. 10–12 miles E of Tecolotlan, 21 July 1951; cut-over oak woodland.

Gentry 10950, US. 8–10 miles SW of Autlan, 23 July 1951; cut-over oak woodland.

Gentry 23525, US, INIF, DES, MICH. 1 mile S of Aserradera Guisar, Sierra Manantlan, 5 Feb. 1975 (l, cap).

Gentry et al. 23552, US, MICH, DES, INIF. Sierra Cuale, ca. 7 miles NE of Tuito, 15 Feb. 1975; pine-oak forest with granitic rock, elev. ca. 3,200 feet (l, f).

Gentry 23500, US, MICH, DES, INIF. 5–6 miles by road SW of Chiquilistlan, 30 Jan. 1975; pine-oak slopes, elev. 5,700–6,000 feet (l, f).

McVaugh 26347, MICH. 3–10 km E on road to Mina de Cuale from hwy. junction 5 km NW of Tuito,

Mun., Cabo Corrientes; pine oak forest, elev. 850–1150 m (f, cap).

Nelson 4048, GH. Dry hills between Mascota & San Sebastian, March 14, 1897; alt. 4,000–5,000 feet (l, f).

Ogden & Gilly 51172, US, DES, MEXU. At km 98 along road between Cocula & Tecolotlán, 11.5 miles from Tecolotlán, 2 April 1951.

NAYARIT. *Gentry 23466,* US, MICH, INIF, DES. A few miles E of San Juan Peyotan, Sierra de los Huicholes, 13 Jan. 1975; pine-oak woodland on chalky volcanics, elev. ca. 6,400 feet.

Rose 2196, US, GH. Near Santa Teresa, Aug. 1897 (l, f).

ZACATECAS. *Gentry 23470,* US, MICH, DES, INIF. 11 miles W of San Juan Capistrano, Sierra de los Huicholes, 13 Jan. 1975. Elev. ca. 6,500 feet.

Agave maximiliana katharinae

COLIMA. *Gentry 23535,* US, INIF, DES, MICH. Sierra Ninatitlan, 6 miles E of Ninatitlán, 10 Feb. 1975; oak forest, iron ore slope, elev. 4,000–4,500 feet (l, f).

DURANGO. *Hutchison 2486,* UC. 2.9 miles E of Palmito, 8 June 1962; elev. 6,900 feet.

JALISCO. *Gentry 18309,* US, DES, MEXU. 9 miles W of Autlan, 3 Dec. 1959; oak woodland, alt. ca. 4,600 feet (l, f).

NAYARIT. *Gentry & Gilly 10751,* US, MEXU, MICH, DES. 14–17 miles W of Tepic along road to Jalcocotan, 5 July 1951; mixed tropical forest bordering oak woodland, elev. 3,000–4,000 feet.

Gentry & Gilly 10805, US, MEXU, MICH, DES. 4–8 miles SW of Jalisco, S of Cerro San Juan, 9 July 1951; pine-oak forest over volcanics, elev. 4,500–5,000 feet (l, f, photos).

SINALOA. *Bye 7562,* COLO, DES. W of El Alcazán, ca. 6 miles from Dur. border, 19 July 1977; elev. 1,840 m, pine-oak forest. "Few scattered plants on partially open W slope."

Gentry & Gilly 10530, 10520, US, DES. 3–4 miles E of El Batel along highway Mazatlan to Durango, 28 May 1951; oak & pine forest, south slopes, elev. 6,000–6,500 feet.

Hutchison 2481, 2485, UC. 3–8 miles W of Palmito along road from Mazatlán to Durango, 8 June 1962; elev. 6,600–7,700 feet.

15.

Group *Deserticolae*

Trelease, Missouri Bot. Gard. Rep. 22:45. 1912.

Plants small- to medium-sized, glaucous gray to greenish, freely suckering, or medium-sized to large, green, nonsurculose, the rosettes acaulescent to short caulescent; leaves rigid, coarsely fibered, with thick cuticle, narrowly lanceolate and with weak, easily detached teeth, or broader and with firmer teeth; panicle narrow with short lateral branches, dry scarious peduncular bracts, and small umbellate flower clusters; flowers small with very short open tube, the tepals about equal and 3 to 5 times as long as the tube; spring flowering; capsules small to medium, freely seeding. Sonoran Desert region in southeastern California, Arizona, Baja California, and Sonora.

Typical species: *Agave deserti* Engelm.

The leaf surface in the Deserticolae is generally light colored, glaucous gray or yellowish or light green, minutely sculptured, the cuticle rather thick and rough with variable papillae. The stomata are relatively dense, ranging from 30 to 50 per mm^2 on the upper leaf surface. For further particulars on epidermal morphology, the reader is referred to the companion study (Gentry and Sauck, 1978).

The Deserticolae do not display close morphological relation to any other *Agave* group. Certain other taxa, such as *A. neomexicana* and *A. gracilipes* of the Parryanae, are also small-leaved xerophytes, and the latter has an ideograph reflecting its short tube, much like the Deserticolae. However, the geographic position and other morphological characters of *A. gracilipes* recommend its inclusion in the Parryanae. The Marginatae are also similar to the Deserticolae with their very short tubes and relatively long tepals, but they belong in the subgenus *Littaea* and have other important distinguishing characters, such as their marginate leaves and filament-clasping tepals, while their geographic distribution is distinct.

Agave deserti deserti occupies a centric position in Deserticolae, certainly in physiographic position, and interpretively in morphological relations. Several of the taxa, such as *A. pringlei* and *A. deserti,* are too close to separate specifically, while others, such as *A. sobria* and *A. moranii,* are quite distinct. The latter, along with *A. avellanidens* and *A. gigantensis,* form a distinctive group to themselves, again based mainly on rosette and habit characters. They could well be treated as a subsection, as Berger (1915) and

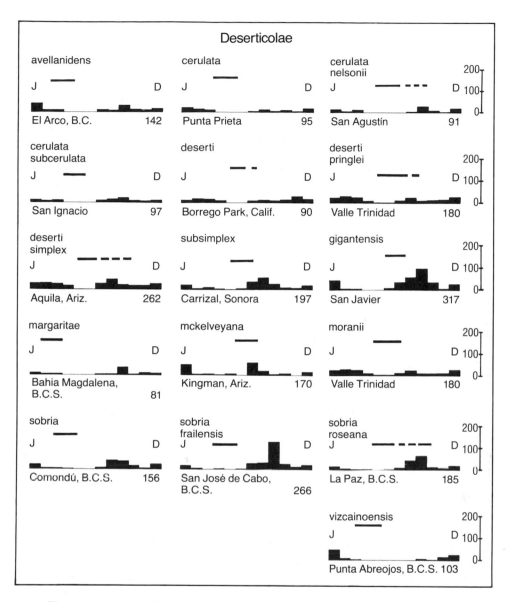

Fig. 15.1. Rainfall (silhouettes) and flowering (bars) perimeters of the Deserticolae. Relevant meteorological stations with average annual rainfall in millimeters. Data from Atlas Meteorológico de Mexico, Servicio Meteorológico Mexicana 1939; Hastings et al. 1959; U. S. Weather Bureau. Flowering periods based on herbarium specimens and field observations, supplemented by plants in cultivation. Uncertainty expressed by broken lines.

most other German taxonomists would probably have done upon cognizance of the groups. However, I prefer to keep the groups simple wherever possible. The Deserticolae forms a natural phylogenic group on both morphological and historical grounds.

Flower structure is quite stable or uniform throughout the Deserticolae (Figs. 15.2, 15.3). The differences or variations in the floral organs are not always critical for defining

Deserticolae

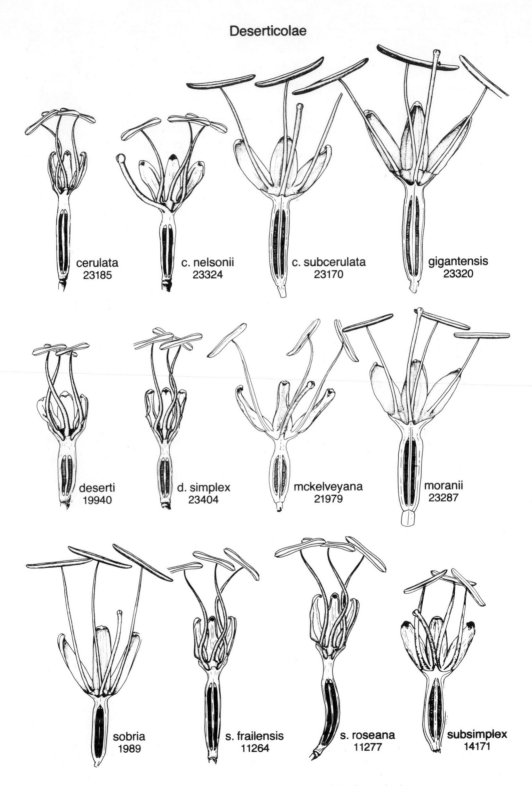

Fig. 15.2. Flower sections representative of the Deserticolae.

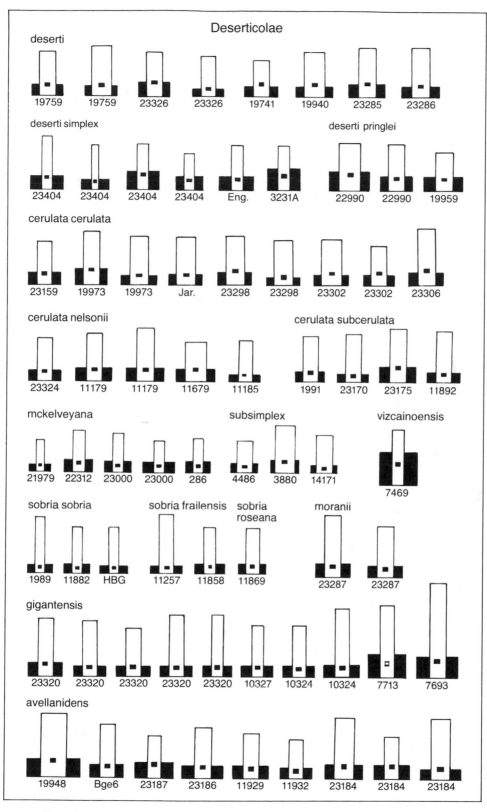

Fig. 15.3. Floral ideographs of the Deserticolae. Measurements listed in Table 15.1.

Table 15.1. Flower Measurements in the Deserticolae (in mm)

Taxon & Locality	Ovary	Tube	Tepal	Filament Insertion & Length		Anther	Total Length	Coll. No.
avellanidens								
Paraíso, BCS	31	6×15	16×4.5	6	38	18	52†	23187
Mesquitál, BCS	{ 34	5×14	19×6	5	46	18	57†	23186
	18	4×12	15×5	4	35	15	37†	11932
Calmallí, BCS	35	6×15	22×7	6	45	23	65†	23184
cerulata ssp. *cerulata*								
Catavinya, BCN	24	4.5×12	16×6	4–4.5	30	14	45†	23159
Laguna Chapala, BCN	31	6×12	20×6	6	32	19	57†	19973
San Luis Gonzaga Bay, BCN	26	3×13	17×7	3–4	35	16	45†	23298
Punta Prieta Valley, BCN	24	4×13	18×8	5	42	18	45†	23302
c. ssp. *nelsoni*								
S of San Miguel, BCN	{ 35	3×12	17×4	3	40	17	55†	23324
	27	5×13	21×7	5	30		54†	11179
c. ssp. *subcerulata*								
W of San Ignacio, BCS	28	2.5×12	18×5	3–2	34	18	48†	23170
E of San Ignacio, BCS	24	5×13	20×6.5	5	36	20	48†	23175
Isla San Marcos, BCS	27	3×12	19	3	33	20	50†	11892
deserti ssp. *deserti*								
Pinyon Flats, Cal.	{ 33	4×12	17×6	5	36	17	53†	19759
	31	3×11	15×4	3	37	16	50†	23326
San Felipe, Cal.	26	4×15	17×8	6	28	15	48†	19940
San Felipe, BCN	38	4.5×12	18×7	4.5	40	18	60†	23286
d. ssp. *pringlei*								
San Matias Pass, BCN	{ 24	6×12	14×5	6.5	20	14	43†	19959
	22	7×15	16×7	7	21	15	44*	19959
d. ssp. *simplex*								
Harquahala Mt., Ariz.	{ 25	6×13	20×3–4	5–6	38	21	50†	23404
	26	7×12.5	17×4.5	5–6	40	20	50†	23404
Providence Mt., Cal.	28	8	15	5 6		17	52†	3231A
gigantensis								
Sierra Palmas, BCS	{ 30	5×13	22×5	5	44	21	56†	23320
	25	4×12	23×5	4	25	24	51†	23320
	27	5×12	18×5	5	37	20	50†	10324
mckelveyana								
Black Mountains, Ariz.	17	3×8	12×3	2–3	28	12	30†	21979
Hualapai Mt., Ariz.	21	4.5–5	15×5	4	25	16	40†	22312
S of Wikieup, Ariz.	21	4×12	12×4	2–3	29	12	36†	23000

Table 15.1. cont.

Taxon & Locality	Ovary	Tube	Tepal	Filament Insertion & Length		Anther	Total Length	Coll. No.
moranii								
Bajada of Sierra San	41	6×13	23×7	6–5	46	21	70†	23287
Pedro Mártir,	40	5×13	24×6	5	43	21	68†	23287
BCN	26	4×13	19×7	4.5	32	17	49†	23287
sobria ssp. *sobria*								
Comondú, BCS	25	3×9	21×3	3	47	22	48†	1989
	29	4×10	17×3	3	35	20	49†	11882
s. ssp. *frailensis*								
Punta Frailes, BCS	37	3.5×12	19×4	3	40	20	58†	11264
	31	2–3×12	24×6	2–3	38	23	57†	11257
	24	4×11	17×4	4	32	18	43†	11858
s. ssp. *roseana*								
Espíritu Santo, BCS	40	5×12	22×5	4–5	42	23	66†	11277
La Paz, BCS	26	3.5×11	17×5	3.5	29	19	46†	11869
subsimplex								
Libertad, Son.	23	5×11	17×8	4–5	30	16	45†	3880
Desemboque, Son.	25	3×10	14×7	2	28	15	43†	14171
vizcainoensis								
Cerro Tordillo, BCS	39	13×15	21×4	9 & 8	50	21	72*	7469
Picachos Santa Clara, BCS	35	8×15	26×5	6	63	25	68*	7713

* Measurements from dried flowers relaxed by boiling.

† Measurements from fresh or pickled flowers.

species, as I had hoped before the respective series of flowering specimens were assembled. Variation in flower structure was sometimes found to be as great within local populations as it was between populations judged to be separate species on other criteria. This is the case, for instance, with *A. deserti* and *A. cerulata*. These two complexes are difficult to separate by morphological characters, although respective populations appear distinct when viewed overall in the field. The chemical compositions of sapogenins appear to substantiate their specific separation (Tables 15.2, 15.3).

However, there are some cases where flower size, proportions, and shapes were correlative with vegetative characters and were useful for separating species, as with the small flowers and narrow tepals of *A. sobria* and *A. mckelveyana*. Leaf form, size, habit, and other variation patterns, combined with geographic distribution, are the principal criteria used in separating the taxa.

The Deserticolae occupy a large area around the Gulf of California, nearly throughout the Sonoran Desert. Species distributions are given in the several maps (Figs. 15.9, 15.19, 15.22, 15.32). They are a strong succulent element characterizing the xerophytic vegetation and have been important hosts to many animals, especially pre-agricultural man. The rainfall silhouettes in Fig. 15.1 show the meager rainfall for all the species and the respective flowering seasons. The flower sections of Fig. 15.2 and the corresponding ideographs, Fig. 15.3, illustrate the close relationships of the group. The measurements of Table 15.1 document the accuracy of the illustrations and the author's conclusions.

Key to the Species of the Deserticolae

1a. Plants surculose; leaves small, usually narrow, less than 10 cm wide, light glaucous gray to yellowish; teeth frequently fragile; panicles short, mostly with only 8–15 umbels   2

1b. Plants not surculose; leaves broad, 10–25 cm wide, green to glaucous gray; teeth firmly attached; panicles elongate with 20–40 lateral umbels   8

2a. Plants of the northern Sonoran Desert; leaves generally gray to light green, broader, less acuminate   3

2b. Plants of peninsular Baja California; leaves generally yellowish to gray, narrower, more acuminate; teeth with a brown ring around base, or leaves larger, frequently cross-zoned   5

3a. Leaves larger, mostly 25–40 x 5–10 cm; teeth very feebly attached; spine decurrent for several cm and confluent with uppermost teeth; flowers 40–60 mm long; filaments inserted on base of tepals or in orifice of tube
deserti, p. 376

3b. Leaves smaller, 3–5 cm broad at mid-blade; teeth feebly or firmly attached; spine very shortly decurrent and not confluent with uppermost teeth; flowers 30–50 mm long; filaments inserted in orifice of tube   4

4a. Leaves 18–30 x 3–5 cm; teeth firmly attached; flowers 30–40 mm long; flowers yellow; tepals 3–4 mm broad in middle, conduplicate, strongly cucullate; ovary constricted below tube. Northwestern Arizona
mckelveyana, p. 390

4b. Leaves 15–25 x 3–6 cm; teeth weakly attached; flowers 40–50 mm long; flowers partly pink or red; tepals 6–8 mm broad in middle, plane or rounded, slightly cucullate; ovary not constricted below tube. Coastal Sonora
subsimplex, p. 404

5a. Leaves small, mostly narrow, long-acuminate, 3–6 cm broad (rarely 7–8 cm); yellowish to light glaucous gray; teeth relatively small, mostly 3–5 mm long, very weakly attached and ringed by a brown margin below base. Mid-peninsular region   *cerulata*, p. 363

5b. Leaves larger, or at least broader, ovate to linear-lanceolate, mostly 5–12 cm broad at mid-blade, green to gray glaucous; teeth usually large, 8–20 mm, variously flexed, firmly attached, not ringed by a brown basal margin   6

6a. Flower tube shallow, 3–5 mm deep; leaves mostly 40–60 cm long, sometimes conspicuously cross-zoned. Sierra de la Giganta region
sobria, p. 396

6b. Flower tube 8–12 mm deep; leaves short, usually less than 50 cm long, not cross-zoned. Outer coastal region   7

7a. Leaves short, less than 25 cm long, ovate to oblanceolate, the margin prominently mammillate and with long (8–15 mm) slender teeth
margaritae, p. 389

7b. Leaves 25–40 cm long, lanceolate, the margin nearly straight to undulate, with shorter, more flattened teeth
vizcainoensis, p. 407

8a. Leaves 70–120 cm long, triangular long-lanceolate, deeply guttered, the margin nearly straight; terminal spine subulate, light gray; peduncle sometimes swollen below the panicle. South and east slopes of the Sierra San Pedro Mártir   *moranii*, p. 392

8b. Leaves 40–70 cm long, broadly lanceolate, plane to concave above, margin lightly to deeply sinuate; spine coarser, dark brown to dark gray, frequently sinuous; peduncle not swollen below panicle   9

9a. Leaves green, broadly linear-lanceolate, not or scarcely narrowed towards the base; panicle generally occupying one-half the shaft with 25–35 laterals; flowers yellow to orange. Central pensinula *avellanidens*, p. 361

9b. Leaves light green to glaucous gray, ovate acuminate to spatulate, conspicuously narrowed toward the base; panicle in upper one-third of shaft with 18–25 laterals; flower buds whitish or pale green, opening pale yellow. Southern peninsula *gigantensis*, p. 386

Agave avellanidens
(Figs. 15.1, 15.3, 15.4, 15.5, 15.22; Table 15.1)

Agave avellanidens Trel. Missouri Bot. Gard. Rep. 22:60, 1912.

Single, medium to large, multi-leaved, green rosettes, 6–12 dm tall, 10–15 dm broad, with stems to 5 dm long; leaves 40–70 x 9–14 cm, broadly linear-lanceolate to ovate, little or unnarrowed base, short-acuminate, thick-fleshy, rigid, smooth, green, the margin straight or undulate and frequently corneous; teeth rather regularly spaced, mostly 1–3 cm apart, but variable in size and curvature, 5–15 mm long, straight or variously flexed, flattened, dusky gray over brown; spine strong, conical, 2.5–4.5 cm long, broadly grooved above, brown to grayish, strongly decurrent as a corneous margin; panicle 4–6 m tall with 25–35 lateral branches of dense, large, globose umbels of small flowers; flowers

Fig. 15.4. *Agave avellanidens* west of Calmalli. The deep narrow panicles and small bracts distinguish it from *Agave shawii goldmaniana*. Photograph by Bruce Gentry.

Fig. 15.5. *Agave avellanidens;* upper, a series of "normal" leaves; lower, a form with bizarre armature of 2-3-cuspid teeth.

pale yellow, drying orange-yellow, 40–70 mm long, slender; ovary 20–40 mm long, fusiform, tapering to base, the neck sometimes constricted, pale yellow; tube 4–6 mm deep, 13–15 mm wide, wide-spreading, ridged within and grooved without from overlapping tepal sinuses; tepals ca. equal, 16–24 mm long, linear-lanceolate, obtuse, the inner with thick keel and sometimes lobate within an inlapping base; filaments slender, 30–45 mm long, inserted on rim of tube; stamens 15–22 mm long, yellow, centric; capsules dark brown, broadly oblong, 20 x 35 mm, not stipitate and scarcely beaked; seeds unknown.

Type: *Brandegee 6,* Paraíso (S of San Borja), Baja California, 1 May 1899, UC (l, f, cap).

Agave avellanidens is a *shawii*-like member of the Deserticolae. The elongated many-leaved green rosettes look like *A. shawii goldmaniana* where the two meet and mingle southeast of Punta Prieta. This relationship has been discussed under *goldmaniana,* group Umbelliflorae. When in flower, *A. avellanidens* is distinguishable by the long, narrow, small-bracted panicles of small flowers with short open tubes. These inflorescence characters align it with the Deserticolae rather than with Umbelliflorae, regardless of the shawiian appearance.

The distribution of *A. avellanidens* is poorly known. So far as known, it is limited to a small area in mid-peninsula between latitudes 20° and 28° N. The type locality, Rancho Paraíso, is several miles south of the Misión San Borja in the mountains and apparently has not been visited by any botanist except Brandegee, who was following mule trails. The species is abundant along the old auto road from Calmallí to Rancho Mesquitál (Fig. 15.5). *Agave shawii goldmaniana* is scattered abundantly along this old road north of Rancho Mesquitál to Punta Prieta, but occasional elongated panicles appear to represent *A. avellanidens*. Certain sightings were also made along the road from San Borja to Rancho Rosalito. *Agave avellanidens* and its associates of *Fouquieria, Yucca,* and *Pachycormus* comprise many exotic and beautiful plantscapes. Perhaps *A. avellanidens* is extensive in that large unknown area about and south of the Sierra Calmallí east of the main road.

Agave cerulata

Agave cerulata, ssp. *cerulata*
(Figs. 15.1, 15.2, 15.3, 15.6, 15.7, 15.8, 15.9; Tables 15.1, 15.2)

Agave cerulata Trel. Missouri Bot. Gard. Rep. 22:55, 1912.

Small yellow to pale green, rarely light glaucous gray, abundantly surculose, few-leaved rosettes 25–50 cm tall; leaves mostly 25–50 x 4–7 cm, long acuminate, narrowly lanceolate to triangular-lanceolate, yellow to light green, sometimes cross-zoned, the margins nearly straight to mildly undulate with low teated teeth; teeth small, 1–4 mm long, irregularly spaced, sometimes lacking through much of the blade, grayish brown, bordered with a brown ring at base, weakly attached; spine 3–6 cm long, acicular, light gray to dark gray, decurrent only to uppermost teeth or less; panicle slender narrow, 2–3.5 m tall, with mostly 6–12 small lateral umbels, the whole white waxy glaucous in bud stage; bracts small scarious triangular; flowers waxy white in bud, opening pale yellow, mostly 45–60 mm long; ovary 22–32 mm long, fusiform, narrowly tapered toward base; tube 3 x 11 mm to 5 x 14 mm, broadly funnelform or discoid with thick nectary and prominent bulges opposite filament insertions; tepals 16–22 mm long, ascending to spreading, equal, elliptic, the inner wide with low broad keel and broadly overlapped at base by outer; filaments inserted at base of tepals on rim of nectary, variable in length,

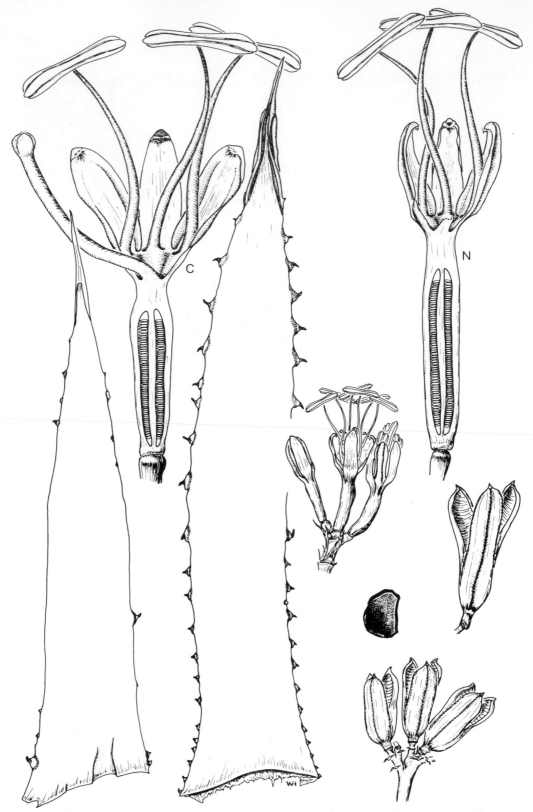

Fig. 15.6. *Agave cerulata* ssp. *cerulata*, C drawn from topotypic *Gentry 23185, 11924;* leaves, flower cluster, capsules x ½, one unusually elongate, flower section x 1.6. Smaller leaf shows a recurrent nearly toothless form. N, flower section of *A. cerulata* ssp. *nelsonii* x 1.6.

Fig. 15.7. Upper, fairy ring of *Agave cerulata* on calcareous pebble pavement in northwest corner of San Andres Valley. Bruce Gentry photograph. Lower, *A. cerulata* on rolling plain near Laguna Seca Chapala, May 1952.

Fig. 15.8. Upper, *Agave cerulata* near Laguna Seca Chapala. The dense clusters of rosettes with long acuminate leaves are characteristic. Lower, east of Punta Prieta along road to Bahía Los Angeles, *A. cerulata*, as a robust, almost white glaucous form with cross-zoned leaves, May 1952.

usually incurved at anthesis, 30–40 mm long; anthers 15–20 mm long, centric to excentric; capsules 3–5 x 1.2–1.3 cm, narrowly oblong, waxy light gray, narrowly stipitate, bluntly apiculate; seed 5 x 3 mm, lunate, sooty black, the hilar notch open or obscure, the marginal wings pronounced on both sides around the curvature.

Type: *Nelson & Goldman 7180,* Calmallí, Baja California, 29 Sept. 1905, US.

Characteristics and Habitat

Most plants of *A. cerulata* are characterized by slender, yellow, long-acuminate, lanceolate leaves with brown eyelets ringing the weakly attached, moderate to small teeth, small narrow panicles whitened from peduncle to capsules with a waxy bloom, and light-yellow spreading tepals. *A. cerulata* has been confused with its near relative *A. deserti,* but the latter is more robust, the leaves light gray-green rather than yellow-green, 4–7 times longer than broad (vs. 5–12 times longer than broad). *A. deserti* generally lacks the distinctive brown eyelets ringing the teeth and the capsules are generally broader and, although glaucous, lack the waxy white bloom. The geographic distributions of the two species are distinct (cf. Figs. 15.9 and 15.19), although they may meet on the peninsula south of Sierra San Pedro Mártir in an area I have not explored.

Agave cerulata is primarily a product of the upland maritime environment where frequent fogs and ocean breezes temper the desert climate, while *A. deserti* is a product of the hot arid continental desert with high insolation and more extreme circadian temperature changes. Figures 15.7 and 15.14 show the two species in their characteristic respective habitats. There are significant chemical differences between these two species, as may be inferred from Tables 15.2 and 15.3.

Chemistry

The data in Table 15.2 on the sapogenin contents of *Agave cerulata* are the most detailed for any *Agave* species and are assembled from U.S.D.A. sources (Wall, 1954; Wall et al., 1954a, 1954b, 1955, 1957). Of the four genins present, hecogenin and manogenin are the more prevalent. Hecogenin occurs in 89 percent, manogenin in 82 percent, tigogenin in 24 percent of the 46 plant samples. These four compounds shift about quite at random, as there is little apparent correlation with time of year, locality, plant part, or stage of plant maturity. The three cases in the table of "leaf, dead" represent dry leaves from rosettes that had flowered and died. However, it is apparent that total sapogenin content is higher in December than during other months. The average content of the 13 leaf samples of December is 1.5 percent, or about double the average content of 0.66 percent of the 28 other seasonal leaf samples.

The average sapogenin content of *Agave cerulata cerulata* is 0.87 percent, of *A. cerulata nelsonii* is 0.96 percent, and of *A. cerulata subcerulata* is 0.98 percent. This is nearly 1 percent and contrasts strongly with 0.25 percent for *A. deserti.* Although samples of the latter are few, they show that there are chemical and physiological differences between *A. deserti* and *A. cerulata* as well as morphological and geographic ones.

Populations

The populations of *Agave cerulata* are vast, extending through northern and central Baja California from just above latitude 30° to ca. 27°N (Fig. 15.9). The clustering clones must number over a million and the individual rosettes many millions more— a wild species population probably exceeded only by *Agave lechuguilla* of the northern desert region of Mexico. *A. cerulata* is polymorphic, showing several forms at many localities. An example of this was recorded on May 17, 1952 near Calmallí, the type locality. The forms noted here were itemized as follows, with Gentry collection numbers in parentheses:

    A. Rosettes large, leaves 45–60 cm long; flg. stalk to 6–7 m (*11919*).

    B. Size medium, glaucous-yellow leaves (*11921*).

    C. Size small, leaves thick (*11922*).

    D. White glaucous with slender narrow leaves (*11923*).

    E. Thick-headed glaucous rosettes with part of the margins toothless (*11924*).

    F. Short, thick-leaved, yellow-green with numerous proximal teeth (*11926*).

Corresponding forms as well as other striking variants were noted in other localities. The extensive collections and detailed study necessary to organize a rational taxonomy of these genetic forms has not been attempted. However, the three geographic subspecies set forth below will clarify some relationships.

    Agave nelsonii Trel. has long been considered conspecific with *A. deserti* (Johnston, 1924; Shreve and Wiggins, 1964) and indeed is very similar in habit, in the gray glaucous leaves, and in other characters. However, the ceriferous inflorescence of *A. nelsonii,* the brown eyelets ringing the teeth, its chemistry and its geographic distribution relate it with the *A. cerulata* complex. I am accordingly treating it as a subspecies of *A. cerulata.* This and two other variants are separable according to the following key and descriptions.

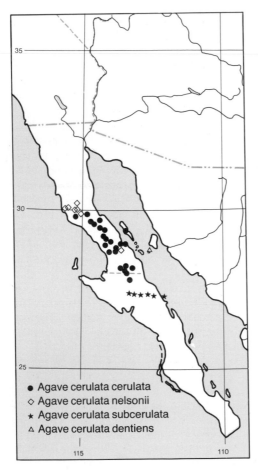

Fig. 15.9. Distribution of *Agave cerulata* and its subspecies, based on herbarium specimens.

Key to Subspecies
of *Agave cerulata*

1a. Leaves yellowish or light glaucous, long-acuminate, mostly 6–12 times as long as broad, the margins nearly straight to mildly undulate; teeth small, fragile; spines acicular 3–6 cm long   2

1b. Leaves mostly light-gray to bluish glaucous over green, short-acuminate, mostly 3–6 times as long as broad, the margins undulate to mammillate; teeth larger, firmer; spines subulate, 2–4 cm long   3

2a. Leaves generally yellowish, 25–50 cm long; panicles narrow, the lateral branches 15–20 cm long on upper ⅓ to ¼ of shaft   *c. cerulata*

2b. Leaves light glaucous gray, 40–55 cm long; panicles broad, the lateral branches 30–40 cm long on upper ½ of shaft. Isla San Esteban   *c. dentiens*

3a. Leaves larger, 25–40 cm long, the margins undulate with small teats or nearly straight. Sierra San Miguel to San Agustín   *c. nelsonii*

3b. Leaves smaller, 15–28 cm long, the margins conspicuously crenate with prominent teats. Vicinity of San Ignacio to San Marcos Island   *c. subcerulata*

Edibility of A. cerulata

Considering the relatively high sapogenin contents, *Agave cerulata* is probably not among the best for eating. However, it was eaten by the Cochimi Indians, a fairly large tribe inhabiting the central part of the peninsula. They ate the thick stem or *cabeza,* which contains less sapogenins than the leaves. Wenceslaus Linck, a Jesuit missionary, founded the Misión de San Francisco Borja during the 5th decade of the 18th century. In 1766 he set out with thirteen soldiers and a band of missionized Indians to explore northern Baja California to the mouth of the Colorado River (Burrus, 1966). He traveled over 480 km through the heartland of *Agave cerulata*. On the second day northwest of San Borja at Vimbet he wrote, "A native chieftain and his entire settlement of Nuestra Señora de Guadalupe had prepared for us thousands of *mescales,* a food which is their daily bread and their favorite sustenance."

Agave cerulata ssp. *dentiens*
(Fig. 15.9)

Agave cerulata ssp. *dentiens* (Trel.) Gentry, Cal. Acad. Sci. Occ. Pap. No. 130: 43, 1978.

   Agave dentiens Trel. Missouri Bot. Gard. Rep. 22:51. 1912.

Medium-sized, acaulescent, surculose, open, few-leaved, green to light glaucous gray rosettes, 50–70 cm tall, 8–15 dm wide, forming dense clumps; leaves 40–70 cm long, 4–7 cm wide in the middle, wider below, triangular-long-lanceolate, thickly crescentic at base, concave above towards apex, rigid, light gray glaucous, sometimes bluish, cross-zoned, the margins straight, usually set with small, weak, friable teeth 1–2 mm long, or margin nearly toothless; spine acicular, 3–5 cm long, brown to gray, narrowly, shortly channeled above, shortly decurrent; panicle 3–4 m tall, open, with 8–18 wide-spreading lateral branches in upper half of shaft; buds ceriferous, pale yellow, in small congested umbels; flowers pale yellow, slender, 49–53 mm long; ovary fusiform, 32–35 mm long with slender neck, constricted at base; tube shallow, open 3 x 10 mm swollen with nectaries below filament insertions, grooved; tepals ca. equal, 15–16 mm long, 4–5 mm wide, ascending-spreading, occasionally recurving, conduplicate and involute with anthesis, rounded and shortly hooded at apex; filaments slender, 28–30 mm long, bulbous at base with nectaries; anthers 15–16 mm long, yellow, centric; capsules glaucous ceriferous, 40–50 x 15–20 cm, oblong, short-stipitate, short-beaked. (Flower description from *Felger s.n.* San Esteban Island, June 20, 1977.)

   Type: *Rose 16819,* San Esteban Island, Baja California 12 Apr. 1911, MO, US.

There is little doubt that *Agave dentiens* is closely related to *A. cerulata*. The old frayed capsules on the type specimen show a waxy bloom like that of *A. cerulata,* while the long, narrow, acuminate leaf-form and small, weak, brown-eyeleted teeth are additional evidence for *cerulata* affinity. The random occurrence of toothless margins is also a *cerulata* character. A long-leaved, light-glaucous variant (*Gentry & Fox 11953*) of *cerulata* (Fig. 15.8) found on the peninsula between Punta Prieta and Bahía de Los Angeles looks like the island *dentiens*. The satellitic-island position of *dentiens* in relation to the peninsular *cerulata* populations is very similar to that of *A. sobria roseana,* here also treated as a subspecies. Moran's color photographs of collection *4079* on San Esteban Island show a generally uniform population with rather consistent characters of a wide-branching panicle and tall, long-leaved, light gray-glaucous, narrow rosettes. This is in obvious distinction from the narrow panicles and usually small, yellowish rosettes of peninsular *cerulata*. But this difference seems to be at the subspecific level. It is possible that the agaves on Angel de la Guarda Island should be aligned with *A. cerulata dentiens,* but until flowering specimens of the respective populations are assembled, the disposition

Table 15.2 Sapogenin Content in *Agave cerulata*
(given in percentages on dry weight basis)

COLL. NO.	BAJA CALIFORNIA SOURCE LOCALITY	MONTH COLL.	PLANT PART	SAPOGENINS*				
				TOTAL	HEC.	GIT.	MAN.	TIG.
cerulata ssp. *cerulata*								
10346	Calmallí	May	leaf (white)	1.2	25		68	7
10346	Calmallí	May	leaf (yellow)	1.1	65		35	
10346	Calmallí	Dec.	leaf (white)	1.1	x	x	x	x
10346	Calmallí	Dec.	leaf (yellow)	0.5	x	x	x	
10346	Calmallí	Dec.	leaf (dead)	1.2	x		x	
11322	Laguna Chapala	Apr.	leaf	1.1	20		45	x
11322	Laguna Chapala	Dec.	leaf	1.0	x	x	x	x
11322	Laguna Chapala	Dec.	leaf (dead)	0.5	x		x	
11322	Laguna Chapala	Dec.	leaf	0.4	50		50	
11322	Laguna Chapala	May	leaf	1.2	35		65	
11324	Laguna Chapala	Apr.	leaf	0.1	x		x	
11188	Laguna Chapala	Sept.	leaf	0.8	69	31		
11740	Sierra San Luis	Apr.	leaf	0.6	80			20
11741	Sierra San Luis	Apr.	leaf	0.72	60		40	
11744	Sierra San Luis	Apr.	leaf	0.9	7			93
11746	Sierra San Luis	Apr.	leaf	0.6	20		70	10
11919	Calmallí	May	leaf	0.6	40		60	
11926	Calmallí	May	leaf	0.6	45		55	
11924	Calmallí	May	leaf	0.8	55		25	20
11962	Laguna Chapala	May	leaf	0.1	100			
11962	Laguna Chapala	Dec.	leaf	3.4	x	x	x	x
11953	E of Punta Prieta	May	leaf	0.4	30		70	
cerulata ssp. *nelsonii*								
10370	San Fernando	Sept.	leaf	1.2			100	
10370	San Fernando	Sept.	stem	0.2	70		15	
10370	San Fernando	Apr.	leaf	0.7	5			95
10370	San Fernando	Apr.	leaf	0.25				100
10370	San Fernando	Dec.	leaf	1.3			100	
10370	San Fernando	Dec.	stem	0.6			100	
10370	San Fernando	Dec.	leaf	1.0	x	x	x	x
10370	San Fernando	May	leaf (dead)	1.1	x	x	x	x
11160	San Fernando	Sept.	leaf	0.77	75		25	
11160	San Fernando	Apr.	leaf	0.94	70		30	
11162	San Fernando	Sept.	leaf	1.0	90		10	
11162	San Fernando	Apr.	leaf	0.6	50		30	20
11164	San Fernando	Apr.	leaf	0.5	75		25	
11170	San Fernando	Dec.	leaf	2.2	x		x	x
11665	San Fernando	Apr.	leaf	0.3			50	50
11665	San Fernando	Dec.	leaf	2.5	x		x	x
11665	San Fernando	Dec.	leaf	1.3	x		x	x
11666	San Fernando	Dec.	leaf	0.5	55			
11669	San Fernando	Dec.	flower	0.25		60		
cerulata ssp. *subcerulata*								
10330	San Ignacio	Apr.	leaf	0.35	64	8	11	17
10330	San Ignacio	May	leaf	0.3	45	30		25
10330	San Ignacio	Dec.	leaf	1.5	x	x	x	x
11892	Isla San Marcos	May	leaf	0.7	x		x	
12393	E of San Ignacio	Dec.	leaf (dead)	2.0	x		x	

SOURCE: Wall (1954) and Wall et al. (1954a, 1954b, 1955, 1957).
* Hec. = hecogenin; Git. = gitogenin; Man. = manogenin; Tig. = tigogenin.

would be uncertain. Altogether, ssp. *dentiens* looks like an isolated genome undergoing evolution and, perhaps, on its way to complete speciation.

The Seri Indians called *Agave cerulata dentiens* ''?emme'' (Felger and Moser 1970). The Seri visited San Esteban Island to collect the plant and found the more fibrous plants the most savory, distinguishing also a white-and-green variety. Large baking pits were observed on the island, and the Seri reported that as many as 100 heads were cooked at a time. They also relieved thirst with juice of the macerated leaves after they were cooked in a fire. After fermenting for several days, the juice was also drunk with warm water added. Felger reports (personal communication) that the green-leaved form generally occupies the higher elevations of the island, where fogs or clouds tend to hang and reduce insolation.

Agave cerulata ssp. *nelsonii*
(Figs. 15.1, 15.2, 15.3, 15.6, 15.9, 15.10, 15.11; Tables 15.1, 15.2)

Agave cerulata Trel. ssp. *nelsonii* (Trel.) Gentry, Cal. Acad. Sci. Occ. Pap. 130: 44, 1978.
   Agave nelsonii Trel. Missouri Bot. Gard. Rep. 22:61. 1912.

Short-stemmed, rather small, cespitose, compact, glaucous green rosettes, 50–75 cm in diameter at maturity; leaves mostly 20–35 x 6–8 cm, rarely larger, lanceolate to triangular-lanceolate, short-acuminate, usually a little narrowed above a broad base, thick rigid, light green with a gray to bluish glaucous bloom; teeth in the mid-blade 3–9 mm long, frequently on small teats, brown-ringed at base, grayish brown, closely regular-spaced 1–2 cm apart, relatively firmly attached; spine 2–4 cm long, strongly sub-ulate, flat or openly grooved above, grayish brown, decurrent to first or second pair of upper teeth; inflorescence slender, 2.5–4 m tall, with 15–20 ascending to arching laterals in upper half of shaft, the yellow flower umbels compact, globose; flowers slender, light yellow, 45–55 mm long; ovary 25–35 mm long, slightly angled, the neck long, grooved, slightly constricted; tube small, 3–5 mm deep, openly funnelform; tepals subequal, linear, erect, incurved at tip, hooded, the inner with prominent kcel; filaments slender, 30–40 mm long, inserted at base of tepals on nectary disk; anthers centric, 15–18 mm long; capsule 4–4.5 x 1.5 cm, oblong or smaller and pyriform, stipitate, with acute beak; seeds 4 x 5 mm with hilar notch at apex and thin marginal wing.

Type: *Nelson & Goldman 7111*, San Fernando (Sierra San Miguel), Baja California, 4 Sept. 1905, US. Goldman's photo (Trelease 1912:pl.65) shows very well the habit of this plant and its habitat on the igneous highlands with open cover of cirio, ocotillo, cactus and low shrubs.

Agave cerulata ssp. *subcerulata*
(Figs. 15.1, 15.2, 15.3, 15.9, 15.11, 15.12, 15.13; Tables 15.1, 15.2)

Agave cerulata ssp. *subcerulata* Gentry, Occ. Pap. Cal. Acad. Sci. No. 130: 44, 1978.

Small, acaulescent, few-leaved, open, glaucous green, cespitose rosettes 15–30 cm tall, 30–50 cm in diameter; leaves 15–30 x 2.5–7 cm, lanceolate to triangular-lanceolate, thickly fleshy, arching upward, guttered at maturity, white glaucous to glaucous green, margins crenate with prominent teats under well developed teeth; teeth at mid-blade 3–8 mm long, variously flexed, weakly attached, grayish brown, 1–3 cm apart; spine 2–4 cm long, subulate, usually sinuous, grayish brown; panicles 2–3 m tall, slender, waxy glaucous, with mostly 8–10 small lateral umbels of light yellow flowers; flowers 44–55 mm long; ovary 23–30 mm long with grooved neck 4–7 mm long; tube 3–6 mm long, shallow, openly spreading; tepals equal, 17–22 x 6–7 mm, smooth, thick, oblong, apex broadly rounded, slightly hooded, the inner with broad low keel; filaments 33–40 mm long, inserted on rim of nectary disk at base of tepals; anthers 18–21 mm long; capsules waxy glaucous, mostly pyriform or obovoid, 3.5–4 x 1 cm.

Fig. 15.10. *Agave cerulata nelsonii* on igneous rocky slope of Sierra San Miguel near San Fernando. John McClure photograph.

Fig. 15.11. Upper, *Agave cerulata nelsonii* detail of rosette showing the typical short-acuminate leaves. Lower, *A. cerulata subcerulata* near San Ignacio; the few-leaved rosettes with short thick leaves are characteristic.

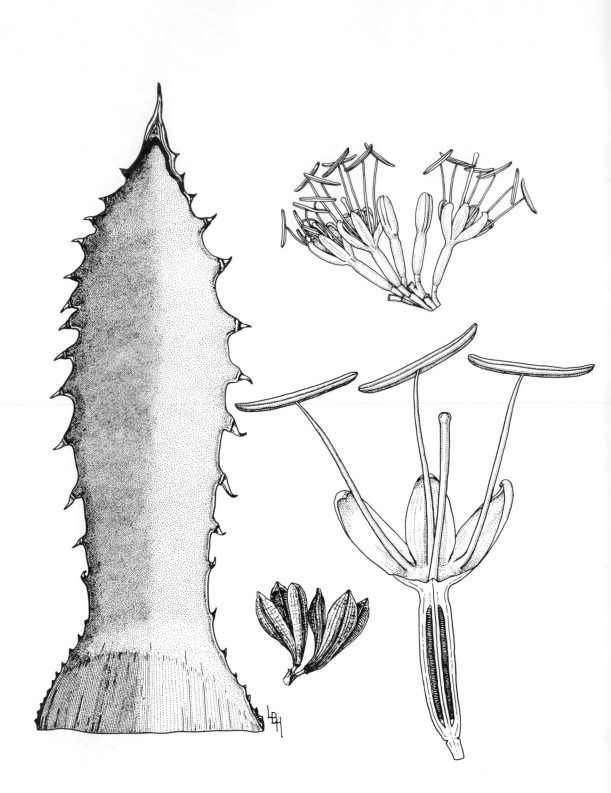

Fig. 15.12. *Agave cerulata* ssp. *subcerulata*. Leaf drawn from type *Gentry 10330;* flowers from topotypic *Gentry 23170*. Leaf, flower cluster, capsule, x ½, flower section x 1.6.

Type: *Gentry 10330,* San Ignacio, Baja California, 3 Apr. 1951, "northern slope with volcanic rocks; Cardon-Fouquieria-Jatropha, etc.," deposited in US. Duplicates in DES, MEXU.

Agave cerulata subcerulata strongly resembles *A. subsimplex* across the Gulf along the Sonoran coast, but the relationship is clearly with *A. cerulata.* The depauperate, gray to yellowish rosettes growing on the gypsiferous island, San Marcos, may reach only 10 cm in stature on open exposures. On northward slopes, the rosettes are larger and the panicles reach 2.5–3 m in height with 5 to 8 lateral branches (Fig. 15.13). Johnston (1924:998) made a similar observation. The ovary and neck of the San Marcos Island flowers collected are more slender than the peninsular ones, but otherwise the flowers are similar, including the waxy bloom on ovaries and pedicels. The short, thick, relatively broad leaves with long teeth on prominent teats provide characters for recognizing this subspecies both in the native habitats and in botanical garden collections. The distribution as far as known is shown in Fig. 15.9, while habit and habitat appear in Figures 15.11 (*lower*) and 15.13.

Fig. 15.13. *Agave cerulata subcerulata* on the island of San Marcos, May 1952.

Agave deserti

Agave deserti ssp. deserti

(Figs. 15.1, 15.2, 15.3, 15.14, 15.15 *top*, 15.17, 15.19; Tables 15.1, 15.3)

Agave deserti Engelm., Trans. Acad. Sci. St. Louis 3:310, 370. 1875.
  Agave consociata Trel. Missouri Bot. Gard. Rep. 22:53. 1912.

Medium-sized, light gray, sparingly or prolifically suckering rosettes mostly 30–50 cm tall, 40–60 cm in diameter; leaves variable, mostly 25–40 x 6–8 cm, lanceolate to linear-lanceolate, scarcely narrowed above the broad clasping base, acuminate, gray glaucous to bluish glaucous, often cross-zoned, thick, rigid, concave above, convex below, usually regularly armed with slender-cusped teeth, from small, the larger 2–3 mm long, to large, the larger 6–8 mm long, gray, loosely attached, mostly 15–30 mm apart; spine strong, generally 2–4 cm long, light brown to grayish, openly grooved above, decurrent to the first or second tooth above; panicles generally 2.5–4 m tall, on slender shafts with scarious triangular bracts 8–15 cm long and 6 to 15 short laterals with small umbels in upper ¼ to ⅕ of shaft; flowers yellow, 40–60 mm long; ovary 22–40 mm long with a slightly narrowed neck 4–6 mm long, greenish; tube shallow, spreading 4–6 mm deep, 12–15 mm wide, lined with a thick nectiferous disk; tepals equal, 14–20 mm long, 6–8 mm wide, light yellow, broadly linear, spreading at anthesis, rounded and abruptly hooked inward at apex; filaments 25–35 mm long, inserted at base to tepals; anthers 13–18 mm long, yellow; capsules ovoid to oblong or obovoid, mostly 3.5–5 cm long, 1.5–1.8 cm wide, thick walled, short-stipitate; seeds black, 4 x 5 mm.

Type: Based on Emory in 1846 and Hitchcock and Palmer in 1875, collected on Rancho San Felipe, San Diego County, California.

Distribution, Variability, and Relationships

Agave deserti is a large variable complex, the limits of which are hard to define. The geographic distribution is mapped in Figure 15.19 and documented in the Exsiccatae. While the extensive population of *A. deserti* at the type locality along San Felipe Creek (Fig. 15.14) is morphologically homogenous, there are other localities showing wide variability in leaf form. One of these localities is Pinyon Flats by the northwest slopes of Santa Rosa Mountain along Route 74 in southern California.

Cave (1964:166) has reported a collection (*Hutchinson 710*) from Baja California with chromosomes $n = 59$. This is a polyploid condition, as the basic number in *Agave* is $n = 30$. The genetic variability in the *Agave* complex indicates a high potential for species evolution, as examples are the populations of *A. deserti pringlei* and *A. cerulata nelsonii* in our present Recent Period.

Ivan Johnston (1924) included several of Trelease's taxa as synonyms under *A. deserti*, viz. *A. pringlei*, *A. nelsonii*, *A. consociata*, and *A. dentiens*, thus forming a kind of superspecies category. After considerable study and vacillation, I am using Trelease's name, *A. pringlei* as subspecies to designate scattered populations bordering the main complex on the south and west of the *deserti* area. Another group in and adjacent to Arizona is segregated as subspecies *simplex,* the name referring to its more simple habit.

A. cerulata Trel. is closely related but appears to be a separate specific complex, difficult to separate morphologically. Reasons and characters for separating it from the *deserti* complex are given under *A. cerulata*. *A. cerulata* does not belong with *A. sobria* Bdge. where Johnston placed it. The following outline separates three main *A. deserti* geographic populations, none of which is in all individuals morphologically distinct.

Fig. 15.14. *Agave deserti* near the type locality along the San Felipe arroyo, San Diego Co., California. Upper photograph shows the desert habitat; lower is detail of drought-shrunken leaves. Otherwise these are robust plants in deep sandy soil.

Fig. 15.15. Upper, *Agave deserti* on plain of the desert southeast of Cerro Borrego, Baja California. The high small panicles are characteristic in this region. Lower, *A. deserti pringlei* in San Matias Pass with green and pale glaucous forms growing side by side.

Key to Subspecies of *Agave deserti*

1a. Leaves mostly 25–40 cm long, 4–7 times longer than broad, moderately acuminate, the margins usually straight; teeth weakly attached; spine decurrent as a corneous margin only to the upper first or second pair of teeth. Tube 3–8 mm deep   2

1b. Leaves mostly 40–70 cm long, 8–12 times longer than broad, long-acuminate, the margin straight; teeth firmly attached; spine conspicuously decurrent in a corneous margin frequently to the mid-blade or even be-

low. Tube 5–8 mm deep. San Matias Pass and vicinity   *pringlei*

2a. Rosettes copiously surculose, forming large clones; flower tube 3–5 mm deep; filaments inserted on base of tepals. Western side of the Gulf of California   *deserti*

2b. Rosettes generally single, rarely with 1–3 offsets; flower tube 5–10 mm deep; filaments inserted in orifice of tube below tepal bases. NE of the Gulf of California   *simplex*

Uses and Chemistry

Agave deserti is among the more edible of the agaves. Castetter et al. (1938) wrote an excellent account of uses of agave by the Indians of the American Southwest. Their map of "mescal finds" (ibid.:37), including "mescal pits," shows that *A. deserti* was eaten and its fibers used throughout the *deserti* area. "Mescal pits" are very common and have been traced as far south as the latitude of Punta San Fermin in the Gulf of California (ibid.:60). This is in Baja California at the southern margin of the known distribution of *A. deserti* (e.g., *Gentry & McGill 23286* on the San Pedro Mártir bajada). Barrows (1967:59) stated that the Coahuilan Indians of Southern California made much use of *A. deserti* for food and fiber. They called the plant "a-mul," sections of the flowering stalk "u-a-sil," the leaves "ya-mil," and the yellow blossoms "amul-sal-em," all of which were cooked in various ways and eaten. Barrows regarded the agaves as one of the principal plant resources of southwestern Indian tribes.

Other tribes known to use *A. deserti,* as well as other species in their respective areas, are the Pimas and Papagos of both Arizona and Sonora, the Yumans, Kamias, Chemehuevis, and Yavapais along the Río Colorado, and the Digueños and Cocopahs of northern Baja California. Other adjacent tribes, such as the Utes and Apaches, received cooked *A. deserti* in trade. Other early reports, such as that of the German missionary Baegert (Rau, 1864), who was stationed near the 25th parallel on the western slope of the Sierra de la Giganta in the 1770s, and Barco (1973), who was stationed at San Javier between 1738 and 1768, corroborate the widespread use of agaves in Baja California. The Indians of the peninsula were hunters and gatherers; they did not practice any agriculture, and the wild agaves were of major importance for their survival in this aridly inhospitable land.

The absence of mescal pits south of Sierra San Pedro Mártir does not mean that agaves were not eaten south of this latitude. The account of Miguel del Barco (1973:123), tells us why there are no mescal pits southward. It just was not the custom of the Cochimi to dig cooking pits. They roasted their mescal upon the ground "en forma de monton." The Cochimi ranged through a large central section of the peninsula and spoke a different language than the northern Digger Indians. The Cochimi guides on Link's expedition of 1766 (Burrus, 1966) found they could not communicate with the San Pedro Mártir people. The southern end of this sierra marks a division margin between the two language-and-cultural groups; so also do the southern limits of mescal pits.

Agave deserti is important to wildlife; birds visit the flowering stalks for nectar and insects. Small rodents live about the plants and are protected by the armed rosettes. The pack rats, *Neotoma,* make nests among the clumps of rosettes, gnaw through the leaves, and eat the flowers and perhaps the seeds. Wild bighorn sheep, like cattle, eat the new-flowering shoots. As a wildlife resource growing in our most arid deserts, *Agave deserti* deserves conservation and perhaps should be established on desert mountains where it is now lacking.

Table 15.3 gives the sapogenin content found in *Agave deserti* by chemists of the Agricultural Research Service in Philadelphia, Pennsylvania (Wall, 1954; Wall et al., 1954a, 1954b, 1955, 1957). Hecogenin is the leading steroid. The steroid content is erratic to absent in leaves, absent in the edible stems, and appears to run about 0.6 to 1 percent in the fruits.

Agave deserti ssp. *pringlei*
(Figs. 15.1, 15.3, 15.15 *bottom,* 15.16, 15.19; Tables 15.1, 15.3)

Agave deserti Engelm. ssp. *pringlei* (Engelm. ex Baker) Gentry, Cal. Acad. Sci. Occ. Pap. 130: 20, 1978.
> *Agave pringlei* Engelm. ex Baker, Handbook Amarillid. 182. 1888.

Green or whitish, cespitose, rather strict rosettes, offsetting closely from root crown, 4–7 dm tall, 5–8 dm wide; leaves narrowly triangular lanceolate, very long-acuminate, mostly 40–70 x 5–7 cm, thick and widest at base, deeply crescentic in cross section, guttered and thinner above, green to yellowish green or light glaucous gray; spine 3–4 cm long, aciculate, reddish brown to light gray, usually narrowly grooved for short distance at base, decurrent as a corneous margin to upper teeth or to the mid-blade; teeth rather regularly spaced 1–2 cm apart, mostly 5–10 mm long, slender and little curved, more rarely smaller or more broadly flattened and irregularly flexed; panicle 3–6 m tall with 10–15 lateral branches, very narrow or more ample with longer laterals and larger umbels; flowers 40–60 mm long, yellow; ovary 20–35 mm long, fusiform, roundly angulate, with furrowed neck; tube ample, 5–8 mm deep, bulging, strongly furrowed from tepal sinuses; tepals nearly equal, 15–20 x 5–7 mm, spreading, linear, rounded over finely cucullate apex, the inner wider than outer and with broad low keel; stamens small, filaments inserted on rim of nectary in orifice of tube; capsules 3.5–5.5 x 1.2–1.5 cm, mostly oblong, beaked, short-stipitate; seeds as for species.

Type: "Central mountains of Lower California, alt. 6,000 feet, Orcutt! Described from a dried specimen sent to Kew by Mr. C. G. Pringle." Probably from the Sierra Juárez in 1882, when Orcutt collected there.

The type, perforce, is only nomenclatorial in scope. The type specimens are but a pitifully small expression of the peninsular highland populations. In the description above, I have drawn this taxon's outline from the listed specimens (see Exsiccatae) to include the apparent introgressions with *A. deserti deserti* and with *A. moranii* as well. Without including some morphological characters of *A. moranii, A. deserti pringlei* would appear as a green, highland ecotype of *A. deserti.*

The variable population in and about San Matias Pass shows relation to *A. deserti deserti* in the tall, very narrow panicle and the shorter white glaucous leaf, which are present in differing combinations in some of the clones. Relation to *A. moranii* appears in the denser broader panicles and yellowish-green leaves of other clones, as well as leaf elongation in both green and whitish forms. Other specimens on the western slopes of the Juárez-Mártir mountain axis are very similar to *A. deserti deserti* in leaf size and form but have atypically green leaves. The type collections of *A. pringlei* appear to be of this green form. The population in and about San Matias Pass appears to be a "hybrid swarm," and individual genotypes here, as well as elsewhere, reflect genetic factors of one or the other

Fig. 15.16. *Agave deserti pringlei* in San Matias Pass, Baja California. Upper, showing detail of rosette and cross section of leaf bases where specimens were cut. Lower, rosette of green clone in flower, April 1963.

"parent" in varying degree and combinations. Subspecies *pringlei*, with its surculose habit, slender panicles, and smaller size is more closely related to *deserti* than it is to the larger, nonsurculose, and diffusely paniculate *moranii*. The deeper tube with its deeper insertion of filaments is perhaps the most unique character distinguishing *A. deserti pringlei* from its neighbors, but incidence of this character appears to be sporadic, judging from the limited flower specimens available.

Agave deserti ssp. *simplex*
(Figs. 15.1, 15.2, 15.3, 15.17, 15.18, 15.19; Tables 15.1, 15.3)

Agave deserti ssp. *simplex* Gentry, Occas. Pap. Cal. Acad. Sci. No. 130: 22, 1978.

Small- to medium-sized, mostly single, light green to light glaucous gray, compact, short-stemmed rosettes; leaves mostly 25–40 x 6.5–10 cm, rarely to 50 cm, lanceolate, firm, rigid, mildly concave above, with undulate to crenate margins, broadest in mid-blade; spine stout, subulate, 3–4 cm long, broadly grooved above, dark brown to light gray, decurrent for several centimeters to upper second or third pair of teeth; teeth in mid-blade mostly 5–8 mm long, 1–3 cm apart, brown to pruinose gray, frequently brown and gray-ringed about base, variously curved or reflexed; panicles 4–6 m tall with 8 to 15 short umbellate branches in upper third of shaft, the peduncles stout, large reflexed bracts below rapidly decreasing in size upward and persisting erect, scarious; flower clusters small, congested, pale yellow to ferruginous in bud; flowers 40–60 mm long, yellow with

Table 15.3. Sapogenin Content in *Agave deserti*
(given in percentages on dry weight basis)

COLL. NO.	SOURCE LOCALITY	MONTH COLL.	PLANT PART	SAPOGENINS* TOTAL	HEC.	GIT.	MAN.	TIG.
deserti ssp. *deserti*								
	California:							
10041	San Felipe Narrows	Dec.	leaf	0				
10041	San Felipe Narrows	Dec.	stem	0				
10044	San Felipe Narrows	Dec.	whole pl.	0.1	50		50	
10051	Pinyon Flats	Jan.	flower	0				
10051	Pinyon Flats	Jan.	fruit	0.6	x			
10051	Pinyon Flats	Jan.	seed	0				
10051	Pinyon Flats	Jan.	fruit	0.6	70	5	5	20
11650	Pinyon Flats	Apr.	leaf	0				
11652	Pinyon Flats	Apr.	leaf	0.75	20	14		66
12436	Pinyon Flats	Dec.	leaf (dead)	0.6	x	x	x	
deserti ssp. *pringlei*								
	Baja California:							
10287	S of San Pedro Mártir	Mar.	leaf	0				
10287	S of San Pedro Mártir	Mar.	stem	0				
deserti ssp. *simplex*								
	Arizona:							
9947	Harquahala Mt.	Nov.	leaf	tr.				
9947	Harquahala Mt.	Nov.	stem	0				

SOURCE: Wall (1954) and Wall et al. (1954a, 1954b, 1955, 1957).
* Hec. = hecogenin; Git. = gitogenin; Man. = manogenin; Tig. = tigogenin.

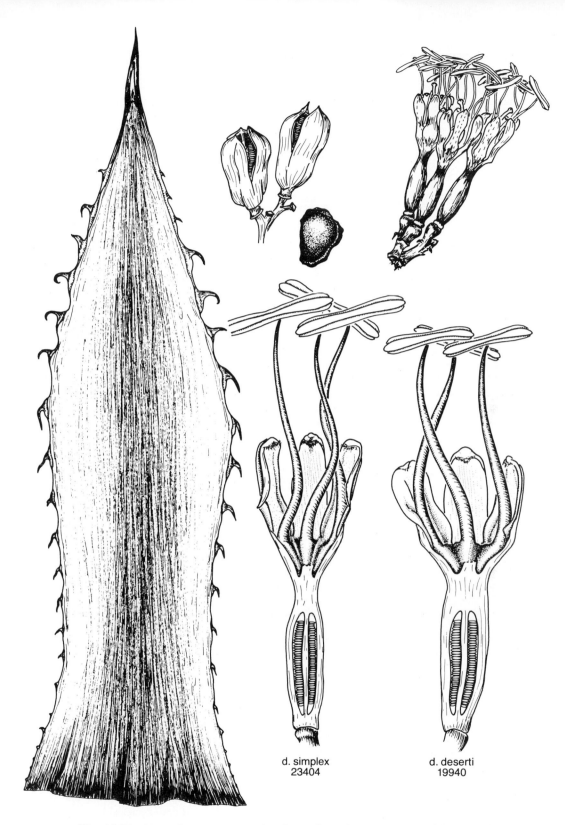

d. simplex
23404

d. deserti
19940

Fig. 15.17. *Agave deserti* ssp. *simplex* drawn from the type, *Gentry 23404;* leaf, flower cluster, capsules x ½, flower section x 1.7, seed x 2. Flower section of ssp. *deserti* for comparison x 1.7.

Fig. 15.18. *Agave deserti simplex* in the Silverbell Mountains, Pima Co., Arizona.
The plants are frequently depauperate, resemble *A. mckelveyana* with small flowers,
33–46 mm long, but the tube is characteristically deep, 5–7 mm.

pale-green ovary; ovary 20–30 mm long, fusiform, with constricted grooved neck 4–6
mm long; tube short, 5–8 mm long, open funnelform, grooved, angular; tepals subequal,
15–20 x 4–5 mm, erect to spreading, rather thin, linear, conduplicate wilting and incurved, apiculate-cucullate, the outer slightly longer, flat to rounded on back, the inner
with broad low keel; filaments 30–42 mm long, unequally inserted below tepal bases high
in tube, slender, elliptic in cross-section, pale yellow; anthers 15–21 mm long, yellow,
regular, centric; capsules 3.5–4.2 x 1.5–2 cm, oblong to obovoid, stipitate, rounded to

rostrate at apex, rather thin-walled; seeds 5–6 x 4–4.5 mm, lunate to lacrimiform, dull or shiny black, the margin with a raised, irregularly fluted wing.

Type: *Gentry 23404,* N slope of Harquahala Mountain, 12 miles (ca. 19 km) W of Aguila, Yuma Co., Arizona, 12 June 1974; deposited in US, isotypes in DES, MEXU, ARIZ.

Relationships and Comparisons

Agave deserti simplex is distinguished from *A. deserti deserti* by its predominantly solitary habit, deeper flower tube, tepals being only ca. 2–3 times longer than the tube (vs. 4–5 times longer), filaments being inserted in the tube (vs. inserted on the base of the tepals), and the disjunct distributions of the two taxa (Fig. 15.17). It is not possible to separate the two taxa on leaf specimens alone, even when accompanied by photographs. However, if labels state the origin of the specimen and whether the population from which it was selected consists of cespitose plants or generally of singles, identification of leaf specimens becomes relatively certain. Large plants frequently offset, especially at maturity, but the offsets may wither and die.

Another close relative of *A. deserti simplex* is *A. mckelveyana,* which occupies higher montane elevations northeast of the *simplex* area. As in *A. deserti simplex,* the filaments of *A. mckelveyana* are inserted in the orifice of the tube, but the latter is a smaller plant with small narrow leaves, slender wandlike inflorescence, and small flowers. It is possible that these two taxa will be found in contact, and if so, it will be interesting to see if there is introgression.

In Arizona the scattered populations of *simplex* rosettes sucker much more sparingly than the California colonies, and the large cespitose California clones are absent. In the Lechuguilla Desert of southwest Arizona, one of our most arid regions, the subspecies is represented by small drought-depauperized, scattered individuals, with leaves commonly only 20–25 cm in length. Other than for their small size, the characters of the plants are quite like those of the species—rather broad, gray, cross-banded, thick rigid leaves and slender small panicles. However, also found in the region is a long-leaved robust form in the Mohawk Mountains *(Wiggins 8643),* which I have not seen in habitat.

The native uses of *Agave deserti simplex* are essentially what has been reported above for the species.

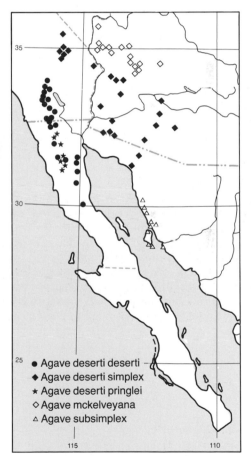

Fig. 15.19. Distribution of *Agave deserti, A. mckelveyana,* and *A. subsimplex,* based on herbarium specimens listed in the Exsiccatae.

Agave gigantensis

(Figs. 15.1, 15.2, 15.3, 15.20, 15.21, 15.22; Table 15.1)

Agave gigantensis Gentry, Cal. Acad. Sci. Occas. Pap. No. 130: 63, 1978.

Single, acaulescent, green to glaucous green, rather open, few-leaved, medium-sized rosettes 5–10 dm tall, 8–12 dm broad; leaves 40–75 x 11–16 cm, plane, rigid, thick-fleshy, sclerophyllous, smooth, broadly lanceolate, acuminate, widest in the middle, markedly narrowed at base, green to glaucous, turning red to purplish with flowering, and depressed, the margin undulate to prominently mammillate; teeth large, sometimes grotesque, mostly 10–20 mm and even longer, the bases thick, frequently 2–3 cuspidate and confluent along upper leaf margins, brown to light grayish, variously flexed and curved, generally remote, up to 6–8 cm apart; spine 3–6 cm long, strongly subulate, deeply sulcate above, straight or sinuous, gray, long decurrent as a heavy corneous margin, sometimes to mid-leaf; inflorescence 4–5 m tall, slender, the peduncular bracts narrowly lanceolate, thickly chartaceous; panicles rather narrow, in upper ⅓ to ¼ of shaft, with 15 to 25 rather small umbels of bright pale-yellow flowers; buds waxy white; flower 48–60 mm long, slender, with spreading tepals; ovary slender, fusiform, with short, constricted, slightly grooved neck; tube short discoid, spreading, bulbous with tepal bases, 4–5 mm deep, 11–13 mm broad; tepals 18–25 x 5–6 mm, linear, rounded and briefly hooded at apex, the inner with prominent keel, grooved within and strongly overlapped at base by outer; filaments 30–45 mm long, variable in length, slender, inserted on rim of tube with base of tepals; anthers 18–25 mm long, somewhat excentric; capsules 3.5–4 x 1.2–1.5 cm, oblong, non-stipitate, non-beaked; seeds unknown.

Type: *Gentry & McGill 23320,* above Rancho San Sebastián, Sierra de las Palmas, 31 miles (ca. 50 km) by road W of San Bruno, Baja California, elev. 3,800–5,000 feet (1,150–1,520 m), 20 June 1973, US. Isotypes DES, SD, MEXU.

The small waxy pale flowers with short spreading tube in a slender panicle places *A. gigantensis* in the Deserticolae. As with other members of this alliance, the habit and leaves provide specific distinction. The leaves are remarkable in their broad form above a narrowed base, their singular quality of color, bud-printing, and firm sclerophyllous smoothness, and bizarre variable large teeth on mammillate margins (Figs. 15.20, 15.21). The nearest phyletic relatives appear to be the disjunct *A. avellanidens* and *A. moranii* to the north. The name *gigantensis* is given because the distribution is coincident with the Sierra de la Giganta range of the southern peninsula (Fig. 15.22).

On Sierra de las Palmas (also known locally as Sierra Campana), *Agave gigantensis* is closely associated with an oak woodland community that mantles the craggy tops of these volcanic mountains. *Nolina beldingii* is a conspicuous member of this community. Southward, *A. gigantensis* has been observed and collected on the brushy slopes west of the Cerro Giganta itself and along the scarp-rim country above Loreto (see Exsiccatae citations). Altogether, it is known to range from elevations between 2,000 and 5,000 feet (600 and 1,520 m) between latitudes 27° and 25° N. This is a small area, and *A. gigantensis* shapes up as another rare endemic plant. Fortunately, it occupies a very rugged and largely inaccessible habitat, which in this age of destructive man will favor its survival. The rancher at San Sebastián, who guided us upon the mountain, stated that this agave is more abundant on "Sierra del Potrero," northward in this same mountain range. It is a symmetrical, beautiful plant that could well be valued and cultivated by the sophisticated agave fanciers. It does not reproduce vegetatively, however, and seeds would be needed to keep it generating.

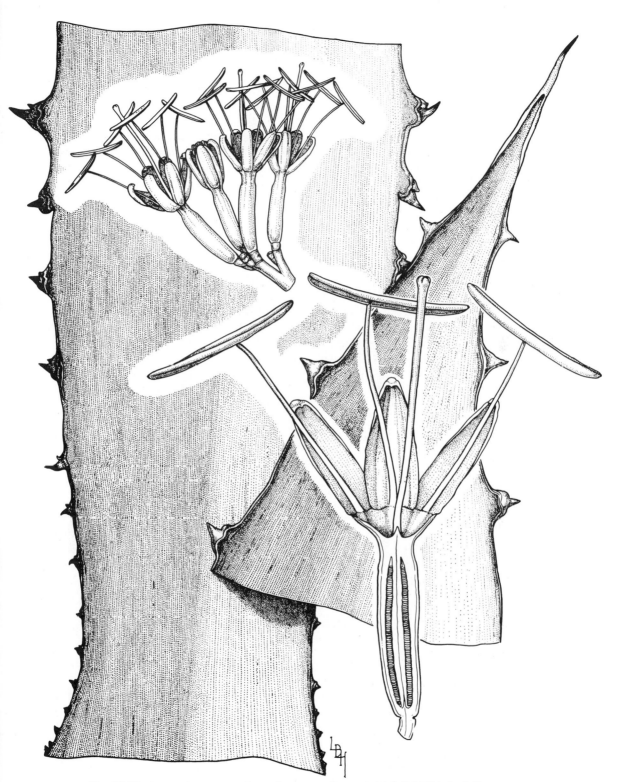

Fig. 15.20. *Agave gigantensis,* drawn from type *Gentry & McGill 23320;* leaf, flower cluster x ca. ½, flower section x 1.5.

Fig. 15.21. *Agave gigantensis;* upper, a transplant flowering on the Gentry ranch near Murrieta, California, with geotropic flowers and a hummingbird. Lower, a mature rosette on Sierra de las Palmas, Baja California Sur.

The name "lechuguilla" is applied to this agave by the people of Comondú. It is considered as one of the best for distilling into mescal. Of the five samples of leaves sent in for analysis, only two showed sapogenin in minor amounts, 0.3 percent hecogenin. Undoubtedly, this is one of the more edible agaves to which Miguel del Barco referred in his early account (quoted in the Introduction).

Agave margaritae
(Figs. 15.1, 15.22)

Agave margaritae Brandegee. Proc. Calif. Acad. Sci. ser. 2. 2: 206. 1889.
   Agave connochaetodon Trel. Missouri Bot. Gard. Rep. 22.58, 1912.

Small, compact, cespitose, glaucous gray to yellowish green rosettes with 40–50 leaves; leaves 25 x 10 cm to 12 x 7 cm, ovate to broadly lanceolate, short-acuminate, thick, fleshy rigid, concave above, narrowed above the base, the margin crenate with moderate to prominent teats; teeth at mid-blade 4 –5 mm or to 8–15 mm long, variously curved or flexed, 1–1.5 cm apart, weakly attached, reddish brown to grayish; spine 2–3 cm long, subulate, shortly and shallowly grooved above, shortly decurrent; panicles 2–3.5 m tall, slender, with 6 to 12 short lateral branches in upper ⅓ of shaft; flowers light yellow, 45–50 mm long; ovary 25–30 mm long, fusiform; tube ca. 10 mm deep; tepals 15 x 5 mm, attenuate; filaments 25 mm long, adnate to the base of tepals (Brandegee) or inserted in the throat of tube (Trelease); capsules 30–50 x 15–20 mm, not stipitate, oblong or pyriform; "seeds 3–4 mm in diameter, smooth" (Brandegee).

Type: *Brandegee s.n.,* 14 Jan. 1889, Magdalena Island, Baja California, UC. The type consists of two leaves, flowers, and a section of a lateral branch with capsules.

Trelease separated his species *A. connochaetodon* from *A. margaritae* on the basis of longer, more flexuous teeth on larger teats, as he wrote, "with considerable hesitancy that, because of the very different arming of the specimens collected, a second species is recognized for these islands." The population on Isla Margarita shows considerable variation in the size, curvature, teating, and the color of the teeth. Trelease's segregate, in the writer's opinion, is only one of the variants of *A. margaritae* and is accordingly treated as a synonym here. The deep flower tube, the short broad leaves, together with other leaf characters described above distinguish this species from all other peninsular taxa. The relatively deep tube of *A. margaritae* indicates relationship with *A. vizcainoensis,* but on leaf characters they appear specifically distinct, as discussed in the account of *A. vizcainoensis. Agave margaritae* is known with certainty only from Magdalena Island, just off the outer coast in the southern part of the peninsula.

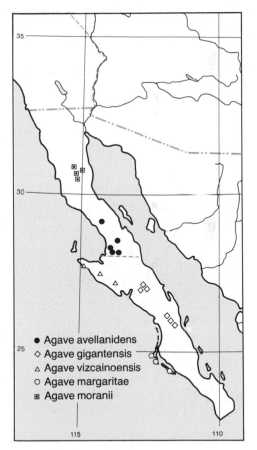

● Agave avellanidens
◇ Agave gigantensis
△ Agave vizcainoensis
○ Agave margaritae
⊡ Agave moranii

Fig. 15.22. Distribution of *Agave avellanidens, A. gigantensis, A. margaritae, A. moranii,* and *A. vizcainoensis,* based on herbarium specimens listed in the Exsiccatae.

Agave mckelveyana

(Figs. 15.1, 15.2, 15.3, 15.19, 15.23, 15.24; Table 15.1)

Agave mckelveyana Gentry, Cactus Succulent J. (U.S.) 42:225. 1970.

Small, single or suckering, rather few-leaved rosettes 20–40 cm tall; leaves 20–35 x 3–5 cm, linear or lanceolate and broadest in the middle, light glaucous green or yellowish green, firmly spreading, the margin nearly straight or undulate with low teats; spine 1.5–4.0 cm long, shortly decurrent, subulate, rounded except for shallow groove above at base, castaneous to gray; teeth small to medium, the larger at mid-blade 4–8 mm long, mostly 1–3 cm apart and downflexed, grayish with reddish tips, rather friable; panicle small, narrow 2–3 m tall with 10–19 laterals in upper half of shaft with small compact umbels; pedicels short-bracteolate; flowers small, 30–40 mm long, the perianth and stamens openly spreading, yellow; ovary light green, 16–22 mm long including constricted neck faintly grooved below tepal sinuses, fusiform or cylindric; tube shallow, open, 3.0–4.5 mm deep, 8–9 mm broad; tepals 12–13 x 3–4 mm spreading, thin, with open sinuses, linear conduplicate, the outer longer and narrower than the round-keeled inner, both abruptly hooded at tips; filaments 25–30 mm long, inserted in orifice of tube, flattened toward base; anthers 12–16 mm long, yellow; capsules 30–45 x 10–14 mm, oblong, narrowly stipitate, obtuse to apiculate, thin-walled, striate; seeds 5.0–6.5 x 4.0–4.5 mm, obovate from apical hilum or half-moon, the faces rugose, completely margined with a low wing.

Type: *Gentry 21979,* Sitgreave Pass in Black Mountains, ca. 4 miles (ca. 6 km) NE of Oatman, Arizona, 26 June 1966, US. It occurs here as a small scattered population on rocky volcanic slopes between 3,000 and 4,000 feet (900 and 1,200 m) elevations.

Agave mckelveyana occupies a central part of western Arizona (Fig. 15.19). Its habitat is with the chaparral and juniper associations on rocky slopes between 3,000 and 6,000 feet (ca. 900 and 1,800 m) elevations. I have not seen it growing with other *Agave* species, but McKelvey collected it in contiguous numbers with *A. utahensis* in the Aquarius Mountains and with *A. parryi* var. *couesii* in the Juniper Mountains. It appears to occupy an ecologic niche of its own, quite distinct from that of its near relative, *A. deserti simplex,* which is confined generally to the Sonoran Desert at lower elevations. The habitat of *A. mckelveyana* is better watered and is cooler than that of *A. deserti.*

Marcus Jones collected a cutting of inflorescence in 1930 near Oatman (*Jones 25167*) and mixed it with his specimens which he described as *A. utahensis* var. *discreta.* Trelease annotated some of McKelvey's earlier collections as *A. aquariensis,* but the name was never published. The slender panicles, small flowers with shallow tubes and spreading tepals place it in the section Deserticolae of Trelease. Its nearest relatives appear to be *A. deserti* and *A. subsimplex* Trel. The small leaves of *A. mckelveyana* with variably flexed teeth resemble those of *A. subsimplex,* from which it can hardly be separated except by the flower differences. The flowers of *A. mckelveyana* differ from these two, and other relatives, in smaller size, in the narrow spreading, thin, linear, conduplicate tepals with sharply inflexed cucullate tips, and small ovary with constricted neck.

Some observers of *A. mckelveyana* have regarded it as a depauperate form of *A. deserti.* The plants in some of the drought-inhibited populations of *A. deserti simplex,* as in the Lechuguilla Desert, are about the size of *A. mckelveyana,* but they still have the essential, if subtle, characters of *A. deserti.* The latter are generally more robust than *A. mckelveyana.* The pattern of variation in *A. mckelveyana* is phenotypically homogenous and does not intermesh with the variability of the polymorphic *A. deserti.* A leaf series of *A. mckelveyana* is shown in Fig. 15.24; flowers are shown in Fig. 15.23. I find no

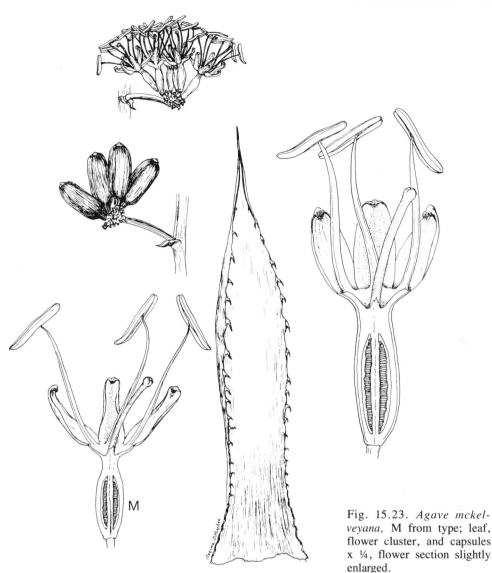

Fig. 15.23. *Agave mckel-veyana*, M from type; leaf, flower cluster, and capsules x ¼, flower section slightly enlarged.

intergrading between the two groups and believe them to be biologically distinct species. They probably separated from a common ancestor long ago.

There is nothing particularly remarkable about this small *Agave mckelveyana,* and its habit of growing among shrubbery keeps it obscure. The flowering season appears to be during May, June, and July. However, flowering time will vary from year to year and according to elevation. The life span is not known. Mature flowering and fruiting rosettes observed had only 30–40 leaves, but as the rate of leaf growth is unknown, the relatively few leaves do not necessarily indicate a short-lived rosette. Suckering offsets were observed on other plants on the Black Mountains and again on Hualapai Mountain.

In the Hualapai Mountains, Abert's tree squirrel, *Sciurus abertii,* has been observed gathering green capsules (word of Rodney Engard). In one instance the squirrel was observed cutting and carrying the fruits down. Other fruit-denuded panicles in the same locality were attributed to the work of this animal. However, this squirrel was introduced into the area in 1941–45.

Fig. 15.24. *Agave mckelveyana* in natural habitat in Sitgreave Pass in the Black Mountains, Arizona, June 1966. Right, a series of mature leaves.

Agave moranii

(Figs. 15.1, 15.2, 15.3, 15.22, 15.25, 15.26, 15.27; Table 15.1)

Agave moranii Gentry, Cal. Acad. Sci. Occ. Pap. No. 130: 58, 1978.

Large, single, short-stemmed, mostly light-green rosettes with symmetrical sword-leaved habit, 1–1.5 m tall, 2 m broad; leaves triangular-long-lanceolate, 70–120 x 8–12 cm, deeply guttered, straightly ascending to spreading, rigid, deeply rounded beneath, light green to yellowish green, sometimes glaucous, strongly armed; spine stout 4–6 cm long, nearly white with castaneous tip, broadly grooved above, decurrent to mid-blade; teeth sinuously flexed or curved, flattened, 6–12 mm long through the mid-blade and below, 2–4 cm apart, reduced and more remote toward apex on the white corneous leaf margin, light gray, the base broad or continuous with the margin; panicle 4–5 m tall, stout, frequently swollen and closely bracteate below base of panicle, with 20–30 closely spaced, compact, large, umbellate branches subtended by prominent, lanceolate, reflexing, strong bracts; flowers 50–70 mm long, bright yellow, on slender bracteolate pedicels 1–3 cm long; ovary 25–40 mm long, fusiform, tapered at base and with short, grooved, thick neck; tube shallow, 4–6 x 12–13 mm, discoid-spreading, the nectary ridged by decurrent filaments; tepals 18–24 x 6–7 mm, erect to ascending, broadly linear, acute, with papillate tufted beak, the inner with prominent bulbous keel and 2 costae within, slightly lobed at base; filaments very slender, 35–46 mm long, inserted on rim of nectary and partly on base of tepals; anthers centrically affixed; capsules 5–7 x 1.6–2 cm, long-oblong, stipitate, short-beaked, yellowish to brown; seeds 7–8 x 5–6 mm, lacriform, shiny, the marginal wing smooth, variable in width.

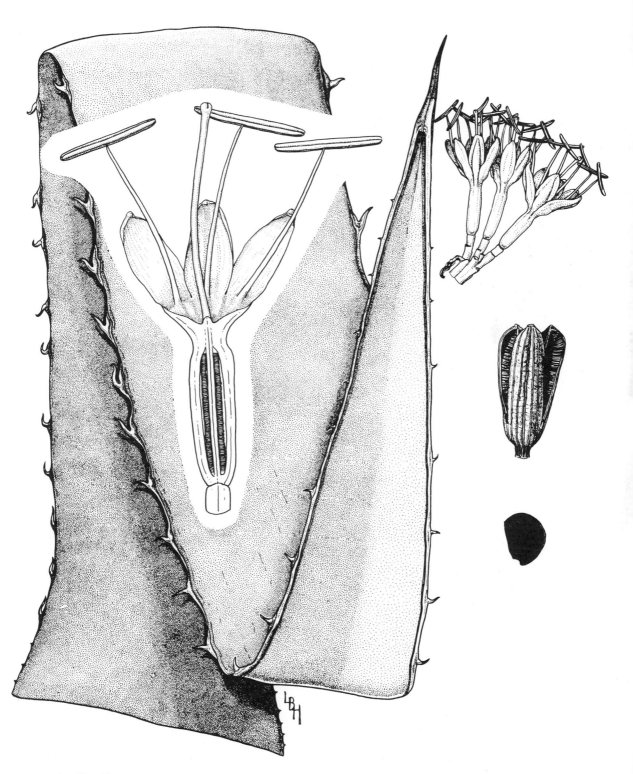

Fig. 15.25. *Agave moranii*, drawn from the type *Gentry & McGill 23287;* leaf, flower cluster, capsule x ca. ½, flower section x 1.5.

Fig. 15.26. *Agave moranii* in fine array on a northeastern bajada of Sierra San Pedro Mártir, near Rancho Parras. Photo by John McClure, June 1973.

Type: *Gentry & McGill 23287,* 2–3 miles (3–5 km) SE of Rancho Agua Caliente on E bajada of Sierra San Pedro Mártir, Baja California, elev. ca. 1,500 feet (450 m), 13 June 1973 (l, f, cap) US, isotypes in DES, SD.

Agave moranii is distinguished from all other Deserticolae by its large single rosettes of large, long, rigid leaves with white corneous margins apically, by its large-bracteate, stout peduncles, sometimes swollen below the panicles, and by its relatively congested panicles. The only agave with which it may be easily confused, especially from leaf specimens alone, is *A. deserti pringlei* in the same region. As discussed in the description of *A. deserti pringlei,* these two taxa appear to hybridize; without complete specimens they may not be satisfactorily identified. This is the case with *Moran 15256* from Rancho San Pedro Mártir on the western slope of Sierra San Pedro Mártir. A photograph of the plant, however, shows that it is probably surculose—the large panicle, small-bracteate with well-spaced laterals, leading me to place it with *A. deserti pringlei.* The respective habitats appear distinct: *A. moranii* is desertic, while *A. deserti pringlei* is at home in the chaparral zone on the Pacific slopes of the mountains. *Agave moranii* occupies a small area of the southern Sierra San Pedro Mártir and its eastern bajadas, between elevations of 1,500 and 5,000 feet (450 and 1,850 m). The specific name is in honor of Reid Moran, the first botanist to make recognizable collections of the species.

Agave moranii appears aligned phylogenetically with *A. avellanidens* and *A. gigantensis,* all having separate areas along the sierran axis of the peninsula (Fig. 15.22). As stated above, the three are distinct from other Deserticolae by their non-surculose habits, and their broader and greener leaves.

Fig. 15.27. A flowering plant of *Agave moranii* and detail of a thickened peduncle with congested bracts.

Near a fork in the road 7 miles (11 km) south of Rancho Agua Caliente near Rancho Parras, there is a fine population of *Agave moranii* (Fig. 15.26). It is most concentrated on low ridges running out from the sierra side, but it is also scattered on sandy slopes and in the arroyos. Leaf color varies from a light green or glaucous green to yellow-green; a few are lightly zoned. Leaf form and armature are fairly constant, but there is some variation in teeth. All healthy plants have outstanding radiating leaf attitudes. Leaf margin is white with a dark brown border along the green. Some peduncles are distinctive with congestion of bracts below the lowest laterals where the peduncle is perceptibly swollen (Fig. 15.27). This suggests the same flowering habit as *A. macroculmis, A. murpheyi,* and *A. parrasana,* which have the peduncle initiated in the fall, arrested over winter, and a second elongation with the warmer days of spring.

The people living at Agua Caliente have no name to distinguish *A. moranii* from *A. deserti,* which is common in the same vicinity. They stated that the flowers were good to eat. The leaf has an abundance of strong coarse fibers and probably was employed by the Cocopah Indians who formerly inhabited the region.

Agave sobria

Agave sobria Bdge. ssp. sobria
(Figs. 15.1, 15.2, 15.3, 15.28, 15.29, 15.32; Tables 15.1, 15.4)

Agave sobria Brandegee, Proc. Calif. Acad. Sci. ser. 2. 2:207. 1889.
    Agave affinis Trel. Missouri Bot. Gard. Rep. 22:56. 1912.
    Agave carminis Trel. Missouri Bot. Gard. Rep. 22:25. 1912.
    Agave slevinii I. M. Johnston. Proc. Calif. Acad. Sci. ser. 4. 12:1000. 1924.

Small- to medium-sized, usually cespitose, open, few-leaved, bright glaucous-gray rosettes with short stem, or appearing stemless, 50–150 cm in diameter; leaves linear to lanceolate, long-acuminate, variable, but mostly 80 x 10 cm to 45 x 5 cm, frequently cross-zoned, straight to curved, sometimes twisted, plane to somewhat concave above, thick and concave below toward base, the margin undulate to mammillate; teeth remote, mostly 5–10 mm long with broad, flattened, gray bases, reddish toward apex, variously flexed or straight, mostly 3–4 cm apart; spine acicular, mostly 3–6 cm long, narrowly grooved above; panicles slender, sometimes arching, 2.5–4 m tall with 12–20 short lateral branches, the umbels of flowers compact, nearly globose; peduncular bracts small, triangular, chartaceous; flowers pale yellow, 45–55 mm long, slender; ovary 25–35 mm long, tapering at base, the neck short, scarcely constricted; tepals about equal, 17–22 x 3–4 mm, very narrow, linear, rather thin, involuting at anthesis, rounded at tip, cucullate, the inner keeled; tube short, spreading, 3–4 mm long, 9–11 mm wide; filaments inserted at base of tepals on nectary disk, 35–47 mm long, slender; anthers 18–23 mm long, centrically attached; capsules 5–6.5 x 1.5–1.8 cm, oblong, thick-walled, short stipitate, apiculate; seeds 7–8 x 5–6 mm, lunate, the margins narrowly winged.

Type: *Brandegee 2*, 28 Mar. 1889, Comondú mesas (Fig. 27), Baja California, UC, DS.

Agave sobria is distinguished by its slender flowers with long narrow tepals and its very light-glaucous, long, lanceolate leaves with remote teeth. It is widely scattered but common on both sides of the Sierra de la Giganta mountain axis, from near sea level to 3,500 feet (1,070 m) elevation (Fig. 15.32). It appears to make its best growth on shadier or northern slopes in canyons, such as below the volcanic rim rock on the talus slope above Comondú. Figure 15.29 shows the plants at home at the type locality. As examined there in 1951, the rosettes commonly throw out one or more branches from above the base, thus forming cespitose plants. When small, these shoots are readily lifted out of the dry trunk tissue as separate plants with young hard roots started downward. Large rosettes develop when the plants are single or only double. The plants are scattered and do not form dense colonies. On the eastern foot of the Sierra de la Giganta the plants are generally smaller and drought depauperate.

Edibility and Use

The people at Comondú call this plant "mescal pardo" or "pardito," saying also that it is good for making mescal. This is contrary to what Brandegee (1889) has reported and contrary to his use of the name *sobria* (for sober, or not given to alcohol); so I questioned the inhabitants closely about its edibility and suitability for mescal. They were unanimous in pronouncing it good, and they added that *Agave aurea,* growing in the vicinity, was the one not suitable for mescal. Hence, I believe that Brandegee confused these two agaves on this point in his notes or memory.

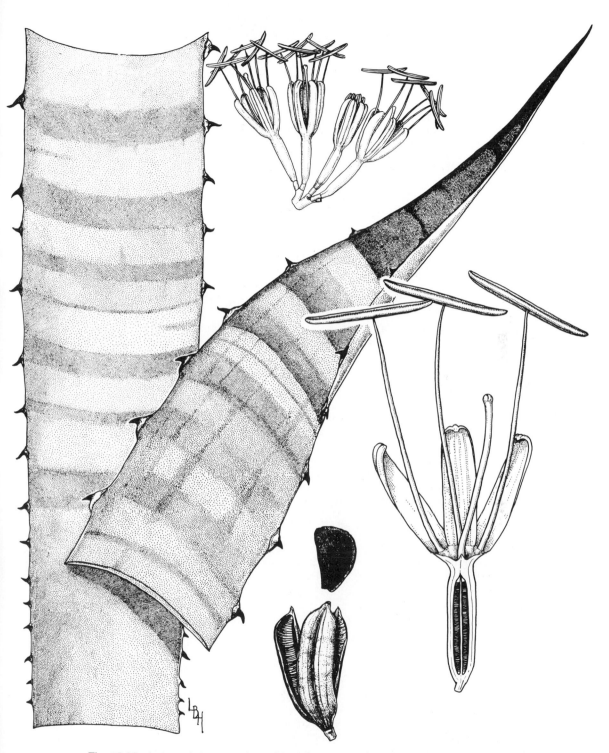

Fig. 15.28. *Agave sobria* ssp. *sobria,* drawn from topotypic materials, *Gentry 11882* x ca. ½, flowers from *Barclay & Arguelles 1989;* cluster and capsules x ca. ½, flower section x 1.5.

Fig. 15.29. *Agave sobria* at the type locality on the lava rocks above San Miguel de Comondú, October 1951.

However, the Comondú informants, consisting of several older men, said that the agave best for eating and for distilling is called "lechuguilla" and grows in the Sierra de la Giganta. A guide was appointed among them, and I dispatched with him Juan Arguelles, my field assistant. They went by horse and returned with several plants (*Gentry & Arguelles 10324, 10327*) collected on the highland near the base of the mountains above the Llano San Julio. The plants were not flowering, but leaf specimens were pressed, young plants kept for planting, and samples sent into the USDA laboratory in Philadelphia for analysis. With this local information and our present series of specimens, it is now possible to orient the detailed description of agave uses made by Miguel del Barco in the 1770s. He was located for about 30 years (1735–68) at the Mission of San Javier in the Sierra de la Giganta, and his remarks apparently apply mainly to *A. sobria* and *A. gigantensis,* both of which grow around the mission and were used extensively by the Cochimi Indians. There was a kind that was not eaten, wrote Barco, probably what is now called *Agave aurea*. Most of his remarks I have translated and quoted before (1978).

Chemistry

Table 15.4 summarizes the sapogenin content of *Agave sobria* as determined by the United States Department of Agriculture on their cortisone-source survey during the 1950s (Wall, 1954; Wall et al., 1954a, 1954b, 1955, 1957). The four genins—hecogenin, tigogenin, manogenin, and gitogenin—continue as in the *cerulata* complex to chemically characterize the Deserticolae, both in quantity and in random distribution. However, manogenin is lacking in the subspecies *frailensis,* nearly so in ssp. *roseana,* but frequent in ssp. *sobria*. Sapogenin content is conspicuously higher in *frailensis,* averaging 1.5 percent in leaves, compared with 0.6 percent in *sobria* and 0.5 percent in *roseana*. Early summer shows a higher average content of sapogenins in leaves (May 1.3 percent) than for winter (Nov.–Dec. 0.7 percent), but the data are scarce and not always strictly comparable. This is the reverse of what was obtained for *A. cerulata,* which is more in the winter-rainfall region, while *A. sobria frailensis* is in the summer-rainfall region. Is higher genin content correlated with dry seasons rather than with temperature? The lack of sapogenin in the edible stems or "cabezas" confirms the statement of the Comondú people that *Agave sobria* is a good source for mescal.

Agave sobria, although it reproduces asexually, is a freely seeding sexual species. It shows a lot of variability among individuals in certain localities, especially in the southern part of its area and on adjacent islands, where isolation has had time to foster genetic divergence. Trelease's *A. roseana,* which Johnston reduced to a variety, appears to represent a natural segregate of *sobria,* and accords with my use of subspecies in this treatise. I am separating it and another geographic variant, hitherto unrecognized, according to the following key and descriptions.

Key to the Subspecies of *A. sobria*

1a. Leaves linear lanceolate, mostly 50–80 cm long, 6–10 times longer than wide, the margin merely undulate; tepals very narrow, 3–4 mm wide   *sobria*

1b. Leaves broadly lanceolate, 25–50 cm long, 4–5 times longer than wide, the margins prominently mammillate; tepals 4–6 mm wide   2

2a. Rosettes few-leaved, openly spreading, yellow-green; leaves larger, 40–50 cm long, with large teeth few and remote on mammillate margins. Isla Espíritu Santo and adjacent peninsula   *roseana*

2b. Rosettes with more leaves, compact, urceolate, bluish-gray glaucous; leaves smaller, 25–30 cm long, with smaller teeth closely spaced along mammillate margin. Punta Frailes and vicinity, Cape District   *frailensis*

Agave sobria ssp. *roseana*
(Figs. 15.1, 15.2, 15.3, 15.30, 15.31, 15.32; Tables 15.1, 15.4)

Agave sobria Bdge. ssp. *roseana* (Trel.), Gentry, Cal. Acad. Sci. Occ. Pap. 130: 54, 1978.

  Agave roseana Trel. Missouri Bot. Gard. Rep. 22:59. 1912.
  Agave sobria var. *roseana* I. M. Johnston. Proc. Calif. Acad. Sci. ser. 4. 12:1002. 1924.

Small, open, few-leaved, cespitose or single, yellow-green rosettes; leaves 35–50 x 7–10 cm, broadly lanceolate, acuminate, concave above, frequently twisted, the margin prominently mammillate, teats 1–1.5 mm high, with remote large flexuous teeth, sometimes bicuspid, the larger 10–25 mm long; spine 5–7 cm long, sinuous to contorted, acicular, narrowly grooved above, castaneous to gray; panicles 2.5–3.5 m tall, slender, with 8 to 12 compact, globose umbels of light yellow flowers in upper third of shaft; peduncular bracts small, triangular; flowers 45–65 mm long; ovary 30–40 mm long, cylindric; tube 4–5 mm long, spreading; tepals 20–22 x 4–5 mm, long-linear, erect-ascending, incurved at apex and cucullate, narrowing with anthesis, inner strongly keeled;

Table 15.4. Sapogenin Content in *Agave sobria*
(given in percentages on dry weight basis)

COLL. No.	BAJA CALIFORNIA SOURCE LOCALITY	MONTH COLL.	PLANT PART	SAPOGENINS*				
				TOTAL	HEC.	GIT.	MAN.	TIG.
sobria ssp. *sobria*								
11291	Comondú	May	leaf	1.6	55			45
11291	Comondú	May	fruit	0.6	100			
11291	Comondú	Nov.	leaf	0.5	x		x	x
11291	Comondú	Nov.	leaf (dead)	0.4	x	x	x	
11811	Sierra de Palmas	May	leaf	1.3			100	
12384	Sierra de la Giganta	Dec.	leaf	0.0				
12384	Sierra de la Giganta	Dec.	leaf	0.6	x		x	x
12387	Bahía Concepción	Dec.	leaf	0.2	100			
12387	Bahía Concepicón	Dec.	stem	0.0				
12387	Bahía Concepción	Dec.	leaf	0.2	x		x	x
sobria roseana								
11274	Islota Gallo	Oct.	leaf	0.7	40	15	10	35
11274	Islota Gallo	Oct.	leaf	0.2	60			40
11277	Espíritu Santo Is.	Oct.	leaf	0.6	57			42
11869	La Paz	May	leaf	0.45	25	10		65
sobria ssp. *frailensis*								
11264	Punta Frailes	May	leaf	1.1	20	10		70
11264	Punta Frailes	Nov.	leaf	1.4	x			x
11257	Punta Frailes	Oct.	leaf	1.7	10			90
11257	Punta Frailes	Nov.	leaf (dead)	1.3	x	x		x
11858	Punta Frailes	May	leaf	2.2				100
12367	Punta Frailes	Nov.	leaf	1.7	x			x
12368	Punta Frailes	Nov.	stem	0.0				
12369	Punta Frailes	Nov.	leaf (dead)	1.3	x	x		x

SOURCE: Wall (1954) and Wall et al. (1954a, 1954b, 1955, 1957).
* Hec. = hecogenin; Git. = gitogenin; Man. = manogenin; Tig. = tigogenin.

Fig. 15.30. Lower left, *Agave sobria* ssp. *frailensis* near Punta Frailes. Right and upper, *Agave sobria* ssp. *roseana* with inflorescence on the hills east of La Paz. The small rounded dense umbels of light yellow flowers are characteristic.

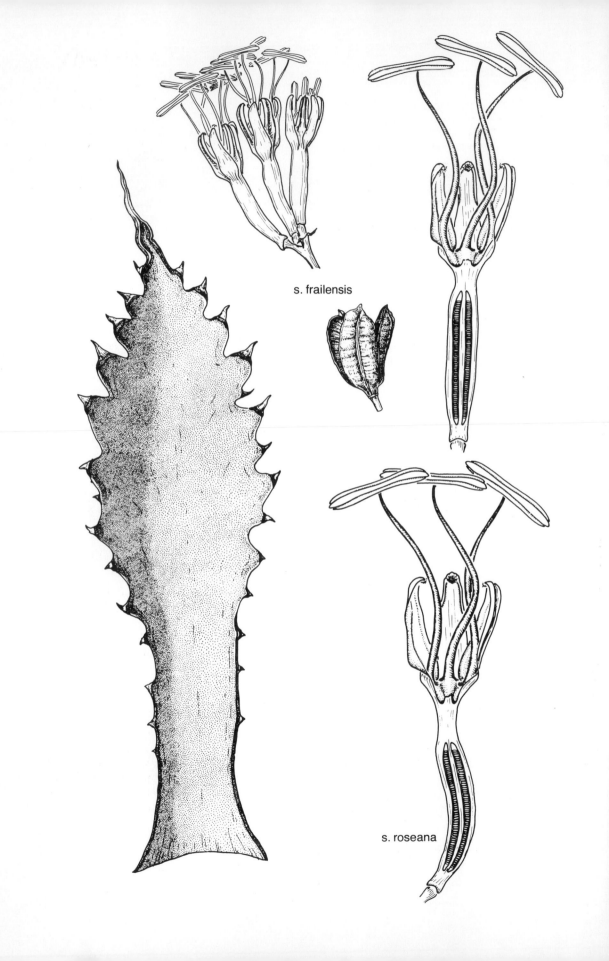

s. frailensis

s. roseana

tepal sinuses at first closed, wilting open; filaments 35–42 mm long, inserted at base of tepals on rim of tube; stamens 18–20 mm long; capsules 4–4.5 x 1.5–1.8 cm, oblong, short-stipitate.

Type: *Rose 16854,* 18 Apr. 1911, Espíritu Santo Island, Baja California, US.

Agave sobria roseana is distinguished by its open, few-leaved rosettes, its short broad leaves with prominent mammillate margins with large irregularly flexed teeth, and its restriction to the southeastern portion of the *sobria* area. It owes its distinction apparently to long isolation on Espíritu Santo Island and the adjacent islets, Islote Gallo and Islote Gallina. It also occurs near La Paz on the peninsula along with plants that are typically *sobria,* but I have had no trouble in separating the few specimens collected there. I did not find evidence of crossing between them, but such may occur. *A. sobria roseana* appears morphologically consistent both on the islands and about La Paz. Variations noted were differences in size of plants and in the size and number of teeth. The population on Islote Gallina showed cespitose plants as common, but all plants on Islote Gallo were singles. Islote Gallo had a rather uniform and dense colony, while elsewhere the plants were widely scattered. As an ornamental this plant is striking because of its bizarre teeth on big teats.

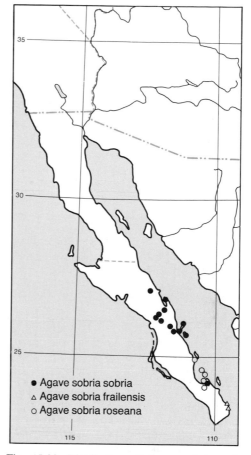

Fig. 15.32. Distribution of *Agave sobria* and subspecies, based on herbarium specimens; see Exsiccatae.

Agave sobria ssp. *frailensis*
(Figs. 15.1, 15.2, 15.3, 15.30, 15.31, 15.32; Tables 15.1, 15.4)

Agave sobria ssp. *frailensis* Gentry, Cal. Acad. Sci. Occ. Pap. No. 130: 54, 1978.

Small, compact, sometimes urceolate, glaucous green to bluish glaucous, sparingly cespitose rosettes; leaves mostly 20–35 x 6–8 cm, conduplicate, lanceolate, acuminate, markedly narrowed below, the margin mammillate and with numerous, flexuous, castaneous to graying teeth mostly 6–10 mm long; spine subulate 3–4 cm long, frequently sinuous or contorted, decurrent to upper teeth; panicle slender, 3–4 m tall, with deltoid chartaceous peduncular bracts and 10–15 compact yellow umbels in upper ⅓ to ¼ of

◄ Fig. 15.31. *(opposite page) Agave sobria* ssp. *frailensis,* from type *Gentry & Cech 11264,* leaf and flower cluster, capsule x ⅔, flower section x 1½, and *A. sobria* ssp. *roseana* flower section x 1½, from *Gentry & Cech 11277.*

shaft; flowers 45–63 mm long, slender; ovary 25–40 mm long, cylindric, with short unconstricted neck; tube 2.5–4 mm deep, small, sulcate from tepal sinuses; tepals 17–23 x 4–6 mm, linear-lanceolate, incurved and cucullate at apex, the outer dark-tinged at tip, the inner with prominent keel; filaments slender 30–40 mm long inserted on rim of tube; anthers 18–22 mm long, centric or excentric; capsules short oblong, 3–4 x 1.5–1.7 cm, sessile to short-stipitate, rounded at apex.

Type: Gentry & Cech 11264, 4 miles (ca. 6 km) N of Punta Frailes, Baja California, 7 Oct. 1951, on granitic slopes, US. Transplant flowered in Murrieta, California, July 1962.

Agave sobria frailensis differs from its near relative *A. sobria roseana* in its smaller size, more numerous leaves, lighter gray-glaucous color, softer flesh of the leaves, and more numerous, more regular, and smaller teeth. In 1951 and 1952 it was found in typical form scattered upon the granitic coastal cerros 2 to 8 miles (3–13 km) northwest of Punta Frailes in the Cape District on the peninsula. This is the only area in which it was observed.

Because of its compact urceolate rosette form and the glaucous, sometimes bluish, color of the leaves, it makes an attractive ornamental. A transplant (*Gentry & Cech 11264*) flowered in Murrieta after 11 years, but numerous offsets from the base of the plant subsequently failed to root.

Agave subsimplex

(Figs. 15.1, 15.2, 15.3, 15.19, 15.33, 15.34; Table 15.1)

Agave subsimplex Trel., Missouri Bot. Gard. Rep. 22:60, 1912.

Small, single or cespitose, glaucous, colorful, low-spreading rosettes, 20–35 cm tall, 50–70 cm broad; leaves variable, 12–35 x 3–5 cm, lanceolate to ovate, long-acuminate to short-acuminate, but little narrowed toward the base, thick, rigid, rounded on back, hollowed in inner face, gray glaucous or light yellow-green, or sometimes purple-tinged, the margin nearly straight or strongly teated; teeth variable, friable, the larger 3–15 mm long, straight or variously flexed, rarely bicuspid, brown or more often yellowish gray; spine subulate, 2–4 cm long, not or but little decurrent, frequently sinuous, shallowly grooved above, glaucous gray; panicle slender, narrow, 2–3.5 m tall, with 5–8 short laterals; flowers in small umbels, yellow to pink 40–45 mm long; ovary about 25 mm long with unconstricted long (5 mm) neck; tube shallow, spreading, 3–4 mm deep, 10 mm wide; tepals 12–15 x 6–7 mm, equal, ascending, elliptic, plane, widest at the middle, apiculate and scarcely hooded; filaments 25–28 mm long, round in cross section, inserted below base of tepals 3 mm above bottom of tube; anthers 13–15 mm long, centrically attached; capsules 40 x 15 mm to 35 x 10 mm, variable, oblong, sometimes narrowly so, light glaucous (*Gentry 10221*), bluntly apiculate, narrowly or broadly stipitate, the valves thick and striate-nerved; seeds (*10221*) mostly 4.5 x 3 mm, sooty black, roughly lunate, hilum notch narrow and the opposite corner frequently apiculate, edges rimmed with a sharp, winglike flange.

Type: Sonora: Seal Island, just off Tiburón Island, 13 Apr. 1911, *Rose 16811,* US. Known only from coastal Sonora, where it is thinly scattered in small colonies on the outwash slopes of the granitic and volcanic mountains and the adjacent islands.

Reid Moran in 1966 found *Agave subsimplex* common on Turner's Island, or Isla Dátil, as it is also called, just south of Tiburón Island in the Gulf. Moran's (1967) account provides a short botanic history of the plant, a diagnostic description, and his field observations.

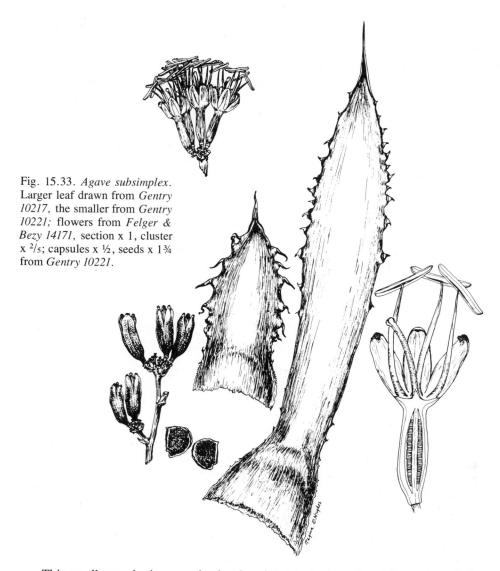

Fig. 15.33. *Agave subsimplex.* Larger leaf drawn from *Gentry 10217,* the smaller from *Gentry 10221;* flowers from *Felger & Bezy 14171,* section x 1, cluster x $^2/_5$; capsules x ½, seeds x 1¾ from *Gentry 10221.*

This small xerophytic agave is closely related to *A. deserti* and *A. cerulata* Trel., as expressed by their common characters of small variable leaves, nearly tubeless flowers, and narrow small panicles, which characterize the group Deserticolae. I think the characters outlined in the key to species should be sufficient to distinguish *A. subsimplex* from *A. deserti,* but depauperate plants of the latter in the Lechuguilla Desert of Arizona resemble *A. subsimplex.* The two species are not known to cohabit, but it is possible that they may be found together on some unexplored mountains in northwestern Sonora, such as the coastal Sierra del Alamo. *Agave subsimplex* may be more closely related to the polymorphic *A. cerulata* of Baja California, as indicated by the small, variable leaves and the narrow, oblong, waxy capsules. However, the Sonoran plants have more spreading rosettes, gray not yellow, with thicker, wider, less acuminate leaves, and the filaments are inserted below the base of the tepals. In *A. subsimplex* the nectarious interliner of the tube, from the rim of which the filaments ascend, does not completely fill the tube, as it does in *A. cerulata.* The flowers of *A. subsimplex* are smaller than those

Fig. 15.34. *Agave subsimplex* along the arid Sonoran coast, February 1951.

of *A. cerulata,* 40–45 mm vs. 45–60 mm. The pink to red color frequently occurring in the pistil, filaments, and corolla (see notes of collectors in Exsiccatae) is distinctive of *A. subsimplex.* Such color is not known in the flowers of *A. deserti* and *A. cerulata.*

The Seri Indians, a maritime hunting-and-gathering people of the Sonoran coast, called this agave "ahmmo," as nearly as I could render the sound, and stated that it was gathered for cooking and eating. Whiting noted the Seri name as "den;kl." Small amounts of sapogenin (0.07–0.14 percent) were found in the leaves. Felger and Moser (1970) render the name phonetically as "?aamXW." They reported that the pit-baked stems are eaten in several ways: cubed and cooked with sea turtle, or made into flat cakes or patties, which may be stored for later use or taken on long trips. These cakes could also be soaked in water and consumed as a sweet drink. The various agave viands all had Seri names and indicate an habitual, important resource for survival by a people living on the outer edge of human environment.

Agave vizcainoensis
(Figs. 15.1, 15.3, 15.22, 15.35; Table 15.1)

Agave vizcainoensis Gentry, Cal. Acad. Sci. Occ. Pap. No. 130: 67, 1978.

Plants acaulescent, small to medium, surculose or single, the rosettes 30–50 cm tall, 50–90 cm wide, few-leaved and open in habit; leaves lanceolate, 25–40 x 6–10 cm, glaucous gray to green, sometimes reddish, broadest in middle, narrowed above base, thick-fleshy, rather rigid, the margin undulate, corneous above with decurrent spine; teeth largest along mid-blade, 5–10 mm long, 1–3 cm apart, slender or broadly flattened, dark brown to grayish, nearly straight or curved, spine 2.5–4 cm long, stoutly subulate, mostly rather straight, brown to grayish, shallowly grooved above, long decurrent; panicle 2–3 m tall with 8 to 15 spreading lateral umbellate branches in upper half of shaft; flowers yellow 65–75 mm long, on stout small-bracteolate pedicels 8–12 mm long; ovary green, 36–41 mm long, in fresh flower neck scarcely constricted but drying very narrow, 6–8 mm long; tube funnelform, 8–12 mm deep, 15 mm broad, bulging at filament insertion; tepals 21–26 x 4–5 mm, linear-lanceolate, involute above spreading base, obtuse, cucullate, strictly erect at anthesis; filaments 55–70 mm long, inserted somewhat unequally 7–9 mm above base of tube; stamens 21–26 mm long, excentric; pistil to 70–85 mm long with broad lobate stigma; capsules 5–7 x 2 cm, oblong, stipitate, rostrate, striate, brown; seeds 6–7 x 4.5 mm, lacriform, black, marginal wing narrow.

Type: *Gentry 7469,* Cerro Tordillo and vicinity, Sierra Vizcaíno, Baja California, 12–13 March 1947; elev. 400–800 feet (120–250 m); desert of dispersed succulent trees and suffrutescent shrubs, UC; isotypes. DS, ARIZ, MEXU, DES.

The relatively deep flower tubes of this species suggests affinity with *A. margaritae* of the same geographic region, but the latter, with its short, broad, long-toothed leaves and small size, appears distinct. Neither species is well known; the collections are few and incomplete. The species do not show close relation with other members of the Deserticolae, but they are assigned here for want of a more suitable section to receive them. Their small size, surculose habit, shape of leaf and armature, and the small scarious bracts of the peduncle exclude them from the *A. shawii* complex. The deepening tube of subspecies *simplex* in the pivotal complex of *A. deserti,* at almost opposite ends of the Deserticolae area, seems also to permit their inclusion in the Deserticolae.

The plants growing about the Pichachos de Santa Clara are more robust and less surculose than those in the Sierra Vizcaíno proper and resemble *A. gigantensis. Gentry 7713* agrees well with the type in leaf morphology, but the tepals are relatively long in relation to the shorter tube (Table 15.1). Another specimen (*Gentry 7693*), also from the Picachos de Santa Clara, is more extreme with tepals to 35 mm long.

Fig. 15.35. Type collection of *Agave vizcainoensis, Gentry 7469,* from an isotype sheet in the Gentry Herbarium.

Deserticolae Exsiccatae

Agave avellanidens

BAJA CALIFORNIA. *Brandegee s.n.,* UC. Type. Paraíso, 1 May 1899 (upper ½–⅔ of leaf, flowers, and old caps. All similar to what I have seen and collected between Calmallí and Mesquitál Rancho).

Gentry & Fox 11944, DES, MEXU, US. Ca. 9 miles E of Punta Prieta, 19 May 1952.

Gentry & Fox 11933, 11932, DES, MEXU, US. 6 miles (ca. 10 km) W of Calmallí, 17 May 1952.

Gentry & Fox 11929, DES, MEXU, US. 6 miles (ca. 10 km) W of Calmallí, 17 May 1952 (l, f, photo).

Gentry & Gentry 23184, DES, MEXU, US. 5½ miles (ca. 9 km) W of Calmallí, 12 Apr. 1973 (l, f, photo).

Gentry & Gentry 23186, DES, MEXU, US. 15–17 miles (ca. 24–27 km) SE of Mesquitál, 12 Apr. 1973 (l, f).

Gentry & Gentry 23187, DES, MEXU, US. 7 miles (ca. 11 km) SE of Mesquitál along road to El Arco, 12 Apr. 1973 (l, f).

Hammerly 69, CAS. 29 miles (ca. 47 km) N of Mesquitál, 27 Sept. 1941 (doubtfully referred here).

Wiggins 5726, DS. Between Calmallí and Mesquital Rancho on mesas, 31 May 1931 (l,f).

Agave cerulata cerulata

BAJA CALIFORNIA. *Brandegee s.n.,* DS. Cardón Grande (between San Ignacio & Calmallí), 22 Apr. 1889 (l, f, cap).

Carter & Kellogg 2953, UC. Isolated red hill in sandy plain, 18.2 km W of Misión Santa Gertrudis, 18 Dec. 1950.

Ferris 8576A, DS. 1 mile (ca. 1.6 km) S of Laguna Seca Chapala, 6 Mar. 1934 (l, cap).

Gentry 10346, DES, MEXU, US. Calmallí, between mine and houses of town, 6 Apr. 1951, type locality.

Gentry 10359, DES, MEXU, US. Ca. 7 miles (ca. 11 km) S of Tinaja Yubay and 15 miles (ca. 24 km) NE of Punta Prieta; 8 Apr. 1951 (l, cap).

Gentry 10369, MEXU, US. Rancho Jaraguay, 9 Apr. 1961 (l, early f, eaten by cattle).

Gentry 11188, DES, MEXU, US. 4 miles (ca. 6 km) NW of Laguna Seca Chapala, 21 Sept. 1951 (l, cap).

Gentry 19973, DES, MEXU, US. 10 miles (ca. 16 km) S of Laguna Seca Chapala, 29 Apr. 1963 (l, f, photo).

Gentry & Cech 11322, DES, MEXU, US. 4 miles (ca. 6 km) NW of Laguna Seca Chapala, 24 Oct. 1951 (l, photo).

Gentry & Fox 11919, 11921, 11924, DES, MEXU, US. 6 miles (ca. 10 km) W of Calmallí, 17 May 1952.

Gentry & Fox 11953, DES, MEXU, US. 21 miles (ca. 34 km) E of Punta Prieta on road to Bahía de Los Ángeles, 20 May 1952 (l, photo).

Gentry & Fox 11961, 11962, DES, MEXU, US. 4 miles (ca. 6 km) NW of Laguna Seca Chapala, May 1952 (l, photo).

Gentry & Gentry 23159, DES, MEXU, US. 20 miles (ca. 32 km) SE of San Agustín, elev. 2,100 feet (ca. 640 m), 5 Apr. 1973 (l, early f).

Gentry & Gentry 23185, DES, MEXU, US. 6½ miles (ca. 10 km) W of Calmallí, 12 Apr. 1973.

Gentry & McGill 23298, DES, MEXU, US. 10 miles (ca. 16 km) S of San Luis Gonzaga Bay along road to Laguna Chapala, 17 June 1973 (l, f).

Gentry & McGill 23302, DES, MEXU, US. W. of Sierra Calamajué, ca. 30 miles (ca. 48 km) N of Punta Prieta, 17 June 1973 (l, f, photo).

Gentry & McGill 23306, DES, MEXU, US. 16–20 miles (ca. 26–32 km) NE of Punta Prieta along road to Bahía de Los Ángeles, 17 June 1973 (l, f).

Gentry & McGill 23307, DES, MEXU, US. Ca. 40 miles (ca. 65 km) NE of Punta Prieta along road to Bahía de Los Ángeles, 17 June 1973 (l, f).

Gentry & McGill 23314, DES, MEXU, US. 6–8 miles (ca. 10–13 km) S of road fork from Los Angeles Bay along road to San Borja, 18 June 1973 (l, f).

Harbison s.n., SD. 12 miles (ca. 19 km) E of Calmallí, 8 Apr. 1947 (l, f).

Harbison s.n., SD. 20 miles (ca. 32 km) S of Punta Prieta, 9 Apr. 1947 (l, inflo).

Harbison s.n., SD. Agua Amarga, ca. 15 miles (ca. 24 km) W of Los Angeles Bay, 15 Apr. 1947 (l, f).

Johnston 3487, 3489, CAS, GII, SD, UC, US. Los Angeles Bay, 6 May 1921, "in small groups on rocky mountainside," (yellow leaves with brown-ringed teeth, l, f, cap).

Johnston 3405 a-g, CAS, US. Angel de la Guarda Island, Palm Canyon, 3 May 1921 (a series of yellow leaves, mostly with reduced teeth, brown-ringed, and some toothless), "gregarious on hillside."

Moran 4106, BH, DS. Motherless Island, Los Angeles Bay, 10 May 1952 (l, f).

Moran 2007, DS, UC. 16 miles (ca. 26 km) N of Punta Prieta, elev. ca. 1,700 feet (ca. 520 m), 22 Apr. 1946 (l, f).

Moran 8167, DS, SD, UC. 3 miles (ca. 5 km) E of El Arco, elev. ca. 250 m, 5 Apr. 1960 (l, f).

Moran 8185, 26318, DES, SD. Arroyo Estaton, Isla Ángel de la Guarda, 15 Apr. 1960; elev. ca. 25 m (f, cap).

Nelson & Goldman 7180, US. Type. Calmallí, elev. 800 feet (ca. 240 m), 29 Sept. 1905. (Leaves and flowers characteristic of what I have seen through middle Baja California; no problem here.)

Raven et al. 12631, UC. 1.6 km S of Rancho Santo Ignacito, elev. 560 m, 21 Apr. 1958.

Stover & Harbison s.n., SD. 35 miles (ca. 56 km) N of Punta Prieta, 5 May 1939 (ca. 1.6 km).

Thomas 7972, SD, US. 1 mile (ca. 1.6 km) NW of Pozo Alemán, elev. ca. 800 feet (ca. 240 m), 26 May 1959 (l, bud).

Wiggins 5721, DS. 15 miles (ca. 24 km) NW of San Ignacio, 30 May 1931 (l, f).

Wiggins 5724, DS, UC, US. Calmallí, 31 May 1931 (l, f).

Wiggins 5734, DS. 5 miles (ca. 8 km) N of Punta Prieta, 1 June 1931 (l, f).

Wiggins & Wiggins 14882, CAS, DS, S end of Isla Ventana, Bahía de los Ángeles, near beach, 18 May 1959 (l, cap).

Agave cerulata dentiens

BAJA CALIFORNIA. *Bostic s.n.,* SD. San Esteban Island, sandy arroyo near SE corner, 21 June 1965.

Johnston 3194, CAS, US. San Esteban Island, 20 Apr. 1921, "common in small colonies on hillsides" (has brown-ringed teeth like *cerulata,* l, cap).

Moran 4079, SD. San Esteban Island, 6 May 1952.

Moran 21748, SD. San Esteban Island, arroyo near E side, Apr. 1975, "large colonies on hillsides and in arroyo" (in bud).

Rose 16819, US. Type. San Esteban Island, 12 Apr. 1911 (l, cap).

Agave cerulata nelsonii

BAJA CALIFORNIA. *Gentry et al. 10370, 11155, 11162, 11164, 11165, 11665, 11666,* DES, MEXU, MICH, US. 2–3 miles (ca. 3–5 km) N of San Fernando, Sierra San Miguel, type locality, 9 Apr. 1951, 10 Sept. 1951, 10 Apr. 1952 (l, cap).

Gentry et al. 10376, DES, MEXU, US. 18 miles (ca. 30 km) E of Rosario, 10 Apr. 1951.

Gentry et al. 11178, DES, MEXU, US. 28 miles (ca. 45 km) E of Rosario on Sierra San Miguel, 13 Sept. 1951.

Gentry et al. 11179, DES, MEXU, US. 17 miles (ca. 27 km) E of Rosario, 14 Sept. 1951 (l, f).

Gentry et al. 11185, DES, MEXU, US. Ca. 4 miles (ca. 6 km) SE of San Agustín, elev. ca. 2,000 feet (ca. 3,200 m), 21 Sept. 1951 (l, f).

Gentry & McGill 23311, DES, MEXU, US. 4 miles (ca. 6 km) S of Los Angeles Bay road fork on road to San Borja, elev. 900 feet (ca. 275 m) 18 June 1973 (l, f).

Gentry & McGill 23315, DES, MEXU, US. 10 miles (ca. 16 km) N of San Borja, elev. ca. 1,700 feet (ca. 500 m), 18 June 1973 (l, f).

Gentry & McGill 23322, DES, MEXU, US. Ca. 5 miles (ca. 8 km) N of San Fernando, Sierra San Miguel, elev. ca. 1,750 feet (ca. 530 m), 22 June 1973 (l, f).

Gentry & McGill 23324, DES, MEXU, US. 11 miles (ca. 18 km) E of Rosario, Rancho Porvenir, elev. 450 feet (ca. 140 m), 24 June 1973 (l, f).

Harbison s.n., SD. 1 mile (ca. 1.5 km) W of Rancho Arenoso; elev. ca. 500 m (?).

Moran 22643, SD. S of Rancho San Miguel (Sierra San Miguel Range), elev. ca. 900 m, 9 Aug. 1975 (f).

Moran & Reveal 22064, SD. Ridge 3 miles (ca. 5 km) SW of San Isidro, 30°44′N, 115°34′W, elev. ca. 1,120 m, 20 July 1975, "occasional in chaparral" (l, f).

Nelson & Goldman 7111, US. Type. San Fernando (Sierra San Miguel), alt. 1,400 feet (ca. 425 m) 4 Sept. 1905 (l, f).

Agave cerulata subcerulata

BAJA CALIFORNIA. *Barclay & Arguelles 1991,* DES, MEXU, US. Ca. 5 miles (ca. 8 km) W of San Ignacio, 29 Apr. 1966 (l, f).

Gentry 10330, DES, MEXU, **US.** Type. San Ignacio, 3 Apr. 1951 (l, f, cap, inflo).

Gentry 11892, DES, MEXU, US. Arroyo de la Teneria, Isla San Marcos, 13 May 1952 (l, f, cap, photo).

Gentry & Fox 11926, DES, MEXU, US. 6 miles (ca. 10 km) W of Calmallí, 17 May 1952.

Gentry & Gentry 23170, DES, MEXU, US. 10 miles (ca. 16 km) W of San Ignacio, 8 Apr. 1973 (l, f, photo).

Gentry & Gentry 23175, DES, MEXU, US. 24 miles (ca. 40 km) E of San Ignacio along road to Santa Rosalia, 9 Apr. 1973 (l, f).

Harbison No. P, DES. Cuesta de las Vírgenes, 1972.

Johnson 3649, 3650, CAS, GH, SD, UC, US. San Marcos Island, on gypsum, 12 May 1921 (l, f, cap).

Pinkava & McGill P12287, ASU, DES. Ca. 70 miles (ca. 110 km) W of Santa Rosalia, Route 1. 30 May 1974, Butte base (l, f).

Agave deserti deserti

BAJA CALIFORNIA. *Gentry s.n.,* DES. Granitic sandy highland on Mex. Rt. 2, S. of Jacumba, Apr. 1963 (photo).

Gentry & Arguelles 22990, DES, MEXU, US. Valle de Trinidad, elev. ca. 2,600 feet (ca. 800 m), 3 May 1972 (l, f).

Gentry & McGill 23285, DES, MEXU, US. 17–18 miles (ca. 27–29 km) W of San Felipe along road to Valle Trinidad, elev. ca. 1,950 feet (ca. 600 m), 13 June 1973 (l, f).

Gentry & McGill 23286, ASU, DES, MEXU, US. 2.6 miles (ca. 4 km) SE of Rancho Agua Caliente on E bajada of Sierra San Pedro Mártir, elev. ca. 1,450 feet (ca. 440 m), 13 June 1973 (l, f).

Hastings & Turner 66–6. ARIZ. 12.6 miles (ca. 20 km) W of turnoff toward San Matias Pass, elev. 1,300 feet (ca. 400 m), 4 Oct. 1966 (l, cap).

Hutchison 710, UC. Between Alaska and Mexicali, km 140, E slopes of the sierra, 31 Dec. 1952 (flowered UC Botanical Garden, Aug. 1963; karyotype n=59, Cave).

Pinkava & McGill 8648, ASU, DES. Along Rte. 2, 38.5 miles (ca. 62 km) W of junction with main route to Mexicali, 7 June 1971 (l, f, br).

Wiggins & Wiggins 16044, DS. Granitic sandy bajada 16 miles (ca. 26 km) W of San Felipe Hwy. along road to San Matias Pass, 2 Apr. 1960 (l, f).

CALIFORNIA. *Abrams 3976,* DS. Between San Felipe & Carisso Creek, San Diego Co., 4 July 1903 (l, f).

Ball & Everett 22777, UC. Ca. 3 miles (ca. 5 km) NE of Banner Trading Post, State Hwy. 78, 6 Nov. 1957 (l, cap).

Ball & Everett 22671, UC. 6.6 miles (ca. 11 km) SW of Palm Village Hwy. junction 111 along Hwy. 74, elev. 1,650 feet (ca. 500 m), 20 Aug. 1957 (toothless l, and cap).

Barr 67-211, 67-210, ARIZ. 40 miles (ca. 65 km) W of El Centro near Ocotillo, 13 May 1967; elev. 1,700 feet (ca. 520 m) (l, f).

Cleveland 7218, SD. Jacumba, 25 June 1885 (f).

Crovello 295, UC. 2.6 miles (ca. 4 km) E of Vallecito Stage Station, San Diego Co., elev. 2,300 feet (ca. 700 m), 15 Dec. 1963 (f).

Eastwood 18639, CAS. Road to Mountain Springs, San Diego Co., 25 Apr. 1932.

Ferris & Rossbach 9685, DS. Borrego Springs, P.O. road and State Hwy. 78, San Diego Co., May 1938 (l, f).

Ferris 7060, DS. Between Jacumba & Mt. Springs, San Diego Co., 18 Apr. 1928 (l, cap).

Flemming 753, SD. Foot of Banner Grade, Sentenac Canyon (San Diego Co.), Apr. 1926.

Gander 1325, SD. Carriso Creek (San Diego Co.), 3 Jan. 1936 (l, f).

Gander 4810, SD. Bull Willow Canyon, San Diego Co., 15 Dec. 1937 (l, f).

Gentry 10034, 10034a, DES, MEXU, MICH, US. 3–4 miles (ca. 5–6 km) S of Palm Desert, Santa Rosa Mt., 20 Dec. 1950.

Gentry 10041, 10044, DES, MEXU, MICH, US. Yaqui Wells near San Felipe Creek, San Diego Co., 29 Dec. 1950 (l, cap).

Gentry 10051, DES, MEXU, US. Pinyon Flats, Hwy. 74, Riverside Co., elev. ca. 4,000 feet (ca. 1,200 m), 5 Jan. 1951 (l, f, cap).

Gentry 17759, DES, MEXU, US. Pinyon Flats (Riverside Co.), 7 July 1959.

Gentry 19741, DES, MEXU, US. San Felipe Ranch near Rt. 789 and road to Warner's Ranch, San Diego Co., 25 May 1962 (l, f).

Gentry et al. 11650, 11652, DES, MEXU, MICH, US. Pinyon Flats (Riverside Co.), 3 Apr. 1952.

Gentry & McGill 23326, DES, MEXU, US. Pinyon Flats (Riverside Co.), 26 June 1973.

Hall 2117, DS, UC, US, E base of San Jacinto Mts., Colorado Desert, June 1901 (l, f).

Hitchcock & Palmer in 1875, **Emory** in 1846, **MO.** "The types." E of San Felipe Ranch, San Diego Co. (Probably along San Felipe Creek, above or below The Narrows.)

Howell 3243, CAS. San Felipe Wash, halfway from Borrego Valley to Yaqui Wells, San Diego Co., 26 Nov. 1927 (l, f, cap).

Huey s.n., SD. Borrego Narrows (San Diego Co.), Apr. 1931 (f).

Lester s.n., SD. San Isidro Mt., N of Borrego, May 1926.

Mearns 2972, US. E base of Coast Range, edge of Colorado Desert, 7 May 1894.

Mearns 3109, 3025, 3147, DS, US. Mountain Springs, San Diego Co., May 1894 (l, inflo).

Palmer 462, 88, GH. Type collection in part. San Felipe Canyon, San Diego Co., 1875 (inflo).

Vasey 626, US. Mountain Springs grade (San Diego Co.), June 1880 (l, f).

Woglum 156, SD. Mountain Springs grade (San Diego Co.), 7 May 1936.

Wolf 9455, UC. Palms to Pines Hwy. 74, 2.7 miles (ca. 4 km) below Dos Palmos Spring, Riverside Co., 29 Sept. 1939 (l, f).

Woodcock s.n., SD. Foot of Banner Grade, Sentenac Canyon (San Diego Co.), 10 Apr. 1929 (f).

Yates 5459, UC. 5 miles (ca. 8 km) W of San Felipe Canyon, San Diego Co., elev. 1,500 feet (ca. 460 m), 8 Apr. 1936 (l, f, cap).

Agave deserti pringlei

BAJA CALIFORNIA. *Border 547, 473,* DS. 3 miles (ca. 5 km) WNW of Santa Catarina, 25 May 1961, elev. ca. 4,000 feet (ca. 1,200 m) (l, f. In Sierra Juárez).

Gentry 10287, DES, MEXU, US. Northwest end of Sierra San Pedro Mártir, 23 Mar. 1951.

Gentry 16723, DES, MEXU, US. Near San Matias Pass, 22 June 1957 (l, f, photo).

Gentry 19959, DES, MEXU, US. San Matias Pass, 23 Apr. 1963 (l, f, photo).

Harbison s.n., SD. Near El Progreso, Sierra Juárez, elev. ca. 1,500 m; 32°18′ N, 115°54′ W, 1 Aug. 1965 (l, f).

Hastings & Turner 66–17, ARIZ. San Matias Pass, 19.1 miles (ca. 31 km) E of Valle Trinidád, 5 Oct. 1966, elev. 1,950 feet (ca. 600 m) (l, cap).

Moran 9838, SD, UC. 5 miles (ca. 8 km) SE of Las Filipenas, Sierra Juárez, elev. ca. 1,620 m, 30 June 1962 (l, f).

Moran 9849, SD. Just E of San Matias Pass, elev. ca. 1,020 m, 30 June 1962 (l, f, caps). "Rosettes clustered, to 1.2 m, of ca. 50 leaves; leaves green, channeled, curved, to 7 dm long, floral stem 3–5 m tall. . . . ''

Moran 15256, SD. Rancho San Pedro Mártir, Sierra San Pedro Mártir, elev. ca. 1,700 m, 5 July 1968 (l, f). "Rosettes 1–1½ m, of ca. 50 green or slightly glaucous leaves 4–8 dm long. Occasional in chaparral."

Moran 18639, SD. 2 miles (ca. 3 km) NE of Alamito, Sierra Juárez, elev. ca. 1,150 m, 3 Oct. 1971 (l, f).

Moran 21983, SD. Sierra San Pedro Mártir, 30°57′N, 115°36′W, elev. ca. 1,600 m, "on metamorphic rock in small arroyo, mile NW of oak pasture" (l, cap).

Moulis & McGill 555, ASU, DES. Sierra San Pedro Mártir, ca. 18.5 road miles (ca. 30 km) E of Meling Ranch, 21 Aug. 1972 (l, f, cap).

Orcutt s.n., UC. Hanson's Ranch, 29 July 1883.

Orcutt s.n., DS. Lower California, 1892, ex. herb. Pringle. (Sierra Juárez.)

Orcutt s.n. K, MEXU. Type. Sierra Centrales, elev. 6,000 feet (ca. 1,830 m), 7 Oct. 1882 (l, cap).

Agave deserti simplex

ARIZONA. *Engard s.n.*, DES. Near Indian Springs, Harcuvar Mt., Yuma Co., June 1974 (f only).

Gentry 23404, DES, SD, **US**. Type. N slope of Harquahala Mt., 12 miles (ca. 19 km) W of Águila, Yuma Co., 12 June 1974 (l of 5 & f of 4 pls.).

Gentry 20590, DES, MEXU, US. Near a sheep tank in Cabeza Prieta Mts. (Yuma Co.), 3 May 1964.

Gentry & Engard 23562, ASU, DES, US, S end of Silverbell Mts., Pima Co., 17 June 1975. Limestone, elev. 3,000–4,500 feet (ca. 900–1,070 m) (l, f).

Gentry & Ogden 9947, MEXU, US. N slope Harquahala Mt., elev. 3,500–5,000 feet (ca. 1,000–1,500 m), 10 Nov. 1950 (l, f).

Gentry & Ogden Photo, DES. Cunningham Pass, Harcuvar Mts., Yuma Co., 14 Nov. 1950.

Gentry & Weber 23410, DES. San Tank Mt., Yuma Co., 27 July 1974; rocky volcanic slope, elev. 2,800–3,000 feet (ca. 850–900 m) (l, cap, 2 live).

Goldman 2310, US. Tinaja Altas, Yuma Co., 20 Nov. 1913.

Harbison 4312, ARIZ, SD. Canyon leading to Del Oso Pass, Kofa Mts., Yuma Co., 8 June 1943 (The largest leaf seen of this species, with large teeth) (l, f).

Harrison et al. 7303, ARIZ. Table Top Mt., 16 Aug. 1930 (l, cap). (Doubtfully referred here.)

Mearns 305, US. Tule Mts., Mexican boundary line, 11 Feb. 1894 (l looks like *A. zebra*).

Peebles & Smith 14415, ARIZ, GH, US. Sierra Estrella, Maricopa Co., elev. 2,500 feet (ca. 760 m), 12 July 1939 (l, cap).

Peebles & Smith 13873, 13878, 13881, ARIZ, GH, US. Cunningham Pass, Harcuvar Mts., Yuma Co., 17 May 1938 (l, f, cap).

Van Devender s.n., ARIZ. Along Hwy. I 8 in Telegraph Pass, Gila Mts., N slope, elev. ca. 800 feet (ca. 250 m), 31 Dec. 1972 (l, cap).

Weber Photo, DES. Summit of Little Horn Mts., May 1970.

Weber & McGill 2549, ASU, DES. ½ mile (ca. 0.8 km) S of Sheep Tank Mine, Yuma Co., Ariz., 1 June 1970 (l, f).

West s.n., ARIZ. 18-Mile Drive, Organ Pipe Cactus National Monument, Pima Co., 27 May 1962 (l, f).

Wiggins 8643, ARIZ, DS, GH, UC, US. Rocky canyon near small tanks ca. 10 miles (ca. 16 km) S of Hwy. 80 on E foot of Mohawk Mts., Yuma Co., 1 Mar. 1937 (l, cap—Atypical elongate leaf; flowers should be collected).

CALIFORNIA. *Blakley 3231A*, DES. Above Mitchell's Caverns, Providence Mts. (flowered Santa Barbara Botanic Garden, 22 June 1972).

Brandegee s.n., UC. Providence Mts., 31 May 1902, "plants solitary."

Ferris & Bacigalupi 8161, DS. Canyon above Bonanza King Mine, Providence Mts., San Bernardino Co., 27 Apr. 1932.

Jaeger s.n., SD. 20 miles (ca. 32 km) NE of Amboy, E end of Granite Mts. (San Bernardino Co.), 26 Mar. 1929.

Munz et al. 4302, US. Vicinity of Bonanza King Mine, E slope of Providence Mts., Mohave Desert, alt. 3,500–5,000 feet (ca. 1,070–1,500 m), 21–24 May 1920 (cap).

Shaw s.n., DES. Ivanpah Mt. (Riley's claim), San Bernardino Co., Spring, 1974 (l, cap).

Van Devender et al. 74–29, ARIZ. S slopes of Whipple Mts., San Bernardino Co., elev. 1,600 feet (ca. 500 m), 17 Feb. 1974 (l, f).

Wolf 10183, DS, US. N side of Old Dad-Granite Mt. Range, Snake Spring, Mohave Desert, elev. 4,500 feet (ca. 1,400 m), 30 Apr. 1941.

Wolf 3165, CAS, DS, US. From Copper Basin on road to Parker Bridge, Whipple Mts., E Colorado

Desert, San Bernardino Co., elev. 1,400 feet (ca. 400 m), 30 Apr. 1932 (l, cap, buds).

Wolf & Everett 9023, CAS, DS, UC, US. Same locality as *Wolf 3165,* 29 July 1937 (l, cap).

SONORA. *Gentry 21203,* DES, MEXU, US. Km 2509 in mountain pass SE of Quítovac, 4 Sept. 1965, ele. 1,800–2,000 feet (ca. 550–600 m) (l, cap).

Agave gigantensis

BAJA CALIFORNIA. *Barclay & Arguelles 1990,* MEXU, US. Mountains above Rancho San Sebastián, 28 Apr. 1966.

Carter & Reese 4552, UC. Cuesta de las Parras just above Rancho de las Parras, road between Loreto & San Javier, 5 June 1963, alt. ca. 350 m.

Gentry 7693, 7713 (pars.) ARIZ, DES, MEXU, SD, UC. Picachos de Santa Clara, 5–10 Nov. 1947 (l, f).

Gentry 10339, 10342, DES, MEXU, US. Picachos de Santa Clara, Vizcaino Desert, 4–5 Apr. 1951 (doubtfully assigned here).

Gentry & Arguelles 10327, 10324, MEXU, US. Between Llano San Julio & Sierra de la Giganta, 2 Apr. 1951 (l, f, photo).

Gentry & Fox 11778, DES, MEXU, US. Sierra de las Palmas above Rancho San Sebastián, 27–29 Apr. 1952 (l, cap).

Gentry & McGill 23320, DES, MEXU, **US.** Type. Sierra de las Palmas, above Rancho San Sebastián, 20 June 1973, elev. ca. 4,000 feet (ca. 1,200 m).

Johnston 3843, CAS, US. Puerto Escondido, 29 May 1921 (l, f), "single plant in wash."

Agave margaritae

BAJA CALIFORNIA. *Beauchamp 2149,* SD. Magdalena Island, on slopes along arroyos N of Punta Magdalena, elev. ca. 5 m, Apr. 1971 (l, cap).

Beauchamp 2109, SD. West end of N side of Margarita Island, elev. ca. 50 m, 6 Apr. 1971.

Brandegee s.n., UC. Type. Magdalena Island, 14 Jan. 1889 (l, f, cap).

Gentry, Fox, Arguelles 11903, 11905, DES, US. Cienegita, Isla Santa Margarita, on sandy bajada or plain, 15 May 1952.

Moran 3540, BH. Santa Maria Bay, 31 Mar. 1952 (l, f).

Moran 4187, BH. Man-of-War Cove, Magdalena Bay, 21 May 1952 (l, cap).

Rose 16261, US. Santa Maria Bay, Isla Santa Magdalena, 18 Mar. 1911 (type of *Agave connochaetodon* Trel.).

Agave mckelveyana

ARIZONA. *Bezy 286,* ARIZ. 8.3 miles (ca. 13 km)

W of Hillside (Yavapai Co.), elev. 3,900 feet (ca. 1,200 m), 8 June 1964 (f).

Braem s.n., DS. Dean Mountain Rd., Hualapai Mts., 6 Oct. 1935 (l, caps).

Breitung 18156, DS. McClout Mts., ca. 18 miles (ca. 13 km) N of Congress (Yavapai Co.), 23 Aug. 1959.

Eastwood 18387, CAS. Aquarius Mts. (Mohave Co.), 13 May 1931 (l, bud).

Eastwood s.n., CAS. Coyote Pass between Oatman and Kingman, May 1931.

Gentry 21979, ARIZ, DES, **US.** Type. Sitgreave Pass in Black Mts., ca. 4 miles (ca. 6 km) NE of Oatman, 26 June 1966.

Gentry 23000, DES, MEXU, US. 26 miles (ca. 43 km) SE of Wikieup along Route 93, Mohave Co., elev. 3,400 feet (ca. 1,040 m), 10 June 1972 (l, f).

Gentry 23002, DES, MEXU, US. 6 miles (ca. 10 km) E of Route 93 along road to Burro Creek, Mohave Co., elev. 3,600 feet (ca. 1,000 m), 10 June 1972.

Gentry & Ogden 9981, DES, MEXU, US. Near Oatman, elev. 3,000–3,200 feet (ca. 900–975 m), low bush cover with scattered juniper and yucca in igneous mountains (Mohave Co.), 15 Nov. 1950 (l, cap).

Gentry & Whitehead 22312, ARIZ, DES, US. Hualapai Mts., along road to Recreation Park from Kingman, elev. ca. 5,000 feet (ca. 1,500 m), 1 July 1967 (l, first f).

Harrison & Kearney 7303, ARIZ. Table Top Mt., Mohave Co., 16 Aug. 1930 (l, cap).

Haskell s.n., ARIZ. Near Gold Road (Mohave Co.), 6 Dec. 1941 (l, cap).

Jones 25167, POM. Near Oatman (Mohave Co.), 1930 (in part, cap, mixed with leaf of *A. utahensis).*

Kearney & Peebles 12570, ARIZ. Rocky hills S of Hillside, Yavapai Co., elev. 3,600 feet (ca. 1,100 m), 17 Sept. 1935, "scape 8 ft. high (other plants with scapes twice as tall, but leaves always small)" (l, cap).

McKelvey 2235, 1515, 1653, A. West of Burro Creek, Aquarius Mts. (Mohave Co.), 14 May 1931 (l, f, cap).

McKelvey 2250, A. Black Mts., between Kingman and Oatman (Mohave Co.), 16 May 1931 (f in fluid).

McKelvey 4056A, A. Camp Verde region, 23 Apr. 1934 (l, cap).

McKelvey 4078, 4080, A. Juniper Mts. (Yavapai Co.), Apr. 1934.

McKelvey 1648, 1649, 4071, A. Between Kirkland and Hillside (Yavapai Co.), 29 Mar. 1930 (whole plant with cap).

Stephens s.n., UC Hualapai Mts., 5 July 1903 (l, f).

Agave moranii

BAJA CALIFORNIA. *Chambers 629,* DS, Canyon del Diablo to N and W of Picacho del Diablo, E flank of Sierra San Pedro Mártir, 6 miles (ca. 10 km) from Canyon mouth, elev. 4,500 feet (ca. 1,400 m), 17 June 1954 (f, cap, bract).

Gentry & McGill 23287, DES, MEXU, **US.** Type. 2–3 miles (ca. 3–5 km) SE of Agua Caliente, on E plain of Sierra San Pedro Mártir, 13 June 1973 (l, f).

Moran 18295, DES, MEXU, US. SE side of San Felipe Valley (Sierra Santa Rosa?) elev. ca. 530 m, 9 Mar. 1971.

Moran Photo, SD. Cerro Chato, S side of Sierra San Pedro Mártir, elev. ca. 1,800 m, 3 June 1963.

Moran 21562, SD. Sierra San Pedro Mártir, Arroyo del Cajón ca. 2 miles (ca. 3 km) from the mouth, near 30°51′N, 115°16′W, elev. ca. 810 m, 28 Dec. 1974.

Moran 24434, SD. La Vibora, Arroyo la Grulla 4 km SW of La Grulla, 9 Aug. 1977, 30°52′N; 115°30½′W; elev. ca. 1,900 m, south slope in chaparral (l, cap).

Moran 25580, SD, DES. SW slope above Campo Noche, Canyon del Diablo, Sierra San Pedro Mártir, 4 May 1978; elev. ca. 1,975, 30°59′N, 115°23½′W (l, cap).

Moran 25628, SD, DES. Common on west slope, Canyon del Diablo, Sierra San Pedro Mártir, 6 May 1978; elev. ca. 1,650 m (l, cap).

Moran 24604, SD. Arroyo el Alamar, Sierra San Pedro Mártir, 31°12′N, 115°37½′W, 23 Aug. 1977; elev. ca. 1,300 m (l, cap. Appears to be *moranii* X *pringlei,* the caps. being very small).

Agave sobria frailensis

BAJA CALIFORNIA. *Gentry & Cech 11264,* MEXU, US. Type. 4 miles (ca. 6 km) northward of Punta Frailes, 7 Oct. 1951 (l, f).

Gentry & Cech 11257, MEXU, US. 5–8 miles (ca. 8–13 km) N of Punta Frailes, 7 Oct. 1951 (l, photo).

Gentry & Fox 11858, MEXU, US. 5–8 miles (ca. 8–13 km) N of Punta Frailes, 6 May 1952 (l, f, cap).

Agave sobria roseana

BAJA CALIFORNIA. *Brandegee s.n.,* UC. La Paz, 14 Apr. 1892.

Gentry 11274, MEXU, US. Islote Gallo, 10 Oct. 1951 (l, f, cap).

Gentry & Cech 11277, MEXU, US. A W ridge of Isla Espíritu Santo, Golfo de California, 10 Oct. 1951 (l, photo).

Gentry & Fox 11869, MEXU, US. Ca. 3 miles (ca. 5 km) E of La Paz, 7 May 1952 (l, f, photo).

Hastings & Turner 64–154b, SD. 6 miles (ca. 10 km) N of La Paz.

Johnston 3989, 4001, UC, US. The Isthmus, Espíritu Santo Island, 31 May 1921.

Johnston 4002, 4003, UC, US. San Gabriel Bay, Espíritu Santo Island, 1 June 1921 (l, f).

Rose 16854, US. Type. Espíritu Santo Island, 18 Apr. 1911.

Wiggins 17828, DS. Isla Partida, just N of Isla Espíritu Santo, 20 Apr. 1962.

Agave sobria sobria

BAJA CALIFORNIA. *Barclay & Arguelles 1989,* MEXU, US. Vicinity of San Miguel de Comondú, 26 Apr. 1966, topotype (l, f).

Brandegee s.n., DS, UC. Type. Comondú Mesas, 23 Mar. 1889 (l, margins, br, cap, f).

Brandegee s.n., MO, UC. Cape region mountains, 20 Sept. 1899 (f, type of *A. brandegeei,* mixed with *A. aurea*).

Carter 5486, UC. Cumbre de la Cuesta de Las Parras, north of road, alt. ca. 1,750 m, road from Loreto to San Javier, 5 July 1970.

Carter 5487, UC. Cuesta S of Arroyo Liguí, south of Loreto, 8 July 1970.

Carter & Sharsmith 4940, UC. Along trail from San José de Agua Verde to Bahía Agua Verde, on Gulf drainage, alt. ca. 360 m, 4 June 1965.

Gentry 10304, MEXU, US. Arroyo Purísima several miles above Purísima, 31 Mar. 1951.

Gentry 10308, MEXU, US. Arroyo Purísima above Purísima, 1 Apr. 1951.

Gentry 11303, DES, Ca. 10 miles (ca. 16 km) W of Canipolé (photo).

Gentry 12382, MEXU, US. Ca. 3 miles (ca. 5 km) N of San Javier, Sierra Giganta, 2 Dec. 1952.

Gentry 12387, MEXU, US. E side of Bahía Concepción, Rancho Salto, 3 Dec. 1952.

Gentry & Cech 11291, DES, MEXU, US. Comondú, N slope of volcanic rim, 15 Oct. 1951 (l, photo).

Gentry & Fox 11811, MEXU, US. Rancho San Andreas, Sierra de las Palmas, S of Santa Rosalía, 27–29 Apr. 1952.

Gentry et al. 11876, MEXU, US. 2–3 miles (ca. 3–5 km) NE of La Paz, 7 May 1952.

Gentry et al. 11882, DES, MEXU, US. Comondú, 10 May 1952 (cap).

Harbison s.n., SD. Danzante Island, 7 Apr. 1962 (l, f, cap, very long ovaries).

Harbison s.n., SD. Top of grade on road from Comondú to Loreto, 6 Oct. 1967.

Hutchison 7399, 7473, SD. 13 miles (ca. 21 m) S of El Coyote, Concepcion Bay (½ mile N of km 85) May & Jan. 1975.

Johnston 3857, CAS, US. Ballenas Bay, Danzante Island, 24 May 1921 (l, f, cap—leaf and its armor similar to *A. roseana*).

Johnston 3887, CAS, UC, US. Agua Verde Bay, 26 May 1921.

Moran 3936, BH, DS. Canyon S of Balandra Bay, Carmen Island, 18 Apr. 1952 (l, f).

Purpus s.n., MO, US. San José del Cabo, Jan.–Mar. 1901 (f only, mixed with *A. aurea).*

Wiggins 11446, UC. 6 miles (ca. 10 km) W of Canipolé, 17 Nov. 1946 (l, different form).

Agave subsimplex

SONORA. *Felger 2721,* ARIZ. Isla Cholludo (Roca Fuca, Seal Island), 30 Oct. 1958, "common and widespread, suckering, leaves bluish; inflo. branched, only one plant in fl.; anthers yellow, filaments and tepals dusty rose-colored; base of tube pinkish white" (l, f).

Felger & Bezy 14171, ARIZ. Foothills at N side of Cerro Tepopa, 4 miles (ca. 6 km) by road W of 9 miles (ca. 14 km) by road S of Desemboque, 15 May 1966 (f).

Felger & Bezy 14182, ARIZ. 6.3 miles (ca. 10 km) S of Desemboque San Ignacio, middle bajada with sandy soil, 14 May 1966, "few scattered colonies, leaves faintly cross-banded, tepals white to pinkish" (f).

Gentry 4486, DES, ARIZ. Puerto Libertad, 22 May 1939 (f).

Gentry 10217, DES, MEXU, US. N end of Sierra Coloral, 28 Feb. 1951 (l, cap, pl).

Gentry 10221, 10222, 10223, DES, MEXU, US. Cerro Punta Tepopa, 1 Mar. 1939 (l, cap, pl).

Hastings et al. 63–8, ARIZ. Punto Cirio, about 7 miles (ca. 11 km) S of Puerto Libertad, 3 Jan. 1963 (l, cap—small leaf with big teats).

MacDougal & Shreve s.n., ARIZ. Kino Point, 17 Nov. 1923.

Moran 13022, SD, UC. Turner's Island, elev. ca. 25 m, "common on rocky slopes." Tepals light yellow or reddish, filaments pink to dark red, 25 Apr. 1966 (l, f).

Rose 16811, US. Type. Seal Island, just off Tiburón Island, 13 Apr. 1911.

Whiting 9047, ARIZ. Tiburón Island, vicinity of Tecomate, 14–22 June 1951, "abundant on rocky mountain slopes on W side of island. This one from NW corner."

Agave vizcainoensis

BAJA CALIFORNIA. *Gentry 7469,* ARIZ, DES, DS, MEXU, MICH, UC. Type. Cerro Tordillo, Sierra Vizcaíno, 12–13 Mar. 1947, elev. 400–800 feet (ca. 120–240 m) (l, f).

Gentry 7713, 7693, ARIZ, DES, DS, MICH, UC, UM. Picachos de Santa Clara, Vizcaino Desert, 5–10 Nov. 1947 (l, cap. Doubtfully assigned here, in part.).

Howell 10660, CAS. San Bartolomé Bay (= Bahía Tortuga), 14 Aug. 1932 (l, f, s).

16.

Group Ditepalae

Plants small to large with mostly light glaucous rosettes, freely seeding, cross-pollinating outbreeders, single or sparingly surculose; leaves always well armed, firm to rigid; panicles generally open, with scarious reflexing persisting bracts, only one species regularly bulbiferous; flowers generally reddish in bud and yellow at anthesis; flower tubes deep, equaling or longer than the tepals; tepals dimorphic, short, drying leathery, persisting erect, the outer larger and overlapping the inner, usually red-tipped, conspicuously corneous at apex; filaments inserted unequally in mid-tube, the three opposite the outer tepals inserted higher; pistil slender with clavate trilobate glandular stigma, over-reaching the stamens in post-anthesis; capsules strong, woody, generally long oblong. Northern Sierra Madre Occidental and outliers.

> Typical species: *Agave shrevei* Gentry, Carn. Inst. Wash. Publ. 527:95, 1942. (Ditepalae=dimorphic tepals.)

Trelease (1912) set up 10 species to form his group Applanatae. I find it more systematic to separate his group into two sections: the Ditepalae, separable primarily on flower structure, i.e., long deep tube, short leathery, persistently erect, unequal tepals, and the filaments inserted at two levels; and the Parryanae with shorter, stiffer, more numerous leaves, and longer, narrower, wilt-curving subequal tepals. Trelease in his time knew only two species with marked dimorphic tepals, *A. applanata* and *A. palmeri*. Berger (1915) placed the latter in his Reiche Americanae, but otherwise retained Trelease's Applanatae, adding his new species, *A. parrasana*. With the specimens and species added during the last 50 years, the two groups are much better discernible. Rather than subvert the term Applanatae away from its original context, it is preferable to abandon the name in favor of two new definable groups, the Ditepalae and Parryanae.

The leaf color in the Ditepalae is commonly a glaucous gray, but this varies from the nearly white forms found in *Agave shrevei* and *A. applanata* to the silvery sheen on some forms of *A. palmeri* and the glaucous yellow-green color of this species and several others. *A. wocomahi* is a distinctive green, while *A. colorata* has a zoned leaf, frequently reddish or pink alternating with light gray on dry rocky sites. The colors are imparted by the anthocyanins and other pigments in cells underlying the cuticle, which is a highly light-reflective covering, colorless or light gray to nearly white. The epidermis has a rather thick over-lying cuticle, minutely sculptured, as in all *Agave*. To the touch, *A. applanata*, *A. shrevei*, and *A. colorata* are slightly asperous, while *A. wocomahi* is quite smooth, the other species ranging imperceptibly between these extremes. Under the microscope, a rather regular stomatal pattern emerges with only occasional confluences. The stomatal complex is elaborate with 5 planes in surface view. The margins of the surface pore are irregular. The 2 polar lips are broad, the 2 lateral lips narrow; rim of the suprastomatal chamber is absent (for terminology, see Gentry and Sauck, 1978).

The Ditepalae are among the more highly modified or advanced in the genus *Agave*. The strong erect flowers are sturdy red and yellow receptacles providing nectar for birds, bats, and insects, and are specializations common to many of the advanced seed plants of the Tertiary Period. The outer tepals are conspicuously larger than the inner, turn leathery with drying, and the cucullate apex is hardened with a dark, glandular, corneous cap. The dimorphic tepals are unique among the monocotyledons, having nearly progressed from a single ranked 6-merous corolla to a 2-ranked stage of 3-merous sepals and petals, like the dicotyledons. Other sections of the genus show similar evolutionary flower specialization, but none in such advanced degree. With the exception of *A. murpheyi,* all the Ditepalae are freely seeding out-breeders, several appearing self-sterile, relying little on vegetative reproduction, and are altogether recognizable as Tertiary modern seed plants. Introgressive intermediates, as appear to exist in *A. palmeri* and *A. shrevei,* are, therefore, not surprising.

Pollination and a Nectar Flyway

Recent studies have established that bats are regular and important pollinators of agaves, especially *Agave palmeri*. The bats of the genus *Leptonycteris,* subfamily Glossophaginae, are anatomically structured for nectar-lapping and pollen feeding. They are of migratory habit and have been observed to feed in small flocks during the seasonal flowering of *A. palmeri* (Howell, 1979). Anthers dehisce during night hours. These bats feed on flowers other than agave, but certain structures of the latter are notably co-adaptive with bats; e.g., abundance of nectar in a strongly scented mass in individual cuplets held erect by geotropic flowers. The tough short leathery tepals of the Ditepalae appear unusually well structured to support the clambering bats and protective of the nectar-holding tubes. Such structures and functions in disparate organisms can develop only over long periods of time and indicate adaptive evolution. Geologic time is an appropriate term in the case of agaves, where generation spans require 15 to 20 or more years. If you ever asked yourself, how and when did agaves get this way?, this co-adaptation is one clue to consider. I would call this bio-relationship another case of symbiosis, rather than a "syndrome." There is nothing psychopathic about it; nectar-feeding, rather than nectar-phobia.

Table 16.1. Stomata Counts per square mm of Representative Species in the Ditepalae

SPECIES	COLL. NO.	SURFACE		LOCALITY
		UPPER	LOWER	
applanata	{ 12597	42	30	Cadereyta, Quer.
	{ 20444	33	22	Nochistlán, Oax.
colorata	3050	23	23	Aquibiquichi, Son.
durangensis	10576	22		Sierra Registro, Dur.
fortiflora	{ 19808	25		Sierra Jojoba, Son.
	{ 11630	22		
palmeri	21990	32	31	Texas Canyon, Ariz.
shrevei	11385	31	30	Sierra Saguaribo, Son.
s. ssp. *magna*	23360	34		Barranca Batopilas, Chih.

Presumably, other members of the Ditepalae participate in the bat-agave symbiosis, because their flowers are structurally similar. The flowering seasons in Fig. 16.1 indicate much more. There is a wave of nectar flow from spring to winter, north to south. Starting with the spring flowering *A. colorata* in Sonora followed by *A. palmeri* in Arizona, the taxa provide a seasonal supply of bat food through the Sierra Madre Occidental to the south: viz., *A. shrevei* of southern Sonora and Chihuahua to *A. durangensis* and *A. wocomahi* through Durango and Zacatecas, which flower up to December. South of there the winter bloomers of the Group Crenatae are in flower production and could lead the bats via the Group Hiemiflorae to Central America. Is this the migratory pathway of *Leptonycteris,* a nectar flyway mutually evolved?

The Ditepalae give distinction to and characterize the succulent element in the Sierra Madrean flora (Figs. 16.15, 16.17, 16.32). Rainfall and the breeding seasons are outlined in Fig. 16.1. Outside this region there is one disjunct member, *A. applanata,* which, although it shows considerable variance in rosette ontogeny and fibrous leaf morphology, is typically Ditepalae in floral structure. Before man's apparent carriage of this plant to the Sierra Madre region, it was probably isolated from its congeners for considerable time in the Tertiary Period.

The Ditepalae are all low in sapogenin content (see Table 16.2). All the species have

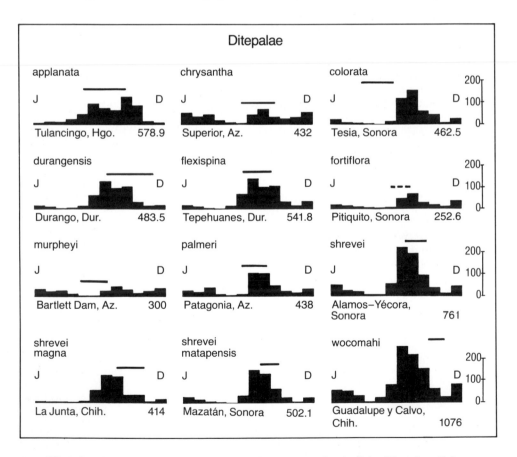

Fig. 16.1. Rainfall (silhouettes) and flowering seasons (bars) of the Ditepalae. Relevant meteorological stations with an average annual rainfall in millimeters. Data from U. S. Weather Bureau and Servicio Meteorológico Mexicana (1939); see also Hastings et al. (1964, 1965, 1969). Flowering periods based on herbarium specimens and field observations; uncertainty expressed by broken bars.

generally been acceptable and useful as food and beverage to the Amerindians, as noted in the following accounts of species. Historical accounts record widespread use by the Tepehuan, Tarahumara, Warihio, Mayo, Yaqui, Opato, Lipan and other tribal groups of the Apaches, the Pimas, Papagos, and Maricopas; all tribes having access to the edible Ditepalae *Agave*. They are the predominant agave group in the northern Sierra Madre Occidental.

Key to Species of the Ditepalae

1. Leaves of mature plants linear to long-lanceolate, 5 to 12 times longer than wide; widest at or below the middle, margin straight or nearly so   2
1. Leaves of mature plants ovate to lanceolate, 2 to 6 times longer than wide; widest at or above the middle, margin undulate or prominently teated   6
2. Leaves very rigid and tough, the latest on mature plants much longer than on younger stages, mostly 50–130 cm long; teeth remote, mostly 2 to 5 cm apart   *applanata*, p. 421
2. Leaves less rigid and tough, the latest on mature plants not much longer than on middle stages, mostly 40–70 cm long; teeth proximal, mostly 0.5 to 2 cm apart (except *chrysantha*)   3
3. Leaves linear; spines conical, short, to 2 cm long; panicle regularly bulbiferous; flowers 70 mm long   *murpheyi*, p. 440
3. Leaves lanceolate; spines acicular, 2.5–6 cm long; panicle normally not bulbiferous   4
4. Flowers larger with broad tubes, 75–82 mm long; tepals ca. 2 times the length of tube   *fortiflora*, p. 438
4. Flowers smaller with narrow tubes, 45–65 mm long; tepals equaling or shorter than tubes   5
5. Leaf edge relatively straight with closely set teeth, mostly 0.5 cm apart; tepals conspicuously tipped with red or reddish brown   *palmeri*, p. 443
5. Leaf edge undulate; larger teeth mostly 1–3 cm apart; tepals yellow   *chrysantha*, p. 426
6. Rosettes medium to large, generally not suckering, green to glaucous gray; panicles large, deep, 5–8 m tall, with 20 to 30 umbels   7
6. Rosettes small to medium, surculose, glaucous gray; panicles smaller, 2–5 m tall with 8 to 20 umbels   9

7. Medium to large, dark green to light green rosettes; leaves smooth, the margins straight to undulate; tepals equaling or exceeding tube in length   *wocomahi*, p. 456
7. Large glaucous gray rosettes; leaves finely asperous, the margins frequently deeply crenate and prominently teated; tepals shorter than the tube   8
8. Rosettes large, mostly 1.4–1.7 m tall, 2.5 m broad; leaves long-lanceolate, 1.2–1.5 m long, flexible; larger teeth remote, 3–5 cm apart; filaments long, 55–70 mm   *shrevei* ssp. *magna*, p. 451
8. Rosettes shorter, mostly 0.8–1.2 m tall, 1.2–1.8 m wide; leaves short-lanceolate, 0.4–0.9 m long, rigid; larger teeth proximal, 1–2 cm apart; filaments shorter, 48–57 mm long   *durangensis*, p. 433
9. Rosettes small, 25–30 cm tall; leaves 35–40 in number, 16–30 cm long; panicles 2–3 m tall with 6–12 few-flowered umbels; filaments not conspicuously 2-ranked in insertion   *flexispina*, p. 436
9. Rosettes larger, mostly 40–70 cm tall; leaves more than 40, 30–60 cm long; panicles more than 3.5 m tall, or, if shorter, with 15 to 20 densely flowered umbels; filaments inserted at two distinct levels in tube   10
10. Leaves uniformly gray; marginal teats moderate in size; panicle over 3 m tall, small, open with few umbels in upper ⅓ of shaft; tepals shorter than the deep tube   *shrevei*, p. 447
10. Leaves conspicuously cross-zoned and sometimes tinged with red; teats prominent; panicle deep, relatively dense with many umbels in upper ½ to ⅔ of shaft; tepals about equaling tube in length   *colorata*, p. 431

Table 16.2. Sapogenin Content in the *Ditepalae*
(given in percentages on dry weight basis)

COLL. NO.	SOURCE LOCALITY	MONTH COLL.	PLANT PART	SAPOGENINS*			
				TOTAL	HEC.	MAN.	TIG.
applanata							
12597	Cadereyta, Quer.	May	leaf	0.1			
12579	Onemile, Puebla	May	leaf	0.0			
12580	Tulancingo, Hidalgo	May	leaf	0.2			
chrysantha							
9984	Queen Creek, Ariz.	Dec.	leaf	0.5			
9984	Queen Creek, Ariz.	Dec.	stem	0.0			
9991	Globe, Ariz.	Dec.	leaf	0.0			
11625	Superior, Ariz.	Mar.	leaf	0.3	90	10	
?	Superior, Ariz.	July	whole plant	0.25	90	10	
colorata							
10270	Aquibiquichi, Son.	Mar.	leaf	0.0			
10270	Aquibiquichi, Son.	Mar.	inflorescence	0.0			
10270	Aquibiquichi, Son.	Mar.	flower	0.0			
11368	Bachaca, Son.	Dec.	leaf	0.1			
11641	Sierra Bojihuacame, Son.	Apr.	base	0.0			
durangensis							
10576	Sierra Registro, Dur.	June	leaf	0.0			
10589	Valle de Mezquital, Dur.	June	leaf	0.0			
fortiflora							
11618	Sierrita de Lopez, Son.	Feb.	leaf	0.0			
11630	Sierra Jojoba, Son.	Mar.	leaf	0.2			
murpheyi							
9978	B. T. Sw. Arboretum, Ariz.	Nov.	leaf	0.0			
9978	B. T. Sw. Arboretum, Ariz.	Nov.	flower bud	0.0			
10028	Sells, Ariz.	Dec.	flower bud	0.02			
palmeri							
9915	Pyramid Mt., N.M.	Oct.	leaf	0.5			
9915	Pyramid Mt., N.M.	Oct.	inflorescence	0.0			
9932	Sonoita, Ariz.	Dec.	leaf	0.0			
11627	Nogales, Ariz.	Mar.	leaf	0.0			
?	Nogales, Ariz.	July	stem	0.0			
shrevei							
10235	Sierra Charuco, Son.	Mar.	leaf	0.0			
10235	Sierra Charuco, Son.	Mar.	inflorescence	0.0			
11385	Sierra Saguaribo, Son.	Dec.	leaf	0.44		c.30	61
11401	Sierra Canelo, Chih.	Jan.	leaf	0.0			
11404	Sierra Charuco, Son.	Jan.	leaf	0.0			
11580	Guajaray Scarp, Son.	Feb.	leaf	0.0			
11605	Navojoa, Son.	Feb.	leaf	0.0			
s. ssp. *matapensis*							
11612	Nacori, Son.	Feb.	leaf	0.0			
11607	Sierra de Matape, Son.	Feb.	leaf	0.0			
wocomahi							
11586	Guajaray Scarp, Son.	Feb.	leaf	0.0			
11588	Guajaray Scarp, Son.	Feb.	leaf	0.0			
11397	Guicorichi, Chih.	Dec.	leaf	0.0			

SOURCE: Wall et al. (1954–57).

* Hec. = hecogenin; Man. = manogenin; Tig. = tigogenin; c= chlorogenin.

Agave applanata
(Figs. 16.1, 16.2, 16.3, 16.6, 16.7, Tables 16.1, 16.2, 16.3)

Agave applanata Koch ex Jacobi, Hamb. Gart. Blumenztg. 20:550, 1864.

Rosettes medium to large, mostly glaucous gray, multi-leaved, single, strongly armed, radial in the wild forms 5–10 dm tall, 10–20 dm broad, in the cultivars 12–20 dm tall, 25–30 dm broad; leaves on mature plants much longer than earlier stages, 40–60 x 7–10 cm (wild) or 100–140 cm long (cult.), linear lanceolate, usually widest at or near the base, very rigid, hard-fleshed with coarse, strong, abundant fiber, the margins corneous throughout or margin lacking in the mid-blade; teeth very strong, sharp, the larger through the mid-blade 8–15 mm long, nearly straight or frequently flexed downward, dark brown becoming light waxy pruinose, mostly 4–6 cm apart, smaller and more proximal on wild forms; spine very strong, 3–7 cm long, 0.8–1.4 cm wide at base, flat or broadly hollowed above, dark reddish brown aging grayish, decurrent with the corneous margin; panicle 4–8 m tall, narrow and deep with numerous laterals; peduncular bracts short, scarious, reflexed; flowers yellow with greenish ovary, 55–80 mm long; ovary 35–38 mm long, angular cylindric, tapering toward base; tube cylindric, 15–22 mm deep, thick-walled, deeply furrowed from tepal sinuses; tepals unequal, thick, linear-lanceolate, brown-tipped, cucullate, the outer 15–22 mm long, the inner shorter, narrower with a narrow prominent keel; filaments 45–55 mm long,

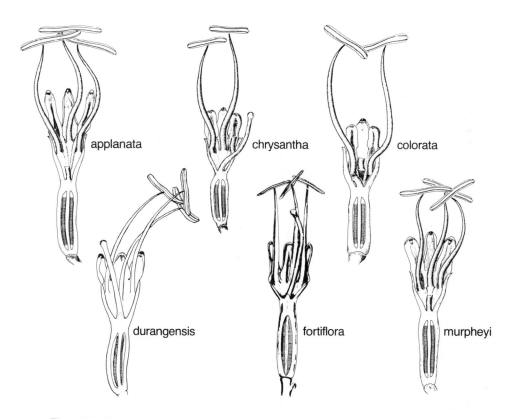

Fig. 16.2. Long sections of Ditepalae flowers: *Agave applanata*, *A. chrysantha*, *A. colorata*, *A. durangensis*, *A. fortiflora*, and *A. murpheyi*.

inserted 2-ranked in mid-tube, flattened; anthers 23–30 mm long, centric or excentric; capsules and seed not seen.

Type: None ever cited; *Trelease No. 1,* MO, ''Limon, on the Interoceanic R. R. above Jalapa,'' Veracruz, may be taken as neotype.

In this locality the species grows wild upon the weathering lava beds which extend for several miles northward and westward at elevations of 7,600 feet (2,300 m) and above. It appears to be naturally endemic in this semiarid cool highland of Veracruz and adjacent

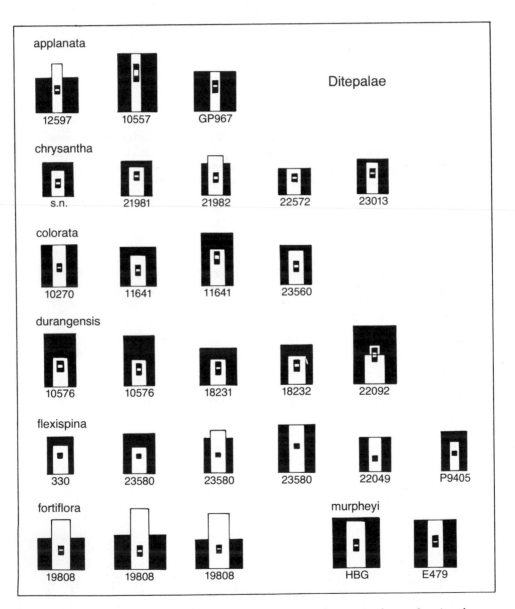

Fig. 16.3. Ideographs of Ditepalae flowers: *Agave applanata, A. chrysantha, A. colorata, A. durangensis, A. flexispina, A. fortiflora,* and *A. murpheyi.* Depicted are relative proportions of tubes (black) to outer tepals (white column) and filament insertion in tubes. Specific series show intraspecific variation.

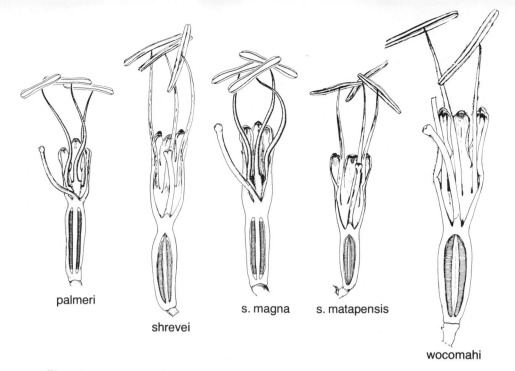

Fig. 16.4. Long sections of Ditepalae flowers: *Agave palmeri, A. shrevei, A. shrevei* ssp. *magna, A. shrevei* ssp. *matapensis,* and *A. wocomahi.*

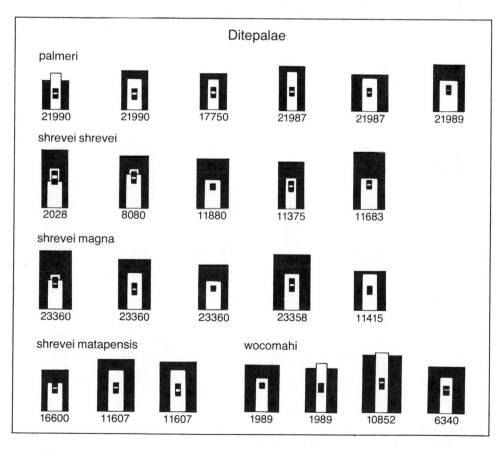

Fig. 16.5. Ideographs of Ditepalae flowers: *Agave palmeri, A. shrevei* ssp. *shrevei, A. shrevei* ssp. *magna, A. shrevei* ssp. *matapensis,* and *A. wocomahi.*

Fig. 16.6. *Agave applanata*. A submature plant on the lava bed near El Limón in highland Veracruz and a fruiting plant south of Creel, Chihuahua, October 1973.

Puebla. Elsewhere it is found widely as a cottage cultivate with local adventives over the Mexican plateau from central Chihuahua to Oaxaca (see Fig. 16.7).

The short, leathery, dimorphic tepals and 2-ranked filament insertions definitely align this polymorphic species with other members of the Ditepalae. The short stiff compact leaves in the juvenile and middle stages of growth resemble leaves of the Parryanae. The long ensiform leaves that finally appear with the maturing plants are quite distinct from any near relatives. The large sharp teeth and spines held rigidly on the tough leaves make it one of the most formidable agaves, and it draws blood from the most experienced collectors. The largest rosettes, which reach 2 m in height, are among the most arresting in the genus, especially when appearing in light glaucous forms. No rhizomatous or bulbiferous offspring have been observed. Numerous seedlings have been noted among old colonies and the species appears to be a free seeder, although no capsules or seeds have been collected. Introgression between *Agave applanata* and *A. durangensis* may explain variants showing leaf characters of both these species, as observed 8 to 10 miles (12 to 16 km) south of Boca de Mezquital in central Durango, *Gentry & Gilly 10589* (see Exsiccatae). Near Nochistlán, Oaxaca, many plants were found dead or dying from an unrecognizable cause.

Trelease (1911) gave numerous bibliographic citations, many of which, however, are horticultural catalogue listings. These, as well as others, may not really represent this species. He wrote, ''Long cultivated, but of doubtful origin, and greatly misunderstood because of the difference between juvenile, moderately developed, and mature plants.'' Jacobi noted that it was in a Berlin garden by 1856. Laurentius listed its source as ''Mirador'' which is misleading (see p. 57).

The use of the very strong fibers and the distribution of *Agave applanata* appear to be intertwined. At Magueyitos in highland Veracruz it was called "maguey de ixtle," where it was cultivated in the yard of an Indian dwelling. The Indian woman owner showed us how the leaves on a mature plant (*Gentry et al. 20420*) could be pulled manually right off the drying trunk. This was done with a short sidewise jerk and proved much easier than cutting them. It may have been more generally cultivated in the central highlands of Mexico before the modern development of henequen plantations, the fiber products of which are now cheaply available in markets everywhere. The long-leaved, strong-fibered, large forms, occurring spontaneously near old settlements, as near Nochistlán in Oaxaca, may be survivals of earlier cottage plantings and selections.

In Chihuahua the plant has been called "maguey de la casa" (field note on *Gentry & Bye 23363*) and the Tarahumara name was given as "socolume." In Durango and Chihuahua the species is found sporadically along the old Indian trading trail between central Mexico and Casas Grandes, Chihuahua. The plants observed and collected near Cusarare, Chihuahua, were spontaneous (Fig. 16.6), but not far away are two old Jesuit missions. It has been observed in old cemeteries in Durango and the plant may have religious or ceremonial significance not yet reported. All of these observations suggest that *A. applanata* has been disseminated far to the north by man's hand in historic or prehistoric times.

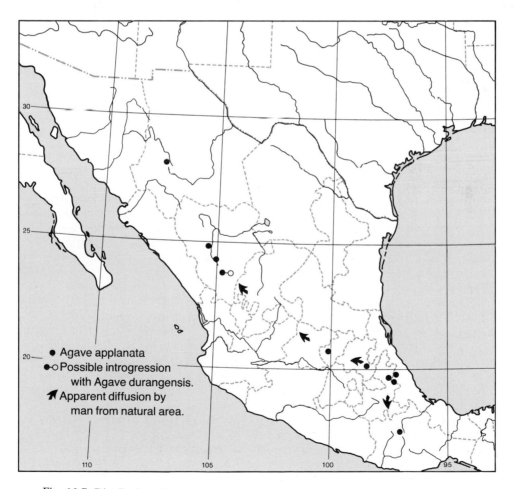

Fig. 16.7. Distribution of *Agave applanata,* based on herbarium specimens, sightings, and photographs.

Agave chrysantha
(Figs. 16.1, 16.2, 16.3, 16.8, 16.9, 16.10; Tables 16.2, 16.3)

Agave chrysantha Peebles, Biol. Soc. Wash. Proc. 48: 139, 1935.

> *Agave palmeri* var. *chrysantha* (Peebles) Little ex Benson, Amer. J. Bot. 30: 235, 1943.
> *Agave repanda* Trel., in Arnold Herb. only.

Rosettes small compact to rather large and open, grayish to yellowish green, 5–10 dm tall, 8–18 dm wide, usually single, suckering infrequently; leaves mostly 40–75 x 8–10 cm, linear-lancolate to lanceolate, usually but little narrowed below the middle, widest in the middle, straight, deeply guttered, the margin nearly straight to repand; larger teeth mostly 1–3 cm apart, 5–10 mm long, straight or flexed, the smaller teeth toward the base and interstitial; spine slender, 25–45 mm long, brown or castaneous to gray in age, openly grooved above, decurrent for 5–15 cm to upper teeth; panicles 4–7 m tall, small, narrow, with 8 to 18 small congested umbels in upper ¼ to ⅓ of shaft; flowers yellow, 40–55 mm long; ovary 22–30 mm long, slender, with short constricted neck; tube 8–13 mm deep, 10–12 mm wide, with bulging angles, narrowly grooved, thick-walled; tepals dimorphic, thick, strictly erect, the outer 9–15 mm long, a little shorter to little longer than the tube,

Fig. 16.8. *Agave chrysantha* in the type locality in Queen Creek Canyon near Superior, Arizona, with fruiting panicles, Dec. 1951.

Fig. 16.9. A mature fruiting plant of *A. chrysantha* with collapsing leaves.

rarely red-tipped; filaments inserted 2-ranked, 3–9 mm above base of tube, 35–48 mm long; anthers 16–20 mm long, yellow; capsules congested, 35–50 x 13–15 mm, linear oblong to obovoid, apiculate, stipitate or sessile, the walls rather thin; seeds 6–7 x 4.5–5 mm, mostly crescentic, punctulate rugose, with strong marginal wing.

Type: *Peebles & Harrison 5543*, US Queen Canyon, Pinal Mountains, Arizona, flowering, June 21, 1928 (2 sheets, l, f).

Agave chrysantha has a small range and is endemic in central Arizona on the granitic and volcanic mountain slopes between 3,000 and 6,000 feet (900 and 1,800 m) elevations (Fig. 16.10). The main populations are on and about the Santa Catalina, Pinal, and Mazatzal ranges with an eastern segment on the southern slopes of the Sierra Ancha, and northwestern outliers on the New River Mountains and northeastern slopes of the Bradshaw Mountains. It is frequently associated with the chaparral and juniper communities and barely enters the pine woods of the Transition Zone.

The nearest relative of *Agave chrysantha* is quite clearly *A. palmeri*. The flower structures are very similar, agreeing generally in size and organ proportions. As summary of the flower measurements (Table 16.3), the following averages provide these floral formulas:

   palmeri   o=28,   t=14.0,   s=12.3,   f at 6–8,   45, a=20,   tl=53.3
   chrysantha   o=25,   t=11.4,   s=11.4,   f at 5–6,   40, a=19,   tl=47.6

A. chrysantha has a smaller flower and a more shallow tube with proportionally longer tepals. While this is significant, it is too small a difference for distinguishing many individuals, considering the variability in most populations of both species. *A. chrysantha* inflorescence differs consistently in its clear yellow perianth, in the more congested umbels, and the shorter panicles. In the central area of its range, leaves of *A. chrysantha* are distinct in their broader, frequently shorter, more lanceolate form and remote large

Table 16.3. Flower Measurements in the Ditepalae (in mm)

TAXON & LOCALITY	OVARY	TUBE	TEPAL	FILAMENT INSERTION & LENGTH		ANTHER	TOTAL LENGTH	COLL. NO.
applanata								
Querétaro, Quer.	43	13×17	18×4	10&7	41	25	74†	12597
Durango, Dur.	38	22×17	22×5	18&12	53	30	80†	10557
S Papasquiaro, Dur.	35	15×16	15×4&13	11&9	50	25	65†	GP967
chrysantha								
Queen Creek, Ariz.	25	9×10	12×5&11	4–3	35	18	45†	s.n.
	27	13×12	10×5	6&5	38	21	51†	s.n.
N of Globe, Ariz.	24	13×12	11×6&11	9–8	47	17	49†	21981
	30	12×11	15×6&13	8&7	48	20	56†	21982
New River Mts., Ariz.	25	10×13	10×5½×9	7&6	35	11	44†	22572
Black Canyon Mesa, Ariz.	23	13×12	12×5&10	9&7	35	18	48†	23013
colorata								
Aquibíquichi, Son.	29	16×14	16×5&14	8&6	48	26	61†	10270
Sierra Bojihuacame, Son.	27	15×14	12×7&10	7&5	54	21	53†	11641
	40	20×13	14×5&12	12&9	65	24	73†	11641
B. San Carlos, Son.	33	15×12	13×5&12	7&6	57	24	60†	23560
durangensis								
	34	20×12	11×5&9	9&7	55	23	65†	10576
	30	19×12	10×6&8	7&6	52	24	59†	10576
Sierra Registro, Dur.	28	14×14	10×6&8	8&5	48	20	54†	18231
	21	15×13	11×7&10	9&7	45	19	48†	18232
	40	22×17	11×8&9	13&11	57	23	73†	22092
flexispina								
	22	15×13	14×6&11	7–6	45	18	50†	23580
S of Parral, Chih.	29	15×13	10×6&8	7–6	42	15	54†	23580
	31	13×11	16×6&14	7&7	50	23	59†	23580
	37	18×14	18×7&16	10	40	21	73†	23580
Río Florida, Dur.	30	13×12	13×5&12	5&5	42	21	55†	22049
SE of Tepehuanes, Dur.	25	15×11	11×3	6–7	42	22	52†	P9405
	33	15×11	13×4	6–7	39	23	60†	P9405
Tepehuanes, Dur. (type)	24	14×10	11×6&9	8&6	42	20	49*	330
fortiflora								
	44	12×18	19×7	8&6	60	20	75†	19808
Sierra Jojoba, Son.	45	13×20	23×7	8&6	45	22	79†	19808
	45	11×18	21×7	8	60	21	77†	19808
	48	11×20	23×8	8–9	60	22	82†	19808
murpheyi								
Hunt. Bot. Gard., Cal.	32	17×15	18×7&17	10&8	50	25	68†	s.n.
	35	19×19	18×7&17	11–10	45	26	71†	s.n.
Bartlett Dam, Ariz.	29	18×16	18×6&16	11–9	45	25	65†	E479
palmeri								
	31	11×11	14×5&9	7–5	40	20	56†	21990
Texas Canyon, Ariz.	25	10×12	12×6&10	5–4	50	16	46†	21990
	36	15×11	12×6&10	7–5	47	21	63†	21990
NW of Nogales, Ariz.	28	14×10	12×5&10	8&6	42	17	52†	17750

Table 16.3. cont.

TAXON & LOCALITY	OVARY	TUBE	TEPAL	FILAMENT INSERTION & LENGTH		ANTHER	TOTAL LENGTH	COLL. NO.
Sierra Jojoba, Son.	27	17 × 11	15 × 5 & 13	8 & 7	46	25	59†	21987
	36	14 × 15	13 × 7 & 11	8 & 7	43	23	62†	21987
Sierra Baviso, Son.	22	18 × 12	12 × 6.5	9 & 8	40	21	52†	21989
shrevei								
Sierra Canelo, Chih.	31	23	10 × 6 & 8	14 & 10	50	22	64*	2028
	30	21	10 × 6 & 8	13 & 9	45	22	62*	2028
Sierra Charuco, Chih.	33	20 × 11	13 × 7 & 11	12 & 10	58	26	65*	8080
Guajaray Scarp, Son.	30	19 × 12	11 × 7 & 10	8 & 8	50	27	61†	11580
Sierra Potrero, Son.	36	18	12 × 5 & 10	10 & 8	50	23	66*	11378
Sierra Saguaribo, Son.	37	22 × 12	12 × 7 & 11	11 & 9	40		71†	11633
s. ssp. *magna*								
Barranca Batopilas, Chih.	33	22 × 13	11 × 6 & 9	11 & 9	48	25	65†	23360
	34	19 × 13	14 × 6 & 12	9 & 6	65	27	66†	23360
	27	17 × 11	11 × 6 & 9	9 & 7	55	23	55†	23360
Quirare, Chih.	30	21 × 14	14 × 6 & 12	11 & 8	58	28	65†	23358
Cerro Babuyo, Sin.	34	15 × 13	14 × 6 & 12	9 & 7	58	23	62†	11415
s. ssp. *matapensis*								
Matape, Son.	23	16 × 11	11 × 7 & 9	10 & 8	45	20	51†	16600
	41	20 × 14	16 × 7 & 14	10 & 8	52	30	77†	11607
	34	19 × 14	16 × 7 & 14	10 & 7	55	28	68†	11607
wocomahi								
Guicorichi, Chih.	32	18	13 & 11	11 & 9	55	22	62*	1989
	34	17	19 × 5 & 17	11	55	27	70*	1989
Sierra Charuco, Chih.	35	22 × 15	23 × 5 & 18	13 & 10	62	33	78†	10252
Surutato, Sin.	33	18	15 × 5 & 13	11 & 10	65	25	65*	6340
	34	18	14 × 5 & 13	10 & 8	63	26	67*	6340

* Measurements from dried flowers relaxed by boiling.

† Measurements from fresh or pickled flowers.

teeth. Unlike *A. palmeri* the margin is frequently undulate to repand. Both of them frequently show small teeth interstitial to the large ones. On the eastward and southern margins of the *A. chrysantha* area, the *A. palmeri* side, the variable forms of the two taxa become too intermingled to identify individual plants as one or the other. Peeble's note on the Chimney Creek population expresses this condition: "Typical plants of neither *A. palmeri* or its golden flowered variety could be found at Chimney Creek. Presumably, the variations in size, color, & texture of the flowers and the size and form of the fruits—is the result of hybridization between typical purple-flowered *A. palmeri*—with the golden flowered variety found in the Pinal and Santa Catalina Mts."

The coherent leaf form of *Agave chrysantha* also breaks down in its area of contact with the distantly related *A. parryi*, as about Sierra Ancha and in the Mazatzal Range. The leaf form of the former in such areas shortens, broadens, and thickens like the *A. parryi* leaf. I found it very difficult to identify leaf specimens of McKelvey from near Coolidge

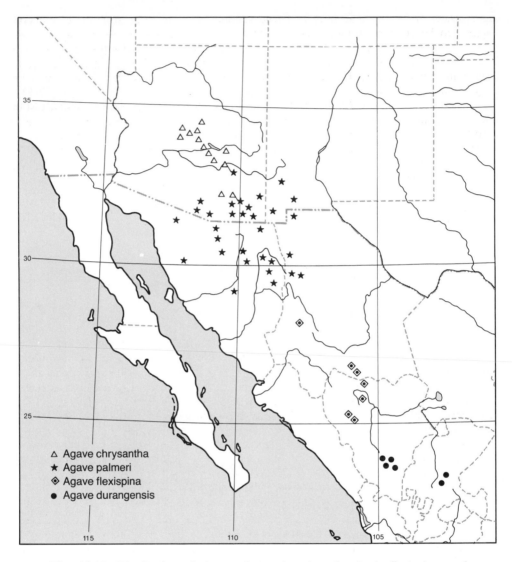

Fig. 16.10. Distribution of *Agave chrysantha, A. palmeri, A. flexispina,* and *A. durangensis,* based on herbarium specimens.

Dam, as they had the shorter lanceolate form of *A. chrysantha* with the small proximal teeth of *A. palmeri.* Leaf specimens alone may be difficult or impossible to identify with confidence, but when accompanied by flowers and photos are usually easily placed. The series of flowers studied from the central part of the *A. chrysantha* area are all of the Ditepalae kind. However, they show instability in the tube to tepal ratio (Table 16.3) and in the insertion of the filaments; a criteria spread quite as unsatisfactory to taxonomic clarity as that found in the leaves.

In the core of its area it has good morphological weight. It may be a geologically young species which has not yet reached a stable or isolated condition. It appears to be mixing genes with its neighbors and may even have had its origin through introgression with *A. palmeri* and *A. parryi,* which because of their larger species areas are theoretically older. The Ditepalae have advanced considerably in their modified perianth, as well as spatially and environmentally from the generic center of origins. Some of the taxa appear to be adapting frontier plants naturally generating new forms and species; *A. chrysantha* among them. Such speculations on origins and species definitions point out the need for intensive studies on *Agave* populations. Until such studies are made, I prefer to maintain Peebles' binomial rather than accept Little's trinomial, which was made without any explanation whatever.

There is a thriving massive population of typical *Agave chrysantha* above and east of Superior along a spectacular mountain section of route 70. Peebles wrote of this (loc. cit.): "When in flower *A. chrysantha* is a striking and attractive feature of its native mountains. It is especially fine and abundant in the canyons above Superior, Pinal County. The golden-yellow flowers are copiously nectariferous and emit a strong odor resembling that of coconut meat. Reproduction is by means of the numerous and highly fertile seeds and by offsets, rarely by bulbils developed in abnormal flower clusters." A young offset brought to the Desert Botanical Garden in the winter of 1939–40 did not bloom until June 1979—40 years!

Agave colorata
(Figs. 16.1, 16.2, 16.3, 16.11, 16.12, 16.13, 16.30; Tables 16.1, 16.2, 16.3)

Agave colorata Gentry, Carn. Inst. Wash. Publ. 527:93, 1942.

Plants short-caulescent, small to medium, few-leaved, compact, tapered at base or hemispherical, light gray, suckering sparingly from base; leaves 25–60 x 12–18 cm, ovate, short-acuminate to lanceolate, thick, firm, convex below toward base, plane to concave above, frequently cross-zoned and red-tinted, glaucous, asperous, the margins prominently crenate or mammillate; teeth mostly 5–10 mm long in the mid-blade, 1.5–3 cm apart, smaller below, straight or flexuous, brown to grayish; spine mostly 30–50 mm long, subulate, straight or flexuous, narrowly grooved above in lower half, brown to gray; panicle 2–3 m tall, narrow, with 15 to 20 densely flowered umbels in upper $^1/_2$ to $^2/_3$ of shaft; lower bracts leaf-like, the others soon drying appressed at base, reflexed, reddish, 15–10 cm long; flowers 50–70 mm long, reddish in bud and on tepals, opening yellow, thick, in dense umbels; ovary pale green, 25–40 mm long with short unconstricted neck; tube 15–20 mm long, 12–14 mm wide, cylindric, thick-walled, narrow within, narrowly grooved; tepals unequal, linear, thickly fleshy, strictly erect, the outer larger, overlapping, 12–16 mm long, usually reddish on cucullate apex, the inner shorter, thickly keeled; filaments 45–65 mm long, stout, inserted at two levels in mid-tube, yellow or pink, flattened, tapered; stamens 22–26 mm long, yellow; pistil thick, sometimes pink, exceeding stamens after anthesis; capsules 4.5–5.5 x 1.5–1.7 cm, oblong to clavate, thickly short-stipitate or sessile; seeds 6 x 5 mm, lachrymiform, the hilar notch frequently obscure, marginal wing narrow.

Type: *Gentry 3050*, Peñasco Blanco, near Aquibiquichi, Sonora, Feb. 1937, in Dudley Herb., CAS, isotype in ARIZ.

Agave colorata is a foothill or coastal plant growing on open rocky sites in the Thorn Forest (Fig. 16.12). With its dimorphic tepals and large deep tube, it fits nicely into the Section Ditepalae. Its nearest relative, morphologically and geographically, is *A. shrevei*, from which it differs in the broader cross-zoned leaf with deeply crenate to undulate margins, the more densely flowered panicles with closely set umbels, and the shorter, wider tube and thicker ovary. *A. shrevei* grows in the higher elevations of the Sierra Madre. Recent collection of flowering specimens near Bahía San Carlos (*Gentry & Engard 23560*) definitely place that population with *A. colorata*, not with *A. fortiflora* as was previously reported by Gentry (1972). Plants with atypical linear-lanceolate leaves

Fig. 16.11. *Agave colorata* with inflorescence, cultivated near Murrieta, California, June 1965.

Fig. 16.13. A long-leaved form of *Agave colorata* in a palm canyon north of San Carlos Bay, Sonora.

Fig. 16.12. *Agave colorata* at the type locality near Aquibiquichi, Sonora, ca. 1938, with pink cross bands.

from a shady canyon site, N of Bahía San Carlos, have recently flowered in the writer's garden; the leaf zonation and flower morphology are that of *A. colorata* (Fig. 16.13).

Solitary flowering plants in the Desert Botanical Garden in Phoenix, Arizona, have seeded abundantly, but plants from the same clone grown at Murrieta, California, have produced little or no seed. The lack of fruit set and seed at Murrieta may be attributable to the climate there with dry summers, which does not support the inherent seasonal physiological rhythms of *A. colorata* (Fig. 16.1). Hummingbirds habitually visit the flowering agaves at Murrieta, but they obviously do not guarantee fruit and seed set. Small live plants collected in March 1952 flowered at Murrieta in June 1965, indicating a generation period of about 15 years.

Agave colorata was reported by the natives as suitable for pit-baking and was called "mescal ceniza." It was perhaps formerly decimated as a sugar source by the Mayo and Yaqui Indians who occupied its area rather densely in precolonial times. It is uncommon in nature and on my very intensive travels through its area I do not recall it from any localities other than those listed in the Exsiccatae. Because of its compact size and bright glaucous leaves with pinkish cross zones, it makes an attractive ornamental, especially plants from the type locality (Fig. 16.12). It is found in southwestern gardens (Breitung, 1968).

Agave durangensis
(Figs. 16.1, 16.2, 16.3, 16.10, 16.14, 16.16, 16.17; Tables 16.1, 16.2, 16.3)

Agave durangensis Gentry sp. nov.

Medium to large, single or cespitose, short-stemmed, glaucous gray, broad-leaved, heavily armed rosettes 8–12 dm tall, 12–18 dm broad; leaves mostly 40–90 x 14–22 cm, broadly lanceolate, widest in the middle, narrowed above the broad base, fairly straight to outcurving, plane to concave especially toward apex, thick and convex toward the base, asperous, pruinose, the margins deeply crenate-mammillate; teeth prominent, 1–2 cm long, broadly flattened, variously flexed, generally 1–2 cm apart; spine strong, 4–6 cm long, broadly channeled above, pruinose gray over brown; panicle 7–8 m tall, deep, open, with 18–30 sinuously spreading, trifurcate laterals in upper ¾ of a zigzag shaft; peduncular bracts 15–25 cm long, scarious, remote, reflexed; flowers in small umbels, 60–80 mm long, closely, persistently erect, yellow; ovary 30–45 mm long including unconstricted neck; tube cylindric, 15–22 mm deep, broad and fleshy, lightly grooved; tepals unequal, strictly appressed to filaments, becoming leathery, the outer larger, 10–12 mm long, thickly rounded on the back and over-lapping the inner, with apex conspicuously papillate, almost corneous, reddish, the inner tepal smaller, sharply keeled; filaments 48–60 mm long, somewhat flattened, inserted 2-ranked 8–12 and 6–10 mm above base of tube; anthers 18–25 mm long; capsules 4.5–6 x 1.6–1.8 cm, oblong, obscurely stipitate, rounded and shortbeaked at apex; seeds small, 4.5–6 x 3.5–4.5 mm, lunate to obovate, with broad but little raised rim wing.

Type: *Gentry & Gilly 10576*, Sierra Registro, 18 miles SE of Cd. Durango along highway to Nombre de Dios, Durango, alt. ca. 6,000 feet, June 8, 1951 (l., plants); flowers from same produced from transplant at Murrieta, California, Aug. 1962. Deposited in US.

Planta plerumque caule singulo vel infrequenter caespitosula, 8–12 dm alta, 12–18 dm lata. *Folia* lati-lanceolata, 40–90 cm longa, 14–22 cm lata, pallidi-viridia plana vel conduplicata, arcuata crassa rigida; *margine* valde crenato, *dentibus* supra mamilas plerumque 1–2 cm longis, curvatis vel reflexis, plerumque 1–2 cm separatis; *spina terminali* 4–6 cm longa robusta pallide brunneis. *Inflorescentia* paniculata, scapo incluso 7–8 m alta, rhachidi sinuosa 18–30 ramis; *floribus* luteis 60–80 mm longis erectis; ovarium cylindricum apice incluso 30–45 mm longum; *perianthii tubo* cylindrico 15–22 mm longo lato carnoso; *segmentis*

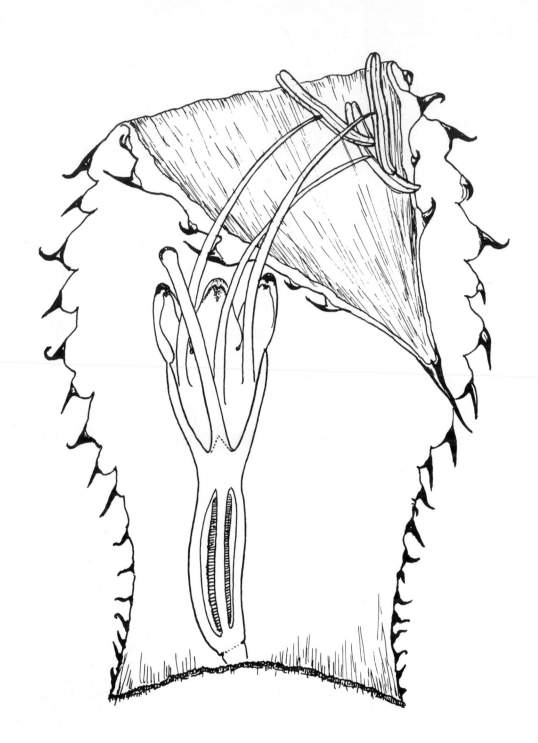

Fig. 16.14. *Agave durangensis*. A short broad-leaved form drawn from *Gentry 18230* and flower from *Gentry 18231*, Sierra Registro, Durango.

Fig. 16.15. *Agave durangensis*. A small colony on sun slope atop of Sierra Zacatecas, Zacatecas.

Fig. 16.16. A large specimen of *A. durangensis* with over-sized deep panicle on Sierra Registro, Durango, 1966.
▼

inaequalibus, exterioribus 1–12 mm longis, 6–8 mm latis, persistentibus erectis, apice rubro paene corneo, interioribus 8–10 mm longis. *Filamenta* 48–60 mm longa lutea duo-ordinata ad medium tubi inserta; *antheris* luteis 18–25 mm longis. *Capsulae* oblongae 4.5–6 x 1.6–1.8 cm, brevi-rostratae. *Semina* parva 4.5–6 x 3.5–4.5 mm, lunata vel obovata, margine brevi-alato.

Agave durangensis occupies the rocky slopes and gravelly bajadas in grama grassland between elevations of 5,500 and 8,500 feet (1,700 and 2,600 m). It grows in scattered colonies on the open slopes of southern Durango and Zacatecas (Fig. 16.15).

While *Agave durangensis* clearly belongs in the Ditepalae (see Figs. 16.2, 16.3, 16.4, 16.5), it does not show a particularly close relationship with other members of the section. It is distinguished by its large, rigid, broadly lanceolate leaves heavily armed with flexuous teeth on a deeply crenate margin, and its large, deep, open panicle with a flexuous or zigzag shaft. The dimorphic tepals are broad, especially the outer, and short. The cuticle is rather asperous. When not flowering it may be confused with *A. scabra* of the Section Americanae growing in the same region.

In November of 1959, *A. durangensis* was abundant along the eastern slopes of Sierra Registro between Cd. Durango and Nombre de Dios, where it grew singly or in close clusters. It varied greatly in size, color of leaf, and size of teeth, some of which were hooked and 2 cm long. The flowers were light clear yellow tipped with a strong red. Most of the panicles were maturing fruit at that time. However, the entire population appeared to constitute but one species, and the wide foliar and size variation indicated a fertile, free-seeding, outbreeding system. On the western slopes of Sierra Registro, below the Boca de Mezquital, the population appeared to be infused with genes of *A. applanata*. Some of the variables, such as the narrower lanceolate leaf and the remote teeth, looked like characters of *A. applanata*, but this aspect was not thoroughly explored.

On Sierra Registro it was being cut for making mescal. Some of the burros carried "cabezas" as large as those of the cultivated tequila. Transplants cultivated in gardens in southern California and the Desert Botanical Garden in Arizona have proved to be hardy and responsive in these two climates.

Agave flexispina

(Figs. 16.1, 16.3, 16.10, 16.17, 16.18; Table 16.3)

Agave flexispina Trel., Contr. U. S. Nat. Herb. 23: 133, 1920.

Small, single or cespitose, generally glaucous green, few-leaved, open rosettes 25–35 cm tall and twice as broad; leaves 16–30 x 6–8 cm, ovate, acuminate, nearly plane to concave above, glaucous to yellowish green, the margins undulate to crenate with teated teeth, the larger teeth mostly 5–8 mm long, mostly with down-flexed tips, brown to pruinose, 1–1.5 cm apart, sometimes with small interstitial teeth; spine 2.5–3.5 cm long, acicular, usually flexuous, flat to openly grooved near the base, brown to pruinose gray, decurrent to the upper small teeth; panicles 2.5–3.5 m tall, slender, rather open and frequently narrow with 6 to 12 small, few-flowered umbels, the shaft with small, dry, appressed, hardly persistent bracts; flowers 50–70 mm long, greenish yellow with red tinge on buds, tepals and filaments; ovary 22–35 mm long, cylindric, slightly angulate with short obscure neck; tube 13–18 mm deep, cylindric to urceolate and bulging in middle, slightly or narrowly grooved; tepals broad, lanceolate to linear, broadly rounded at apex, the outer larger, strongly over-lapping inner, 10–18 mm long, persisting erect and drying leathery, the inner with a high narrow keel; filaments 40–50 mm long, sometimes broadly flattened, inserted at one level, sometimes irregularly so; anthers 17–23 mm long, yellow or bronze colored; capsules 35–45 mm long, 15–17 mm broad, oblong, strongly trigonous, tardily dehiscent, abruptly stipitate, round at apex; seeds 5.5–7 x 4–5 mm, obovate to lunate, the rim wing low.

Fig. 16.17. *Agave flexispina* on the Chihuahua grassland about 20 miles (32 km) south of Parral, Chihuahua.

Fig. 16.18. The rosettes of *Agave flexispina* at maturity are only 20–30 cm tall with about 40 leaves. July 1975.

Type: *Palmer 330* in 1906, Tepehuanes, Durango, in US National Herbarium, consists of 2 sheets; 2 leaves, a bract-leaf, and flowering branch.

The habitat of *Agave flexispina* is the grama grassland and oak woodland of the eastern bajadas and mesas of the Sierra Madre Occidental from southern Chihuahua south through highland Durango between 1,300 and 2,300 m elevations (Fig. 16.10). So far as observed, individuals within populations are widely dispersed and so are the populations.

This species appears like a small edition of *Agave shrevei* or *A. palmeri*. However, *A. flexispina* also differs in having the filaments inserted on a single level in the tube, while in *A. shrevei* and *A. palmeri*, and most other members of the Ditepalae, the insertion is on two distinct levels. The rosettes and inflorescences in *A. flexispina* are consistently smaller throughout the populations than are normally found in other species of the Ditepalae. Plants of *A. flexispina* in the garden may grow larger than wild ones, as the transplant, *Gentry 22049,* in the Desert Botanical Garden with leaves 40 x 9 cm at flowering time.

Agave fortiflora

(Figs. 16.1, 16.2, 16.3, 16.19, 16.20, 16.21; Tables 16.1, 16.2, 16.3)

Agave fortiflora Gentry, U. S. Dept. Agr. Handb. 399: 122, 1972.

Light-gray, open rosettes up to 1 x 1.8 m, mostly single but occasionally suckering; leaves 50 x 8 to 100 x 12 cm, straightly ascending or outcurving and conduplicate, light gray glaucous, usually cross-zoned, widest in the middle, gradually narrowed above the dilated base, long-acuminate, the margin straight or small-teated under the mid-teeth; teeth in the mid-blade 5–10 mm long, 1–3 cm apart, generally flexed downward but also erect, smaller and variously flexed above and below the middle, with smaller interstitial teeth irregularly occurring; spine 3–5 cm long, subulate, chestnut to light gray, rounded below, narrowly grooved above, decurrent with a narrow corneous margin to the upper teeth; surface of leaf, spine, and teeth finely tuberculate rugose; panicle 4–6 m tall with 12–18 laterals, open and deeply ovoid in outline; bracts short, triangular, chartaceous, 5–6 cm long at the lower laterals; pedicels 4–20 mm long, bracteolate; flowers in dense yellow umbels, long-persisting, erect, 72–82 mm long; ovary pale green, 45–50 mm long, conspicuously angled below outer tepals; tube pale yellow, broadly bulging, 11–13 mm deep, 18–20 mm wide, the sinuses overlapping and deeply grooved below; tepals erect, yellow, 20–23 mm long, 7–8 mm wide, the outer rounded, linear-lanceolate, with an involute margin, obtuse, the inner with a high narrow keel and involute hyaline margins, broadly hooded at tip; filaments strong, persisting erect after anthesis, 60 mm long, elliptic in cross section; anthers 20–22 mm long, yellow, centrally affixed; pistil stout, triquetrously knobbed at the stigma; capsules and seeds not seen.

Type: *Gentry 19808,* Huntington Botanical Garden, San Marino, Calif., July 2, 1962, deposited in U.S. Natl. Herbarium, No. 2549708 and No. 2549709 believed to have been collected originally on Sierra Jojoba, ca. 15 miles northeast of El Datil, Dist. Altar, Sonora, March 23, 1952, *Gentry & Fox 11630* (l, & pl.). Also assigned here are leaf specimens "on valley plain, 3 miles NE of Sierrita de Lopez, Sonora, Feb. 1952, *Gentry 11618*" (Fig. 16.20).

This light-gray glaucous xerophyte is distinct in floral morphology; the large flowers are strong (whence the name, *fortiflora*), and long lasting, persisting erect in close-set umbels long after anthesis. The ovary is thick, roundly 3-angulate, with the grooves from the tepal sinuses furrowing the thick neck. The tube is broad and bulging at the attachment of the filaments and 3-angulate, with the corners descending from the outer tepal bulges.

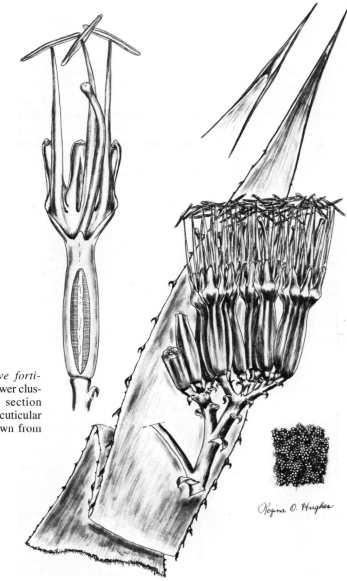

Fig. 16.19. *Agave forti-flora*. Leaf and flower cluster x ⅜; flower section slightly reduced; cuticular pattern x 18. Drawn from *Gentry 19808*.

The flattening of the filaments is extreme in *A. fortiflora,* making them almost strap-shaped and quite the broadest observed in the genus.

Agave fortiflora does not show relation to other members of the Ditepalae. The flowers are distinct with their long tepals and relatively short broad bulging tube. However, the tepals are erect and dry that way, as in the Ditepalae. The flower proportions are similar to some of the *A. wocomahi,* but the latter is very different in leaf characters. On the whole, the variability in the Ditepalae appears broad enough to admit *A. fortiflora* without distortion of the section.

Some doubts about the origin of the type specimen were expressed with its initial description (Gentry, 1972), because the garden tag in the Huntington Botanical Garden was lost. Recently we have compared in detail the leaf anatomy of my original leaf

Fig. 16.20. *Agave fortiflora (Gentry 11618)* in a valley near Sierrita de Lopez, Sonora.

collection at Sierra Jojoba with the Huntington grown plant (Gentry and Sauck, 1978). The cuticle, epidermis, and stomata of the two specimens are almost identical and leave no doubt that the two specimens are specifically alike. *A. fortiflora* is definitely of Sonoran origin.

The flowering season appears to be in summer, since the Huntington Botanical Garden specimen flowered during June and July. The rosettes observed produced offsets sparingly, and old or extensive colonies are rare. More recent attempts to recollect the species on Sierra Jojoba have failed to relocate the species. Locally, the plant is called "lechuguilla" and the Papago Indian guide said it was good for pit-baking and eating.

Agave murpheyi
(Figs. 16.1, 16.2, 16.3, 16.21, 16.22, 16.23; Tables 16.2, 16.3)

Agave murpheyi F. Gibson, Boyce Thompson Inst. Contrib. 7: 83, fig. 1. 1935.

Medium size, light green, compact, freely suckering rosettes, 60–80 cm tall, ca. 1 m broad, and richly bulbiferous, rarely seeding inflorescence. Leaves 50–65 x 6–8 cm, linear, short-acuminate, light glaucous green to yellowish green, frequently lightly cross-zoned, firm, straight, bud-printing clear, the margin undulate with small regular teeth 3–4 mm long, mostly 1–2 cm apart, bases brown, cusps graying; spines short, 12–20 mm long, conic, very shortly grooved or flattened above, dark brown becoming

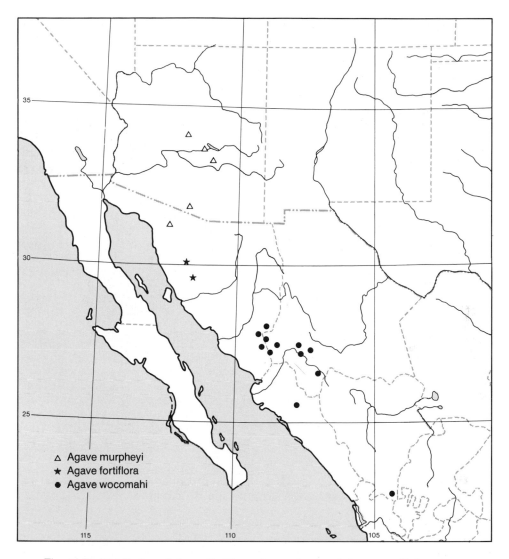

Fig. 16.21. Distribution of *Agave fortiflora, A. murpheyi,* and *A. wocomahi,* based on herbarium specimens (see Exsiccatae).

grayish; panicle bulbiferous, 3–4 m tall on a thick shaft with scarious, triangular bracts 10–15 cm long and 10–15 short, compact laterals in upper ¼ of shaft; flowers pale waxy green with purplish to brownish tips, 65–75 mm long; ovary thick, rounded, 32–40 mm long, scarcely narrowed at neck; tube deep, urceolate, 16–20 mm deep, 16–19 m wide, 6-grooved from the tepal sinuses; tepals unequal, the outer longer, 15–19 mm long, 7 mm wide in the middle, overlapping the inner, linear, rounded on back, strictly erect, thick, the tip thickly hooded, brownish, the inner strongly keeled, pale yellow; filaments 45–50 mm long, unequally inserted 8–12 mm above bottom of tube, yellow; anthers 25–26 m long, yellow, slightly excentric; capsules obovate to oblong or ovate, 2–2.5 x 5–6 cm, stipitate, beaked; seeds "dull black, 6 x 9 mm, thin" (loc. cit).

Type: From along Queen Creek near Boyce Thompson Southwestern Arboretum, Superior, Arizona; "Deposited and cotypes cultivated at The Boyce Thompson Southwestern Arboretum, Superior" (Gibson, loc. cit., not seen). The cultivated clone is still living there.

Fig. 16.22. *Agave murpheyi* in Opuntia orchard at Rancho San Luisito, Sonora, reported by the owner to have been brought in from nearby desert.

Affinity and Habitat

The flower structure clearly aligns this agave with other members of the Ditepalae. In addition to the separating characters given in the key to species, I may add here its prolifically bulbiferous inflorescence, which is short, compact, and limited to the top ¼ or ⅓ of the shaft. It blooms earlier than other agaves in its area. Its habit of starting the flowering shoot in winter means that the succulent, turgid tissues are somewhat resistant to the light winter freezes regularly occurring in its area. However, a severe winter freeze, of 20°F (−7°C), or lower, killed the shoot apices in the Verde River population in 1976–77. No further resurgence of inflorescence occurred subsequently in this case (word of Rodney Engard). All plants I have seen show relatively little variation and, once the species is observed, it is readily recognized again. The aspect and essential characters of the species are plainly there, including the prolific reproduction with bulbils and rhizomatous offsets.

Agave murpheyi is indigenous to what Shreve called "the arborescent desert," ranging from the lower slopes of the Tonto Rim chaparral to southwestern Arizona and

Fig. 16.23. A nearly mature rosette of *Agave murpheyi* in Opuntia orchard at Rancho San Luisito, Sonora.

adjacent Sonora, at elevations of 1,500 to 3,000 feet (460 to 930 m) on mountainous slopes, or bajadas. Annual rainfall ranges from 200 to 450 mm (8–18 inches). Local particulars were reported earlier (Gentry, 1972) and collections are listed in the Exsiccatae. However, *A. murpheyi* has never been observed in extensive or dense populations. Some of the clones appear to have been associated with old Indian living sites. The propagules are easily transported and transplanted.

Agave palmeri
(Figs. 16.1, 16.4, 16.5, 16.10, 16.24, 16.25, 16.26; Tables 16.1, 16.2, 16.3)

Agave palmeri Engelm., Acad. Sci. St. Louis Trans. 3: 313, 1875.

Rosettes generally single, rarely late-suckering, 5–12 dm tall, 10–12 dm broad, rather open about the conal bud; leaves mostly 35–75 x 7–10 cm usually narrowed above the base, lanceolate, long-acuminate, rather rigid, somewhat guttered, thick at base, convex below, pale green to light glaucous green or reddish-tinged, the margins nearly straight or undulate with or without small teatlike bases under closely set, rather regular, slender teeth, which are variously flexed, smaller teeth sometimes between the larger; spine strong, 3–6 cm long, shortly and openly grooved above the base, castaneous or brown to aging gray, acicular; panicle deep, broad, open, 3–5 m tall, with triangular bracts and 8–12 horizontal laterals in upper ⅓ of shaft; flowers 45–55 mm

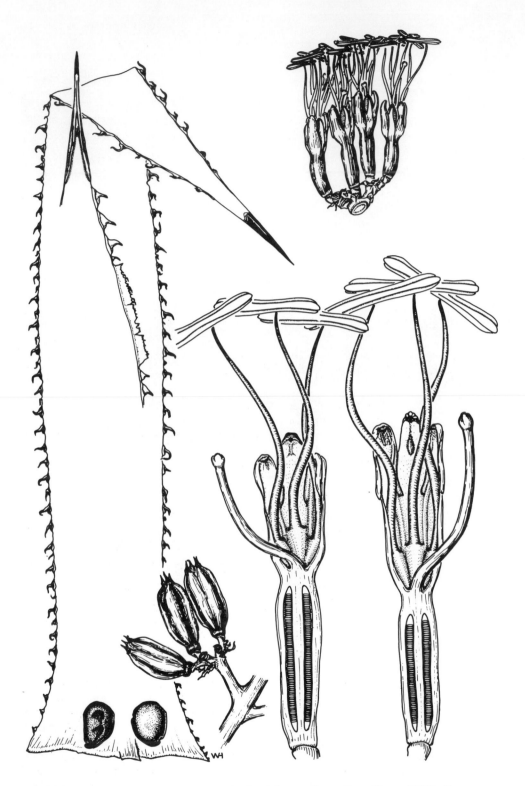

Fig. 16.24. *Agave palmeri*. Typical leaf and flowers drawn from *Gentry 21990*, Texas Canyon, Arizona. For comparison are also depicted a coxcomb teeth inset and a long section of flower with corrugated apex of outer tepal of plant from Sierra Jojoba, Sonora, *Gentry & Weber 21987*.

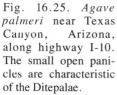

Fig. 16.25. *Agave palmeri* near Texas Canyon, Arizona, along highway I-10. The small open panicles are characteristic of the Ditepalae.

long, narrow, pale greenish yellow to waxy white, reddish in bud; ovary shiny green, 25–30 mm long including short neck; 12–14 mm deep, 10 mm wide; tepals persistently erect, leathery, yellow to pink below, conspicuously red to brownish on calloused tip, dimorphic, the outer 10–13 mm long, longer than and overlapping the inner, the inner 3 mm shorter and narrowly keeled on back; filaments 40–45 mm long, reddish or pale yellow, inserted midway in tube with those opposite the outer tepals inserted higher than the others; anthers 15–18 mm long, yellow; capsule oblong to pyriform, 3.5–6 x 1.8–2 cm, the dried walls thick and strong, short to long apiculate; seeds sooty black, 5–7 mm along the straight edge, 4–5 mm broad, thin, flat.

Type: Not specified; from Schott, Palmer, & Rothrock who made scattered collections in southeastern Arizona.

Fig. 16.26. Detail of a rosette of the small form of *A. palmeri* in its northwestern area, this one on Sierra del Humo, Sonora.

Distribution and Taxonomic Notes

A. palmeri is a widely scattered but characteristic plant of the oak woodland and grama grassland communities at altitudes between 3,000 and 6,000 feet (930 and 1,850 m). It shows good development on the limestone slopes at altitudes between 4,000 and 5,000 feet (1,240 and 1,550 m) on the Patagonia Hills and along highway 86 where it passes through the granite of Texas Canyon in Arizona. A smaller form occurs in northern Sonora in somewhat lower elevations on the rocky brush slopes and comprises the western segment of the species (Fig. 16.26).

Characteristics of the species are the long lanceolate leaf, typically narrow in its northern range, its close-set, slender teeth, sometimes with smaller ones between, and the reddish tepals about equaling the tube. About the margins of its considerable area it shows what is taken for introgressive variation: on the north in the Rincon Mountains with *A. chrysantha* Peebles, on the south in east central Sonora with *A. shrevei*. I have separated the southern representatives of *A. palmeri* from *A. shrevei* on leaf and flower characters; i.e., the broader leaves and the much deeper tube and relatively shorter tepals of the latter (Fig. 16.4, 16.5). The broad, glaucous green leaves in populations in the Río Bavispe watershed resemble those of *A. shrevei,* but flowers collected by Felger (*3595, 3525, 3343*) show the smaller flowers with shallower tubes of *A. palmeri* and led me to cite these collections, along with *Gentry 16641,* under this species. An introgressive aspect in these eastern Sonoran populations extends as far south as the upper Guajaráy country.

The scattered population along the main highway in Sonora between Nogales and Magdalena consists of small plants, reflecting the drier and lower elevations. Similar small plants were encountered on the Sierra del Humo (Fig. 16.26) and on the northern slopes of the Baboquivari Mountains, along the grade up Kitt Peak. These populations are apparently *A. palmeri* and perhaps their small apartness should be nominated sub-specifically. However, they are a small part of a large polymorphic group, including, besides the species mentioned, *A. colorata, A. flexispina,* and other nondesignated forms ranging from central Arizona and New Mexico into Durango, Mexico.

Agave palmeri flowers in June and July (Fig. 16.1). Hummingbirds, other wild fowl, and many insects feed on the flower nectar. Some of these may effect chance pollination, but night-flying migratory bats of the genus *Leptonycteris* are now known to be persistent specialized pollinators. The sturdy erect flowers are structurally well suited for cooperation with the bat visitors, whose flock feeding habits appear to determine in some degree the scattered colonial nature of the *A. palmeri* distribution.

Native Uses

In northeastern Sonora, the people call *A. palmeri* "lechuguilla" and said it was collected for eating and for making mescal. It was much used for food, fiber, and beverage by the Indian tribes inhabiting its region. Castetter, Bell and Grove (1938) reported its use by the Papago, Pima, and Apache tribes. The young tender flowering shoots are still eaten by the Mexicans. They are also eaten by cattle, rodents, and other animals. It is among the sweeter kinds with little or no sapogenin reported in assays (Table 16.2); 0.5 percent hecogenin was found in leaves of *Gentry & Ogden 9915* of Pyramid Mountain, New Mexico. Several other assays of plants in other localities were negative. It is easy to grow and is frost hardy in our gardens.

Agave shrevei
(Figs. 16.1, 16.4, 16.5, 16.27, 16.28, 16.29, 16.30 ; Tables 16.1, 16.2, 16.3)

Agave shrevei ssp. *shrevei* Gentry, U.S. Dept. Agriculture Handbook No. 399: 111, 1972.

Agave shrevei Gentry, Carnegie Inst. Wash. Pub. 527: 95, 1942.

Rosettes small to medium, light gray, glaucous, closely suckering with maturity; leaves ovate, short-acuminate, 20–35 x 8–10 cm, plane to conduplicate, lanceolate, acuminate, 50–60 x 12–18 cm, generally narrowed above the base, firm, thick, straight or outcurving near the tip; teeth variable, on small to prominent teats, the larger teeth in mid-blade 5–10 mm long, straight or flexed upward or downward, dark brown to gray; spine stout, mostly 25–50 mm long, brown, acicular, with a narrow or open groove from base to above the middle; panicle 2.5–5 m tall with 8–16 distal, ascending laterals in upper ⅓; bracts deltoid, long-acuminate, drying early; umbels small; flowers persisting erect, slender, 60–70 mm long, light green to pale yellow, buds red-tipped; ovary 25–35 mm long including constricted neck; tube cylindric or urceolate, 18–23 mm deep, 10–12 mm wide; tepals strictly erect, leathery, pea green to light yellow, unequal, the outer 10–12 mm long with red to purplish calloused tips, plane, widely overlapping the shorter, keeled inner; filaments 40–50 mm long, yellow or red, inserted unequally in mid-tube, flattened; anthers large, 22–26 mm long, yellow; capsules oblong, 4.5–7 x 1.5–2.5 cm, short-stipitate and shortly beaked; seeds 6–7 x 4.5–5 mm, hilar notch small, the marginal wing prominent.

Type: *Gentry 2028,* Sierra Canelo, Chihuahua, Oct. 8, 1935 (old dried flowers) deposited in Dudley Herbarium, Stanford Univ.; now in *CAS.*

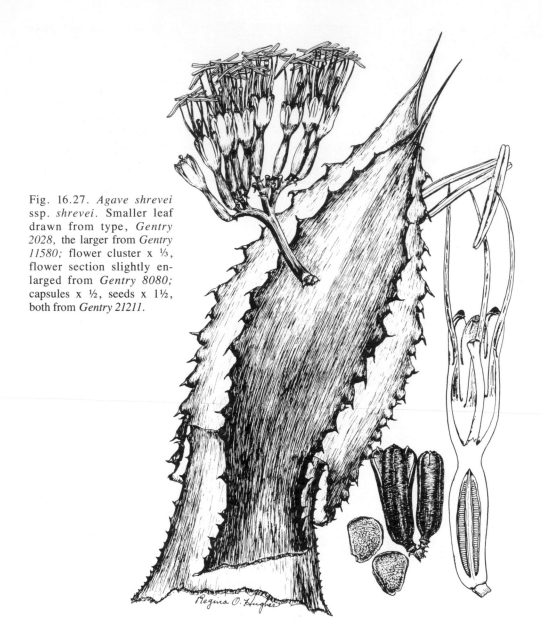

Fig. 16.27. *Agave shrevei* ssp. *shrevei*. Smaller leaf drawn from type, *Gentry 2028,* the larger from *Gentry 11580;* flower cluster x ⅓, flower section slightly enlarged from *Gentry 8080;* capsules x ½, seeds x 1½, both from *Gentry 21211.*

Common and widely scattered from middle Sonora southward on both sides of the Chihuahua-Sonora boundary, from 3,000- to 6,000-foot (930 to 1,850 m) altitudes of the northern Sierra Madre Occidental, in open, rocky, calcareous slopes of the oak woodland and pine-oak forests. Flowers in summer. (Figs. 16.28, 16.29.)

Good distinguishing features of this agave are its light glaucous gray, broad leaves with teated margins under well-developed brown teeth and its leathery perianth with deep tube, short, persistently erect tepals, much shorter than the tube. Some of the flower forms are quite striking, as the red filaments in a pea-green perianth. The typical or smaller leaved *A. shrevei* shows leaf resemblance to *A. flexispina* Trel., but the latter's flowers appear closer to *A. palmeri,* the tube and tepals being about equal. The Guajaráy population with its larger leaf and variable forms shows some *A. palmeri* characters and there appears to be some introgression. The obscure relative *Gentry 11605,* cultivated in Navojoa, was reported as from Tepahue, which is adjacent to the Guajaráy country. The

Fig. 16.28. *Agave shrevei shrevei,* an inflorescence with red filaments in Gentry garden near Murrieta, California.

Fig. 16.29. A colony of *A. shrevei shrevei* in the pine-oak forest zone on Sierra Tecurahui, southeastern Sonora.

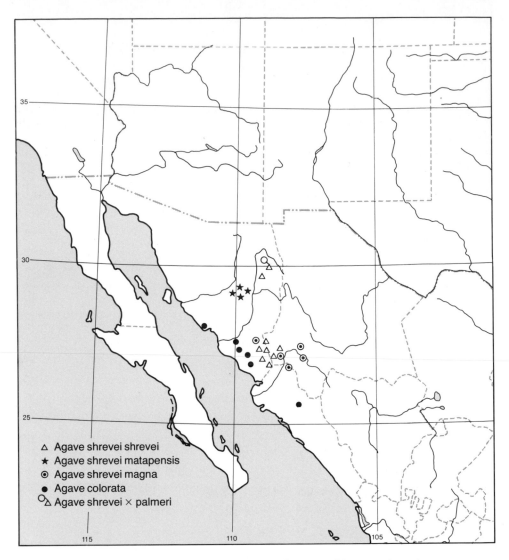

Fig. 16.30. Distribution of *Agave colorata, A. shrevei,* and its two subspecies *magna* and *matapensis,* based on herbarium specimens.

Mayo Indian growing the plants in his garden in the ''barrio'' known as Chijuca also stated that he grew them to eat, as the pit-baked heads were very sweet.

In Sonora and adjacent Chihuahua it is known as ''lechuguilla,'' ''ceniza,'' or as ''totosali'' by the Warihio Indians, who pit-baked the heads for eating. Bye et al. (1975) report the Tarahumara call it ''o'tosa'' and pit-bake the ''hearts'' for eating and making ''mescal bread'' or a fermented drink. The Mexican mountaineer mescal distillers value the plants and no doubt have made depletions in the sierran populations along the remote Sonora-Chihuahua border country. The numerous reports of its use and its general lack of sapogenins (Table 16.2) all indicate that it has been a valuable food resource for indigenous peoples of the Sierra Madre Region. See also Trias and Blight (1972).

With recent collections, a large form of *Agave shrevei* has appeared in the barranca region of southwestern Chihuahua, and it is described below.

Agave shrevei ssp. *magna*
(Figs. 16.1, 16.4, 16.5, 16.30, 16.31, 16.32; Tables 16.1, 16.3)

Agave shrevei Gentry ssp. *magna* Gentry ssp. nov.

Mature rosettes mostly single, 1.4–1.7 m tall, broadly spreading to 2.5 m with open crown; mature leaves mostly 1.2–1.5 m x 15–25 cm, lanceolate, long acuminate, light glaucous gray, outcurving, guttered, thickly fleshy and crescentic toward base, finely asperous, the margins remotely crenate; teeth large along most of blade, 6–10 (–15) mm long, mostly 3–5 cm apart on prominent teats, flexed or curved upward or downward, small interstitial teeth frequently present; spine 3.5–6 cm long, acicular, narrowly grooved for ⅓–½ the length, usually decurrent to uppermost teeth, lustrous brown to pruinose gray; panicles 6–7 m tall, narrow, deep, the shaft frequently sinuous, with 20–30 small umbels of yellow flowers on short branches; peduncular bracts remote, drying reflexed; flowers 55–70 mm long; ovary small, cylindric, roundly angled, neck slightly constricted and lightly grooved or unconstricted and ungrooved; tube deep cylindric, 17–22 mm deep, 11–13 mm broad, thick-walled, slightly narrowed in middle; tepals unequal, closely appressed, erect, thick, linear-ovate, the outer 11–13 mm long, reddish brown and calloused at apex, broadly over-lapping the narrowly keeled inner, the keel not reaching to apex; filaments 50–70 mm long, broad toward base, narrowly tapering above, inserted in two ranks 2–3 mm apart; anthers 23–27 mm long, yellow, centric; capsules 4.5–6 x 1.7–2 cm, oblong, broadly short-stipitate, thick-walled, shortly apiculate, dark brown to black; seeds 7 x 6 mm, deeply lunate, with a low marginal wing and obscure hilar notch.

Type: *Gentry & Bye 23360*, 3–4½ road miles NE of bridge over Rio Batopilas, Barranca de Batopilas, Chihuahua, volcanic sun slope, elev. 3,500–4,500 feet, 10 Oct. 1973 (l, f, cap, photo). Holotype US, isotypes DES, MEXU.

Planta caule plerumque singulo. *Folia* pallida lanceolata longi-acuminata, 1.2–1.4 m longa, 15–25 cm lata; *dentibus marginalibus* validis, pro parte maxima 6–10 mm longis, subulatis, lexuosis, in mamilis remotis; *spina terminali* 3.5–6 cm longa, aciculari decurrente, supra anguste canaliculata. *Inflorescentia* 6–7 m alta angusta paniculata ramis 20–30 lateralibus; *floribus* luteis 55–70 mm longis; *ovarium* 30–35 mm longum cylindricum; *perianthii tubo* 17–22 mm longo cylindrico; *segmentis* inaequalibus, exterioribus 11–13 mm longis, 6 mm latis, apice rubro-brunneolo calloso. *Filamenta* 50–70 mm longa duo-ordinata ad medium tubi inserta; *antheris* luteis 23–27 mm longis. *Capsulae* oblongae 4.5–6 cm longae, 1.7–2 cm latae brevi-stipitatae. *Semina* 7 x 6 mm lati-lunata.

The name *magna* (from Latin, great or large) I have given because of the plant's large size and splendid proportions. The flower structure is very similar to *Agave shrevei*, as the flower drawings (Fig. 16.4) and the ideographs (Fig. 16.5) clearly show. *A. shrevei magna* differs in having very long filaments, those of *Felger 800* being up to 70 mm long. But the conspicuous difference is the larger size of the plants (Fig. 16.32). The large leaves arch outward, probably owing to their weight, giving a flat-topped, open crown appearance to the rosettes. The plants were abundant, when observed in 1973, on a south-facing slope of the canyon that many years previously had been cleared of trees, probably for planting maize. It appeared that opening of the forest and mixing of the soil had provided favorable opportunity for increase of a local agave population.

The Tarahumara Indians of the Barranca Region call this agave "o'tosa," while more eastward on the plateau they call it "me" (Bye et al., 1975). They pit-bake the stems to eat and for making mescal bread and a fermented drink. The flowering stalks are also cooked for food and small plants are used in curing and in death ceremonies. The Mexicans call this agave "mescal blanco" or "mescal ceniza" and have used it for making the distilled mescal drink.

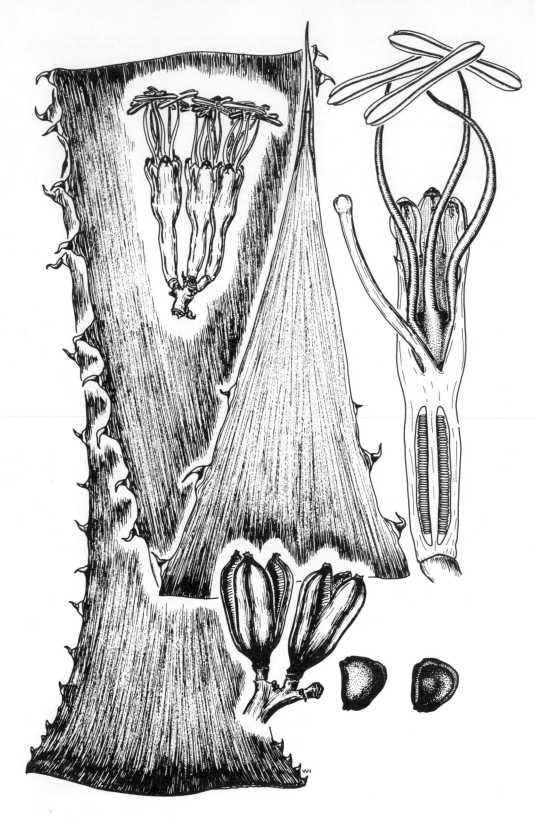

Fig. 16.31. *Agave shrevei* ssp. *magna* drawn from the type, *Gentry & Bye 23360*. The corneous outer tepal tip is dark red in life.

Fig. 16.32. *Agave shrevei magna* on rocky slope in oak woodland, Barranca de Batopilas, Chihuahua, October 1973.

An earlier collection of robust plants, *Gentry 11415,* from Cerro Babuyo in extreme NE Sinaloa appears to belong in this subspecies. The limestone on top of this mountain is covered with a palm-oak savanna with an abundance of *Nolina matapensis* and some *Dasylirion.* Cave-dwelling Indians, probably Warihios, living on this mountain, pit-baked this agave. The open pit, near a cleared flat space for their ceremonial dance "tuwuri" (Fig. 16.33), showed it was recently in use at the time of my visit in 1952. The breech-clouted Indian who walked by escaped me before I could talk with him. The guide, who lived by the road at the foot of the mountain, had planted this agave in an enclosure near his house. Some of his plants flowered nine years later and the ones I observed at his house were larger than the wild ones upon the mountain.

Fig. 16.33. Indian mescal pit on Cerro Babuya, Sinaloa, with cracked stone and charcoal near rim; trees of *Ipomoea arborescens pachylutea* above men.

Agave shrevei ssp. *matapensis*
(Figs. 16.1, 16.4, 16.5, 16.30, 16.34; Tables 16.2, 16.3)

Agave shrevei ssp. *matapensis* Gentry, U.S. Dept. Agri. Handb. No. 399: 115, 1972.

Small to medium, light gray glaucous, rather open rosettes, suckering late and sparsely; leaves short-lanceolate, 30–45 x 8–14 cm, mostly plane to conduplicate toward apex, the larger teeth in the mid-blade down-flexed, the margin nearly straight to undulate; panicles 4–5 m tall with 8–14 laterals in upper half of shaft; flowers waxy yellow-green with tips of outer tepals reddish; filaments greenish to pink, anthers yellow; ovary 22–40 mm long including the short unconstricted neck 2–3 mm long; tube cylindric, 15–20 mm long, 11–14 mm broad; outer tepals 11–16 x 7 mm, rounded on back, thick; filaments inserted 8–10 mm above base of tube; filaments 45–55 mm long, slightly flattened; anthers 20–28 mm long; capsules (*Gentry 17623*) short, 35–40 x 20 mm, short-oblong, apiculate, the valves strong and woody; seeds smoky black, mostly 5–7 mm along straight edge, 5–6 mm broad, rugose, scarcely winged and unfluted along margin.

Type: *Gentry 11607*, Sierra Batuc, North of Matapé, Sonora, Feb. 16, 1952, deposited in U.S. Natl. Herbarium. No. 2540344. Widely scattered in sparse colonies in the savanillas of *Lysiloma*-oak-*Nolina*

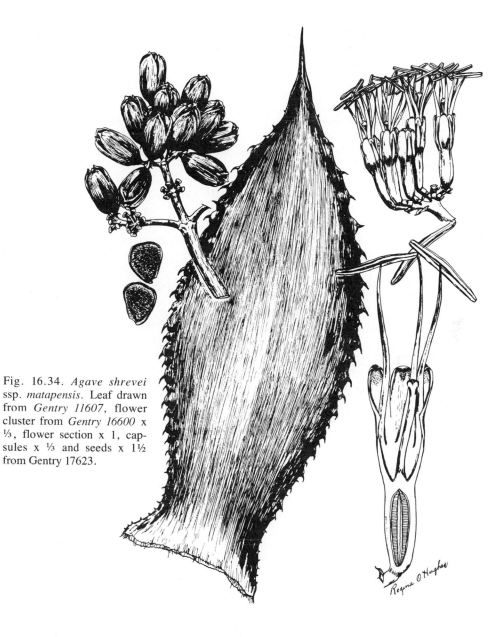

Fig. 16.34. *Agave shrevei* ssp. *matapensis*. Leaf drawn from *Gentry 11607,* flower cluster from *Gentry 16600* x ⅓, flower section x 1, capsules x ⅓ and seeds x 1½ from Gentry 17623.

woodland on granitics and deformed limestones, alt. 2,000 to 3,000 feet (620 to 930 m). The largest colony observed was in and about the mountain pass 6–8 miles (10–13 km) northeast of Matapé along the road to Sahuaripa.

These small to medium, light-gray glaucous rosettes resemble some forms of *A. shrevei,* from which it is distinguished by its deflexed teeth, smaller flowers with shorter ovary unconstricted in the very short neck, the broader and longer tepals in proportion to the shorter tube, and the short, broad capsules. It produces offsets late and seeds abundantly. The variability observed about Matapé indicates an outbreeding population. The flower description is drawn mainly from pickled flowers from plants that flowered in my Murrieta planting.

Agave wocomahi
(Figs. 16.1, 16.4, 16.5, 16.21, 16.35, 16.36, 16.37; Tables 16.2, 16.3)

Agave wocomahi Gentry, Carnegie Inst. Wash. Pub. 527, p. 96. 1942. Figure 13.33.

Rosettes medium to large, 0.8–1.3 m tall, 1.5–2 m broad, non-suckering, eventually depressed and open at maturity; leaves 30 x 9 cm to 90 x 25 cm, mostly lanceolate to linear-lanceolate, rarely ovate, somewhat narrowed toward the base, plane, thick-fleshy, rather rigid, ascending to depressed in age, dark green to glaucous green, smooth, the margins straight to undulate with large teeth, 1–2 cm long, dark brown to glaucous brown, variously flexed, these below the mid-blade frequently down-flexed, and smaller interstitial teeth irregularly occurring; spine stout, 3–6 cm long, usually sinuous, flattened or hollowed in a broad groove above, short or long-decurrent; panicle 3–5 m tall, open, with 8–15 laterals in upper ⅓ of shaft; bracts in the mid-shaft 15–20 cm long, scarious, appressed or reflexed; flowers yellow, in small umbels, 65–85 mm long, erect; ovary cylindrical, 34–40 mm long including the neck 2–5 mm long, light green; tube deeply funnelform, 18–22 mm deep, 15 mm broad, light yellow, narrowly grooved below the tepal sinuses; tepals dimorphic, 15–23 mm long, 5 mm wide, yellow, erect, thick, linear, conduplicate, rounded and deeply hooded at tip, the outer with papillate pubescence well below apex and sometimes red-tipped, the inner shorter and prominently keeled; filaments 60–65 mm long, inserted unequally 9–14 mm above base of tube; anthers yellow, large, 26–34 mm long, excentric; pistil with a large stigmatic head; capsules short-stipitate, oblong, 50–60 mm long, 15 mm or more in diameter; seeds 7 x 4.5–5 mm , shiny black, oblique, flattened on the end opposite the hilar notch, finely punctulate, wavy corrugate, with a narrow margin partly upcurved.

Type: *Gentry 1989,* Huicorichi (or Guicorichi), Sierra Huicorichi, Rio Mayo, Chihuahua, 7 Oct. 1935 (f, l); holotype *CAS,* isotypes ARIZ, DES.

Agave wocomahi is native to the rocky, calcareous, open mountain slopes through the pine-oak forests at elevations between 4,500 and 8,100 feet (1,400 and 2,500 m), from southeastern Sonora and adjacent Chihuahua to southern Durango (Fig. 16.21). As a montane mesophyte, one might expect it to be abundant through the Sierra Madre, but it is an uncommon plant, the populations consisting of small colonies with widely scattered individuals.

On the basis of floral morphology *A. wocomahi* belongs in the Section Ditepalae, as indicated by the ideographs (Figs. 16.3, 16.5) and illustrations (Figs. 16.2, 16.4). It sometimes grows with *A. shrevei,* from which it is usually easily distinguished by its dark green color and more remote teeth. It is more easily confused with *A. bovicornuta* of the Section Crenatae, but the latter generally has lighter yellowish green leaves, conspicuously narrowed just above the base, and the flowers have different structural features (cf. ideographs, Figs. 14.3 and 16.5). An unusual variant with very short broad leaves is represented by *Gentry 11588,* found along the Guajaráy Scarp, a striking topographic feature described in Río Mayo Plants (1942). I have noted no intergrading of *A. wocomahi* with its neighbors. A more recent leaf collection has extended the distribution to southern Durango (Fig. 16.21).

The Warihio Indians and their Mexican neighbors call this plant "wocomahi," stating that it is sweet and suitable for eating and making mescal. Both these peoples and the mountaineers of Sinaloa stated that the flowers were cooked and eaten like squash (*Cucurbita* spp.). The Tarahumara Indians call it "ojcome" and are reported to eat the cooked heads. They are eaten as fresh chunks out of hand or as pounded dried cakes,

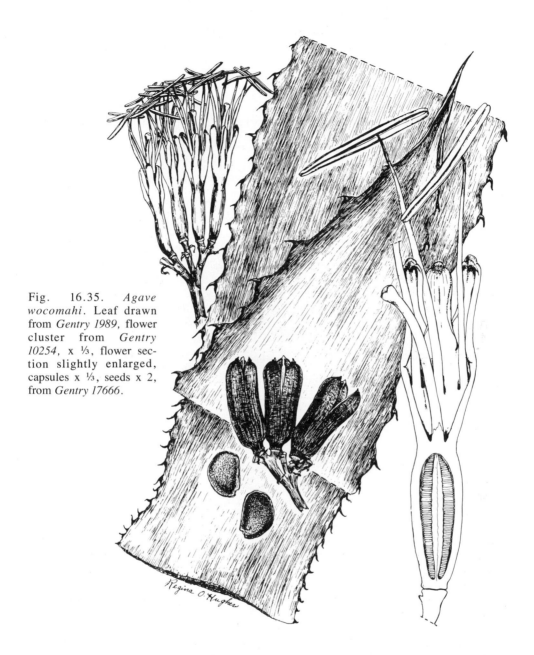

Fig. 16.35. *Agave wocomahi*. Leaf drawn from *Gentry 1989*, flower cluster from *Gentry 10254*, x ⅓, flower section slightly enlarged, capsules x ⅓, seeds x 2, from *Gentry 17666*.

"mesagoli," or made into a fermented drink, "sugui," or "tesguino" as it is generally called through the Sierra Madre country (Bye et al., 1975). The fiber of the leaves is used locally for cord, rope, and packsaddle pads.

A single plant that flowered in my Murrieta garden in 1962 did not set viable seed, indicating that it required cross-fertilization or a different climatic rhythm. This maturation required 11 years from the small plant collected, *Gentry 10252*, in 1951. No vegetative reproduction has been observed in this species and it appears to be a confirmed seeder and outbreeder.

Fig. 16.37. Detail of leaves and teeth of a submature plant of *Agave wocomahi* at the same locality as above, October 1973.

Fig. 16.36. *Agave wocomahi* on the sierran highlands south of Creel, Chihuahua, with Robert Bye beside a flowering plant.

Ditepalae Exsiccatae

Agave applanata

CHIHUAHUA. *Gentry & Bye 23363*, DES, US, MEXU. 3 miles S of Cusarare, pine-juniper, rocky sun slope, elev. 7,000–7,500 ft., Oct. 11, 1973 (l, photo).

DURANGO. *Gentry & Gilly 10557*, DES, US, MEXU. In old cemetery, Tapias Coloradas, 17½ miles N of El Pino in Rio Chico Basin, June 6, 1951 (transplant fl'ed. at Murrieta, Calif. July 18, 1971) (l, f).

Gentry & Gilly 10589, DES, US, MEXU. 8–10 miles S of Boca de Mezquital, along road from Cd. Durango to Mezquital, semi-arid, open, dispersed, deciduous woodland (thorn forest?) with *Bursera, Ipomoea, Fouquieria*, etc., elev. ca. 5,500 ft., with steep gravelly slopes, June 11, 1951. (Perhaps hybrid X *A. durangensis*).

Gomez-Pompa 967, MEXU. In cemetery, Santiago Papasquiaro (f).

HIDALGO. *Gentry & Arguelles 12580*, 3 miles E of Tulancingo, 21 April 1953; open grassy slope in grama grassland (4 live pl., photos).

OAXACA. *Gentry, Barclay, Arguelles 20444*, DES, US, MEXU. 1 mile SE of Nochixtlan along highway, open sedimentary hills, elev. ca. 6,000 ft., Sep. 11, 1963 (l, photos).

PUEBLA. *Gentry & Arguelles 12579*, "Onemile" (sign post), E of Zacatepec, along road to Jalapa. 17 April 1953 (5 live pl. & anal. sample).

Trelease l, MO. Lava bed. Limon, 2/26/06 (lvs).

QUERETARO. *Gentry & Arguelles 12597*, DES, US, MEXU. ca. 15 miles W of Cadercyta along road to Queretaro, May 1, 1953 (4 live pl) (transplant fl'ed. at Murrieta, Calif., July 6, 1971) (l, f).

VERACRUZ. *Gentry, Barclay, Arguelles* 20415 DES, US, MEXU. Lava beds by Limon, weathered lava, elev. ca. 7,800 ft., Sep. 4, 1963 (l, photos).

Gentry, Barclay, Arguelles 20420 DES, US, MEXU. Magueyitos, cult. for fiber as "Maguey de Ixtle," Sept. 5, 1963 (l, f, photos).

Trelease 1 Mo, Neotype. Limon, "on the Interoceanic R.R. above Jalapa."

ZACATECAS. *Foster & Glass 1961*, HNT. W of Jerez in mountainous area, 6 Nov. 1969 (cult. HNT, fl'ed. Oct. 1976).

UCBG 61-1518-1, University of California Botanical Garden, Berkeley, Calif. Source of Cave karyotype, 12 Oct. 1962. (Large plant with very heavy apical margin and strong teeth almost to spine.)

Agave chrysantha

ARIZONA. *Collom 112*, U.S. Collom Camp, Mazatzal Mts., Gila Co., elev. 1,200 m, July 15, 1933.

Darrow s.n., ARIZ, US. Limestone slope 15 miles S of Cutler on road to Winkelman, Pinal Mts.; June 6, 1943 (l, f).

Eastwood 17358, 17359, CAS. Apache Trail Summit, May 21, 1929.

Eastwood 8767, CAS. Fish Creek, May 19, 1919.

Eastwood s.n., CAS. Road to Winkleman, Pinal Mts., May, 1929.

Ferris 1071, CAS, DS. Sabino Canyon, Santa Catalina Mts., June 26, 1918 (l, f).

Gentry 23013, DES, US, MEXU. Ca. 10 miles N of Black Canyon City along rt. I. 17, elev. ca. 3,400 ft., volcanic rocks & clay soil on open plain, June 15, 1972 (l, f).

Gentry 23444, DES. Black Canyon mesa along rt. I. 17, near power line, elev. ca. 3,400 ft., clay soil from volcanic rocks, Oct. 13, 1974 (cap).

Gentry 22568, DES, US, MEXU. Near Camp Creek, Maricopa Co., open chaparral, elev. 3,000–4,000 ft., June 10, 1968 (l, cap).

Gentry 22572, DES. 11 miles W of Magazine Spring on side road to Brushy Basin, New River Mts., grassy sun slope, elev. ca. 4,200 ft., June 11, 1968 (l, f).

Gentry 17756, DES, US, MEXU. Queen Creek Canyon above Superior, rocky slopes, elev. 3,000–3,500 ft., July 7, 1959 (l, f).

Gentry 21981, DES, US, MEXU. Mile 266, along route 60, 15 miles N of Globe, steep, red rocky, sun slope with scrub oak, elev. ca. 4,000 ft., July 18, 1966 (l, f).

Gentry 21982, DES, US, MEXU. Mile 264 along route 60, 13 miles N of Globe, scattered on rocky slopes, July 19, 1966 (l, f, photos).

Gentry 3994, DES, ARIZ, UC. Pine above Payson, pine-oak forests to chaparral, Oct. 1938 (l, cap).

Gentry 23417, DES. Ca. 1 mile E of Cleator, Bradshaw Mts., Yavapai Mts., elev. ca. 3,400 ft., igneous, palo verde-ocotillo, Aug. 18, 1974.

Gentry 23418, DES, US. 5 miles W of Cleator along road to Crown King, Bradshaw Mts., elev. ca. 4,350 ft., igneous, chaparral, Aug. 18, 1974 (l, f, cap).

Gentry & Ogden 9984, US, MEXU, DES. Mountains 1 mile E of Superior, limestone, ocotillo-*Simmondsia-Agave*, Nov. 24, 1950 (l, photos).

Gentry & Ogden 9990, DES, US, MEXU. 25 miles NE of Globe along Hwy. 60, hill & mt. slopes with grama grasslands or open chaparral, Nov. 25, 1950.

Gentry & Ogden 9991, DES, US, MEXU. 16 miles NE of Globe along Hwy. 60, hill & mt. slopes with grama grasslands or rather open chaparral, Nov. 25, 1950 (l, photos).

Gentry & Sauck 23411, DES. 4 miles N of Superior near Ridgeline T. 1 N, R12E, SW ½ Sec. 27, N exp., juniper, elev. 4,720 ft., July 24, 1974 (l, f).

Harris C16284, US. The Basin, Santa Catalina Mts., July 11, 1916.

Harrison 2083, ARIZ, US. Cliffs above Superior, June 6, 1926 (l, f).

Harrison, Kearney, Hope 7981, 7972, ARIZ, US. Along Chimney Creek, Rincon Mts.

Higgins 7850, OSH. 3 miles south of Pine along hwy. 87; sandy to gravelly soil; oak-juniper woodland, 19 July 1973 (l, f).

Kearney & Peebles 8732, 8733, 8734, 8735, ARIZ, US. Along Chimney Creek, Rincon Mts., July 31, 1932 (f).

Kearney & Peebles 12036, ARIZ, US. 14 miles NE of Globe, Gila Co., elev. 4,700 ft., July 21, 1935 (f).

Lemmon s.n., UC. Santa Catalina Mts., 1880s.

McKelvey 1621, 1622, 1624, A. Mt. Lemmon Road on Santa Catalina Mts., March 21, 1930 (l, cap).

Peebles 14344, ARIZ, US. Summit of Bush Hwy., Mazatzal Mts., elev. ca. 4,000 ft., June 19, 1939 (l, f).

Peebles & Harrison 5543, Type. **US.** Queen Canyon, Pinal Mountains, June 21, 1928 (l, f, 2 sheets). July 16, 1931. Also here in 1940.

Price s.n., DS. Santa Catalina Mts., Pima Co., Jan. 17, 1894.

Pringle 467, US. Santa Catalina Mts., June 1882 (l, f).

Toumey s.n., UC, US. Chiricahua Mts., Aug. 15, 1895 (f only—hardly from the Chiricahuas!).

Agave colorata

SINALOA. *Gentry 7237,* DES. Los Pucheros, Sierra Surutato, open rocky slope, pine-oak forest zone, March 17, 1945 (l, f, photos); cult. from small plant at Murrieta, Calif., fl'ed. May 1957.

SONORA. *Boutin & Kimnach 3273,* HNT. Nacapuli Canyon, NW of Guaymas, Feb. 1970 (cult. HNT, fl'ed. July 1979).

Gentry 3050, DES, ARIZ, **DS** Type: Peñasco Blanco, near Aquibiquichi, Feb. 1937.

Gentry 10270, DES, US, MEXU, TEX. Aquibiquichi, white calcareous ridge in thorn forest, March 9, 1951 (type locality) (f).

Gentry 11368, DES, US, MEXU. Rancho Bachaca, ca. 15 miles SE of Navojoa, thorn forest, on limey tuffaceous cliff of N exposure, Dec. 19, 1951.

Gentry, Arguelles 11641, US, DES, MEXU. Sierra Bojihuacame, volcanic foothill, elev. 500–700 ft., March 1952, fl'ed. at Murrieta, Calif., June 1965, June 1975 (l, f, cap, photo).

Gentry & Arguelles 19881, US, DES. Cerros 4–5 miles N of Bahia San Carlos in palm canyon, 29 March 1963.

Gentry & Engard 23560, US, DES, MEXU. Same, 17 June 1975 (l, f).

Phillips et al. 75-161. ARIZ. 3 miles N of San Carlos Bay, 26 Oct. 1975.

Agave durangensis

DURANGO. Gentry & Gilly 10576, Type DES, **US.** 18 miles SE of Cd. Durango along hwy. to Nombre de Dios, Sierra Registro, rocky slope with scattered trees & shrub in grama grassland, elev. ca. 6,000 ft., June 8, 1951, transpl. fl'ed. at Murrieta, Calif., Aug., 1962 (l, f, cap, photos).

Gentry & Gilly 10591, DES, US, 8–10 miles S of Boca de Mezquital, along road from Cd. Durango to Mezquital, steep gravelly slopes, elev. ca. 5,500 ft., June 11, 1951 (f).

Gentry & Arguelles 18230, 18231, 18232, 18233 DES, US, MEXU. Sierra Registro, 21 miles SE of Cd. Durango, rocky grassland slope, elev. ca. 6,400 ft., Nov. 19, 1959 (l, f).

Gentry 22089, DES, US. 25 miles S of Cd. Durango along road to Mesquital, Sierra Registro, sun slope in grassland, elev. 6,500 ft., Nov. 3, 1966.

Gentry 22092, DES, US., MEXU. 32 miles S of Cd. Durango along road to Mezquitál, grassland, elev. 5,500 ft., Nov. 3, 1966 (l, f, photos).

Pinkava et al. 9601, ASU 11.5 miles NW of Nombre de Dios along rt. 45 to Durango, July 29, 1972 (l, f).

ZACATECAS. *Gentry 23395,* DES, US, MEXU. Sierra de Zacatecas, 5 miles N of Cd. Zacatecas, open slope with Opuntia on volcanic rock, elev. 8,400 ft., March 18, 1974 (l, photos).

Gentry 23396, DES, MEXU, US. N slope of Sierra Zacatecas, 8 miles N of Cd. Zacatecas, elev. 7,850 ft., March 18, 1974 (l, photo).

Gentry & Engard 23606, US, MEXU, DES, INIF. 12–15 miles NE of Cd. Zacatecas along rt. 54. 4 Aug. 1975; *Acacia-Opuntia* grassland, elev. ca. 7,200 feet.

Agave flexispina

CHIHUAHUA. *Gentry & Arguelles 17922,* DES. 4 miles S of Matamoros (Cuevas), oak woodland and grama grassland, elev. 1,950–2,100 m, 4 Oct. 1959 (l, f, cap.).

Gentry & Engard 23580, US, MEXU, DES. 20 miles S of Parral along rt. 45, open volcanic rocky slope in grassland, elev. ca. 5,900 ft, 28 July 1975 (l, f, cap.).

LeSueur 245, GH. Majalca, Aug. 20, 1935 (l, f of a different species).

Palmer 138, GH. Vicinity of Santa Eulalia, elev. ca. 1,300 m. 28–29 April 1908.

DURANGO. *Gentry 22049,* DES, US. Crossroad to Canutillo along rt. 45 by Rio Florida, grama grassland, elev. 6,000 ft., Oct. 25, 1966 (l, f, cap, photos).

Gentry & Arguelles 18142, DES, MEXU, US. 15–18 miles NE of La Zarca, 11 Nov. 1959; grama grassland, elev. ca. 6,500 ft. (l, f).

Kimnach & Brandt 1093, HNT. 2 miles W of Zarca on road to Palmito, 18 Nov. 1967; alt. 7,000 ft. (cult HNT, fl'ed. Aug. 76).

Palmer 330 in 1906, Type. K, UC, **US.** Tepehuanes (l, f, 2 sheets).

Pinkava et al. 9405, DES, ASU. Along rt. 39, ca. 23 miles NW of Santiago Papasquiaro, July 25, 1972; rocky hillside. Topotypic as this is only a few miles from Tepehuanes (l, f).

ZACATECAS. *Gentry & Gilly 10575,* ? DES, US. Sierra Papanton, ca. 14 miles W of Sombrerete, oak-juniper grassland, open rocky slope of southern exposure, elev. ca. 7,500 ft., June 8, 1951 (lvs. from transpl. Murrieta, Calif., Fall, 1958).

Agave fortiflora

SONORA. Gentry 1980, Type: **US,** DES, MEXU. Hunt. Bot. Gard., San Marino, Calif., 2 July 1962 (l, f); presumably originally from Sierra Jojoba, No. 11630 below.

Gentry & Fox 11630, US, MEXU, DES. Sierra Jojoba, 15 miles NE of El Datil, Altar Dist., 23 March 1952; limestone ridge with dispersed shrub of *Cercidium, Jatropha* and *Bursera.*

Gentry 11618, US, MEXU, DES. On valley plain ca. 3 miles NE of Sierrita de Lopez, 19 Feb. 1952; desert shrub with *Olneya, Cercidium, Larrea,* cactus etc.

Agave murpheyi

ARIZONA. *Eastwood 17068a,* CAS. Road to Packard, May 19, 1929.

Eastwood 17350, CAS. Corn Creek, Apache Trail, May 20, 1929 (l, bulblets).

Engard et al. 479, DES. 4 miles S of Bartlett Dam on mesa near Verde River, Maricopa Co., 18 May 1975 (l, f, bulbil).

Gentry & Ogden 9978, US, DES, MICH. Topotype clone, Boyce Thompson Southwest Arboretum, Nov. 24, 1950 (l, photo).

Gentry, Weber, Engard 22981, DES. Ca. 5 miles S of Bartlett Dam and 2 miles E of Verde River, elev. ca. 2,000 ft., April, 1972.

Gibson s.n., Type: "Deposited and cotypes cultivated at the Boyce Thompson Southwest Arboretum, Superior."

McKelvey 1319, A, CAS. Apache Lodge, Roosevelt Dam, June 13, 1929 (photos, bulblets).

HORTICULTURE. *Gentry & Koch 19860* US, MICH, DES. Huntington Botanical Garden, July 31, 1962 (l, f).

SONORA. *Engard s.n.,* DES. Km post 204 between Hermosillo and Magdalena, 4 July 1975.

Gentry 21204, US, DES. San Luisito along road to Quitovac from Caborca, 4 Sep. 1965 (l, photos).

Agave palmeri

ARIZONA. *Barkely 14A574,* UC. Cave Creek Canyon, Cochise Co., Sep. 27, 1944 (l, cap).

Benson 10263, ARIZ. Limestone hill between Bisbee and Douglas, Cochise Co., elev. 3,500 ft., April, 1940 (l, cap.).

Benson 10247, ARIZ. 1 mile E of Canelo, Huachuca Mts., Santa Cruz Co., elev. 5,000 ft., April 15, 1940 (l, cap).

Benson 11111, DS. 2 miles E of Lowell, Mule Mts., Cochise Co., elev. 5,000 ft., April 14, 1942 (l, cap).

Brandegee s.n., UC. Santa Rita Mts., Nov. 1891.

Coville 1950, US. Double Circle Ranch, Graham Co., June 28, 1904 (f).

Eggleston 10956, US. About Portal, Cochise Co., Chiricahua Nat. For., elev. 1,600 m, 26–29 Sept. 1914.

Engard & Squire 146, DES. Whetstone Mts., Cochise Co., elev. ca. 4,800 ft., 30 Sep. 1973 (l, f, cap.).

Gentry 21990, DES, US. Texas Canyon along rt. 80, 9–10 miles E of Benson, open granitic slopes with scattered oaks, 25 July 1966 (l, f).

Gentry 17750, DES, US. 8 miles W of Nogales hwy. along road to Ruby, oak woodland, July 4, 1959 (f).

Gentry & Ogden 9932, DES, US. Ca. 5 miles W of Sonoita in the Patagonia Hills, oak woodland with grama grassland with coarse calcareous soils, elev. 3,500–4,000 ft.

Gentry 17749, DES, US. MEXU. 5 miles W of Nogales hwy. along road to Ruby, terrace slope in oak woodland, July 4, 1959.

Goldman 2341, US. Graham Mts., elev. 6,000 ft., July 26, 1914 (l, f).

Lehto & McGill 120494, ASU, DES. Peña Blanca Rd. (Az. 289) 6.5 miles W of Jct. with I-19, Santa Cruz Co. Oak grassland.

Lemmon Hb., UC. Huachuca Mts., Oct., 1882.

Lemmon & wife s.n., US. Huachuca Mts., Sep., 1883 (l, f).

McKelvey 615, 616, A. Mule Mts., 9 Feb. 1929 (lvs., large, long-leaved form).

McKelvey 937, 938, 1015, 1116, A. Sierra Ancha, road to Carr's Ranch and Pleasant Valley 7-21 May 1929.

McKelvey 1080, A. Road from Globe to Coolidge Dam, 1 May 1929 (3 lvs., 3 forms).

McKelvey 1323, A. Near Salt River on road to Carr's Ranch in Sierra Ancha, June 16, 1929 (l, infl).

McKelvey 1555, A. Sabino Canyon, Santa Catalina Mts., 14 March 1930 (l, cap).

McKelvey 1568, 1569, A. Between Vail and Sonoita on road to Patagonia, March, 1930 (l, cap).

McKelvey 1588, 1586, 1629, A. Ranger Station road in Rincon Mts., March, 1930 (large strong caps).

McKelvey 1636, 1638, A. Pajarito Mts., along road from Nogales to Ruby, March, 1930 (cap).

Nealley 56, US. Steins Pass, AR-NM.

Peebles 14214, US. Dragoon to Benson, 3 Nov. 1938 (l, cap).

Peebles, Harrison, Kearney 4302, 4303, US, ARIZ. Graham Mts., July 22, 1927 (l, infl).

Peebles, Harrison, Kearney 4528, 4539, US, ARIZ. Santa Rita Mts., July, 1927 (l, infl).

Peebles, Harrison, Kearney 5600, ARIZ. Patagonia Mts., 18 Aug. 1928 (f).

Pringle in 1884, US. Santa Rita and Mustang Mts., 20–23 July 1894 (l, f).

Purpus, 1903, US. Vicinity of Benson, Cochise Co.

Rose, Standley, Russell 12325, US. Vicinity of Benson, Cochise Co., 2 March 1910 (l, cap.).

Toumey s.n., US. Chiricahua Mts., 20–23 July 1894 (f).

Vasey 579, US, A. Vicinity of Benson, Cochise Co., 1881.

Wooten s.n., US. Santa Rita Mts., 10 Sep. 1914 (f, photo).

Wooten, s.n., US. Range Reserve near Tucson, 11 July 1911 (l, f).

CHIHUAHUA. *Borja L. B-408, B-409*, ENCB. 25 km SWS of Palomas.

Engard & Getz 237, DES. Cerro Huerfano. 36 miles N of San Buenaventura along road to Nueva Casas Grandes near KM post 171, rocky outcrop, 4 July 1974 (l, photo).

Gentry & Arguelles 22947, DES, US, MEXU. Near Col. Juarez, rocky hill slopes in grassland, elev. ca. 5,000 ft., 7 June 1971 (l, photo).

Nelson 6469, US. Near Sierra en Media, plains, 24–26 Sep. 1899 (l, f).

NEW MEXICO. *Gentry & Ogden 9915*, DES, US. Pyramid Mountains, about 7 miles S of Lordsburg, acaulescent, big-butted *Agave*, suckering at inflorescence in large plants; smaller flowering stalks not suckering, 5–7 m, 30 Oct. 1950.

Goldman 1402, US. Animas Mts., alt. 5,800 ft., 7 Aug. 1908.

Goldman 1442, US. San Luis Mts., Grant Co., alt. 5,500 ft., 14 Aug. 1908 (l, fl).

Goldman 1477, 1478, US. Florida Mts., alt. 6,000 ft., 8 Sep. 1908 (l, fl).

McKelvey 2088, 2089, A. Burro Mts., May, 1931 (strong, elongate, oblong caps. 50–60 x 16–18 mm [valves]).

McKelvey 4032, A. Tyrone, 17 April 1934 (l. narrow 45 x 5.5 cm with many small teeth; caps. 5–6 x 1.6 cm [valves]).

Mearns 506, US. Cloverdale, Las Animas Valley, 17 July 1892 (f).

Mearns 545, US. San Luis Mts., Grant Co., near summit, 19 July 1892 (l, f).

Worthington s.n., DES. N slope of hill at old Mahoney Mine, Luna Co., Florida Mts., 25 March 1978 (l, cap).

SONORA. *Barclay & Arguelles 2013*, US, DES. 16 miles southeast of Magdalena, 13 May 1966.

Felger 3457, ARIZ. 15.4 miles southeast of Magdalena along road to Cucurpe, alt. ca. 3,400 ft., 16 July 1960 (l, f).

Felger & Beasly 3525, ARIZ. El Coyote northeast of Moctezuma on north-facing slopes with lower oaks, alt. ca. 3,400 ft., 19 July 1960 ("now in full flower—flowers white with callous tip, ovary green").

Felger & Marshall 3343, ARIZ. 6.3 miles northwest of Nacori Chico, "with lower oaks ca. 3,400 ft. alt., common, non-suckering" (l, buds).

Felger 3781, ARIZ. 8.3 miles north of Magdalena, alt. ca. 3,100 ft., 19 Aug. 1960 (l, fl).

Felger 4026, ARIZ. 15 miles north of Fronteras, alt. ca. 4,100 ft., 7 Sep. 1960 (l, f).

Felger 3595, ARIZ. 16.1 km north of Nacozari, oak-grassland, 22 July 1960 (l, f).

Felger 3525, ARIZ. El Coyote (NE Sonora), 19 July 1960; elev. ca. 3,400 ft. with lower oaks. "Leaves gray-blue green; fls. white with callous tips, ovary green (l broad, f).

Gentry 16641, DES, US. Mountain above & E of Guasabas on volcanics in oak woodland, alt. 4,000 ft., 21 May 1957 (pl).

Gentry 10274, DES, US, MEXU. Sierra de Pajaritos, about 12 miles east of Ures, March 11, 1951 (l, cap, photo).

Gentry 19888, DES. 16 miles southeast of Magdalena along road to Cucurpe, April, 1963.

Gentry 21200, DES. Sierra del Humo, Dist. Caborca, near Rancho Cimarron, 3 Sep. 1965 (l, cap, live pl.).

Gentry & Arguelles 22943, DES, US, MEXU. Ca. 10 miles NE of Aribabi along road to Huachinera, alt. ca. 4,000 ft., oak-juniper grassland, 4 June 1971.

Gentry & Arguelles 22942, DES, US, MEXU. 3 miles E of Aribabi along road to Huachinera, alt. ca. 4,000 ft., oak grassland, 4 June 1971.

Gentry & Arguelles 22931, DES, MEXU, US. Ca. 15 miles NW of Nacozari along road from Bocoachic, alt. ca. 3,500 ft., rocky sun slope, 2 June 1971.

Gentry 19986, DES, US, MEXU. Sierra Jojoba, rocky igneous slope, elev. 2,500–3,000 ft., 20 May 1963.

Gentry & Weber 21989, DES, US, MEXU. Pass by Sierra Baviso along road from Magdalena to

Cucurpe, Sonora, gravelly slopes with *Yucca,* Ocotillo, etc., elev. 3,000–3,500 ft., 23 July 1966 (l, f, photo).

Gentry & Weber 21987, DES, MEXU, US. Near summit of Sierra Jojoba, brushy slopes, 24 July 1966 (l, f, photo).

Nabhan et al X517, ARIZ. 5 miles S of Cananea on Rio Sonora tributary, 26 Aug. 1976; *Prosopis–Quercus* woodland (l, cap).

Nabhan X-480, ARIZ. 37 miles S of Colonia Morelos, Rio Bavispe, 20 Aug. 1976 (l, f, cap).

Nelson 6469, US. Near Sierra del Media?

White 3049, MICH, US. Arroyo de la Galera (loop of the Bavispe River), 27 July 1940 (l, f). Mesquite grassland.

White 517, MICH. Santa Rosa canyon near Bavispe, 15 July 1938 (l, f).

Wiggins s.n., DS. 43 miles S of Nogales along hwy. to Magdalena, 21 Feb. 1933 (l, seedlings).

White 3022, US. Cañon de Agua Amarga, Rio de Bavispe loop, 30–31 Aug. 1940; oak grassland (l, cap.).

White 3920, ARIZ, US. Rancho de la Nacha, R. de Bavispe loop, 15 Aug. 1941; oak grassland (l, f).

White 4580, ARIZ, US. Colonia Morelos, R. de Bavispe loop, Sept.-Oct. 1941; elev. 2,600 ft. (l, f).

Agave palmeri × *chrysantha*

ARIZONA. *Kearney & Peebles 8732, 8733, 8735,* ARIZ. Near Ranger Station, Rincon Mts., 31 July 1932. Also at Chimney Creek, *8753.*

Agave shrevei magna

CHIHUAHUA. *Felger 8080,* ARIZ. Barranca de Batopilas, ca. 5 miles above La Bufa bridge, 9 June 1963; elev. ca. 3,400 ft., leaves blue green, flowers yellow above green ovary (l, f).

Gentry & Bye 23360, type, DES, **US,** MEXU. 3–4 ½ road miles NE of bridge over Rio Batopilas, Barranca Batopilas, volcanic sun slope, elev. 3,500–4,500 ft., 10 Oct. 1973 (l, f, cap, photos).

Gentry & Bye 23358, DES, US, MEXU. 2 miles SW of Quirare along road to la Bufa, rocky sun slope, elev. ca. 6,000 ft., 9 Oct. 1973 (l, f, photos).

Gentry & Bye 23356, DES, US, MEXU. 2.5 miles NE of Quirare along road to la Bufa, volcanic rocky slope with oaks and pines, elev. ca. 6,500 ft., 8 Oct. 1973 (l, photos).

Kimnach & Brandt 901, HNT. Top of canyon of Rio Batopilas, NE of La Bufa mine, 3 Oct. 1967; alt. 6,000 ft. Also by same collectors *934* near bridge to La Bufa mine (cult H.B.G., fl'ed. Sep. 1975).

Kimnach & Brandt 990, HNT, ¼ mile W of Parajes on Chihuahua–Mochis RR by Rio Santiago, 20 Mar. 1967; alts. 4,500 ft.

Pennington 33, TEX. Yepachic, Chih., 10 July 1970. (l, f shoot and buds eaten.)

SINALOA. *Gentry 11415,* DES, US, MEXU. Cerro Babuyo, ca. 16 miles northeast of Choix, oak-palm-*Nolina* grassland on limestone mountain top, elev. ca. 4,000 ft., 3 Jan. 1952 (l, f, cap, photos).

SONORA. *Gentry & Arguelles 17680,* DES. Cedros Range, Nov., 1958; oak savanna on volcanics, elev. ca. 3,500 feet. (l, f, br, photo. Transplant flowered in Des. Bot. Gard., Aug.–Sep., 1976).

Agave shrevei matapensis

SONORA. *Boutin & Kimnach 3736,* DES. Above Rancho La Puerta, Sierra Batuc., limestone slopes, elev. ca. 3,600 ft., 7 Dec. 1972.

Gentry 11612, DES, US, MEXU. 2 miles E of Nacori, elev. ca. 2,000 ft., 16 Feb. 1952 (l, cap).

Gentry 17622, 17623, DES. Mountain pass 6–8 miles NE of Matape 4 Nov. 1958 (cap, seed).

Gentry & Arguelles 11607, DES, type, **US,** MEXU. Sierra de Matape, Tepehuaje-oak-*Nolina* woodland, on granitics and deformed limestone, elev. ca. 3,000 ft., live plants collected 16 Feb. 1952 (flowered at Murrieta, Calif. Aug., 1965), (l, f, cap, photos).

Gentry & Arguelles 16600, DES, US. Between Nacori Grande and Matape, live plants collected 12 May 1957, flowered at Murrieta, Calif., Sept., 1961 (l, f).

Agave shrevei shrevei

CHIHUAHUA. *Bye 7818,* COLO, DES. Mpio. de Guachochic, on Creel-Guachochi road between Basiguare and Cusarare above Arroyo Colorado, 1 Aug. 1977; elev. ca. 1,980 m, pine-oak forest (l, f).

Gentry 2028, type, DES, ARIZ, **DS.** Sierra Canelo, Rio Mayo, rocky slopes in oak and pine country, 8 Oct. 1935 (l, f, cap).

Gentry 8080, DES. Arroyo Hondo, Sierra Charuco, 16–30 April 1948 (flowered at Murrieta, Calif. summer 1965), (live pl, l, f).

Pennington 34, TEX. Yepachic, 10 July 1970. Lvs., f. shoot and buds eaten.

SONORA. *Boutin & Kimnach 3684B,* HNT. Along road to Bermudas from Santa Ana, 2 Dec. 1972; elev. 3,000 ft. (cult. H.B.G., fl'ed Aug. 1979).

Gentry 11385, Sierra Saguaribo, Sonora. (Live plants only.)

Gentry 11580, 19865, DES, US, MEXU. Guajaray Scarp, north of Curopaco and Arroyo Guajaray, oak forest on precipitous volcanic slopes, elev. 4,500–5,000 ft., 13 Feb. 1952 (l, f, photo), fl'ed. Hunt. Bot. Gard., 14 Aug. 1962.

Gentry & Arguelles 22941, DES. 2 miles E of Aribabi along road to Huachinera, alt. ca. 4,800 ft., oak woodland on volcanics, 4 June 1971.

Gentry 11582, DES, US. Guajaray Scarp, north of Curopaco and Arroyo Guajaray, oak forest on precipitous volcanic slopes, elev. 4,500–5,000 ft., 13 Feb. 1952 (l, photo).

Gentry & Arguelles 11598, DES, US. Rancho Santa Barbara del Agua Blanca, northwest of Curopaco, extensive oak forest savanna over broad foothills; compact calcareous soils, 14 Feb. 1952 (l, cap, photo).

Gentry 16641, DES. Mountain above (E) Guasabas, oak woodland, volcanics, alt. ca. 4,000 ft., 21 May 1957 (l, cap, photo).

Gentry 11378, DES, US. Sierra Potrero, Rio Mayo, Col. winter 1951–52 (flowered at Murrieta, Calif. 15 Aug.–10 Sep. 1957), (l, f).

Gentry 21211, DES. Sierra Charuco, Sonora-Chihuahua border, pine-oak woodland, elev. ca. 4,000 ft., 10 Sep. 1965 (cap).

Gentry 21187, DES. Sierra de la Ventana, Rio Mayo, oak woodland, alt. ca. 4,000 ft., 28 May 1965 (photo).

Gentry, Barclay, Arguelles s.n., DES. Sierra Tecurahui, Dist. Alamos, rocky volcanic slopes of the pine-oak forest, elev. ca. 4,500 ft. (photo).

Gentry 11605, DES. Chihucu near Navojoa, 15 Feb. 1952 (l, photos).

Gentry 21211, DES. Sierra Charuco, Son.-Chih. border, Alt. ca. 4,000 feet, 10 Sep. 1965 (cap.).

Agave wocomahi

CHIHUAHUA. Gentry 1989, DS, ARIZ, DES. Type: Sierra Huicorichi, Rio Mayo, 7 Oct. 1935 (l, f).

Gentry & Arguelles 11397, DES, US. Sierra Huicorichi, Rio Mayo.

Gentry & Arguelles 10252, 10254, DES, US. High rocky rims around Arroyo Hondo, Sierra Charuco, 6–7 Mar. 1951 (flowered at Murrieta, Calif., 25 July–10 Aug. 1962).

Gentry, Correll, Arguelles 17988, DES, US. Sierra Chinatu, Dist. Guadalupe y Calvo, 9 Oct. 1959.

Gentry & Bye 23352, DES, US, MEXU. Ca. 8 miles SE of Creel, elev. ca. 8,000 ft., 7 Oct. 1973 (l, photo).

Gentry & Bye 23354, DES, US, MEXU. Ca. 25 miles S of Creel along new road to Canyon Urique, volcanic sun slope, elev. ca. 7,000 ft., 8 Oct. 1973 (l, f, photo).

Kimnach & Brandt 878, HNT, 4.5 miles from Basihuare direction of Creel, 9 Mar. 1967; 7,000 ft. (cult H.B.G., fl'ed. Aug. '75.)

Pennington 535, TEX. Near Rio Conchos, N of Hararachic, alt. ca. 4,800 ft., 10 July 1955. "Fish stupefaction plant; crush leaves."

DURANGO. *Gentry 23467*, US, DES, INIF, MICH. 2½ miles S of Tepetates lumber mill. Sierra de los Huicholes, 13 Jan. 1975; elev. ca. 8,150 ft., pine-oak-madroño, rocky sun slope.

SINALOA. *Gentry 6340*, DES, ARIZ, GH, US, DS, MICH. Ocurahui, Sierra Surutato, 1–10 Sep. 1941 (l, f).

Gentry & Arguelles 18379, DES. Sierra Surutato, elev. ca. 6,000 ft., pine-oak, in rocks, 13 Dec. 1959 (f).

SONORA. *Felger & Stronack 3052*, ARIZ. 6 miles E of Reparo along road from Nuri to Yécora, 11 Dec. 1959 (l, cap.).

Gentry 10225 in part (l), DES, US. Sierra Charuco, Rio Mayo, 6 Mar. 1951.

Gentry 11586, 11588, DES, US, MEXU. Guajaráy Scarp, N of Curopaco and Arroyo Guajaráy, 13 Feb. 1952.

Gentry 17666, DES, US. Sierra de Yecora, 15 Nov. 1958.

Pennington 143, TEX. Santana, 6 Aug. 1970 (l, f).

17.

Group Hiemiflorae

Plants small to medium, rarely gigantic, generally nonsurculose, freely seeding, winter flowering; leaves broadly ovate to lanceolate, frequently light glaucous or pruinose in symmetrical, multi-leaved, short-stemmed rosettes; inflorescence paniculate to racemose; flowers in relatively small clusters, either in loose sparsely bracteolate umbels, or more often in tightly balled, densely bracteolate clusters on very short to medium length lateral peduncles; flowers small to medium, variously colored, the tepals exceeding the tube in length; filaments inserted above the mid-tube; capsules ovate to oblong, thin-walled or thick-walled. Southeastern Mexico to Nicaragua.

Typical species: *Agave hiemiflora* Gentry. (Hiemiflorae = winter flowering.) Guatemalenses Trelease (in part) Trans. Acad. Sci. St. Louis 23, No. 3, 1915.

The Hiemiflorae constitute a rather natural group with some morphological features that make them readily recognizable in their native region. The conspicuous feature is the tightly balled flower clusters borne in narrow deep panicles. In some species, as *hiemiflora, congesta,* and *potatorum,* the lateral peduncles are often very short, only a few centimeters, giving a racemose appearance. In other species, as *atrovirens, hurteri,* and *thomasae,* the lateral peduncles are longer, but the flowers are tightly set in ball-like clusters. In all of these the pedicels bear rather large dark-colored bractlets, characteristically tufting the flower clusters; a feature also appearing in Chiapan members of the Polycephaleae of the subgenus Littaea.

There is a subgroup of the Hiemiflorae, represented by *A. pachycentra, A. wercklei,* and *A. seemanniana* that have broader, shorter, more open panicles with flowers loosely clustered, and the pedicellar bractlets are reduced in number and size and are rather colorless. However, the other characters of leaves and flowers admit them in this section. Altogether they form a natural geographic group in tropical Mesoamerica and constitute the majority of species in Central America.

With review of flower structure, as represented in the statistics of Table 17.1 and the ideographs of Fig. 17.3, a large range of variability is evident within several species. Variation is especially marked in *Agave potatorum,* somewhat less so but still strong in *A. pachycentra, A. hiemiflora,* and *A. hurteri.* Such series appear to represent unsolved taxonomic problems. Considering the long distance from the United States of most Hiemiflorae, there are a surprising number of collections, made mostly by a handful of collectors: Elzada Clover (1947), Dennis Breedlove (1964–74), William Trelease (1915), and Howard Scott Gentry (1953–75) (see Exsiccatae). These represent collecting forays, rather than careful field study of the agave populations. The group needs the sustained attention of a resident agaveologist the year or years around, who can supplement his collection by developing more criteria in cytology, epidermal anatomy, chemical taxonomy, and biotic factors in the environments, including pollinators. Trelease's treatment of the Guatemalan agaves was quite cursory and Berger's even more limited. There are significant additions and changes of concepts in my revision, but I am not satisfied with it. There are a number of species proposed by Trelease that I was unable to locate in the field, such as *A. minarum, A. kellermaniana,* and *A. deamiana;* they remain *incertae sedis.*

Apparently, relatively little use has been made of this group of plants by the indigenous Americans. The few uses I have observed have been annotated below under accounts of species. It is probable that an inquiring ethnobotanist could add a great deal to the written knowledge of the uses of these agaves.

Key to Species of the Hiemiflorae

1. Small to medium-sized plants; leaves at plant maturity mostly less than 60 cm long; spines shorter, mostly 2–4 cm long, with wide shallow groove or flat above   2
1. Large to medium-sized plants; leaves of mature plants mostly over 60 cm long; spines long, mostly 4–6 cm, and with deep narrow grooves   5
2. Plants polymorphic in leaves and inflorescence; leaf margins straight or deeply crenate; panicle narrow with subsessile flower clusters, or wider with longer lateral peduncles; flower tube deep or shallow; tepals longer or shorter than tube. Puebla & Oaxaca   *potatorum,* p. 490
2. Plants more constant in leaf and inflorescence; leaf margins generally all crenate to undulate; panicle narrow to broader but never with subsessile umbels; tepals all longer than tube. Chiapas-Nicaragua   3
3. Base of flower clusters congested with numerous dark-colored, persistent bractlets 1–2 cm long; panicles narrow and deep with 20–40 umbels, the lateral peduncles frequently short (10–15 cm long); leaves lanceolate, mostly 3–6 times longer than wide, little nar-

rowed at base   *hiemiflora,* p. 480
3. Flower clusters not congested at base with bractlets; bractlets pale, small, 0.5–1 cm long; panicles broader, the lateral peduncles 20–30 cm long; leaves short, ovate to spatulate, only 2–3 times longer than wide, markedly narrowed towards base   4
4. Rosettes medium-sized, single, leaves 30–50 cm long; panicles 3–4 m tall with 18–30 umbels; flowers 50–70 mm long   *seemanniana,* p. 494
4. Rosettes very small, cespitose; leaves 12–30 cm long; panicles 2–3 m tall with 8–12 umbels; flowers 30–35 mm long   *pygmae,* p. 494
5. Panicles deep, narrow, with numerous compact rounded flower clusters, thickly bracteolate at base   6
5. Panicles relatively short, or broad and open, with small, loosely clustered flowers, and with fewer smaller bractlets   10
6. Plants very large; leaves 1.5–2 m long, 25–40 cm wide; panicles 8–10 m tall, shorter than the supporting peduncle. Sierra Madre Oriental, Puebla, Veracruz, Oaxaca   *atrovirens,* p. 468

6. Plants smaller; leaves 0.6–1.5 m long; panicles deep, longer than the supporting peduncles   7
7. Inflorescence very narrow or racemose; flower clusters 40–50, sessile to short-pedunculate. Chiapas
    congesta, p. 476
7. Inflorescence wider, the flower clusters generally 30–40, the lateral peduncles 20–30 cm long. Guatemala   8
8. Rosettes generally large with numerous variable leaves (100 or more); teeth variable, frequently large, sometimes lacking; flowers slender, not markedly triquetrous   *hurteri,* p. 482
8. Rosettes medium-sized with fewer and less variable leaves; teeth small, sometimes with denticles between the larger, or if larger, more remotely spaced; flowers triquetrous or not   9
9. Leaves wider (15–19 cm); teeth smaller, only 1–2 mm long, sometimes with intervening small denticles; umbels 30–40; flowers markedly tri-

quetrous with angulate tubes and ovaries   *thomasae,* p. 500
9. Leaves narrower (10–15 cm wide); teeth larger without intervening denticles; umbels 15–20; flowers slender, not triquetrous   *lagunae,* p. 485
10. Leaves with more prominent teeth, mostly 5–10 mm long in mid-blades; panicles on long peduncles, the branches borne in upper half of axis
    pachycentra, p. 486
10. Leaves with small teeth along the mid-blades, 2–4 mm long; panicles on short peduncles, equaling or but little exceeding the leaves   11
11. Panicles short (2.5–3 m), the peduncle not exceeding the leaves; flowers smaller (40–50 mm long), the tube 4–5 mm deep, the tepals 3–4 times as long   *parvidentata,* p. 488
11. Panicles tall (to 8 m), the peduncle longer than the leaves; flowers larger (60–65 mm long), the tube 8–9 mm deep, the tepals 2 times as long
    wercklei, p. 500

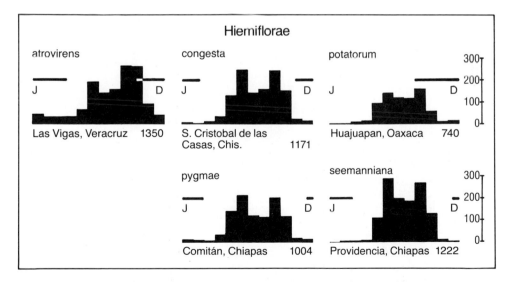

Fig. 17.1. Rainfall (silhouettes) and flowering (bars) perimeters of the Mexican species of Hiemiflorae. Relevant meteorological stations with annual rainfall in millimeters. Data from Atlas Climatológico de México, 1939, Sec. Agri. y Fom., México, D. F., and Boletín Meteorológico No. 2, 1968, Comm. Fed. Elect., México. Flowering periods are based on herbarium specimens and field observations; uncertainty expressed by dotted lines.

Agave atrovirens
(Figs. 17.1, 17.2, 17.3, 17.4, 17.5, 17.7; Table 17.1)

Agave atrovirens Karw. ex Salm, Hort. Dyck. 7: 302, 1834.
> *Agave latissima* Jacobi, Hamb. Gard. Zeit. 20: 499, 1864.
> *Agave coccinea* Roelz ex Jacobi, Hamb. Gartenz. 21: 61, 1865.
> *Agave mirabilis* Trel., Contr. U.S. Nat. Herb. 23: 131, 1920.

Rosettes large to very large, single, openly spreading, 1.5 x 3 m to 2 x 4 m, non-surculose; leaves mostly 150 x 25 cm to 200 x 40 cm, lanceolate, green to light glaucous or glaucous variegated, succulent, thick at base (to 25 cm), usually narrowed below the mid-blade, openly concave, sometimes valleculate, the edges about straight, convex below, surface smooth; teeth moderate, regular, the larger along the mid-blade

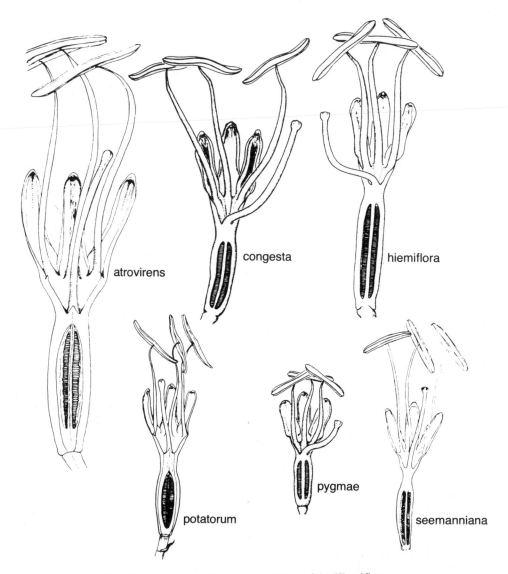

atrovirens

congesta

hiemiflora

potatorum

pygmae

seemanniana

Fig. 17.2. Flower sections representative of the Hiemiflorae.

mostly 4–7 mm long with low broad bases, 1–4 cm apart, brown to grayish brown; spine strong, 3–5 cm long, straight or sinuous, broad at base, widely openly grooved above, the keel rounded below and markedly produced into the leaf apex; panicle narrow, 8–12 m tall, with mostly 18 to 30 lateral branches in upper ½–⅓ of the shaft, the congested flowers in ball-like clusters; flowers thick fleshy, red to purple in bud, 70–100 mm long, opening yellowish within; ovary 30–50 mm long, cylindric, tapered at base, 3–6-angulate, the thick neck furrowed, 4–7 mm long; tube 11–15 mm deep, deeply furrowed, bulging by filament insertions, thick-walled; tepals unequal, the outer larger, 30–34 x 8–9 mm, red with margins paler, thick-fleshy, linear, erect, incurved at tips, conduplicate-involute, the inner with very thick keel and thin margins; filaments large, flattened, 70–80 mm long, purplish or red-spotted, inserted above middle of tube; anthers 34–37 mm long, straight to sinuous, yellow to bronze, excentric; pistil large overreaching anthers in post-anthesis; capsules ovoid, 4–5 x 2–2.5 cm, valves rather thin-walled, the pedicels 5–12 mm long, tapering from apex; seeds mostly 8–9 x 6–7 mm, thinly compressed, deeply lunate, shiny black.

Neotype: *Gentry 22377*, ca. 20 miles SE of Miahuatlan along road to Pochutla, on Sierra Madre del Sur, Oaxaca, Oct. 26, 1967, alt. ca. 7,500 feet on open mountain slope, in US, dups. in DES, MEXU (l, f, photo). The known distribution of *A. atrovirens* is shown in Fig. 17.7. It is strictly a high montane species between 6,000 and 11,000 feet (1,850 and 3,400 m) elevations in the Sierra Madre Oriental from northern Puebla, adjacent Veracruz, and eastern Oaxaca to the Sierra Madre del Sur in Oaxaca. For further details, see Exsiccatae.

The natural affinities of *Agave atrovirens* are with the Chiapan and Guatemalan agaves forming the Hiemiflorae. Like them it is a winter bloomer, develops a straight strong shaft, bears its flowers in tightly compacted, bracteolate, ball-shaped, purplish clusters in deep narrow panicles. The flowers of *A. atrovirens* are particularly fleshy succulent, and dimorphism is well developed between the conspicuously larger outer tepals and the high-keeled smaller inner tepals. It shares this latter character and its giant size with the cultivated magueys of the Section Salmianae, but it does not have the broad pyramidal panicles and thick fleshy peduncular bracts, and its smooth leaves are less fibrous and of quite different texture than the Salmianae. *A. atrovirensis* is further distinguished by the broad-based end spine with flat or open groove above and a ventral corneous spur intrudes into the flesh of the leaf apex. Berger, in remarks under *A. latissima* (1915, p. 131), called this protrusion a "zungenformig verlangert," or a ligulate prolongation, and concluded, "by the very broadly grooved endspine (this agave) is distinguished from all other large agaves." An unusually long intrusion is shown in Fig. 17.4, taken from the neotype specimen. Other sheets, however, show this same character, but in lesser degree, and a Berger specimen in US also shows it.

None of the preceding students of *Agave* have understood *A. atrovirens*. Being a non-surculose plant reproducing by seed, the early introduction of Karwinsky went out of European horticulture unnoticed a couple of decades later. It was reintroduced later unrecognized and assigned other names. Baker, Berger, and Trelease confused it with the giant maguey, *A. salmiana,* cultivated extensively over the central Mexican plateau.

Karwinsky, through Salm-Dyck in the original account (1834), reported, "crecit in Imperio Mexicano in summo monte Tanga." This was the same mountain where Karwinsky obtained *Furcraea longaeva,* which Karwinsky and Zuccarini (1933) further identified as "summo monte Tanga, provinciae Oaxaca, 10,000 pedes supra oceanum, in declivibus Quercebus et Arbustis consitis." I have been unable to locate Monte Tanga on any map. The Oaxacans of whom I inquired did not know, nor did Thomas MacDougal or Boone Halberg, both modern Americans who have hiked and lived for many years in the Oaxacan mountains. The modern Mexican botanists do not know. However, the 10,000-foot (3,100-m) habitat tells us more than Salm-Dyck's short

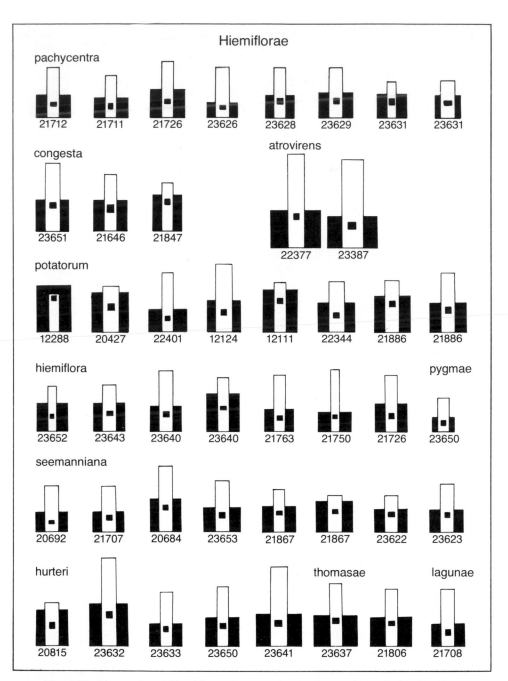

Fig. 17.3. Ideographs of Hiemiflorae flowers, showing proportions of tube (black) to outer tepal (white) and locus of filament insertion (black squares) in the tube. From measurements in Table 17.1.

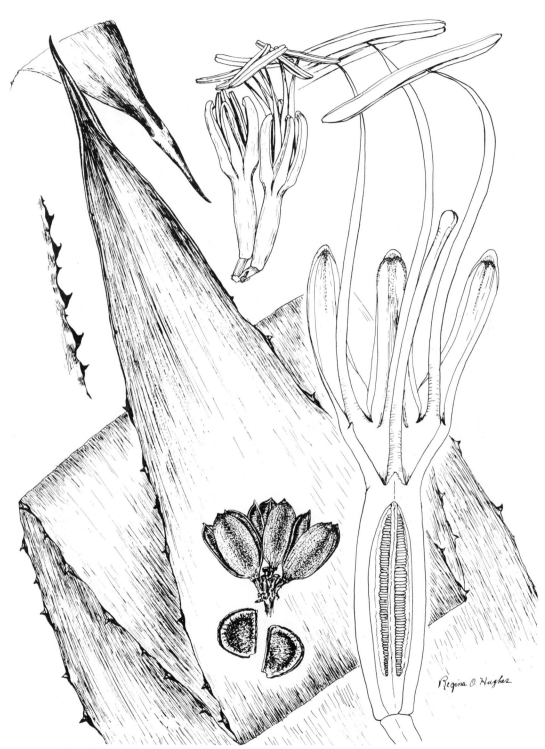

Fig. 17.4. *Agave atrovirens*. Part of leaf with detail of spine and leaf margin, reduced; flower and capsule cluster reduced ca. $^2/_5$; flower section x 1.35; seeds x 1.8. Drawn from *Gentry 22377*.

Fig. 17.5. *Agave atrovirens*. Upper, ca. 20 miles south of Miahuatlán, Oaxaca, on slope of the Sierra Madre del Sur. Lower, some handsome giants in the colony on Sierra de Ixtlán, Oaxaca.

description. The *Agave* here accepted as the real *atrovirens* is the only one of the large central Mexican magueys that fit both the habitat and the description. It does not grow on the drier central Mexican plains with the cultivated magueys. However, a light glaucous form of it is cultivated for pulque, fences, and terraces about Las Vigas in company with *A. salmiana,* where the road to Jalapa crosses the Sierra Madre Oriental. Trelease called this form *A. mirabilis,* which is recognized as a variety here below.

The population about the neotype locality on the Sierra Madre del Sur is a large one. Both wild and cultivated row plants can be observed along the crests of the mountains for a distance of 15 to 20 miles (24 to 32 km). The symmetrical glaucous rosettes made a showy display with their tall inflorescences during October of 1967. However, the largest and most beautiful plants observed formed the colony on the high wet mountain about 14 miles (22 km) northeast of Ixtlán de Juarez (Fig. 17.5 lower). The largest individual rosettes were about 2 m tall and 4 m broad with 80 to 100 leaves up to 2 m long and 40 cm broad, ascending curvaceously, the margins finely and closely toothed, green to glaucous or light-glaucous variegated, very thick at base. Some of the horizontal hollows on the leaves held pools of rain water. The broad stems including leaf bases were nearly 1 m in diameter. Other observations in eastern Oaxaca indicate a distribution all along the high slopes of northeastern Oaxaca.

Local names have been reported as "maguey del cumbre" (Oaxaca), "maguey javalin" (Sierra Madre del Sur), "tepeme" (Tehuacán), and "tuah" (Acatepec). The most common use is for fence plants and for holding terraces on steep cultivated slopes. In northeastern Puebla and adjacent Veracruz the plants are tapped for pulque, but this practice was not found in Oaxaca. Apparently such tribes as the Mixes and Zapotecs did not practice the pulque culture. In the high country about Sierra Zempoaltepec, the Mixes peel off large sections of the cuticle and use it as a wrapping to carry and keep their tortilla sandwiches fresh, when they go off to their fields. While we have marshaled chemistry and industry to devise a wrapper, the Indian finds his ready-made by his host the maguey.

Propagation stock is still taken from wild stands and transplanted to cultivated lands. The natives cut off the young flowering shoots which causes prolific offsets to form in the axils of the upper leaves and bracts. After a season of growth these young rosettes form calluses at their bases and are easily separated from the old drying parent plant, where they can strike root and become a genetic replica of the parent. The black seeds collected from old capsules appear to be viable and are doubtless the generative source forming the variable individuals noted in wild populations. Additional illustrations of this remarkable plant can be found in a short account of this agave written years ago (Gentry, 1964). Schultze (1942), as *A. coccinea,* features a photograph of the inflorescence with some of the lateral peduncles broken down by the weight of the large tightly packed umbels. I have also observed this weakness in the native habitat.

Agave atrovirens var. *mirabilis*
(Fig. 17.6)

Agave atrovirens Karw., var. *mirabilis* (Trel.) Gentry, stat. nov.
  Agave mirabilis Trel., Contr. U.S. Nat. Herb. 23: 131, 1920.

Large to gigantic single rosettes with massive, soft-succulent, upcurving-ascending, broadly lanceolate leaves up to 2 m long. Differs from the species only in having the leaves consistently light gray glaucous (Fig. 17.6).

Type: *Trelease 7,* Las Vigas, Veracruz, 3/2/05. MO, DES.

Known certainly only from Las Vigas and vicinity at elevations from 7,000 to 8,000 feet (2,150 to 2,480 m). It probably also occurs through these elevations in adjacent

Table 17.1. Flower Measurements in the Hiemiflorae (in mm)

TAXON & LOCALITY	OVARY	TUBE	TEPAL	FILAMENT INSERTION & LENGTH		ANTHER	TOTAL LENGTH	COLL. NO.
atrovirens								
Miahuatlán, Oax.	50	13 × 18	33 × 7 & 31	11 & 10	65	37	96†	22377
	46	14 × 18	32 × 7 & 30	11 – 10	80	37	91†	22377
Tehuacán, Pue.	28	11 × 19	31 × 9 & 28	8 – 7	55	34	71†	23387
	29	11 × 18	31 × 9 & 27	8	65	34	72†	23387
congesta								
S San Cristobal, Chis.	33	11 × 14	25 × 6 & 23	10 – 9	45	28	68†	23651
	28	11 × 13	24 × 6 & 22	10 – 9	54	26	62†	23651
Comitán to S Cristobal, Chis.	26	12 × 12	19 × 4 & 17	8 & 6	45	25	55†	21846
	29	10 × 10	21 × 4 & 19	8 & 7	45	24	60†	21846
	36	13 × 10	17 × 3 & 16	10 – 11	30	25	65†	21847
hiemiflora								
Sierra San Cristobal, Chis.	23	10 × 11	16 × 3.5 & 15	6	32	22	48†	23652
Huehuetenango, Guat.	28	10 × 11	17 × 5 & 16	7 – 6	40	22	54†	23643
	41	9 × 12	22 × 5 & 20	6 – 7	45	27	72†	23640
	33	14 × 12	19 × 4 & 17	9	42	27	65†	23640
	39	11 × 12	20 × 4 & 17	8 – 9	40	19	70†	21426
Cunen-Zacapulas, Guat.	30	8 × 10	20 × 3.5 & 19	5	37	23	58†	21765
Uspantán, Guat.	30	6 × 12	23 × 4 & 22	5	38	23	60†	21750
	34	8 × 14	22 × 3 & 20	7	42	24	64†	21750
Salama, Guat.	27	9 × 11	15 × 5 & 13	6	35	19	51†	21726
	32	10 × 13	20 × 5	6 – 5	40	21	61†	21726
hurteri								
	34	13 × 10	16 × 5	7 – 8	40	23	63†	20815
	42	15 × 14	32 × 5 & 30	12 & 10	70	32	88†	23632
	27	10 × 14	18 × 4 & 16	8 – 7	46	20	55†	23632
Zunil, Guat.	31	11 × 12	22 × 4 & 21	9 – 7	55	23	64†	23632
	27	8 × 13	19 × 4 & 19	7 – 6	47	21	55†	23633
	34	9 × 13	23 × 5 & 21	7 – 5	45	23	66†	23634
	32	10 × 11	18 × 3 & 16	8 & 7	53	18	59†	23634
Solola, Guat.	31	10 × 12	20 × 4 & 19	8 – 7	41	22	61†	23630
	33	10 × 13	21 × 4 & 19	8 – 7	43	23	63†	23630
Huehuetenango, Guat.	42	12 × 17	28 × 6 & 26	9 – 8	44	29	84†	23641
	40	11 × 16	30 × 5 & 28	9 – 7	45	33	80†	23641
hurteri x *hiemiflora*								
Huehuetenango, Guat.	23	11 × 10	18 × 3.5 & 16	9 – 8	33	20	52†	23642
	22	14 × 11	15 × 3.5 & 13	10 – 9	31	20	51†	23642
lagunae								
Amatitlán, Guat.	39	8 × 12	21 × 5 & 20	6 – 5	46	20	67†	21708
	40	7 × 12	20 × 5 & 19	5 – 4	48	21	67†	21708
pachycentra								
Aldea de Barril, Guat.	25	8 × 11	17 × 5 & 16	5 – 4	35	21	50†	21712
	26	8 × 14	19 × 4	5 – 4	42	19	52†	21712
El Progresso, Guat.	34	9 × 12	19 × 5 & 18	5 – 6	46	22	62†	23629
	34	8 × 12	17 × 4 & 16	4 – 6	44		60†	23629
Sanarate, Guat.	27	7 × 12	15 × 4 & 14	3	38	20	48†	21711
	29	10 × 12	18 × 4 & 17	6 – 7	32	22	56†	21711
	29	8 × 10	14 × 5 & 13	5 – 6	30	18	51†	23631

Table 17.1. Flower Measurements in the Hiemiflorae (in mm)

Taxon & Locality	Ovary	Tube	Tepal	Filament Insertion & Length		Anther	Total Length	Coll. No.
El Rancho, Guat.	28	6×12	18×4.5&17	3–4	35	20	53*	23626
	24	8×10	20×4&19	6&5	42	22	51*	23628
Ipala, Guat.	33	11×10	11×4.5&10	6	29	20	56†	23621
	34	9×10	13×4&12	5	36	21	52†	23621
Cruz, Guat.	30	10	16	6–5	36	20	56*	Tre.5
Jalapa, Guat.	25	10	15	6–5	52	21	50*	10025
potatorum								
Tehuacán, Pue.	27	17×12	14×3&12	13&12	32	21	58†	12288
	26	16×12	13×3&11	13&12	35	20	54†	12288
	32	15×13	17×6&16	10	45	25	63†	20427
Huajuapán, Oax.	37	7×14	21×5.5&19	5	34	23	65†	22401
	33	9×13	21×3	5	38	26	62†	22401
Tejupán, Oax.	35	11×12	24×5&22	8&6	50	32	71†	12124
Yanhuitlán, Oax.	23	14×12	17×4&15	9–10	45	22	55†	12111
	21	16×12	17×4	12&11	50		56†	12111
Oaxaca, Oax.	31	10×11	18×4&16	8	37	24	58†	12057
Mitla, Oax.	35	11×14	18×6&17	6–7	37	25	63†	22344
	33	10×13	18×5&16	6	38	25	60†	22344
Cameron, Oax.	50	13×14	19×5.5&18	10–11	36	29	82†	21886
	34	10×13	21×5.5&19	7–9	38	26	64†	21886
pygmae								
Jocote-Trinitaria, Chis.	16.5	6×9	13×4&13	4–4.5	20	16	35†	23650
	16	4×8	12×3.5&11	3	20	14	32†	23650
seemanniana								
Danli, Hond.	27	7×12	16×5&16	4	38	20	49†	20692
	28	7×12	17×5&16	3–4	38	20	52†	20692
Puerto del Diablo, Hond.	31	7×11	17×5&17	5	40	20	55†	21707
	28	8×11	16×5&15	6&5	40	19	52†	21707
Tegucigalpa, Hond.	30	11×12	22×5	7–8	40	25	63†	20684
	34	12×12	24×5	8–10	55	25	71†	20684
Cardenez, Chris.	26	6	14	5	36	18	47*	16435
	33	9×14	19×6&18	7	37	22	61†	23653
	34	7.5×14	17×7&16	4–6	35	18	57†	23653
La Cruz, Chis.	36	10×12	17×5&16	8–6	37	19	63†	21867
	25	11×13	13×5&12	9–8	23	19	48†	21867
	26	9×12	15×5&13	7–6	35	18	50†	21867
Olopa road, Guat.	26	10×12	11×4.5&10	8–9	33	18	47†	23622
	30	8×12	17×4.5&16	6–7	33	18	55†	23623
thomasae								
Chonimacanac, Guat.	38	11×14	23×4&20	10&9	55	22	73†	23637
Cristobal-Huehuete., Guat.	29	10×14	19×4&19	8–9	48	20	57†	21806
	35	10×14	21×4&19	8–9	44	21	65†	21806
Patzicia, Guat.	36	6×12	29×7&27	5	56	25	70†	20862
	36	6×14	25×6&24	5	65	24	66†	20862

* Measurements from dried flowers relaxed by boiling.

† Measurements from fresh or pickled flowers.

Fig. 17.6. *Agave atrovirens* var. *mirabilis* near Las Vigas, Veracruz. Arguelles after leaf specimen with Barclay contemplating.

Puebla. This is a cool montane climate with adequate moisture throughout the year, adjacent to the wetter Huasteca Region on the eastern slopes of the Sierra Madre Oriental.

In Las Vigas it is called "maguey blanco" and "maguey cenizo," and is tapped for "agua miel" and "pulque." In row plantings it may be mixed with *A. salmiana* and *A. mapisaga,* which are also tapped for pulque.

Agave congesta
(Figs. 17.1, 17.2, 17.3, 17.7, 17.8, 17.9; Table 17.1)

Agave congesta Gentry, sp. nov.

Rosettes medium size, 1–2 m broad, to 1 m tall, non-surculose, compact, with numerous ascending to spreading, thickly succulent leaves on short stem; leaves lanceolate, 50–120 x 12–22 cm, acuminate to short-acuminate, at first curved-ascending, then horizontally extending, plane, green to yellow-green, sometimes faintly glaucous or pruinose, leaf margins undulate to crenate, variously mammillate; teeth moderate to rather large, mostly with cusps 5–10 mm long from broad low base, usually remote, 3–5 cm apart, straight to variously curved, dark brown to grayish brown; spine stout, 3–7 cm long, gray to castaneous, very broad at base, widely, flatly grooved above, sharply decurrent to upper teeth; inflorescence racemose, 6–8 m tall, straight, prominently bracteate along peduncle, with 40 to 50 congested, rounded flower clusters, sub-sessile or on short lateral peduncles, along upper half or more of shaft; pedicels densely subtended by purplish bractlets, 1–2 cm long; flowers dark-colored (orange to reddish or purplish) in bud and tepals, opening yellow, 55–70 mm long, soft fleshy; ovary 30–40 mm long, angular-fusiform, with short neck; tube 10–13 mm deep, deeply funnelform, thick-walled, grooved, knobby; tepals 17–25 mm long, linear lanceolate, involute, thick rounded cucullate, the outer larger, dark colored apically, the inner shorter, with thick prominent keel; filaments 40–55 mm long, slender, purplish, inserted at two levels near orifice of tube; anthers large, 24–28 mm long, irregularly curved to sinuous, bronze-colored, ca. centric; pistil stout, lengthening in post-anthesis.

Type: *Howard & Marie Gentry 23651,* ca. 14 miles W of San Cristóbal de las Casas, Sierra San Cristobal, Chiapas, on wooded limestone slope, ca. 8,000 feet elevation, 3 Jan. 1976 (l,f); holotype US, isotypes DES, MEXU.

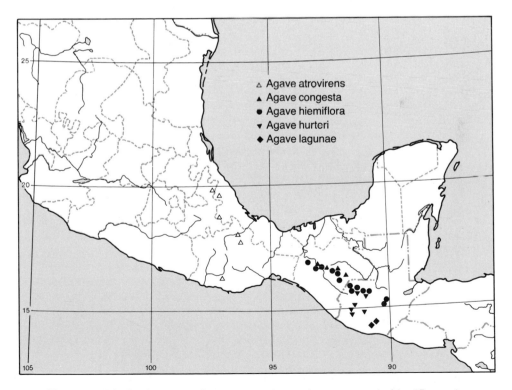

Fig. 17.7. Distribution map of *Agave atrovirens, A. congesta, A. hiemiflora, A. hurteri,* and *A. lagunae.*

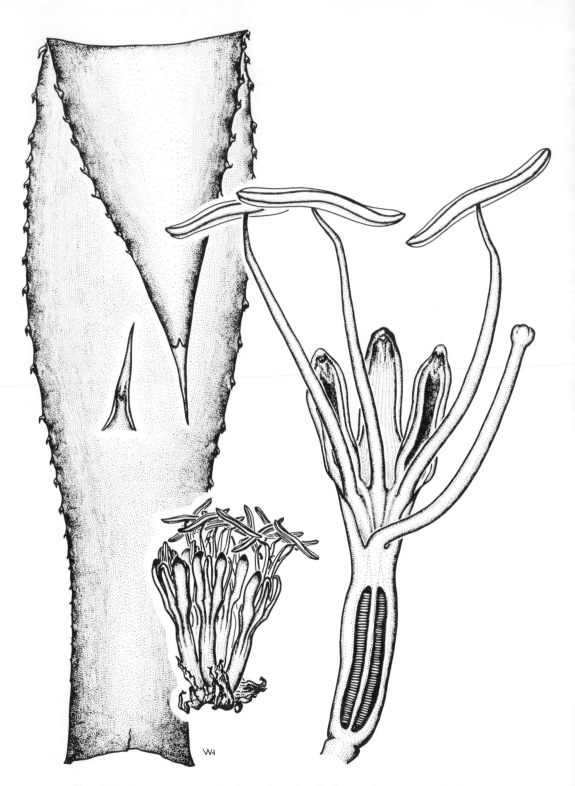

Fig. 17.8. *Agave congesta*. Leaf greatly reduced, flower cluster x ca. ½, flower section x 1.8. From type collection, *Gentry 23651*.

Fig. 17.9. *Agave congesta*. Flowering plant in natural situation on Sierra de San Cristóbal de las Casas, Chiapas.

Planta non-surculosa 1–2 m alta, 1 m alta. *Folia* viridia, numerosa, 50–120 cm longa, 12–22 cm lata, lanceolata horizontaliter curvata; *margine* undulata vel crenata, *dentibus* plerumque 5–10 mm longis, 3–5 cm separatis rectis vel curvatis, brunneis vel griseobrunneis; *spina terminali* robusta, 3–7 cm longa, griseo-castanea supra late canaliculata decurrente. *Inflorescentia* scapo incluso 6–8 m alto, recta racemosa umbellis 40–50 subessilis congestis; *bracteis* siccis triangulis reflexis, 25–10 cm longis; *floribus* 55–70 mm longis, versicoloris purpureis vel luteis. *Ovarium* angulato-fusiforme apine incluso 30–40 mm longum. *Perianthii tubo* infundibuliformi 10–13 mm longo; *segmentis* 17–25 mm longis, lineari-lanceolatis crassis involutis, apice obtuse cuculatis. *Filamenta* 40–55 mm longa, gracilia purpurea ad apicem tubi inserta; *antheris* 24–28 mm longis, irregulariter curvatis vel sinuosis, ca. centric.

Agave congesta is so-named for the congested condition of the flowers in the tightly balled bracteose clusters. It is further distinguished by the dimorphic dark colored tepals and the 2-level insertion of the filaments. *A. hiemiflora* has these characters also, but in less degree; *A. congesta* is a much larger plant, with more numerous leaves, and darker colored flowers.

A. congesta is widely scattered at 7,000 to 8,000 feet (2,150 to 2,480 m) elevations in the pine woodland on Sierra San Cristóbal (see Fig. 17.9). The series of leaves collected south of Las Casas (see Exsiccatae) show the usual variability in size, shape, and armature of *Agave* species. *A. congesta* will probably be found eventually on neighboring sierras to the north and east of San Cristóbal de las Casas.

Agave hiemiflora
(Figs. 17.2, 17.3, 17.7, 17.10, 17.11; Table 17.1)

Agave hiemiflora Gentry sp. nov.

Plants single, non-surculose, acaulescent, compact, with 50–90 openly spreading, glaucous green to yellowish leaves; leaves mostly 30–55 x 10–15 cm, lanceolate, acuminate, gradually narrowed and thickened toward the base, rather soft fleshy, light gray glaucous to pale green, plane to slightly hollowed above, margins undulate to deeply crenate with prominent mammae under the teeth; teeth moderate, the larger through the mid-blade commonly 5–8 mm long, the slender cusps variously flexed up or down, 1–3 cm apart, light brown to dark brown, or teeth smaller and more distal on undulate margins; spines generally 2–4 cm long, slender or thick, sinuous to contorted to straight, but always openly grooved to flat above; panicles 4–5 m tall, slender, narrow, with 20 to 30 small umbellate branches in upper ⅔ to ½ of shaft, the peduncular bracts remote, triangular, quickly drying; the lateral peduncles very short (4–5 cm) or longer (10–20 cm); flower clusters small, dense, densely tufted with scarious, frequently dark-colored bractlets subtending the slender elongate pedicels; flowers generally 50–70 mm long, sometimes red in bud, opening yellow, slender; ovary angulate-cylindric, 25–40 mm long, with short, smooth, little-constricted neck; tube 8–13 mm long, funnel-form, angulate to rather knobby, lightly grooved; tepals unequal, 16–23 mm long, much longer than the tube, involuting, the outer linear-lanceolate, obtuse, frequently reddish toward tip, erect-ascending, the inner smaller, narrower, but widened at base, in-lapping the outer, keel narrow; filaments 35–45 mm long, slender, flattened adaxially, inserted rather unevenly just above mid-tube; anthers 18–27 mm long, yellow, straight, regular, centric; pistil at length longer than the stamens, with knobby triquetrous stigma. Capsules small, ovoid to oblong, 3–3.5 x 1.4–1.7 cm, thin-walled, scarcely beaked; seed 5–6 x 3.5–4 mm.

Type: *Howard & Marie Gentry 23640,* 7 miles NE of Huehuetenango along road to Aguacatan, Guatemala, Jan. 1, 1976; on limestone slope in pine-oak forest 6,500–7,000 feet elevation. Holotype in US, isotypes in DES, MEXU.

Planta caule singulo. *Folia* pallida glauco-viridia, lanceolata 30–55 cm longa, 10–15 cm lata, plerumque planta; *margine* crenata dentibus supra mamilas plerumque 5–8 mm longis, flexuosis brunneis vel pallide brunneis; *spina terminali* 2–4 cm longa, gracili vel robusta, sinuosa vel contorta vel recta, semper supra laticanaliculata. *Inflorescentia* angusta paniculata scapo incluso 4–5 m alta, ramis 20–30 proximis multi-bracteolis ad pedicelae; *floribus* 50–70 mm longis purpurei vel rubris vel luteis; *ovarium* apice incluso 25–40 longum, cylindrico-angulatum. *Perianthii tubo* 8–13 mm longo infundibuliformi; *segmentis* inequalibus 16–23 mm longis lineari-lanceolatis involutis apice saepe roseis ad basim imbricatis. *Filamenta* 35–45 longa, supra medio tubo insertis; *antheris* 18–27 mm longis luteis. *Capsulae* ovatae vel oblongae 30–35 mm longae 14–17 mm latae. *Semina* 6–7 x 4–5 mm.

Agave hiemiflora is a highland relative of *A. seemanniana* and *A. congesta,* sharing with them a freely seeding outbreeding habit, moderate leaf variability, and the openly grooved to flat spine. It differs from *A. seemanniana* in the more regular form and size of leaf, the less narrowed haft of leaf, and the densely tufted bracteolate apex of the lateral peduncles. It shares this latter character with *A. congesta* but differs from this taxon in its small rosette, less robust and more ramified inflorescence, and more evenly inserted filaments at a lower level in the tube (see ideographs, Fig. 17.3). The three species form an evolutionary complex, each appearing more stable morphologically than their polymorphic congener, *A. potatorum,* to the north of them which occupied a generally more arid and perhaps more changeful habitat in prehistoric times.

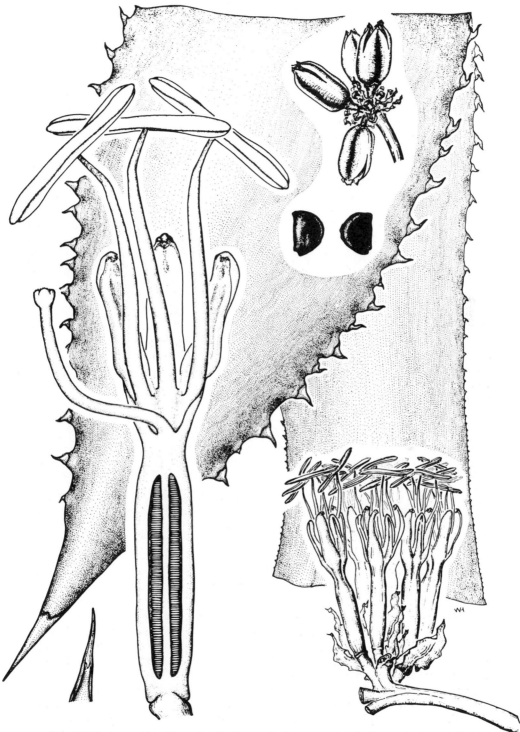

Fig. 17.10. *Agave hiemiflora*. Leaf and capsule clusters, reduced; flower section x 1.7, seeds x 1.7. Drawn from the type, *Gentry 23640*.

Fig. 17.11. *Agave hiemiflora* in natural habitat on a southern slope of Sierra de Cuchumatanes, Guatemala. Compare inflorescence with *Agave seemanniana* (Fig. 17.24).

Agave hurteri
(Figs. 17.3, 17.7, 17.12, 17.13, 17.14; Table 17.1)

Agave hurteri Trel., St. Louis Acad. Sci. Trans. 23: 136, 1915.
  Agave samalana, Trel., St. Louis Acad. Sci. Trans. 23: 142, 1915.

Rosettes medium to large, non-surculose, rather open with outcurving leaves to compact with numerous ascending leaves, 1–1.8 m tall x 2–3 m broad, with thick short stem; leaves mostly 70–130 x 15–22 cm, rarely narrower, lanceolate, broadest at or above the middle, acuminate, thick-fleshy, plane to slightly hollowed above, thickly rounded below, generally ascending to spreading, rarely reflexed, light glaucous to pale green and yellow-green, surface slightly rough above, more asperous below, margins about straight; teeth small to moderate, the larger along the mid-blade 3–8 mm long, 1–3 cm apart, rarely smaller and closer together or quite toothless, straight to curved, dark brown; spine

Fig. 17.12. *Agave hurteri* Flowering plant at the type locality by Zunil, Guatemala, April 1965.

generally subulate, 4–6 cm long, broad at base, openly grooved above, dark brown to grayish brown; panicles 5–7 m tall, stout, deep, narrow, with 30–45 lateral peduncles in upper ⅔ of shaft; flowers 55–85 mm long, in tightly balled clusters, greenish yellow to yellow with purple-tinged tepals and bronze or purplish or yellow stamens, densely tufted at pedicellate bases with dark-colored bractlets 1–2 cm long; ovary 30–45 mm long, cylindric, usually rather thick, with short neck; tube 9–15 mm deep, funnelform to angulate-cylindric, thick-walled, grooved; tepals unequal, linear-lanceolate, the outer 16–28 mm long, ca. twice as long as tube, rounded to obtuse at tip, purple-tinged, erect-ascending, the inner conduplicate-involute, with thin margins; filaments 45–70 mm

Fig. 17.13. *Agave hurteri*. Variation in leaves in the population at Zunil, Guatemala, the type locality.

Fig. 17.14. Part of the large stand of *Agave hurteri* in the cliff-rimmed canyon above San Cristóbal, Depto. Totonicapán, Guatemala.

long, stout to slender, flattened adaxially, inserted somewhat irregularly below orifice of tube; anthers 18–30 mm long, regular, straight, centric; capsules oblong, 20 x 50 mm, neither stipitate nor beaked; seeds 5 x 8 mm.

Type: *Trelease 3,* Zunil on the Samala River, Guatemala, 10 April 1915, ILL (l, f, 3 sheets).

Agave hurteri is a variable complex in the mountains of northwestern Guatemala between 6,000 and 8,500 feet (1,850 and 2,630 m) above sea level. It forms a thriving colony on the steep rocky southeastern slope above the town of Zunil along the Río Samala. An even larger population is conspicuous farther up this same valley about the crossroads town of San Cristóbal east of Quezaltenango (Figs. 17.12, 17.14). Such populations are widely scattered through the pine forest zone northeast to the Sierra Cuchumatanes and at least as far east as Cunen, Departamento de Quiche.

In 1965 and again in 1975 I paid particular heed to finding *A. samalana* that Trelease described from the Samala Valley below Zunil. There were plenty of flowering plants available in 1975, and a series of plants were collected and the whole population of the valley reconnoitered. The plants were found to be very variable in leaf and flower. I was unable to find any consistent set of correlating characters for separating the population into two species. As *A. hurteri* was described with flowers, it is nominally selected to represent the species. Although closely related to other members of the Hiemiflorae with tightly round-clustered umbels, *A. hurteri* is usually recognizable by its larger many-leaved rosettes and consistently longer lateral peduncles in the deep large panicles.

Very little use is made of this handsome agave in Guatemala. However, it enhances the beautiful mountain slopes, too rocky and rough for cultivation, where it still survives. Near Zunil I was given the name ''gi'' for the plant.

Agave lagunae
(Figs. 17.3, 17.7; Table 17.1)

Agave lagunae Trel., St. Louis Acad. Sci. Trans. 23: 143, 1915.

Rosettes medium-sized, single, openly spreading, rather few-leaved, acaulescent or nearly so; leaves 40–70 x 8–12 cm, linear to lanceolate, acuminate, glaucous green, plane to concave, finely asperous, with nearly straight margins; teeth moderate, the larger through the mid-blade 4–6 mm long, slender, mostly curved, with slightly elevated bases on type, 1–3 cm apart, dark brown; spine 3–4 cm long, subulate, the groove openly channeled to narrow, dark brown; panicle 3–4 m tall, rather open, with 15–20 lateral peduncles in upper half of shaft, with rather small umbels, pedicellate bractlets 1–1.5 cm long; flowers 60–70 mm long, slender, the tepals and anthers yellow; ovary 39–45 mm long, slender, cylindric, with long slender constricted, grooved neck; tube 7–10 mm deep, funnelform; tepals 18–21 mm long, 5 mm broad, thin, lanceolate, faintly striate, the outer involute toward apex, apiculate, the inner with flattened keel and inner costae not converging at tip; filaments 45–50 mm long, somewhat flattened, inserted about middle of tube; anthers 20–21 mm long, regular, centric; pistil extended in post-anthesis with clavate trilobate stigma.

Type: *Trelease 10,* above Amatitlán near the lake, Guatemala, 11 April 1915, ILL (l).

Known only from the type locality and vicinity. Trelease cites it also from the ''flanks of the Volcán de Agua,'' which is near Antigua, but I searched for it there without success. He must have meant the lower outlying slopes toward Palin and Escuintla (see Exsiccatae). It is apparently a lowland species, poorly known. The filaments are set unusually low in the tube for this section.

Agave pachycentra
(Figs. 17.3, 17.15, 17.16, 17.18; Table 17.1)

Agave pachycentra Trel., St. Louis Acad. Sci. Trans. 23: 135, 1915.
   Agave tenuispina Trel. ibid. p. 140.
   Agave opacidens Trel. ibid. p. 140.
   Agave eichlami Berger, Die Agaven 1915; 200, 1951.
   Agave weingartii Berge, ibid. p. 200.

Rosettes generally single, rarely surculose, medium size, rather open, up to 1 m tall, 1.5–2 m broad, on short stems; leaves variable, mostly 60–100 x 12–18 cm, broadly lanceolate, acuminate, rarely linear-lanceolate, plane to guttered, thick fleshy toward the gradually narrowed base, glaucous white to yellowish green or pale green, asperous above, rougher or scabrous below, margins generally undulate with hollows between teeth; teeth variable, small to medium, closely spaced or remote, mostly 5–10 mm long through the mid-blade, 1–3 cm apart, the cusps straight or variously curved above low broad bases, brown; spines generally 4–6 cm long, finely subulate to nearly conic from broad base, broadly to narrowly grooved above, keeled to rounded below, scabrous, long decurrent to upper teeth; panicles 4–6 m tall, open, rather irregular, the shaft usually crooked, white pruinose when young, with 20 to 30 small umbels on rather long lateral

Fig. 17.15. *Agave pachycentra* in natural habitat near Laguna Retana, Depto. Jutiapa, Guatemala. Photo by Hugh Iltis.

Fig. 17.16. *Agave pachycentra* near Laguna Retana, Guatemala. Photo by Hugh Iltis.

peduncles, the pedicelar bractlets few, small, 5–10 mm long; flowers 45–62 mm long, yellow above green ovary, uncrowded on slender pedicels 5–12 cm long; ovary 25–35 mm long, thick-fusiform to slender cylindric, with constricted grooved neck; tube 6–11 mm deep, funnelform to urceolate, knobby, grooved; tepals subequal, 13–20 x 4–5 mm, linear-lanceolate, erect to ascending, incurved, involute, the outer frequently reddish at apex, closely over-lapping inner at base, even below sinus, the inner with narrow thick keel; filaments slender, 35–45 mm long, inserted in mid-tube or just above; anthers 17–22 mm long, regular, straight, centric. Capsules (*20910, 20873*) 40–45 x 15–17 mm, oblong to ovoid, stipitate, apex rounded, nearly beakless; seeds 6 x 4–4.5 mm, lunate to lachrymiform, very finely punctate, marginal wing raised to ca. thickness of seed.

Type: *Trelease 2,* Cruz (between El Rancho and El Progreso, Depto. de Progreso), Guatemala, 20 March 1915, ILL (leaf only).

La Cruz is a railroad station in the lowland valley of Río Motagua at 1,000–2,000 feet (310–375 m) elevation. The vegetation about there is Thorn Forest, semi-arid with a long dry hot spring. *A. pachycentra* is widely scattered in the hilly country around La Cruz and is common from there south and eastward below 4,000 feet (1,240 m), especially on volcanic rock outcrops, to El Salvador and Pacific Honduras (see Fig. 17.18).

Agave pachycentra is generally recognizable by its rather open medium-sized to rather large rosettes with pale green to nearly white, broadly lanceolate leaves and crooked, loosely flowered, open panicles with small yellow umbels. It is highly variable in leaf size, form, armature, and coloring. Trelease described three "species" from around La Cruz, separated primarily on spine differences; variants of what I regard as forms of a polymorphic complex. Two of the four taxa that Berger described from potted plants grown by the firm of Haage & Smidt (Germany?), I have interpreted as belonging to this complex. As Trelease's account of the Guatemalan agaves antedates that of Berger, Trelease names have priority.

As with other non-fiber agaves, this plant is little used by the Guatemalans. I found it planted along wire fences east of El Rancho. Like other wild agaves they call it simply "maguey." No doubt, some natural stands have been cleared off cultivated lands, and it owes its continued existence in large part to the rocky volcanic sites it inhabits, which are unsuitable to agriculture.

Agave parvidentata
(Figs. 17.17, 17.18)

Agave parvidentata Trelease, J. Wash. Acad. Sci. 15: 395, 1925.
  Agave compacta Trelease, J. Wash. Acad. Sci. 17: 161, 1927.

Single, subcaulescent, medium-sized, dense rosettes to 1 m tall, 1.7 m wide with numerous leaves and dense deep, short-peduncled, bulbiferous panicle. Leaves pale green to light gray glaucous, ovate-lanceolate, acuminate, 80–100 x 15–25 cm, contracted in thick base, thickly fleshy, smooth, plane, ascending to outcurving or incurved and concave above middle, sometimes plicate, the margin straight; teeth 3 (–5) mm long through the mid-leaf, 10–20 mm apart, reduced upward and downward, deltoid from lenticular bases, nearly straight; spine ca. 5 cm long, acicular, smooth, dull light brown, involutely grooved to above the middle, decurrent for more than its length; panicle deeply oblong, ca. 2.5 m tall, with over 30 ball-shaped umbels; peduncle shorter than the leaves; lateral branches closely ramified, the tertiary branchlets conspicuously small-bracteolate; flowers yellow 40–50 mm long, slender; ovary 20–25 mm long, fusiform; tube openly conical, ca. 5 mm deep; tepals 15–20 mm long, linear, subequal, spreading, involute, tip acute with small hood; filaments slender, ascending-spreading, inserted in orifice of tube; anthers 16–20 mm long, yellow, centric. Fruit unknown.

Type: *Calderon 2085, US* 1169884-5. Cult. in San Salvador, Salvador, 1924 (l, f). "Maguey."

This plant, originally known only as a cultivated ornamental, has since been found spontaneous on Salvadoran volcanoes, as the Volcán de Santa Ana in northern Salvador (Fig. 17.17), where it grew with dispersed shrubbery on a rocky slope at 7,500 feet (2,300 m) elevation.

A later collection of this species in San José (type of *Agave compacta*), *Calderon 2251*, US, carries a good photograph and the followng revealing note by the collector:

"2251. This is a maguey cultivated in the Patio de Ensayos in San Jacinto, where I took the specimens, although the same species is planted also in the parks. It is called "pulque" and I have seen it only as an ornamental plant, it being probably imported. The flowers collected are the first that opened on the first branches, about Nov. 27, 1924. At the present time, March 21, 1925, there are still flowers, but no capsules have developed, only bulblets. I sent you a photo taken Nov. 27.... The scape is about 2.5 m high. The leaves are very thick and weigh on the average NINE POUNDS. I gave orders that the fibers should be separated as a sample here in the herbarium, but the mozo informed me that it was impossible because the juice of the leaves was very irritant and burned the skin severely."

This agave is perhaps nearer to *A. pachycentra* than to other members of the Group, because of the small flowers with shallow tubes and the diffuse small-bracteolated umbels. Some white-leaved forms with low deep panicles observed as *A. pachycentra* in the Valle de Monjas, Guatemala (*Gentry 20910*), now appear to me much like this much later acquaintance, *A. parvidentata*. The panicles and the rosettes of *A. parvidentata* are better developed than *A. pachycentra*, and in general conformation resemble its other Central America relative, *A. wercklei*. The very short-peduncled panicle of *A. parvidentata*, branching from the top level of the rosette, is apparently a consistent character.

Fig. 17.17. *Agave parvidentata* high on the volcanic mountain, Volcán de Santa Ana, Salvador, February 1959; Dorothy Allen as witness. Photo by Paul Allen.

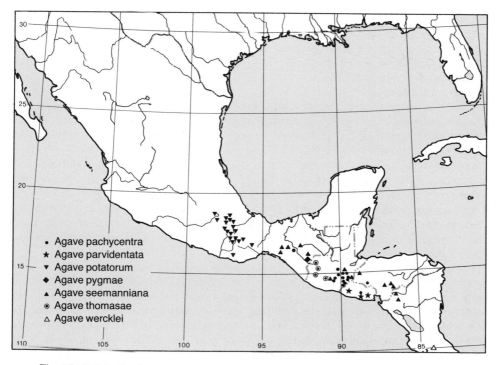

Fig. 17.18. Distribution map of *Agave pachycentra, A. parvidentata, A. potatorum, A. pygmae, A. seemanniana, A. thomasae,* and *A. wercklei.*

Agave potatorum

(Figs. 17.1, 17.2, 17.3, 17.18, 17.19, 17.20, 17.21; Table 17.1)

Agave potatorum Zucc., Nov. Act. Caes. Leop.-Carol. 16 (2): 674, 1833.
    Agave scolymus Karw. ex Hort Dyck. 307, 1834.
    Agave saundersii Hook. f., Bot. Mag. 1865: t. 5493, 1865.
    Agave verschaffeltii Lem., Ill Hort. 15: t. 564, 1868.
    Agave potatorum var. verschaffletii (Lem.) Berger, Die Agaven 186, 1915.

Plants polymorphic, generally non-surculose; rosettes small, compact to openly spreading, generally with 50–80 leaves and acaulescent or infrequently short-stemmed with 100 or more leaves; leaves glaucous white to green, mostly 25–40 x 9–18 cm, ovate to short-lanceolate, plane to somewhat hollowed above, thickened and narrowed toward base, soft fleshy but rather rigid, the margins undulate to deeply crenate with mammillate prominences, especially above mid-blade; teeth prominent, 5–10 mm long or more, mainly 1–3 cm apart, the slender variously curved cusps from low broad bases, castaneous to grayish brown; spines generally 3–4.5 cm long, broad at base, sharply pointed, sinuous, broadly grooved to flat above, sharply decurrent as a ridge to proximal upper teeth, roundly keeled and flesh intrusive below, castaneous to grayish brown; inflorescence 3–6 m tall, racemose with subsessile flower clusters or paniculate with lateral peduncles, the shaft conspicuously bracteate with red to purplish bracts, with 15 to 30 flower clusters in upper ½ to ¼ of shaft; flower polymorphic, in small compact umbels, light green to yellowish, frequently tinged red or purplish on buds, tepals and filaments, 55–80 mm long, rather thickly succulent; ovary 25–50 mm long, generally thick with short smooth neck; tube 10–17 mm deep, cylindric to funnelform, thick-walled, grooved; tepals unequal, 13–24 mm long, linear to broadly lanceolate, erect-ascending to incurved, the outer larger, thickly rounded, the inner with thick keel not reaching to tip; filaments rather short, 35–50 mm long, rather broad below, flattened adaxially, sometimes colored; stamens rather large, 20–30 mm long, straight to sinuous, generally centric, yellow; pistil large, over-reaching stamens in post-anthesis, with large globular triquetrous stigma; capsules large, 4–5.5 x 2 cm, ovoid to oblong, subsessile to short stipitate, shortly beaked; seeds 6–7 x 5–6 mm, lachrymiform, black and shining, the rim slightly winged, hilar notch shallow.

A type was never specified for *A. potatorum*. In 1969 I found 2 specimens in the Munich Herbarium, consisting of parts of inflorescence only. The older "1867, Hortus Monacensis" represented a small form with short lateral branches, as I had observed in Oaxaca. The other "7.3.79, H.b.M." is from a larger inflorescence with lateral peduncles 8 cm long. Since these specimens date from 33 and 46 years after the publication of Zuccarini's species, their suitability as types is doubtful. However, they can be regarded as neotypes.

Saunders in his Refugium Botanicum, Tab. 328, 1872, published one of the earliest illustrations of *A. potatorum* under the name *A. scolymus* Karw. It depicts quite clearly the common form found around Tehuacán with straight leaf margins and umbellate panicle.

Agave potatorum is an attractive ornamental and since Lemaire's publication of Stroobant's fine lithographic illustration in 1868 (Fig. 17.20), it had been widely known in horticulture as *A. vershaffeltii*, which Berger (1915) reduced to a variety. The only perceptible difference between *potatorum* and *verschaffeltii* is the more prominent mammillate margins of the latter. However, the two forms occur at random, side by side in some of the populations, and I can find no other correlative characters for separating them. The conspicuous structural differences in leaves, in inflorescence, racemose or paniculate (Fig. 17.21), short flower tubes or long flower tubes, variations in tepals, and so on, none of them correlative consistently, have thwarted my efforts to segregate them

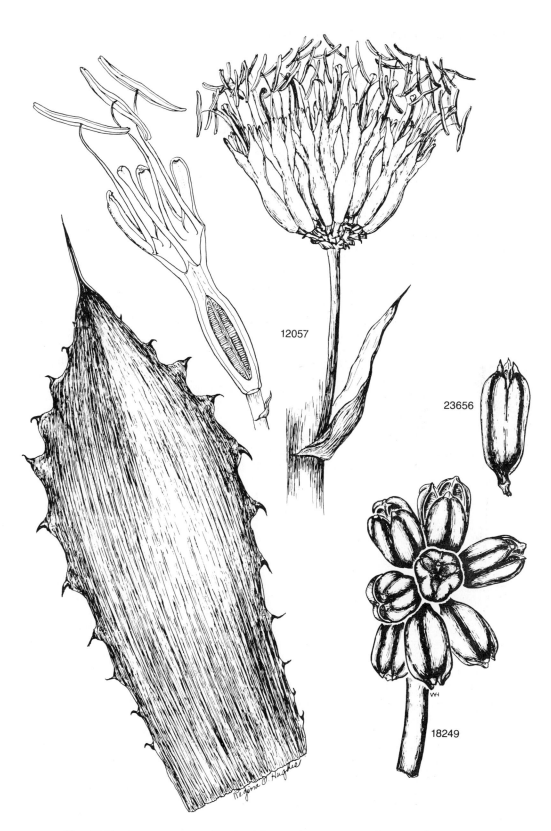

Fig. 17.19. *Agave potatorum*. Leaf and flower cluster x ca. ½, flower section x ca. 1, from *Gentry 12057;* capsules x ½–³/₅ from different plants.

AGAVE VERSCHAFFELTII *Serie-fioide (Mexique).*

Fig. 17.20. Lemaire's illustration of *Agave potatorum* in Ill. Hort. 15: taf. 564, 1868.

as specific or subspecific taxa. Because of its polymorphic nature, the species is taxonomically frustrating.

Agave grandibracteata Ross, *A. littaeoides* Pampanini, *A. galeottei* Baker, and *A. multiflora* Todaro are obviously related to *A. potatorum*. Whether they should be treated as synonyms, varieties, or otherwise, I leave to later students. Careful study of the field populations and their analysis by objective criteria would doubtless put them into better taxonomic perspective.

The range of *A. potatorum* is shown in Fig. 17.18. Its habitat is the semi-arid highlands of Puebla and Oaxaca between 4,000 and 7,500 feet (1,240 and 2,300 m) elevations, where it is associated on the grassy slopes with remnants of pine and oak vegetation. On the denuded sedimentary slopes along the Puebla-Oaxaca highway near Yanhuitlan is usually visible the form with small rosettes and racemose inflorescences. This form occurs again just northwest of Huajuapán in the palm thicket vegetation on Mixtecan limestone. Eight to 14 miles (13 to 22 km) southeast of Camerón along the same highway to Tehuantepec is a scattered population in oak woodland, outstanding with large individual rosettes. The general flowering season is fall, September to December. A Nahuatl name for *A. potatorum* is ''papalometl'' (butterfly agave), a beautiful name for a beautiful plant.

Fig. 17.21. *Agave potatorum*. Upper, comparison of umbellate and racemose forms of infructescences. Lower, the large, mammillate leaf form found south of Camerón in southern Oaxaca.

Agave pygmae

(Figs. 17.1, 17.2, 17.3, 17.18, 17.22; Table 17.1)

Agave pygmae Gentry sp. nov.

Rosettes small to very small, cespitose, short-stemmed, with few remote leaves; leaves 12–28 x 7–12 cm, ovate, short-acuminate or truncate, widest above the middle, strongly constricted near the base, the margins sharply crenate, bluish glaucous to dull gray green; teeth small, 3–5 mm long, 1 cm apart, the short bases on marginal teats, cusps straight to slightly curved, grayish brown; spine 1.5–2 cm long, subulate, broadly grooved above, short-decurrent, grayish brown; panicles small, 2–3 m tall, slender, open, with 8 to 12 small umbels in upper half of shaft, the lateral peduncles slender, 15–20 cm long; flowers 31–35 mm long, yellow, uncrowded, on small pedicels 3–8 mm long, the pale bractlets few, 3–8 mm long; ovary 15–17 mm long, fusiform, with very short neck; tube 4–6 mm deep, 8–9 mm broad, funnelform, lightly grooved; tepals subequal, 11–13 x 3–4 mm, spreading, broadly linear, rounded-cucullate at apex, the inner with low keel not reaching apex, sublobed at base; filaments ca. 20 mm long, flattened adaxially, the elbow near apex, inserted near mid-tube; anthers yellow, 19–20 mm long, regular, straight, centric (Fig. 17.22).

Planta parva caule caespitosula. *Folia* pauca ovata, 12–28 cm longa, 7–12 cm lata, pallidi-viridia brevi-acuminata ad basim angustata; *margine* crenata dentibus supra mamilas plerumque 3–5 mm longis cinereo-brunneis; *spina terminali* subulate 1.5–2 cm longa, supra late-canaliculata brevi-decurrente. *Inflorescentia* panaliculata 2–3 m alta, diffusa, ramis ca. 8–12 lateralibus; *floribus* parvis 31–35 mm longis flavis; *ovarium* parvum fusiforme 15–17 mm longum. *Perianthii tubo* 4–6 mm longo, 8–9 mm lato; *segmentis* subequalibus 11–13 mm longis, 3–4 mm latis, lati-linearibus apice rotundi-cuculatis. *Filamenta* ca. 20 mm longa, prope apicem flexis ad medium tubo insertis; *antheris* luteis 19–20 mm longis regularibus rectis.

Type: *Gentry & McClure 23650,* ca. 28 miles N of Mexico-Guatemala border along route 190, S of Comitán, Chiapas, on white limestone, elev. ca. 3,000 feet, 2 Jan. 1976. Holotype US, dups. MEXU, DES. Known only from the type locality.

Agave pygmae is closely related to *A. seemanniana* and *A. hiemiflora.* It is easily distinguished from either by its small size. It may only be a depauperate form generated by the arid limestone on which it was found. However, the small, distinct form of the normal appearing flowers, differing in their broad linear tepals, and the lobing at the base of the inner tepal, together with the cespitose habit and short-stemmed rosettes, appear to warrant its recognition on the species level. Further study of the plants in their native habitat may reduce it to subspecific rank. In any event, it is an interesting addition to the small agaves of special interest to the succulent fanciers. It is a great contrast to its more distant relative in the section, the giant *A. atrovirens.* The two leaves in Fig. 17.22 both represent mature flowering plants.

Agave seemanniana

(Figs. 17.1, 17.2, 17.3, 17.11 (lower), 17.18, 17.23, 17.24, 17.25; Table 17.1)

Agave seemanniana Jacobi, Nachtrage I in Abh. Schles. Ges. Vaterl. Cult. Abth. Natur Wiss. 1868: 154, 1868.

Agave tortispina Trel. St. Louis Acad. Sci. Trans. 23: 135, 1915.
Agave carol-schmidtii Berger, Die Agaven 1915: 199, 1915.
Agave guatemalensis Berger, ibid. p. 201?

Rosettes compact, small to medium, single, non-surculose, acaulescent; leaves generally 30–50 x 12–20 cm, or 2 to 3 times as long as wide, rarely more slender, ovate to broadly lanceolate or spatulate, plane to hollow-upcurved, thickly succulent, thickened

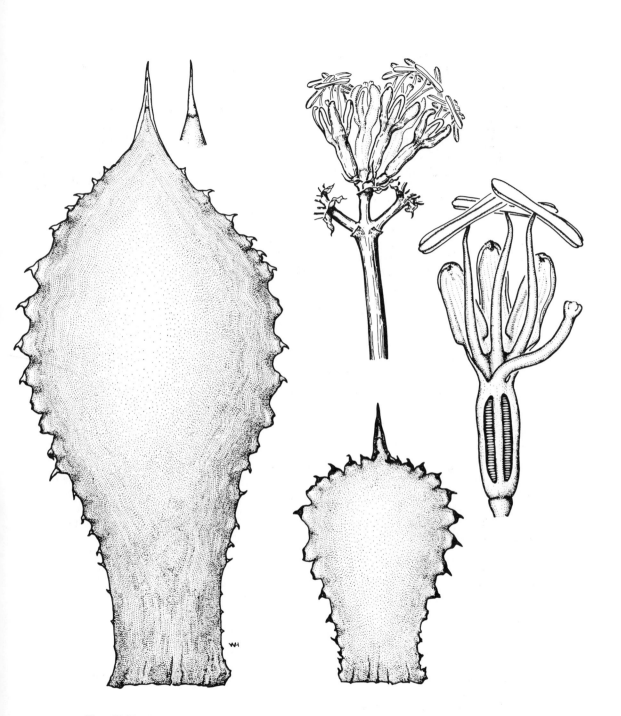

Fig. 17.22. *Agave pygmae*. Mature leaves and flower cluster ca. ⅔; flower section x 1.8. Drawn from the type, *Gentry & McClure 23650,* by Wendy Hodgson.

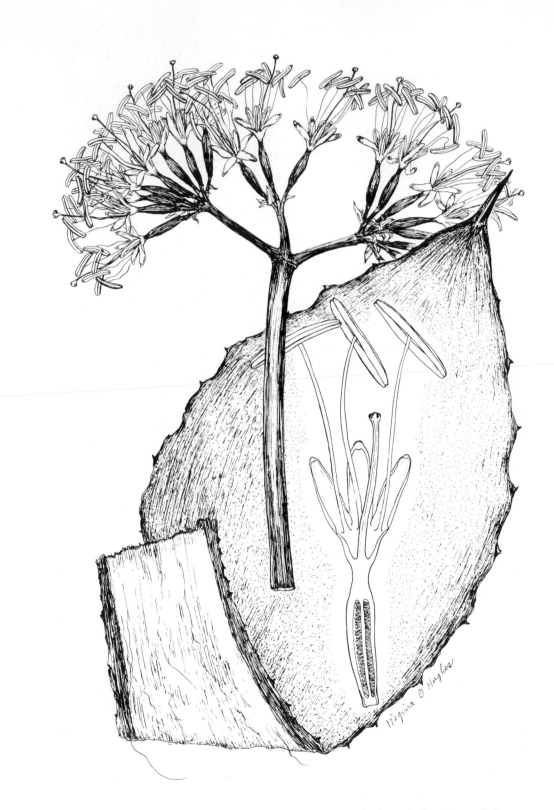

Fig. 17.23. *Agave seemanniana*. A form with undulate leaf margin found near Col. Lazaro Cardenez, Chiapas. Drawn from *Gentry 16435*, reduced by ½.

Fig. 17.24. *Agave seemanniana* on a pine-topped hill south of La Cruz, along road to Arriaga, Chiapas. December 1965.

and strongly narrowed at base, light glaucous green to yellowish green, the margins undulate to sharply crenate; teeth deltoid, mostly straight or some curved, 5–10 mm long, rarely much larger, 1 to 2 or 3 cm apart, on usually prominent marginal teats, dark brown to grayish brown; spine 2 to 4 cm long, subulate, very broad at base, conspicuously decurrent as a sharp ridge to upper teeth, broadly grooved above, dark brown to grayish; panicles 3–4 m tall, rather open, deep oval in outline, with 18 to 30 spreading umbels on upper half of shaft, the peduncle with rather small triangular remote bracts; flower clusters diffuse on slender pedicels with sparse pale bractlets; flowers generally 50–70 mm long, slender, yellow above green ovary; ovary slender, 25–38 mm long, fusiform to cylindric, with lightly furrowed neck; tube 7–11 mm deep, broadly funnelform, thick-walled, deeply grooved; tepals 13–24 mm long, linear, obtuse, erect to ascending, over-lapping at base, thick, the outer slightly larger, somewhat involute, sometimes red-tipped, the inner with thick keel; filaments 35–50 mm long, slender, flattened on adaxial side, inserted unevenly below orifice of tube; anthers 18–25 mm long, yellow, regular, straight, mostly centric. Capsules (*16435*) 3–4 x 1.5–1.7 cm, oblong, short stipitate or sessile, beak short, stout; seeds 5.5–7 x 3.5–4.5 mm, lunate, hilar notch open and frequently notched or flattened at opposite corner, the testa lightly rugose semireticulate.

Neotype: *Gentry 20684*, 8–12 km SE of Tegucigalpa along road to Zamarano, Honduras, April 1965 (l,f) in US, dupl. DES, MEXU. The only specimen under this name found in European herbaria was an unidentifiable leaf from Shaw of the Mo. Bot. Gard. received by Kew in 1885.

Agave seemanniana is generally recognizable in its area by its small compact rosettes with broad plane glaucous leaves, markedly narrowed at the base. It shows considerable variation in form and size of leaf and in leaf margins and teeth. The small population near Somoto in northern Nicaragua is remarkable for its very large teeth (Fig. 17.24). The terminal spine, widened at the base and sharply decurrent, with its broad flat groove of the upper surface, is also consistently characteristic, but it shares this character with its near relative, *A. hiemiflora*.

A. seemanniana represents an aggregate of very scattered populations from Chiapas to northern Nicaragua (Fig. 17.18). Its area is wholly tropical from ca. 1,000 to 5,000 feet (300 to 1,550 m) elevations in the rather arid pinelands of central Honduras. The most extensive population observed is scattered over the low hills in Chiapas from Colonio Cardenez to Las Cruces and along the road to Arriaga (Fig. 17.24). The species does not ordinarily reproduce vegetatively, and it appears to have run out of European cultivation before the end of the last century. Like other members of the Hiemiflorae, it flowers in winter and is probably seeding by late spring. Seeds are needed to reintroduce this handsome plant to the many fanciers of the exotic succulents.

The history of this plant is better known than most of the agaves introduced to Europe during the 19th century. As an example of how occidental man handled these succulent acquisitions, it is of fitting interest to outline the taxonomic history a bit further here.

The plant was originally introduced by Berthold Seemann, naturalist on the voyage of the H.M.S. Herald, 1849–51, who collected seeds in Nuevo Segovia (now northern Nicaragua or adjacent Honduras). Some of these seeds were propagated by a Mr. Bull of England, and through him plants reached the Royal Botanic Gardens in Kew and perhaps other gardens. Jacobi describes this species from immature living plants with leaves 4¼ inches (10.8 cm) long, 3 inches (7.6 cm) wide, broadly spatulate, short acuminate, narrowed at the base to 1½–2 inches (3.8–5 cm) wide, the margin crenate with variously hooked castaneous teeth on fleshy prominences and smaller teeth between.

Trelease in his account of the Agaveae of Guatemala (1915) interpreted the species as belonging in the section Rigidae, as indicated by his annotation of *Deam 6154*. Berger placed the species in his section Scolymoides (1915) and remarked that there was a plant in the Kew Gardens "perhaps rightly named." Standley, the ace of Mesoamerican botanists, regarded the glaucous green single plant agaves with broad crenate leaves common on the central mountains of Honduras all as *Agave seemanniana* (personal communication, 1951). Miranda, who worked intensively with the Chiapan flora 1945–60, regarded a similar broad-leaved agave near Tuxtla Gutierrez as *A. seemanniana*. However, no critical account of the characteristics of these plants was ever made. No type specimen of *A. seemanniana* was ever designated and the original seedlings of Mr. Bull have long been extinct.

In 1965, while engaged in bean work, I visited Nicaragua and made a special effort to find and collect *A. seemanniana* in its natural habitat. In northern Nicaragua 2 miles (3–4 km) E of Somoto I found a small population of a beautiful toothy agave (Fig. 17.25 lower) (*Gentry 21631*), and again 8–9 miles (13–15 km) south of Condega (*Gentry 21630*). None were observed south of Condega. Unfortunately at this time, Nov. 10, there were no flowering plants available, the inflorescences being in early bud. Five young plants were collected at Somoto and several other collections of similar live plants were made in adjacent Honduras. These were sent to the Plant Introduction Station of Agricultural Research Service in Miami, Florida, the Huntington Botanical Gardens in California, and to the Desert Botanical Garden in Phoenix, Arizona. However, there is no record of any of these blooming. They are tropical plants and may have been killed by the occasional freezes that visit all these gardens. The herbarium specimens, however, remain secure, and on them I have based my concept of *Agave seemanniana*. They testify to what has been written and will serve as source for other taxonomic circles, not yet born.

Fig. 17.25. *Agave seemanniana*. Upper, rosettes in a colony along road to Arriaga, south of La Cruz, Chiapas. Lower, large-leaved forms with big teeth near Somoto, Nicaragua.

Agave thomasae

(Figs. 17.3, 17.18, 17.26, 17.27; Table 17.1)

Agave thomasae Trel., St. Louis Acad. Sci. Trans. 23: 138, 1915.

Rosettes single or moderately suckering, medium-sized, openly spreading, light gray, acaulescent; leaves 60–120 x 12–17 cm, broadly lanceolate, acuminate, narrowed and thickened toward base, plane to mildly guttered, soft succulent, pliable, pruinose or light glaucous to pale green, smooth to slightly asperous below; teeth minute, 1–2 mm long, 1–2 cm apart, reduced to denticles below; spine subulate to acicular, 3–4.5 cm long, dark brown, shallowly grooved above; panicle deep, narrow, 5–8 m tall, with 30 to 60 congested balled umbels on upper ½ to ⅚ of shaft; flower clusters purple to yellow, tufted at base with dark-colored bractlets; flower 60–70 mm long; ovary 30–38 mm long, strongly 3-angulate, tapering from tube or with a short grooved neck; tube 6–11 mm deep, funnelform, knobby, triangulate, rather thin-walled, sulcate; tepals unequal, 19–29 mm long, erect to incurved, long linear, 2 to 4 times as long as tube, bluntly acute, cucullate, the inner smaller with prominent keel; filaments 50–60 mm long, purple or yellow, inserted near orifice of tube; anthers 20–25 mm long, regular, excentric, yellow; capsule not seen.

Type: *Trelease 19,* cultivated in Hotel Thomas, Quetzaltenango, Guatemala, 6 April 1915, *NI.* (Named for Señora Dora Thomas, owner of the Hotel Thomas.)

Agave thomasae has been observed and collected along the highway from Quezaltenango to Huehuetenango, spontaneous near houses and cultivated about the native houses. It is one of the few suckering agaves in Guatemala and hence lends itself readily to local cultivation. However, it is not abundant. It is a high altitude species of the pine woods zone between 6,500 and 9,000 feet (2,000 and 2,800 m).

Agave thomasae is distinguished by its soft gray pruinose leaves with minute teeth and the strongly triangulate form of the variously colored flowers.

Agave wercklei

(Figs. 17.18, 17.28)

Agave wercklei Weber ex Berger, Agaven 1915: 201, 1915.

Rosettes medium to large, non-surculose, compact, 1–2 m tall with heavy broad short-acuminate leaves and large profuse bulbiferous panicles. Leaves variable, ovate to lanceolate, short-acuminate with inrolled apex, 70–150 cm long, green to white glaucous, thickly fleshy, narrowed at the base, plane to concave toward apex, ascending to outcurved above, bud-printed, the young leaves with rough surface below apex; teeth small, 3–4 mm long, deltoid, black, on smooth rounded, straight to undulate margins; spine conical at base to an acicular point, 2–3 cm long, dark brown or black, narrowly grooved above, finely short-decurrent; panicle 8 m tall with short deltoid-bracted peduncle, with ca. 45 short, ramified, umbellate, horizontally spreading, thickly flowered branches; flowers golden yellow, 62 mm long, on 5 mm long pedicels; ovary elongate, narrowed at both ends, 40 x 7 mm; tube openly funnelform, 8–9 mm deep; tepals 17 x 4 mm, elliptic or oblong blunt; filaments to 90 mm long, inserted above mid-tube; anthers 16 mm long, golden yellow; pistil to 10 cm long; panicle apparently bulbiferous.

Fig. 17.26. *Agave thomasae*. Flowering plant 21 miles (34 km) north of San Cristóbal road fork along route CA-1, Dec. 1965, and a rosette with minute teeth near Patzizia, Depto. Chimaltenango, Guatemala.

Fig. 17.27. *Agave thomasae*. A pale glaucous rosette with minute teeth near Patzizia, Depto. Chimaltenango, Guatemala.

Lectotype: *Berger s.n.,* Cult. La Mórtola Garden, Veintimiglia, Italy, 19-IX-1910, *US*. Plant originally from Carlos Werckle collected at Peña Negra, near Rio Grande westward of San Ignacio de Candellaria at 600–800 m elev., Costa Rica.

The description above is based largely on Berger's (loc. cit.). Dried flowers of the lectotype appear to have been very succulent in life and were borne in large diffusive umbels with sparse small bractlets. The large thick leaves, deep panicles, and flowers show relationship to the Salvadoran *A. parvidentata*.

Trelease also published the name *wercklei* as sp. nov. (Contr. U.S. Nat. Herb. 23: 132, 1920), based on *Alfaro & Tonduz' 17553,* cult. in San José, Costa Rica, MO, which consists of six sheets. The leaves of these specimens were damaged and poorly prepared,

Fig. 17.28. An immature plant of *Agave wercklei* cultivated as a yard ornamental in San José, Costa Rica. April 1965.

but together with the flowers appear to represent a quite different species than the Werckle material. But the name is untenable. Trelease's specimens are of a slender-leaved, graceful, white-glaucous plant, apparently belonging with the Americanae Group. I observed it in cultivation in Costa Rica, where I collected offsets in 1965; one is established in the Desert Botanical Garden, where we await its flowering.

A. wercklei is also a beautiful plant, but sensitive to frost. There are apparently both green and whitish forms of it. The ones I saw, cultivated as yard ornamentals in San José, Costa Rica, were all green, forming compact handsome rosettes (Fig. 17.28). Werckle reported it as native to the region of the Río Grande on the Pacific slope of Costa Rica, where it grew in hot country in sparse grassland, also the home of the tropical giant Ceiba trees and vanilla orchids. When provided good soil and watered, the cultivated plants exceed the wild ones in size and beauty, the rosettes reaching 2 m tall. The panicle is apparently very tall and deep.

Hiemiflorae Exsiccatae

Agave atrovirens

OAXACA. Gentry 22377, US, MEXU, DES. Neotype. Ca. 20 miles SE of Miahuatlán along road to Pochutla on Sierra Madre del Sur, 26 Oct. 1967, elev. ca. 7,500 feet (l, f, photo).

Gentry 22353, DES. 6–8 miles W of Ayutla, Dist. Tamazulapán, 19 Oct. 1967, elev. ca. 6,200 feet (photo).

Gentry 12062, DES. 14–18 miles NE of Cd. Oaxaca along road to Ixtlan de Juarez, 29 Aug. 1952 (l, photo).

Gentry et al. 20273, 20423, US, MEXU, DES. Ca. 14 miles NE of Ixtlan de Juarez along road to Tuxtepec, Aug. & Sep. 1963, elev. 9,000–9,500 feet in pine-oak forest (l, pl, cap, photo).

Ernst 2377, US. Between Mitla and Cuesta, 30 Jan. 1966 (l, f, cap).

VERACRUZ. *Gentry 23387,* US, MEXU, DES. 18 miles NE of Tehuacán along road to Orizaba, 13 March 1974, elev. ca. 7,300 feet (l, f).

Ogden et al. 5159, US, DES, MEXU. Azumbilla Canyon near Cordoba highway, 16 miles N of Tehuacán road, 3 Feb. 1951, "edge of cloud forest at 3,000 m elev." (l, f).

Agave atrovirens mirabilis

VERACRUZ. *Gentry & Arguelles 12577.* 16 miles NW of Jalapa along road to Puebla, 17 April 1953 (anal. pl).

Gentry et al. 20413, DES. Las Vigas, 4 Sep. 1963, elev. ca. 7,500 feet (cap, f, photo).

Trelease s.n., MO, DES. Type. Las Vigas, above Jalapa, V.C. "Maguey Blanco" 8/26/1903 (3 sheets l, cap, buds, photos).

Trelease s.n., MO. 3/2/05 (f).

Agave congesta

CHIAPAS. *Breedlove 33512,* CAS, DES. 6–10 km N-NE of La Soledad along logging road from Las Margaritas to Camp Alegro, Mun. de Independencia, 17 Feb. 1973; elev. 1,600 m, slope with *Pinus & Quercus* (l, f).

Gentry 23651, US, DES, MEXU. Type. 14 miles W of San Cristobal Casas along hwy. to Tuxtla Gutierrez, 8,000 feet elev., wooded limestone slopes, 3 Jan. 1976 (l, f).

Gentry 21846, US, DES, MEXU. 18 miles NW of Comitan along road to San Cristobal Casas, elev. 6,500–7,000 feet, pine & oak limestone slope, 3 Dec. 1965 (l, f).

Gentry 21847, US, MEXU, DES. 21–22 miles NW of Comitan along road to San Cristobal Casas, alt. 6,500–7,000 feet, 3 Dec. 1965 (l, f).

Gentry 12179, 12180. 14 miles S of San Cristobal Casas, pine-oak forest, elev. ca. 6,500 feet, 18 Sep. 1952.

Gilley et al. 7114, US, DES, MEXU. 13 km W of San Cristobal de las Casas, along road to Tuxtla Gutierrez, 14 August 1951.

Agave hiemiflora

CHIAPAS. *Breedlove 23852,* CAS, DES. 18–20 km N of Ocozocoautla along road to Mal Paso, Mun. Ocozocoautla, 26 Jan. 1972; elev. 800 m, Montano rain forest (l, f, cap).

Gentry 23652, US, MEXU, DES. 25 miles W of San Cristobal Casas along road to Tuxtla Gutierrez; limestone slope, elev. ca. 5,200 feet (l, f) 3 Jan. 1976.

Gentry 12203, US, MEXU, DES. Near Piedra Parada along road to San Cristobal Casas, elev. ca. 5,500 feet, 21 Sep. 1952 (l, cap).

Gentry 12179, US, MEXU, DES. 14 miles S of San Cristobal Casas along road to Comitan; pine & oak zone, elev. ca. 6,500 feet, 18 Sep. 1952.

Gentry 12184, US, DES, MEXU. 15 miles S of Comitan along highway, elev. ca. 4,000 feet, 20 Sep. 1952.

Gilly et al. 7119, US. 20–25 km W of San Cristobal de la Casas along road to Tuxtla Gutierrez, 14 Aug. 1951.

Raven & Breedlove 20071, CAS. 3 miles NW of Teopisca along hwy. 190, Mun. de Teopisca, 28 Jan. 1965; elev. 6,100 feet, slope with *Quercus & Pinus* (l, f, cap).

GUATEMALA. *Clover 9623,* MICH. 17 miles N of Salama, Depto. de Baja Verapaz along road to Coban, alt. 4,000 feet, 17 Dec. 1946 (f).

Clover 9656, MICH. 3 miles W of Salama on road to San Miguel, alt. 3,500 feet, 15 Dec. 1946.

Gentry 23640, US, DES. Type. 7 miles E of Huehuetenango along road to Aguacatan; pine-oak forest on limestone, elev. ca. 7,000 feet, 1 Jan. 1976 (l, f).

Gentry & McClure 23643, US, DES. 9 miles NE of Huehuetenango along highway CA-1; grassy mt. slope, elev. ca. 6,000 feet, 2 Jan. 1976 (l, f).

Gentry 21726, US, DES. 3–4 miles N of Salama, Depto. Baja Verapaz, elev. 3,800–5,000 feet, 26 Nov. 1965.

Gentry 21426, US, DES. 9.5 miles E of Huehuetenango along road to Aguacatan, alt. ca. 6,000 feet, 14 Oct. 1965 (l, f).

Gentry 21750, US, DES. Ca. 2 miles W of Uspantan, Depto. Quiche, alt. ca. 5,000 feet, 27 Nov. 1965 (l, f).

Gentry 21765, US, DES. Between Zacapulas & Cunen, Depto. Quiche, alt. ca. 6,500 feet, 27 Nov. 1965 (l, f).

Agave hurteri

GUATEMALA. *Gentry 20815, Gentry & McClure 23632, 23635A,* US, DES. Rocky slopes above Zunil along road to Rio Salama, elev. 7,000–7,500 feet; 21 April 1965 & 29 Dec. 1975 (l, f, photos, pl).

Gentry & McClure 23633, 23634, 23634A, US, DES. Near Asunción above tunnel along road to Santa Maria below Zunil, elev. ca. 6,500 feet, 29 Dec. 1976 (l, f).

Gentry & McClure 23636, DES. 1.5 miles E of San Cristobal CA-1 crossroad, Depto. de Totonicapán; rocky slope below cliff, elev. ca. 8,000 feet, 30 Dec. 1975 (photo, pl).

Gentry 23630, US, DES. Just below Solola above Lago Atitlán, rocky volcanic slopes & cliffs, elev. ca. 6,500 feet, 22 Dec. 1975 (l, f).

Gentry 23641, 23642, US, DES. 7 miles NE of Huehuetenango along road to Aguacatan; pine-oak forest on limestone, elev. ca. 7,000 feet; 1 Jan. 1976 (l, f).

Gentry s.n., DES. Mt. grade S of Cunen, Depto. Quiche, alt. ca. 6,000 feet, 6 May 1965 (photo only).

Gentry 21408, DES. River gorge above San Cristobal, Depto. de Totonicapan, 13 Oct. 1965; alt. 7,500–8,000 feet, volcanics (cap, photos).

Trelease 3, ILL. Type. Zunil, on the Samala River, 10 April 1915 (2 sheets, l, f).

Agave lagunae

GUATEMALA. *DeHam s.n.,* MO. Palin to Escuintla, 12/3/1908.

Gentry 21708, US, DES. In pass below Amatitlan, on shady cliff, 24 Nov. 1965 (l, f).

Kradolfer s.n., MO. Palin to Escuintla, 1908.

Trelease 10, ILL. Type. Above Amatitlán near the lake, April 1915.

Agave pachycentra

CHIAPAS. *Gentry & Arguelles 12227,* US, MEXU, DES, MICH. Mesa de Copoya, 7–9 miles S of Tuxtla Gutierrez, 28 Sep. 1952; grassland in old cleared field (l, photo).

GUATEMALA. *Clover 9657, 9576,* MICH. On mts. near Salama, alt. 3,500 feet, 10 Dec. 1946.

Clover 10025, 10025A, MICH. E of Volcán Jumay, near Jalapa, Depto. de Jalapa, alt. 5,500 feet, 22 Jan. 1947.

Clover 9138, 9139, MICH. Near Linda Vista, near Chiquimula, alt. 1,400 feet, 19 Sep. 1946.

Clover 9181, 9182, MICH. Along road to Montanas de las Minas, near Chiquimula, alt. 2,500 feet, 26 Sep. 1946.

Clover 9203, 9204, 9210, 9663, MICH. Near Chiquimula, Depto. Chiquimula, 26 Sep. to 26 Dec. 1946, alt. 720–800 feet.

Gentry 20973, US, DES. Near Sanarate along new highway, Depto. de Progreso; thorn shrub, alt. 2,000–2,500 feet, 4 May 1965 (l, cap).

Gentry 12444, US, DES. Km 60 along road from Guatemala to Zacapa; elev. 2,000–3,000 feet, 16 Jan. 1953.

Gentry 21711, US, DES. Near Sanarate, Depto. Progreso, alt, ca. 1,800 feet, 25 Nov. 1965 (l, f).

Gentry 21712, US, DES. Aldea de Barril, Depto. Progreso, alt. ca. 1,200 feet, 25 Nov. 1965 (l, f).

Gentry 23631, US, DES. 2.5 miles W of Sanarate, Depto. Progreso, along highway CA-9, elev. ca. 3,000 feet, 27 Dec. 1975 (l, f).

Gentry 23629, US, DES. El Progreso, Depto. Progreso, along highway, elev. ca. 1,500 feet, 17 Dec. 1975 (l, f).

Gentry 23628, US, DES. 5 miles W of El Rancho, Depto. Progreso, elev. ca. 1,750 feet, 17 Dec. 1975 (l, f).

Gentry 23628, US, DES. 5 miles W of El Rancho, along highway to Puerto Barrios, 16 Dec. 1975 (l, f).

Gentry 23621, US, DES. 1 mile E of Ipala, Depto. Chiquimula, thorn forest, elev. ca. 3,000 feet, 14 Dec. 1975 (l, f).

Gentry 20910, US, DES. Valle de Monjas, Depto. de Jalapa, 29 April 1965.

Gentry 23620, US, DES. 5 miles NE of Jutiapa, along route CA-1, elev. ca. 3,200 feet, 14 Dec. 1975.

Iltis G-23, ILL. Laguna Retana, 5–7–9 km N of El Progreso, Dept. Jutiapa, alt. 1,150 m, 29 Dec. 1975 (l, cap, bud).

Kellerman 7066, MICH, UC. Jalapa; El Rancho, 18 Jan. 1908 (l, f, cap).

Trelease 2, ILL. Type. La Cruz, Depto. Progreso, 2 March 1915 (l, & l snips).

Trelease 4, ILL. Cruz, Depto. Progreso, 20 March 1915 (l, & l snips). Type of *A. tenuispina.*

Trelease 5, ILL. Cruz, Depto. Progreso, 20 March, 1915 (2 sheets, l, f, cap). Type of *A. opacidens.*

HONDURAS. *Gentry 21689,* DES. Valley below Jesus Otoro, Depto. Intibuca, alt. 1,700–2,000 feet, 17 Nov. 1965 (photo).

SALVADOR. *Gentry 21707,* US, DES. Puerto del Diablo near San Salvador, alt. 3,000 feet, 22 Nov. 1965 (l, f).

Agave parvidentata

SALVADOR. *Allen, s.n.,* DES. Volcán de Santa Ana, Feb. 1959; elev. 7,500 feet (photos).

Calderon 2085, Type, **US.** Cult in garden, San Salvador, 1924 (2 sheets, l, f).

Calderon 2251, US. Patio de Ensayos in San Jacinto, San Salvador, 1924–25 (l, f, photo, 6 sheets, type of *A. compacta* Trel.).

Rohweder 1075, MO. Volcán de San Vicente, Depto. San Vicente, 7 March 1951; elev. 2,160 m (f, large, 65–70 mm long).

Rohweder 1076, 1077, MO. Cerro Chulo, Los Planos Renderos, Depto. San Salvador, 7 Dec. 1950; elev. 1,050 m (l, f, 45 mm long).

Agave potatorum

OAXACA. *Conzatti 4099,* US. Cerro de el Rosario, Dist. del Centro, 18 Dec. 1920; elev. 1,700 m (l, f).

Gentry 22401, 23656, US, MEXU, DES. 6 miles NW of Huajuapan along highway, limestone hills with palm & sotol, alt. ca. 6,000 feet, 6 Nov. 1967 (l, f); Jan. 1976 (cap).

Gentry 12124, US, MEXU, DES, MICH. Ca. 15 miles NW of Tejupán along highway; limey slope with shrub & agave, 7 Sep. 1952 (l, f).

Gentry 12111, 22399, US, MEXU, DES, MICH. Ca. 2 miles NW of Yanhuitlan along highway, sedimentary hill, alt. ca. 7,500 feet; 11 Sep. 1952–4 Nov. 1967 (l, f, cap).

Gentry & Halberg 12057, 12079, US, MEXU, MICH, DES. 6–8 miles NE of Cd. Oaxaca along road to Ixtlan, elev. 5,000–6,000 feet, Aug. 1952 (l, f).

Gentry 22344, US, MEXU, DES. 8 miles E of Mitla along road to Ayutla, volcanic rocky slope with oak, elev. ca. 6,800 feet (l, f).

Gentry 12106, US, MEXU, MICH, DES. 7 to 8 miles SE of Miahuatlán along new road over Sierra Madre del Sur, oak-pine forests, elev. 6,500–7,000 feet, 2 Sep. 1952.

Gentry 21886, 23655, US, MEXU, DES. 8–10 miles SE of Camerón along Oaxaca-Tehuantepec highway; oak woodland, elev. 4,200–4,300 feet, 6 Dec. 1965 & 7 Jan. 1976 (l, f, cap).

Iltis et al. 1147, ILL, MEXU, US. Near Yanhuitlan, km 420 Panam. highway, alt. 2,500 m, 20 Aug. 1960.

Ogden & Gilly 51193, US, MEXU, DES, MICH. 1 mile N of La Joya along highway from Cd. Oaxaca to Acatlan; oak & pine, 18 April 1951.

F. Miranda 4619, MEXU. Barranca del Rio Chico E of Cuicatlan, 14 Sep. 1948.

Rose 11527, US. Tomellin Canyon, 7 Sep. 1906.

Rose & Hough 4592, US. Near Cd. Oaxaca, June 16–21, 1899.

Schoenwelter JSOX-78, US. Near Cueva Blanca, 22 July 1966 (infl., l).

Trelease s.n., MO. Tomellin, 2/10/05 (l, f).

PUEBLA. *Endlich 1909,* MO. Tehuacan, Apr. 1907.

Gentry 12288, US, MICH, MEXU, DES. 22–24 km NE of Tehuacán along highway to Veracruz; limestone sun slope, 16 Oct. 1952 (l, f).

Gentry & Barclay 20427, US, MEXU, DES. 4–10 miles SW of Tehuacán along road to Zapotitlán; arid thorn forest over limestone hills, elev. 5,600–6,000 feet, Aug. 1963 (l, f).

Griffiths 9/6/09, MO. Tehuacán.

F. Miranda 2157, MEXU. Matamoros, 27 July 1942.

Ogden & Gilly 51201, US, MEXU, MICH, DES. 10 miles S of Izucar de Matamoros along highway to Acatlán, 19 April 1951.

Rose et al. 5950, US. Near Techuacán, 1–2 Aug. 1901 (l, cap).

Rose et al. 10024, US. Near Tehuacán, Aug.–Sept. 1905 (l, f).

Rose & Rose 11261, 11269, 11421, 11426, US. Near Tehuacán, Sept. 1906.

Trelease 8/12/03, MO. Tehuacán.

Agave pygmae

CHIAPAS. Gentry & McClure 23650, US, DES, MEXU. Type. White limestone dome 28 miles N of Guatemala-Mexico border along highway, elev. ca. 3,100 feet, 2 Jan. 1976 (l, f).

Agave seemanniana

CHIAPAS. *Gentry 16435, 23653,* US, MEXU, DES. Near Col. Lazaro Cardenez along Panam. highway; upland savannah hill slopes, elev. ca. 2,000 feet; 22 March 1957 & 6 Jan. 1976 (l, f).

Gentry 21867, US, MEXU, DES. 6–10 miles S of La Cruz fork along road to Arriaga; granitic slope with grassland and pines; elev. 2,300 feet, 5 Dec. 1965 (l, f).

Harriman & Jensen 14673, DES. Ca. 2 miles N of km 200 on rt. 190 S of Comitán, 11 Jan. 1978; limestone, arid rocky roadside (l, f).

Laughlin 1528, CAS. Paraje Navenchauk along hwy. 190, Mun. de Zinacatán, 17 Aug. 1966; elev. 7,200 feet, slope with *Quercus* & *Pinus* (l, f).

GUATEMALA. *Deam 6154,* MO, US. Fiscal, 3 June 1909; elev. 3,700 feet, sides of mountain ravine.

Gentry 12474, DES. 3–5 miles W of Estor, Depto. de Izabal, on open sun slope with short woods, elev. ca. 250 feet, 27 Jan. 1953 (photo, pl).

Gentry 23622, 23623, US, DES. 5–6 miles E of Chiquimula highway along road to Olopa; limestone rocks & cliffs, elev. ca. 2,850 feet, 15 Dec. 1975 (l, f).

Trelease 1, ILL. La Cruz, between El Rancho and Sanarate, Depto. Progreso, March 1915 (l, type of *A. tortispina* Trel.)

HONDURAS. Gentry 20684, US, DES. Neotype. 8–12 km SE of Tegucigalpa along road to Zamorano; open rocky slope with pine, April 1965 (l, f).

Gentry 20785, DES, US. 5–10 km W of Cucuyagua along road to Ocotopeque; sedimentary bluff, 14 April 1965.

Gentry 20692, US, DES. Ca. 7 km S of Danli along road to Paraiso; grassland on volcanic rocky slope, elev. ca. 2,500 ft., 4 April 1965 (f, photo).

Kimnach 445, HNT. 14 km from Tegucigalpa along road to Zamorano, 18 Aug. 1962 (cult. & fl'ed. at H.B.G. July 1977).

Molina 8619, US. Drenaje del Rio Choluteca cerca del pueblo de Oropoli, Depto. El Paraiso, 9 Mar. 1958; alt. 400 m.

Molino 25930, US. Rocky cliff on Rio Hondo, near La Venta on road to Talanga, 14 Feb. 1971; alt. 700 m (l, f).

NICARAGUA. *Gentry 21631,* US, DES. 2 miles E of Somoto, Old Segovia Province; open hill slope, elev. ca. 2,500 feet, 10 Nov. 1965 (l, photo).

Gentry 21630, US, DES. 8–9 miles S of Condega, Depto. de Estil; volcanic cliffs, elev. ca. 2,300 feet, 10 Nov. 1965.

Agave thomasae

GUATEMALA. *Gentry 23637,* US, DES. Chonimacanac or Tierra Blanca, ca. 23 miles N of Huehuetenango along highway CA-1; pine & oak forest zone, elev. ca. 9,000 feet, 30 Dec. 1975 (l, f).

Gentry 21806, US, DES. 21 miles N of San Cristobal road fork along road to Huehuetenango; grassy hill slope, elev. ca. 8,500 feet, 2 Dec. 1965 (l, f).

Gentry 20862, US, DES. Near Patzicia, Depto. de Chimaltenango, alt. ca. 6,500 feet, 23 April 1965 (l, f).

Harmon & Fuentes 4870, MO. 11 km S of Chemal, Depto. de Huehuetenango, 12 Nov. 1970; elev. 10,000 feet.

Trelease 19, ILL. Type. Cultivated at Hotel Thomas, Quezaltenango, April 1915 (l, photo).

Agave wercklei

COSTA RICA. Berger s.n., Lectotype, **US.** Cult. in La Mortola Garden; original from Carlos Werckle collected near Peña Negra, Rio Grande, westward of San Ignacio de Candelaria at 600–800 m elev.

18.

Group Marmoratae

Die Agaven 1915: 270.

Plants small to large, multiannuals, rarely surculose, freely seeding, with few leaves; leaves thickly succulent, linear to ovate lanceolate, mostly gray green, frequently cross-zoned and roughly scabrous, the margins mammillate-crenate with small to large teeth; terminal spines short, not exceeding 3 cm, not decurrent, and with little or no dorsal grooving (*A. zebra* excepted); inflorescence paniculate, the panicles in upper half of shaft with 10–25 diffusely spreading, flat-topped umbels of small bright yellow protandrous flowers, 30–50 mm long, with short tubes.

Typical species: *Agave marmorata* Roezl, Belg. Hort. 33:238, 1883.

The species of this group are all readily characterized by several or all of the following characters: the grayish scabrous leaves with crenate margins and small terminal spines, the small bright yellow flowers with small tubes ¼–½ as long as the tepals (Fig. 18.2). In the form of terminal spine, especially through *A. marmorata* and *A. gypsophila*, and in flower structure they resemble the Inaguense and other groups of the Caribbean *Agave*. Each of the four species here allocated to the Marmoratae are distinct and occupy disjunct geographic areas. *A. marmorata* is the only one among them of any abundance. The others are relatively or actually rare and limited in areas; see map of distributions (Fig. 18.3).

There is wide divergence in the habitats of the species, each habitat being distinct, as indicated in Fig. 18.1. *Agave nayaritensis* is the mesophyte among them, occupying the well watered cliffs and rocks of tropical northwestern Nayarit with nearly continuous soil moisture through summer to early spring. *A. zebra* is at the arid extreme of this, sitting upon a few arid mountains in the Sonoran Desert with protracted annual droughts and light winter frosts. The habitats of both *A. marmorata* and *A. gypsophila* are semi-arid; the latter is drier than the yearly rainfall indicates because of the poor moisture holding capacity of the gypsiferous substrate and limestone it occupies. Further notes on respective habitats and uses will be found in the species accounts following.

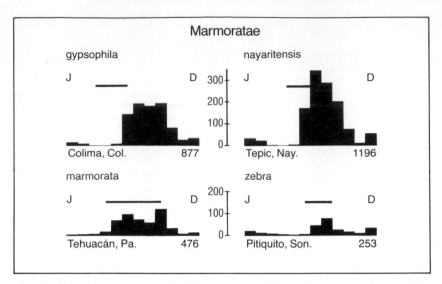

Fig. 18.1. Perimeters of rainfall (silhouettes) and flowering seasons (bars) of the Marmoratae; relevant meteorological stations with annual rainfall in millimeters. Data from Atlas Climatológico de México, 1939. Flowering periods based on herbarium specimens and field observations.

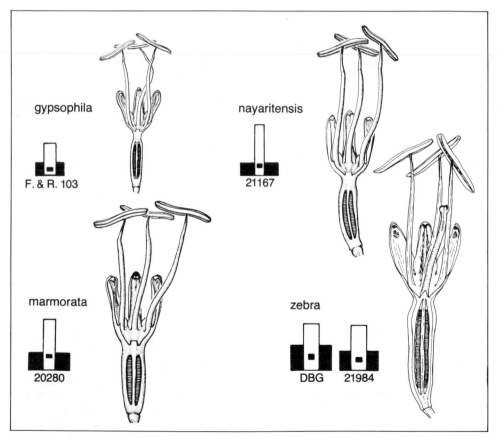

Fig. 18.2. Long sections of flowers of the Marmoratae and their ideographs; based on measurements in Table 18.1. The relative smallness of *Agave gypsophila* is due in part to dry flowers, which never regain full size even when relaxed in hot water.

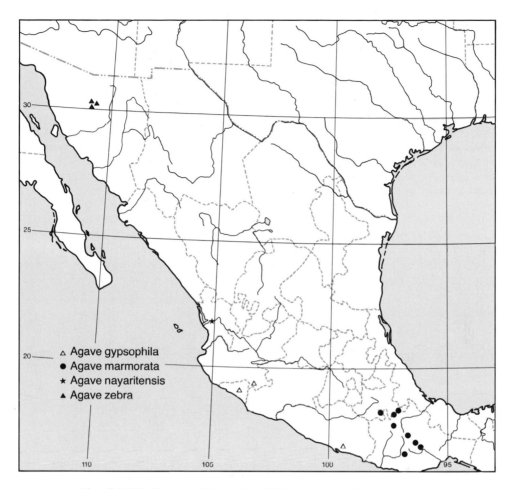

Fig. 18.3. Distribution of the species of Marmoratae, so far as is known.

Key to Species of the Marmoratae

1. Leaves relatively weak, fragile, narrower, 6 to 10 times longer than wide; teeth small, not exceeding 4 mm in length   2
1. Leaves strong, rigid, wider, only 4 to 5 times longer than wide; teeth large, 6–20 mm long   3
2. Leaves fragile, brittle, gray, scabrous, not exceeding 8–10 cm wide; margins distinctly crenate   *gypsophila*, p. 510

2. Leaves not fragile, green, nearly smooth, 12–15 cm wide; margins straight to undulate
                        nayaritensis, p. 515
3. Leaves larger, 80–135 cm long; spines short, conical, 1.5–3 cm long
                          marmorata, p. 512
3. Leaves shorter, usually 50–80 cm long; spines elongate, 3.5–7.5 cm long, decurrent   *zebra*, p. 516

Agave gypsophila
(Figs. 18.1, 18.2, 18.3, 18.4, 18.5; Table 18.1)

Agave gypsophila Gentry, sp. nov.

Rosettes multiannuals, nonsurculose, small, openly spreading, few-leaved, maturing with 20–30 leaves; leaves 45–100 (–110) x 7–12 cm, linear lanceolate, weak, brittle, with little fiber, slightly narrowed near the base, thick, deeply convex below, flat above, the sides above distinctly undulate, the whole generally arching, glaucous gray, asperous; margins closely dentate with small mammae, involute at apex, weak teeth 1–2 mm long, and interstitial denticles; spines very small, 5–15 mm long, conic, dark brown, nondecurrent; panicles 2–3 m tall with relatively few, wide-spreading few-flowered umbels in upper half of slender shaft; flowers yellow, 30–35 mm long (measurements from dried relaxed fls.), on slender small-bracteolate pedicels; ovary 18–20 mm long, fusiform, with furrowed neck; tube 4–5 mm deep, 9 mm wide, openly funnelform, grooved; tepals 10–11 x 3–4 mm, ca. equal, spreading, linear, galeate-apiculate; filaments 20–25 mm long, inserted near base of tube; anthers 11–12 mm long, yellow, nearly centric; pistil over-reaching stamens in post-anthesis; capsules and seeds not known.

Type: *Floyed & Ryan 103, MICH*. 4 miles SE of Acahuizatla, Guerrero (along highway to Acapulco), 14 June 1954; rocky clay loam; elev. 3,000 feet (portions l & infl.). "Yellow corolla, up to 3 ft. high."

Planta parva brevicaule singula paucifolia. *Folia* linearia vel lanceolata 45–100 (–110) cm longa, 7–12 cm lata, arcuatia sinuata fragilia grisea scabrosa, apice involuto, *margine* crenata, dentibus supra mamillas parva 1–2 mm longis; *spina terminali* minima, 5–15 mm longa conica spadicea nondecurrent. *Inflorescentia*

Table 18.1. Flower Measurements in the Marmoratae (in mm)

TAXON & LOCALITY	OVARY	TUBE	TEPAL	FILAMENT INSERTION & LENGTH		ANTHER	TOTAL LENGTH	COLL. NO.
gypsophila								
	18	4×9	10×4&10	2–1	20	12	31*	103
Acahuizotla, Gro.	19	5×9	10×3	2	22	11	34*	6193
	19	4×9	11×3	2	23	11	34*	6193
marmorata								
	20	6×13	16×4	5	34	18	41†	20280
Huajuapán, Oax.	22	6×13	16×4	5	37	18	44†	20280
	23	5×13	16	4.5	36	18	45†	20280
nayaritensis								
Mirador del Aguila, Nay.	20	4×11	16×4&15	3–4	35	15	40†	21167
	24	4×12	17×4	3–4	37	16	44†	21167
	25	4×12	16×3	3–4	36	15	45†	21167
zebra								
	30	7–8×8	12×4–5	3–4			48*	10205
	23	7×7	10×4	4–5			40*	21207
Sierra del Viejo, Son.	30	6×10	14×5&13	3	35	17	50†	21984
	25	5×10	11×5.5	3	30	12	40†	21984
	24	6×13	14×5&13	4	35	16	43†	21984
	30	6×12	15×5&14	3–4	38	18	53†	21984

* Measurements from dried flowers relaxed by boiling.

† Measurements from fresh or pickled flowers.

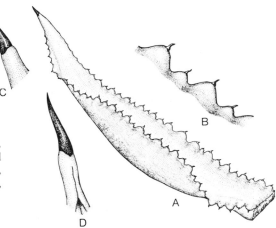

Fig. 18.4. *Agave gypsophila*. A, terminal third of leaf x ½; B, detail of leaf margin from mid-leaf x 2; C, ventral view of terminal spine; D, dorsal view of terminal spine.

Fig. 18.5. *Agave gypsophila* in litter of forest floor in natural habitat southwest of Cd. Colima.

panaliculata scapo incluso 2–3 m alta, ramis remotis paucis; *floribus* parvis luteis 30–35 mm longis; *ovarium* fusiforme, apice incluso 18–20 mm longum. *Perianthii tubo* 4–5 mm longo, 9 mm lato, sulcato; *segmentis* ca. equalibus 10–11 mm longis, 3–4 mm latis, linearibus adscendentibus galeati-apiculatis. *Filamenta* 20–25 mm longa prope basine tube inserta; *antheris* luteis 11–12 mm longis. *Capsula* et semina non vidi.

Agave gypsophila is a small sprawling plant, easily distinguished by its linear, brittle, brindled, gray leaves with close-set mammillate, small teeth and spines (Fig. 18.4). On the gypsiferous hills, 13–15 miles (21–24 km) southwest of Cd. Colima at about 400 m elevation, the plants were very weakly and shallowly rooted in the debris over the rocky forest floor, like a saprophyte. The population there is small, hard to locate, and grows scattered under the thorn forest with shallow roots over a gypsum formation (whence the name). It also occurs on limestone, 20–25 miles (32–40 km) east of Cd. Colima along road to Pihuamo in Jalisco. A third locality is near Acahuitzotla, Guerrero, along the highway to Acapulco, where three different parties collected it between 1951 and 1954 (see Exsiccatae). The species is sensitive to frost, introductions at Phoenix, Arizona, and Murrieta, California, having been killed or severely damaged by lows of 20°–25°F (−4°– −8°C). It is a distinctive oddity among agaves rather than ornamental.

Agave marmorata

(Figs. 18.1, 18.2, 18.3, 18.6, 18.7, 18.8; Table 18.1)

Agave marmorata Roezl, Belg. Hort. 33: 238, 1883.
  Agave todaroi Baker, Handb. Amaryll. 195, 1888.

Rosettes large, short-stemmed, single, rarely surculose, openly spreading with 30–50 leaves, 1.2–1.3 m tall, 2 m broad; mature leaves 100–135 x 20–30 cm, broadly lanceolate, in various attitudes, frequently undulate, thick at base, flat above, convex below, infolding along mid-blade, involute at base of spine, glaucous gray to light green, sometimes zonate, generally roughly scabrous, the margin crenate with fleshy mammae; teeth mostly 6–12 mm long, 2–5 cm apart, flattened, the cusps from very broad bases, mostly straight, castaneous to dark brown, interstitial teeth few or none; spine usually short conic, 1.5–3 cm long, rarely short-decurrent; panicle 5–6.5 m tall with stout peduncle and 20–25 large diffuse decompound umbels in upper half of shaft; flowers small, brilliant yellow, 40–48 mm long; ovary 20–25 mm long, light green, cylindric, the neck unconstricted and scarcely grooved; tube 5–6 mm deep, 12 mm wide, shallowly funnelform, grooved, the sinuses soon spreading; tepals equal, 14–16 mm long, 4 mm broad, linear, fleshy, erect, involute, the apex galeate, dark colored; filaments 34–37 mm long, inserted in orifice of tube or near base of tube, slender, yellow; anthers 18 mm long, centric, yellow; capsules oblong, 4 x 1.6–1.8 cm slender stipitate, short-beaked, thin-walled (*Gentry 22357*, dehisced); seeds not seen.

Neotype: Tab. 8442, Curtis Bot. Mag. 1912.

Berger states that "*Agave marmorata* was collected by Roezl in the Province of Tehuacan, Mexico." Trelease adds (1920) "unquestionably the Cerro Colorado near Tehuacan." However there are about three Cerro Colorados near Tehuacán, but the plant is still common around Tehuacán.

The name *marmorata* was given for the grayish marble-like hue and texture of the leaves. The short conic spines, the coarse rough, grayish, highly mammillate leaves, and the small golden flowers in large diffusive panicles distinguish this species. However, it exhibits considerable variability in leaf characters and the marked difference found in high or low filament insertion on the tube walls is unusual. The high insertion was found in *Gentry, Barclay & Arguelles 20280,* a plant also atypical in its relatively smooth leaves (Fig. 18.7). The plant otherwise closely conforms to *A. marmorata.* It may be another

Fig. 18.6. *Agave marmorata* neotype. Tab. 8442 Curtis Botanical Magazine, July 1912.

instance of gene incursion from another species in past generations. No likely cosponsor was found nearby.

The names "pisomel" (corrupted from Nahuatl, pitzometl), "huiscole," and "maguey curandero" are reported for this agave. The last name indicates it may have been used in folk medicine. This agave has proved to be frost sensitive in southwestern U.S. gardens. At Murrieta, California, and the Desert Botanical Garden in Phoenix, Arizona, night temperatures of 25°–20°F (−4°−−8°C) have killed small to medium plants. It is recommended for culture only in frostless sunny situations.

Fig. 18.7. *Agave marmorata* in golden flower on September 2, 1952, near Huajuapán, Oaxaca.

Fig. 18.8. *Agave marmorata* southwest of Tehuacán, Puebla. A rather typically canted rosette in full sun and submature.

Agave nayaritensis
(Figs. 18.1, 18.2, 18.3, 18.9, 18.10; Table 18.1)

Agave nayaritensis Gentry, sp. nov.

Rosettes medium size, acaulescent, single, rarely surculose, few-leaved, open, light green. Leaves lanceolate, 85–115 x 12–15 cm, rather floppy, etiolated because of tree shade, widest above middle, narrowed toward base, long acuminate, the cuticle somewhat asperulous, light green, the margin undulate to straight, with small teeth 1–3 mm long, regularly spaced 1–1.5 cm apart, castaneous or darker, with scattered intervening denticles; spine 9–15 mm long, conical, with short narrow groove above, dark brown, nondecurrent; panicle 3–4 m tall, diffusive, deeply oval in outline, with 14–15 widely spreading decompound umbels of bright yellow flowers in upper half of shaft; peduncle stout, with large triangular-lanceolate chartaceous bracts; flowers small, 40–45 mm long, on small bracteolate pedicels; ovary 20–25 mm long, rounded trigonal, with short furrowed neck; tube 4 mm deep, openly spreading, short-grooved; tepals subequal, 15–17

Fig. 18.9. *Agave nayaritensis* at Mirador del Aguila, north of Tepic, Nayarit. The small-flowered inflorescence became part of the type, May 1965.

Fig. 18.10. A younger plant of *A. nayaritensis* with full length leaves growing at the type locality along the edge of a cliff.

mm long, 3–4 mm wide, linear from spreading sinuses, erect to slightly incurved, involute, galeate at tip, rather thin, the inner with small keel; filaments 35–37 mm long, rounded, attenuately slender, inserted near orifice of tube; anthers 15–16 mm long, yellow, centric; capsules and seeds unknown.

Type: *Gentry 21167,* Mirador del Aguila, N of Tepic along rt. 15, Nayarit, May 21, 1965; altitude ca. 2,200 feet, mixed tropical forest, on cliff edges (l, f), *US,* dups. MEXU, DES.

The small scattered population on the volcanic cliff edges about Mirador del Aguila is the only known locality (Fig. 18.9).

Planta plerumque caule singulo paucifolia recta diffusa. *Folia* lanceolata 85–115 cm longa, supra medium 12–15 cm lata, flaccida viridia, parce scabrosa, margine rectiuscula; *dentibus* parvis 1–3 mm longis, 1–1.5 cm distantis, castaneis vel obscurioribus, interjacentibus denticulatis minutis; *spina terminali* 9–15 mm longa angusti-conica, brunnea non-decurrente, supra brevi-canaliculata. *Inflorescentia* panaliculata scapo incluso 3–4 m alta, ramis 14–15 diffusis remotis umbelliformipartitis; *floribus* luteis 40–50 mm longis; *ovarium* 20–25 mm longum apice incluso, trigone cylindricum. *Perianthii tubo* 4 mm longo vadoso infundibuliforme; *segmentis* subequalibus 15–17 mm longis, 3–4 mm latis, tenuibus linearibus rectis vel incurvatis involutis, apice galeatis. *Filamenta* 35–37 mm longa ad apicem tubi inserta; *antheris* 15–16 mm longis luteis. *Capsula* et semina non vidi.

Agave nayaritensis is distinguished by its broad diffusive panicle of small bright yellow flowers with shallow open tubes and slender green leaves with small teeth and spine. These characters readily admit its grouping in the Marmoratae, but its relationship with other species is not close.

No local name or uses were obtained for this plant, there being no local residents in the vicinity of the type locality. Young plants were collected (*Gentry 10706*) and sent to Florida or California for propagation, but I have no further record of them. They apparently did not survive.

Agave zebra
(Figs. 18.1, 18.2, 18.3, 18.11, 18.12, 18.13; Table 18.1)

Agave zebra Gentry, U.S. Dept. Agri. AH No. 399: 126, 1972.

Rosettes mostly single, medium-size, low spreading, rather open, light gray, xerophytic, strongly armed, 4–6 x 10–16 dm leaves generally 50–80 x 12–17 cm lanceolate, broadest near the middle, narrowed above the base, deeply guttered, arcuate, thick, rigid, light gray glaucous, scabrous, conspicuously cross-zoned, with strongly undulate margins; spine 3.5–7.5 cm long, acicular, mostly very narrowly grooved above, scabrously decurrent for 5–10 cm to upper teeth, yellowish brown to light gray; teeth large, 25–40 on a side, on prominent teats, those on the mid-blade mostly 10–20 mm long, with broad, low, scabrous bases, 1–3 cm apart, flattened, variously curved and flexed, gray with castaneous tips; panicles to 6–8 m tall, narrow, with large bracts toward the shaft base and 7–14 small laterals in upper ¼ to ⅕ of shaft; flowers small, yellow, 40–55 mm long; ovary 25–32 mm long, slender, cylindro-angulate, straight from angled very short pedicels, neck slightly constricted and 6-sulcate; tube funnelform, 6–7 mm deep, bulging, deeply sulcate; tepals about equal, 12–15 mm long, 5–6 mm wide, erect to ascending, lanceolate, revolute, cucullate, the inner prominently keeled; filaments broad, inserted in mid-tube, 35–40 mm long; anthers small, 15–18 mm long; capsules small, mostly 4–5 x 1.2–1.5 cm, oblong to obovate, the walls strong and conspicuously lined, stipitate, apex round, apiculate; seeds 4.5–5 x 4–4.5 mm obliquely rounded, concave on one side or both faces, the marginal wing very low, hilar notch shallow.

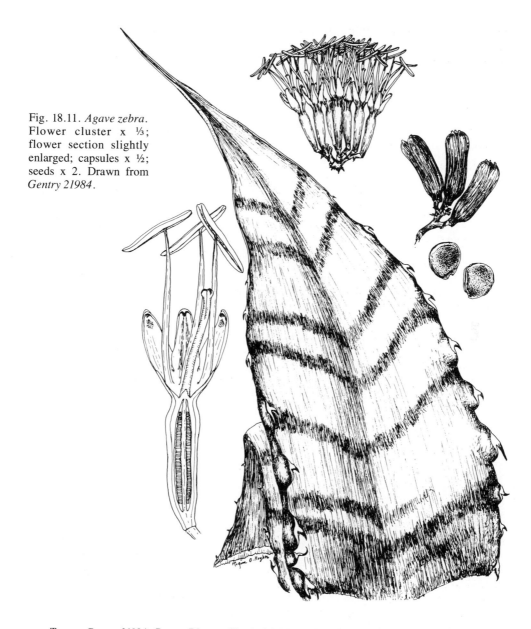

Fig. 18.11. *Agave zebra.* Flower cluster x ⅓; flower section slightly enlarged; capsules x ½; seeds x 2. Drawn from *Gentry 21984.*

Type: *Gentry 21984,* Puerto Blanco, Sierra del Viejo, about 20 miles south of Caborca, Sonora, July 24, 1966, open rocky slopes of limestone mountain, alt. 2,000 to 2,500 feet, deposited in U.S. Natl. Herbarium No. 2549706 and 2549707.

A. zebra is certainly known from only the Sierra del Viejo, and Cerro Quituni, about 26 miles south of Caborca.

A. zebra was used for a short while for making mescal, but no such activity has been observed since 1950. The leaves have a rather weird aspect: marbled crossbanding on deeply guttered, outcurving leaves, flagrantly armed with prominent teeth. If over-watered in the garden it becomes rather monstrous. It has proven cold hardy in our southwestern gardens.

Fig. 18.12. *Agave zebra* with *Fouquieria splendens* and *Jatropha cuneata* on lime-stone rubble on bajada of Sierra del Viejo, Sonora.

Fig. 18.13. *Agave zebra*, showing detail of leaf and its armature.

Marmoratae Exsiccatae

Agave gypsophila

COLIMA. *Gentry 19563, 22193, 23533,* DES, INF, MEXU, MICH, US. 13–14 miles SW of Cd. Colima crossroad along road to Manzanilla, 5 Dec. 1959, 22 Nov. 1966, 9 Feb. 1975; elev. ca. 1,200 feet; Thorn Forest on gypsiferous substrate.

GUERRERO. Floyed & Ryan 103, UC, **MICH.** Type. 4 miles SE of Acahuizatla, 14 June 1954; rocky clay loam, elev. ca. 3,000 feet (l, f).

Moore & Valiente 6193, BH, MEXU. Km 338 on road to Acapulco ca. 3 km beyond Acahuizotla, 28 March 1952 (l, f).

Ogden 5104, MEXU, US. Highway between Chilpancingo and Acapulco, 3 Jan. 1951; crevices of perpendicular road cut, soil pH 7.81.

JALISCO. *Gentry 23532,* DES, MEXU, MICH. Ca. 25 miles NE of Cd. Colima along road to Pihuamo, 8 Feb. 1975; elev. ca. 2,100 feet, dog-toothed limestone slope with sarcophytic vegetation.

Agave marmorata

HORTICULTURE. *Berger s.n.,* US. Cult. La Mórtola Botanic Garden, 1909 (l, f, photo).

OAXACA. *Gentry 12108,* DES, MEXU, MICH, US. 7–8 miles N of Miahuatlán along road to Cd. Oaxaca, 2 Sep. 1952; steep sedimentary hillside with gravel.

Gentry 20280, DES, MEXU, US. 24 miles NE of Huajuapán along road to Tehuacán, 11 Aug. 1963; alt. ca. 6,000 feet, volcanics and sedimentaries (l, f, photo).

Gentry & Tejeda 22367, DES. Ca. 12 miles S of Mitla road junction along hwy. to Tehuantepec; elev. ca. 5,600 feet, oak woodland and short-tree forest (cap).

Ogden & Gilly 51188, DES, MEXU, US. Along Tehuantepec-Oaxaca hwy. 5 miles SE of Totolapán, 14 April 1952.

PUEBLA. *Gentry 12121,* DES. Between Tehuitzingo and Acatlán along hwy. to Oaxaca, 6 Sep. 1952 (photo, pl.).

Gentry et al. 20225, DES. 4–10 miles SW of Tehuacán along road to Zapotitlán, Aug. 1963 (photo, pl).

Iltis et al. 1113, MICH, WIS. 12 km NW of Petlalcingo along road to Acatlán (Ca. 18° N; 97° 56′ W), Aug. 19, 1960; alt. ca. 1,350 m (l, cap).

F. Miranda 2863, 3091, MEXU. Acatlan, 16 July 1943.

Ramirez L. s.n., MEXU. Cerro Colorado near Tehuacán.

Rose & Rose 11265, US. Near Tehuacán, 1 Sep. 1906.

Trelease s.n., ARIZ, MO. Tehuacán, 11 Aug. 1903 (leaf clippings, some smooth, others scabrous).

Agave nayaritensis

NAYARIT. Gentry 21167, US, DES, MEXU. Type: Mirador del Aguila, ca. 15 miles N of Tepic, May 21, 1965; alt. ca. 2,200 feet, mixed tropical forest, cliffs (l, f, cap, pl).

Gentry & Gilly 10706, DES, MEXU, MICH, US. Mirador del Aguila, 28 June, 1951 (l, pl).

Agave zebra

SONORA. Gentry & Weber 21984, US, DES, MEXU. Type: Puerto Blanco, Sierra del Viejo, 24 July 1966; rocky limestone slope with ocotillo and desert shrubs (l, f, cap).

Gentry 10205, DES, MEXU, US. Sierra del Viejo, 25 Feb. 1951 (l, cap); elev. ca. 2,000 feet, open rocky slopes of limestone mt.

Gentry & Arguelles 19897, DES, MEXU, US. Cerro del Viejo ca. 25 miles S of Caborca, 5 April 1963.

Gentry & Arguelles 19899, DES, US. Cerro Quituni ca. 26 miles S of Caborca, 7 April 1963; elev. 2,000–3,000 feet, arid limestone mountain.

Gentry 21207, US, DES, MEXU. Puerto Blanco, Sierra del Viejo, 2 Sep. 1965 (l, cap).

19.

Group Parryanae

Small to medium, globose, compact, glaucous gray to green rosettes, suckering sparingly or prolifically with vigorous rhizomes. Leaves short, broad, rigid, mostly closely imbricated, the teeth conspicuously larger toward the leaf apex; scape of the panicle strong, with large lanceolate bracts irregularly reflexing; panicles mostly deep with 15–40 laterals in upper half or more of shaft, copiously flowered; flowers red to purplish in bud; tubes well developed but shorter than the tepals; tepals at anthesis linear-lanceolate, conduplicate, over-lapping at base, narrowing and incurving in post-anthesis wilt; filaments inserted above the mid-tube usually at uniform level; capsules rather small, ovoid to oblong, stipitate, strong-walled, freely dehiscing and freely seeding. Central Arizona to Big Bend region of Texas and southward in Mexican highlands to Guanajuato.

Typical species: *Agave parryi* Engelm., Acad. Sci. St. Louis Trans. 3: 310, 1875.

Trelease (1912) included most of these species in his group Applanatae along with *A. applanata* Koch. Berger (1915) maintained Trelease's grouping. *Agave applanata* in flower and leaf structure, however, shows no close affinity with the species here included in the section Parryanae. It is aligned rather with other members of my section Ditepalae. *A. havardiana,* however, with its deep flower tube and sometimes 2-level insertion of filaments does show affinity with the Ditepalae; a sectional bridging strand.

Most members of the Parryanae can be quickly recognized to section by the short, broad, closely imbricated leaves with the largest teeth along the upper one-quarter of the leaf. The longer more slender tepals distinguish all the Parryanae from the short-tepaled Ditepalae, while their short leaves exclude them from confusion with most of the Americanae. They are mesophytes rather than xerophytes and do not inhabit the dry desert habitats. While some occur in the Chihuahuan Desert, as *A. parrasana,* they are restricted to the larger mountains, where more moisture occurs than over the lower surrounding desert plains. Some small marginal populations, as in *A. gracilipes,* on lower dry desert heights are depauperate by drought, surviving as it were, by the skin of their leaves. Rainfall silhouettes and flowering periods are indicated in Fig. 19.1.

The Apache-Agave Connection

The Parryanae, a seed and sucker group, constituted a vital renewable resource of food, fiber, and drink for many southwestern Indian tribes. Since the time of Cabeza de Vaca (1542), travelers and botanists have provided observations of Indian uses. More recently, archaeologists have added more information from studies of the refuse heaps of campsites, caves, and ruins. Castetter, Bell, and Grove gathered together a great deal of this information from 150 bibliographic titles (1938).

There is a remarkable coincidence between the distribution of this group of agaves and the tribes and clans of the wide-ranging Apaches. It is as though agave attracted and held these wild hunting people. The western Apache in central Arizona made habitual use of *Agave parryi* and its variety *couesii*. The Mescalero Apache extended the use activity to *A. parryi*, *A. neomexicana*, and *A. gracilipes* across southern New Mexico to western Texas. The Lipan Apache made use of *A. havardiana* in western Texas and adjacent Mexico, while the Chiricahua Apache followed *A. parryi* southward deep into the Chihuahua highlands. It is apparent that the habitual annual use of agaves, during the 500 to 600 years the Apaches are thought to have spent in Parryanae country, was sufficient to ingrain agave into their eating habits and ceremonies.

The Apaches must have reduced many agave populations and in places may have threatened agave survival, but it is difficult to make any precise assessment of this biotic

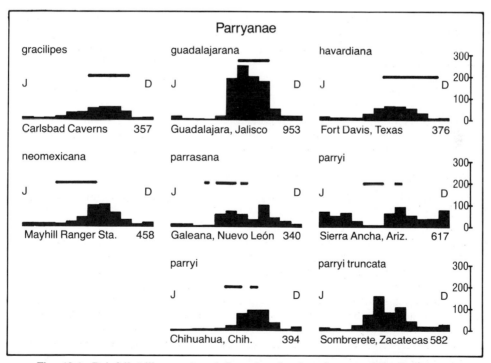

Fig. 19.1. Rainfall (silhouettes) and flowering (bars) perimeters of the Parryanae. Relevant meteorological stations with average annual rainfall given in millimeters. Data from U. S. Weather Bureau and Servicio Meteorológico Mexicana. Flowering periods based on field observations, herbarium specimens, supplemented by plants in cultivation; uncertainty expressed by dotted line.

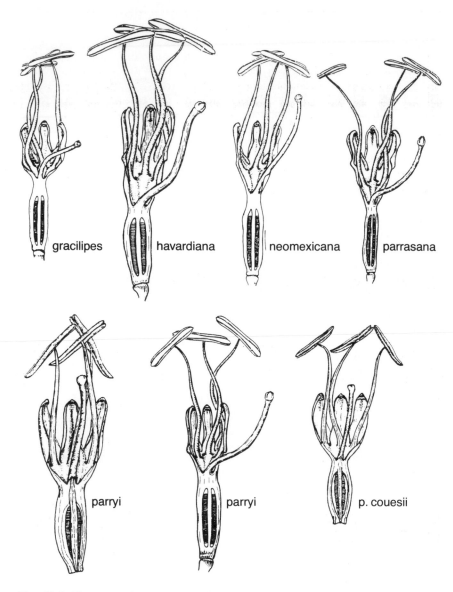

Fig. 19.2. Flower sections representative of the Parryanae. The two of *parryi* show the range of variability in that species.

pressure. The Indian harvested only the maturing individual plants. The young plants were left to sucker and increase. The mescal eaters were generally non-agricultural and did not plant mescal. However, there is some evidence that their living habits coincidentally favored agave dissemination. In gathering agave poles, for instance, they may have scattered seed and covered it by foot along their paths. Seed may have found its way to their refuse dumps, a microhabitat favorable to seedling establishment. Minnis and Plog (1976) found a strong correlation between Apache campsites and *Agave parryi* clones outside the natural agave habitat in the Sitgreaves-Apache National Forest in northern Arizona. The seed and sucker breeding habit of *Agave* has *ipso facto* operated throughout the biome, before and with man, for agave survival.

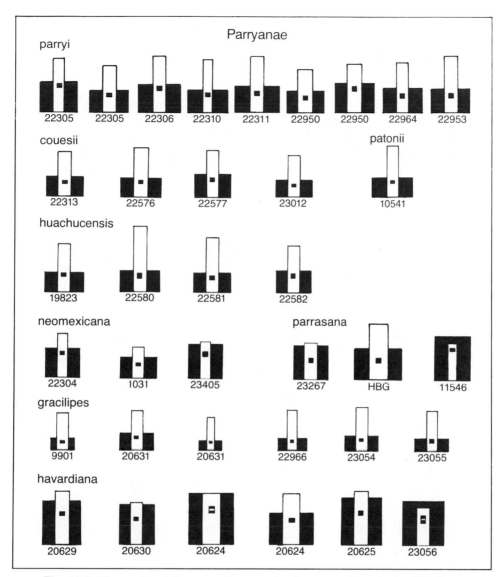

Fig. 19.3. Ideographs of Parryanae flowers, showing proportions of tube (black) to outer tepals (white) and locus of filament insertion (black squares) in tube. From measurements in Table 19.2.

The few chemical analyses available, Table 19.1, show little or no steroids present in the Parryanae. The negative results early discouraged further sampling for cortisone precursors in this group of agaves.

The flower measurements of Table 19.2 document the floral ideographs of Figs. 19.2 and 19.3, provide a statistical comparison of flower parts and supplement the following taxonomic account.

As in other groups of *Agave,* the capsules in the Parryanae have developed little distinction among species. The study represented by Fig. 19.4 shows as much differentiation within species as between species.

Table 19.1. Sapogenin Content in the Parryanae

Coll. No.	Source Locality	Month Coll.	Plant Part	Sapogenin Total%
gracilipes				
9901	Guadalupe Mt., Tex.	Oct.	infl.	0.
9901	Guadalupe Mt., Tex.	Oct.	stem	0.
9901	Guadalupe Mt., Tex.	Oct.	seed	0.
havardiana				
9885	Arroyo Limpio, Tex.	Oct.	leaf	0.
9885	Arroyo Limpio, Tex.	Oct.	infl.	0.
neomexicana				
9905	Guadalupe Mt., Tex.	Oct.	leaf	0.
9908	Mayhill, N.M.	Oct.	leaf	0.
parrasana				
11546	Sierra Parras, Coah.	Feb.	leaf	0.
parryi				
9987	Seneca, Ariz.	Nov.	leaf	0.
9987	Seneca, Ariz.	Nov.	infl.	0.
	Flagstaff, Ariz.		leaf	0.2 hecogenin
10551	Chinacates, Dur.	June	leaf	0.
10571	Sierra Papanton, Zac.	June	leaf	0.
parryi var. *couesii*				
9972	Ash Fork, Ariz.	Nov.	leaf	0.
9972	Ash Fork, Ariz.	Nov.	infl.	0.
parryi vari. *huachucensis*				
9920	Sonoita, Ariz.	Nov.	leaf	0.02
9920	Sonoita, Ariz.	Nov.	infl.	0.1 hecogenin
parryi var. *truncata*				
10566	Sierra Papanton, Zac.	June	leaf	0.
10568	Sierra Papanton, Zac.	June	leaf	0.

Source: Wall (1954); Wall et al. (1954a).

Table 19.2. Flower Measurements Representative of the Parryanae (in mm)*

Taxon & Locality	Ovary	Tube	Tepal	Filament Insertion & Length		Anther	Total Length	Coll. No.
gracilipes								
Guadalupe Mt., N.M.	29	6 × 14	18 × 5 & 17	5	45	20	52	23054
	31	6 × 15	18 × 5 & 17	5	45	21	54	23054
	25	6 × 14	18 × 5 & 17	4–5	40	21	49	23057
El Paso Gap, N.M.	29	6 × 14	16 × 4	5	38	19	50	23055
	27	5 × 14	17 × 4	4	40	19	49	23055
Cerro Presidio, Chih.	28	3 × 10	12 × 3	2.5	20	12	42	22970
	25	4 × 10	12 × 3.5	3	·18	12	46	22970
W of Sueco, Chih.	33	5 × 13	17 × 4	4	25	17	53	22966
	32	5 × 13	15 × 4	4	20	17	52	22966
Hueco Hills, Tex.	27	5 × 11	13 × 3	3–4	34	15	45	23060
	23	5 × 11	15 × 3	3–4	38	18	43	23060
	20	5 × 11	15 × 3.5	4	30	16	39	23061
gracilipes x *lechuguilla*								
N of Allamore, Tex.	33	7 × 14	17 × 5	6	28	19	58	20631
	24	4 × 9	14 × 3	4	23	17	42	20631
havardiana								
Guadalupe Mt. bajada, Tex.	34	17 × 17	15 × 5 & 15	11 & 9	43	23	66	23056
	35	19 × 16	14 × 5 & 13	13 & 11	45	23	68	23056

Table 19.2. cont.

Taxon & Locality	Ovary	Tube	Tepal	Filament Insertion & Length		Anther	Total Length	Coll. No.
Arroyo Limpio, Tex.	31	18×16	22×6	13–14	60	30	71	20629
	35	20×15	25×5	12–12	60	30	80	20630
	28	16×14	17×5	10–11	50	22	61	20630
	37	22×17	22×6	12&14	64	29	80	20630
Marathon, Tex.	34	21×18	21×7	16&14	65	30	75	20624
	36	13×18	21×7	9&10	54	26	68	20624
Chisos Mts., Tex.	41	19×17	22×6	13–14	55	28	82	20625
Sierra Carmen, Coah.	45	12×16	30×6&27	8&6	45	25	86	23132
	47	14×18	27×5	8&7	45	24	88	23132
Marrion, Coah.	37	20×16	24×6&23	15–14	53	28	80	23571
Los Huerfanos, Coah.	38	20×15	21×6&20	12–11	60	27	77	23572
neomexicana								
Organ Mts., N.M.	36	12×14	19×4&18	10	38	23	67	22305
	33	10×13	18×4&17	8–9	35	22	61	22304
Mayhill, N.M.	37	8×15	14×4	6	40	15*	58	1031
	32	10×15	11×5	6	36	15	52	1031
Guadalupe Mt., N.M.	36	14×12	15×4&13	10&9	43	22	64	23405
	32	13×13	15×4	10–9	40	22	61	23405
	22	9×9	13×3	6	35	16	42	22407
parrasana								
Hunt. Bot. Gard., Cal.	46	13×20	23×8	8–7	55	22	83	
	54	15×21	23×8	8–9	65	22	91	
Sierra Parras, Coah.	28	14×14	15×6	8	49	19	58	23267
	26	13×14	15×5	8–9	48	18	54	23267
parryi var. *parryi*								
Sierra Ancha, Ariz.	47	10×17	21×4&20	7	50	26	77	22310
	43	9×15	24×4&22	7–8	48	26	75	22310
	36	10×18	24×6&22	8	45	28	71	22311
Pinos Altos, N.M.	33	12×15	23×4.5&22	11	42	23	67	22305
	33	9×16	19×5	8	40	23	60	22305
Wild Horse Mesa, N.M.	30	11×18	23×6&21	10	46	24	64	22306
Col. Juarez, Chih.	32	9×16	19×7	6–7	46	19	60	22950
	37	12×17	20×6	9–11	53	20	63	22950
Buenaventura, Chih.	35	9×16	23×5	7–8	54	25	67	22953
	31	11×14	17×5	6–7	42	24	60	22953
Majalca, Chih.	34	10×17	20×6	7–8	44	25	63	22964
p. var. *couesii*								
S of Ash Fork, Ariz.	23	8×15	17×6&16	6	44	18	46	22313
Jerome, Ariz.	32	8×16	20×6&18	6&5	40	21	59	22576
SW of Prescott, Ariz.	26	9×14	20×5&18	8–6	42	22	54	22577
S of Flagstaff, Ariz.	34	7×14	15×4	4–5	35	19	54	23012
p. var. *huachucensis*								
Huachuca Mt., Ariz.	47	9×17	27×5&25	7–8	58	27	81	22580
Sonoita, Ariz.	37	8×14	22×5&21	7–6	52	24	68	22581
Hunt. Bot. Gard., Cal.	34	8×16	20×4	8	46	20	62	19823
patonii								
Río Chico, Dur.	38	8×17	21×5&19	7&6	42	19	66	10541
	45	7×17	17×4&16	6	41	23	70	10541

* Measurements from fresh or pickled flowers.

Key to the Species and Varieties of Parryanae

1. Leaves relatively short, broad, short-acuminate, mostly 2 to 3½ times longer than broad; spines 2–3 cm long; flowers relatively large, mostly 60–90 mm long   2
1. Leaves relatively narrow, more acuminate, mostly 3½ to 6 times longer than broad; spines more variable, 2–5 cm long; flowers smaller, mostly 40–60 mm long   5
2. Rosettes globose, suckering copiously with long rhizomes forming large spreading clones; panicles deep with 20–40 umbels; flowers mostly 60–80 mm long   3
2. Rosettes not globose, more openly flat-topped, with few or no suckers; panicles rather wide and open with 10–20 large umbels; flowers larger, mostly 70–90 mm long   7
3. Rosettes large with many leaves; mature leaves 20–40 cm long, acuminate
                          parryi, p. 538
3. Rosettes small with few leaves; mature leaves 10–30 cm long, truncate to short-acuminate   4
4. Leaves of mature wild plants 10–20 cm long, not mammillate, the margins straight or nearly so; inflorescence more compact with 20 or more branches   *parryi* var. *truncata*, p. 543
4. Leaves of mature wild plants 20–30 cm long, conspicuously mammillate toward apex; inflorescence laxly elongate with 15 to 20 small remote umbels
                    guadalajarana, p. 531
5. Leaves ovate lanceolate, short-acuminate, 20–30 cm long, with short spines 2–3 cm long; flowers with well developed tubes, ca. ½ as long as tepals. Central Arizona
                parryi var. *couesii*, p. 542
5. Leaves lanceolate, acuminate, 20–35 cm long, frequently with subulate spines 3–5 cm long; flowers with deep or shallow tubes   6
6. Flowers larger, mostly 55–67 mm long, with deeper tubes (10–14 mm); spring flowering
                    neomexicana, p. 535
6. Flowers smaller, mostly 40–55 mm long, with shallow tubes (5–7 mm); fall flowering   *gracilipes*, p. 526
7. Leaves larger, 30–70 x 15–25 cm, long-acuminate; teeth reflexed along middle and lower leaf margins; tepals about equaling to slightly longer than tube. Big Bend Region and adjacent Mexico   *havardiana*, p. 531
7. Leaves smaller, 20–30 x 10–12 cm, the apex abruptly acute; teeth not reflexed along margins; tepals about twice as long as tube. Sierras of southern Coahuila   *parrasana*, p. 537

Agave gracilipes

(Figs. 19.1, 19.2, 19.3, 19.4, 19.5 G, 19.6, 19.7, 19.11; Tables 19.1, 19.2)

Agave gracilipes Trel., Mo. Bot. Gard. Rep. 22:95, 1912.

Small, acaulescent, compact, pale yellow-green to glaucous green rosettes with 60–100 leaves at maturity, sparingly and closely surculose; leaves 18–30 x 5–7 cm, lanceolate, mostly glaucous green, rigid, thick at base, concave above, convex below, acuminate, the margin nearly straight to undulate with teeth bases, slightly narrowed above base, the cuticle smooth; spine 2.5–5 cm long, dark reddish brown to light gray, sturdy, usually straight, narrowly to openly grooved above; teeth regular, the larger along the mid-blade commonly 4–8 mm long, mostly rather straight or merely curved, regularly spaced 1.5–2.5 cm apart, light gray, 8–12 per side; inflorescence 1.8–2.5 m tall, the shaft slender and with scarious lanceolate bracts, the panicle frequently narrow, deep, with

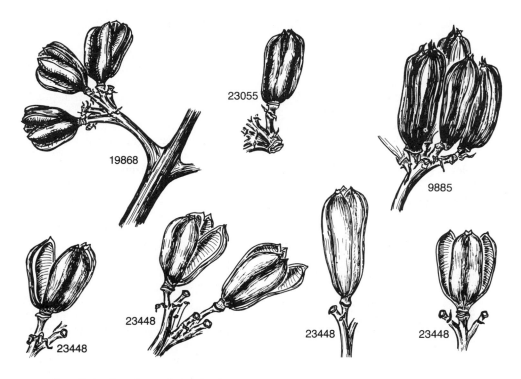

Fig. 19.4. Capsular variation in the Parryanae: 19868, 23055 = *Agave gracilipes;* 9885 = *A. havardiana;* 23448 = *A. parryi couesii.*

30–40 laterals in upper ⅓–½ of shaft, or panicle short, open with only 10–12 laterals in upper ¼ of shaft; flowers small, mostly 40–55 mm long, red in the bud, the tepals yellow when open, anthers yellow, filaments and pistil sometimes red; ovary slender to thick cylindric to fusiform, 25–32 mm long; tube shallow, 4–7 mm deep, knobby by base of filaments, grooved below tepal sinuses; tepals 14–18 mm long, subequal, spreading to ascending at anthesis, overlapping at base with thin margins which become involute above, soon narrowing, withering, and sinuses spreading; the margin of the inner tepals very wide, filaments 30–45 mm long, inserted just below rim of tube, slender; anthers 16–20 mm long; capsules oblong to ovoid, 3.5–4 x 2 cm, sessile or short stipitate; seeds not seen.

Type: *Mulford 293, 293a,* 1895. Sierra Blanca, Texas.

The range of *Agave gracilipes* extends from the Guadalupe Mountains of southeastern New Mexico through the western Big Bend of Texas southward in Chihuahua to the latitude of Sueco and San Buenaventura (Gentry & Arguelles 22966). It is a calciphile of the highland grama grasslands found between elevations of ca. 1,250 and 1,850 m. The various plant communities where *A. gracilipes* has been collected were noted as: juniper-pinyon grassland (top of Sierra Guadalupe), sotol grassland, *Acacia*-grassland, *Larrea-Yucca* grassland, open treeless grama grassland, and ocotillo low bush desert. Near Samalayuca, Chihuahua, the population was drought-pauperized and near its limit of tolerance to aridity. It is paradoxical that a plant so narrowly limited in range and habitat should exhibit such a wide range of genetic variability, as expressed through its divergent morphology and as inferable from the species description above. Just as a statesman may

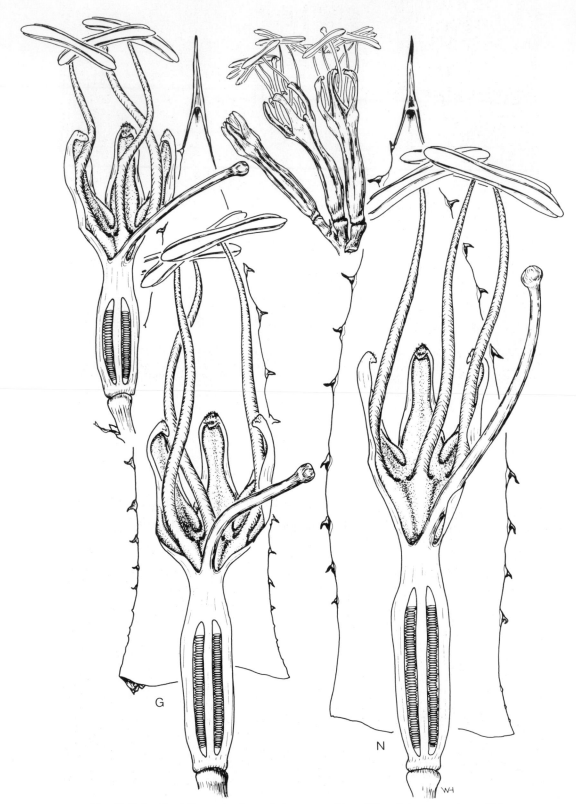

Fig. 19.5. Leaf and flower sections of *Agave gracilipes* (G) and *A. neomexicana* (N).
Flower sections x 1.8, leaves x ³/₅.

Fig. 19.6. *Agaves gracilipes* on the Hueco Hills of western Texas: Upper, hybridizing with *A. lechuguilla,* the latter with two leaves and infructescence on the right. Lower, a typical rosette.

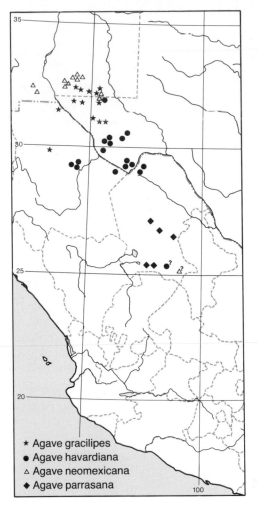

* Agave gracilipes
● Agave havardiana
△ Agave neomexicana
◆ Agave parrasana

Fig. 19.7. Distribution map of *Agave gracilipes, A. havardiana, A. neomexicana,* and *A. parrasana.* Based on herbarium specimens (see Exsiccatae) and observations. Map modified from M. Johnston, outlining the Chihuahuan Desert and marginal vegetation (dotted).

learn much from a revolutionary, so may a biologist gain by a study of the radical *gracilipes* gene pool.

This smallest member of the Parryanae appears similar to *A. neomexicana* in leaf and rosette. It is separable from *A. neomexicana,* however, by its smaller flowers with shallow open tubes and its fall blooming habit. Some of the populations of *A. gracilipes* are also distinguished by a narrow deep panicle with 30–40 short-peduncled umbels, but other individuals or populations may have relatively few inflorescence branches and a wider more open terminal panicle, much like those of *A. neomexicana.* This similarity and dimorphic variation of *A. gracilipes* is perhaps explained by the following observations.

Agave gracilipes is frequently found growing with *A. lechuguilla,* as 11–12 miles (18–20 km) north of Allamore, in the Hueco Hills, and about the Guadalupe Mountains. Such populations of *A. gracilipes* have numerous individuals with variable leaves, many of which show progressive narrowness toward leaves of *A. lechuguilla,* some rosettes even have leaves with a fragile corneous margin like *A. lechugilla* (Fig. 19.6). The inflorescences of such plants are paniculate, but deep and narrow varying toward the narrow spike of *A. lechuguilla.* The flowers of such plants are frequently very red, not just with the red flush common on buds of the Parryanae, but also with red filaments and red pistils, which is common on *A. lechuguilla* in the region. The obvious conclusion is that *A. gracilipes* is strongly infused with genes of *A. lechuguilla.* Nowhere, however, has a stable self-perpetuating population of *A. gracilipes* x *lechuguilla* been observed. Unlike *A. chisosensis,* this cross has apparently not yet achieved an isolated self-perpetuating status.

On second thought, however, *A. gracilipes* may be a specific product of geologically past introgression between *A. neomexicana* and *A. lechuguilla,* the habitats of which now are generally separated by about 2,000 feet (610 m) elevation. Perhaps we should regard *A. gracilipes* as a product of the combined genes of *A. neomexicana* x *lechuguilla,* but which has not yet achieved genetic isolation from *A. lechuguilla.* Cytological investigation might throw considerable light on the relationships of these nominal populations and their natural evolution. Further comments on the relation of *A. gracilipes* and *A. neomexicana* are given under the latter (p. 537).

Agave guadalajarana
(Figs. 19.1, 19.8)

Agave guadalajarana Trel., Contr. U.S. Nat. Herb. 23: 123, 1920.

Rosettes small, compact, light grayish green, single rarely suckering, 25–35 cm tall, broader than tall; leaves numerous, obovate to oblong, obtuse, 20–30 x 8–12 cm, the inner shiny glaucous, the outer dull gray, closely imbricate, rigid, plane to incurved, the upper teeth 8–10 mm long, remote prominently mammillate, reddish brown to dusty gray, those from the mid-leaf to base much smaller, 3–4 mm long, 5 to 10 mm apart on nearly straight margin; spine 2.5 cm long, subulate, grayish, straight to sinuous, flat to shallowly hollowed above, roundly keeled below; panicles with slender shaft, 4–5 m tall with 15 to 20 small umbels in upper half of shaft; "flowers 60 mm long" (Trelease); tepals slender, much longer than tube; capsules 4.5 x 1.8 cm, stipitate, shortly beaked, thick-walled, freely seeding; seeds lunate, 4 x 6 mm.

Type: *Pringle 4473,* on rocky summit of hills ca. 8 miles W of Guadalajara, Jalisco, 22 June 1893, *MO,* isotypes K, US, MEXU.

The native habitat of *Agave guadalajarana* is the grassy slopes of oak woodland around 5,000 feet (1,500 m) elevations on volcanic rocky soils. The climate is an open one below the Tropic of Cancer with a protracted dry season from January to May (see Fig. 19.1). The species is known only from the environs of Guadalajara (see Exsiccatae).

Agave guadalajarana was distributed by Pringle as *A. megalacantha* Hemsl. In most of the large herbaria reviewed it was still disposed under that name until my annotations. I have not seen any good duplication of Bourgeau's type collection, *No. 1020 bis* or *1020b* of *A. megalacantha.* Trelease looked for it in 1903 at the type locality, lava beds near Mexico City, and I have also looked for it. The only agave resembling the Bourgeau specimen that I have found on the Mexico City lavas is *A. inaequidens* Koch. They are all much large than *Bourgeau 1020b,* and Bourgeau, apparently being intimidated by the size and succulence of typical *A. inaequidens,* selected a depauperate plant for pressing, just as other botanists have done in many other cases.

The relationship of *A. guadalajarana* is with the Parryanae. It resembles other members of this section in the compact rosettes with closely imbricated, short, broad, rigid leaves with the largest teeth confined to the upper third of leaf, and the proportions of tube and tepals, the latter being twice or more longer than the tube. This latter character precludes it from the Ditepalae, although it bears some resemblance to that group in the long lax panicles with relatively small umbels and the mammillate leaf margins. The short obtuse leaves resemble those of *A. truncata,* but the mammillate leaf margins and lax panicles distinguish *A. guadalajarana* from *A. truncata* and all other Parryanae.

The compact size and gray coloring of *Agave guadalajarana* should make an interesting addition in our agave gardens. Seeds of the species were obtained in 1975; the seedlings are being cultivated in the writer's garden and will be widely distributed.

Agave havardiana
(Figs. 19.1, 19.2, 19.3, 19.4, 19.7, 19.9 (H), 19.10; Tables 19.1, 19.2)

Agave havardiana Trel., Mo. Bot. Gard. Rep. 22: 91, 1912.

Rosettes medium to large, rather open, glaucous gray, mostly single, suckering sparingly, 5–8 dm high and twice as broad; leaves 30–60 x 15–20 cm, rarely larger, thick, rigid, ovate-acuminate, glaucous gray to light green, occasionally yellowish,

Fig. 19.8. *Agave guadala-jarana* in natural habitat about 20 miles north of Guadalajara, Jalisco, in oak woodland, January 1975.

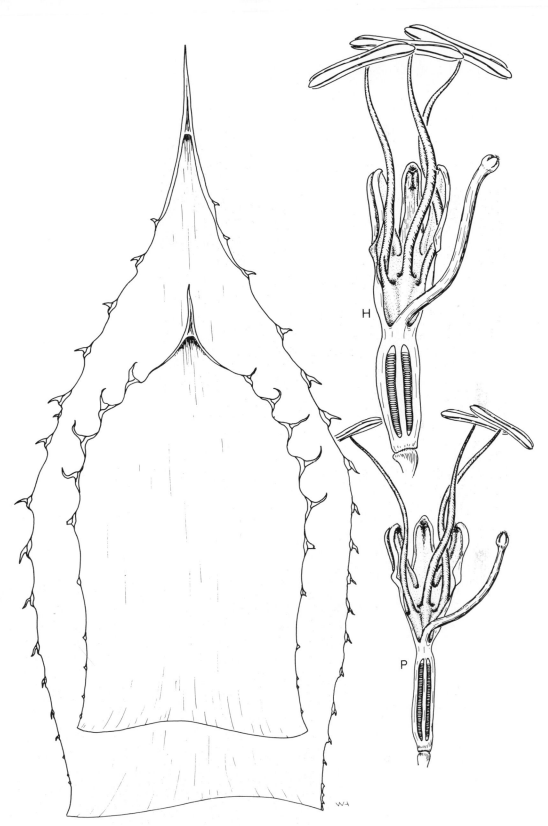

Fig. 19.9. *Agave havardiana*. H, leaf and flower section to compare with smaller leaf and flower section of *A. parrasana*, P; leaves x ca. ½, flowers slightly enlarged.

Fig. 19.10. *Agave havardiana* in natural habitat along the Arroyo Limpio near Fort Davis, western Texas, January 1975.

broadest at clasping base, slightly narrowed above base, the blade proper widest below middle, concave above, rounded below; teeth numerous, the larger toward leaf apex, mostly 7–10 mm long, gradually diminishing downward, 1.5–2 cm apart, the uppermost about straight, the others reflexed; spine stout, mostly 3–5 cm long (–8–10 cm), straight or sinuous, dark brown to grayish, smooth, broadly grooved above, round-keeled below, long decurrent, sometimes as a complete corneous margin; panicle 2–4 m tall, broad, open, with 12 to 20 lateral peduncles with large umbels of yellow flowers; flowers 68–88 mm long; ovary 30–40 mm long, green, with short thick neck; tube 14–22 mm deep, 15–18 mm wide, deep funnelform, thickly walled below filament insertion, finely grooved; tepals 18–24 mm long, narrowing to 4–5 mm at anthesis, the inner shorter, narrower, involute-conduplicate, wilting early and inward; filaments 50–65 mm long, long-tapered to apex, inserted rather irregularly above mid-tube; anthers 25–30 mm long, yellow, excentric; capsules oblong 4–5.5 x 1.4–2 cm, broadly short-stipitate, thick-walled, on strong pedicels; seeds 6–7 x 4–5 mm, smoothly black, marginal wing low, hilar notch small.

Type: *"Havard s.n., Oct. 1881,* Guadalupe Mountains,'' Texas and other later Havard collections, as Havard flowers in 1884 MO. (plate 85, Trelease loc cit.).

Agave havardiana ranges through the Big Bend region of Texas and the adjacent mountains in Coahuila and Chihuahua (Fig. 19.7). The habitats include the rocky grass-land slopes of the mountains, frequently on limestone and with scattered oak and juniper, between elevations of 4,000 and 6,500 feet (1,240 and 2,000 m). In the lower elevations it is associated with *Prosopis, Acacia,* and *Larrea* shrub desert. In eastern Chihuahua along the Chihuahua-Ojinaga road at 4,500 to 5,000 feet (1,400 to 1,550 m) elevation there is a scattered population in *Larrea-Yucca* desert shrub. Although the limey soil here is very shallow, the plants are exceptionally robust. In 1974 many of the flowering plants at this location were blown over and uprooted by a windy storm.

Among the Parryanae, *Agave havardiana* is distinguished by the deep tube and relatively short tepals, and by the very broadly based but acuminate leaves with reflexed teeth. The tube-tepal ratio is variable (see ideographs, Fig. 19.3), and there is a tendency to a 2-level insertion of the filaments, found so commonly in the Ditepalae. However, the tepals are atypically slender and long for the Ditepalae, rapidly involute and wilt incurved after anthesis. Trelease's leaf type (plate 85, loc. cit.) from the Guadalupe Mountains is atypically narrow for the species and may carry some genes of *A. neomexicana,* which is more abundant there, but his description, illustrated flowers and leaf (plate 86) are clearly applicable to *A. havardiana. Gentry & Engard 23056* collected on the Guadalupe Mountain bajada certainly agrees with the main stream of *havardiana* in the rest of the Big Bend mountains (Fig. 19.10).

Agave neomexicana
(Figs. 19.1, 19.2, 19.3, 19.5 N., 19.7, 19.11, 19.15 (left); Tables 19.1, 19.2)

Agave neomexicana Wooton & Standley, Contr. U.S. Nat. Herb: 16: 115, pl. 48, 1913.

Small to medium, light glaucous gray to light green, freely suckering, few to many-leaved, rather flat-topped rosettes; leaves 20–45 x 5–12 cm, lanceolate, usually broadest near middle, mostly rather short-acuminate, rigid, thickly rounded below, concave above; teeth above the mid-blade mostly 5–7 mm long, 1–3 cm apart, nearly straight or flexed, slender, dark brown to grayish; spine 2.5–4 cm long, subulate to acicular, flat above in a broad shallow groove, decurrent for 1 to several teeth; panicle 3–4 m tall with rather remote scarious reflexed triangular bracts, with mostly 10–17 compact umbellate branches in upper half of shaft; flowers red to orange in bud, yellow at anthesis, 55–67 mm long;

Fig. 19.11. *Agave neomexicana* in the type locality in the Organ Mountains, New Mexico, and a flowering plant of *A. gracilipes* (upper left) on the Hueco Hills for comparison.

ovary 32–38 mm long, slender, fusiform, furrowed in a constricted neck 4–7 mm long; tube 12–14 mm deep, and about as wide, funnelform, deeply furrowed and bulging with filament insertions; tepals nearly equal, 15–20 mm long, 3–4 mm wide, erect to ascending, narrowly linear, involute, the outer reddish on the cucullate apex, the inner all yellow, conduplicate, involute; filaments slender 35–45 mm long, sometimes irregularly inserted 9–11 mm above base of tube; anthers 20–24 mm long, yellow, centric; capsule oblong-elliptic, light brown. Seeds not seen.

Type: *Standley 541*, col. in Organ Mts., Dona Ana Co., New Mexico, June 1906, in *US* No. 498333.

Agave neomexicana suckers abundantly, much like *A. parryi,* but the rosettes and clones are generally smaller , the leaves more slender, and the inflorescence smaller with fewer branches. *A. neomexicana* and *A. gracilipes* do grow sympatrically in the Guadalupe Mountains and perhaps elsewhere. Without inflorescence characters most leaf specimens of these two species are not distinguishable. Their relationship is discussed under *A. gracilipes* (p. 530). *A. neomexicana* is distinguished by its spring blooming habit and the deep flower tube, the tepals being only 1 to 2 times as long as the tube, while the tepals of *A. gracilipes* are 2.5–4 times as long as the tubes.

Although these two agaves grow together, separate flowering seasons tend to preclude gene exchange. Even though exceptional instances of togetherness flowering may occur and some gene exchange happen, back-crossing of such offspring would tend to "swamp" out continuance of such crosses. No morphological intermediates of the two species have been recognized.

The post-conquest name of *A. neomexicana* was "mescal," like many other agaves in the deep extensive frontier of New Spain. The Mescalero Apaches received their subtribal name because of their habit of eating these "mescales," and they ranged particularly over the area of *A. neomexicana*. The charred cooked "heads" and leaf bases, sweet with sticky syrup, stuck to their faces, hands, and clothing, making them look dark and dirty. They probably made large depletions in the agave populations. They warred with the plains Comanches, who are thought to have confined the Apaches mainly to the mountains. This may have been a factor in helping conserve the extensive stand of agaves along the eastern bajada of the Sierra Guadalupe, where the Apache and the Comanche ranges met—a tribal no-man's-land.

Agave parrasana
(Figs. 19.1, 19.2, 19.3, 19.7, 19.9(F), 19.12; Tables 19.1, 19.2)

Agave parrasana Berger, Notizblatt Bot. Gard. Berlin 4: 250, 1906.
    Agave wislizeni Engelm.,* St. Louis Acad. Sci. Trans. 3: 320, 1875.

Rosettes small, compact, single, few or no suckers, glaucous gray green to green, acaulescent; leaves generally 20–30 x 10–12 cm, closely imbricated, ovate, short-acuminate to merely acute, plane to concave above, thick, rigid, frequently light gray to bluish glaucous; teeth largest near leaf apex, 5–10 (–15) mm long, diminishing rapidly below, slender from small low bases, straight to curved, 1–2.5 cm apart, grayish brown; spine 2–3 (–4) cm long, slender from broad base, flat to openly grooved above, sharply decurrent to upper teeth, dark brown to grayish; panicle 3–4 m tall, with 12–15 lateral peduncles, subtended by large reddish to purplish bracts, also on peduncle; umbels compact, bracteolate; flowers flushed red or purple, opening yellow, 50–60 mm long; ovary 25–30 mm long, with short unconstricted neck; tube cylindric, 13–14 mm deep and as broad, lightly grooved; tepals 15 x 5–6 mm, lanceolate, rather thick, pale yellow, the

---

*An illegitimate name. See Gentry, Cact. & Succ. Journ. US. 48: 102–4, 259, 1975.

Fig. 19.12. *Agave parrasana.* Left: At right, large imbricated bracts closely imbricated due to winter growth repression; at left, a panicle of *A. parryi,* Gentry garden near Murrieta, California. Right: Leaves of *A. parrasana.*

outer more acute, 2 mm longer than inner, red-flushed at apex; filaments 45–50 mm long, slender, inserted near mid-tube; anthers 18–19 mm long, yellow, centric. Capsules and seeds not seen.

Type: *Berger s.n.* La Mórtola Botanic Garden, Italy 1915. Originally from Sierra de Parras, Coahuila, col. by Purpus in 1905, *US.* Description drawn from wild plants. Cultivated plants in garden and greenhouse are more robust.

Agave parrasana is known only from a few limestone mountains in southeastern Coahuila (Fig. 19.7). It grows between elevations of 4,500 and 8,000 feet (1,400 and 2,480 m), mainly associated with chaparral shrub and the pine and oak communities above the desert proper. Flowers in summer.

This species is easily distinguished by its short, broad, abruptly short-acuminate leaves. It differs from all other Parryanae by the purplish colored, large, succulent bracts on the peduncle and sheathing the budding umbels (Fig. 19.12, *top*). In its native region it is called "noah." Because of its natural cold-hardiness, moderate size, distinctive leaf form, and coloring, it is a fine ornamental succulent for the garden.

Agave parryi
(Figs. 19.1, 19.2, 19.3, 19.12, 19.13 (P), 19.14, 19.15 (left), 19.18; Tables 19.1, 19.2)

Agave parryi Engelm., Acad. Sci. St. Louis Trans. 3: 310, 1875.
   Agave chihuahuana Trel., Mo. Bot. Gard. Rep. 22: 1911.
   Agave patonii Trel., ibid. p. 92.

Rosettes compact, globose, medium-sized, glaucous gray to light green, freely suckering, 40–50 x 60–75 cm, with 100 to 160 closely imbricated leaves; leaves mostly 25–40 x 8–12 cm, linear-ovate, short acuminate, rigid, thick, nearly plane to concave above, rounded below; teeth mostly 1–2 cm apart, small, the largest above mid-blade 3–7 mm long, mostly rather straight on a nearly straight margin, dark brown to grayish; spine 15–30 mm long, nearly flat above, dark brown aging to gray, decurrent to the first or second teeth; panicles stout, 4–6 m tall, with large reflexing bracts on peduncle and with 20–36 stout lateral peduncles on upper half of shaft; flowers in bud pink to red, opening yellow, mostly 60–75 mm long; ovary 30–45 mm long with long neck (6–9 mm), mildly constricted and grooved; tube 8–12 mm deep, 15–18 mm wide, fleshy, thickly angled, deeply grooved; tepals subequal, 18–24 x 4–6 mm, thick, ascending to erect, soon involute, linear above the rather open sinuses, papillate within well below the hooded tip, the inner thickly keeled, 2-costate within; filaments broad, 40–55 mm long, inserted near orifice of tube 7–11 mm above base of tube; anthers excentric to centric, 20–24 mm long, yellow; pistil eventually exceeding stamens with capitate 3-lobed stigma; capsules on stout pedicels, 3.5–5 x 1.5–2 cm, short-stipitate, beaked, strong-walled; seeds 7–8 x 5–6 mm, half-moon in outline, with low, thick rim and shallow hilar notch.

Type: *Rothrock 274*, in 1874, Rocky Canyon (probably in Graham County, Arizona), MO ?, as specified by Trelease, following Engelmann who wrote (loc. cit.). "The first who, collecting foliage, flowers, and fruits, enabled me to connect all those scattered fragments, was Dr. J. T. Rothrock, Surgeon and Naturalist of Lieut. Wheeler's Southwestern Expedition of 1874. He met with the plant in 'Rocky Canyon' and as far north as Camp Apache in northeast Arizona."

Agave parryi has the most extensive range of any species in this section, being found from central Arizona and southwestern New Mexico southward through the highlands of Mexico to Durango (Fig. 19.18). Its habitats include the rocky slopes of the grama grasslands, the oak woodland, the pine and oak forests, and the Arizona chaparral. The elevations of these zones range generally from 1,500 to 2,500 m. In the north along the slopes of the Mogollon Rim and the Mogollon Mountains of New Mexico, the *parryi* area outlines the cold fringe of Agaveland. The clonal clusters and populations are widely scattered, and nowhere has the species been observed as a community codominant. The annual precipitation through this region ranges between 20 and 40 inches (500 and 1,000 mm), but there is a long spring dry season. Summer temperatures are mild but winter lows in the north reach down to about −20°C.

Agave parryi is distinguished by its compact, multi-leaved, freely suckering, light green to grayish rosettes. The paniculate inflorescence is consistently stout, straight, and regular in outline. The tepals are generally about twice as long as the broad thick tubes (see ideographs, Fig. 19.3), this character prevailing in the varieties as well as in the matrix of the species itself. The series of specimens now available (see Exsiccatae) from central Arizona to central Mexico do not lend themselves to subspecific segregation. There are no correlative breaks in the subtle variability of populations that one can observe widely in the field; as the more open and smaller rosettes in the higher Chihuahuan mountains, or the paler forms represented by *A. patonii* of Durango, the satiny pruinose, light-reflective forms of *A. chihuahuana* on the grassland plains of central Chihuahua. A competent taxonomic segregation of these forms requires a detailed analytic field study beyond the limits of this work. The varieties and forms already recognized in this complex are annotated below.

Throughout historic times the common name of *A. parryi* has been "mescal," a general country term of Indian origin for northern Agaveland. *A. parryi* and its varieties constituted a vital resource for the Apache tribes and clans (see discussion above). The compact attractiveness of *A. parryi*, in or out of pots, together with its natural cold-hardiness, makes it suitable for outdoor garden culture over much of temperate United States and Europe. The colorful red and yellow blooms are frequently seen on color postal cards in the Southwest.

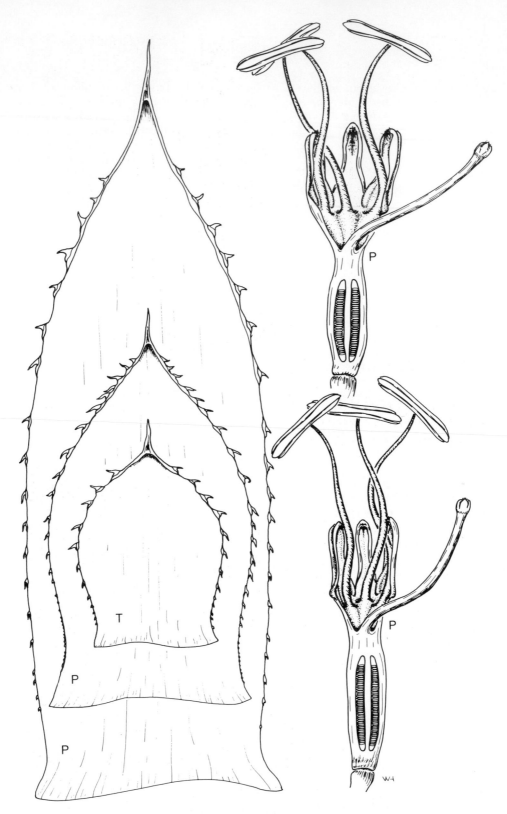

Fig. 19.13. *Agave parryi*. P, a large leaf and flower from near Silver City, New Mexico *(Gentry 22306)*, and small leaf and flower from near San Buenaventura, Chihuahua *(Gentry & Arguelles 22953)*. T, smallest leaf, *A. parryi truncata (Gentry & Gilly 10566)*, Sierra Papanton, Zacatecas. All from submature rosettes.

Fig. 19.14. *Agave parryi* in Chihuahua: Upper, in juniper grassland near Col. Juarez with Juan Arguelles. Lower, on the Cumbres de Majalca in oak woodland, June 1971.

Fig. 19.15. Left, an array of leaves of *Agave neomexicana* and *A. parryi* from New Mexico and Arizona. Right, *Agave parryi* var. *couesii* in natural habitat of dispersed chaparral near Prescott, Ariz.

Agave parryi var. *huachucensis*
(Figs. 19.3, 19.18; Tables 19.1, 19.2)

Agave parryi var. *huachucensis* (Baker) Little ex Benson, Am. Journ. Bot. 30: 235, 1953.
  Agave huachucensis Baker, Handb. Amaryll. 1888: 172, 1888.

Differs from typical *parryi* in being more robust, the leaves larger, up to 50 x 20 cm, and the panicle broader with larger flowers.

Type: *Pringle s.n.,* Huachuca Mountains, Arizona, June 30, 1884, alt. 5,000–8,000 feet (1,550–2,480 m).

It is primarily an oak woodland species occupying open slope sites between 5,000 and 7,000 feet (1,550 and 2,150 m) elevations, but it also occurs in the higher pine forest zones. It certainly extends into Sonora as far south as the Sierra Huachinera, and populations in adjacent northeastern Chihuahua are probably to be assigned here too.

Agave parryi var. *couesii*
(Figs. 19.2, 19.3, 19.4, 19.15, 19.18; Tables 19.1, 19.2)

Agave parryi var. *couesii* (Engelm. ex Trel.) Kearney & Peebles, Journ. Wash. Acad. Sci. 29: 474, 1939.

This variety is separable from typical *parryi* by its smaller flowers, 30–55 mm long vs. 55–75 mm long, in combination with its smaller leaves. The apex of the tepals is more densely papillate in *couesii* than in the species, judging from the depiction in Fig. 19.2. It appears to be limited to the highland slopes of the Agua Fria and Verde River watersheds in central Arizona (see Fig. 19.18). Small-leaved forms of *A. parryi,* however, occur at

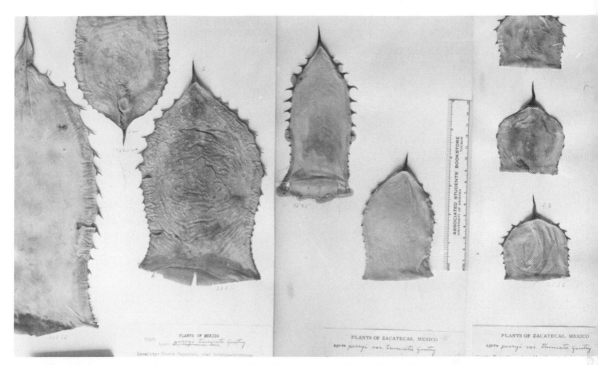

Fig. 19.16. *Agave parryi* var. *truncata* collected in natural habitat on Sierra Papanton, September 1963 and June 1951. All small leaves are from old stemmy plants at the type locality.

random elsewhere, as on the Sierra Ancha of Arizona (*Gentry 2231*) and again in central Chihuahua (*Gentry & Arguelles 22953*). Some of the latter are compact, attractive, pale, nearly white, glaucous plants.

Brietung described and illustrated a toothless form, *Agave parryi* f. *integrifolia* (Cact. & Succ. Journ. Amer. 35: 76, 1963) 10 miles (16 km) SW of Prescott, Arizona along hwy. 89; type *CAS*. This form was found growing in association with *A. parryi couesii*. The suppression of teeth appears to be a homologous character widely distributed in the genus *Agave*.

Agave parryi var. *truncata*
(Fig. 19.1, 19.13, 19.16, 19.17, 19.18; Table 19.1)

Agave parryi Engelm., var. *truncata*, var. nov.

Small, light gray, short-stemmed plants, copiously surculose with long rhizomes, infrequently flowering. Leaves very short and broad, 10–30 x 7–12 cm, short oblong, plane, the apex concave, short acuminate to truncate, margin repand to undulate; teeth variable, the larger 4–8 mm long, commonly flexed downward, dark brown, spine 2–2.5 cm long, openly grooved, sinuous, dark brown. Inflorescence as in *A. parryi;* tepals 2–2.5 times longer than the tube (10 mm deep).

Type: *Gentry & Gilly 10566*, Sierra Papanton, ca. 14 miles W of Sombrerete, near Zacatecas-Durango border, along highway 45; oak-juniper grassland, elev. ca. 7,500 feet, June 8, 1951, holotype *US;* isotypes DES, MICH, MEXU. Collected here also on same data are leaf variants Nos. *10568, 10570*.

Ab *A. parryi* var. *parryi* rosulis minoribus et foliis truncatis multo brevioribus differt.

This dimunitive variety is distinguished by its very small broad leaves with acute to truncate apex (whence the name). The more mammillate margined forms resemble the small-leaved *Agave guadalajarana,* but the latter has a more slender inflorescence with more remote, fewer, and smaller umbels. The wild plants appeared rather depauperate, the leaves of some only 7–15 cm long (Fig. 19.16). However, the leaves collected were from

Fig. 19.17. *Agave parryi truncata* flowering in Huntington Botanical Garden, July 1977, with numerous rhizomatous offsets. The mother plant is sitting upon a gravelly mound of rock, not long-stemmed as may appear. Heavily watered, the rosette is gigantic compared with the wild plants.

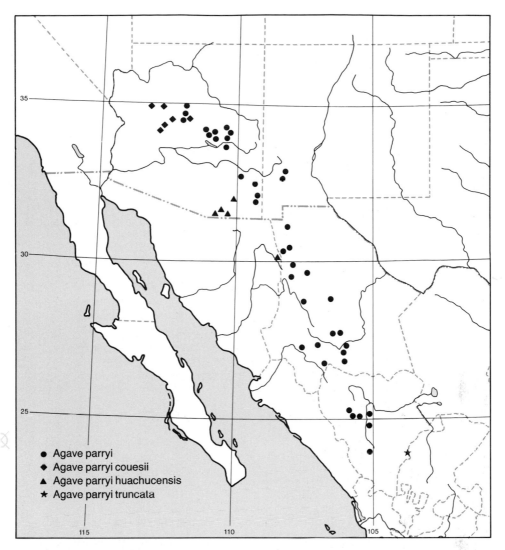

Fig. 19.18. Distribution map of *Agave parryi* and varieties. Based on herbarium specimens (see Exsiccatae) and observations.

old plants with dried frayed old leaf remains on the short stems (Fig. 19.16). The three leaf forms of the above cited numbers represent different clones growing together in a site with a radius of only 40–50 feet (12–15 m). Beginning in 1951 repeated visits were made to this locality for locating inflorescence, but only shoots decapitated by cattle were ever found. Several small populations are scattered on the western slope of Sierra Papanton, some with larger plants, all generally short-acuminate, but none as truncate as the original clone of *10570*. Other collectors have found this variety, and it may be offered in the succulent trade as *A. patonii*, as it has been called at Huntington Botanical Gardens, following my suggestion. It is a very handsome little agave well suited to pot culture in patios, window ledges, and outdoors in rock gardens. Keep it depauperate and call it a "bonsai"!

The flowers of var. *truncata* are known only from a relatively large specimen, which flowered in Huntington Botanical Gardens in June 1977. The garden tag was lost on this specimen, but Gentry's records show that plants of the above numbered series were sent to Huntington in 1951. The conditions at Huntington are ideal for many agaves. The rich soil in an open climate with a regular watering regime are reflected in the large vigorous plants found there.

Parryanae Exsiccatae

Agave gracilipes

CHIHUAHUA. *Gentry & Arguelles 22970,* DES. Ca. 32 miles S of Cd. Juarez and 1.8 mile SE of rt. 45 on road to Microhondas tower, limestone mt., elev. ca. 4,500 ft., small colony on E slope, June 15, 1971 (l, f, photo).

Gentry & Arguelles 22966, DES. 32 miles W of Sueco along road to Buenaventura, treeless grassland on rolling hills, elev. ca. 5,400 ft., June 14, 1971 (l, f).

NEW MEXICO. *Gentry & Engard 23054,* DES. 40 miles SW of Hwy. 285 along rt. 137 on Guadalupe Mountain, juniper-pinyon grassland, limestone, elev. ca. 5,750 ft., Sep. 25, 1972 (l, f, photos).

Gentry & Engard 23055, DES. 1 mile W of El Paso Gap, Guadalupe Mts., sotol-grassland, limestone, elev. ca. 5,350 ft., Sep. 25, 1972 (l, f).

Gentry & Hughes 23408, DES. Ca. 20 miles S of Pinyon, Otero Co., open, nearly bare, rocky limestone slope, elev. ca. 5,300 ft., June 27, 1974 (l, cap).

McKechnie s.n., DES. Near Queen, Eddy County, along rt. 137, Lincoln Nat. park in Guadalupe Mts., limestone rocky bajada slope in highland desert, Sep. 13, 1960 (l, f, photo).

Mulford 397a, MO. Humbolt Mt., July 1895.

Reid 1342, TEX (at El Paso). 0.7 km SE of Anthony's Nose, Dona Ana Co. 32°02'N, 106°32'W, 29 Oct. 1978 (l only; need infl. for certain iden.).

Gentry & Correll 20631, DES. 12 miles N of Allamore turn-off of rt. 80, Hudspeth Co., limestone hill by Keeling house, elev. ca. 5,500 ft., June 19, 1964 (l, f, photos).

Gentry & Engard 23060, DES. 5 miles E of Hueco along rt. 180, *Larrea-Yucca* grassland, limestone, elev. ca. 5,350 ft., Sep. 27, 1972 (l, f, photos).

Gentry & Engard 23061, DES. Hueco Hills, 5 miles E of Hueco along rt. 180, *Larrea-Yucca* grassland, limestone, elev. ca. 5,350 ft., Sep. 27, 1972 (l, f, photos).

Gentry & Engard 23057, DES. E bajada of Guadalupe Mts., 42–43 miles SW of Carlsbad along rt. 180, acacia grassland, elev. ca. 4,350 ft., Sep. 26, 1972 (l, f, cap).

Gentry & Engard 23058, DES. 20 miles E of Hueco along rt. 180, *Larrea* grassland, elev. ca. 4,650 ft., Sep. 27, 1972.

Gentry & Engard, DES. 5 miles E of Hueco along rt. 180, *Larrea-Yucca* grassland, limestone, elev. ca. 5,350 ft., Sep. 27, 1972 (l, cap, photo).

Gentry & Ogden 9901, DES. Eastern piedmont of Sierra Guadalupe, grama grassland, coarse calcareous soil over limestone, Oct. 25, 1950 (l, f, photos).

Havard s.n., GH. Guadalupe Mts., 1881 (l, cap).

Waterfall 5398, GH. 1 mile W of McAdoo ranchhouse, near head of Victoria Canyon, Sierra Diablo Plateau, Hudspeth Co., July 28, 1943 (l, f).

Mulford 293, 293a, MO. Sierra Blanca, 1895. Type.

Rose et al. 12224, US. Vicinity of Sierra Blanca, 24 Feb. 1910.

Agave gracilipes × lechuguilla

TEXAS. *Gentry 19868,* DES. Ca. 26 miles E of El Paso along rt. 180, by the last (E) big turn thru the Hueco Mts., rocky slope, March 21, 1963.

Gentry & Correll 20632, DES. 12 miles N of Allamore turn-off rt. 80, Hudspeth Co., limestone hill by Keeling house, elev. ca. 5,500 ft., June 19, 1964.

Agave guadalajarana

HORTICULTURE. *Berger s.n.,* US. 1904, 1912. Cult. La Mórtola (received from Max. Herb., Naples, 1904 as *A. megalacantha*).

JALISCO. *Gentry 23498,* US, INIF, MICH, DES. 13 miles N of junction along new road to Colotlan, ca. 20 miles N of Guadalajara, 27 Jan, 1975; oak grassland, elev. ca. 5,000 feet (l, cap, pl).

Pringle 4473, MO, K, US, MEXU. Type. On rocky summit of hills ca. 8 miles W of Guadalajara, 22 June 1893.

Rose & Hough 4804, US. Guadalajara, 7 July 1899.

NAYARIT. *Norris & Taranto 14380,* MICH. 4 miles E of La Cienaga ca. 5 miles NW of Mesa del Nayar, 29 July, 1970; pine and oak forest ridge (l bud; doubtfully assigned here).

Agave havardiana

CHIHUAHUA. *Gentry & Engard 23206,* DES. N base of Sierra Rica, "Rancho Consolación," on brushy limestone slope, elev. 4,200–4,500 ft., May 3, 1973 (photos).

Gentry & Engard 23082, US, DES, MEXU. 35 miles NE of Aldama along road to Coyame, elev. 4,850 ft., 2 Oct. 1972 (l, cap).

Gentry & Engard 23571, US, MEXU, DES. Ca. 3 miles E of Marrion along rt. 16 to Coyame, elev. ca. 4,500 feet, 26 July, 1975 (l, f).

Gentry & Engard 23572, US, MEXU, DES. 3 miles E of Los Huerfanos, off rt. 16, elev. ca. 4,675 feet, 26 July 1975 (l, f).

COAHUILA. *Gentry & Engard 23132,* DES. Los Cojos Minas, SW slope of Sierra del Carmen, oak woodland, limestone, elev. 6,000–6,500 ft., Oct. 12, 1972 (l, photos).

Henrickson 11394, DES. Ca. 22 air miles E of Boquillas, near Puerto Boquillas on road to El Jardin on rocky limestone alluvium, elev. 4,500 feet, 27 July 1972 (l, cap).

Henrickson 11459, DES. Ibid., elev. 5,500 feet (l, f).

Johnston & Muller 656, TEC. Sierra del Pino, ca. 4 miles NE of camp at La Noria, 23 Aug. 1940 (l, f, cap).

Nelson 6139, US. General Cepeda, 20 April 1902 (l, cap).

TEXAS. *Bailey 413,* US. Davis Mountains, 12 July 1901 (l, f, cap, photo).

Bailey 390, US. Chisos Mountains, 15 June 1901 (l, f, cap, photo).

Ferris & Duncan 3577, MO. Chisos Mts., July 15–18, 1921.

Gentry & Correll 20625A, DES. SE of the Basin, Chisos Mts., June 17, 1964 (f, photos).

Gentry & Correll 20629, DES. Arroyo Limpio ca. 5 miles NE of Fort Davis, oak-juniper grassland, rocky hill slope, June 18, 1964 (l, f, photos).

Gentry & Correll 20630, DES. Along Arroyo Limpio 3½ miles NE of Fort Davis, oak-juniper grassland, elev. ca. 5,000 ft., rocky mesa, June 18, 1964 (l, f, photo).

Gentry & Correll 20623, DES. 25–26 miles S of Fort Stockton, Madera Mts., limestone slope with juniper and shrub, elev. ca. 4,500 ft., June 15, 1964 (l, photos).

Gentry & Correll 20624, DES. Ca. 10 miles NE of Marathon, limestone hill in open grassland on upland plain, June 15, 1964 (l, f, photos).

Gentry & Correll 20625, DES. Mouth of Juniper Canyon, Chisos Mts., along arroyo in rocky outwash slope, June 16, 1964.

Gentry & Engard 23056, DES. E bajada of Guadalupe Mts., 42–43 miles SW of Carlsbad along rt. 180, *Acacia* grassland, elev. ca. 4,350 ft., Sep. 26, 1972 (l, f).

Havard s.n., MO. Type, Guadalupe Mts., Oct. 1881. (3 l, f, infruct. One f. o=25, t=15, s=14, f @ 10–11, tl=55 mm. 2nd. f. t=15, s=14, f @ 10, tl=64 mm. One leaf very large, but lanceolate rather than deltoid. One leaf is probably *A. neomexicana,* but the deep flower tube and the large broad leaf points it toward the rest of *havardiana* as described by Trelease and represented so abundantly in the Davis Mts., and other mountains in western Texas. In another type folder are several other leaves definitely belonging to *A. neomexicana.* Even though they were collected at the same locality and same date, they should be excluded from the type folder.)

Havard, s.n., US. Chisos Mountains, Aug. 1883 (f).

Havard s.n., MO. Chinacate Mts. in 1880.

McKelvey 1957, A. Limpia Creek (near Fort Davis) 23 April 1931.

Mueller 7959, US. Chisos Mts., 20 June 1931 (l, f).

Mueller s.n., US. Boot Springs, Chisos Mts., 5 July 1932 (l, f).

Sperry T729, T730, US. 17 & 19 miles from Fort Davis on Kent Road, Davis Mts., Jeff Davis Co., 12 June 1939 (l, f, photo).

Sperry 1618, US. Green Gulch, Chisos Mts., Brewster Co., 18 June 1939 (l, f, photo).

Sperry 1619, 1620, US. Basin toward saddle, Chisos Mts., Brewster Co., 18 June 1939 (l, f, photo).

Warnock Lot 9, US. Glass Mts., Victor Pierce Ranch, Brewster Co., 7 Oct. 1950.

Warnock Lot 49, US. Green Gulch, Chisos Mts., Brewster Co., 20 Nov. 1950; "single acaulescent plants."

Wright 1906, GH. Wild Rose Pass (Davis Mts.), 12 June 1851; "common on mountain side, Field No. 426" (f).

Agave neomexicana

COAHUILA. *Palmer 1310,* US. 12–14 leagues S of Saltillo in 1880 (l, f, cap). Rather doubtfully assigned here, but it agrees generally with this species.

NEW MEXICO. *Barclay & Thompson 1031,* DES. 5 miles W of junction of State Highways 83 & 24, Chaves Co., dry, rocky caliche soil with *Yuccas* and chollas, June 27, 1961 (l, f).

Gant s.n., US. San Andres Mountains, Dec. 1902.

Gentry 22304, DES. Dripping Springs, Organ Mts., 13–14 miles E of Las Cruces, volcanic rocky slopes, elev. 5,800–6,000 ft., June 18, 1967 (l, f, photos).

Gentry & Hughes 23406, DES. 1–2 miles SW of Guadalupe Administration Center, Guadalupe Mountain, pinyon and juniper mesa, limestone, elev. ca. 6,000 ft., June 27, 1974 (l, f).

Gentry & Hughes 23407, DES. 3 miles SW of Queens, Guadalupe Mountain, dry rocky limestone S slope in pinyon-juniper, elev. ca. 6,250 ft., June 27, 1974 (l, f).

Gentry & Hughes 23409, DES. 6 miles E of Mayhill along rt. 82, pinyon-juniper, S exposure, limestone, elev. ca. 6,100 ft., June 28, 1974 (l, f, cap).

Gentry & Ogden 9908, DES. Mountain about 4 miles E of Mayhill, Sacramento Mts., open rocky limestone slope, elev. ca. 7,000 ft., Oct. 26, 1950 (l, photos).

Rehn & Viereck s.n., US. Alamogordo, Otero Co., 6 June 1902 (l, f, cap).

Reid 1332, DES, TEX-P. 3 miles N of Cooks Peak, Luna Co., 32°33′N, 107°43′W, 8 Oct. 1978, elev. 1,830 m.

Standley 541, US. Type. Organ Mts., Dona Ana Co., June 1906; elev. ca. 5,600 ft. (l, f).

Standley 643, US. On and near Tortugas Mountains, Dona Ana Co., 30 June 1911; alt. 1,270–1,400 feet.

Wooten s.n., US. Organ Mts., 28 May 1912 (l, cap).

Wooten s.n., US. San Andres Mts., 6 June 1914 (l, f).

Wooten s.n., US. Jornado Range Reserve, June 1914 (photo).

Worthington 2865, DES, TEX-P. E side top of Baldy Peak, Florida Mts., Luna Co., 32°05'20"N, 107°37'15"W, 27 May 1978; elev. 6,700 ft., limestone, relatively flat area on top (l, cap).

TEXAS. *Crawford s.n.,* TEX-P. Franklin Mts., S of South Mt. at the 6,053 ft. peak, 6 April 1976; alt. 6,050 ft. (l, cap).

Gentry & Hughes 23405, DES. McKittrick Canyon, Guadalupe Mts., 25 June 1974; elev. ca. 5,200 ft., rocky limestone slope with Juniper and oak (l, f, photo).

Gentry & Ogden 9905, DES, US. Eastern piedmont of Sierra Guadalupe, 25 Oct. 1950; grama grassland, coarse calcareous soil over limestone (l, cap, photo).

Hinckley 4506, US. Mouth of McKittrick Canyon, Culberson Co., 12 July (l, f).

Worthington s.n., June 18, July 16, 31, DES, TEX-P. Franklin Mts., N of El Paso, elev. 6,000–7,000 ft., igneous & limestone rocky slopes (l, f, cap).

Agave parrasana

COAHUILA. Berger s.n., US. K. Type. Cult. L. Mórtola, VIII, 1912, "Blatter von importtierten pflanzen!'' (f, 2 l with old rotten bases & 2 younger l).

Gentry 11546, DES. Sierra de Parras, northern limestone slope with rocky rubble and low bushy dispersed cover, elev. 4,000–4,500 ft., Feb. 1–3, 1952.

Gentry s.n., DES. Hunt. Bot. Gard., San Marino, Calif., cultivated—may be from col. of *Gentry 11546* from Sierra de Parras, Feb. 1–3, 1952, May 1962 (l, f, photos).

Gentry & Engard 23250, DES. Canyon de la Hacienda, Sierra de los Maderos, limestone, forested canyon, elev. 6,000–6,500 ft., May 10, 1973 (l, photo).

Gentry & Engard 23245, DES. N end of Sierra San Marcos, S of Cuatro Cienegas, elev. ca. 6,500 ft., May 9, 1973 (l, buds, photos).

Gentry & Engard 23267, DES. Top ridge of Sierra de Parras, elev. ca. 8,000 ft., May 14, 1973 (l, f).

Gentry & Engard 23100, DES. Sierra de Parras, 3 miles SW of Parras, brushy N slope with limestone rocks, elev. 6,000–6,500 ft., Oct. 7, 1962.

Henrickson 11839, DES. Canyon de la Gavia, Sierra de la Gavia, ca. 35 miles S of Monclova; limestone, with oak, *Prunus, Ceanothus,* etc., alt. 5,500 feet, 3 Aug. 1973.

Minckley s.n., DES. Top of Sierra de San Marcos, NE-facing slope, Aug. 20, 1968.

Pinkava et al. 5980, DES. Sierra de San Marcos, opposite Los Fresnos, NE-facing slopes, April 4–5, 1969 (l, photo).

Purpus s.n., MO. Near Parras in 1905 (l, typically small truncate leaf with large teeth). One sheet annotated by Trelease as *A. wislizeni.* All in type folder. Appears to be of same origin as Berger's type material.

Agave parryi parryi

ARIZONA, *Allen 939,* UC. 30 miles N of Globe, Gila Co., June 24, 1937 (l, f).

Barr 65-234, CAS. Summit above Rustler Park, Chiricahua Mts., elev. 9,200 ft., Cochise Co., July 14, 1962 (l, f).

Clark 8154, AZ. Mesa at Balanced Rock, Chiricahua Nat. Mon., June 19, 1939 (l, f).

Eastwood 17597, 17560, CAS. Sierra Ancha, May 1929.

Eastwood 17528, CAS. Oxbow Hill, Apache Trail and vicinity, May 27, 1929 (X *chrysantha,* l).

Gentry 4518, DS, UC, CAS, DES. Along Tonto Rim from Pleasant Valley to Prescott Mts., elev. 3,500–6,500 ft., June 1939 (l, f, br).

Gentry & Weber 22310, DES, US, MEXU. Ca. 1 mile N of Juniper Flat, Sierra Ancha, Gila Co., June 20, 1967 (l, f, photo).

Gentry & Weber 22311, DES, US, MEXU. On ridge above, 1 mile W of Parker Creek Exp. Station, Sierra Ancha, June 20, 1967 (l, f, photos).

Gentry & Ogden 9987, DES, US, MEXU. Ca. 5 miles S of Seneca along rt. 60, Nov. 25, 1950 (l, photos).

Harrison, Kearney, Fulton 5974, 5975, ARIZ. Near summit of Sierra Ancha highway, June 23, 1930 (l, f).

Jones 5M-11-26, DS. Pine, VI-2-1890 (f).

Kearney & Peebles 12053, ARIZ. 33 miles NE of Globe, Gila Co., elev. 5,500 ft., July 21, 1935 (f, cap).

Kusche s.n., CAS. Cave Creek Canyon, Chiricahua Mts., elev. 6,000–8,000 ft., July 29–Aug. 9, 1927 (f).

Lemmon s.n., UC. Santa Rita Mts.

McKelvey 939, 1074, 1114, 1115, 1116, A, 1320, CAS, A. Sierra Ancha along road to Pleasant Valley, May & June 1929 (l, f).

McKelvey 1169, A. Juniper Flat, Sierra Ancha, May 1929.

McKelvey 1316, 1317, A, CAS. Sycamore Creek near Payson, June 12, 1929 (f & l like *A. couesii*).

McKelvey 1154, 1156, A. Osborn Hill near Payson, May 27, 1929.

McKelvey 1318, A. Between Payson & Pine, May 13, 1929.

McKelvey 985, A. Road to Amethyst Mine in Mazatzal Range (as *A. parryi* X *chrysantha*).

McKelvey 1327, A. Pinal Range on road from Globe to Winkelman, June 19, 1929 (l, f).

McKelvey 2073, A. Santa Rita Mts.

Rehder 1. Oak Creek Canyon near Flagstaff, July 13, 1914 (l, f).

Rothrock in 1874, MO. Type: Rocky Canyon (Prob. Graham Co.).

Thornber & Shreve 7930, ARIZ. Graham Mts., near sawmill, Sep. 7–13, 1914.

Wetherill 1046, 4073, ARIZ. Oak Creek Canyon 11½ miles S of Sedona, June 11, 1940 (cap).

Wolf 2459, DS, CAS. 3 miles NW of Pine, Gila Co., elev. 6,000 ft., July 1928.

CHIHUAHUA. *Bye 7634,* COLO, DES. W of Parral on road to Balleza in small llano with juniper-piñon and *Opuntia,* 22 July 1977; elev. ca. 1,980 m.

Engard & Getz 251A, DES. 9 miles S of San Lorenzo along the Rio Santa Clara, rocky *Acacia-Fouquieria* grassland slope above the flood plain of Rio Santa Clara, July 5, 1974 (l, photo).

Engard & Getz 236, DES. 51 miles S of Casas Grandes between Ascension and Janos on Mexico Hwy 2 near KM post 212, lower bajada of west facing slope on Hwy margin, July 4, 1974 (l, photo).

Gentry 18252, DES. Las Tapias, Sierra Campana, oak woodland, elev. 5,000–5,500 ft., Nov. 25, 1959 (l, cap).

Gentry & Arguelles 22946, DES. 5 miles SW of Casas Grandes along road to Col. Juarez, elev. ca. 5,000 ft., on rocky slope in open grassland, June 7, 1971 (l, photos).

Gentry & Arguelles 22962, 22964, DES, ARIZ, MEXU, US. 17 miles W of rt. 45 along road to Cumbres de Majalca, elev. ca. 6,200 ft., sunny canyon slope, volcanic, June 14, 1971 (l, f, photos).

Gentry & Arguelles 22953, DES. 7 miles SW of Buenaventura along new road to Ignacio Zaragosa, elev. 6,400 ft., rocky volcanic slope with cholla, June 9, 1971 (l, f, photos).

Gentry & Arguelles 22957, DES. 2 miles N of Peña Blanca, elev. ca. 7,000 ft., talus slopes of volcanic cliffs, June 10, 1971 (l, f, photo).

Gentry & Arguelles 22950, DES. 11 miles SW of Col. Juarez, elev. ca. 5,500 ft., rocky mountain grass slope, June 7, 1971 (l, f, photos).

Gentry & Bye 23350, DES. 1 mile S of Creel, elev. ca. 8,000 ft., in soil-filled crevices of tufaceous rim rock, south exposure, Oct. 6, 1973 (l, photos).

Gentry 22145, DES. 20–25 miles W of Parral along road to Huachochic, elev. 6,000–6,500 ft., open hill slopes with grass and second growth oak, Oct. 30, 1966 (photos).

Gentry, Correll, Arguelles 18019, DES. 15 miles NE of El Vergel along road to Parral, elev. 1,950–2,200 m, oak woodland and grama grassland, Oct. 21, 1959 (l, photos).

LeSueur 245, UC. Majalca, Aug. 20, 1935 (l, f).

Palmer 138, MO, US. Vicinity of Santa Eulalia, 28–29 April 1908; alt. ca. 1,300 m (l, f, 3 sheets).

Pringle s.n., US. Santa Eulalia Mts., 5 June 1885 (l, f).

Pringle 958, MEXU, MO. Hills near Cd. Chihuahua, 8 Sep. 1886 (l, f, cap. type of *A. chihuahuana* Trel.).

Rose 11654, US. Vicinity of Cusihuiriachic, 2–3 April 1908.

Rose 11671, US. Santa Eulalia Mts., 4 April 1908 (ann. by Trelease as *A. chihuahuana*).

Townsend & Barber 73, GH, MO, US. Near Colonia Garcia, Sierra Madre, 28 June 1899; elev. 7,400 ft. (l, f, short-leaved form).

Wooton s.n., MO. 5 miles SE of Col. Garcia, 23 June 1899, rocky places (l, f, —more likely from T. & B.).

DURANGO. *Gentry & Gilly 10541,* DES. Rio Chico ca. 20 miles W of Cd. Durango along hwy. to El Salto, June 1951 (transplant fl'ed. at Murrieta, Cal., May 1972) (l, f, cap), Grama grassland, elev. 6,500–7,000 ft.

Gentry & Gilly 10553, 10551, 10554, US, MEXU, DES. 6 miles S of Chincacates along road to Patos and Guatinape, near railroad, coarse gravelly soil along arroyo, grama grassland, elev. 6,500 ft., June 4, 1951 (topotypic).

Gomez-Pompa s.n., DES. Near Santiago Papasquiaro, Nov. 1961 (photo).

Gonzales Medrana—Quero 1381, MEXU. 35 km. W of Cd. Durango on road to Mazatlan, km. 995, elev. 2,250 m, Sep. 22, 1966.

Palmer 228, GH, MEXU, UC, MO. Tobar, May 28–31, 1906 (l, f).

Paton 158/11/3, MO. Chinacates (along Tepehuanes RR) April 1911 (l —Type of *A. patonii* Trel.).

GUANAJUATO. *Gentry, Barclay, Arguelles 20086,* DES. 6–7 miles N of rt. 110 along rt. 57, open gravelly desert plain, elev. 6,300 ft., June 27, 1963 (l, photo).

NEW MEXICO. *Gentry 22305,* DES. Ca. 3 miles N of Pinos Altos, Grant Co., elev. 6,500 ft., tuffaceous volcanic rocks with pine and juniper, June 18, 1967 (l, f, photos).

Gentry 22306, DES. Wild Horse Mesa, ca. 16 miles N of Pinos Altos along road to Gila Cliff Dwellings, Grant Co., elev. 6,750 ft., open volcanic slopes with pine and juniper, June 18, 1967 (l, f, photos).

Greene s.n., Near Silver City, June 16, 1880 (l, f).

Greene s.n., Near Santa Rita del Cobre, June 30, 1877 (f).

McKelvey 4017, 4019, 4020, A. Pinos Altos Mts. (l, cap, br).

McKelvey 4023, A. Silver City.

McKelvey 4031, A. Tyrone (l, cap).

McKelvey & Kellogg 4949, A. Silver city, June 3, 1934 (f, br).

Metcalfe 262, UC, K, GH. Mogollon Creek, Mogollon Mts., Socorro Co., elev. ca. 8,000 ft., July 18, 1903 (l, f).

Rusby 411, 10186, UC, MO. Bear Mts., May 1881 (f).

Wolf 2608, DS. Bear Mt., 5 miles from Silver City, Grant Co., July 12, 1928.

Agave parryi couesii

ARIZONA. *Earle s.n.,* DES. 10 miles E of Oak Creek Canyon, Oct. 12, 1969 (cap).

Eastwood s.n., CAS. Old wood road, Granite Mt., near Prescott, June 2, 1929.

Eastwood 17717, CAS. Near Prescott, June 3, 1929 (l, bud).

Eastwood 16817, CAS. On road to Phoenix from Prescott, Nov. 11, 1928 (l, old f).

Eastwood 17663, CAS. Senator Mine near Prescott, June 2, 1929.

Fulton 9687 ?, ARIZ. Upper Oak Creek, Coconino Co., June 18, 1934 (f).

Fulton 5935. 5 miles W of Altman P.O., Yavapai Co., June 2, 1929.

Gentry 23448, DES. Ca. 3 miles SE of Stoneman Lake, Coconino Co., pinyon-juniper, clay soil from volcanic rocks, elev. 6,000–6,500 ft., Oct. 13, 1974 (l, cap).

Gentry 23012, DES. 1 mile above Stoneman Lake Rd. along hwy. 79, Yavapai Co., S slope above arroyo, juniper and sparse chaparral, elev. ca. 5,800 ft., June 15, 1972 (l, f).

Gentry 22575, DES. E city limit of Jerome, Yavapai Co., limestone grassy open hill slopes, elev. 4,300 ft., June 14, 1968 (l, f, photos, cap).

Gentry 22313, DES. Ca. 20 miles S of Ash Fork along road to Prescott, open limestone slope, elev. ca. 4,600 ft., July 2, 1967 (l, f, photos).

Gentry 22576, DES. 11 miles W of Jerome along road to Jerome, pinyon-juniper slope on igneous rock, elev. 5,700 ft., June 14, 1968 (l, f of one pl).

Gentry 22577, DES. 11 miles SW of Prescott along road to Congress, brushy granitic slope, elev. 5,400 ft., June 15, 1968 (l, f, photos).

Gentry & Ogden 9972, DES. 11 miles S of Ashfork, juniper-grassland, Nov. 16, 1950 (l, photos).

Kearney & Peebles 13976, ARIZ. 5 miles SW of Prescott, elev. 5,800 ft., July 16, 1938 (l, f).

McDougal s.n., US. Flagstaff, July 1891 (l, f).

McKelvey 1228, 1227A, 1238, 1238A, 1238B, A. Region of Prescott, April-June, 1929.

McKelvey 4077, A. Walnut Creek, Yavapai Co., April 26, 1934.

McKelvey 4079, A. Juniper Mts., (l, cap).

McKelvey 1227, 1229, 1230, A. Senator Highway near Prescott, June 2, 1929.

Peebles 13882, US. Near Prescott, Yavapai Co., elev. 5,800 ft., June 10, 1938.

Peebles, Harrison, Kearney 4247, US. Between Congress Junction and Prescott, July 17, 1927 (f, cap).

Agave parryi huachucensis

ARIZONA. *Benson 10243,* ARIZ. Canelo Hills, Santa Cruz Co., oak woodland, elev. 5,000 ft., April 15, 1940 (l, cap).

Gentry 22580, DES. Miller Canyon, Huachuca Mts., rocky forested sun slope, elev. 6,500 ft., June 16, 1968 (l, f, photos).

Gentry 22581, DES. 8 miles SE of Sonoita on road to Canelo, grassland, elev. 5,000 ft., June 17, 1968 (l, f, photos).

Gentry 22582, DES. Ca. 6 miles SE of Sonoita along road to Canelo, grama grassland, elev. ca. 5,000 ft., June 17, 1968 (l, f, photos).

Gentry & Ogden 9920, DES. Ca. 3 miles W of Sonoita, low hill in grama grassland, Nov. 5, 1950 (l, photos).

Haskell s.n., SE of Sonoita, Santa Cruz Co., June 20, 1942 (f).

Lemmon s.n., UC. Huachuca Mts., Sep. 1882.

McKelvey 532, A, CAS. Near Patagonia on road to Nogales, June 18, 1929 (l, f).

McKelvey 1326, A. 5 miles S of Sonoita near Nogales, June 18, 1929 (l, infl).

Pringle s.n., K. Type. Huachuca Mountains, Arizona, June 30, 1884, alt. 5,000–8,000 feet.

Toumey s.n., ARIZ. Huachuca Mts., July 19, 1894 (l, f).

CALIFORNIA. *Gentry 19829,* DES. Hunt. Bot. Gard., San Marino, cultivated, July 6, 1962 (l, f, photo).

SONORA. *Gentry & Arguelles 16642,* DES. Sierra de Huachinera, pines, elev. 7,500 ft. (live pl, photo).

Agave parryi forma *integrifolia*

ARIZONA. Breitung 18157, CAS. Type. 10 miles SW of Prescott, Coconino Co., along Hwy. 89, elev. 5,000 ft., 1956 and herbarium leaf in 1959.

Agave parryi truncata

DURANGO-ZACATECAS. Gentry & Gilly 10566, Type **US,** dups. DES, MEXU, MICH. Sierra Papanton, ca. 14 miles W of Sombrerete (near Durango-Zacatecas border), 8 June 1951; elev ca. 7,500 ft., open rocky slope of S exposure in oak-juniper grassland.

Gentry & Gilly 10568, 10570, DES, MEXU, MICH, US. Ibid., other clones, all within 50 ft. radius.

Gentry et al., probably *10566,* DES, HNT, US. Cult. Hunt. Bot. Gard. (garden tag lost) fl'ed. July 1977; see Fig. 19.17.

Gentry et al. 20466, DES, MEXU, US. Sierra Papanton, near Durango-Zacatecas border, 23 Sep. 1963; open rocky limestone slope along route 45.

20.

Group Rigidae

Berger, Agaven 1915: 226 (in part).

Plants small to large multiannuals, surculose, with short to elongate stems; leaves ensiform, firmly fleshed, strongly fibrous, usually patulous, regularly armed with small to medium size teeth on nearly straight margins, the terminal spines small to medium, variable, conic to subulate, the upper surface openly grooved to merely flattened; inflorescence an open panicle with relatively few spreading umbellate branches, frequently bulbiferous; flowers protandrous, small to medium, greenish to yellow; tepals generally ca. twice as long as the tube, rapidly conduplicating, reflexing sharply along tube with drying; filaments inserted ca. in mid-tube; seeding capsules generally present, broadly ovoid, sometimes absent. Wide ranging in Mexico and Central America, absent in southwestern U.S.

Typical species: *Agave angustifolia* Haw. (Fig. 20.6).

This is the sword-leaved group of agaves recognizable by their narrow, mostly rigid, outstanding leaves deployed in a radiate spiral. They are further characterized by the rather small, open, small-bracteate panicles with relatively few branches; greenish to pale yellow, weak flowers with quickly wilting tepals that reflex with drying, and the broad ovoid capsules with large seeds. Although there is a general homogeneity of flower part proportions (Figs. 20.2, 20.3), there are puzzling variations. Some of the variations are not consistent between species, and considerable interspecific variation occurs. When such variations are not correlated with other characters in leaf and inflorescence, there is little support for specific separations.

Some of the species are distinct, easily characterized and recognized, as *A. aktites, A. karwinskii, A. decipiens, A. macroacantha,* and *A. datylio,* while others are a polyglot of varying forms, as, for instance, the extensive *A. angustifolia* complex. I attempted to separate from this the common extensive Pacific complex, reported earlier as *A. pacifica* Trel. (Gentry, 1972). However, so blending and extensive are the variants in both the Atlantic and Pacific branches of this complex, that I am unable to separate them even as subspecies. Other groups that are difficult to separate and define as species are the cultivated groups (fiber and beverage), most of which are clones and some are sterile hybrids, as *A. fourcroydes.* Flowering specimens are very difficult to obtain in these groups, as agricultural practices do not permit normal flowering. Some of the cases in the following account are maintained as species more for taxonomic convenience than founded on basic morphological and biological knowledge. By eliminating many unseparable specific names here-to-fore recognized, I hope the present treatment will permit other botanists to identify the species and varieties accounted for below. This has hardly been possible before.

There are several older names, as *A. rigida* and *A. vera-cruz* of Miller, which lack types or any historically founded specimens, variously interpreted by botanists for the last 125 years, and it appears impossible definitely to connect these vaguely described names with any living forms. For instance, a specimen in US herbarium annotated as "Ex Herb. Hort Reg. Kew 6375, cult. Kew," with no date, was annotated as "*A. rigida* Mill." It consists of leaf segment and flowers but appears to belong rather with *A. lurida* Aiton. Trelease reviewed the problems with older names in detail (1908) but did not succeed in re-establishing the identities of these names, nor did he cite any specimens to represent

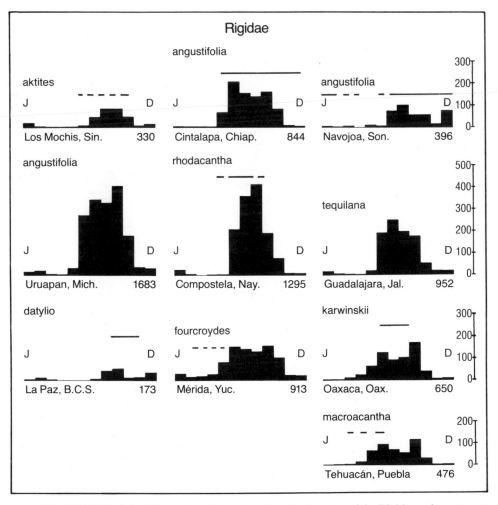

Fig. 20.1. Rainfall (silhouttes) and flowering (bars) perimeters of the Rigidae; relevant meteorological stations with average annual rainfall in millimeters. Data from Atlas Climatológico de México, 1939. Broken bars indicate uncertainty in flowering.

them. Because of their obscurity, I am not employing such names. Several cases of uncertain status are annotated at the end of this section under Nomina Incertae Sedis.

Several taxa of the Rigidae are of economic importance for their production of excellent fibers and liquors. Brief accounts of these productions will be found under the accounts of species, especially the two principal ones, *A. fourcroydes* and *A. tequilana*. The group is tropical, as the distribution maps show (Figs. 20.12 and 20.20), nourished by the convective storms of summer and fall (Fig. 20.1).

Key to Species of the Rigidae

1. Small plants, the leaves generally less than 60 cm long   2
1. Plants larger, the leaves generally more than 70 cm long, and if smaller, then not bluish gray   7
2. Leaves 1–2 cm wide, thin, recurving, with cartilaginous margins
    stringens, p. 582
2. Leaves 2–5 cm wide, not thin and recurving, without cartilaginous margins   3
3. Leaves bluish or gray glaucous, 25–60 cm long, except for shorter leaves of *A. breedlovei*; flowers 50–70 mm long   4
3. Leaves green to yellowish green, 50–80 cm long; flowers 40–60 mm long   5
4. Leaves 25–35 cm long; flowers 50–55 mm long, tepals about equaling tube
    macroacantha, p. 579
4. Leaves 40–60 cm long; flowers 60–70 mm long, tepals much longer than tube
    aktites, p. 556
5. Teeth, except near the base, coarsely deltoid, blunt, 3–5 mm long, 2–4 cm apart; flowers 40–50 mm long. Baja California   *datylio*, p. 571
5. Teeth slender, sharp, upcurving, 1–3 mm long, 1–1.5 cm apart; flowers 50–60 mm long. Central America   6
6. Leaves linear, 60–70 cm long; spine to 1 cm long   *panamana*, p. 580
6. Leaves linear lanceolate, 30–50 cm long; spine 2–3 cm long
    breedlovei, p. 567
7. Plants arborescent, the trunks 1.5 m or more tall   8
7. Stems generally short, less than 1 m tall   10

8. Trunks very thick with enlarged leaf bases, 15–25 cm, up to 1.5–1.6 m tall; flowers 70–80 mm long, the ovaries large and thick, 40–48 x 12–14 mm   *decipiens*, p. 573
8. Trunks thinner, sometimes to 2 m or more tall; flowers 45–70 mm long; ovaries smaller, 20–40 mm long   9
9. Leaves short, 45–65 cm long; flowers small, 45–57 mm long; ovaries 20–30 mm long   *karwinskii*, p. 577
9. Leaves elongate, 1.3–2 m long; flowers larger, 60–70 mm long; ovaries 35–40 mm long   *fourcroydes*, p. 573
10. Mature plants with leaves generally 1.5–2.5 m long   11
10. Mature plants with leaves generally 0.7–1.5 m long   12
11. Leaves rigid, unreflexed, to 2–2.6 m long; teeth large, 5–6 mm long through mid-blade, closely spaced, 1–3 cm apart; flowers with small tubes, 10–11 mm broad. Wild
    rhodacantha, p. 580
11. Leaves reflexed, 1.5–2 m long with small teeth 2–4 mm long, remote; flowers larger with deep tubes, 14–16 mm broad, cult.   *cantala*, p. 568
12. Leaves at maturity mostly 8–12 cm wide, light green to white, or bluish gray glaucous; inflorescence (so far as known) large, diffuse, with 20–25 densely flowered umbels. Mostly cultivates   *tequilana*, p. 582
12. Leaves generally smaller, narrower than above, white glaucous to green; inflorescence sparsely flowered, generally with 10–18 small umbels. Wild and cultivated   *angustifolia*, p. 559

Table 20.1. Flower Measurements in the Rigidae (in mm)

Taxon & Locality	Ovary	Tube	Tepal	Filament Insertion & Length		Anther	Total Length	Coll. No.
aktites								
Yavaros, Son.	26	15×15	24×5&22	9&8	34	25	65†	22015
	31	15×12	25×6&24	9–10	45	29	70†	22015
angustifolia								
Huajuapán, Oax.	31	12×10	27×3&26	7–6	44	26	60†	20261
Oaxaca, Oax.	25	10×10	21×3	6	45	25	56†	12081
Candelario, Oax.	27	15×10	22×2&21	10–11	48	26	63†	22380
Taxco, Mor.	24	13×12	23×4&21	9&8	34	28	60†	5865
Tehuantepec, Oax.	20	10×10	22×4	6–7	22	20	52†	12235
Perspire, Hond.	28	8×10	20×4	6	35	19	57†	12519
Pers.-Savanna, Hond.	25	15×12	23×3&21	9–11	35	31	63†	21464
Huigalpa, Hond.	26	16×12	19×3&17	12	45	25	61†	21578
Managua, Nicarag.	22	12×11	23×3&21	9–10	35	26	57†	21577
Puente Nacional, Ver.	27	15×9	19×4&18	8–9	38	26	61†	15879
(Pacific)								
NW Son.	26	8×12	18×4	4–5	45		52†	19885
Navojoa, Son.	30	11×14	19×5&18	5–6	40	24	61†	19875
S Tecurahui, Son.	25	11×10	19×4&17	6	35	24	57†	19359
	28	14×13	24×5&22	10	45	30	67†	19368
Son.-Sin. border	20	9×11	22×4	5–6	37	28	52†	19656
Fuerte, Sin.	29	11×10	22×4&20	8	42	25	63†	19655
Topolobampo, Sin.	27	11×11	19×4	6	45	26	57†	11425
Durango, Dur.	26	10×12	17×3&15	5–6	38	23	54†	22091
a. var. *rubescens*								
Cerro Colorado, Sin.	20	9×9	19×3&18	7&5	45	18	48*	5215
cantala								
Hunt. Bot. Gard., Cal.	32	15×15	28×4&27	11–9	50	32	74†	23676
	41	17×15	28×4&26	13–11	54	34	85†	23676
c. var. *acuispina*								
Coyoles, Hond.	24	14×12	21×4	9	48	23	59†	20694
Tulanga, Hond.	25	16×15	20×4&20	11&13	55	26	60†	21467
	27	16×14	20×3&19	12–13	60	25	63†	21467
datylio								
La Paz, B.C.S.	20	7×8	14×3.5	5	40	18	41*	11200
	19	7×8	14×3	5	40	16	39*	11200
d. var. *vexans*								
Comondú, B.C.S.	26	4×8	21×3	3	40	23	51*	4322
	29	4.5×8	18	3–4	36	14	52*	4322
decipiens								
Hunt. Bot. Gard., Cal.	47	12×17	18×5	6&5	44	24	78†	19749
	44	12×18	22×5	9&6	48	25	79†	19749
fourcroydes								
Hunt. Bot. Gard., Cal.	34	12×13	18×4&16	6–7	44	22	64†	23680
	39	12×14	18×5&17	7–8	47	24	69†	23680
	30	10×12	19×5&18	6–5	42	22	59†	4008

Table 20.1. cont.

Taxon & Locality	Ovary	Tube	Tepal	Filament Insertion & Length		Total Anther Length*	Coll. No.
karwinskii							
Mitla, Oax.	20	11 × 8	15 × 4	5–6	30	22	46†	12049
Zapotitlán, Pue.	27	11 × 11	17 × 3 & 15	7 & 5	40	21	55†	20357
macrocantha							
Zapotoande, Pue.	25	14 × 9	16 × 3.5	7–8	35	21	54†	DBG
	25	14 × 10	15 × 2.5	7–8	40	20	54†	DBG
rhodacantha							
Compostela to Mazatán, Nay.	28	8 × 10	22 × 4	5	46	22	56	10704
	31	8 × 10	23 × 4	6	44	28	62	10806
	25	10 × 9	16 × 4	6–7	40	13	54*	10783
tequilana							
Hunt. Bot. Gard., Cal.	32	10 × 13	26 × 4	6	45	25	68†	19827
	34	10 × 12	28 × 4	7 & 5	50	26	72†	19827

* Measurements from dried flowers relaxed by boiling.

† Measurements from fresh or pickled flowers.

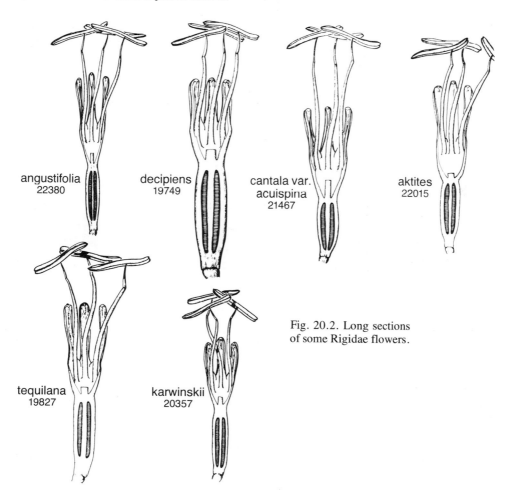

angustifolia
22380

decipiens
19749

cantala var.
acuispina
21467

aktites
22015

tequilana
19827

karwinskii
20357

Fig. 20.2. Long sections of some Rigidae flowers.

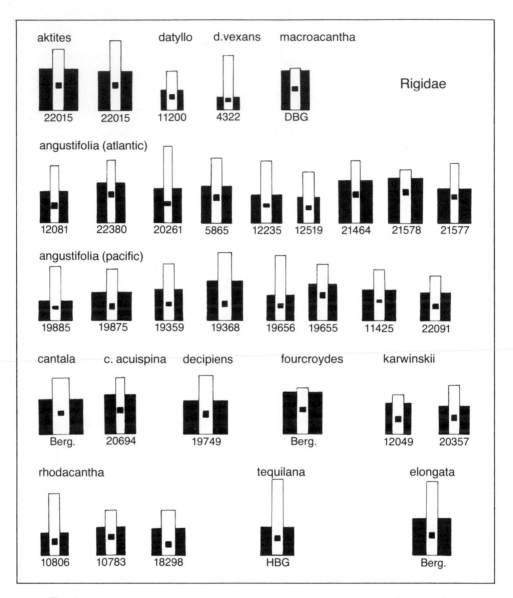

Fig. 20.3. Ideographs of Rigidae flowers showing relative proportions of tubes (black) to outer tepals (white columns) and filament insertion in tubes. Specific series show intraspecific variation.

Agave aktites
(Figs. 20.1, 20.2, 20.3, 20.4, 20.5, 20.20; Table 20.1)

Agave aktites Gentry, Agave Family in Sonora, U.S. Dept. Agri. Handb. 399: 148, 1972.

Small glaucous, surculose rosettes with broadly globose stems, 4–7 dm tall, 6–11 dm wide; leaves 40–60 x 2–4 cm, linear, straight, patulous, unequal in the rosette, nearly flat above, convex below, broadly clasping at base, bluish glaucous gray, sometimes cross-zoned, smooth or asperous; spine short, 12–20 mm long, usually broad at base and

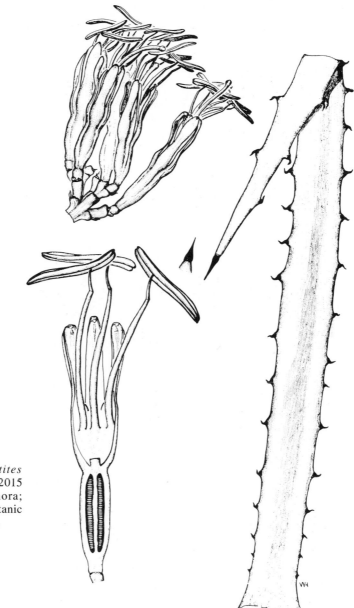

Fig. 20.4. *Agave aktites* drawn from Gentry 22015 from Yavaros, Sonora; flowered in Desert Botanic Garden in 1974 and 1976.

flattened above, abruptly subulate, dark brown to grayish; teeth 3–5 mm long, generally upcurved with slender flexuous tips, irregularly spaced, 1–3 cm or 4–5 cm apart; panicle 3–4 m tall, narrow, with 10–15 short branches in upper third or quarter of shaft, the flowering umbels small; flowers pale greenish, 64–70 mm long, not opening well, the tepals adhering, quickly wilting; ovary 26–31 mm long, angulate-cylindric, neckless, tapering to pedicel; tube 14–16 mm deep, 12–16 mm broad, cylindric to globose, narrowly grooved, thick-walled below filaments; tepals unequal, 21–25 mm long, long linear, soon infolding becoming spatulate with broadly rounded hooded tips, the outer larger, closely over-lapping inner, which is non-costate within; filaments 40–50 mm long,

Fig. 20.5. *Agave aktites* in native habitat on sandy island of Lechuguilla, Sinaloa.
Part of the leaves removed for analysis.

inserted at wide angle 8–10 mm above base of tube; anthers 23–29 mm long, slender,
extremely excentric; pistil over-reaching stamens in post-anthesis.

Type: *Gentry 11470*, Isla Lechuguilla, north of Topolobampo, Sinaloa, Mexico, Jan. 7, 1952, *US*,
dups. in MEXU, DES (l, photo).

On sand dunes with coastal thorn forest. Flower description here provided for the first
time is drawn from two specimens that flowered in the Desert Botanical Garden, summers
of 1974, 1976, both having been collected as small plants on the beach sands near Yvaros,
Sonora, Oct. 1966. (See Exsiccatae.)

Agave aktites is the only agave known to grow naturally and regularly within the
maritime zone of beach dunes. It is apparently nearly limited to this habitat. However,
long before the species was recognized, small bluish gray rosettes were observed several
miles inland, southeast of Los Mochis, Sinaloa, *Gentry* 11423. But this was beach area in
the late Recent geologic period. In any case, the species is a narrow endemic, being a
coastal or beach dweller, as the name implies, in northern Sinaloa and adjacent Sonora.
The wide-ranging *A. angustifolia* is common in the same region, but *A. aktites* is usually
recognizable by its smaller size and narrow bluish gray leaves.

My Mayo Indian guide to the Island Lechuguilla, Carlos Preciado, reported that the
thick stems of *A. aktites* were sweet and good eating. His people had pit-baked as many as
40–50 "cabezas" at a time, gathered on the island. He considered them superior to other
agaves growing on the mainland. He said that the neighboring island of Santa María
had many more plants of this species than had Lechuguilla Island. The round thick stem
(Fig. 20.5) indicates a large starch buildup in the meristem, morphologically confirm-

ing Preciado's report. However, the plants grown in the relatively dry stony soil in the Desert Botanical Garden did not develop the thick stems. This suggests that *A. aktites* is closely adapted to its habitat of deep sand and for best development needs to draw upon the deep soil moisture, which may persist into dry periods in its relatively humid native atmosphere. The dry hot atmosphere in our Phoenix garden may be also largely responsible for the lack of complete flower development and total lack of fruit development. Scattered bulbils did follow flowers upon both panicles in the garden.

Analysis of the leaves of both island and mainland samples showed only negligible amounts of sapogenins.

Agave angustifolia
(Figs. 20.1, 20.2, 20.3, 20.6, 20.7, 20.8, 20.9, 20.12; Table 20.1)

Agave angustifolia Haw., Syn. Pl. Succul. 72, 1812, 78, 1819; Revis. Pl. Succul. 35, 1821.

Agave jacquiniana Schultes, Syst. 7: 727, 1829.
Agave jacquiniana Schultes ex Hook., Curtis Bot. Mag. Tab. 5097, 1859.
Agave ixtli Karw. ex Salm, Hort. Dyck. 304, 1837.
Agave elongata Jacobi, Hamb. Gartenz. und Blumen. 168, 1865.
Agave ixtloides Hook., Curtis Bot. Mag. 3: 27, pl. 5893, 1871.
Agave exselsa Baker, Gard. Chron. n.s. 8: 397, 1877.
Agave spectabilis Todaro, Hort. Bot. Pan. II: 4, t. 25, 1878–92.
Agave wightii Drum . & Prain, Agri. Ledger 1906: 78, 91, 102, 105, 128, 139, 140, 147, 1907.
Agave wrightii Drum., Mo. Bot. Gard. Rep. 18: 27, 1907.
Agave lespinassei Trel., St. Louis Acad. Sci. Trans. 18: 33, 1907.
Agave endlichiana Trel., ibid. p. 34.
Agave aboriginum Trel., ibid. p. 34.
Agave zapupe Trel., ibid., p. 32.
Agave prainiana Berger, Agaven, 246, 1915.
Agave bergeri Trel. ex Berger, ibid., p. 250.
Agave kirchneriana Berger, ibid., p. 252.
Agave sicaefolia Trel., St. Louis Acad. Sci. Trans. 23: 141, 1915.
Agave donnell-smithii Trel., ibid., p. 144.
Agave pacifica Trel., Contr. U.S. Nat. Herb. 23: 118, 1920.
Agave yaquiana Trel., ibid. p. 120.
Agave owenii I. M. Jtn., Calif. Acad. Sci. Proc. ser. 4, 12: 999, 1924.

Rosettes surculose, radiately spreading, with stems 2–6 (–9) dm long; mature leaves generally 60–120 x 3.5–10 cm (cult. larger), linear to lanceolate, mostly rigid, hard fleshy, fibrous, ascending to horizontal, light green to glaucous gray, flat to concave above, convex below, narrowed and thickened toward base, margin straight to undulate, sometimes thinly cartilaginous; teeth generally small, 2–5 mm long, rarely longer, evenly and closely spaced or remote, commonly reddish brown or dark brown, from low narrow bases, the cusps slender, curved or variously flexed; spine variable, 1.5–3.5 cm long, conical to subulate, dark brown, graying in age, flat to shallowly grooved above, non-decurrent or thinly decurrent; panicle 3–5 m tall, open, sometimes bulbiferous, the peduncle usually longer than the panicle, bracteate with quick-drying narrow triangular bracts; umbellate branches 10–20, horizontally spreading; flowers green to yellow, 50–65 mm long, quickly wilting, the tepals drying reflexed along tube; ovary small 20–30 mm long, angulate-cylindric, somewhat ribbed, tapered at base, neck short, lightly grooved; tube funnelform to slightly urceolate, 8–16 mm deep, grooved; tepals unequal, 18–24 mm long, 3–5 mm wide, rapidly infolding, at first erect in incurved, soon wilting, drying reflexed against tube, the outer broadly over-lapping inner at base, obtuse to rounded and small-hooded at apex; filaments 35–45 mm long, slender, flattened, inserted in mid-tube;

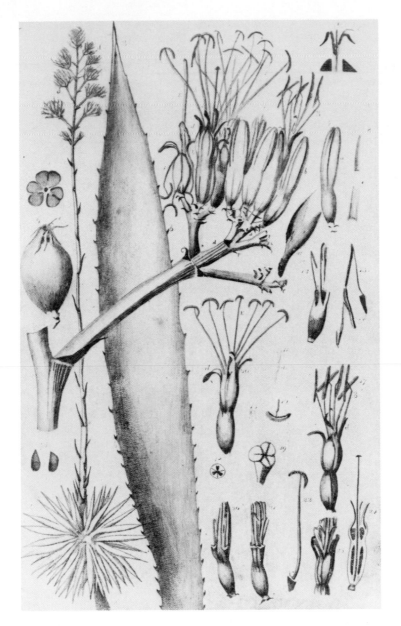

Fig. 20.6. *Agave angustifolia*. Retouched drawing of Tozzetti's 1810 drawing; here elected as lectotype.

anthers yellow, 20–30 mm long, centric or excentric; pistil over-reaching stamens after anthesis; capsules broadly ovoid, large, 3 x 5 cm, dark brown, woody, short stipitate, raggedly beaked; seeds large, dull black, 9–12 x 7–8 mm, the hilar notch long cleft.

Lectotype: pl. 6, T. Tozzetti, Ann. Mus. Imp. Firenze 2 (2): 25, 31–5, 1810. Here reproduced (Fig. 20.7) after Trelease, Mo. Bot. Gard. Rep. 19, pl. 30, 1908.

Trelease gave us a clarifying account of the complex and incomplete history of this name. The plants figured by Tozzetti were growing on the island of St. Helena. ''It was at about this time that the plants were received from St. Helena at Chelsea (England), which Haworth recognized as being his own *A. angustifolia*.'' (Trelease 1908: 283)

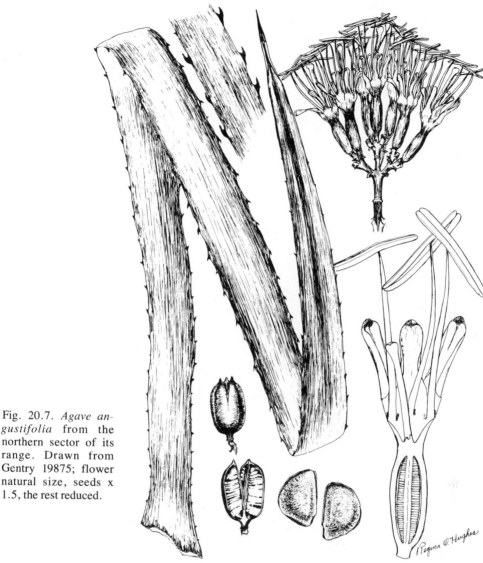

Fig. 20.7. *Agave angustifolia* from the northern sector of its range. Drawn from Gentry 19875; flower natural size, seeds x 1.5, the rest reduced.

The Habitats

This *Agave* has the most wide-ranging distribution of agaves in North America (Fig. 20.12). It is mainly a plant of the *tierra caliente* ranging from Costa Rica, on both Atlantic and Pacific coasts, to Tamaulipas and northwestern Sonora. The common major vegetation formations in which it occurs are tropical savanna, thorn forest, and the drought-deciduous tropical forests in both low and middle elevations: sea level to 1,500 m or, rarely, more. As a heliophyte it keeps to the more open sites, but frequently it will persist with etiolated growth under the light shade of trees and shrubs. Its more extreme habitats include the arid Sonoran Desert, where it survives with about 250 mm (10 inches) of average annual rainfall, to the montane pine-oak forest zone, as near Uruapan, Michoacan (Fig. 20.9), with 1,680 mm (56 inches) of average annual rain. The northern forms survive without damage the light winter freezes, but tropical

Fig. 20.8. *Agave angustifolia* in natural habitat on the thorn forest grassland near Tecolotlán, Jalisco. February 1975.

forms are frost sensitive. Three rainfall silhouettes are given in Fig. 20.1 as general representations. A detailed study of the environments combined with close taxonomic study could contribute much toward understanding the evolution of this complex and other *Agave* species.

Variability of the Complex

As though true to its cosmopolitan nature, it shows itself in many forms. Previous students, being exact and careful men, have coped in their own way by naming many of the variants as species. The aggregate result, at this time of my taxonomic turn, violates my sense of species, because I cannot separate them consistently with any combination of characters. I had hoped to maintain the northwest Pacific populations from the Atlantic ones, at least subspecifically, but I cannot. It is at last very simplifying to accept the

Fig. 20.9. *Agave angustifolia* in montane habitat of pine and oak near Uruapán, Michoacán, November 1966. Although the site is well watered, growth appears moderate and genetically controlled.

complex for what it appears to be: an extensive variable species. As a species it is taxonomically more satisfactory than some others, as, for example, *A. potatorum*.

In passing, I will mention some of the variations that commonly occur in this species complex. Most of them fail to correlate well in any place or to have population characteristics. Some occur at random, segregating like Mendelian factors. Stems may be short or long, in or out of cultivation. Suckering is very prevalent, but it may be early or late in the life cycle. While bulbils have been observed, vivipary is generally uncommon, even on panicles where seed does not develop. Leaves may be few or many as on stemmy plants. They vary in color from green to yellowish green to nearly white glaucous gray. A bluish gray color predominates on many of the mescal plantings in Oaxaca and appears much like the "mescal azul" filling the fields of *Agave tequilana* in Jalisco, but there are spine and teeth differences to show these respective cultivates are not of the same clone. Generally they are narrow outstanding ensiform leaves, but an atypical broadened form appears out of nowhere, especially in Oaxaca, quite similar to that of Fig. 20.6. Teeth are generally small and upcurved in *A. angustifolia,* but they do vary in size and spacing. There is more variation in terminal spines, which may be relatively long and acicular or very short and broad, decurrent or non-decurrent. The ideographs (Fig. 20.3) indicate variations in size and proportions of flowers. On the whole they are weak flowers, quickly passing, the tepals drying reflexed.

Cultivation and industrial use has also made clear that some forms mature earlier than others, some have relatively high starch content, and that still others develop superior fiber and in greater quantity. Altogether, these variations amply illustrate we are dealing with a freely seeding outbreeding complex, that has been widely assorted by circumstances of habitat, changing climates over a long period of time, and man's interventions. Out of this *Agave* gene reservoir have appeared the following selections in taxonomy and agriculture.

Key to Some Varieties of *Agave angustifolia*

1. Leaves longer, 8–17 dm   2
1. Leaves shorter, 3–5 dm long   5
2. Leaves narrow, 3–4 cm wide, wild plants   *rubescens*
2. Leaves wider, 8–14 cm wide, cultivates   3
3. Leaves whitish gray glaucous, relatively broad, 10–14 cm wide; spine short, stout, 1–2 cm long, short-decurrent   *letonae*
3. Leaves bluish gray glaucous to pale green, mostly 7–10 cm wide; spines acicular, 2–3 cm long, non-decurrent   4

4. Leaves bluish gray glaucous to green; teeth larger, 3–4 mm long through mid-blade, raised on small teats   *nivea*
4. Leaves pale green, teeth mainly 3–5 mm long, not raised on marginal teats   *deweyana*
5. Medium sized plants on well developed trunks 5 dm or more tall; leaves with white to pale yellow margins   *marginata*
5. Dwarf plants on short stems with small, rather thin, light green leaves, 30–35 x 2–3 cm   *sargentii*

Agave angustifolia var. *deweyana*
(Fig. 20.12)

Agave angustifolia Haw., var. *deweyana* (Trel.) Gentry, stat. nov.
  Agave deweyana Trel., St. Louis Acad. Sci. Trans. 18: 35, 1909.

Type: *Dewey 649*, cultivated, Victoria, Tamaulipas, Mexico, Feb. 2, 1907, US.

"Zapupe verde." The variety is not well marked, various clones apparently being cultivated under this name. The type is a narrow leaf 5–6 cm wide, while later collections in the vicinity, *Ogden Gilley & Hernandez 51115, Gentry 12265,* have wider leaves, 100–115 x 7–10 cm, and more remote teeth. The plant was named after Lyster H. Dewey, fiber specialist with the U.S. Department of Agriculture early in the century. Cultivated in fiber plantations, mainly in the state of Tamaulipas, but Trelease also lists it from Veracruz.

Agave angustifolia var. *letonae*
(Figs. 20.11, 20.12)

Agave angustifolia Haw., var. *letonae* (Taylor) Gentry, stat. nov.
  Agave letonae Taylor ex Trel., Wash. Acad. Sci. J. 15: 593, 1925.

Type: *Milner Oct. Nov. 1923, MO.* Cultivated at Sucesión Letona, San Miguel, El Salvador.

This is a robust, nearly white-leaved plant, developing a broad trunk with several years of leaf cutting (Fig. 20.11). It has been of considerable economic importance for fiber production in Salvador and has been noted in minor cultivation in Guatemala. A fungal disease has been reported to cause serious losses in El Salvador, the humid summer climate there favoring fungal growth.

Agave angustifolia var. *marginata*
(Fig. 20.10)

Agave angustifolia Haw., var. *marginata* Hort.

This is a handsome, neatly growing plant developing a trunk 3–6 dm tall with numerous leaves margined with white or yellow (Fig. 20.10). It is widely distributed as an ornamental around the world. A closely allied form, *A. angustifolia* var. *variegata* Trel.

Fig. 20.10. *Agave angustifolia* var. *marginata* at a private home in San José, Costa Rica.

Fig. 20.11. *Agave angustifolia* var. *letonae* near Laguna Retana, Depto. de Jutiapa, Guatemala. The trunks indicate several years of cutting for fiber production.

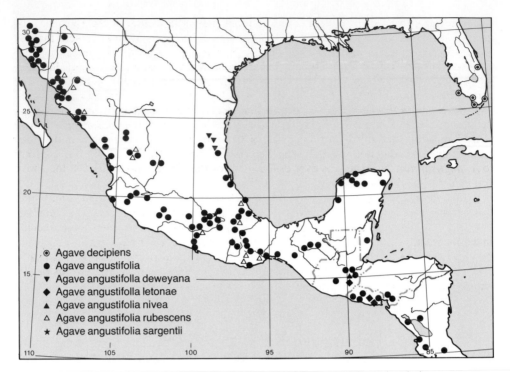

Fig. 20.12. The extensive distribution of *Agave angustifolia* and varieties.

(Mo. Bot. Gard. Ann. Rep. 19: 287, 1908) has the marginal white in unusual width and the remainder of the leaf silvery gray or milky. It is reported to have arisen in the botanical garden of the College of Science at Poona, India, about 1895.

Agave angustifolia var. *nivea*
(Fig. 20.12)

Agave angustifolia Haw., var. *nivea* (Trel.) Gentry, stat. nov.
   Agave nivea Trel., St. Louis Acad. Sci. Trans. 23: 143, 1915.

   Type: *Trelease 11,* About El Rancho, Dept. de Progreso, Guatemala, April 1915, MO.

This is a long-leaved, short-stemmed plant, the leaves characteristically a dull bluish gray, 130–140 x 9–10 cm. I collected the plant from hedges in the type locality in 1965 and again in 1975 and the leaves match those of the type like a clone. The plant is still known only from this locality, where it is used locally for fiber and fence plants. The inflorescence is still unknown, but the natives reported it to flower in April. Offsets are now growing in the Desert Botanical Garden.

Agave angustifolia var. *rubescens*
(Fig. 20.12; Table 20.1)

Agave angustifolia var. *rubescens* (Salm.) Gentry, stat. nov.
   Agave rubescens Salm., Hortus Dyckensis 8, 306, 1834.

   Trelease's plates 32–24, Mo. Bot. Gard. Ann. Rep. 1907, can be taken as Neotype.

The variety *rubescens* differs from the more prevailing forms of the species in its narrow, less rigid leaves. The leaf specimens reviewed range from 80 to 130 cm long by 3–4 (–5) cm wide, as arbitrarily segregated. In some populations the gradation from narrow to broader than 5 cm is clinal. Trelease reviewed this taxon (1907: 254), identifying with it some specimens from the dry highlands of southern Puebla and Oaxaca. The cartilaginous margin of the leaves he and Salm-Dyck mention is very thin and inconspicuous. On some specimens I find there are minute blunt serrulations on this margin between the teeth, lacking on others. In herbarium MO Trelease left a note saying, "There is no indication that any writer since Jacobi has seen authentic *rubescens,* and it may be that even he did not see it." The flowers are small, *Gentry 5215,* but conform to *A. angustifolia.* No type was ever established for this name, and so I leave the above neotypic designation to anchor my interpretation.

This varietal recognition may be of horticultural value, and it does draw attention to this common form of *A. angustifolia.* Typical examples have been observed from the Isthmus of Tehuantepec to southern Sonora, but my segregation comes too late to include in the Exsiccatae the collections reviewed in most herbaria other than the DES. It occurs in the drier drought deciduous forests and bordering grasslands, especially on thin rocky soils where tree canopy is light and virtually lacking during the long dry seasons of late winter and spring. It is apparently a late fall-winter bloomer.

Agave angustifolia var. *sargentii*
(Fig. 20.12)

Agave angustifolia Haw., var. *sargentii* Trel., Mo. Bot. Gard. Ann. Rep. 22: 99, 1912.

Type: Grown at Missouri Botanical Garden from C. S. Sargent of Brookline, Massachusetts, pls. 100, 101.

Plant dwarf with trunk ca. 25 cm high with numerous leaves, spreading, straight, smooth, grayish green narrowly oblong-lanceolate, 2.5–3 x 25–30 cm, with dark gray, minutely roughened, non-decurrent spine 3 x 20–25 cm; teeth nearly black, glossy, 1–2 mm long, 10–15 mm apart with slender cusps; inflorescence bulbiferous, ca. 1 m tall, with few simple branches at top; flowers yellowish green, ca. 40 mm long; ovary lightly glaucous 7 x 18 mm, obconical; tube deeply urceolate-conical, 8 mm deep; segments 4 x 15–17 mm quickly drying; filaments inserted nearly in throat of tube, 25 mm long, somewhat maroon-dotted like the style. A leaf collection, *Ogden 5134,* 15 miles from Puebla, on Puebla-Tlaxcala road, 12 Jan. 1951, in open field, appears to represent this variety.

Agave breedlovei

Agave breedlovei Gentry, sp. nov.

Small single or cespitose light green plants with spreading rosettes 60–90 cm in diameter and narrow umbelliform panicles sometimes bulbiferous. Leaves linear-lanceolate, 30–50 x 4–5 cm, rather rigid, patulous, light green, long acuminate, slightly narrowed at base; margin straight to undulate, without corneous edges; teeth 2–3 mm long through the mid-blade, 5–15 mm apart, the slender cusps flexed or straight, reddish to brown, with short low bases; spine 1.7–3 cm long, conic-acicular, openly grooved above, finely decurrent to upper teeth, rounded below, dark brown; panicle to ca. 3 m tall, slender, with several slender umbellate, few-flowered branches 18–25 cm long, sparsely bracteolate; flowers protandrous, greenish yellow with reddish filaments, 50–60 mm long, the tepals drying reflexed after anthesis; ovary 25–30 mm long with neck; tube ample, funnelform, 15–16 mm deep, 12 mm diameter at orifice, apparently thin-walled; tepals 16–18 mm long, 4 mm wide, linear, rounded-mucronate, erectly ascending, the outer and inner nearly equal; filaments 55–60 mm long, reddish, slender, diverging-ascending, inserted subequally in mid-tube, 8–10 mm above tube bottom; anthers 24 mm long, yellow, centric.

Type: *Breedlove 37904, CAS.* 8 km E of Las Margaritas along road to La Soledad, Mun. Las Margaritas, Chiapas, 15 Sep. 1974 (l, f; the latter in various stages including the dry deflexed tepals of post-anthesis).

Planta, parva, singula vel caespitosula, 60–90 cm diametro. *Folio* pauca, lineari-lanceolata, 30–50 cm longa, 4–5 cm lata, viridia, patulosa; *margine* recta vel undulata, *dentibus* parvulis 2–3 mm longis, 5–15 mm separatis, flexuosis vel rectis, brunneis; *spina terminali* conica vel acicula, 1.7–3 cm longa, brunnea, supra lata canaliculata. *Inflorescentia* panaliculata, scapo incluso 3 m alta; *floribus* viridi-luteis, 50–60 mm longis; *ovarium* apice incluso 25–30 mm longum. *Perianthii tubo* infundibuliformi 15–16 mm longo, 12 mm diametro; *segmentis* 16–18 mm longis, aequalibus. *Filamenta* medio tubi inserta, 55–60 mm longa; *antheris* 24 mm longis, luteis.

Agave breedlovei is distinguished by its short linear-lanceolate green leaves and the deep flower tubes with relatively short tepals, which about equal the tube. Flower ideographs of *A. macroacantha* and *A. karwinskii* show very similar proportions (Fig. 20.3), but the short leaves of *A. breedlovei* with close-set teeth are quite distinctive. The species appears closest morphologically to *A. panamana,* as reflected in the contiguous positions in the foregoing key to species. However, the longer, linear, more flexible leaves of *A. panamana* and its maritime habitat suggest that Breedlove's Chiapan collections should at least be proposed as a different species. I feel handicapped by never having seen either of these taxa in natural habitat.

Breedlove & Thorne 30509 has a leaf much like *A. breedlovei,* while the flowers and infructescence are those of *A. seemanniana.* Perhaps the efforts of the two collectors were not well synchronized?? Another sheet tentatively assigned here is *Breedlove 36756* with short lanceolate leaf and a bulbiferous infructescence like that of *A. angustifolia,* which helps align this taxa with the Rigidae. Seeds of this latter number were planted at the Desert Botanical Garden in 1980 for further study as the plants grow, provided the six-year-old seeds are viable.

The distinction of this taxa is based primarily on the short narrow leaves, while the dimorphic flowers show a taxonomic weakness in lack of correlations in the several collections available. The species is therefore intended as a provisional name. I have not seen *A. breedlovei* in life and cannot judge how truly representative the leaf specimens are. If they are the uppermost bract-like leaves, not the mature full grown leaves of the rosette proper, then this proposed species should be submerged in the extensive variable complex represented by *A. angustifolia.* The specific name is given after the collector, Denis Breedlove, who has reaped a fine harvest of Chiapan specimens during the past two decades.

Agave cantala
(Figs. 20.3, 20.13, 20.14; Table 20.1)

Agave cantala Roxb., Hort. Beng. 25, 1814.
    Agave cantula Roxb., Fl. Ind. 2: 167, 1832.
    Agave candalabrum Todaro, Hort. Bot. Pan. II: 4, t. 15, 1878–92.
    Agave rumphii Jacobi, Hamb. Gartenz. und Blumen. 261, 1865.

Tall, slender, loose-leaved, surculose, green rosettes, 2–2.5 m tall; stems 3–6 dm long; leaves 1.5–2 m long, 7–9 cm wide, linear, long-acuminate, thin, frequently reflexing, concavo-convex, roundly keeled below toward base, light or dark green, smooth above, rough below; teeth small, brown, mostly 2–3 cm apart, the larger 3–4 mm long, antrorsely curved, reduced or lacking toward apex, margin straight; spine very small, 5–15 mm long. Panicle sometimes bulbiferous, 6–8 m tall with ca. 20 loosely diffusive umbels in upper half of slender shaft; bracts narrow, chartaceous, soon reflexed; lateral peduncles slender, ascending to recurved; flowers greenish tinged with purple or reddish, 70–85 mm long, slender; ovary 32–42 mm long, fusiform, tapering below to a basal rim, shortly constricted at union with tube but virtually neckless, roundly 6-ribbed; tube 14–17 mm deep, 15 mm wide, cylindric-funnelform, grooved; tepals sub-equal, 25–28 mm long, rapidly involute-conduplicating, linear-spatulate, apex rounded-cucullate, inner with narrow keel; filaments 50–55 mm long, stout, abruptly outcurving from tube wall, inserted 11–14 mm above bottom of tube; anthers slender, slatey purple, slightly curved, excentric; pistil stout.

Type: ?   Neotype: ?

Agave cantala can be recognized by its thin long narrow leaves, weak and frequently reflexed above the middle, small teeth, and large purplish green flowers in broad, diffusive, slender-peduncled panicles. It seldom flowers and has long gone unrecognized in the Americas. Roxburg named it in India, whence it was doubtless carried from North America by European traders. Trelease apparently did not know the species by sight, but he listed it in Standley's Trees and Shrubs of Mexico. His *A. acuispina,* treated here following, appears assignable to *A. cantala.* Trelease (1920) cites as synonyms several names of early 19th-century authors presumably applicable to this plant, as *A. vivipara, A. flaccida* Haw., *A. madagascariensis* Spreng., *Furcraea cantala* Haw., and *F. madagascariensis* Haw. Berger added more, but they are hardly worth reprinting here.

My acquaintance with the species began with a submature specimen plant in the Huntington Botanical Gardens (Fig. 20.13) obtained from New York Botanical Garden in 1932 as *A. franceschiana.* In Mexico, two offsets were collected from a fairly mature plant, looking very much like *A. cantala,* about 5 miles (8 km) east of Villa Guerrero, Jalisco. The owner stated that the plant was spontaneous, and it may be native there but certainly not in abundance, as that is my only recollection of the plant in Jalisco. Considering its long good-quality fiber, the plant was probably long used by the native Mexicans and may have been carried about and traded in preconquest times. Its specific origin may always remain unknown. I photographed the plant in 1955 in Bam, Fars Province, Iran, but it is uncommon in that country.

Apparently it has long been cultivated for fiber in southeast Asia from the Philippines to India. Its fiber is reported of fine quality, commanding a higher price than sisal, but the plant is more subject to disease, the leaf is harder to decorticate, and cutters do not like to handle the prickly leaves (Lock, 1969).

Agave cantala var. *acuispina*
(Figs. 20.2, 20.3; Table 20.1)

Agave cantala Roxb., var. *acuispina* (Trel.) Gentry, stat. nov.
   Agave acuispina Trel., Wash. Acad. Sci. J. 15: 393, 1925.

Rosettes tall, sparsely suckering, forming trunk 5–7 cm tall; leaves narrowly long linear, at maturity 1.4–1.7 m x 6–8 cm, green, erect-ascending, occasionally reflexed, margin straight to undulate; teeth small, 2–4 mm long, curved finely from small low bases, 1–3 cm apart, chestnut-colored to dark brown, lacking on apex of leaf for upper 15–20 cm; spine 1.5 cm or more long, 3–5 mm broad at base, slender conical, shiny dark brown to graying, ungrooved, non-decurrent; inflorescence 6–7 m tall, bulbiferous, with broad diffuse open panicle in upper half of shaft, with 20–35 broadly spreading lateral branches; flowers green with tepal lobes light greenish yellow, filaments and anthers

Fig. 20.13. The long, narrow, weak leaves of *Agave cantala* grown at Huntington Botanical Gardens, San Marino, California, 1951.

Fig. 20.14. *Agave cantala* in flower at Huntington ➤ Botanical Gardens, July 19, 1977.

reddish or spotted, or anthers bronze-colored; flowers 57–63 mm long; ovary short, 25–30 mm long, constricted above with short neck and below; tube broadly funnelform, 14–16 mm deep, 12–15 mm wide, with narrow grooves; tepals subequal, 19–21 mm long, the outer broad at base, thin, lanceolate, involute, conduplicate, narrowing early and beginning to reflex, the hood small; filaments 55–60 mm long, flattened, inserted subequally 11–13 mm above base of tube; anthers regular, centric; capsules broadly ellipsoid, 2.5–3.5 x 3 cm, short stipitate.

Type: *Calderon 2084*, Hacienda El Platinar, Dept. de San Miguel, El Salvador, Jan. 1924, *US*, ''maguey''; in hedge, wild, not used for fiber.''

Trelease did not have flowers to describe. My description is drawn from *Gentry 20694* and *21467,* both from Honduras (see Exsiccatae). The tall open panicle of *A. c. acuispina* with its greenish flowers with large deep tube, and the narrow long leaf form with small teeth, appear to align this cultivar with *A. cantala*. It differs from *A. cantala* in its shorter sturdier leaves with heavier end spine and the shorter flowers. In Honduras it was found used as a hedge plant, but not in fiber plantations. However, the natives doubtless used the fine fiber locally before the advent of agave plantation agriculture.

Agave datylio
(Figs. 20.1, 20.3, 20.15, 20.20; Table 20.1)

Agave datylio Simon ex Weber, Bull. Mus. Hist. Nat. Paris 8: 224, 1902.

Rosettes small to medium size, 6–10 dm tall, 10–15 dm wide, suckering freely, the rhizomes frequently elongate; leaves lance-linear 50–80 cm long, 3–4 cm wide, green to yellowish green, somewhat glaucous in youth, canaliculate above, rounded below, radiately spreading, rather rigid nearly straight; margin nearly straight; teeth mostly 3–5 mm long, deltoid, flattened, rather blunt, dark brown, usually remote or 3–6 cm apart, more closely spaced below; terminal spine large, conical to subulate, 2.5–4 cm long, dark brown to grayish, shortly decurrent, scarcely or flatly grooved above; panicles 3–5 m tall with 8 to 15 branches of small umbels in upper half of shaft; flower greenish yellow, 40–55 mm long; ovary 20–30 mm long; tube funnelform, 5–10 mm deep; tepals 15–20 mm long, erect to ascending; filaments 35–45 mm long, inserted 4–6 mm above base of tube; anthers 15–20 mm long, yellow; capsules oblong to pyriform, 1.5–2 cm in diameter, 3.5–4 cm long, short-stipitate; seeds 7 x 6 mm.

Fig. 20.15. *Agave datylio* near La Paz, Cape District, Baja California, September 1951.

Fig. 20.16. *Agave datylio* var. *vexans* in the sandy valley below Comondú, Baja California, October 1951.

Neotype: *Gentry & Arguelles 11200, US,* DES, MEXU, MICH. About 4 miles E of La Paz, Cape District, Baja California, 29 Sep. 1951. (l, f, photo, fig. 20.15.)

Agave datylio is widely scattered in the Cape Region at lower elevations on granitic sandy soils. It has no close relatives on the peninsula. It belongs with the widespread sword-leaved group, whose nearest geographic member is *A. aktites* Gentry of the Sonoran-Sinaloan coast. The latter, however, is amply distinct with its narrower bluish leaves, sharply cuspid teeth, larger rotund stem, and larger flowers with deep bulging flower tube. Trelease set off *A. datylio* in a monotypic group, Datyliones, but it fits well in Berger's Rigidae along with the *A. angustifolia* alliance.

A. datylio is locally called "datiliyo" or "datilillo," while its variety *vexans* at Comondú is called "mescaliyo."

Agave datylio var. vexans
(Figs. 20.3, 20.16, 20.20; Table 20.1)

Agave datylio var. *vexans* (Trel.), I. M. Jtn., Proc. Calif. Acad. Sci. 12: 1003, 1924.
Agave vexans Trel., Mo. Bot. Gard. Rpt. 22: 62, 1912.

This variety differs from the species in the smaller size of rosettes and leaves, ranging from 30 to 50 cm long, and a more glaucous or yellowish color. Teeth are usually remote, but specimens from Arroyo Purísima have close-set small teeth below the mid-leaf, the larger being only 2–3 mm long. Johnston separated the variety as having shorter filaments (20–30 mm), but this does not hold as I have specimens from near Comondú with filaments 35–40 mm long, as for the species. The variety occurs on sandy soils in the lower elevations on both sides of the Sierra Giganta.

It appears to be a xerophytic ecotype of the species. I had them growing for many years in Murrieta, but they did not flower there, perhaps owing to their shady situation.

Agave decipiens
(Figs. 20.2, 20.3, 20.17; Table 20.1)

Agave decipiens Baker, Kew Bull. Misc. Inf. 1892: 183, 1892.

Agave laxifolia Baker, Curtis Bot. Mag. 122: pl. 7477, 1896.

Arborescent with trunk 1–3 m long with deep crown of green patulous leaves, the large bases thickening the stem; leaves mostly 75–100 x 7–10 cm, narrowly lanceolate, fleshy, concave, rigidly spreading to recurving, green, long-acuminate, narrowed at thickened base, the margins repand; teeth through the mid-blade 2–3 mm long on low prominences, 1–2 cm apart, with a few interstitial smaller teeth, the slender cusps upcurving, dark brown; spine 1–2 cm long, conical, ungrooved, non-decurrent, dark brown; panicle 3–5 m tall, often bulbiferous, with 10–12 or more umbellate branches in upper half of shaft; flowers greenish yellow, 60–80 mm long, fetid; ovary large and thick, 40–48 mm long, 12–14 mm broad, neckless, tapered below, grooved nearly to base; tube 11–13 mm deep, funnelform, thick fleshy, grooved; tepals subequal, 18–22 x 6–4 mm, linear, thick, conduplicate, incurved, cucullate; filaments slender, 40–50 mm long, inserted on 2 levels at mid-tube or slightly above; anthers 22–25 mm long, yellow, excentric; capsules ellipsoid to oblong, 3.5–5 cm long, stipitate.

Type: Unknown. The illustration in Curtis Bot. Mag. 122: t. 7477, 1896 can be taken as lectotype, regardless of Baker calling it *A. laxifolia*.

The elongated stems of large plants appear very thick owing to the thick bulbous bases of the clasping leaves. In fertile soil on the Florida peninsula the plants are said to reach 4 m in height with leaves exceeding 2 m in length. Small (1933) reported it grows on coastal sands and on old Indian village sites, "kitchen-middens." Some of the leaf collections from the Yucatán peninsula resemble this agave, but the flowers relate such leaves to *A. angustifolia,* as *Lundell 8110. Agave decipiens* is still known to grow spontaneously only in Florida. The only flowering material I have seen was growing in the Huntington Botanical Gardens, San Marino, California, *Gentry 19749,* from which the above description is drawn.

Agave fourcroydes
(Figs. 20.1, 20.3, 20.18, 20.19; Table 20.1)

Agave fourcroydes Lem., Ill. Hort. 11:65, 1864.

Agave sullivani Trel., U.S. Nat. Herb. Contr. 23: 119, 1920.

Large, suckering, sword-leaved rosettes developing a thick stem 1–1.7 m tall. Leaves straight, rigid, 1.2–1.8 m long, 8–12 cm wide, linear, thickly rounded at base, guttered, acuminate, the margins straight with regularly spaced, slender, dark brown teeth 3–6 mm long; spine stout, conical, mostly 2–3 cm long, dark brown, openly short-grooved above at base; panicle bulbiferous, 5–6 m tall with 10–18 lateral umbels in upper half of shaft; flower greenish yellow, 60–70 mm long; ovary 35–40 mm long, fusiform, roundly trigonous, tapered toward base, neck briefly constricted; tube 12–16 mm deep, urceolate, thick-walled, grooved, 12–15 mm wide; segments subequal, 18–16 x 3–4 mm, linear, closely over-lapping, at first erect then sharply reflexing, the outer flat-backed, inner keeled; filaments stout, 45–60 mm long, inserted 6–8 mm above bottom of tube, tapered toward apex; anthers 20–24 mm long, pale yellow, slender, excentric; pistil stout.

Neotype: ?

Fig. 20.17. *Agave decipiens*, showing the broad stem with bulging leaf bases. Huntington Botanical Gardens, summer 1977.

Fig. 20.18. Decorticating mills and fiber drying racks for *Agave fourcroydes* at Pericos, Sinaloa, December 1951.

This is the plantation fiber agave of eastern Mexico, being of primary economic importance in Yucatán, Veracruz, and southern Tamaulipas. The plants are widespread in botanical gardens and as cottage plantings in tropical countries. They are generally recognizable by their tall vigorous habit, the long narrow light green leaves, rigid and unyielding. As the plants grow, the lower more mature leaves are cut for fiber production, and over a period of ten years or more, develop conspicuous trunks 1 m or more tall. They are not hardy, a few degrees of frost frequently damaging the leaves. They can endure considerable drought, as in the long hot rainless springs in monsoon type of climate, but good leaf growth and quality fiber production require a minimum of ca. 30 inches (750 mm) of annual rainfall in frostless climates. The Obregón plantation south of Ciudad Obregón, Sonora, for instance, with less than 20 inches (500 mm) of annual rainfall, never produced fiber of sufficient length and quality to meet trade standards.

The fiber is coarse, strong, excellent for rope and coarse twine, as bailing twine, is resistant to seawater, but for finer fiber artifacts is less suitable than sisal and some other agave fibers. The production of henequen for 1965 is given in Table 22.2 as 160,000 metric tons, 147,150 of which was produced by Mexico. A by-product of fiber mills in the past two decades has been the recovery of the sapogenin, mainly as hecogenin, from the waste bagasse. Cortisone and some of the sex hormones can easily be synthesized from hecogenin, which occurs in something under 1 percent dry weight basis in the leaves.

Edward Thompson in a consular report (1899) lists some figures on henequen exports from Yucatán. For the ten years of 1889–98 exports averaged 49,147 metric tons per

Fig. 20.19. *Agave fourcroydes* in Pericos plantation showing several years of cutting. December 1951. Planting stock was brought from Yucatán.

annum at an average price of about 7¢ US per pound. His account gives other details of historical interest on this "sisal grass" industry.

> This plant has been in use among the ancient inhabitants of Yucatan from the earliest times. The writer has found it imbedded in the form of cord in the stucco figures that ornamented the facades of the mysterious ruined cities of Yucatán. There are two wild varieties of henequen, called by the native "cahum" and "chelem." The fiber of these wild plants is used to some extent by the native in the making of cordage for domestic use, and some claim that hammocks made from the fiber of the cahum are the best.

> Between the years 1750–1780, quite a furor was created in commercial countries of the Old World by the discovery that the fiber of a plant found in Yucatán was good for ship cordage. Spain sent over a royal commission to report upon the discovery, and in a few years many of Spain's commercial and war vessels were using cordage made from henequen. For some reason, probably because of the primitive method of preparing it, the use of the fiber gradually declined, until at the commencement of this century the former trade had been forgotten.

Agave fourcroydes, like *A. sisalana*, is apparently a sterile hybrid. The ovaries never develop and produce seed. Bulbils are often produced in the axils of bracts and pedicels, following the flowers. They may persist for a year or more on the standing panicles. These bulbils will strike root and grow with soil coverage, but they are not generally utilized in plantation operation, since they require more time and expense than the rhizomatous suckers, vigorously produced by all young plants. These must be kept from growing in the harvest lanes between the rows of plants, so the special weeding and propagation stock are convenienced in one operation. The prickly leaves are hard to handle by the leaf cutters, but *A. fourcroydes*, in spite of the ups and downs of the fiber market, continues to be among the leading hard fiber producers of the world.

Agave karwinskii
(Figs. 20.1, 20.2, 20.3, 20.20, 20.21; Table 20.1)

Agave karwinskii Zucc., Act. Acad. Caes. Leop.-Carol. 16(2): 677, 1833.
  Agave corderoyi Baker, Gard, Chron. n. ser. 8: 398, 1877.
  Agave bakeri Ross, Boll. Soc. Sci. Nat. ed. Econom. Palermo 1894.

Plants arborescent with stems to 2–3 m tall, apparently forming clonal colonies by spreading rhizomes, the leaves forming deep crowns, reflexing along stems with age; leaves 65–40 x 7–3 cm, linear-lanceolate, ascending to radiately spreading, green, guttered or concave above, convex below, narrowed and thickened toward base, acuminate, involute toward base of spine, the margin straight; teeth through mid-blade 3–5 mm long, pyramidal and nearly straight to finely cuspidate and flexuous, dark brown, 2–4 cm apart; spine variable, 1.5–4 cm long, subulate or conical with thickened base, decurrent or nondecurrent, dark brown to grayish and corroding at base; panicle ca. 3–3.5 m long, openly diffuse in oval outline, with 10–15 branches in upper third of shaft; flowers small, greenish to pale yellow, with ferruginous tinge on tepals, 45–57 mm long; ovary 20–30

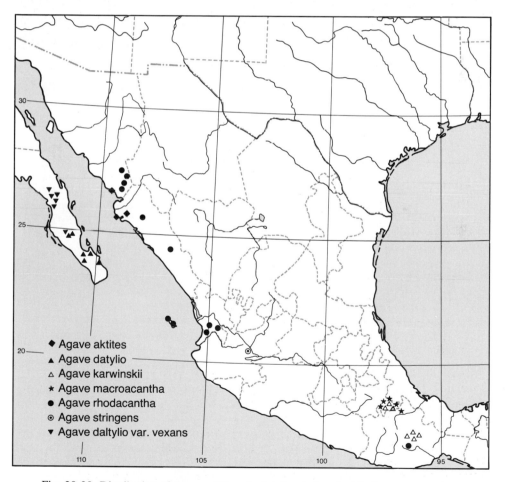

Fig. 20.20. Distribution of *Agave aktites, A. datylio, A. karwinskii, A. macroacantha, A. rhodacantha,* and *A. stringens.*

Fig. 20.21. *Agave karwinskii* in the arid hills near Zapotitlán, Puebla, August 1963.

Fig. 20.22. *Agave macroacantha* in native habitat near Techuacán, Puebla, August 1963.

mm long, angular cylindric, lightly 6-grooved in short neck; tube 10–11 mm deep, 8–11 mm broad, bulging between the grooves, rather thin-walled; tepals unequal 11–19 mm long, linear spatulate, involute, incurving before anther dehiscence, quickly wilting; filaments 35–40 mm long, inserted irregularly in mid-tube or slightly above; anthers 20–22 mm long, yellow, regular, centric; capsule broadly ovoid to oblong, 4–5 x 2.5–3 cm, woody.

Neotype: *Gentry 12049,* Mitla vicinity, Aug. 1952, *US,* dups. DES, MEXU, MICH (l, f). ''Cultivated as a fence plant along roads and fields,'' also wild in Oaxaca Valley.

Agave karwinskii is easily recognized by its tall habit and relatively small leaves and small flowers. It is a true xerophyte growing spontaneously only in the more arid regions of southern Puebla and Oaxaca between 5,000 and 6,000 feet (1,550 and 1,850 m) elevations (Fig. 20.21). Berger reported garden plants to reach 4 m in height, but in native habitat it usually appears in open clonal clumps 2–3 m tall, frequently in shallow stony soil. When closely planted in rows it forms excellent enduring fences. The tall long-lived trunks appear as polypodial plants, but I have no observations on this point. Specimen leaves resemble those of *A. decipiens,* but *A. karwinskii* lacks the small interstitial teeth commonly notable on *A. decipiens,* and the flowers of the latter are larger with disproportionally large ovaries. Leaf specimens from the Tehuacán area show the terminal spine distinctly decurrent, while Mitla leaves are scarcely or not at all decurrent.

Agave macroacantha
(Figs. 20.1, 20.3, 20.20, 20.22; Table 20.1)

Agave macroacantha Zucc., Act. Acad. Caes. Leop.-Carol. 16(2): 676, 1833.

Rosettes small, commonly cespitose, bluish gray green, eventually 25–40 cm tall on short stems with numerous patulous leaves; leaves 25–35 x 2.5–3 cm, linear, acuminate, rigid, radiately spreading, bluish gray glaucous, flat above, convex below, the margin straight or undulate; teeth irregularly spaced 1–3 cm apart, the larger 3–4 mm long, dark brown, with slender cusps mostly curved from small low bases; spine 3–3.5 cm long, subulate, straight to sinuous, dark brown to gray, flat above, rounded below, not decurrent; panicles slender, sometimes bulbiferous, about 2 m tall with 10–14 shortly spreading branches in upper half of shaft; flowers pruinose green with purple tinge on tepals, 50–56 mm long; ovary 25–30 mm long, angular-fusiform, with constricted deeply grooved neck; tube broadest in middle or urceolate, 14 mm long, grooved; tepals ca. equal, 13–16 mm long, 2.5–3.5 mm wide, quickly wilting at anthesis, the inner with narrow keel but soon receding, sinuses open at anthesis, filaments 35–40 mm long, sharply bent from regular insertion 7–8 mm above base of tube, spotted; anthers spotted, 20–21 mm long, slightly excentric; capsules oblong 4.5 x 2 cm, dark brown, stipitate, with constricted prominent beak; seeds 6–7 x 4–5 mm, triangular, dull black.

Neotype: *Gentry, Barclay, Arguelles 20242,* 21 miles SE of Tehuacán along road to Teotitlan, Puebla, Aug. 5–22, 1963; alt. 3,800–4,500 feet, in open dry deciduous forest, US (l, infruct.).

So far as observed about Tehuacán and northern Oaxaca, *Agave macroacantha* presents a regular form of rosette and leaf of a bluish gray color, quite stiffly armed. As a free-seeding and suckering species, it manifests little genetic variation. Plants on dry rocky soils are depauperate, somewhat smaller than described. It is common on the dry sedimentary slopes about Tehuacán.

Trelease (Mo. Bot. Gard. Rep. 18: 231–56, 1907) wrote a detailed review of the complicated taxonomic history of this plant through European literature, listing many synonyms and misapplications of Zuccarini's original name. Several atypical variants were reported, some doubtless due to artificial growing in greenhouses, but of these Trelease listed only two that appeared assignable to Zuccarini's name: *A. macroacantha* var. *integrifolia* Baker, leaves of the normal form and size, and entire; and *A. macroacantha* var. *latifolia* Trel., with leaves broad and repand. He was doubtful that the latter belonged in this species, and did not cite it in his later account (1920) of the agaves of Mexico. His photos (loc. cit. pls. 27, 28) show a robust plant more like *A. peacockii,* which grows west of Tehuacán, where his photos were taken.

Agave panamana

Agave panamana Trel., U.S. Nat. Herb. Contr. 23: 114, 1920.

Small, few-leaved, suckering, short-stemmed, open rosettes with bulbiferous inflorescence; leaves narrowly linear-lanceolate, 50–65 x 4.5–5 cm, thin, apparently deployed irregularly; teeth 1–2 mm long, ca. 15 mm apart, slender, upcurving; spine small, ca. 1 cm long, 2 mm wide at base, dark brown; panicle 2–3 m tall with small lateral umbels; flowers 60 mm long with segments equaling tube; filaments inserted in upper third of tube.

Type: *Howe s.n.* Urava Island, Panama, Pacific coast, 1909, *NY.*

This plant was subsequently observed and collected by other botanists (see Exsiccatae). I. M. Johnston in 1949 wrote (Sargentia 8: 96), "Known only from some small barren islets in Bodega Bay off Long Beach and from islets and large rocks forming the point of land at the south end of Plaza Grande." The plant appears to be sea salt tolerant like *A. aktites* and *A. colimana,* which also grow within reach of salt spray at high tides. The specimens indicate a weak plant battling for survival on nearly soilless rocks, the only sites available where it will not be shaded out by jungle forest. It can be expected to form more robust and symmetrical plants when grown in frostless gardens.

Agave rhodacantha
(Figs. 20.1, 20.3, 20.20, 20.23; Table 20.1)

Agave rhodacantha Trel., U.S. Natl. Herbarium Contr. 23: 117. 1920.

Large, single or cespitose, multileaved, radial rosettes 2–3 m tall, 3–5 m in diameter, truncate, 5–9 dm, or acaulescent; leaves 1.4–2.5 m x 8–15 cm linear, hard-fibrous, rigid, straight, much thickened and scarcely narrowed at base, green to faintly glaucous green, smooth, the margin straight to undulate; teeth regular, mostly 4–8 mm long, firm, slender, very sharp, flexed upward, dark brown, mostly 1–3 cm apart; spine 1–2.5 cm long, conical but frequently with a subulate tip, dark brown, and with a short open groove above; panicle 7–9 m tall, deep, broad, with 30–45 large remote laterals; flowers green with tepals yellowing at anthesis, 55–65 mm long (dried and relaxed); ovary 25–35 mm long, including short neck, fusiform; tube urceolate, 8–10 mm long, 9–10 mm wide; tepals subequal, 16–23 mm long, 4 mm wide, linear, strongly involute, rounded and strongly hooded at apex, wilting at anthesis and soon reflexing; filaments 40–45 mm long, inserted 5–7 mm above bottom of tube; anthers 14–28 mm long; capsules 7–8 cm long, 2.5–3 cm wide, oblong, shortly beaked long-stipitate, the stipe 10–15 mm long; seeds large, oblique, 8 x 10 mm, the curved margin not winglike, the hilum notch shallow.

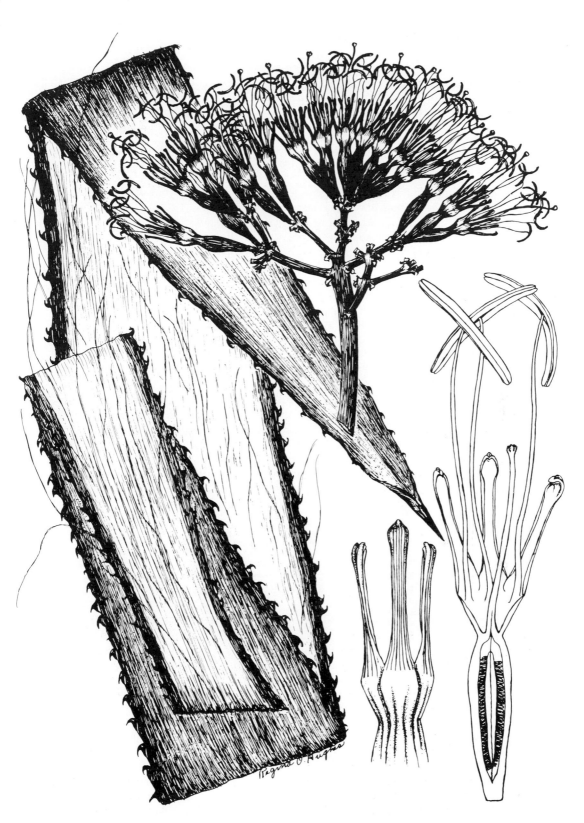

Fig. 20.23. *Agave rhodacantha* drawn from *Gentry & Gilly 10704;* leaf and flower umbel x ½; flower section x 1.7.

Type: Lundstrom in 1909, Mocorito, Sinaloa, *MO*. (As reported by Trelease, but the collection is no longer there.)

Trelease's description is very brief. The above description is based primarily upon 3 flowering specimens collected in Nayarit: *Gentry & Gilly 10806, 10786, 10704* (see Exsiccatae). Widely scattered along the more moist mountain slopes at elevations from 50–1,000 m from southern Sonora at least to about Bahía Banderas, Nayarit, and apparently SE to Oaxaca.

The very long, rigid leaves and large inflorescences with large, long-stipitate capsules distinguish this agave from its near relative, *A. angustifolia*. However, leaf size and armature of *A. rhodacantha* can become quite variable and some of the forms with small distal teeth are inseparable in themselves from *A. angustifolia*, like *Gentry & Gilly 10756*, 1 mile (1.6 km) east of Mazatán along road from Compostela, Nayarit, but this collection has the long-stipitate capsules of *A. rhodacantha*. It is not known if the two species hybridize, but some of the variants suggest they do. Some young plants received from the Tres Marías Islands, *Gentry 11565, 11567*, appeared to represent *A. rhodacantha* and are to be identified with the "giant Agave" referred to by Nelson (North Amer. Fauna 14: 7–13. 1899): María Madre Island, May 3–25, 1897, *Nelson 4264* (cap. & l.). During the 1940s and 1950s it was being exploited on María Madre Island for its long fibers with labor of the convicts incarcerated there.

Agave stringens
(Fig. 20.20)

Agave stringens Trel., U.S. Nat. Herb. Contr. 23: 114, 1920.

Leaves concave, thin and recurving, very glaucous, 1–2 cm wide, 60 cm long or more, with a dark conical spine about 2 mm wide and 8 mm long; teeth curved scarcely 5 mm apart, 1–2 mm long, very sharp and slender, red or brown, the intervening cartilaginous margin nearly straight.

Type: *Trelease in 1904*, Río Blanco barranca, near Guadalajara, Jalisco, *MO* (leaf only, infl. unknown).

The type and a specimen from a plant grown at Missouri Botanical Garden show thin narrow leaves with numerous slender teeth, both appearing immature. These with Trelease's brief description appear to represent a rare and distinctive species. However, it cannot be properly placed, even to section, until mature leaves and inflorescence are known.

Agave tequilana
(Figs. 20.1, 20.2, 20.3, 20.24, 20.25; Table 20.1)

Agave tequilana Weber, Mus. Nat. D'Hist. Nat. Bull. 8: 220, 1902.
  Agave palmaris Trel., Contr. U.S. Nat. Herb. 23: 116, 1920.
  Agave pedrosana Trel., ibid. p. 116.
  Agave pes-mulae Trel., ibid. p. 117.
  Agave pseudotequilana Trel., ibid. p. 119.
  Agave subtilis Trel., ibid. p. 116.

Plants surculose, radiately spreading, 1.2–1.8 m tall with short thick stems 3–5 dm tall at maturity; leaves 90–120 x 8–12 cm, lanceolate, acuminate, firm fibrous, mostly rigidly outstretched, concave, ascending to horizontal, widest through the middle, narrowed and thickened toward base, generally glaucous bluish to gray green, sometimes

Fig. 20.24. Cultivated *Agave tequilana* near Tequila, Jalisco. The leaves are generally cropped about twice during the last three to four years, which increases growth of stem and yield of sugars, the growers say. Photo by J. Fitz-Randolf.

cross-zoned, the margin straight to undulate or repand; teeth generally regular in size and spacing or rarely irregular, mostly 3–6 mm long through mid-blade, the slender cusps curved or flexed from low pyramidal bases, light brown to dark brown, 1–2 cm apart, rarely remote and longer; spine generally short, 1–2 cm long, rarely longer, flattened or openly grooved above, the base broad, dark brown, decurrent or not decurrent; panicle 5–6 m tall, large densely branched with 20–25 large diffusive decompound umbels of green flowers with roseate stamens; flowers 68–75 mm long on small bracteolate pedicels 3–8 mm long; ovary 32–38 mm long, cylindric, 6-ridged, with unconstricted short neck, slightly tapered at base; tube 10 mm deep, 12 mm wide, funnelform, grooved; tepals subequal, 25–28 mm long, 4 mm wide, linear, erect but withering quickly in anthesis, turning brownish and dry; filaments 45–50 mm long, bent inward against pistil, inserted at 7 and 5 mm above base of tube; anthers 25 mm long; "capsula ovata breviter cuspidata; seminibus semi-orbicularibus maximis; hilo sub-ventrali" (Weber loc. cit.).

Type: Weber's Fig. 1 (loc. cit.) may be taken to represent the type, a photograph taken by M. Diguet at Tequila, Jalisco. It shows a light glaucous plant with cross-zonations and regular evenly spaced teeth, typical of plants cultivated for tequila in Jalisco. Apparently, there are no specimens of this agave in European herbaria.

Agave tequilana is distinguished from its close relatives in *A. angustifolia* by its larger leaves, thicker stems, and heavier more diffusive panicles of relatively large flowers with tepals long in proportion to the relatively short tube (Figs. 20.2, 20.3). Since these differences are of degree rather than of distinct contrast, their separation as a species is nominal, but appears tenable for the Rigidae, where species are so hard to define. Certainly, the commercial trade with this important economic plant will profit by the maintenance of a simple binomial.

Plants closely related to *A. tequilana* grow wild on the semi-arid slopes west and south of Tequila, as along the road from Cocula to Tecolotlán, Jalisco. There is a large variable population in the oak woodland, some of which closely resemble the cultivated forms of *A. tequilana*. One would expect that the tequila growers made their original selections for planting from among these wild stocks. Unfortunately, there is no flowering material of such populations for study. This is also true of several cultivated forms described by Trelease. These were all described from leaf specimens from the cultivated tequila fields of Jalisco, are based on minor characters of leaf armature, and are here treated as synonyms of *A. tequilana*. There may be one or two exceptions to this allocation when flowering specimens can be obtained; for the present, however, I view them as selected forms of *A. tequilana*.

Some of the cultivated mescal plants growing in the state of Oaxaca resemble *A. tequilana* and I have identified two leaf collections from there as *A. tequilana* (see Exsiccatae). These forms appeared there in the 1960s, contrasting with the forms I had observed in the 1950s. When I inquired of a local Oaxacan mescal distiller, if his "mescal azul" plants had not been brought from Jalisco, he replied No, it was local stock. A man may be responsive but dissembling, or honest but ignorant, whereas a plant has a constant image. I rest with the morphological evidence, i.e., the broad bluish or whitish glaucous leaves, sturdy tapered spine, etc. My investigation of many agave questions have perforce been superficial.

Agave tequilana is the source of the famous distilled liquor, "tequila," the name taken from the town of Tequila, presumably a place name of Indian origin. The town, about 40 miles (65 km) northwest of the city of Guadalajara, has been the hub of the tequila industry for over 100 years. The tequila plantations cover thousands of acres over the Jaliscan Plateau region of western Mexico. The bluish gray variety, "mescal azul," now the productive favorite, lends a unique color aspect to the cultivated fields.

Table 20.2 outlines the production of tequila for 1963 through 1968. It was presented in a "brief" to the Alcohol, Tobacco, and Firearms Division, Internal Revenue Service, Washington, D.C. (O'Neill, 1969). It estimated that tequila imports into the United States would reach one million gallons by the end of 1971. The brief stated that it required about 15 pounds of agave to produce 1 liter of tequila. It was estimated that it required about 4.8 million agave plants to provide the 25 million liters estimated for 1968.

Figure 20.25 shows the trimmed stems or "cabezas," which are ready for transport to the distillery. These heads, which contain a meristematic mass of starchy tissue, are first steam-cooked in large ovens, which converts the starch to sugar. It is then macerated

Table 20.2. Tequila Production in Jalisco, 1963–68.

CALENDAR YEARS	MILLIONS OF LITERS	INCREASE OVER PREVIOUS YEAR
1963	19.7	5%
1964	20.0	1%
1965	20.5	3%
1966	22.1	8%
1967	23.5	6%
1968	25.0 (Est.)	7%

SOURCE: O'Neill (1969). National Association of Alcoholic Beverages, Inc.

Fig. 20.25. Cabezas of *Agave tequilana* on their way to the distillery; near Tequila, Jalisco, October 1952.

and the juice is placed in large vats, where anaerobic bacteria convert the sugar to alcohol. The fermented liquor is then distilled and the product, issuing as a colorless liquor with an alcoholic content of 37 degrees G. L., is fresh or raw tequila. It is consumable directly, but is generally held for further refinement in large oaken vats. Post-distillation modification may be done by fortifying with cane alcohol, coloring and flavoring with crude brown sugar, or simply aging in oak kegs. The industry is modernized, conducted by large companies with standard brands. It is economically important as a dollar earning product, not to mention the calories and stimulation, good or bad, that come to the consumers. The invention of the "margarita" cocktail and its popularity in the U.S. appears mainly responsible for the upsurge of tequila exports and the large expansion of tequila plantings in Jalisco and adjacent states. It gives one pause to consider the large effect of a civilized but relatively trivial refinement of a common class of product.

Nomina Incertae Sedis

Agave amaniensis Trel. & Nowell, Kew Bull. Misc. Info. 1933: 465.

A toothless form of vigorous growth producing superior fiber. Said to be related to Trelease's zapupe group. The circumstances of its origin are unknown, but it may be a hybrid between species grown in the East African Agricultural Experiment Station, Amani, Tanganyika (now Tanzania). It may belong with the Sisalanae. This was the last *Agave* species named by Trelease.

Agave collina Greenm., Proc. Amer. Acad. 32: 296, 1897.

This is a stout, short-stemmed plant with nearly white glaucous to pale green leaves with nearly black teeth and spine. It might well be treated as a variety of *A. angustifolia*, but it intergrades morphologically with other adjacent forms or is congenetical with more distant forms. It is frequent in middle elevations from Morelos to Guerrero. The leaves of the type collection are small, cut well above the leaf bases and, as Trelease noted (Herb. Mo), "evidently leaf from scape." My measurements of the dried flowers of the type are: o=25, t=16, s=19, tl=60 mm.

Agave costaricana Gentry, Allan Hancock Pacific Exped. 13: 195, 1949.

Known only from the type collection made in Salinas Bay, Costa Rica. The description indicates a plant much like *A. panamana* Trel., with which it should be compared. A series of specimens should be taken along the Pacific Coast of Central America. Elmore found it flowering in March, viz the type: *Elmore E18, AHFH*, MICH. Along rocky bank of dry stream bed, at or near Port Parker, Salinas Bay, Costa Rica, March 18, 1939.

Agave gutierreziana Trel., Contr. U.S. Nat. Herb. 23: 116, 1920.

This long-leaved agave apparently belongs in this group, but the sterile type leaves its affinities uncertain.

Agave longisepala Todaro, Hort. Bot. Pan. II : 34, t.31, 1878–92.

The sectional place of this taxon is uncertain. Berger described it under his Scolymoides (1915) but suggested it might be better belong with the Americanae. I know it only by Todaro's excellent plate. This shows a symmetrical plant with linear-lanceolate patulous leaves, and more diagnostic, long tepals (−40 mm) and long filaments inserted toward base of a short tube. Until found and recognized in nature, the taxonomic place of this plant will remain obscure.

Rigidae Exsiccatae

Agave aktites

SINALOA. *Gentry 11423,* DES. 8 mi. SE of Los Mochis, 4 January 1952 (l, photo). Coastal Thorn Forest, clay loam soil.

Gentry 11470, DES, **US.** Type:—Isla Lechuguilla, N of Topolobampo, 7 January 1952 (l, photo). Sand dunes captured by Coastal Thorn Forest.

Gentry 11573, DES. Isla Santa María off Topolobampo, January 1952 (l). Sandy island.

SONORA. *Gentry & Arguelles 22015,* DES. Near El Faro in Yavaros, 19 October 1966 (l, f). Coastal sand dunes with *Jatropha, Croton, Opuntia.* Transplants flowered in Des. Bot. Gard. 1974 & 1976.

Agave angustifolia

AGUASCALIENTES. *Rzedowski 14075,* ENCB. 6 km NE of Ojocaliente, Mncpo. Calvillo, 25 August 1960. 1,800 m. alt.

Rzedowski 14170, ENCB. Sierra del Laurel, near Calvillo. 28 Aug. 1960 (cap.) 2,200 m.

BELICE. *Bartlett 11547,* US. Pine Ridge, Duck River, 17 Feb. 1931 (l, cap).

CAMPECHE. *Schubert 1672,* DES, NA. Paraiso between road & ocean, Campeche–Champoton Rd., 1958 (l). "Small plants scattered along shore."

CHIAPAS. *Breedlove & Thorne 30357,* CAS, DES. Choreadero near Derna, Río de la Venta, Mun. de Ocozocuatla, 16 Dec. 1972. Tropical deciduous forest, elev. 800–1,000 m.

Gentry 12232, DES. 26 mi. W of Cintalapa, Chiapas along Hwy. to Oaxaca, 29 Sept. 1952 (l, photo). Edge of Pinal on open hilltop with grass.

Miranda 2725, MEXU. Tuxtla Gutierrez, 21 April 1943.

Miranda 6491, MEXU. Above Chacona Hwy. S. of Fernando, N. of Tuxtla Gutierrez; 6 August 1950 (l, cap.).

CHIHUAHUA. *LeSeur 1172,* ARIZ, UC, TEX. Río Bonito, Dec. 1936 (l, f).

COSTA RICA. *Gentry s.n.,* DES. Alajuela & San Jose, April 1965 (photo). Planted as an ornamental about houses & perhaps used as fiber.

Elmore E-18, AHFH, MICH, DES. Port Parker, Salinas Bay, 24 March 1939 (f, l; type collection of *A. costaricana* Gentry, see Nomina Incertae Sedis).

DURANGO. *Gentry 22091,* DES, MEXU, US. 34–35 mi. S. of Cd. Durango, along road to Mesquital, 3 Nov. 1966 (l, f, photo). Desert chaparillo, alt. 5,000 ft.

Gentry & Gilly 10585, DES, MEXU, MICH, US. 8–10 mi. S. of Boca Mesquital, along road from Cd. Durango to Mesquital, 11 June 1951 (l, f). Semiarid, open, dispersed deciduous woodland with *Bursera, Ipomaea, Fouquieria,* etc., ca. 5,500 ft. elev.

Soule 2086, MO. Camp 2, Inferno, on Río Mezquital, Dur., 17 Nov. 1970; elev. 4,500 ft.

Soule 2124, MO. Tecomates along Río Mezquital, Dur., 20 Nov. 1970; elev. 4,200 ft.

GUATEMALA. *Deam 6240,* MO, US. Fiscal, 7 June 1909; elev. 3,700 ft., ledge in canyon (l, annot. Trel. as *A. donnell-smithii).*

Gentry & Gentry 23624, DES, US. 4½ miles E. of Chiquimula Hwy. along side road to Olopa, 15 Dec. 1975 (l). Cult. in small fiber-maize field with wild & cult. Furcraea, elev. ca. 2,800 ft.

Smith 2085, US. Escuintla, Depto. Escuintla, 1890; alt. 1,100 ft. (l, f). Type of *A. donnell-smithii* Trel.

Trelease 6, OSH. Hillside at El Rancho, April 1915 (l). Type of *A. sicaefolia* Trel.

GUERRERO. *Gentry et al. 5865,* DES. Ca 3 mi. N of Taxco, along road to Cuernavaca, Morelos, 11 August 1953 (l, f). Cult. at Huntington Botanical Garden, flowered 31 July 1962.

Gentry & Fox 11996, DES. 2–3 mi. S of Ixcateopan, 15 Aug. 1952 (photo). Open woody grassy hillslope; 5,000–6,000 ft.

Gentry & Fox 11999, DES. 3–4 mi. N of Pachivia, 15 Aug. 1952 (l, photo). Open grassy slope on limey shale.

Iltis & Cochrane 154a, DES, WIS. Along Mexican Hwy. 51 at km. 55, Rancho Viejo, ca. 5 km. E of Teloloapan, 18° 20′ N, 99° 49′ W, 11 April 1975; alt. 1,490 m. (l, cap., photo).

Moore 5265, BH, MEXU, US. Canyón de Zopilote, vicinity of km. 264 near Venta Vieja, 4 Oct. 1949. Alt. ca 2,300 ft. (l, f).

Moore 5519, BH, MEXU, US. Ca. 16 km. from Iguala on road Teloloapan, 5 Nov. 1949 (l, f).

Ogden 5114, DES, MEXU, NA. 12 km. N. of Chilpancingo, 4 Jan. 1951 (l).

Palmer 340, US. Acapulco and vicinity. 1895.

HONDURAS. *Gentry 12519,* DES. Perspire, Depto. Choluteca, 27 Feb. 1953 (l, f). Cult. at Huntington Botanical Garden, San Marino, California, flowered 3 July 1962.

Gentry 21464, DES. Between Perspire & Savannagrande along Hwy. 22 Oct. 1965 (l, f, photo). Tropical foothills: alt. 2,000–2,800 ft.

JALISCO. *Gentry 10464,* DES, MEXU, MICH, US. Tierra Blanco between Villa Corona & Tecolot-

lan, 17 May 1951 (l). Cut-over woodland with secondary "monte" on volcanic terrain. Elev. 4,500–5,000 ft.

Gentry & Gentry 23526, DES, MEXU, MICH, US. 4 mi. SW. of Tecolotlan along road to Autlan, 6 Feb. 1975 (l, cap, photo). Grassland on sedimentary hills, elev. ca. 3,900 ft.

Gentry & Gentry 23527, DES, MEXU, MICH, US. 11 mi. E. of Tecolotlan along road to Autlan, 6 Feb. 1975 (l, cap). Short oak forest on volcanics. Elev. ca. 5,600 ft.

Ogden & Gilly 51177, DES, MEXU, MICH, NA. At km. 98 along the road between Cocula & Tecolotlan, 11.5 mi. from Tecolotlán, 2 April 1951 (l, photo).

Nelson 6525, US. Hills fronting Lake Chapala, Jal. 5 Jan. 1903.

STATE OF MEXICO. *Hinton 6374,* U.S. Acatitlan, Dist. Temascaltepec, Mex. 8/1/34 (l, f).

MICHOACAN. *Gentry s.n.,* DES. Ca. 8 mi. E. of Uruapán, Nov. 1966 (photo). Volcanic hillside in open pine woods; alt. ca. 4,500 ft. Infrequent at this altitude.

Nelson 6930, MEXU, US. In canyon 20 mi. S. of Uruapán, 22 March 1903.

Nelson 6935, MEXU, US. Base to summit of Volcano Jorulla, 28 March 1903 Mich.

MORELOS. *Gentry et al. s.n.,* DES. Near Cuernavaca, Nov. 1961 (photo). In open *Ipomoea*—deciduous tropical forest & along fences & roads.

Matuda 37696, MEXU. On road between Cuernavaca & Tepoztlan. 15 April 1964 (l, f, cap) Alt. 1,600 m.

Miranda 1188, MEXU, Oaxtepec, 23 March 1941.

Miranda 1466, MEXU, Jojutla, 5 July 1941.

Miranda 1514, MEXU, Miacatlan, 3 Aug. 1941.

Ogden 5136, DES, MEXU, NA. Near Hwy. to Atotonilco, 12 Jan. 1951 (l). Rocky open hillside.

Pringle 6349, MEXU, MO, US. Type of *A. collina* Greenm. Hillsides near Cuernavaca, 5,000 ft., fls. 17 June 1896. (Young leaf, as Trelease noted "evidently from near scape." Fl. o=25, y=16, s=19, tl=60.)

Rose & Hough 4356, US. Near Cuernavaca.

Rose & Painter 8050, US. State of Morelos, 27 Aug. 1903.

Trelease 4, 5, MO. Cuernavaca, 8 Aug. 1903 (l, 4 sheets).

Vasquez 298, MEXU, Yantepec, Canyón de Lobos, 20 Aug. 1972. Alt. 1,250 m.

Vasquez 2406, DES. Cerro de Gallo, km. 5 on Hwy. to Chinameca, 25 March 1970 (l, f).

NAYARIT. *Rose s.n.,* US. Acaponeto, July 30, 1897. Cult.

NICARAGUA. *Gentry 21577,* DES. Managua. 6 Nov. 1965 (l, f). Cult. as an ornamental in front yard.

Gentry 21578, DES. 35 mi. NW of Juigalpa, Depto. Chontales, 7 Nov. 1965 (l, f). Sparsely scattered in region, wild. Open pasture lands, alt. ca. 500 ft.

OAXACA. *Conzatti 3987,* US. Cerro Fahame, Santa Catarina, Dist. de Nochixtlan, Oax., 15 June 1920, 1,700 m (l, f).

Gentry 12235, DES, MEXU, MICH, US. Pan American Hwy. & Trans-Isthmus Hwy., 41 km. E of Tehuantepec, 29 Sept. 1952 (l, f). Thorn forest near sea level.

Gentry 12245, DES, MEXU, MICH, US. 45 mi. NW of Tehuantepec along Pan American Hwy., 30 Sept. 1952 (l). Foothills with short-tree forest. Plentiful on open hills of savanillas.

Gentry & Arguelles 21334, DES. Ca. 2½ mi. SW of Cameron along side road, 29 Sept. 1965 (l, photo). Rocky volcanic slope with short-tree forest; extensive population on open wooded rocky slopes; alt. 2,800 ft.

Gentry 22380, DES, MEXU, US. 6 mi. S of Candelario, Dist. Pochutla, 27 Oct. 1967 (l, f). Tropical savanna; alt. ca. 1,000 ft.

Gentry et al. 20261, DES, MEXU, US. 29 mi. NE of Huajuapán, along road to Tehuacán. 5 Aug. 1963 (l, f). Savanilla on volcanics; alt. ca. 6,000 ft.

Gentry 12081, DES, MEXU, MICH, US. 6–8 mi. NE of Cd. Oaxaca along road to Ixtlan, 28 Aug.–1 Sept. 1952 (l, f, photo). Steep rocky slopes with disturbed, mixed deciduous forest, 5,000–6,000 ft.

Gentry et al. 20265, DES. 3–4 mi. E of Cd. Oaxaca, 6 Aug. 1963 (photo). Cultivated; alt. 5,000 ft.

Iltis et al. 1226a, WIS. Ca. 20 km. NE of Oaxaca along road to Ixtlan, 23 Aug. 1960 (l, cap).

Miranda 1188, MEXU. Oaxtepec, 23 March 1941.

Miranda 1514, MEXU. Miacatlán, 3 Aug. 1941.

Miranda 1466, MEXU. Jojutla, 5 July, 1941.

Miranda 4651, MEXU. Road between Cuicatlan & Reyes Papalo, 15 Sept. 1948.

Nelson 2554, US. Between Totalapa and San Carlos, April 1895; alt. 3,000–3,800 ft.

Nelson 2573, US. From Huilotepec, Oax., 4–11 May 1895; alt. 100 ft. (f, bulb).

Ogden & Gilly 51186, DES, MEXU, US. At km. 721 on Tehuantepec–Oaxaca de Juarez Hwy. 47 mi. NW from Tehuantepec; 14 April 1951 (l, cap, photo).

Ogden & Gilly 51184, DES, US. 13 mi. from Tehuantepec along Hwy. to Cd. Oaxaca, 14 April 1951 (l, bulbil, photo). Dry hills in mixed shrub growth.

Ramirez L. s.n., MEXU. Monte Alban.

Rose & Hough 4591, US. San Felipe & Monte Alban, Oax. June 1899.

Rose & Hough 4651, US. Near City of Oaxaca, June 1899.

PUEBLA. *Gentry et al. 20349-A,* DES. Near Acatepec along road to Huajuapán, 19–27 Aug. 1963 (l). Arid thorn forest over limestone hills, alt. 5,500–6,000 ft.

Martinez 5414, MEXU. Cercanias de Axtlico, 2 Oct. 1956.

Miranda 1397, MEXU. Matamoros, 22 May 1941.

Miranda 2073, MEXU. Cd. Tlaacoctli, 23 July 1942.

Ogden 5134, MEXU, US. 15 mi. from Puebla along road to Tlazcala, 12 Jan. 1951.

Nelson 7087, US. Atlixco, Puebla. June 1903 (l, f).

QUINTANA ROO. *Lundell 7817,* MEXU. Coba, E of ruins, 5 July 1938 (l, cap). In advanced deciduous forest.

SALVADOR. *Calderon 2082,* US. Hac. El Platanar, Dept. San Miguel, Salvador, Jan. 1924, "wild, not cult. for fiber."

Calderon 2083, US. Finca El Platinar, Dept. San Miguel, Salvador, Jan. 1924 (toothless form; probably cultivated).

Calderon 2148, US. Nueva Concepción, Dept. Chalatenango, Salvador, 1924 (4 unpressed lvs & infl).

Rohweder 1078, MO. Cerro San Jacinto, San Salvador, 12 Dec., 1950.

Rohweder 1074, MO. Ostrand Acajutla, Depto. Sonsonate, 10 Aug. 1951.

SINALOA. *Gentry 11425,* DES, MEXU, MICH, US. Topolobampo, 5 Jan. 1952 (l, f, photo). Volcanic cerro; thorn forest.

Gentry 11425-A, DES. Vicinity of Los Mochis, Jan. 1952 (f). Coastal plain thorn forest; scattered over coastal plain among shrubbery and/or upon adjacent cerro slopes.

Gentry 11481, DES, MEXU, US. Isla Piedron, Mazatlán, 15 Jan. 1952 (l). Littoral sand hills.

Gentry 11485, DES, US. Isla Creston, Mazatlán, 20 Jan. 1952 (l). Rocky maritime slope.

Gentry 19655, DES. S. bajadas of Cerro de Fuerte; 18–24 mi. N of Los Mochis, 15 Dec. 1961 (l). Rocky volcanic slopes with coastal thorn forest; elev. 200–1,000 ft.

Gentry & Arguelles 18385, DES. 3 mi. SE of Los Mochis, 15 Dec. 1959 (l, f). Coastal thorn forest; alt. ca. 30 ft.

Gentry & Preciado 11571, DES. Near Jipón by Bahía Topolobampo, Jan. 1952 (l).

Howell 10533, DS. Creston Island, Mazatlán, 1 Aug. 1932 (l).

M. E. Jones 23463, MO. San Blas, 1 Feb. 1927.

Mason 1756, CAS. Maria Madre Isl., Tres Marías Islands, 18 May 1925 (l, f). Abundant along the shore and hilltops.

Ogden & Gilly 51214, 51216, DES. Ca. 5 km. NW of village of Pericos, 26 May 1951 (l). Scattered in very dry scrub growth.

Ortega 6688, US. Vicinity of Culiacán, Sin. (f).

Rose s.n., MO. Mazatlán, Apr. 1910 (l).

Rose 3154, US. Concepción, Sin. 4 July 1897.

Rose 1372, US. Near Palmito, Sin., 12 July 1897.

Rose et al. 13323, US. Vicinity of Topolobampo, 23 Mar. 1910 (l, cap).

Rose et al. 13627, US. Vicinity of San Blas, 28 Mar. 1910.

Rose et al. 13870, US. Vicinity of Mazatlán, 1 April 1910.

Trelease s.n., MO. Creston Isla, Mazatlán 5 March 1904 (3 sheets l, buds). Type of *A. pacifica* Trel.

Trelease s.n., MO. Topolobampo 3 March 1904 (l, photos).

SONORA. *Abrams 13315*, DS. 20 mi. NW of Hermosillo, 13 April 1932 (l).

Brandegee s.n., UC. Hermosillo, 16 May 1892.

Gentry 10212, DES, MEXU, MICH, US. 4 mi. N of Rancho Tinaja, Altar Dist., 27 Feb. 1951 (l, photo). In sandy arroyo of outwash bajada; only patch known in the area.

Gentry 10258, DES, MEXU, MICH, US. San Bernardo, 8 March 1951 (l). Widely scattered in short-tree forest.

Gentry 10280, DES, MEXU, MICH, US. 8 mi. E of Ures, 11 March 1951 (l, photo).

Gentry 11346, DES, MEXU, MICH, US. Islota Almagre Grande, Bahía Guaymas, 15 Dec. 1951 (l). Coastal desert shrub with *Pachycereus* forest. Occurring on the shade slopes of both islands in the harbor, but lacking on sun slopes.

Gentry 11364, DES, Alamos, 17 Dec. 1951. Previously cultivated for mescal.

Gentry 16644, DES, US. 6–8 mi. NW of Bacadehuachi, 21 May 1957 (l, photo). Rocky slope of open semi-arid woodland; alt. ca. 3,000 ft.

Gentry & Arguelles 19875, DES, MEXU, US. Near Navojoa, 28 March 1963 (l, f). Coastal plain; thorn forest.

Gentry et al. 19359, DES, US. Sierra Tecurahui, ca. 20 mi. E of Alamos, 26–28 Oct. 1961. Short-tree forest; alt. 2,000 ft.

Goldman 302, US. Camoa, Rio Mayo, 25 Jan. 1899 (l, f, photo).

Johnston 3085, CAS. Guaymas Harbor, 14 April 1921. On scoria-covered islet in harbor, plants frequent, in groups.

Jones 23462, UC, MO. Guaymas, 26 Jan. 1927.

Lindsay 1148, DS. Hill in Guaymas Bay, 24 March 1937 (l, f).

Rose 1232, US. Guaymas, Son. 5–11 June 1897.

Rose et al. 12430, US. Hermosillo, 5 Mar. 1910 (l, bud).

Rose et al. 12620, US. Vicinity of Empalme, 11 March 1910 (l, cap).

Rose et al. 12847, US. Vicinity of Alamos, 14 March 1910 (l, cap).

Rose et al. 12954, US. Vicinity of Alamos, 16 March 1910; cult.

Trelease 391, MO. Ures-Hermosillo, 18 Aug. 1900 (l, photo). The form with heavy black teeth, glaucous. Photo shows a zoned glaucous leaf. Type of *A. yaquiana* Trel.

Wiggins 6463, DS. 1 mi. S of Peón between Río Yaqui & Empalme, 7 March 1933.

Wiggins s.n., DS. 7 mi. N of La Colorada, Mina San Fernando Road, 5 May 1948 (l, f).

Wiggins & Rollins 317, DS. 4 mi. NE of Colorado on road to Mazatlán, 6 Sept. 1941.

TAMAULIPAS. *Dewey 585*, UC. Jaumave, 30 May 1903.

Wooten s.n., US. Cult. at Saladito, 23 June 1919 (f).

VERACRUZ. *Dewey 161*, MO. US. Ilda. la Soledad, 72 km. W. of Tuxpam, Feb. 11, 1907. Cult. for fiber, "zapupe azul" (l with dark brown remote teeth, dark spine). Type of *A. zapupe* Trel.

Endlich 1160B, MO. Huatusca, Finca El Mirador, 10 March 1906 (l, just ordinary *angustifolia!*) Type of *A. endlichiana* Trel.

Endlich s.n., MO. Cult. Tuxpam, 31 Oct. 1906.

Fox & Arguelles 5820, DES, US. Puente Nacional; 1 mi. E on road to Cd. Vera Cruz. 14 July 1952. Partly cleared land, densely brushy in part.

Gentry 12287, DES. Above Acultzingan, along hwy., 16 Oct. 1952 (photo). Open rolling abandoned fields.

Gentry 15879, DES. Near Puente Nacional, 11 Nov. 1955 (l, f). Thorn forest. Common through the brushy disturbed low forest. Alt. below 1,000 ft.

Gentry & Dorantes 23374, DES, MEXU, US. Villa Rica, where Cortez landed, on rocky point 8–10 m above sea, 28 Feb. 1974 (f, cap).

Gilly et al. 7031, DES. Near Salinas, N of Alvarado along hwy. from Cd. Vera Cruz, 5 Aug. 1951 (l, f).

Lespinasse June 1908, MO. Tuxpan, "zapupe silvestre" (l, 2 sheets, rather broad sword-leaf with deltoid teeth with curved cusps. Numerous spines on another sheet). Type of *A. aboriginum* Trel.

Lespinasse s.n., MO. Tuxpam, V. C., June 1908 (l, 6 sheets), "zapupe verde and zapupe azul."

Lot 795, 796, CAS, MEXU. Isla Verde, 22 Nov. 1970. Coastal dunes.

Lot 1370, MEXU. Isla de Enmedio, Frente a Anton Lizardo, 17 June 1971.

Ogden & Gilly 51158, US. Between Paso de Ovejas and Puente Nacional along Jalapa–C. Veracruz hwy, 18 Mar. 1951 (l, photo).

Vincent-Palmer; *Dewey 11*, MO, US. Isla de Juana Ramirez, 10 March 1910.

YUCATAN. *Gaumer 950*, US. Isamal.

Gaumer 23164, MO. San Pedro, Jan. 1915.

Goldman 608, US. Progreso, Feb. 24–Mar. 5, 1901 (l, f, det. by Trelease as *A. ixtli*). Short broad f.

Lundell 8110, MEXU. Near Telchac, May–Aug. 1938. In scrub on low sand dunes.

Matuda 37507, 37509, MEXU. From the coast of Campeche to Progreso, 11 Jan. 1966. Near the ocean.

Miranda s.n., MEXU. Between Merida and Progreso, 15 Sept. 1955. In low deciduous forest, on rocks.

Purpus 8884, UC. Los Conejos. Feb. 1923 (l, cap).

A. Schott 201, MO, US. Merida, Yucatan, 5 Feb. 1865, "The common wild plant, Chelem." (l, f).

ZACATECAS. *Rose 2755*, US. Near Plateado, 2 Sept. 1897 (l, f).

Rose 3537, US. Near San Juan Capistrano, 18 Aug. 1897 (l, cap).

Agave angustifolia var. *deweyana*

TAMAULIPAS. Type: **Dewey 648, MO, 649, US.** Victoria, Tamps., 2 Feb. 1907. "Cultivated for production of fiber, Zapupe verde."

Gentry 12265, DES, MEXU, MICH, US. Carrisal, N. of Manuel, transplanted from Cd. Vict., 10 Oct. 1952 (l, photo). Cultivated on heavy alluvial coastal plain.

Ogden et al. 51115, DES, US. Rancho La Negra, ca. 10 mi. S of Cd. Victoria on the hwy. to Valles and Cd. Mexico, 17 Feb. 1951 (l, photo). Cultivated for fiber.

VERACRUZ. *Dewey 648*, US. Victoria, 2 Feb. 1907.

Lespinasse A B C, ARIZ, MO. Tuxpan, V.C., June 1908. "Tontoyuca zapupe."

Palmer s.n., US. Isla Juana Ramirez S. of Tampico, V.C., 10 Mar. 1910.

Agave angustifolia var. *letonae*

GUATEMALA. *Gentry s.n.*, DES. Near Laguna Retana, Depto. de Jutiapa, April 1965 (photo). Observed infrequently, cult. for fiber.

SALVADOR. *Calderon 2079, 2080*, US. Haciendas San Antonio & El Platinar, Depto. San Miguel, Jan. 1924 (l, 5 sheets, show the light gray glaucous color).

Gentry 21706, DES, US. 7 miles W of San Miguel along hwy., 21 Nov. 1965 (l, photo); alt. ca. 1,000 ft. Cult. for fiber.

Milner s.n., MO, US. Type. Cult at Sucesión Letona, San Miguel, Oct.–Nov., 1923. (l margins only & spines. A small-toothed leaf of the *angustifolia* complex. US., annotated by Dewey as "cotype.")

Agave angustifolia var. *marginata*

CALIFORNIA. *Gentry 10169*, DES. Huntington Botanical Gardens, San Marino. 9–15 Jan. 1951 (l). Cultivated.

COSTA RICA. *Gentry s.n.*, DES. San José, Mesa Central, April 1965 (photo). Common as an ornamental.

INDIA. *Oza s.n.*, DES. Motibana Palace Compound, Baroda, 25 Aug. 1972 (l, cap). Growing wild.

Agave angustifolia var. *nivea*

GUATEMALA. *Gentry 21366*, DES. El Rancho, Dist. Progreso, 8 Oct. 1965 (l). Cultivated for fiber, mostly as fencelines.

Gentry 23627, DES. El Rancho, Depto. de El Progreso, 16 Dec. 1975. Cult. in hedgerows along houses; elev. ca. 900 ft.

Trelease 11, OSH. Type: About El Rancho, Depto. Progreso, April 1915.

Agave angustifolia var. *rubescens*

CALIFORNIA. *Gentry 10092*, DES. Huntington Botanical Gardens, San Marino, Jan. 1951 (photo). Cult.

GUERRERO. *Ogden 5118*, DES, US. Along road between Iguala and Chilpancingo near km. 261, 4 Jan. 1951 (l). Open dry soil.

JALISCO. *Gentry & Gentry 23458*, DES, MEXU, MICH, US. 21 mi. SW of Valparaiso along road to Huejuquilla, 11 Jan. 1975 (l). Limey volcanic rocks on sun slope; elev. ca. 6,500 ft. Common about Huejuquilla valley.

OAXACA. *Conzatti 4312 ½*, US. De San Pablo a Ayoquezco, Dist. de Zimatlan, 25 Nov. 1921, 1,400 m. (l, f).

Gentry 22380, DES. 6 mi. S of Candelario, Dist. Pochutla, 27 Oct. 1967 (l). Tropical Savanna, alt. ca. 1,000 ft.

Gentry & Gentry 23654, DES. 10 mi. W. of Tehuantepec along hwy. to Oaxaca, 7 Jan. 1976 (l, f). Sparse short-tree forest, elev. ca. 500 ft.

Agave angustifolia var. *sargentii*

PUEBLA. *Ogden 5134*, DES. 15 mi. from Puebla, on Puebla-Tlaxcala Rd., 12 Jan. 1951 (l). Open field, perhaps escape from cultivation.

MORELOS. *Camp 30*, MICH. 20 km. NE of Cuautla, 29 July 1950 (l, f). "Open hillside with volcanic soil; flower white."

Agave breedlovei

CHIAPAS. Breedlove 37904, CAS. Type. 8 km E. of Las Margaritas along road to La Soledad, Mun. de Margaritas, 15 Sep. 1974 (f panicle, l. Panicle slender 18 cm long, sparsely fl'ed. with few small bractlets).

Breedlove 36756, CAS. 5 km W. of Rizo de Oro, Mun. de Cintalapa, 26 Aug. 1974; elev. 900 m. (l, infruct., bulbiferous & broad caps. like *angustifolia*).

Breedlove & Thorne 30509, CAS. 3–5 km N of Cintalapa, Mun. de Cintalapa, 22 Dec. 1972; elev. 900 m. (l of *breedlovei*, f of *seemanniana*).

Breedlove 11764. CAS. Hwy. 190 in Paraje Granadillo, Mun. de Zinacatan, 14 Aug. 1965; elev. 4,500 ft. (Small rosette with abnormal f. shoots. Questionably referred here.)

Agave cantala

HORTICULTURE. *Dewey in 1906*, MO. Manila, Philippines (fiber, bulbils, buds. Other sheets from Palmero, Curracao, etc. Todaro had trouble recognizing *A. cantala* and *A. rumphii* Jacobi.).

Franseschi s.n., MO. Sta. Barbara, Cal. 22 Oct. 1908 (l, f).

Gentry 10167, 23676, DES. US. Huntington Botanical Gardens, San Marino, Calif., 9–15 Jan. 1951 & July–Aug., 1977 (l, f, bul., photos). Cult. Block 20, in partial shade.

Gentry s.n., DES, Fars Province, Iran, June 1955 (photo).

Thompson 59, MO. Los Angeles, Calif. 21 July 1909 (l, photo).

Yan 11715, UC. Canton Christian College, Kwangtung, China, 20 June 1964 (l, f).

Agave cantala var. acuispina

HONDURAS. *Gentry 20694*, DES, US. Coyoles, Depto. de Yoro, 7 April 1965 (l, f, bulbil, photo). In town compound, alt. 1,000–1,500 ft.

Gentry & Freytag 21467, DES, US. 5½ mi. E of road fork by Tulanga along road to Juticalpa, 26 Oct. 1965 (l, f). Cultivated along fence, alt. 2,900 ft.

SALVADOR. Calderon 2084, US. Type: Hacienda El Platanar, San Miguel, "in hedges, maguey; not used for fiber."

Calderon 2082, same loc., Jan. 1924.

Agave datylio

BAJA CALIFORNIA. *Brandegee s.n.* CAS. La Paz, Cape Dist., 4 Nov. 1891 (l, f).

Brandegee 581, UC. San Pedro, 29 Oct. 1891 (l, f, cap).

Brandegee s.n. UC. Paso de los Dolores to Lake Ramon, 4 April 1889 (l, cap).

Hastings & Turner 64-373, ARIZ. 1 mile N of Rancho El Obispo, 20 Oct. 1964; elev. 450 ft. (l, bud).

Gentry 11200, DES, MEXU, US. About 4 miles E of La Paz, Cape Dist., 29 Sept. 1951 (l, f, photo).

Moran 3553, SD. Ensenada de los Muertos, Cape Dist., 1 April 1952 (l, cap).

Nabhan et al. s.n. ARIZ, DES. 10 miles S of La Paz, 7 Nov. 1978 (l, f).

Peters 124, UC. Los Planes, Cape Dist., arroyo bottom, elev. 300 ft., 27 Mar. 1948. "mescal."

Rose 1302, US. La Paz, 14 June 1897.

Wiggins 11501, CAS. 9 miles W of La Paz.

Wiggins 15475, CAS, DS, MEXU. Rancho del Obispo (Magdalena Plain), alt. ca. 150 m, 15 Nov. 1959 (l with long blade of *A. datylio* and remote teeth of *vexans*).

Agave datylio var. vexans

BAJA CALIFORNIA. *Brandegee s.n.* UC. Purisima ?, 1899 (infl).

Gentry 10302, DES. Rancho Panales, Arroyo Purisima, 31 March 1951 (l, f, cap). Rocky sedimentary slope; cardon-pitaya-*Bursera-Larrea-Opuntia*-etc.

Gentry 4322, DES. Comondú, 10 Mar. 1939 (f). Hill slope, common over coastal plain & foothills.

Gentry & Cech 11292, DES. 23 miles SW of Comondú, 16 Oct. 1951 (l, photo). Sandy valley bottomland.

Hastings & Turner 64-377, ARIZ, CAS. 6 miles by road W of San Luis Gonzaga (NW of La Paz), elev. 400 ft., 21 Oct. 1964, flowers yellow with green tinge (l, f).

M. E. Jones 23751, MO. La Paz, 15 Nov. 1925 (f).

Nelson & Goldman 7237, US, Type. About 5 miles SW of El Potrero, 31 Oct. 1905; alt. 1,000–2,400 ft. (l, cap).

Rose 1302, US. La Paz, 14 June 1897.

Rose 16540, US. La Paz, 29 March 1911.

Wiggins 15205, CAS. Mesa 2 miles N of Arroyo San Gregorio, 27 Oct. 1959 (l, f).

Agave decipiens

FLORIDA. *Britton 544*, MO, NY. Border of Hammock, Boot Key, 7–12 April 1909.

Britton 66, NY. Top of cliffs, 2 miles S of Miami, 19 Nov. 1904.

Curtiss 2836, US. Rocks on shore of Bay Biscayne, March 1882 (l, f).

Dewey 1513, MO. Harbor Keys, E. of Kemps Channel, 14 June 1910.

Killip 31427, US. Big Pine Key, N. of railroad near coast, 12–18 Feb. 1935 (l, f, photo).

Killip 41053, US. Big Pine Key, Monroe Co., 23 March 1951.

Killip 41901, US. Big Pine Key, 8 Feb. 1952; shell beach, coastal hammock. "Lvs 60–90 cm long when developed (young ones collected)."

McCarty s.n., MO. 19 April 1895, Ankona.

Moldenke 5833, NY. Roadside, Lower Matecumbe Key, Monroe Co., 18 March 1930.

Small 7367, US, MO, NY. Hammocks, Meigs Key, Monroe Co., 18 Jan. 1916.

Small & Carter s.n., NY. Bull Key, opposite Lemon City, 6 Nov. 1903.

Small & Carter 985, NY. Between Cutler & Black Point, 13 & 16 Nov. 1903.

Small & Small 4625, NY. Hammocks between Miami & Cocoanut Grove, Nov. 26–Dec. 20, 1913 (f).

Standley 12812, US. Vicinity of Marao, Lee Co., 25 Feb. 1916; "shell mound."

HORTICULTURE. *Gentry 19749,* DES. Huntington Botanical Gardens, San Marino, Calif. 1 June 1962, cult. (l, f).

Agave fourcroydes

BAJA CALIFORNIA. *Gentry & Cech 11268,* DES, US. El Triumfo, Cape District, 8 Oct. 1951 (l). Cultivated. Said to have probably come from the plantation W of town, now long abandoned.

Rose 1301, US. La Paz, 14 June 1897, in yard (l, f).

COSTA RICA. *Tonduz 17552,* US. San Jose, Costa Rica—from Paris, France. Jan. 1911.

GUATEMALA. *Gentry s.n.,* DES. Along new hwy. a few mi. E of Sanarate, Depto. de Progreso, May 1965 (photo).

SINALOA. *Gentry 11477,* DES, US. Pericos, 9 Jan. 1952 (l, photo). Cultivated in a large fiber plantation. This plantation started from bulbils brought from Yucatan about 1890.

Rose 1811, US. Rosario, 21 July 1897 (l, f).

SONORA. *Gentry 11361,* DES, US. Obregon plantation, 12 mi. SE of Cd. Obregon, 17 Dec. 1951 (l). In plantation on argillaceous plain.

YUCATAN. *Goldman 612,* US. Progreso, Yucatan, Feb.–Mar. 1901 (l, f).

HORTICULTURE. *Berger s.n.,* US. 6 Aug. 1909. La Mórtola (l, f, 2 sheets, large, 70 mm long, with deep ample tube; fils. inserted midway in tube).

Gentry s.n., DES. Cult. Desert Botanical Garden, Phoenix, Ariz. April–May, 1978; poled through roof of lath house.

Gentry & Kimnach, 23680, DES. Cult. Huntington Botanical Gardens, San Marino, Calif., 7 June 1977; large old clone with several rosettes 2 m tall & others younger.

Agave karwinskii

OAXACA. *Ernst 2439,* US. On San Lorenzo Road N of Mitla. 3 Feb. 1966.

Gentry 12049, DES, MEXU, MICH, US. Near Mitla, 27 Aug. 1952 (l, f, photo). Cultivated as a fence plant along roads and fields.

Gentry 22354, DES, MEXU, US. Mitla, 20 Oct. 1967 (cap). Cultivated on rocky limestone slope along margin of field. Much used as a fence plant.

Kirkby 2722, US. Barranca de Río Grande, near Mitla, 11 Feb. 1966 (l, cap).

Roever s.n., DES, Mitla (l).

Rose & Rose 11302, 11310, US. Mitla, 5 Sept. 1906 (l, infl. photo).

PUEBLA. *Gentry et al. 20357,* DES, US. 4–10 mi. SW of Tehuacán along road to Zapotitlán, Aug. 1963 (l, photo). Arid thorn forest over limestone hills; alt. 5,600–6,000 ft. Scattered populations in arid thorn forest; also cultivated for fences.

Smith et al. 4063, US. Near Cerro Colorado beyond La Cruz, July 1961; alt. ca. 1,000–1,800 m.

Trelease s.n., MO. Cerro Colorado, La Huerta, Tehuacán, 13 Aug. 1903.

Trelease 19, MO. Tehuacán, 5 Feb. 1905.

Agave macroacantha

OAXACA. *Purpus s.n.,* US. Cerro Verde, 1908, Cult. Bot. Gard. Darmstadt. 1913 (Berger leaf).

Rose & Hough 4686, US. Cholula, 25 June 1899.

PUEBLA. *Gentry & Engard s.n.,* DES. Cultivated in Desert Botanical Garden, Phoenix, Arizona, June, 1973, from Zapotlán de las Salinas (l, f).

Gentry et al. 20242, DES, MEXU, **US.** Neotype. 21 mi. SE of Tehuacán along road to Teotitlán, 5–22 Aug. 1963 (l, cap, photo). Short-tree forest; alt. 3,800–4,500 ft.

Endlich 1929, MO. Tehuacán, Apr. 1907 (var. *latifolia*).

Rose & Rose 11263, US. Tehuacán, 1 Sept. 1906.

Smith et al. 3553, US. Near Coxcatlán, Tehuacán area, July 1961; alt. 1,000–1,800 m. (l, infl).

Smith et al. 3663, US. W of Río Salado near Petlanco, 12 July 1961; alt. 1,000–1,500 m. (l, infl).

Trelease s.n., MO. Tehuacán, 11 Aug. 1903.

Trelease s.n., MO. Cerro Colorado, Tehuacán, 13 Aug. 1903.

Trelease 8, MO. Tehuacán hills SW of Riego Hotel (var. *latifolia*) 5 Feb. 1905. = *A. peacockii.*

Agave panamana

PANAMA. *Campbell 1,* GH, US. Playa Grande, San José Isl., Gulf of Panama (ca. 55 miles SSE of Balboa), 20 Mar. 1946 (l, f).

D'Arcy & Croat 4202, MO. Salinas de Chitre, 14 April 1970 (l, f).

Erlanson 549, US. San José Isl., Pearl Archipelago, 30 July 1945 (l, infl, bulb).

Johnston 597, A, US. Panama. Johnston wrote "Known only from some small barren rocky islets in Bodego Bay off Long Beach and from islets and large rocks forming the point of land at the south end of Plaza Grande." (San José Island, Panama. Sargentia 8: 96, 1949.)

Lewis et al. 62, MO, US. Farfan Beach, from Thatcher to Palo Seco, 4 Dec. 1966 (l, f).

Tyson 1809, MO. Farfan Beach area, 8 June 1965.

Agave rhodacantha

NAYARIT. *Gentry 11020*, DES, US. Santa Maria del Oro, 2 Aug. 1951 (l). Along margins of open fields and along stone fences of town.

Gentry & Gilly 10756, 10759, 10771, DES, MEXU, MICH, US. 1 mi. E of Mazatan along road to Compostela, 6 July 1951 (l, f). Highly mixed, short-tree, deciduous, disturbed tropical forest on volcanic mesa and slopes.

Gentry & Gilly 10806, 10783, DES, MEXU, MICH, US. 1–2 mi. SW of Jalisco along road around S shoulder of Cerro San Juan, 1 July 1951 (l, f). Mixed pine-oak forest on mountain slopes with clay soils, generally derived from volcanics. 4,500–5,000 ft. elev.

Maltby 126, US. María Madre Island, May 16, 1897.

Mason 1756, US. María Madre Island, 18 May 1925 (l, f). "Abundant along the shore and hill tops."

Nelson 4264, US. María Madre Island, 3–25 May 1897 (l, cap).

Solis s.n., US. Isla María Madre, 18 May 1925 (l, f).

OAXACA. *Gentry 12109*, DES, US. Ca. 10 mi. N of Miahuatlán along road to Cd. Oaxaca, 2 Sept. 1952 (l). Gravelly sedimentary in Indian-decimated country.

SINALOA. *Gentry 5630*, ARIZ, DES, GH, MO, NA. Capadero, Sierra Tacuichamona, 13 Feb. 1940 (l, cap). Open, rocky, sunny hill slope in oak forest, 3,000 ft.

SONORA. *Arguelles s.n.*, DES. San Bernardo, 10 Aug. 1958 (l). "En las panelas"; cultivated.

Gentry 11376, DES, US. Along Arroyo Guajaray N of San Bernardo, 22–25 Dec. 1951 (l). Savanilla on broken volcanic terrain with scattered *Lysiloma, Quercus, Acacia*, etc.; elev. 2,500–3,500 ft. Growing alone on top of hill with oaks, palms, etc. Only one observed; said to be rare.

Gentry 11407, DES, MEXU, MICH, US. Rancho Yocojigua, ca. 40 mi. SE of Navojoa, 31 Dec. 1951 (l). Cultivated in plantation, without irrigation, for mescal.

Gentry 20479, DES, MEXU, US. Sierra Calabasa (S. Charuco), September 1963 (l). Now cultivated in Tupeyeca, Río Mayo.

Agave stringens

JALISCO. Type: **Trelease in 1904**, MO. Rio Blanco Barranca near Guadalajara, 12 March 1904 (l only, infl unknown).

Trelease 136/4, MO. Cult. Mo. Bot. Gard. (l, photo).

Agave tequilana

HORTICULTURE. *Gentry 19827*, DES. Huntington Botanical Gardens, San Marino, 6 June 1976 (l, f, photo). Cultivated.

JALISCO. *Gentry 11494*, DES. Refugio de Paradón, 25 Jan. 1952 (photo). Cultivated on rocky volcanic soil.

Gentry et al. s.n., DES. Between Tecolotlán and Union de Tula, November 1961 (photo). Rocky volcanic mountain top with secondary scrub oak and shrubs. An extensive, scattered, sword-leaved poly-morphic wild population with both green and glaucous forms. Appears like source of tequila selections.

Griffiths B, MO. El Llano near Tequila, 21 Sept. 1909 (l), "pato de mula." Type of *A. pes-mulae* Trel.

Medrano 1487, MEXU. Alrededores de Tequila, 24 Sept. 1966. Cultivated, alt. 1,560 m.

Ogden & Gilly 51165, 51168, DES. El Arenal, at km. 717 on hwy. between Guadalajara and Tequila, 31 March 1951 (l, photo). Cultivated.

Ogden & Gilly 51175, DES. Ca. 10 mi. from Tecolotlán, along road to Cocula, 2 April 1951 (l, photo).

Rose 4757, US. Near Tequila, 5–6 July 1899 (l, f, photos). Also leaf of wild *A. angustifolia* with small teeth.

Rose & Hough 4756, US. Near Tequila, 5–6 July 1899 (l, f). o=20–21, t =8–10, s=23–24, f = 7–8, 40–45, tl=53–55.

MICHOACAN. *Ogden & Gilly 51181*, DES, US. Ca. 11 miles W. of Jiquilpán along hwy. to Guadalajara, 3 Apr. 1951; cult. for tequila as "maguey blanco" or "maguey manso."

OAXACA. *Gentry & Arguelles 21335*, US, DES. 35 miles SE of Camerón along hwy. to Tehuantepec, 29 Sept. 1965; cult. "light glaucous gray." Cult.; selected clone planted for mescal and reported to mature in 7 years.

Ogden & Gilly 51190, DES, US, MEXU. Ca. 3 miles N of Matatlan along highway to Cd. Oaxaca, 16 April 1951. "Acaulescent to short-stemmed; leaves to 1.3 m, gray-green." "Espadin" (l, photo).

SINALOA. *Gentry 11475*, DES, US. Los Aguajitos between Mocorito and Pericos, 9 Jan. 1952 (l, photo. Cult. in yards and fields).

Gentry 11483, DES, US. Concordia, 20 Jan. 1952 (l, cult. in abandoned plantation).

SONORA. *Gentry 11364*, US. Alamos. 17 Dec. 1951; cult. previously for mescal. (Form with small spine and teeth on light glaucous green leaf.)

21.

Group Salmianae

Berger, Die Agaven 1915: 128.

Plants large, massively succulent multiannuals with thick short stems, usually closely surculose and also seeding, without bulbils; leaves generally green, mostly very large, fleshy, very thick toward base, the margins variously armed; inflorescence very large, the peduncles with large, fleshy, appressed, imbricate bracts; panicles pyramidal to ovoid in outline, with wide-branching, decompound umbels of large succulent proterandrous flowers, with broad thick-walled tubes, the tepals longer than the tubes, dimorphic, thick-fleshy, conduplicating and/or involute, wilting incurved with anthesis; filaments stout, inserted at or above mid-tube, frequently on two levels, those with the outer tepals being 1–4 mm higher; pistil over-reaching stamens in post-anthesis.

Type species: *Agave salmiana* Otto ex Salm.

The recognition characters of this group are the large green thickly succulent rosettes, the thick, imbricated, peduncular bracts, the broad base of the panicle often giving it a pyramidal outline, and the dimorphically modified flowers with unequal tepals. The dimorphism of the tepals recalls the Ditepalae, but the tepals of the Salmianae lack the scleroid rigidity of the Ditepalae, are soft fleshy, wilt and curl quickly, and are much folded with conduplication and involution. These structural characters are not so well set as in the Ditepalae and the 2-level insertion of the filaments is inconstant. Altogether, however, the Salmianae show a high degree of *Agave* specialization and phylogenetically can be regarded as among the most advanced or modern. Their great variability, obviously abetted by man, is a part of their modern modification, a situation of unpredictable eventuation.

This group, as defined and restructured here, has little taxonomic history. Berger founded his ''Reiche Salmianae'' in 1915, including 19 species. His group included a part of Jacobi's ''Semi-marginatae'' (1864) and Baker's ''Submarginatae'' (1888). The range of specimens now available and my long years of population studies lead me to exclude the majority of the plants included by former authors and to reduce the number of species. As stated before, many names were given for horticultural convenience, not for systematic need, and I have been unable to find specimens, living or dead, for many names. It may be

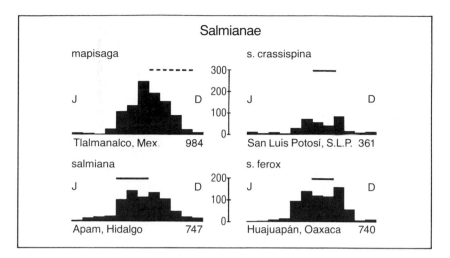

Fig. 21.1. Rainfall (silhouettes) and flowering perimeters of the Salmianae; relevant meteorological stations with annual rainfall in millimeters. Data from the Atlas Climatológico de México, 1939. Flowering periods are based on herbarium specimens and field observations; uncertainty expressed by broken bars.

Fig. 21.2. *Agave mapisaga* (left) and *A. salmiana* in a row planting ca. 20 miles (32 km) west of Cholula, Puebla, along the old Mexico–Puebla highway, October 1952. The longer linear leaves and small teeth of *A. mapisaga* are diagnostic.

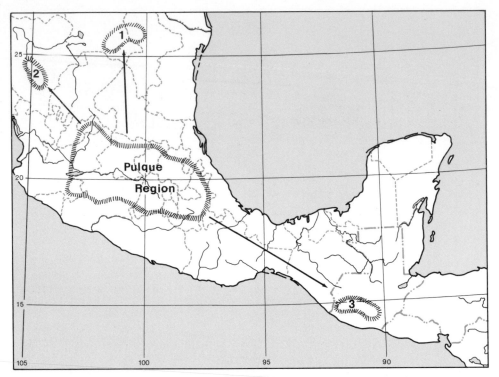

Fig. 21.3. The primary region of pulque industry in Mexico and secondary centers developed with Spanish colonialism: 1. Saltillo-Monterrey; 2, Durango; 3, Guatemala.

that the four species, one subspecies, and four varieties I now conceive as composing this section are inadequate. The highly variable complex of *A. salmiana* has been intensified by man during many hundreds, if not thousands, of years. The number of varieties or forms outstrip the perimeters of this work. No doubt they warrant a detailed taxonomic study, which I hope will be done; such a study, together with still unrecognized plants, can extend knowledge of this group.

The Salmianae are native to the highlands of central Mexico and there are numerous wild and cultivated forms, some of which have rather narrow environmental tolerances. Generally, they occupy elevations between 5,000 and 8,000 feet (1,230–2,460 m). Rainfall ranges from 15 to 36 inches (360–1,000 mm) annual average (Fig. 21.1), and light frosts are common through the drier winter months. Seventy to eighty percent of the rain falls during May through October. Apam (or Apan), Hidalgo, a center of extensive pulque cultivation, has annual rainfall of ca. 750 mm. The graphs of Fig. 21.1 indicate the rainfall for both the wild and cultivated forms of *A. salmiana* and the cultivars of *A. mapisaga*. *A. macroculmis* is a purely wild species of the high montane Sierra Madre Oriental, where both rainfall and low temperatures exceed those shown in the graphs but have not been recorded by meteorological stations.

Agave salmiana and *A. mapisaga* are the No. 1 and No. 2 pulque producers of Mexico. With a little practice, they are easily recognized, even though they often grow side by side (Fig. 21.2). The map in Fig. 21.3 outlines the indigenous area of pulque agriculture in central Mexico. Besides members of the Salmianae, the area includes *A. americana* and a few other species rarely cultivated and tapped for their juice. A brief description of this indigenous industry is given under the account on *A. salmiana*.

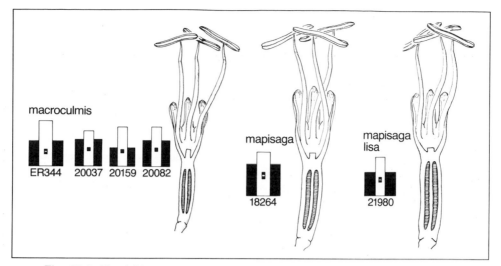

Fig. 21.4. Floral ideographs and long sections of flowers of *Agave macroculmis*, *A. mapisaga*, and *A. mapisaga* var. *lisa*.

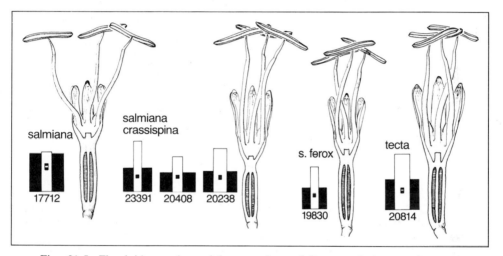

Fig. 21.5. Floral ideographs and long sections of flowers of *Agave salmiana*, *A. salmiana* ssp. *crassispina*, *A. salmiana* var. *ferox*, and *A. tecta*.

Key to Species of the Salmianae

1. Rosettes relatively small, usually not exceeding 1 m in height; leaves triangular lanceolate, less than 1 m long *macroculmis*, p. 598
1. Rosettes usually much larger, 1 m or more tall; leaves usually 1–2 m or more long, linear to lanceolate 2
2. Leaves linear, mostly 1.8–3 m long, the teeth relatively small, 4–5 mm long; spines 3–5 cm long *mapisaga*, p. 602
2. Leaves broadly lanceolate, 0.7–2 m long, the teeth larger, mostly 5–10 mm long; spines longer, 5–10 cm 3
3. Tepals large, 32–33 x 8–9 mm; leaves massive, thick, broad, relatively short, usually not over 1.5 m long, the apex not markedly sigmoidally curved. Guatemala *tecta*, p. 612
3. Tepals smaller, 21–30 x 5–8 mm, leaves massive also but frequently over 1.5 m long (except ssp. *crassispina*), the apex sigmoidally curved. Mexico *salmiana*, p. 605

Agave macroculmis

(Figs. 21.4, 21.6, 21.7, 21.8, 21.9; Table 21.1)

Agave macroculmis Tódaro, Hort. Bot. Panorm. 2: 51, Tab. 37, 38, 1878–92.

Plants medium to large, rigid, multiannual, nonsurculose, green, with thick short stems; leaves 45–85 x 17–26 cm, very broad and thick at base, sometimes slightly narrowed above base, generally triangular long-acuminate, deeply concave above, thickly rounded below, dark green to light green, sometimes faintly glaucous, the margins partly or entirely corneous, stoutly armed; teeth along mid-blade commonly 8–12 mm long, 2–4 cm apart, the cusps from broad well-rounded bases, bent or flexed, castaneous to grayish brown; terminal spines very strong, 4.5–6.5 cm long, subulate, broadly channeled above for ⅔ of its length, rounded or keeled below, long decurrent, dark brown to grayish; panicles 3–5 m tall, on stout peduncles with large fleshy bracts closely imbricate at base of panicles; panicles pyramidal in outline with 10–28 strong densely flowered umbellate branches; flowers 70–90 mm long, reddish in bud, opening yellow, ovaries green, thickly succulent; ovary 35–55 mm long, trigonous-cylindric, with grooved neck; tube 11–16 mm deep, 16–22 mm broad, funnelform, thick-walled, deeply grooved; tepals, 20–28 x 6–8 mm, linear lanceolate, erect to incurving, conduplicate involute, thickly fleshy, bluntly apiculate, cucullate, papillate within, the outer longer, the inner with thick keel; filaments 50–65 mm long, yellow, flattened-angulate, tapering, inserted 8–12 mm above tube bottom; anthers yellow, 25–30 mm long, regular, centric or excentric; pistil over-reaching stamens in post-anthesis, the stigma globose, trilobate; capsules 6–7.5 x 2–2.5 cm, oblong, woody, short-stipitate, short-beaked, chocolate brown; seeds lachrymiform, 7–8 x 6 mm, shiny black.

Lectotype: Tab. XXVII and XXVIII, Hort. Bot. Panorm. 2, 1878–92. From plant grown in the university botanic garden, Palermo, Sicily.

Agave macroculmis is easily recognized in the field by its large fleshy peduncular bracts congested below the panicles, which are massive structures above the formidable extremely broad-based, rigid, long pointed, green leaves. These characters together with the thick succulent flowers qualify it for inclusion in the Salmianae. In certain localities, as upon the high mountains east of Saltillo, there is a marked variability in the inflorescence. The umbels may be very short pedunculate, scarcely emerging from the large clasping bracts, with resulting linear or clavate form of panicle (Fig. 21.8).

Tódaro's plate (Figs. 21.6, 21.7), which represents the type, differs in some respects from the wild plants I am assigning to this name. In his figure the bracts are not bunched below the base of the panicle, as they are in nature, and the filaments are depicted as inserted in the base of the tube. These differences I reconcile as follows. On the Tódaro specimen the bracts were not bunched because of the open winter climate in Palermo. There is depicted, however, some bunching of bracts in the lower branches of the panicle. The flowering shoot could continue growing through the Palermo winter, while in the high mountains of northern Mexico the cold winters enforce cessation of growth. Elongation begins again with warmer winter temperatures. This same habit is also exhibited in *A. parrasana* and *A. moranii,* which are also winter sprouters. The bunching of bracts appears to be an adaptation preventing cold injury to the growing bud initial. *A. murpheyi* is also a winter sprouter, but it does not exhibit this concentration of bracts, and its flower shoots are sometimes injured by winter frosts (Gibson, 1935). The low insertion of the filaments I attribute to faulty drawing. A similar case is shown in the illustration in the same publication of *A. haynaldii* of the Section Marginatae, a group which almost universally has the filaments inserted in the orifice of the tube. Otherwise, Tódaro's illustration

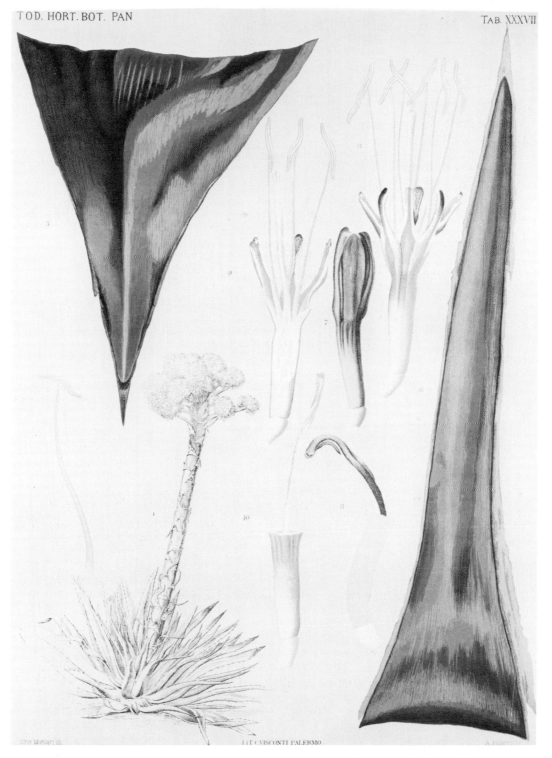

Fig. 21.6. Todaro's illustration of *Agave macroculmis* (loc. cit.), showing habit, bracts, and flowers.

Fig. 21.7. Todaro's Tab. XXXVIII (loc. cit.), *A. macroculmis,* sections of leaf.

Fig. 21.8. *Agave macroculmis* on a mountain above Carneros Pass, south of Saltillo, Coahuila, June 12, 1963. Associates are pine, juniper, and *Yucca carnerosana.*

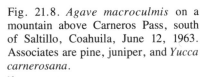

Fig. 21.9. *Agave macroculmis* near Carneros Pass; detail of a rosette with large formidable teeth, quite rigid.

appears to represent quite well the species here described. If this explanation is not tenable, then the *Agave* assigned here may require a new name, of which we already have far too many in the genus.

Agave macroculmis is a high montane species, living generally between 7,000 and 9,000 feet (2,150 and 2,800 m) elevations on limestones in oak and pine forest or with chaparral. Snow may fall in these elevations. Occasional plants are found at lower elevations, especially in northward canyons and slopes, as on the lower slopes near Saltillo at about 6,000 feet (1,850 m) elevation. For *Agave,* it is a mesophyte, requiring relatively high rainfall, cold hardy, and tolerant of some forest shade (Figs. 21.1, 21.8). It has a long range of over 500 miles (800 km) along the Sierra Madre Oriental. It has no natural type locality but was probably first collected from the southern part of its range in the state of Hidalgo, where men were moving along roads between the mines in the mountains.

In the field, I found no special name for this *Agave*. The people called it simply "maguey" or "maguey verde." Near Parras, and probably elsewhere, the flowering shoot in the asparagus stage is much cut for eating. The heavy shoots are packed down off the Sierra de Parras by mule and steam-cooked on the outskirts of Parras. In 1972, I saw them hauled to market in a light truck. They were sold in one-half inch cross-slices for a few cents along the street near the Parras market, and elsewhere. It provides a crude sweet to chew out of the fiber, which is spit upon the ground.

Agave mapisaga

(Figs. 21.1, 21.2 (left), 21.4, 21.10; Table 21.1)

Agave mapisaga Trel., var. *mapisaga*.
> *Agave mapisaga*, U.S. Nat. Herb. Contr. 23: 130, 1920.

Plants large with short massive stems, surculose, 2–2.4 m tall, nearly twice as wide, forming openly spreading rosettes; leaves linear, 185–250 x 19–25 cm, spreading to ascending, sometimes reflexed or inflexed, green or pale glaucous or zonate, at base very

Table 21.1. Flower Measurements in the Salmianae (in mm)*

Taxon & Locality	Ovary	Tube	Tepal	Filament Insertion & Length		Anther	Total Length	Coll. No.
macroculmis								
Sierra San Marcos, Coah.	45	15×23	28×8&24	8–10	46	36	86	ER344
	45	15×20	27×8&25	8–10	53	37	86	ER344
	50	15×21	28×7&26	8–9	62	38	92	ER344
Sierra de Parras, Coah.	55	16×18	27×27	11–10	60	29	98	23267
	56	15×16	26×7	10–9	60	28	95	23267
Carneros Pass, Coah.	37	16×16	19×5&18	10–11	60	27	72	20037
	42	16×16	21×5	10–11	60	30	81	20037
Los Alpes, Coah.	42	16×16	23×6&21	8–9	55	27	76	20074
	44	11×16	22×6&20	7–8	55	27	76	20074
Sierra Potosí, S.L.P.	33	11×15	24×6&23	8–9	48	26	67	20159
	34	11×15	23×6&	9–10	50	27	69	20159
Zimapán, Hgo.	30	15×17	23×6&20	10–11	65	30	68	20082
	35	16×16	24×6&22	10–11	70	30	75	20082
mapisaga								
Zacatecas, Zac.	42	17×20	24×8&20	11–9	58	32	82	18264
	50	21×20	27×7&23	15–12	66	35	98	18264
m. var. lisa								
Hunt. Bot. Gard., Cal.	47	14×19	22×6&20	8–10	57	30	82	21980
	52	14×18	23×5&21	8–10	60	32	89	21980
salmiana								
Hunt. Bot. Gard., Cal.	61	22	25&19	16&14	58	32	108	17712
	48	23×20	23×6	16&14	60	31	95	17712
	51	24×20	22×6	17&15	62	32	98	17712
Guanajuato, Guan.	37	14×17	30×5&28	8	70	35	81	23391
	47	14×17	30×4&28	10&9	65	35	90	23391
Zacatepec, Pue.	39	12×21	21×6&18	10&8	54	31	72	20408
	44	11×21	21×6&19	9&8	52	29	75	20408
	42	11×17	23×7&20	9&7	55	30	73	20204
Tehuacán, Pue.	42	12×18	25×6&21	8–9	63	30	78	20204
	45	13×20	25×8&22	8–9	60	32	85	20238
	43	12×20	27×8&23	8–9	65	33	82	20238
s. var. ferox								
Hunt. Bot. Gard., Cal.	37	12×15	23×5	7&8	50	30	72	19830
	45	12×15	25×5	8–9	55	32	82	19830
tecta								
Quezaltenango, Guat.	38	18×20	32×9&29	13&10	63	30	88	20814
	43	17×20	33×9&31	12&10	64	33	92	20814

* Measurements from fresh or pickled flowers.

thick fleshy, flat above, convex below, the leaf upward guttered, long-acuminate, margin straight to repand with small brown teeth, the cusps 2–5 mm long from low bases, mostly 4–6 cm apart; spine 3–5 cm long, conic-subulate, narrowly grooved above, long decurrent, dark brown to grayish brown. Inflorescence massive, 7–8 m or more tall; peduncle closely set with large succulent bracts; panicle wide-spreading with 20–25 heavy dense decompound umbels; flowers large, succulent, 80–100 mm long, tepals frequently reddish in bud, opening yellow over green ovaries; ovary 40–52 mm long, thick, roundly 3-6-angled, with short unconstricted neck; tube 14–21 mm deep, fleshy thick-walled, funnelform, deeply grooved; tepals unequal, linear, erect to incurving, the outer 22–27 mm long, 7–8 mm wide, but conduplicating and narrowing to 5 mm, the apex bluntly galeate, the inner 2–3 mm shorter, with thick fleshy keel, involute and 2-costate within; filaments 55–70 mm long, inserted at 2 levels 11–15 mm above base of tube, thick but flattened; anthers yellow, 30–35 mm long, excentric; pistil eventually over-reaching stamens, stout; capsules 6–6.5 x 2 cm, oblong, stipitate, short-beaked, brown, thick-walled; seeds 7–8 x 5–6 mm, lacrimiform, black, wavy-sculptured, hilar notch near apex, marginal wing upcurving.

Type: *Trelease 147,* Tacubaya, 16 April, 1900 D. F., Mexico, MO, "maguey mapisaga; planted for pulque." Tacubaya is now engulfed in Mexico City.

Agave mapisaga is obviously related to *A. salmiana,* but is easily distinguished by its longer linear leaves which lack the sigmoid flex on the apex, so characteristic of *A. salmiana* (Figs. 21.10, 21.2). *A. mapisaga* is a cultivar widely scattered over the central Mexican highlands. It is frequently found in cultivation with *A. salmiana,* seldom planted alone. It suckers less than *A. salmiana,* which may largely account for its general minority in plantings. It has the same uses in the maguey culture as *A. salmiana,* described above and under *A. salmiana* following.

Fig. 21.10. *Agave mapisaga* near Cholula, Puebla. Arguelles is cutting leaf sample for analysis.

Agave mapisaga var. *lisa*
(Figs. 21.4, 21.11, Table 21.1)

Agave mapisaga Trel. var. *lisa,* Gentry var. nov.

Plants gigantic, 2–2.5 m tall, forming open rather sprawling rosettes, sparsely suckering; leaves 200–275 cm long, 25–30 cm wide, linear-lanceolate, outspreading to ascending, sometimes incurved or reflexed, toward the base massively thick, little narrowed, flat above, convex below, through the mid-blade concave, green or faintly cross-zoned; margins straight; teeth small, 1–3 mm long, remote; spine subulate; inflorescence massive ca. 8 m tall, with large shaft-clasping bracts with 20–25 wide-spreading lateral peduncles with umbels of yellow flowers 80–90 mm long; ovary 47–55 mm long, thick fleshy, with short furrowed neck; tube 14 mm deep, funnelform, thick-walled, deeply grooved, with open sinuses; tepals unequal, thick fleshy, linear, incurved, the outer 19–21 mm long, involute, the apex thickly galeate, the inner shorter, thickly keeled; filaments relatively short, 57–60 mm long, inserted above mid-tube; anthers yellow, 30–32 mm long, ca. centric.

Type: Gentry 21980, Huntington Botanical Garden (block 20), San Marino, California, July 7, 1966, US, isotype DES (f, photo).

Planta grandissima, 2.5–3 alta, 5–6 m lata, caulis crassis stoloniferis; differt a var. *mapisaga* statura grandiore et tubo floribus breviore.

This gigantic succulent is unexceeded in size by any other agave (Fig. 21.11). It differs from *A. mapisaga mapisaga* in its larger rosettes, shorter flower tube. Hertrich, past superintendent of the Huntington Botanical Gardens, once informed me that the plant had originally been received from the plant dealer Schmoll of Querétaro, Mexico, as "maguey lisa," without other specific information. Subsequently, I was unable to locate it in Mexico. It appears to represent a clonal variety from some obscure locality.

Fig. 21.11. *Agave mapisaga* var. *lisa* in block 20 of the Huntington Botanical Gardens, San Marino, California.

Agave salmiana

(Figs. 21.1, 21.2 (right), 21.5, 21.12, 21.13; Table 21.1)

Agave salmiana Otto ex Salm, var. *salmiana*.

> *Agave salmiana* Bonplandia 7: 88, 1859.
> *Agave cochlearis* Jacobi, Nachtr. II in Abh. Schles. Ges. Vaterl. Cult. Abth. Naturwiss. 1870: 151.
> *Agave coarctata* Jacobi, Nachtr. I in Abh. Schles. Ges. Vaterl. Cult. Abth. Naturwiss. 1868: 147.
> *Agave lehmannii* Jacobi, l. c., Nachtr. I: 146, 1868.
> *Agave mitriformis* Jacobi, l. c. Nachtr. I: 145, 1868.
> *Agave tehuacanensis* Karw. ex Salm, l. c. 89, 1859.
> *Agave jacobiana* Salm, l. c. 88, 1859.
> *Agave atrovirens* var. *sigmatophylla* Berger, Die Agaven: 143, 1915.
> *Agave quiotifera* Trel. ex Ochoterena, Mem. Soc. Alzate 33: 102, 1913.
> *Agave compluviata* Trel. Bailey Stand. Cycl. Hort. l: 234, 1914.

Plants medium size to large, with short thick stems, closely surculose, forming massive rosettes 1.5–2 m tall and ca. twice as broad; leaves 100–200 x 20–35 cm, broadly linear lanceolate, acuminate, thick-fleshy, green to glaucous grayish, deeply convex below at base, concave to guttered upward, the apex sigmoidally curved; margin repand, sometimes mammillate; teeth largest along the mid-blade, mostly 5–10 mm long, 3–5 cm apart, brown to grayish brown, the cusps straight to flexed or curved from low broad bases; spine long, stout, subulate, 5–10 cm long, dark brown, grooved above for over half its length, long decurrent, sometimes to mid-blade as a heavy corneous margin. Inflorescence stout, the peduncle closely imbricate with large fleshy bracts, 7–8 m tall; panicle broad, with 15–20 large decompound umbels in upper half of shaft; flowers 80–110 mm long, thick fleshy, yellow above green ovary; ovary 50–60 mm long, thick, cylindric, with unconstricted neck; tube large funnelform, 21–24 mm deep, 20 mm broad, thick-walled between deep sinal grooves; tepals unequal, lanceolate, curling inward with anthesis, the outer 21–25 x 6 mm and bulging at base, narrowed above, with thin involute margin, bluntly galeate, the inner 2–3 mm shorter, with broad keel; filaments 55–70 mm long, inserted just above mid-tube, those by the outer tepals frequently 1–3 mm higher than others; anthers 30–35 mm long, yellow, excentric; pistil over-reaching stamens in post-anthesis; capsule (*Gentry 22416*) 5.5–7 x 2–2.2 cm, stipitate, beaked, woody, brown; seed 8–9 x 6–7 mm, black, lacrimiform, hilar notch shallow, apical.

Neotype: (Nothing suitable at hand to designate as type representative.)

Taxonomic Notes

See Trelease (1915: 130) and Berger (1915: 130–45) for still more horticultural names that have been applied to this complex. Trelease did not recognize *Agave salmiana* Salm, but misapplied *A. atrovirens* Karw. ex Salm, a distinct species from the high montane habitat, which is described and discussed above in group Hiemiflorae.

Agave atrovirens Karw. as figured by Berger (1915: 143) is obviously a broad-leaved form of *A. salmiana* Otto ex Salm. *Agave latissima* Jacobi, figured by Berger (1915: 130), I interpret as a form of *A. atrovirens* Otto ex Salm. These cases again illustrate the inaccuracies that arise from lack of specimens and information on the natural populations of the variable agaves.

Agave salmiana consists of both wild and cultivated populations and the forms of each are numerous. However, the species is generally recognizable by its broad, heavy, well-armed, green leaves with long-acuminate sigmoid apices and the large peduncular bracts subtending broad, pyramidal, large panicles. The flowers are coarsely fleshy, the tepals dimorphic, narrow, involute, the inner with high fleshy keel bordered by thin, hyaline, inrolling margins. The flowers vary in size but are usually between 8 and 11 cm in length, frequently flushed reddish in the bud but opening yellow. Late March to June appears to be the principal flowering season.

Fig. 21.12. *Agave salmiana* with valleculate leaves photographed near Tepatitlán, Jalisco.

Many of the fields show uniform plants as though from a single clone, variation being in size of plants, depending upon age, crowding, or on individual sites. Other fields show genetic variation among the plants as reflected in differences of teeth size and their spacing, or in size and form of leaf. Trelease (1920) mentions a treatise on maguey written by P. & I. Blasquez, "Tratado de Maguey," published in Puebla, in which 32 forms of maguey are enumerated and illustrated with spines and teeth. I have been unable to find a copy of this work. Trelease dismisses the many binomials listed as not properly used in a taxonomic or botanical sense. Apparently, the 32 names listed reflect the variable forms found in *A. salmiana* and other pulque magueys of the region. On the whole, the culti- vated forms are larger than the wild plants, most of them being polyploids (Cave, 1964; Granick, 1944). (See also Fig. 12.2 of this work.) No doubt the maguey planters gave preference to the larger, more vigorous, and sweeter forms, as they found them in their cultivated fields, or encountered them in wild forms adjacent to their farms. Suckering is common throughout the species. Hence, vegetative propagation is very simple and the usual method for increase.

There are many wild populations of *Agave salmiana* adjacent to its cultivation, especially in the northern parts of its area in San Luis Potosí and Hidalgo. On the semi-arid stony uplands, it grows with *Acacia* and *Opuntia* in a pervasive plant commu- nity that can appropriately be designated as the *Acacia-Opuntia-Agave* association. There is much seedling or true sexual variability in these wild populations, and it is difficult to identify some of them with accuracy. For instance, there is a short broad-leaved form found about the Cerros Derumbadas in northern Puebla that appears much like *A. macroculmis* Tod. In the Tehuacán area and even toward Puebla, short-acuminate leaf forms appear to belong with the taxon called *A. ferox,* the familiar horticultural form, which is readily recognized by its massive short-acuminate leaf form. In view of the polymorphic nature of *A. salmiana,* and its clinal merging with other taxa, in the writer's opinion, *A. ferox* is best treated as a variety.

Agave salmiana is usually separable from its pulque companion, *A. americana,* by the latter's pale glaucous narrower leaves, and from its other common pulque companion, *A. mapisaga,* by the latter's very long narrow, less prominently armed leaves.

The Pulque Industry

Agave salmiana is the great "maguey de pulque" growing so conspicuously over large tracts of land in the central plateau region of Mexico. It is frequent along all roads leading out of Mexico City into the states of Michoacán, Guanajuato, Querétaro, San Luis Potosí, Hidalgo, Tlaxcala, Puebla, and Morelos. This agave area (see Fig. 21.3) is approximately 350 km wide E–W, and 170 km W–E, Jalisco to Tehuacán, comprising about 60,000 km (20,000 sq. miles). It is most pervasive on the high, sandy plains about Tlaxcala and northward. It is generally planted in rows on borders, used as property line fences or in widely spaced rows within fields where other crops are interplanted.

On the high desertic plain of northern Puebla, the rows follow loosely the contour of the land and the borders are huge, as much as 6 or 7 feet (1.8 or 2.1 m) high and 15 to 20 feet (4.5 to 6 m) broad, with the plants spaced in double or triple rows upon the borders. At first, I could see no explanation for such large borders, or why the farmers, working with limited hand tools, had found it necessary to form such immense borders. But then, I remembered the winds that sweep dust and sand across these highlands. It appears logical that these maguey rows are windbreaks causing the sand to settle about these plants, a self-building and self-sustaining land form of linear semi-dunes. When the mature plants

Fig. 21.13. Weaving cloth with fine fiber from the buds of *Agave salmiana;* Valle de Mesquital, Hidalgo.

are tapped and dry out, new offsets are replanted on the same borders, a practice that is doubtless hundreds of years old. The size of the borders are a clue to the longevity of maguey culture, a moot question in the field of Mexican archaeology. By measuring border bulk and the annual rate of sand fall, an enterprising "ingeniero-biólogo" could make an interesting estimate of the age of these agave borders. On mountain slopes, maguey is still planted across slopes, where they hold soil and tend in time to build handy terraces. In such situations, they may or may not be tapped for juice.

Agave salmiana, as a guess, probably supplies ca. 75 percent of the fermented liquor, "pulque," produced and consumed in this region. This is a nourishing mildly alcoholic brew that has balanced out the diet of the farming folk since prehistory (see earlier chapter, pp. 8–14). A fine, strong twillable fiber is also extracted from the conal bud, which is cut out for sap-tapping purposes. From this bud fiber, cord, nets, carrying bags, and even cloth is still woven by surviving Aztec, Otomie, and other indigenous peoples in their villages (Fig. 21.13). Such handicrafts also can still be found in the larger markets of the region, as in the cities of Tehuacán, Puebla, and México. To the uses of sap, fiber, and soil formation, there are still many desultory uses to be added; e.g., property enclosures or fences, poles and leaf thatching for temporary structures, flavoring in "pan de pulque," leaf pulp as poultices on bruises and sores of farm animals, and doubtless other uses that have not come to my attention.

Around the city of San Luis Potosí, a modern use of this maguey has developed. The leaves of both wild and cultivated plants are cut, hauled, and fed as forage to the dairy cattle. The cattle convert the leaves into milk for that city. These xeric plants are capable of productive growth, with only 360–400 mm of annual rainfall (14–16 inches), on the poor arid stony soils, which are less suitable for grass and other conventional forage crops. This is a local innovation in agriculture, suggesting that modern research may still find additional uses for the remarkable maguey plants. As fossil plant resources of coal and oil become scarce and dearer, man must turn more and more to the culture of live plants, but there is no need to wait until the need is dire to study and develop the agave potential. Mexico and other countries with suitable arid land climates might well profit by an economic feasibility analysis of the maguey agave potential.

For vigorous growth, *A. salmiana* needs well-drained, good quality soil, with ample water in the warm summer. Common varieties in southern California and Arizona have withstood 20°F (−7°C) without serious damage. It does not respond well in the Desert Botanical Garden at Phoenix, Arizona, where the hot, dry air of summer commonly reaches 110–15 degrees Fahrenheit in the shade. It does better in the milder summers of southern California. In rich soil with plenty of water this species and *A. mapisaga* var. *lisa* and *A. tecta* reach gigantic sizes.

Key to Varieties and Subspecies of *A. salmiana*

1. Plants massive; stems thick; leaves 1–2 m long; flowers larger, 90–110 mm long   2
1. Plants smaller with narrower trunks; leaves usually not over 1 m long; flowers smaller, 70–90 mm long   3
2. Leaves green, broadly lanceolate
    var. *salmiana,* p. 605
2. Leaves light gray glaucous, linear
    var. *angustifolia,* p. 609
3. Rosettes urceolate, leaves bright green, outcurving, short acuminate, the margins crenate mammillate, with large, broad-based teeth; flowers relatively slender, the tube ca. 15 mm broad. Cultivated   var. *ferox,* p. 611
    see also *A. ragusae,* p. 615
3. Plants variable, the rosettes not typically urceolate; leaves dull green to glaucous green, erect-ascending, rarely outcurved, usually more acuminate, the margins and teeth variable; tube 17–21 mm broad. Wild
    ssp. *crassispina,* p. 609

Agave salmiana var. *angustifolia*

Agave salmiana var. *angustifolia* Berger, Die Agaven 1915: 135–37.

Type: *Berger s.n.,* Cult. La Mortola, 6 VIII 1909, U.S. (flowers only).

This giant agave with its long linear leaves looks like a white variety of *A. mapisaga,* but the type flowers indicate relationship with *A. salmiana,* where Berger placed it. He stated it was frequent in culture along the Mediterranean Riviera. I have not seen it anywhere.

Agave salmiana ssp. *crassispina*
(Figs. 21.1, 21.5, 21.14)

Agave salmiana Otto ex Salm, ssp. *crassispina* (Trel.) Gentry, stat. nov.
   Agave crassispina Trel.in Standley, Contr. U.S. Nat. Herb. 23: 129, 1920; in Bailey, Stand. Cycl. Hort. 1: 234, 1914.

Plants relatively few-leaved, smaller than var. *salmiana,* wild, variable, the rosettes 8–12 dm tall, compact to spreading; leaves broadly lanceolate, 60–90 x 16–25 cm, rarely larger, thickly rigid, narrowed toward the thick base convex below, thinner and concave through the mid-blade, acuminate, green to grayish, the margins undulate to crenate with firm, broadly based teeth, mostly 7–12 mm long, dark brown graying in age, 1–3 cm apart; spines stoutly subulate, 5–9 cm long, openly grooved above

Fig. 21.14. *Agave salmiana* ssp. *crassispina* growing wild about 100 km north of Querétaro along route 57, Guanajuato, March 16, 1974; flowering.

Fig. 21.15. Cultivated varieties of *Agave salmiana* photographed near Saltillo, Coahuila, 1952.

for ¾ of its length, long decurrent; dark brown; inflorescence as in the species; flowers generally yellow above green ovary, 70–90 mm long, thickly succulent; ovary 38–50 mm long, cylindric, neck not constricted; tube 11–14 mm long, 17–20 mm broad, thick-walled, rather finely grooved; tepals unequal, 21–30 mm long; filaments 52–70 mm long, inserted nearly on one level slightly above mid-tube; anthers 30–35 mm long, yellow.

Type: *Trelease* ?, "about San Luis Potosí." (There may be a specimen at MO, but I have no note of it.)

Trelease's name is applicable here to designate the extensive wild populations of *A. salmiana,* occurring from south San Luis Potosí state southward to Tehuacán, Puebla and adjacent northern Oaxaca. They seed and sucker freely, and there is considerable natural variation throughout its range. Generally, they are smaller plants with less thick stems and leaves than the cultivates (Fig. 21.14). The broad leaves with sigmoid tips, long spines, and large succulent bracts of the shafts are indicators of this taxon, just as it is for the rest of the Salmianae species. Some of the short-acuminate leaves in the southern populations resemble those of *A. salmiana* var. *ferox,* and it becomes obvious that there is no distinct morphological break between *salmiana* taxa. However, the cultivated *ferox* of horticulture with bright green leaves is a clone and, as such, is easily recognized.

Agave salmiana crassispina has obviously been a natural resource for man since prehistoric times (MacNeish, 1967). Like *A. americana protamericana,* its wild forms were a reservoir of easily propagated variants for cultivated selections for food, beverage, and fiber. Occasionally, one can still see the small, wild forms of *crassispina* within fields of the large cultivated forms of *salmiana.* Vice versa, there are also large forms of the latter growing spontaneously in open country, whether from seed of cultivar, or wildling, it is usually impossible for the transient botanist to know.

Agave salmiana var. *ferox*
(Figs. 21.1, 21.5, 21.16, 21.17; Table 21.1)

Agave salmiana Otto ex Salm, var. *ferox* (Koch) Gentry, stat. nov.
   Agave ferox Koch, Wochenschr. Ver. Beford, Gartenb. 3: 23, 1860.

Rosettes large, rather compact, urceolate, 1–1.5 m tall, ca. twice as wide, freely offsetting around base; leaves 70–90 x 23–30 cm, broadly oblanceolate, short-acuminate, light bright green, thickly succulent, the margin crenate, heavily armed with castaneous to brown teeth, 10–14 mm long on prominent teats; terminal spine narrowly subulate, 6–7 cm long, castaneous to dark brown, long decurrent; inflorescence as for the species, with heavy dense umbels, but flowers more slender, 70–85 mm long, tube ca. 15 mm broad; capsules (Berger) 58–60–70 mm long, obovoid, stipitate, long-beaked, grayish brown to dark brown; seeds 7 x 5 mm, shining black, with narrow winged margin.

   Neotype: Kew Herb. ''from Mr. Hanbury's garden La Mórtola. Flowers July 2, 1896, leaf July 18, 1896.'' (Sheets I, II, III, IV, flowers, leaf.)

Agave salmiana ferox is a distinguished variety readily recognizable by its thick, graceful, light shiny green, outcurving leaves, with strongly teated margins. Berger (1915) reported it was common in European gardens since the middle of the 19th century. Its regular habit of suckering provides continuous propagation. It is a handsome plant by

Fig. 21.16. *Agave salmiana* var. *ferox* in block 20 of Huntington Botanical Gardens, San Marino, California.

Fig. 21.17. *Agave salmiana* var. *ferox* near Tejupán, Oaxaca.

color and form but is sensitive to frosts below 25°F (−4°C). Berger ascribed it to Mexico, Trelease to the Valley of Mexico, where he may have seen it. My only encounter with the variety in Mexico, but of different clone, was in the Valley of Tejupán in northern Oaxaca (Fig. 21.17). Here, the plant was obviously planted and its leaves were employed in thatching the roofs of summer shelters in the fields. Also, some clones appeared to be spontaneous on open valley slopes. A closely related form was found by C. Earl Smith, *5015,* with small dried flowers, in the Valley of Nochistlán, Oaxaca.

Similar short acuminate-leaved plants are wild and cultivated about Tehuacán, Puebla, but the flowers are somewhat larger with broader tubes, confusing it with ssp. *crassispina,* and the poor botanist feels thwarted in making specific identifications. Its apparent native origin is in the virtually frostless climates of Oaxaca and explains in large part its sensitivity to frosts in more temperate climates, as in Phoenix, Arizona, and southern California.

Agave tecta
(Figs. 21.5, 21.18, 21.19; Table 21.1)

Agave tecta Trel., St. Louis Acad. Sci. Trans. 23: 145, 1915, pls. 26, 27.

Plants massively succulent with very thick broad stem, forming a broad hemispherical open rosette 2 m tall, 4 m broad, freely suckering; leaves broadly lanceolate, 100–160 x 30–40 cm, straightly ascending, deeply convex thick at base, thinning upward, concave

Fig. 21.18. *Agave tecta* near Quezaltenango, Guatemala, April 21, 1965; just beginning to flower.

to guttered, acuminate, margin undulate; teeth largest through the mid-blade, 8–10 mm long, 2–6 cm apart, the cusps flattened, triangular, straight or curved from low bases, dull brown; spines 5–7 cm long, subulate, narrowly, shortly grooved above, decurrent or non-decurrent, dull brown; inflorescence massive, 5–7 m tall; flowers 85–95 mm long, greenish yellow, buds red-tinged; ovary 38–43 mm long, neck grooved, unconstricted; tube funnelform 17–18 mm deep, 20 mm wide, deeply grooved, thick-walled; tepals unequal, linear lanceolate, the outer 32–33 mm long, 8–9 mm wide, thickly apiculate, deeply cucullate, the inner shorter, narrower, thickly keeled; filaments 60–65 mm long, broadly flattened, inserted on 2 levels 10–13 mm above base of tube; capsules (Trelease) immature, oblong, some 30 x 60 mm.

Fig. 21.19. The massive stem and leaves of *Agave tecta* at maturity. Near Quezaltenango, Guatemala, April 1965.

Type: *Trelease 17,* about Quezaltenango, Guatemala, 7 April 1915, *ILL* (leaf margin, spines, flower bud, note).

Large mature specimens of this maguey are clearly portrayed in Figs. 21.18 and 21.19, in the type locality. In 1975, it was still common about the roads and fields of Quezal-tenango, but was not observed being tapped for juice or pulque. Many of the plants were battered or stunted and showed lack of care. Its use there is as a fence plant and yard ornamental. It is isolated by several hundred miles from its sectional relatives and may have been transported to Guatemala in post-conquest times, as it has not been found in the wild and has been observed only in cultivation about Quezaltenango in northwestern Guatemala.

Standley (in Standley & Steyermark, Flora of Guatemala, 24 (III): 119, 1952) pro-vided an interesting historical note on this agave. "It is recorded by Fuentes y Guzman (Recordación Florida I: 289, Madrid, 1882; fide Trelease) that two centuries ago excellent pulque was produced at Almolonga or Ciudad Vieja in Sacatepequez and at San Gaspar. Quite possibly, this was the result of the influence by Mexican mercenaries who took part in the conquest of Guatemala. Trelease is of the opinion that *Agave tecta* must have been the species used for the purpose. Of course, it is conceivable that there may have been large plantations formerly about Antigua and that the plants were destroyed when pulque was no longer demanded or permitted."

Species Incertae Sedis

Agave ragusae
(Fig. 21.20)

Agave ragusae Terr., Palermo Boll. R. Orto Bot. 1: 162, 1897.

This is a large handsome plant, still growing in the Instituto Orto Botanico, Palermo, Sicily, in 1969, where I photographed it (Fig. 21.20). In size, color, and armature of leaf, it resembles *A. salmiana ferox,* but the leaf is more acuminate. The curling leaves are also distinctive, but the half-shade situation may have accentuated the outcurving attitude of the leaves. Terracciono wrote a full description including the inflorescence.

He describes the peduncle as 2–3 m long, closely bracteate, the bracts lanceolate below, becoming smaller and triangular above with smooth margins; panicle lax with divaricate branches, in graceful proportion to the plant; flowers 6–7 cm long, greenish yellow, congested, closely bracteolate, with pedicels longer than the bracteoles, 1.5–2 cm long; ovary 3 cm long, rarely more, cylindric, narrowed at apex; tube 1.5 cm long, rarely more, trigonous, 6-ribbed, the ribs obtuse, apex scarcely dilated; tepals ca. 2 cm long, 0.5–1 cm wide, yellow or greenish yellow, apex lanceolate and little calloused, the exterior plane to convex on back, the interior channeled within, keeled on back; filaments pale green to yellowish, inserted at base of segments or in apex of tube, twice as long as segments; anthers barely 2 cm long; style exserted in maturity ca. 1 cm beyond stamens, apex triquetrous with dilated papillate stigma; capsule large, more or less long-pedicellate, ovate or ovate-oblong, obscurely triquetrous, beaked; seeds semi-orbicular, ribbed, black, large, compressed.

Native country unknown. Named for Henrici Ragusa, in whose garden in Florence it was growing.

Terracciono's description of the closely bracteated peduncle and flowers confirms its placement in the Salmianae. It appears to be another horticultural clonal segregate or variety of *A. salmiana,* like *ferox.*

Fig. 21.20. *Agave ragusae* Terr. in the Botanical Garden of the Istituto del Orto Botanico in Palermo, Sicily, April 1969.

Salmianae Exsiccatae

Agave macroculmis

COAHUILA. *Engard & Getz 344*, DES. Top of Sierra San Marcos, W. of Reynolds Mine Headquarters on Casa Colorada turnoff W. of San Lazaro Pass on Hwy 56. 12 July 1974 (l, f).

Engard et al. 1045, DES. Cerro de la Viga, E of San Antonio, 16 May 1977; elev. ca. 9,000 feet; *Pinus-Quercus* and *Cercocarpus* chaparral.

Gentry 11524, DES. Limestone Mt. 6 miles SE of Saltillo, 1–3 February 1952 (l, photo). Open limestone slope with rocky rubble & low bushy dispersed cover; elev. 5,500 ft.

Gentry et al. 20037, DES. Paso Carneros SW of Saltillo, 12 June 1963 (l, f, photo). Limestone with Pine & Oak; 7,500–8,000 ft.

Gentry et al. 20050, DES. Sierra de la Paila, 8–9 mi. W. of Rancho La Luz, 14 June 1963 (photo). Oak shrub in upland valley; alt. 5,700 ft.

Gentry et al. 20074, DES. Los Alpes & vicinity, ca. 40 mi. E of Saltillo, 15 June–7 July, 1963 (l, f, photo). Montane pine & chaparral limestone slopes; alt. 8,000–9,000 ft.

Gentry & Engard 23109, DES. Sierra San Marcos y Pinos, 35 mi. SW. of Monclova, 10 Oct. 1972 (cap, photo). Pine & shrub grassland; limestone; elev. 7,500–8,000 ft.

Gentry & Engard 23266, DES. Top ridge of Sierra Parras, 14 May 1973 (f, photo). Elev. ca. 8,000 ft.

Henrickson 13190, DES. Sierra de Jimulco, ca 26 (air) mi. SW of Torreón, ca 6 (air) mi, SSW of La Rosita, along trail to summit. 25° 10′N, 10° 15′W, 18 September 1973 (l, f). Open grass slopes with mixture of chaparral, 8,000 ft.

Johnson & Muller 725, A, TEX. Sierra del Pino, S canyon, below oak & pine belts, 26 Aug. 1940 (l, f).

Pinkava et al. P-1356OA, DES. Road past San Antonio, ca 26 mi. E. of jct. Rte 57, near Nuevo León border, 21 June 1976 (l, f). Pine-douglas fir-oak-agave mountain hillside, 8,640 ft. elev.

HIDALGO. *Gentry et al. 20082*, DES. 18 mi. NE. of Zimapán along Rte. 85, 26 June 1963 (l,f, photo). Pine & oak forest, alt. 7,500–8,000 ft.

Ogden 5142, DES. Hill just east of Pachuca, overlooking the town, 18 Jan. 1951 (l). Open hillside of volcanic rock, mostly quartz. Perhaps escaped from cultivation.

Rose & Painter 6757, US. Between Pachuca and Real del Monte, 31 Aug. 1903 (l, br, f).

Rose & Rose 11433, US. Sierra de Pachuca, 24 Sep. 1906 (l, cap, photo).

MEXICO. *Maury 3051*, US. Entre Chimalcoyoc y Santa Ursula, D.F., 4 May, 1890.

Rose & Painter 8041, US. Near Santa Fe, Valley of México, D. F., 23 Aug. 1903 (doubtfully assigned here).

NUEVO LEON. *Gentry 20159*, DES. Sierra Potosí near Galeana, 15 July 1963 (l, f, photo). Limestone mt. top, alt. ca. 9,000 ft.

Mueller & Mueller 6, MEXU. Sierra Madre Mts. by Monterrey (l, f).

PUEBLA. *Ogden et al. 5181*, DES. Near Tehuacán, at San Lorenzo Teotipilco, 4 February 1951 (l). Calcareous hills, with yucca & cacti.

SAN LUIS POTOSÍ. *Gentry & Engard 23597*, DES. Near tunnel to Real de Catorce, Sierra Catorce, 3 August 1975 (l). Rocky limestone slope with little cover; elev. ca 8,900 ft.

TAMAULIPAS. *Meyer & Rogers 2933, 2994*, MO. Dulces Nombres N.L.-Tams. border, 10 Aug. & 18 Aug., 1948 (l, cap).

Nelson 4475, 4497, US. Near Miquihuana, 10 June 1898; alt. 7,000–9,000 ft. (l, f, cap).

Agave mapisaga

HIDALGO. *Gentry & Arguelles 12292, 12294*, DES. Lagunilla NW of Pachuca, 18 Oct. 1952 (photo).

MEXICO. Chimal H. 65, MEXU. Navealpan, 6 June.

Kastelic s.n., MO. Texcoco, Valley of Mexico. 3 Nov. 1910.

Rose & Hough 4503, US. City of Mexico, 3 June 1899 (l, photo).

Trelease 147, MO. Type. Cult. Tacubaya, 16 April 1900 (l, br., photo; teeth small & remote).

OAXACA. *Gentry et al. 20387*, DES. Sierra Mazateca, E of Teotitlan, 21–31 Aug. 1963; alt. 7,500 feet, Short-tree Forest—Cloud Forest.

TAMAULIPAS. *Ogden et al. 5194*, DES. Along road to Tula, ca. 2 km SW of Jaumave, 15 Feb. 1951 (l, photo).

ZACATECAS. *Gentry 18264*, DES. Km. 5 along road to Jalpa from Zacatecas, 28 Nov. 1959 (l, f, photo) Cult., alt. ca. 7,500 ft.

Agave mapisaga var. *lisa*

HORTICULTURE. Gentry 21980, DES, **US**, Type. Huntington Botanical Gardens, San Marino, Calif. 7 July 1966 (f, photo, l). Cult., but reported to have been sent by Schmoll from Mexico as "maguey lisa."

Gentry 10069, DES. Hunt. Bot. Gard., Jan 1951 (photo).

Schmoll, s.n. HNT. 60 seedlings received, as Access. No. 3633, from Querétaro, Mexico, 25 Jan. 1933. (Cult. & fl'ed. June 1978.)

Agave salmiana var. angustifolia

HORTICULTURE. Berger s.n., Type, US. Cult. La Mórtola 6 VII 1909 (f, 2 sheets in type case, 3 sheets outside).

Berger s.n., US. Cult. La Mórtola, 1912 (f), also 13 VII 1905.

MEXICO. *Rose & Painter 8040,* US. Near Santa Fe, Valley of Mexico 23 Aug. 1903.

PUEBLA. *Rose 11264,* US. Near Tehuacán.

SAN LUIS POTOSI. *Eschauzier 1,* US. San Luis Potosí, 1898.

Whiting 776, ARIZ, MO, US. Charcas, July–Aug. 1934 (f).

Agave salmiana ssp. crassispina

COAHUILA. *Kastelic, s.n.* MO. Saltillo, cult., 20 May 1909 (l, other sheets of leaves in 1898).

GUANAJUATO. *Gentry 23391,* DES, MEXU. US. Km 100 on rt. 57, 62 miles N of Querétaro, 16 March 1974 (l, f, photo). Spontaneous population with *Opuntia spp.* on volcanic rocky plains.

HIDALGO. *Moore 2577,* US. Cerro de las Canteras near km 104, Pachuca–Actopan hwy, 24 April, 1947; alt. 2,500–2,700 m (l, f).

PUEBLA. *Gentry et al. 20204,* DES. 18 mi. N of Tehuacán along road to Puebla, 27 July 1963 (l, f, photo). Semi-arid shrubland on limestone, 6,200 ft.

Gentry et al. 20238, DES. 4–10 mi. SW of Tehuacán along road to Zapotitlán, August 1963 (l, f, photo). Arid Thorn Forest over limestone hills; alt. 5,600–6,000 ft.

Ogden et al. 5172, DES. Along side road E. of Córdoba Hwy., 4 mi. NE. of road to Tehuacán, 4 Feb. 1951 (l). Dry hillsides with cacti & *Brahea* palm.

Rose 11264, US. Near Tehuacán.

SAN LUIS POTOSÍ. Eschauzier VIII, MO. Lectotype. San Luis Potosí, 1908.

Eschauzier I, ARIZ, MO. US. San Luis Potosí, Sep. 1908, "maguey cimarron." (Sheet in US dated as 1898.)

Gentry 11500, DES. 20 mi. S. of Ciudad San Luis Potosí, 26–28 January 1952 (l, photo). Highland desert with dispersed low shrub cover.

Gentry 11506, DES. 40 mi. S. of Ciudad San Luis Potosí, 26–28 January 1952 (l, photo). Highland desert with dispersed low shrub cover.

Gentry 23393, DES. 35 mi. NW of San Luis Potosí along Rt. 49 to Zacatecas, 17 March 1974 (photo). Limestone, *Acacia*-Cactus-*Agave* grassland; elev. ca. 6,900 ft.

Griffiths s.n., ARIZ, MO. San Luis Potosí, Sept. 16, 1909.

Palmer 17, US. Near San Luis Potosí, 1902.

Parry & Palmer s.n., MO. Region of San Luis Potosí, 1978, "wild" (l, f).

Rose & Hough 4859, 4860a, US. San Luis Potosí, 3 July 1899.

Whiting 776, ARIZ, MO. Charcas, July to Aug. 1934 (l only).

ZACATECAS. *Lloyd s.n.,* ARIZ. Mazapil. 1908 (l only, cult or wild ?).

Agave salmiana var. ferox

HORTICULTURE. *Gentry 19380,* DES. Huntington Botanical Gardens, 6 July 1962 (l, f, photo). Cult.

Berger s.n., US. Cult. La Mórtola, June 1907 & 1910 (5 sheets f).

Brown s.n. K. Royal Botanic Gardens, Kew, 26 Dec. 1912, cult. (l, f).

MEXICO. *Trelease s.n.,* MO. Santa Fe, suburbs of Cd. Mexico, 23 Aug. 1903.

OAXACA. *Gentry 12126,* DES. Tejupan, 8 Sep. 1952 (l, photo). Cultivated, mostly along fences & trails.

Smith & Kitchen 5015, DES. Yucuita, Nochixtlán Valley, 27 July 1970 (l, f). Hedge rows at margin of village.

PUEBLA. *Miranda 2829,* MEXU. Chila-Zapotitlán, 15 July 1943.

Miranda 4395, MEXU. El Riego–Sta. Ana, Tehuacán, 18 May 1948.

Agave salmiana var. salmiana

COAHUILA. *Gentry 11542, 11544,* DES. 7 miles W of Saltillo, 3 Feb. 1952 (photos; cult. on borders around fields).

COLIMA. *Dodge No. 2,* US. Colima.

DURANGO. *Trelease 145, 149,* MO. Pueblito near Cd. Durango, 11 April 1900 (l, f, photo) "maguey verde." Leaves somewhat glaucous but very green, very broad, long-toothed. Somewhat transverse banded, 3/25/03. (Trelease annotated as *A. compluviata.*)

HIDALGO. *Gentry s.n.,* DES. By Río Tula between Ixmiquilpan & Zimapán, 24 June 1963 (photo). Desert shrub on limey slope; alt. 5,000–5,500 ft.

Pineda R. 729, WIS. 3 km. W. of Tenango del Aire, 3 May 1969 (l, f). Alt. 2,450 m.

Ramirez L. s.n., MEXU. Actopan, March–June 1936.

Ramirez L. s.n., MEXU. Lagunilla, March–June 1936.

Rose & Hough 4494, 4495, US. Near Real del Monte, 2 June 1899.

Rose & Hough 4466, US. Near Pachuca, 1 June 1899.

HORTICULTURE. *Berger, s.n.,* US. Cult. La Mórtola, 13 VII 1905 (3 sheets of fs.).

Gentry 17712, DES. Cult. Huntington Botanical Gardens, San Marino, Caif., 13 June 1959 (l, f).

PUEBLA. *Conzatti 915,* GH. Atlixco, May 1899 (f).

Endlich 1897. MO. Tehuacán.

Trelease s.n., MO. Apam, 13 March 1903 (l cuttings). "The commonest of the Apam Plains pulque species."

Trelease s.n., MO. Tehuacán, 8/14/03.

Weber s.n., MO. Lyon, July 1877. "Flowers from the specimen described by Jacobi."

SAN LUIS POTOSI. *Eschauzier No. 2,* MO, *No. 3,* US. San Luis Potosí, 1898, "maguey verde" (l edge frags.).

ZACATECAS. *Kastelic s.n.,* MO. Aguas Calientes, 8 Aug. 1908.

Rose & Hough 4931, US. Aguas Calientes, 1899.

Agave tecta

GUATEMALA. *Gentry 20814,* DES. Near Quezaltenango, 21 April 1965 (f, photo). Cult. as a border fence plant in volcanic ash soil; alt. 7,500 feet.

22.

Group Sisalanae

Group Sisalanae Trelease (in part). Nat. Acad. Sci. Mem. 11: 16, 47, 1913

Medium to large plants with stems short to elongate, freely suckering, often bulbiferous, sometimes seeding, mostly cultivates; leaves generally narrowly lanceolate, fleshy, mostly green, rarely glaucous or pruinose, radiately spreading, unarmed or irregularly armed, the teeth small, frequently cartilaginous; terminal spines small, shortly grooved above or ungrooved; panicles generally diffuse, decompound, open, frequently bulbiferous; flowers small to medium size, greenish yellow, the tube ample, ca. as wide as long, equaling to ⅔ as long as the tepals; tepals subequal, linear to lanceolate, moderately cucullate, generally involute, conduplicate, soon wilting with anthesis; filaments inserted near mid-tube; pistil over-reaching anthers in post-anthesis.

> Typical species: *Agave sisalana* Perrine.

This group is generally distinguished by its unarmed, or weakly armed, ensiform leaves and mostly small yellow-green flowers with ample tubes in open diffusive panicles, frequently bulbiferous. Trelease included in this group the usually unarmed *Agave sisalana* with the well armed *A. fourcroydes* and *A. angustifolia*. Berger (1915: 230) enlarged the group as "Unterreihe Sisalanae" to include 22 species, which he set off from the rest of his "Reihe Rigidae" as having non-decurrent end spines. This character, when series of specimens are examined, does not hold well. He also included armed and unarmed species.

In reviewing the criteria of both flowers and leaves, I find good grounds for limiting the Sisalanae to a small group of species without teeth or with teeth minute or repressed. There is some support for this grouping in flower morphology as well, but the distinction is not complete. My Sisalanae are mostly cultivates of unnatural or unknown origins. *A. sisalana* is obviously a sterile hybrid, and it is probable that others of the Sisalanae here recognized are hybrid to some degree. I believe the present arrangement will contribute to a more ready placement of the following plants and to recognition of their artificial standings. Several taxa are poorly known, even though they have been in horticulture for many years. *Agave neglecta* and *A. weberi* appear closely related and perhaps should be combined under the latter name. Until more specimens are available to make critical comparisons, I have separated them in the following key, according to authors' descriptions.

The members of the Sisalanae, except *A. kewensis,* have no known natural habitats, but as fiber and ornamental plants have been carried around the world. *A. weberi* will tolerate the light freezes of the lower Rio Grande Valley, but others are more tender. The important fiber plant, *A. sisalana,* requires a warm equitable climate with 40–50 inches (500–600 mm) of annual rain for the production of good quality fiber. The rainfall regimes and flowering seasons of some of the species are given in Fig. 22.1.

Key to Species of the Sisalanae

1. Plants small; leaves arching, recurved, 50–80 cm long, narrowed toward base; panicle relatively short, typically less than 3 m tall, compact; flowers small, 40–60 mm long
   desmettiana, p. 622
1. Plants larger; leaves arching or ascending, 80–150 cm long, frequently not narrowed toward base; panicle tall, diffuse, the branches more remote; flowers 60–80 mm long (*sisalana* 50–60 mm)   2
2. Stems virtually lacking; leaves few (35–60), recurved to curling, sprawling; spines acicular, very narrowly grooved at base; panicle 3–4 m tall
   kewensis, p. 624
2. Stems short to well developed in cultivation; leaves numerous (80 or more per rosette), linear lanceolate, arching or ascending, not sprawling; spines

subulate to slender conical, openly grooved above   3
3. Mature leaves 9–12 cm wide in mid-blade, usually with smooth margins; spines short (1.5–2 cm), nearly black, with short shallow groove above at base; flowers 50–60 mm long
   sisalana, p. 628
3. Mature leaves 15–20 cm wide in mid-blade, usually recurved and with small denticles along parts of leaf margins; spines longer (2.5–5 cm), brown, with long open groove above; flowers 60–80 mm long   4
4. Panicle 8–10 m tall; flowers 55–60 mm long; end spine 2.5 cm long
   neglecta, p. 627
4. Panicle 3–8 m tall; flowers 65–80 mm long; end spine 2.5–5 cm long
   weberi, p. 631

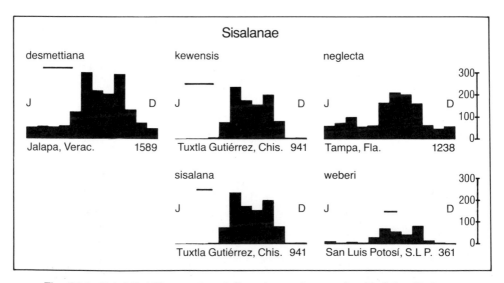

Fig. 22.1. Rainfall (silhouettes) and flowering perimeters (bars) of the Sisalanae; relevant meteorological stations with average annual rainfall in millimeters. Data from Atlas Climatológico de México, 1939. Broken bars indicate uncertainty in flowering seasons.

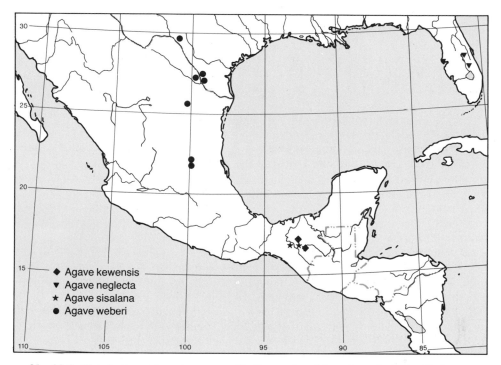

Fig. 22.2. Distributions of spontaneous and indigenously cultivated species in the Sisalanae.

Sisalanae

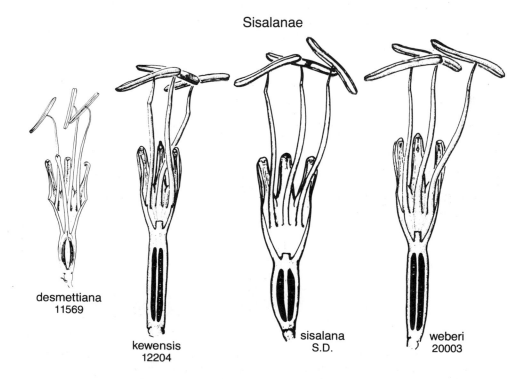

Fig. 22.3. Long sections of Sisalanae flowers.

Agave desmettiana

(Figs. 22.1, 22.3, 22.4, 22.5, 22.6; Table 22.1)

Agave desmettiana Jacobi, Hamburger Garten-Blumenzeitung 22: 217. 1866. Fig. 32.
  Agave regeliana Jacobi, Hamburger Garten-Blumenzeitung 22: 214. 1866.
  Agave miradorensis Jacobi, Abh. Schles. Ges. Vaterl. Cult., Abth. Naturwiss. 1868: 147. 1868.

Medium to small, dark green to glaucous green, symmetrical rosettes 7 x 9 dm, surculose in early years, with graceful arching leaves; leaves 50–60 x 10–12 cm to 70–80 x 7–9 cm, linear-lanceolate, openly ascending, turgidly brittle, abruptly or gradually narrowed toward the base, the base thicker than wide in cross section, finely and sparsely fibrous, unarmed with smooth unmargined edges or armed with small teeth; teeth 1–2 mm high, regular, chestnut brown, 1–2 cm apart or few and irregularly spaced; spine 2–3 cm long, subulate, dark brown to fuscous, shortly and broadly grooved above; inflorescence paniculate, rather deep and narrow in upper ½ to ⅔ of shaft, 2.5–3 m tall; bracts short, deltoid; laterals 20–25 with congested umbels of pale yellow flowers 40–60 mm long, on short pedicels subtended by short scarious bracteoles; ovary short-stipitate, small 15–26 mm long, green, 6-grooved above, with a very short unconstricted neck; tube 10–12 mm long, 14–17 mm broad, prominently flanged within below inner tepals, deeply grooved and ridged outside; tepals 13–15 mm long, green in the bud, yellow at anthesis, drying ferruginous, the outer 1 mm longer than the inner and broadly overlapping at base, the inner sharply involute; filaments 30–40 mm long, inserted 6–8 mm above the tube base, yellow; anthers 13–26 mm long, attached in the middle; capsule and seeds not seen.

  Type: Not specified and origin unknown. Neotype *Gentry 11569,* Guasave, Sinaloa, Mexico, 8 Feb. 1952, cultivated in yard of house (l & f), *US,* DES, MEXU.

I have followed Trelease (1920) in using *desmettiana* as the preferred name for this taxon. Smaller forms or varieties of this species are generally referred to *A. miradorensis* in botanical gardens and horticultural literature. Some of these forms develop light glaucous leaves, which are almost luminous in certain lights and are highly ornamental. Close study of this complex over a few years might well develop a handy subspecific taxonomy and clarify relationships.

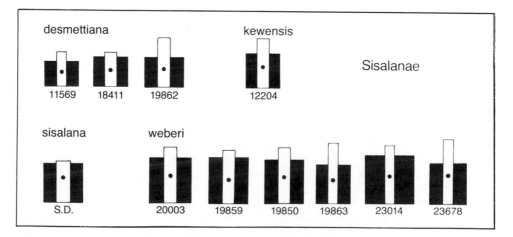

Fig. 22.4. Ideographs of Sisalanae flowers.

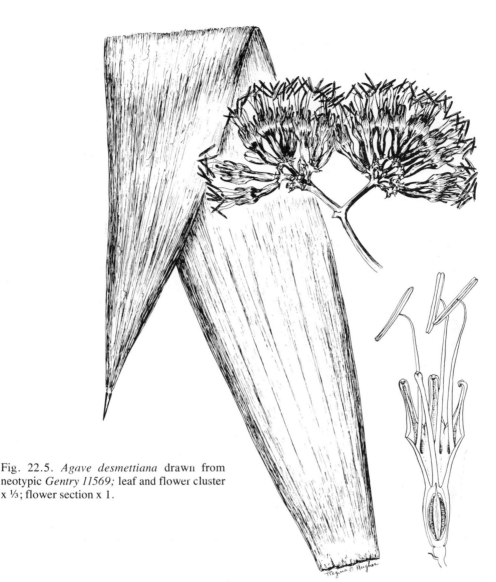

Fig. 22.5. *Agave desmettiana* drawn from neotypic *Gentry 11569;* leaf and flower cluster x ⅓; flower section x 1.

This species is distinguished by its smooth, unarmed, arching leaves, short, compact, paniculate inflorescence, and small flowers with very short ovary and broad, bulging, deeply furrowed tube. The deep exterior furrows are complemented within by conspicuous costae that border the filaments along the inner tepals. These provide structural strength to the thin perianth above the thickened nectariferous tissue of the lower half of the tube. This costate structure is well developed in the Sisalanae. The present taxon differs from all other members of the group reviewed by its smaller, more graceful rosettes, small dense panicles (vs. tall, diffuse), and smaller flowers (35–40 mm vs. 60–80 mm) with a very short, nearly neckless ovary. Like *A. sisalana* and some other members of the group, *A. desmettiana* may have either perfectly smooth leaf margins, or some leaves with small, irregularly spaced teeth, or others with a regular set of small teeth, as in the Ringold Park collection in Brownsville.

Fig. 22.6. A small form of *Agave desmettiana* with lightly armed leaves, suckering prolifically, in the Desert Botanical Garden, Phoenix, Arizona, 1977.

A. desmettiana with its soft, pliant, arching, and frequently glaucous leaves is an attractive ornamental. The glaucous quality and color varies with growing conditions and among individuals. It is, however, sensitive to frost and young plants were killed at Murrieta, California. In the Mexican gardens, it grows rapidly when given plenty of water, throws vigorous offsets early in life, and blooms in the early spring within 8 to 10 years. The fiber is fine and weak. The leaves were found to contain mostly negligible amounts of sapogenin, up to 0.3 percent. Except as an ornamental, no local uses were observed or reported for the plant. Fig. 22.6 is a small-leaved form growing in the Desert Botanical Garden.

Agave kewensis
(Figs. 22.1, 22.2, 22.3, 22.4, 22.7; Table 22.1)

Agave kewensis Jacobi, Hamburger Garten & Blumenzeitung 22: 242, 1866. Curtis Bot. Mag. 1897: t. 7532, 1897.

Rosettes large, single, open, few-leaved, with guttered, arching or sprawling leaves; leaves yellowish green, 120–180 x 12–15 cm, narrowly lanceolate, thickly succulent, pliable, the margin straight or nearly so; teeth small, the larger on upper third of margins 3–4 mm long, 1–3 cm apart, straight, much reduced below or lacking on lower third of leaf; spine 3–4.5 cm long, acicular, narrowly grooved above, non-decurrent; panicles 3–5 m tall, usually deep with 12 to 20 tri-branched umbels in upper half of shaft; pedicels 5–8 mm long, sparsely bracteolate; flowers (*Gentry 12204*) yellow, slender, 60–74 mm long; ovary 30–40 mm long, slender, cylindric, with slightly constricted neck; tube 12–15 mm deep, grooved from open tepal sinuses; tepals unequal, 18–20 x 6–5 mm, lanceolate, falcate, the outer longer and with dark glandular apex; filaments 46–52 mm long, inserted above mid-tube; anthers 26–27 mm long, excentric; capsules (*Gentry 16403*) immature, 3–3.5 cm long, obovate or oblong, stipitate, shortly beaked.

Type: Kew Gardens, Aug. 31, 1895. "From the type plant. Type of Bot. Mag. t. 7532. Mexico." *K*, here reproduced as Fig. 22.7. The small unarmed leaf appears more bract-like than leaf-like, as though the specimen preparer avoided trying to press a thick leaf. The dried tube is only 6 mm deep. However, Jacobi described the leaf with teeth and the tubes of dried flowers are always shrunken.

Berger in his Agaven, p. 219, did not have this taxon among his 274 numbered species, but he gave a postscript account of it under *A. caribaeicola* Trel. Berger had not seen the living plant, its origin was not certainly known, and he obviously was following Trelease, who placed it in the Section Caribaeae. Trelease based his judgment of the plant solely on evidence of the Botanical Magazine illustration.

The reduced teeth, sometimes lacking, together with the deep tube and the near mid-tube insertion of the filaments, places this poorly known agave in the group Sisalanae.

Table 22.1. Flower Measurements in the Sisalanae (in mm)

TAXON & LOCALITY	OVARY	TUBE	TEPAL	FILAMENT INSERTION & LENGTH		ANTHER	TOTAL LENGTH	CULL. NO.
desmettiana								
Guasave, Sin.	13	10×14	14 & 13	6	28–30	15	37	11569
	12	10	14	6	25	15	37	11569
Brownsville, Tex.	15	13×14	14×5	7–8	36	14	41	18411
	15	12×14	12×4	7–8	35	13	39	18411
Hunt. Bot. Gard., Cal.	24	12×16	20×6	7	33	26	56	19862
	26	12×17	20×5	8	42	26	58	19862
	26	13×17	21×5	8	40	27	61	19862
kewensis								
Río Chiapa, Chis.	34	13×14	20×6 & 18	9–10	50	26	67	12204
	38	14×14	19×5 & 17	9	50	26	71	12204
	39	15×14	20×5 & 18	10	52	27	74	12204
sisalana								
San Diego, Cal.	21	17×15	17×6	11–12	39	23	55	s.n.
	24	17×16	17×5	11–12	50	24	58	s.n.
	23	15×16	18×5	10–11	50	24	57	s.n.
weberi								
Webb Co., Tex.	33	20×16	21×5 & 20	11–12	55	30	73	20003
	36	19×16	23×5 & 21	12	60	31	78	20003
Hunt. Bot. Gard., Cal.	38	17×15	27×4 & 26	11–12	54	32	82	23678
	38	16×18	22×5 & 21	11–12	58		75	23678
	37	19×16	22×4 & 20	11	60	33	77	19859
	38	18×16	21×5 & 19	8–10	65	34	76	19859

* Measurements from fresh or pickled flowers.

My two collections, one of which flowered as a transplant in the Huntington Botanical Gardens, both appear close to Jacobi's original description and subsequent portrayal in the Botanical Magazine. I do believe the two collections are assignable to *A. kewensis* and place the true origin of the species as a calciphile long the Río Chiapa. The open sprawling habit of the plant and its appearance on cliffs strongly recall *A. vilmoriniana* or the subgenus **Littaea.**

Fig. 22.7. *Agave kewensis,* Tab. 7532, Curtis Botanical Magazine, 1895. The forked inflorescence is very unusual and attributable to an accident of growth in the Kew greenhouse.

Agave neglecta
(Figs. 22.1, 22.2, 22.8)

Agave neglecta Small, Flora Southwestern U.S. 289, 1903.

Large rosettes 1.3–1.7 m tall, suckering freely, with short stems 3–4 dm tall; leaves 100–150 x 15–25 cm, broadly lanceolate, pale green, glaucous, ascending to arching or reflexed in age, concave, thickened and narrowed toward base, acuminate, the margins nearly straight, with small, fine, closely set teeth below, becoming toothless above; terminal spine 25 x 3 mm, acicular, scarcely decurrent; panicle very tall, to 8–9 or 10 m, open, broad, with 18 to 20 diffuse decompound umbels in upper ⅓ to ½ of shaft, bulbiferous; flowers greenish yellow, 55 mm long; tepals ca. 23 mm long; filaments long exserted, inserted ca. mid-tube or above; capsules oblong to ovoid, ca. 3 cm long; seeds apparently ca. 6 x 4 mm, lunate or crescentic, with deep hilar notch.

Lectotype: *Weber s.n.,* Cult. at Eustis, Florida, July 1895, *MO. NY.* Flowers were sent by Trelease to Dr. Small: "The only flowers of *A. neglecta* that we have are from a cultivated plant at Eustis. Under separate cover I send a few of them—not pretty to look at! The species continues to be appropriately named. Can't you get good full material of it? 6/20/11 Wm. Trelease."

Fig. 22.8. *Agave neglecta* on the beach near Pass-a-Grille, Florida, 1962. A high wind bent the tall peduncle, also a common happening to *A. weberi* at the Desert Botanical Garden.

I am not certainly acquainted with this agave. The best representative material I have seen is a series of photographs of a plant that flowered at Pass-a-Grille Beach, S of St. Petersburg, Florida, sent to me by Gen. Tom Rives, who was attracted by the very tall panicle (Fig. 22.8). The slender scape was bent by a high wind. Rives wrote that the species was quite common in that part of Florida, but that an unusually severe frost of 18°–20°F (−5°–−7°C) during December 12–13, 1963, killed a number of young suckering plants, but the "underground roots of the giant specimen still appears to be very much alive" (letter of Jan. 14, 1963).

Agave neglecta appears closely related to *A. weberi* of the arid Mexican highland and to the variable horticultural *A. desmettiana*. For lack of specimen material there is no way of judging relationships at this writing.

Agave sisalana

(Figs. 22.1, 22.2, 22.3, 22.4, 22.9, 22.10; Table 22.1)

Agave sisalana Perrine, Congr. Doc. 564: 8, 9, 16, 47, 60, 86, Sen. Doc. 300: 36, 105, 140 (2nd Session 25 Cong., U.S.A.), 1838.

Plants forming rosettes 1.5–2 m tall with stems 4–10 dm long, suckering with elongate rhizomes; leaves ensiform, 90–130 x 9–12 cm, green, somewhat lightly zoned in youth, radial, fleshy, finely fibrous, smooth, the margins of mature leaves usually tooth-less, the young leaves with few minute teeth; spine 2–2.5 cm long, subulate, smooth, dark brown, somewhat lustrous, shortly shallowly grooved above, non-decurrent; panicle 5–6 m tall, deeply elliptic outline, with 10–15 (–25) lateral branches of umbellate clusters in upper half of bracteate shaft, bulbiferous after flowering; flowers 55–65 mm long, greenish yellow, malodorous; ovary short fusiform, 20–25 x 8–9 mm, nearly neckless, tube broadly urceolate, 15–18 mm deep, narrowly grooved; tepals equal, 17–18 x 5–6 mm, appressed to erect, conduplicate, involute, linear-lanceolate, the apex obtuse, cucullate, the inner 2-costate within, keeled outside; filaments 50–60 mm long, reddish or dark-spotted, inserted above mid-tube; anthers 23–25 mm long, yellow, centric; capsules and seeds generally lacking.

Neotype: *Gentry 16434*, Ocosocoautla, Chiapas, Mexico, 22 March 1957; cultivated as fence row and fiber plant, *US*, dup. in DES.

A form with well-developed triangular teeth was named by Trelease as var. *armata* (Nat. Acad. Sci. 11: 49, 1913). It occurs sporadically at random in widely scattered localities, as at La Mórtola Botanical Garden, Italy (Berger Die Agaven 1915: 232), in the Caribbean, Ethiopia, *Meyer 8691*.

Agave sisalana is readily recognized by its green unarmed mature leaves with short dark brown, conic to subulate, non-decurrent spine. It is a sexually sterile clone, probably of hybrid origin. This is evidenced by its general inability to produce seed and by its chromosomes. It is a pentaploid: Sato (1935) reported $2n=147$; Doughty (1936) $2n=138$; Granick (1944) $2n=149$, Fig. 22.10. They report lagging chromosomes and other distur-bances in meiosis. The rare cases where *A. sisalana* has produced seed may be due to chance pollen from *A. angustifolia* or other closely related species, but this point is by no means certain. Doughty made artificial crosses between *A. sisalana* and *A. angustifolia* and *A. amaniensis,* using pollen of *A. sisalana*. At least one outstanding fiber producer was achieved by backcrossing his hybrids with *A. amaniensis,* Hybrid No. 11648, which in field trials in Mlingano, Tanzania, yielded about double the fiber yield of *A. sisalana* (Lock 1969).

The origin of *Agave sisalana* is uncertain. Because it was originally exported from Mexico via the port of Sisal in Yucatán, it has long been erroneously reported as of

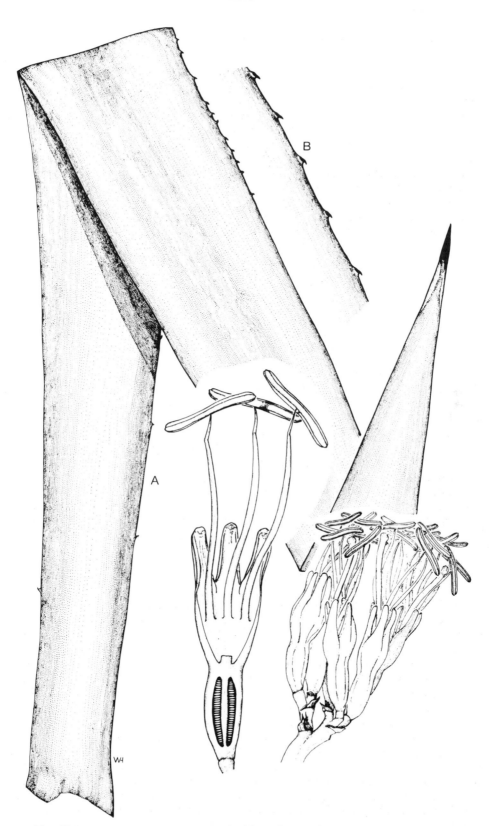

Fig. 22.9. *Agave sisalana*. A, a young leaf from Guatemala, *Gentry 23625;* B, a leaf from Ethiopia, *Meyer 8691*. Adult leaves are usually toothless. Leaves x $^3/_5$; flower section about natural size, from San Diego, California.

Yucatán origin. However, no botanical collections of the plant have ever been made in Yucatán, and botanists who have worked in Yucatán (fide Faustino Miranda, Bernice Schubert, and others) have told me they did not find the plant there. Until recently, all fiber plantation *Agave* there were henequen, *A. fourcroydes* or near relatives with armed leaves. However, *A. sisalana* is cultivated as a native "cottage industry" in the neighboring state of Chiapas. The inhabitants of the small towns from Cintalapa to Chiapa grow it in their yards or as fence rows. They extract the fiber by hand on crude pole or board benches with locally made tools. One I observed was a metal scraper blade mounted on a wooden stick and used like a draw-knife. The leaf was attached to the board and the pulp scraped off the fiber, which was then cleaned and dried. Twisted into twine it could then be employed for making rope, nets, hammocks, and other useful fiber artifacts. The occupation appeared indigenous and, since this is the only area in which *A. sisalana* has been observed and regularly employed in the village complex, I regard the area as a likely place of origin. A possible more specific process in origin may have been via the chance hybridization of *A. angustifolia* with *A. kewensis*. The latter has toothless forms and both species grow in proximity in the state of Chiapas.

From Florida *A. sisalana* was diffused to the Caribbean Islands, to Brazil, to India, and other countries. According to Miranda, the late well-known botanist, it was recently reintroduced to Mexico from Africa for plantation farming (personal communication). Miranda may have had reference to the sisal project of the Banco de México, reported by Halffter (1957) to have initiated plantings with sisal bulbils in the region of Catmis, Yucatán. Another unpredictable happening in the man-agave symbiosis!!

Sisal is one of the leading natural hard fibers in world trade. Table 22.2, taken from Wienk (1969), outlines world production of hard fibers according to the chief producing countries. There is a large literature on *A. sisalana* as an agricultural resource. A useful

Table 22.2. Estimated 1965 World Output of Hard Fibers in Metric Tons

Fiber	Producing Country	Production	Long Agave Fibers, %	Hard Fibers, %
Sisal	Tanganyika	217,650	27.9	24.0
	Mozambique and Angola	97,000	12.4	10.7
	Kenya and Uganda	64,150	8.2	7.1
	Former French Africa and Madagascar	26,950	3.4	2.9
	Brazil	182,900	23.4	20.2
	Haiti	17,300	2.2	1.9
	Others	13,200	1.7	1.5
	Total	619,150	79.2	68.3
Henequen	Mexico	147,150	18.8	16.2
	Cuba	10,150	1.3	1.1
	Others	2,700	0.4	0.3
	Total	160,000	20.5	17.6
Cantala	Philippines	2,500	0.3	0.3
Manila		102,300	—	11.3
Others		22,900	—	2.5
World Total of Long Agave Fibers		781,650	100.0	86.2
World Total of Hard Fibers		906,850	—	100.0

Source: Hard Fibres No. 64 (1967).

Fig. 22.10. Chromosome configurations in *Agave sisalana:* Upper, $2n = 149$, metaphase of root tips (Granick, 1944); lower, two examples of lagging chromosomes, anaphase of first meiotic division (Doughty, 1936).

and well detailed reference is Lock's "Sisal," 2nd edition, 1969. With the waning of our finite petroleum resource and the increased cost of synthetic fibers derived therefrom, it can be expected that the hard natural fibers will increase in economic importance and we will be able to breathe more oxygen more freely.

Agave weberi
(Figs. 22.1, 22.2, 22.3, 22.4, 22.11; Table 22.1)

Agave weberi Cels ex Poisson, Bull. Mus. D'Hist. Natu. 17: 230–232, 1901.
  Agave franceschiana Trel., Trel. ex Berger, Hort. Mortol. 12: 358, 1912.

Plants medium to large, freely suckering, green to grayish, forming open spreading rosettes 1.2–1.4 m tall, 2–3 m wide; leaves 110–160 x 12–18 cm, widest in mid-leaf, narrowed below, lanceolate, rather soft fleshy, pliable, straight to recurving, green or pruinose grayish especially in youth, concave or guttered above, the margin usually toothless along upper ⅓ to ½, denticulate below, the teeth 1–2 mm long, 1 cm or less apart, rarely toothless throughout as on bracts; spine 3–4.5 cm long, subulate, brown to grayish, openly shallowly grooved above in lower half, decurrent for several cm; inflorescence a tall open decompound diffusive panicle 7–8 m tall, sometimes bulbiferous;

Fig. 22.11. *Agave weberi*. Huntington Botanical Gardens, July 1977, budding inflorescence and detail of toothless leaves.

flowers bright yellow above pale green ovaries, 70–80 mm long; ovary 33–40 mm long, cylindric, with short, grooved unconstricted neck; tube 18–20 mm deep, 16 mm wide, rather urceolate, deeply grooved; tepals subequal, 20–24 mm long, erect appressed, involute, obtusely cucullate, the inner narrower, more conduplicate; filaments 55–60 mm long, long-tapering, inserted in mid-tube; anthers 30–31 mm long, yellow, centric or excentric; capsules 55 x 30 mm, stipitate but scarcely beaked.

Neotype: *Gentry, Barclay, & Arguelles 20003,* between Catarina & Laredo along route 83, Webb County, Texas, 4 June 1963; sporadic along road and fence, *US, DES, MEXU.* Reported by Poisson as originally from Moctezuma, a village N of Cd. San Luis Potosí.

I have assigned some related horticultural forms to this species, as found in botanical gardens without specific data on origins. Some forms are green; others are glaucous gray. Usually small teeth or denticles are present to an indeterminate extent on the gracefully arching leaves. The plants are attractive in Phoenix gardens, make good growth, and do not sucker as prolifically as *A. americana expensa,* but the hot summers are too rough on the inflorescence. The gusty strong winds with summer storms may break the high top-heavy panicles or blow the plants over, while the high insolation withers flowers, which may not open and always fail to set seed.

The plant has a desultory local use for pulque and fiber, which is of fine quality. Plantings of it are small and widely scattered in arid northern Mexico, usually about houses and as fences. It appears to be a latecomer in the lower Rio Grande Valley of Texas, where it is planted as an ornamental and the plants appear spontaneously freely suckering along roadsides. The grayer variants may be mistaken at uncritical sight for *A. americana,* but the smooth or nearly unarmed leaf margins distinguish it as *A. weberi.*

Sisalanae Exsiccatae

Agave desmettiana

FLORIDA. *Brumbach 7798,* US. Wulfert, Sanibel Island, 5 Jan. 1972 (l, f); "escaped to palmetto jungles."

Brumbach 8459, US. W Sanibel Island, Lee Co., 13 Oct. 1973 (l, f), 19 March 1976; "roadside escapee."

McCarty s.n., MO. Ankona, 19 May 1895 (l = 98 x 9 cm, with small denticles below the middle).

Meyer & Mazzeo 14294, DES, NA. Univ. Florida, Gainesville, Alachua Co., 4 Oct. 1973 (l, f, 2 sheets).

HORTICULTURE. *Berger s.n.,* US. La Mórtola, cult., 29 Aug. 1909 (l margin, spines, f. tl=55–60, t=15–16 x 18, seg=15–16 mm. Flowers show that it belongs definitely with the Sisalanae and may not be distinct from *A. weberi.*).

Berger copy of Jacobi's original "skisse," US.

Franceschi s.n., MO, DES. Cult. Santa Barbara, California, 24 Sep. 1908 (f, short & broad, like *A. desmettiana*). Type of *franceschiana.*

Gentry 10145, 19862, & s.n. DES, US. Cult Huntington Botanical Gardens, 1951–1977 (l, photos).

SINALOA. *Gentry 11569,* Neotype, DES, MEXU, MICH, *US.* Guasave, 8 Feb. 1952, cult. (l, f. Erroneously as syntype Gentry 1972).

TEXAS. *Gentry & Barclay 18411,* DES, US. Ringold Park, Brownsville, Cameron Co., 14 March 1960 (l, f); cult ornamental.

Agave kewensis

CHIAPAS. *Breedlove 9026,* CAS. El Sumidero de Tuxtla, 8 km. N of Tuxtla Gutierrez, 16 Feb. 1965; elev. ca. 3,400 ft. (l, f, cap).

Breedlove 23920, CAS, DES. At Río Grijalva, 10 km W of Chiapa de Corzo along hwy. 190, Mun. Chiapa de Corzo, 2 Feb. 1972; elev. 500 m. edge of cliff, tropical deciduous forest.

Gentry 12204, DES, US. Highway & Río Chiapa, near Tuxtla Gutierrez, 21 Sep. 1952; limestone cliff, elev. ca. 1,800 ft.

Gentry 16403, DES, MEXU, US. Río Chiapa & Panam. highway, 17 March 1957 (l, infr.).

Matuda 37549, MEXU. En Chacona by Serradero San Fernando, 22 April 1967; alt. 400 m.

F. Miranda 5682, MEXU. N Cahuare 13 Nov. 1949, cerca Chiapa.

F. Miranda 5878, MEXU. Arriba Encanada Chacona–San Fernando, NW of Tuxtla Gutierrez, 8 Jan. 1950 (f).

HORTICULTURE. Anonymous s.n. Type, K. Kew Gardens 31 Aug. 1895. "From the type plant."

Agave neglecta

FLORIDA. *McCarty s.n.,* MO. Ankona, 4/19/95 (l, 98 x 9 cm, with small denticles below middle. Trelease identified as *A. neglecta.* Letters and photos from Franceschi 8/19/02: "photos of *Agave recurvata* which bloomed with me last year." Show a very tall diffuse panicle, much like those in Block 20 of Hunt. Bot. Gard., and like the *Agave* photos of Gen. Rives of Florida. This plant or species differs from *A. weberi* in the smaller leaves, smaller flowers, and taller more profuse panicle with ca. 28 laterals. Leaf more linear than in *A. weberi*).

Rives photos, DES. Pass-a-Grille Beach, S of St. Petersburg, July–Aug., 1962. (Looks close to some hort. forms of *A. desmettiana*.)

Webber s.n., Lectotype, MO, NY. Cult. Eustis, July 1895, "stamens in throat, W. T." (l, f, cap, 3 sheets). The leaf 98 x 8.5 cm, without prickles is very similar to *A. regeliana* in Hunt. Bot. Gard.

Agave sisalana

BAJA CALIFORNIA. *Gentry 11270*, DES, US. El Triumfo, Cape District, 8 Oct. 1951.

CHIAPAS. Gentry 16434, Neotype, DES, **US.** Ocosocautla, 22 March 1957 (l, f).

BRITISH HONDURAS. *Spellman & Stoddart 2189*, US. South Water Cay, on west side of Island, 29 June 1972 (l, bulbils).

FLORIDA. *Curtis 5644*, US. Indian Key, 29 April 1896 (l, f).

Dewey 633, 634, US. Brimstone Hill, Boca Chico near Key West, 3 March 1906 (l, f, bul). "Locally abundant but not cultivated."

Killip 31736, US. Edge of pine woods just back of inn & west, Big Pine Key, 6 March 1936 (l, f).

Pennell 9622, US. Hammock, Stock Island, Monroe Co., 16 May 1917.

Pollard et al. 56, US. Sugar Loaf Key, 12–14 March 1898 (l, f).

Wright et al. 10X, 11X, US. Key West, Jan.–March 1871. "From plants introduced 20 years before by Dr. Perrine." (l, f).

GUATEMALA. *Gentry 21336A*, DES. Several miles W of Sanarate along hwy. to Guatemala City, 8 Oct. 1965 (photo). "Cult. in small field gone to weeds."

Gentry 23625, DES, US. Chiquimula; at highway intersection, 15 Dec. 1975; cult. along roadside.

HORTICULTURE. *Berger s.n.*, US. Mórtola Bot. Gard., Italy. 1911, 1912 (l, f, 6 sheets).

Agave sisalana var. *armata*

ETHIOPIA. *Meyer 8691*, US. By Ras Hotel, Debre Zeit, Shoa Province, 1 Nov. 1964; elev. 6,500 feet.

HONDURAS. *Standley 53829*, US. Vicinity of Tela, Dept. Atlantida, Dec.–Mar. 1928; along beach.

Standley 55993, US. Vicinity of Seguatepeque, Dept. Comayagua, 14–27 Feb., 1928; 1,080–1,400 m.

SALVADOR. *Calderon 1082*, US. San Salvador, Aug. 1922 (l, f).

Agave weberi

HORTICULTURE. *Gentry 10145, 10165, 19850*, DES, US. Hunt. Bot. Gard., San Marino, Calif. 1951, 1962.

Gentry 19859, DES. Hunt. Bot. Gard., 21 July 1962 (l, f).

Gentry 23014, 23675, DES. Desert Bot. Gard., Phoenix, Az. June 28, 1972 & 1979 (l, f, wilting before or with anthesis, sterile. Panicles frequently toppled by summer winds.).

Gentry & Kimnach 23678, DES, US. Hunt. Bot. Gard., 7 June 1977 (l, f, photo).

SAN LUIS POTOSÍ. Eschauzier 6, MO. San Luis Potosí, 1898. "Maguey Huastaco."

TAMAULIPAS. *Kastelic s.n.*, MO. Nuevo Laredo, 28 July 1908.

TEXAS. Gentry, Barclay & Arguelles 20003, Neotype, DES, MEXU, **US.** Between Catarina and Laredo along route 83, Webb Co., 4 June 1963. (l, f) Sporadic along fence and road.

Reed 1207, DES. Near Laredo, 27 July 1951; cult. along highway.

Ten Eyck s.n., MO. Eagle Pass, 7 Feb. 1895.

23.

Group Umbelliflorae

Trelease, Missouri Bot. Gard. Rep. 22:44. 1912

Small to large, freely seeding, long-generation, and frequently long-stemmed plants commonly branching from leaf axils, eventuating in fragmented supine clones; leaves broad, thickly succulent, stiff, well armed, mostly bright green; inflorescence stout, compact, with large, many-flowered, umbellate branches subtended by large, succulent, sheathing bracts; flowers large, fleshy, with ample tubes and strong divergent stamens; ovaries thick; capsules large, thick-walled. Baja California.

Typical species: *Agave shawii* Engelm.

The *shawii* complex is distinguished primarily by perennial plants that flower repeatedly, according to the development of axillary branches. Flowering is not at regular seasons but occurs only as the branching rosettes mature. This appears, from the few records available and from field observations, to require twenty to forty years. Records show that a specimen of ssp. *goldmaniana* (Desert Botanical Garden No. 42) collected as a small plant by George E. Lindsay near San Andrés, Baja California, in the mid-thirties, required more than 35 years to mature and flower in 1972. The trunk at that time was about 2 m long, had declined upon the ground, and is presently being continued by two large branch rosettes. Further exposition of this unique habit is given in the Introduction (Fig. 2.2). The individual rosettes, especially when they have declined upon the ground, rooted, and become separate or independent, may be regarded as multiannuals, a common reproduction habit among agaves. Moran (1964) reported an age of 31 years to maturity for a plant of ssp. *goldmaniana* at the Mina Desengaño in Central Baja California.

It should be noted that this perennial habit is not maintained throughout the various populations. Numerous individuals have been observed that mature and die without leaving any branched rosettes to follow. This is strikingly apparent, for instance, in the *goldmaniana* populations south of Punta Prieta, where single rosettes far outnumber the clustered rosettes (Fig. 23.7). There are no other consistent morphological features to distinguish these single plants from clustered plants, and they are regarded as intrasubspecific. The causes of this variation in habit, whether environmental or genetic, are not known. It may be regarded as a growth-habit potential that is not always individually fulfilled.

The closest group relation of the Umbelliflorae is not clear. The Campaniflorae of the southern half of the peninsula are well separated by their narrower, nonrigid, fine-fibered leaves with regular teeth, and by their shorter, thinner, campanulate flowers in deep, small-bracteate panicles. One species of the Campaniflorae, *A. capensis*, like *A. shawii*,

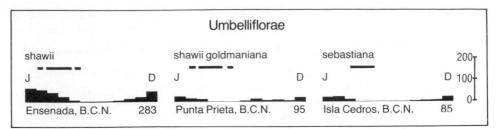

Fig. 23.1. Rainfall (silhouettes) and flowering perimeters (bars) of the Umbelliflorae. Relevant meteorological stations with average annual rainfall given in millimeters. Data from Atlas Meteorológico de México, Servicio Meteorológico Mexicana 1939 and Hastings, 1964. Flowering periods based on herbarium specimens and field observations supplemented by plants in gardens; uncertainty expressed by broken line.

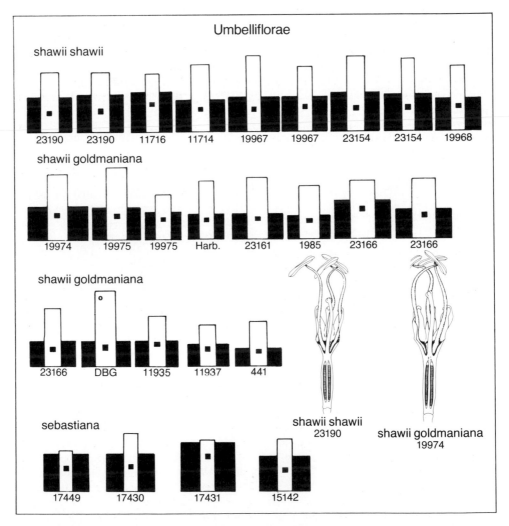

Fig. 23.2. Floral ideographs of the Umbelliflorae, showing relative proportions of the tube (black) to outer tepal (white column), and level of insertion of filament (black square). Measurements are listed in Table 23.1.

branches from the leaf axils. But I do not believe this constitutes close affinity because other, more distant relatives in the subgenus *Littaea* also have this growth habit. The Umbelliflorae and Campaniflorae occupy distinct geographic regions with no sympatric members.

The Umbelliflorae occupy an arid climate, but are unique among agave groups in being confined to a Mediterranean type of climate with winter-spring rainfall and dry summers. The rainfall silhouettes of Fig. 23.1 show only 85–95 mm (3–4 inches) of annual rainfall for ssp. *sebastiana* and *goldmaniana*, respectively, yet they are robust plants with broad leaves. However, the rainfall lack is greatly ameliorated by frequent fogs that condense and bathe the leaves, and reduce insolation and temperatures. Even the more inland populations of *goldmaniana* are regularly visited by inland-moving fogs. The fogs increase and dilute the nectar in the cupulate flowers, which is frequently copious in early morning hours.

The flowering seasons are also indicated in Fig. 23.1. The flowering seasons of *A. shawii* and ssp. *goldmaniana* are relatively long and commonly start with the spring rains, but they appear to vary from year to year and have been observed to flower in the fall.

Agave shawii ssp. *shawii* and *A. sebastiana* look very similar in habit, as they both have elongate, tightly imbricated rosettes, short thick peduncles, and wide congested panicles of very similar, large, succulent, bright-yellow flowers. They are the maritime ecotypes of the species, while *goldmaniana* is the desertic ecotype.

During the early 1950s, samples of leaves, stems, and inflorescences were collected of *Agave shawii* to determine the sapogenin content. The results of this chemical survey of the Agricultural Research Service are brought together in Table 23.2. The sapogenin content was found insufficient as practicable sources for fabricating cortisone and other steroidal drugs. Other economic aspects are noted below.

Key to Taxa
of the Umbelliflorae

1. Panicle about as broad as long, subtended by a cluster of large, succulent, sheathing bracts below lowest laterals; laterals 8 to 15; leaves short, generally only 20–40 cm long, short-acuminate, usually less than 3–3.5 times longer than broad   2
1. Panicle longer than broad, the bracts below lowest laterals smaller and not closely clustered; laterals mostly 20 to 30; leaves longer, generally 40–60 cm long, long-acuminate. Mid-peninsula
        shawii ssp. *goldmaniana*, p. 642
2. Panicle with a rounded crown, the subtending bracts large, succulent, red to purple; leaves bright green. Northern peninsular coast
        shawii ssp. *shawii*, p. 639
2. Panicle with a flat crown, subtending bracts small, scarious, pink to yellow; leaves glaucous gray to green. Cedros Island and vicinity   *sebastiana* p. 645

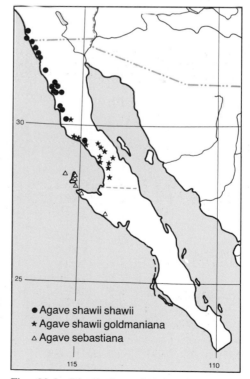

Fig. 23.3. Distribution of the Umbelliflorae, based on herbarium specimens (see Exsiccatae).

Table 23.1. Flower Measurements in the Umbelliflorae (in mm)

Taxon & Locality	Ovary	Tube	Tepal	Filament Insertion & Length		Anther	Total Length	Coll. No.
sebastiana								
San Benito I., BCN	38	15×18	16×5	8–10	42	21	69*	17449
	42	15×19	23×5	10–11	50	20	80*	17430
	35	20×22	20×6	14–15	42	21	75*	17431
	44	15×20	20×6	12	45	20	80*	17431
Natividad I., BCN	52	14×20	21×5	8–9	60	18	86*	15142
shawii ssp. *shawii*								
La Misión, BCN	37	12×19	25×8	8	73	25	74†	23190
	37	16×18	23×7	8–7	72	26	77†	23190
	37	14×18	24×7	9–7	72	25	77†	23190
	42	14×18	26×6	9–10	68	25	81†	23190
	41	16×19	25×7	10–9	70	26	82†	23190
Arroyo Seco, BCN	38	16×16	23×5	11–12	55	21	78†	11716
	29	16×16	24×5	11–12	52	20	67†	11716
	47	12×20	28×7	10		30	88†	11714
	56	14×20	26×7	9	65	30	96†	11714
Socorro, BCN	39	13×20	30×5	7–8	53	27	82†	11967
	53	15×18	25×6	9–10	40	27	92†	19967
	55	13×18	27×5	9–10	60	26	94†	19967
	53	15×20	30×7	10	64	28	97†	23154
	55	16×21	30×7	10–8	70	28	101†	23154
E of Rosario, BCN	50	14×19	28×5	10–8	70	25	92†	23154
	54	14×17	25×7	10	55	26	92†	19968
	55	13×18	27×5	10	60	27	95†	19968
s. ssp. *goldmaniana*								
Punta Prieta, BCN	40	13×24	25×8	9–10	65	26	78†	19974
	51	14×24	27×9	10–11	65	27	92†	19974
	50	13×20	29×8	9–10	53	33	92†	19975
	40	9×14	25×7	8	52	21	76†	19975
	37	11×14	18×6	8	45	22	67†	19975
Tinaja Yubay, BCN	34	9×15	22×6	7	45	26	65†	Harb.
	35	11×15	24×6	8	42	26	71†	Harb.
Punta Prieta, BCN	50	10×21	25×8	8–9	47	24	85†	23161
	49	11×20	25×9	9–10	58	24	84†	23161
	34	9×17	21×8	7	50	26	65†	1985
	34	9×18	22×8	6–8	50	25	65†	1985
W of San Borjas, BCN	57	16×22	24×10	12–13	55	24	96†	23166
	41	12×23	23×9	10–11	70	24	78†	23166
	45	12×23	26×9	8–9	75	26	82†	23166
	36	10×18	24×8	8–6	46	23	70†	23166
San Andrés, BCN	51	12×20	34×9	9–10	70	26	96†	42
	50	11×20	29×8	8–9	70	21	90†	42
	51	10×20	32×8	7–8	58	27	93†	42
Mesquitál, BCN	33	11×17	19×7	9–10	42	20	63†	11935
	34	10×17	21×7	8–10	42	20	64†	11935
	29	9×17	18×7	6–8	45	19	56†	11937
	26	9×17	15×7	5–7	42	18	50†	11937

* Measurements from dried flowers relaxed by boiling.

† Measurements from fresh or pickled flowers.

Agave shawii

Agave shawii ssp. *shawii*

(Figs. 23.1, 23.2, 23.3, 23.4, 23.5, 23.6; Tables 23.1, 23.2)

Agave shawii Engelm., Trans. Acad. Sci. St. Louis 3: 314, 370, 1875.
  Agave orcuttiana Trel. Missouri Bot. Gard. Rep. 22:47, 1912.
  Agave pachyacantha Trel. loc. cit. p. 48.

Single or cespitose, compact, green, small to medium rosettes with short to long stems (to 2 m), erect to decumbent, frequently branching from leaf axils; leaves mostly 20 x 8–10 cm to 50 x 20 cm, ovate to linear-ovate, short acuminate, glossy light to dark green, cuticle slightly asperous, thickly fleshy, rigid, plane to slightly hollowed above, about as wide in mid-blade as at base; spine 2–4 cm long, acicular, broad at base, openly grooved above, dark reddish brown to gray, straight or sinuous, decurrent for 8–10 cm or along entire leaf as corneous margin; teeth very variable in size, shape, in the mid-blade from 5 or 6 to 10–20 mm long, decreasing below, straight or variously flexed, rarely confluent or usually 1–2 cm apart, reddish to dark brown or dark gray; panicle 2–4 m tall,

Table 23.2. Sapogenin Content in *Agave shawii*
(given in percentages on dry weight basis)

COLL. No.	SOURCE LOCALITY	MONTH COLL.	PLANT PART	SAPOGENINS*				
				TOTAL	HEC.	GIT.	MAN.	TIG.
shawii ssp. *shawii*								
	Baja California:							
10282	Km 57 S of Tijuana	Mar.	leaf	0.0				
10283	Km 57 S of Tijuana	Mar.	stem	0.0				
10281	Km 57 S of Tijuana	Mar.	inflorescence	0.0				
10285	San Telmo	Mar.	leaf	0.38				
10379	Socorro	Apr.	fruit	0.35				
10379	Socorro	Apr.	leaf	0.0				
11180	Socorro	Apr.	inflorescence	0.56				
11694	Rosario	Apr.	fruit	0.2				97
11698	Rosario	Apr.	leaf	0.25	55		35	10
11699	Rosario	Apr.	leaf	0.15	100			
10383	Rosario	Apr.	fruit	0.0				
s. ssp. *goldmaniana*								
	Baja California:							
10349	Marmolito	Apr.	leaf	0.0				
10355	Tinaja Yubay	Apr.	inflorescence	0.14			100	
10355	Tinaja Yubay	Apr.	leaf	0.06	50		50	
10355	Tinaja Yubay	Apr.	stem	0.1				
10355	Tinaja Yubay	Apr.	inflorescence	0.15	60	27		
10359	Tinaja Yubay	Apr.	leaf	0.3	63		10	20
11317	Marmolito	Oct.	fruit	0.25	55	28	17	
11318	Punta Prieta	Oct.	leaf	0.3			x	
11319	Punta Prieta	Oct.	leaf	0.0				
11964	Mina Desengano	Oct.	fruit	1.3	70		30	
11936	Mesquital	Oct.	whole plant	0.7	x		x	
12402	Punta Prieta	Dec.	whole plant	0.8			x	
12401	Punta Prieta	Dec.	leaf	0.9	x		x	
12400	Punta Prieta	Dec.	leaf	0.6	x			

SOURCE: Wall (1954) and Wall et al. (1954a, 1954b, 1955, 1957).
* Hec. = hecogenin; Git. = gitogenin; Man. = manogenin; Tig. = tigogenin.

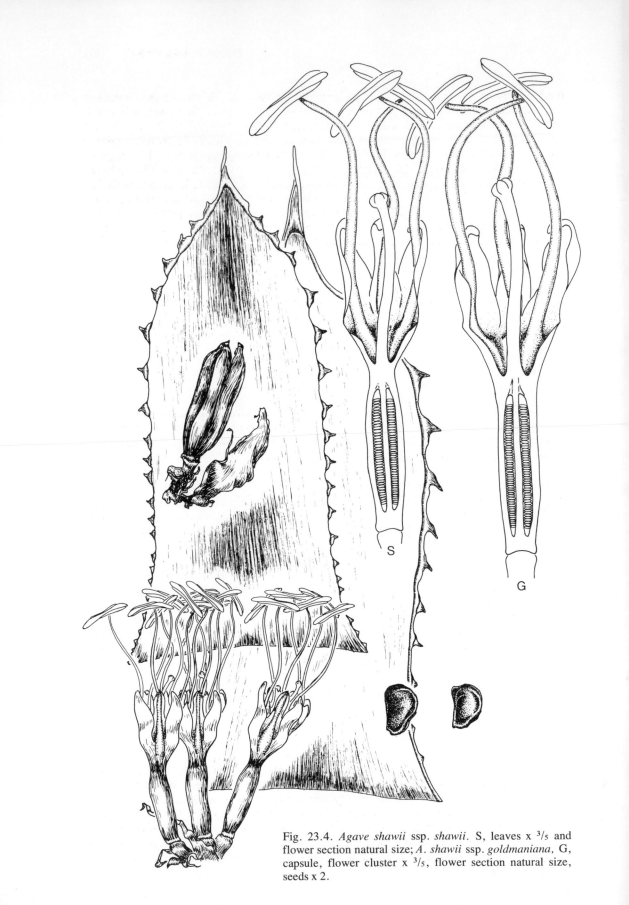

Fig. 23.4. *Agave shawii* ssp. *shawii*. S, leaves x ³/₅ and flower section natural size; *A. shawii* ssp. *goldmaniana*, G, capsule, flower cluster x ³/₅, flower section natural size, seeds x 2.

the peduncle stout, closely imbricated by large, purple, succulent bracts, closely investing the lateral umbellate branches; lateral branches commonly 8 to 14, horizontal to ascending, broad, stout, elliptic to oval in cross section, greenish red to purplish; flowers 75–100 mm long, sessile to short-pedicellate, bracteolate, closely clustered; buds frequently purplish or red; ovary greenish 35–50 mm long, thick-fleshy, cylindric, with grooved, scarcely constricted neck 6–15 mm long, shortly tapered at base; tube amply funnelform, 12–16 mm deep, 16–22 mm wide, thick-fleshy, knobby at filament insertions with furrows between, light yellow, smooth, shiny, sinuses sometimes at unequal levels; tepals unequal, broad and thick at base, becoming linear-lanceolate, thinner, yellow or reddish, soon involute and wilting while stamens extend, the outer tepal longer, 25 x 5 to 38 x 8 mm, rounded to plane on outer face, the inner with fleshy keel and thin involuting margins; filaments thick, strong, 60–70 mm long when extended, inserted near mid-tube 8–12 mm above base of tube, tapered toward apex; anthers large, 20–28 mm long, usually centrically affixed; capsule variable in shape and size, 5.5 x 1.5 cm to 7 x 2.3 cm, obovoid or pyriform to oblong, slightly attenuate to rounded at base, thickly succulent when green and drying as a scurfy exocarp, with a short to long (1.5 cm) beak; seeds 7 x 5–6 mm, lunate to cuneiform, dull black to lustrous black.

Type: *Hitchcock s.n.* 20 miles (32 km) S of San Diego on the mesa, near the ocean, at the initial monument of the Mexican boundary, July 1875, MO (4 1, f, 2 sheets).

Habitat and Characteristics

The habitat of ssp. *shawii* is a narrow one along the maritime northwest coast of the peninsula. Subspecies *shawii* is a frequent component of the coastal sagebrush community with regular associates of *Simmondsia chinensis, Rosa minutifolia, Opuntia, Franseria, Atriplex,* and other endemics like *Aesculus parryi* and *Bergerocactus emoryi.* Ten to twenty miles (16 to 32 km) inland *shawii* also contacts the higher chaparral community, where open slope and dispersed cover allow light for it to grow. Annual average rainfall ranges from 5 to 10 inches (130–250 mm), fogs and dew are frequent, solar intensity is reduced, and temperatures are equable, with mild and infrequent frosts (Fig. 23.1). The bright and gleaming rosettes are especially conspicuous on the low terraces and slopes overlooking the sea (Fig. 23.5).

This habitat contrasts strongly with the more arid and interior habitat of ssp. *goldmaniana* (Fig. 23.7). These two subspecies are regarded as ecotypes of their respective habitats, slowly and presently evolving independently in their respective centers of area. The *goldmaniana* plants around Punta Prieta, for instance, are easily separable from those of *shawii* about La Misión north of Ensenada. They are not treated as species here because of my inability to separate the two along the southern perimeters of the *shawii* area, where the more robust forms of *goldmaniana* mingle with the more compact and dwarflike forms of *shawii*. This blending of forms is viewed as a natural and long-sustained potential of *shawii* to respond robustly to environmental factors in its southern area. The serious student should note, however, that this judgment is based on gross morphological considerations, not on direct genetic evidence. Series of leaves of the two subspecies are shown in Fig. 23.6.

Exploitation of A. shawii

Agave shawii fibers were probably used, and its meristem eaten, by the Indians (Barrows, 1900). During the 1950s there was a brief attempt to use the coarse fibers commercially. This project was featured in a San Diego newspaper, and I later observed where plants had been bulldozed out in piles for hauling to a decorticating machine. The fibers were relatively short and of poor quality; the operation was apparently not commercially successful, and it was discontinued. Before 1950, plants were rarely seen as ornamentals in resident gardens, but since that time they are more commonly planted in the beach developments.

Fig. 23.7. *Agave shawii goldmaniana* in mid-peninsula, left, in branching form; right, in simple rosette form. Compare inflorescence with that of *A. avellanidens*, Fig. 15.4 (photos by Bruce Gentry, April 1972).

Distribution and Habitat

Agave shawii goldmaniana occupies a central part of the peninsula, between latitudes 30°30′N and 28°30′N, or from Laguna Chapala to the vicinity of Mesquitál (Fig. 23.3). Elevations occupied range from 5 to 700 m. Except for a few tributaries that flow into the Bahía de Los Angeles, it is limited to the Pacific slopes. Although there are gaps and irregularities, the population, by virtue of connecting strands, is nearly continuous and very large for agaves generally. It occurs on both igneous and sedimentary rocks. The most concentrated stands and best growth development observed are in the deep granitic soils south of Sierra Calamajué in the highland valley north and east of Punta Prieta, called by Goldman (1951:98) the Valley of San Andrés. Here *goldmaniana* is a co-dominant form in the vegetation along with spectacular forests of *Idria, Pachycereus, Yucca,* and *Pachycormus.* The aridity of this central highland is ameliorated by heavy nocturnal fogs moving in from the southwest from fall to spring, and which are indicated by the epiphytic lichens and *Tillandsia* that festoon the desert trees.

A cryptic member of the Deserticolae, *A. avellanidens,* replaces *A. shawii goldmaniana* south and east of Rancho Mesquitál. The former can scarcely be distinguished from *goldmaniana,* so similar are the green rosettes, unless *A. avellanidens* is in bloom. At that time its smaller, nearly tubeless flowers and narrow, deep panicles are evident. The small peduncular bracts and lack of axil branching in neighboring *goldmaniana* populations further complicate field identification. It is as though the *goldmaniana* population is infused with genes of single-rosette *avellanidens* that inhibit bract growth and axil branching—a speculative inference, of course.

Food for Animals

Leaf bases and hearts are eaten extensively by pack rats (*Neotoma*), who gnaw tunnels through the leaves around the head. *Agave shawii* was frequently noted with the terminal bud undermined and dying. Ravens apparently eat the young flowers. Claw marks, like those of fox or kit fox, on the peduncle showed how a climber had secured two or three young, tender, lateral branches by breaking them from the stalk. Cattle frequently chew the tops off young flowering shoots.

Agave sebastiana

(Figs. 23.1, 23.2, 23.3, 23.8, 23.9; Table 23.1)

Agave sebastiana Greene, Bull. Cal. Acad. Sci. 1:214. 1885.
> *Agave shawii* var. *sebastiana* (Greene) Gentry, Allan Hancock Pac. Exped. 13:49. 1949.
> *?Agave disjuncta* Trel. Missouri Bot. Gard. Rep. 22:51, 1912.

Rosettes medium to rather large, elongate, 6–12 dm tall, closely imbricate, single or cespitose, glaucous gray or green; leaves generally 25–45 x 8–24 cm, broadly linear to ovate, short-acuminate, thick rigid, plane to slightly hollowed above, rounded below, sometimes slightly narrowed toward base, light yellowish to grayish green, bud-printed, the margin usually corneous, dark brown; teeth slender, reddish brown, the larger through mid-blade 5–10 mm long, frequently down-flexed, 1–2 cm apart, or smaller and more numerous; spine stout, black to somewhat gray, 2–3 cm long, rarely shorter, variously grooved above; inflorescence 2–3 m tall, peduncle stout, with deltoid, scarious appressed bracts; panicle short, wide-spreading, rounded to nearly flat, congested with 8 to 12 large umbels of yellow flowers in upper ¼ of shaft; flowers 70–90 mm long, opening yellow, thick-fleshy; buds green; ovary 35–55 mm long, cylindric; tube broadly funnelform,

Fig. 23.8. *Agave sebastiana* on East San Benito Island in the Cedros group; an old clone with reclining trunks (photo by Reid Moran).

Fig. 23.9. *Agave sebastiana* on Cedros Island with light glaucous leaves (photo by Reid Moran).

14–20 mm deep, 18–22 mm broad; tepals 16–25 x 5–7 mm (''17–33 x 11–15''),*
lanceolate, markedly cucullate and glandular floccose at tip, the inner shorter and strongly
keeled; filaments 50–60 mm (''70'')* long, stout, inserted in mid-tube at 8–14 mm above
bottom of tube; anthers 20–21 mm (''24–30'')* long, yellow (measurements from
pressed dry specimens, see Moran numbers in Exsiccatae); capsules large, ca. 30 x 60–80
mm, beaked but scarcely stipitate; seeds large, glossy black, 7 x 11 mm.

> *Type:* *Greene s.n.,* Cedros Island, 1 May 1885, CAS.

Agave sebastiana is closely related to *A. shawii* as shown by the particulars of habit
and flower morphology. However, *sebastiana* differs significantly in the broader flatter
panicle, in the smaller, more remote, scarious peduncular bracts, and in the pale green,
somewhat glaucous leaves with more slender teeth. Bud-printing and the black spines are
more conspicuous on *sebastiana* due to the thin glaucous covering on the leaves. Reid
Moran has provided an excellent series of specimens from San Benito Islands (Fig. 23.8,
23.9) together with a series of color slides. They show that *sebastiana* forms large clones
with new rosettes starting from the bases of old stems. *Moran 15142* from Natividad
Island, southeast of Cedros Island, is a small form without corneous margins, the leaf only
15–20 cm long and pale green, the floral stem only 1 m tall. The flowers, however, are
quite normal for *sebastiana* except for the long stamens. Moran noted it as ''occasional,
near middle of Natividad Island.'' *A. sebastiana* is a very handsome plant which should
respond well in California gardens and perhaps elsewhere.

Agave disjuncta Trel. was based on small live plants collected by J. N. Rose,
reportedly on San Benito Island. Specimens of these plants are no longer in existence.
Reid Moran informs me that he has searched the island for plants of this description
without success.

---

*Moran measurements of fresh flowers.

Umbelliflorae Exsiccatae

Agave sebastiana

BAJA CALIFORNIA. *Anthony 264,* DS, K, US. San Benito Island, Mar.–June 1897 (l, f, fine-toothed form).

Anthony s.n., CAS San Benito Island, Mar.–June 1897 (f).

Beauchamp 2095, SD. West San Benito Island, 4 Apr. 1971, elev. ca. 130 m.

Beauchamp 3193, SD. Isla San Benito Oeste, 28 Feb. 1972. Common on slopes, floral stem 4 feet high (f).

Belding s.n., CAS. May 1881 (f).

Benedict s.n., SBBG. Cedros Island, upper Campo Punta, Norte Canyon, easterly from Cerro del Norte, 15 Mar. 1971.

Benedict s.n., SBBG. West San Benito Island, 200 yards (ca. 180 m) S of fishing village, SE portion of island, 9 Mar. 1971 (side flowering shoot).

Brandegee s.n., UC. San Benito Island, 27 Mar. 1897 (l, f, br).

Brandegee s.n., UC. San Benito Island, 1 Apr. 1897 (l, f, br).

Greene s.n., CAS, DS, UC. Type. Cedros Island, 1 May 1885 (leaves 25–30 cm, lanceolate, broadest in middle with regular teeth mostly curved upwards, 2 cm apart, 5–7 mm long, spine slender 20–30 mm long, openly grooved above, flowers 65–67 mm long, 3 l, f).

Howell 10691, CAS. SE side, Cedros Island, 16 Aug. 1932.

Mason 1986, CAS, K, US. Cedros Island, 3 June 1925 (f, cap).

Moran 2954, CAS. Cedros Island.

Moran 4198, BH, DS. West San Benito Island, 24 May 1952 (l, f—leaf linear-ovate with close-set, slender teeth, reduced below mid-blade, on continuous margin).

Moran 15142, SD. Natividad Island, near middle, elev. ca. 100 m, 24 June 1968 (l, f). "Floral stem 1 m high, plant smaller than usual. Fls. bright yellow, erect, visited by flies."

Moran 17430, 17431, DES, SD. West San Benito Island, elev. ca. 100 m, 19 Apr. 1970 (l, f).

Moran 17449, SD. East San Benito Island, elev. ca. 25 m, 20 Apr. 1970 (l, f). Leaves gray to gray-green.

Moran 21206, DES, SD. Cedros Island, 2 miles (ca. 3 km) from N end, elev. ca. 600 m, 28 Mar. 1974 (l, f). "Fl. stem 3 m, with 10 branches in upper 3 dm."

Moran 19924, SD. East shore of Bahía Tortugas, elev. ca. 10 m, 8 Feb. 1973 (l, f).

Moran 25266, SD. NW of Cerro Prieto, N of San Hipolito, 27°04′N, 114°01′W, 19 Feb. 1978; elev. ca. 200 m (l, f).

Philbrick B75–31, SBBG. West San Benito Island, 20 Jan. 1975. W facing slope, halfway between terrace & summit (l, br, bud).

Philbrich & Benedict B72–77, SBBG. Arroyo Madrid, above Playa Madrid, N of Colorada, SE portion of Cedros Island, 20 Apr. 1972 (l, f).

Solis s.n., MEXU. Isla Cedros, 3 June 1925 (l, f).

Agave shawii goldmaniana

BAJA CALIFORNIA. *Barclay & Arguelles 1985,* MEXU, US. Ca. 25 miles (ca. 40 km) N of Punta Prieta, 6 Apr. 1966 (l, f).

Ferris 8581, DS. W of San Borjas Range (Sierra Calamajué), 12 miles (ca. 19 km) from Laguna Chapala along road to Punta Prieta, 6 Mar. 1934 (l, f, caps).

Gentry 10349, MEXU, US. Near Marmolito, 7 Apr. 1951.

Gentry 10361, 10365, 10366, 10355, MEXU, US. Ca. 7 miles (ca. 11 km) S of Tinaja Yubay and 15 miles (ca. 24 km) NE of Punta Prieta, 8 Apr. 1951 (l, f, cap).

Gentry 10379, 10381, 10382, DES, MEXU, US. Six miles (ca. 10 km) N of Socorro, 11 Apr. 1951 (l, f).

Gentry 11318, 11319, 11320, MEXU, US. Ca. 10 miles (ca. 16 km) N of Punta Prieta, 24 Oct. 1951 (anal., photo).

Gentry 11935, MEXU, US. Ca. 5 miles (ca. 8 km) NW of Mezquitál, Vizcaino Desert, 18 May 1952 (l, f, inflo, photo).

Gentry 19974, MEXU, US. Valley W of Sierra Calamajué, 25 miles (ca. 40 km) N of Punta Prieta, 30 Apr. 1963 (l, f, photo).

Gentry & Fox 11948, DES, MEXU, MICH, US. Ca. 15 miles (ca. 24 km) NE of Punta Prieta, 20 May 1952 (l, f, cap).

Gentry & Gentry 23161, DES, MEXU, US. 16 miles (ca. 26 km) NE of Punta Prieta, 5 Apr. 1973 (f).

Gentry & Gentry 23166, DES, MEXU, US. 8 miles (ca. 13 km) W of Misión San Borja, 7 Apr. 1973 (l, f, photo).

Harbison s.n., CAS. Near Tinaja Yubay, 30 Apr. 1964 (l, f).

Harbison s.n., SD. Four miles (ca. 6 km) S of San Andrés, 25 July 1941 (l, f).

Harbison s.n., SD. Punta Prieta, 29 Apr. 1940 (l, f).

Hastings & Turner 63-231, DS. Arroyo Aguajito, 14.9 miles (ca. 24 km) E of Rosario, elev. 700 ft (ca. 200 m), 19 Oct. 1963 (l, f, cap).

Lindsay DES 42, DES. Near San Andrés, 1930s (l, f, cap).

Moran 17027, DES, SD. 7.6 miles (ca. 12 km) N of Puerto Santa Catarina, 28 Mar. 1970; "Only this one plant seen in the Santa Catarina drainage" (l, f, cap).

Moran 17053, SD. 6 miles (ca. 10 km) E of Punta Canoas, elev. ca. 50 m, 29 Mar. 1970, "Common but only one still flowering" (l, f, cap).

Moran 17121, SD. Puerto San José, elev. ca. 25 m, 30 Mar. 1970 (l, f).

Moran 17191, SD. 8 miles (ca. 13 km) S of Las Palomas, elev. ca. 140 m, 1 Apr. 1970, "Rosettes 13 dm wide, of 125 leaves, 45 x 12 cm" (l, f).

Moran 17204, SD. Boca de Marrón, elev. ca. 5 m, 2 Apr. 1970, "Common on rocky hillside. Rosettes 12–14 dm wide of 200 leaves" (l, f).

Nelson & Goldman 7151, US. Type. Yubay, 30 miles (ca. 48 km) SE of Calamajué, B.C., alt. 2,000 ft. (ca. 600 m), 18 Sept. 1905 (l, cap, photo).

Wiggins 4475, DS. Between El Marmol and Rosario, 40 miles (ca. 64 km) E of Rosario, 12 Mar. 1930 (l, f).

Wiggins & Thomas 171, DS, US. About 14 miles (ca. 23 km) toward the coast from Cerro Blanco (SE of Rosario), alt. ca. 1,200 ft. (ca. 350 m), 8 Feb. 1962 (l, f).

Agave shawii shawii

BAJA CALIFORNIA. *Arnott 12, 21,* UC. 4 miles (ca. 6 km) N of Socorro, 21 Apr. 1955 (f).

Brandegee s.n., UC. Colnett, May 1893 (l, f).

Cox s.n., UC. San Telmo de Abajo, May 1931.

Farmer s.n., SD. Santo Tomás Valley, 24 Dec. 1955 (series of leaves).

Ferris 8524, DS. 16 miles (ca. 26 km) from Colnett Wash on Santo Domingo Road, 2 Mar. 1934 (l, f).

Ferris 8528, 8529, DS, US. 30 miles (ca. 48 km) N of Rosario on road to Santo Domingo, 3 Mar. 1934.

Gentry 4001, UC. Between San Vicente and Hamilton Ranch, 10 Nov. 1938 (l, f).

Gentry 10079, MEXU, US. Huntington Botanical Gardens, 9–15 Jan. 1951 (l, f).

Gentry 10281, MEXU, US. Km 57 S of Tijuana, 21 Mar. 1951 (l, f).

Gentry 10285, MEXU, US. Ca. 120 miles (ca. 16 km) NE of San Telmo, 23 Mar. 1951 (l, f).

Gentry & Arguelles 19968, MEXU, US. 15 miles (ca. 24 km) E of Rosario, 28 Apr. 1963 (f).

Gentry & Gentry 23154, DES, MEXU, US. 12–13 miles (ca. 19–21 km) E of Rosario, 4 Apr. 1973 (l, f, photo).

Gentry & Gentry 23190, DES, MEXU, US. Near La Misión, 27 miles (ca. 43 km) N of Ensenada by power station, old road, 15 Apr. 1973 (l, f, photo).

Harbison 45522, SD. Santa Maria Valley, 31 Aug. 1953.

Harbison s.n., SD. 7 miles (ca. 11 km) SE San Quintín, 21 Apr. 1927 (l, f).

Harbison s.n., SD. San Simón (ca. 10 miles (ca. 16 km) E of San Quintín). 10 Sept. 1955.

Harbison s.n., SD. Camalu Point, 29 Dec. 1949 (l, f, cap June 1939).

Harbison s.n., SD. Arroyo ESE of Rosario, 29 Dec. 1949 (leaves narrow with very fine or small teeth. 1–2 m long & very narrow corneous margin.).

Harbison s.n., SD. Canyon above Hamilton Ranch, 15 Dec. 1953 (l, f).

Harbison s.n., SD. La Misión Point, 30 Dec. 1949 (l, f).

Harbison & Howe s.n., SD. Arroyo 8.4 miles (ca. 14 km) E of Rosario, elev. ca. 200 m. 23 Sept. 1965.

Harding s.n., SD. Halfway between Ensenada and Tijuana, 28 Dec. 1937 (br, f).

Jones s.n., DS, MEXU. West of Tijuana near the sea, 26 Dec. 1924 (f).

Moran 16711, DES, SD, UC. Jatay (S of La Misión), elev. ca. 100 m (f), "Abundant on coastal terrace 1 mile S of Jatay."

Wiggins 21, 442, DS. Valle de San Telmo, 15 miles (ca. 24 km) E of Hwy. No. 1. 17 June 1971 (l, f, insects & hummingbirds).

Wiggins & Gillespie 3911, 4006, CAS, DS, MEXU, US. 37 miles (ca. 60 km) S of Tijuana, 2 miles (ca. 3 km) S of "Halfway House," 8 Sept. 1929.

CALIFORNIA. *Eastwood 2940,* CAS. Boundary Monument, San Diego Co., 24 Apr. 1913 (l, f, topotype).

Gander 494, SD. Near Boundary Monument 258, on shore of Pacific, San Diego Co., 28 Jan. 1936 (l, f).

Hall 3957, UC. SW corner of San Diego Co., Calif., 30 Apr. 1903 (cap).

Harbison s.n., DES. International Monument, San Diego Co., Calif. (topotype).

Hitchcock in 1875, MO. Type. Initial Boundary Monument, S of San Diego, on the mesa near the ocean.

Moran 16707, DES, SD, UC. West side of Point Loma (f), 100 m S of Cal-Western Campus, 18 Dec. 1969. "Perhaps 200 rosettes, brink of sea bluffs."

Moran 21699, SD. Torrey Pines, San Diego Co., 29 Mar. 1975; "uniform spination—may be one clone—it can be considered native."

Stover s.n., SD. Point Loma, San Diego Co., 21 Apr. 1937 (l, f).

Wolf 5412, CAS. 8 ft. N of Mexican Boundary and second small hill from ocean, San Diego Co., elev. 30–40 ft. (ca. 9–12 m), 31 Aug. 1933, topotype.

Wolf 2660, DS. Rancho Santa Ana Botanical Garden from plant brought from SW San Diego Co., 15 Feb. 1932 (l, f, br).

PART IV

Culture of Agaves and Other Addenda

Culture of Agaves

With few exceptions agaves are easy to cultivate in warm climates with well-drained soils. The offsets or basal suckers are the usual propagation stock, as they are of good size, vigorous in growth, and may be planted directly in the site intended for them. Bulbils and seed require nursery care and later transplanting when they are strong enough to stand by themselves. The offsets can be set out directly with fresh roots. They should be planted deep and can be firmed in with water. If the roots are wilted or dry, they should be cut off; then, a drying-off period of one to several weeks is preferable for the healing of damaged tissue and the prevention of rot. Besides preventing rot and molds, this period gives the plant time to readjust for the budding of new roots, which are essential for new growth. This practice is the one generally followed by the fiber and tequila plantation men of Mexico.

In warm weather one or two irrigations per month should be sufficient. The plants do not need irrigation in cold weather.

When the magueys and other large agaves reach a size of three feet (1 m) tall or more, the suckers and lower leaves should be trimmed out, if neat and symmetrical plant form is desired. Snipping off the end spines will make the plants easier to live with. If planted in acid soil, species that are native on limestone or limey soil may require liming for good growth. If cultured in pots, the plants become root bound, and, although they will so endure for many years, they will not achieve normal maturity. However, some agave fanciers purposely maintain the potted dwarfs. As ornamentals outdoors, they should be given adequate spacing, or what they will cover with their radiating leaves at maturity. In the case of big magueys, a space of 10 to 20 feet (3–6 m) in diameter is required.

Generally, agaves need plenty of sunlight, for, if grown in shade, the plants become etiolated with long, weak, floppy leaves and a loose, sloppy rosette. However, in the hot summer sun of Arizona desert, a light shade is helpful for some varieties, particularly for high elevation agaves, such as the large pulque-producing magueys of the central Mexican plateau, e.g., *Agave salmiana, A. mapisaga*. Most forms of these species grow poorly, sunburn, and remain stunted in the Desert Botanical Garden in Phoenix. In Huntington Botanical Garden in southern California in a mild, coastal, foggy climate they thrive and with plenty of water reach gigantic sizes.

Agave tolerance to cold is highly variable in species and varieties and from one freeze to the next. Those from the northern perimeters of agaveland are most resistant. *Agave utahensis, A. parryi, A. neomexicana,* and perhaps a few others can withstand freeze down to zero degree Fahrenheit. Many others from higher elevations in the mountains have withstood temperatures in our gardens down to 15°–20°F, without serious damage; e.g., *A. americana expansa, A. gracilipes, A. havardiana, A. palmeri, A. chrysantha, A. shrevei, A. salmiana, A. scabra, A. parrascana, A. zebra, A. lechuguilla, A. potrerana, A. ocahui, A. pelona.*

Night temperatures of 24°–25°F (−5°C) frequently cause mild to severe leaf "burn" and dieback to many agaves. It seems to be a common turning point in agave tolerance to cold. I have observed the following agaves to suffer damage at 24°–25°F (−5°C) here in the Desert Botanical Garden or at my farm near Murrieta in southern California:

A. salmiana crassispina	A. salmiana	A. murpheyi
A. americana variegata	A. mapisaga	A. angustifolia
A. sisalana	A. promontorii	A. macroacantha
A. fourcroydes	A. karwinskii	A. ghiesbreghtii
A. datylio	A. marmorata	A. triangularis
A. bovicornuta	A. ornithibroma	A. pedunculifera
A. maximiliana	A. shawii	A. polyacantha
A. inaequidens	A. scabra	

Some of the most sensitive species that are damaged by frosts of 2°–3°F (−1°−2°C) include *A. attenuata, A. pedunculifera, A. gypsophila, A. marmorata, A. impressia, A. guiengola,* all of them from tropical climates below the Tropic of Cancer.

The hot summer sun in Arizona also inhibits flower and seed development. In many cases the stamens in flowers fail to expand and dehisce, owing in some cases to insufficient turgor pressure to extend the filaments, e.g., *Agave aktites, A. palmeri.* The anthers remain half caught in the tepals, which also fail to open out giving the anthers room to unfold. There is apparently insufficient soil moisture to feed the leaves and to extend up into the high inflorescence. Heavy watering at the start of the inflorescence shoot will assist flower development in hot climates. Other conditions preventing seed formation are self-sterility and lack of pollinators. Among the species noted as never or rarely setting seed in southwestern gardens are the following:

Agave americana expansa	Agave palmeri	Agave bovicornuta
Agave weberi	Agave chrysantha	Agave maximiliana
Agave desmettiana	Agave wocomahi	Agave inaequidens
Agave murpheyi	Agave flexispina	Agave jaiboli
Agave aktites	Agave marmorata	Agave cerulata dentiens
Agave sisalana	Agave zebra	Agave attenuata
Agave fourcroydes	Agave mckelveyana	Agave victoriae-reginae
Agave shrevei	Agave aurea	Agave pedunculifera

Lack of seed and lack of suckers or bulbils usually results in the ornamental agave going out of culture until new collections are again made in the natural habitat. Although agaves are among the easiest of plants to grow, they all have individual characteristics. Many such characteristics, such as lack of seeding, show that agaves are finely attuned or adapted to their wild native habitats. If there are no bats with *A. palmeri,* it may not be able to set seed at all. Preservation is usually of more concern to the institutional gardener than to the individual, who, if he wants a given agave out of his garden, may be glad to see it flower.

References

Anderson, Edgar. 1949. *Introgressive hybridization*. New York: John Wiley & Sons. 109 pp.

Aschmann, Homer. 1959. *The central desert of California: demography and ecology*. Berkeley & Los Angeles: University of California Press. 315 pp.

Atlas Climatológico de México. 1939. México, D.F.: Servicio Meteorológico Mexicano.

Atlas Geográfico de la República Mexicana. 1942. México, D.F.: Sec. Agri. y Fom.

Bacgert, Johann Jakob. 1972. Mannheim: *Nachrichten von der Amerikanischen halbinsel, Californien*. (Also, Spanish translation by Pedro R. Hendricks Pérez, *Noticias de la Península Americana de California por el Rev. Padre Juan Jacobo Baegert. Con una introducción por Paul Kirchoff*. México: Antigua Librería Robredo de J. Porrúa e Hijos, 1942. A résumé of original published in *Ann. Geogr.* 5: 237–39, 1866. An English translation by M. M. Brandenburg & C. L. Baumann entitled *Observations in Lower California*. Berkeley, 1952.)

Baker, John G. 1888. *Handbook of the Amaryllideae*. London: George Bell & Sons (York St., Covent Garden).

Barco, Miguel Del. 1973. *Historia natural y crónica de la Antigua California*. (Adiciones y correcciones a la noticia de Miguel Venegas.) Edición estudio preliminar, notas y apéndices. By Miguel León-Portilla. Universidad Nacional Autonoma de México, Inst. Invest. Hist. Méx. lxxxv + 464 pp.

Barrows, David Prescott. 1967. *The ethno-botany of the Coahuilla Indians of southern California*. Including a Coahuilla bibliography and introductory essay by Harry W. Lawton, Lowell John Beand (and) William Bright. Banning, Calif.: Malki Museum Press. 82 pp. (Reprint of Ph.D. Thesis, Univ. Chicago, 1897.)

Bennett, W. C., and R. M. Zingg. 1935. *The Tarahumara, An Indian Tribe of Northern Mexico*. 412 pp. Chicago: University of Chicago Press.

Berger, Alwin. 1915. *Die Agaven*. Jena. 288 pp.

Blasquez, P., and I. n.d. *Tratado de Maguey*. Puebla.

Bolton, Herbert Eugene (editor). 1919. *Kino's historical memoir of Pimería Alta*. Cleveland, 2 vols., illus., maps. Arthur H. Clark Co.

Brandegee, T. S. 1889. Plants from Baja California. *Cal. Acad. Sci. Proc. Ser. 2*, 2: 117–217.

Breitung, A. J. 1968. The Agaves. *Cactus & Succulent J. Yearbook.* 107 pp. (Reprint from *Cact. & Succ. J.,* vols. 31–38.) Reseda, Calif.: Abbey Garden Press.

Burrus, Ernest J. (editor and translator). 1966. *Wenceslaus Linck's diary of his 1766 expedition to northern Baja California.* Dawson's Book Shop, Los Angeles. 115 pp.

———. 1967. *Wenceslaus Linck's reports and letters, 1762–1778.* Los Angeles: Dawson's Book Shop. 94 pp.

Bye, Robert A., Don Burgess, and A. B. Tryan. 1975. Ethnobotany of the western Tarahumara of Chihuahua. 1. Notes on the genus *Agave. Bot. Mus. Leafl., Cambridge,* 24: 85–112, illus.

Callen, E. O. 1965. Food habits of some Pre-Columbian Mexican Indians. *Econ. Bot.* 19: 335–43.

Castetter, E. F., W. H. Bell, and A. R. Grove. 1938. The early utilization and distribution of *Agave* in the American Southwest. *Univ. Mexico, Bull.* 6. 92 pp.

Cave, Marion S. 1964. Cytological observations on some genera of the Agavaceae. *Madroño* 17 (5): 163–70.

Chazaro Basañez, Miguel de Jesús. 1981. Nota sobre la tipificación de *Agave obscura* Schiede y su confusión con *Agave xalapensis* Roezl. *Biotica* 6: 435–46. Xalapa, Veracruz.

Christensen, Bodil. 1963. Bark paper and witchcraft in Indian Mexico. *Econ. Bot.* 17: 360–67.

Clavigero, Francisco Xaviero. 1789. *Storia della California.* Venice. 2 vols., map. (Spanish translation by Nicolas García de San Vicente, entitled *Historia de la Antiqua ó Baja California.* Mexico, 1933. 267 pp. English translation by Sara E. Lake & A. A. Gray, *The history of Lower California.* Stanford Univ. Press, Stanford, 1937, 413 pp., maps.)

———. 1807. *The History of Mexico,* Book VII, Sec. Ed. London.

Comisión Federal de Electricidad, Mexico. 1968. Meteorological tables. *Boletin Meteorológico,* No. 2. pp. 1–924.

Cravioto, R. O., G. Massieu, J. Guzman, y J. Calvo de la Torre. 1951. Composición de alimentos mexicanos. *Ciencia* XI (5–6) 153.

Darton, N. H. 1921. Geological reconnaissance in Baja California. *J. Geol.* 29: 720–48.

Davis, Helen B. 1936. *Life and Work of Cyrus Guernsey Pringle.* 750 pp. Burlington: University of Vermont.

Dewey, Lyster H. 1941. *Fibras Vegetales y su Producción en América.* Publ. Union Panamericana, Washington, D.C. 101 pp., fig. 74.

Diguet, Leon. 1902. Etude sur le Maguey de Tequila. *Rev. des Cultures Coloniales* 10: 294–97, 321–26, 357–610.

Doughty, L. R. 1936. Chromosome Behavior in Relation to Genetics of Agave. 1. Seven species of fibre Agaves. *J. Genetics* 33: 197–205.

Ehler, W. L. 1967. Daytime stomatal closure in *Agave americana* etc. *Proc. Seminar on Physiological Systems in Semi-arid Environments.* Albuquerque: University of New Mexico Press.

Engelmann, George. 1875. Notes on agave. *Trans. Acad. St. Louis* 3: 201–322.

———. 1875. Flowering of *Agave shawii. Trans. Acad. St. Louis* 3: 579–82.

Felger, Richard, and Mary B. Moser. 1970. Seri use of *Agave* (century plant). *The Kiva* 35: 159–67.

Gentry, Howard Scott. 1942. Rio Mayo Plants. *Carn. Inst. Wash. Publ.* No. 527.

———. 1949. Land Plants of the California Gulf Region. *Allan Hancock Pacific Expeditions* 13: 81–180.

———. 1963. The Warihio Indians of Sonora-Chihuahua: an Ethnographic Survey. *Bur. Amer. Ethn. Bull.* 186: 61–144, pls. 28–38.

———. 1964. Maguey del Cumbre. *Amer. Hort. Mag.* 43: 158–160.

———. 1967. Putative hybrids in agave. *J. Hered.* 58: 32–36.

———. 1968. *Agave geminiflora* and *Agave colimana* sp. nov. *Cact. Succ. J. Am.* 40: 208–13.

———. 1972. The agave family in Sonora. *U.S. Dep. Agric. Agric. Handbk.* 399. Pp. 1–195, maps, illus.

———. 1978. The agaves of Baja California. *Cal. Acad. Sci. Occ. Pap.* No. 130: 119 pp., illus.

———, and Jane R. Sauck. 1978. The stomatal complex in agave: Groups Deserticolae, Campaniflorae, Umbelliflorae. *Proc. Cal. Acad. Sci.* (4th ser.) 41: 371–87, 51 figs.

Gibson, F., 1935. Agave murpheyi, A New Species. *Boyce Thompson Inst. Contr.* 7: No. 1, 83–85.

Goldman, Edward A. 1951. Biological investigations in Mexico. *Smithson. Miscell. Coll.* 115: 1–476, map, illus.

Goncalves de Lima, Oswaldo. 1956. *El Maguey y el Pulque en los Codices Mexicanos*. Fondo de Cultura Económica, México–Buenos Aires. Pp. 1–275, illus.

Granick, Elsa B. 1944. A karyosystematic study of the genus *Agave. Am. J. Bot.* 31: 283–89.

Hagen, V. W. von. 1943. The Aztec and Maya paper makers. New York: Hacker.

Halffter, O. 1957. Pests affecting the various species of *Agave* cultivated in Mexico (in Spanish). *Rev. App. Ent.* 46, sre. a, Pt. 7: 278.

Harrington, M. R. 1933. Gypsum Cave, Nevada. *Southwest Museum Papers*. Southwest Museum, Los Angeles, California.

Hastings, James R. 1964. Climatological data for Baja California. *Univ. Ariz. Inst. Atmos. Physics Tech. Rep.* No. 14.

———, and Robert R. Humphrey. 1969. Climatological data and statistics for Sonora and northern Sinaloa. *Univ. Ariz. Inst. Atmos. Physics Tech. Rep.* No. 19.

Hemsley, W. B. 1887. *Botany.* Vol. 4. In Godman, F. D., and O. Salvin (eds.) Biologia Centrali-Americana. London: R. H. Porter and Dulau and Co.

Hernandez, Francisco. 1649. Rerum Medicarum Novae Hispaniae Thesaurus. (1651)

Hooker, Joseph. 1871. *Curtis Bot. Mag.* t. 5893, 5940.

Howell, D. J. 1972. Physiological adaptations in the syndrome of chiropterophily with emphasis on the bat Leptonycteris. Ph.D. thesis, Univ. Arizona, xvii + 217 pp.

———. 1979. Flock foraging in nectar-feeding bats. *Am. Nat.* 114: No. 1, 23–49.

Jacobi, G. N. von. 1864–1867. *Versuch zu einer Systematischen der Agaveen.* Hamburger Garten & Blumen-Zeitung, vol. 20–21.

Jacobsen, Hermann. 1960. *Handbook of Succulent Plants.* 1: 64–130.

———, and G. D. Rowley. 1973. Some name changes in succulent plants. *Nat. Cact. & Succ. J.* 28: 4.

Johnston, Ivan M. 1924. Expedition of the California Academy of Sciences to the Gulf of California in 1921. XXX. The botany (the vascular plants). *Proc. Calif. Acad. Sci.,* Ser. 4, 12: 951–1218.

Langman, Ida K. 1964. A selected guide to the literature on the flowering plants of Mexico. Philadelphia: Univ. Penn. Press. 1015 pp.

Larson, R. L., H. W. Menard, and S. M. Smith. 1968. Gulf of California: a result of ocean-floor spreading and transform faulting. *Science* 161: 781–83.

Lenz, Hans. 1950. *El Papel Indigena Mexicano.* Mexico.

Linnaeus, Carl. 1753. *Species Plantarum.*

Lock, G. W. 1969. *Sisal.* Tanganyika Sisal Growers Assoc. London & Harlow: Longmans Green & Co. 2nd Ed. 365 pp., illus.

Lumholtz, Carl S. 1973. *Unknown Mexico.* 2 vols., reprint of 1902 ed. Glorieta, New Mexico: Rio Grande Press.

MacNeish, R. S. 1967. In Byers, D. S.(ed.) et al. *Prehistory of the Tehuacán Valley.* Vol. 1. Austin: Univ. Tex. Press.

Marroquin, A. Sanchez. 1966. Agaves de México en la Industria Alimentario. Cent. Estud. Econ. y Soc. del Tercer Mundo, pp. 526 (prepublication mock-up).

Martin del Campo, Rafael. 1938. El Pulque en el Mexico Precortesiano. *Anales Inst. Biol.* LX: 5–23.

Martinez, Maximino. 1936. *Plants Utiles de Mexico*. México: Edition Botas.

Massieu, G. H., J. Guzman, R. O. Cravioto y J. Calvo. 1948. Determination of some essential amino acids in several uncooked Mexican Foodstuffs. *J. Nutrition* 38: 297.

Mathews, Frank P. 1938. Lechuguilla as a livestock poison on Texas ranges. *Am. Vet. Med. Assoc. J.* 93: 168–75.

McClendon, J. F. 1908. On xerophytic adaptations of leaf structure in *Yucca, Agave,* and *Nolina. Am. Nat.* 42: 308–16.

Minnis, Paul E., and Stephen E. Plog. 1976. A study of the site specific distribution of *Agave parryi* in east central Arizona. *The Kiva* 41: 299–308.

Moran, Reid. 1964. Floración de *Agave goldmaniana* a los 31 años. *Cact. Succ. Mex.* 9: 87–88.

———. 1967. *Agave subsimplex* Trelease. *Cact. Succ. Mex.* 12: 59–61.

Nelson, Edward W. 1922. Lower California and its natural resources. *Mem. Natl. Acad. Sci.* 16 (1st Mem.): 1–194, illus.

O'Neill, Thomas E. 1969. Brief on standard identity for tequila. Nat. Assoc. of Alcoholic Beverages, Inc. Washington D.C. (copy)

Otto, Friederich. 1842. *Arten des Koniglichen Allgemeine Gartenzeitung* No. 7: 49–51. Berlin.

Pennington, C. W. 1958. Tarahumara fish stupefaction plants. *Econ. Bot.* 12: 95–102.

Peterson Frederick. 1961. *Ancient Mexico.* London: Geo. Allen & Unwin Ltd. Pp. 1–314. illus.

Rau, Charles. 1864. An account of the aboriginal inhabitants of the California Peninsula as given by Jacob Baegert. *Annu. Rep. Smithson. Inst.* (1863): 352–69.

Rojas Gonzales, F. 1939. Las Industrias Otomies del Valle del Mezquital. *Rev. Mex. Soc.* 1(1): 88–96.

Rollins, Reed C. 1953. Cytological approaches to the study of genera. In Plant Genera. *Chron. Bot.* 14: 133–39.

Rose, J. N. 1899. Notes on useful plants of Mexico. *Contr. U.S. Nat. Herb.* 209–59.

Salm-Dyck. 1834. *Annotationes Botanicae*. Hortus Dyckensis.

———. 1859. Bermerkungen uber die Gattungen Agave and Fourcroya. Neuen Arten. *Bonplandia* 7: 85–995.

Sanchez Mejorada, F. 1978. Cactaceas y Suculentas de la Barranca de Metztitlan. *Soc. Mex. Cact. Publ.* No. 2. México, D.F. 131 pp.

Sato, D. 1935. Analysis of Karyotypes in *Yucca, Agave* and related Genera. *Jap. J. Genetics* 11: 272–78.

Sauer, Carl. 1965. Cultural factors in plant domestication in the new world. *Euphytica* 4: 301–06.

Schaffer, W. M., and M. V. Schaffer. 1977. The reproductive biology of Agavaceae. *Southw. Nat.* 22: 157–67.

Schiede. 1829. *Linnaea* 4: 581.

Schlechtendal, D. F. L. 1844. *Linnaea* 18: 413, 1844.

Schuchert, Charles. 1935. Historical geology of the Antillean-Caribbean region. New York: John Wiley & Sons. 811 pp.

Schultze, G. M. 1942 (*Agave coccinea*). *Notizb. Bot. Gart. Mus. Berlin-Dahlem* 15: 697.

Scott, W. B. 1937. *A History of Land Mammals in the Western Hemisphere*. Rev. ed. New York: MacMillan Co.

Sharma, A. K., and U. C. Bhattacharyya. 1962. A cytological study of the factors influencing evolution in *Agave. La Cellule* 62: 259–81.

Shreve, Forrest. 1951. Vegetation of the Sonoran Desert. *Carnegie Inst. Washington Publ.* 591. xii + 192 pp, 37 pls.

———, and Ira R. Wiggins. 1964. Vegetation and flora of the Sonoran Desert. Vols. 1 & 2. Stanford, Calif: Stanford Univ. Press. 1740 pp.

Small, J. K. 1933. *Manual of the Southeastern Flora*. New York.

Standley, Paul C. 1920–26. Trees and Shrubs of Mexico. *Contr. U.S. Nat. Herb. 23,* Parts 1–5. Washington, D.C.: Government Printing Office.

Thompson, Edward H. 1899. Sisal grass in Mexico. *U.S. State Dept. Consular Rep.* No. 607: 1–4.

Trelease, William. 1907. *Agave macro-acantha* and Allied Euagaves. *Mo. Bot. Gard. Ann. Rep.* 18: 231–56, illus.

———. 1908. Agave rigida—Furcraea rigida—Agave angustifolia. *Mo. Bot. Gard. Ann. Rep.* 19: 273–87, illus.

———. 1909. The Zapupe agaves. *Trans. Acad. St. Louis* 18: 32–36.

———. 1910. Species in agave. *Proc. Amer. Phil. Soc.* 49: 232–37, illus.

———. 1911. The smallest of the century plants. *Pop. Sci. Monthly* 1911: 5–15.

———. 1912. The agaves of Lower California. *Missouri Bot. Gard. Annu. Rep.* (1911) 22: 37–65, pls. 18–72.

———. 1912. Revision of the agaves of the group Applanatae. *Missouri Bot. Gard. Annu. Rep.* 22: 85–122, illus.

———. 1913. Agave in the West Indies. *Mem. Nat. Acad. Sci.* 11: 55 pp., 116 plates.

———. 1915. The Agaveae of Guatemala. *St. Louis Acad. Sci.* 23 (3): 29–150, 20 plates.

———. 1920. Agave. In Standley, *Trees and Shrubs of Mexico. Contr. U.S. Nat. Herb.* 23: 107–42.

Trias, A. M., and S. Blight. 1972. Hacemos Muchas Cosas con el Mezcal. Publ. Dir. Gen. Educ. Extraescolar Med. Ind., México, D. F. Text in Tarahumara and Spanish, 30 pp., illus.

Wall, M. E. 1954. Steroidal sapogenins XV. *U.S. Dept. Agric. Agric. Res. Serv. AIC-367.* 32 pp. (processed)

——— et al. 1954a. Steroidal sapogenins XII. *J. Am. Pharm. Assoc.* 43: 503–05.

——— et al. 1954b. Steroidal sapogenins VII. Survey of plants for steroidal sapogenins and other constitutents. *J. Am. Pharm. Assoc. (Sci. Ed.)* 43: 1–7.

——— et al. 1955. Steroidal sapogenins XXVI. *U.S. Dept. Agric. Agric. Res. Serv. ARS-73-4.* 30 pp. (processed)

——— et al. 1957. Steroidal sapogenins XLIII. *J. Am. Pharm. Assoc. (Sci. ed.)* 46: 653–84.

——— et al. 1961. Steroidal Sapogenins LX. *J. Pharm. Sci.* 50: 1001–34.

West, R. C. et al. 1964. *Handbook of Middle American Indians.* Vol. 1, *Natural Environment and Early Cultures.* Austin: Univ. Texas Press.

Wienk, J. F. 1969. Breeding long fibre agaves. In *Misc. Papers No. 4,* Landbouwhogeshool, Wageningen, The Netherlands, 511 pp. Ed. Ferwerda & Wit.

Wilson, J. Tuzo et al. 1972. Continents adrift: readings from *Scientific American,* with introduction by J. Tuzo Wilson. San Francisco: W. H. Freeman & Co. 172 pp.

Wolf, Ivan et al. 1960–65. U.S. Dept. Agric. Unpubl. Rep. Inter-departmental from North. Utilization Research & Development Div., Peoria, Ill.

Zuccarini, J. G. 1833. Ober Einige Pflanzen aus den Gattungen Agave und Fourcroya. *Act. Acad. Caes. Leop. Nat. Cur.* 162: 661–80.

Glossary of Special Terms

abaxial, the side of a lateral organ away from the central axis.

acaulescent, stemless or without visible stem below the leaves.

acicular, needle-shaped.

adaxial, the side of a lateral organ next to the central axis.

adventitious buds, those produced in areas without visible bud initials, as from the stem instead of the axils of the leaves.

allopatric, applied to allied species or populations inhabiting separate geographic areas. Compare sympatric.

antrorse, directed forward or towards the apex, as the prickles on a leaf margin.

arcuate, moderately arched or curving.

ascending, of leaves pointed upward and outward at about 20° or more from the horizontal.

attenuate, gradually narrowed or prolonged.

bud-printing, of leaves when margin of one leaf is impressed upon the surface of the next leaf.

bulbiferous, producing bulbils.

bulbil, small plant reproduced vegetatively in the axils of the inflorescence. A form of asexual reproduction.

cabeza, Spanish, head, applied to the thick, short stem of agaves.

campanulate, bell-shaped.

castaneous, chestnut-colored.

caulescent, having a stem or trunk below the leaves.

cespitose, as applied to succulents, growing in clusters by the production of basal branches, suckers, or offsets.

chartaceous, papery, dry, and thin.

circadian (L. *circa*, about, and *dies*, day), relating to biologic variations or rhythms with a cycle of about 24 hours.

clone, a group of individual plants reproduced asexually from a single original parent.

conduplicate, folded together lengthwise.

conic, cone-shaped.

contingent perennial, a plant living more than 2 years and whose flowering is contingent upon the proper climatic conditions, e.g., rainfall and higher temperature.

crenate, applied to leaf margin strongly and abruptly undulate with large teats.

cucullate, hooded at the apex.

cultigen, a plant known to exist only in cultivation. Compare cultivate.

cultivate, as a noun, a cultivated plant with known wild ancestors.

deflexed, bent downward.

descending, of leaves, directed below the horizontal.

ensiform, sword-shaped.

explanate, flattened, spread out flat.

exserted, exceeding the corolla, as the filaments extending beyond the tepals.

filiferae, threadlike structures along the leaf margins.

filferous, having threadlike structures.

friable, said of teeth or leaf margins that are easily brushed or rubbed off.

funnelform, funnel-shaped.

glaucous, whitened with a waxy coating over the epidermis.

guttered, having the sides of the leaves raised to form a trough-shaped leaf, partly conduplicate.

haft, handle, applied to the narrowed lower part of the leaf, which is usually the least prickly part for grasping with the hand.

keel, the fleshy midrib on the tepal or the leaf.

laterals, the main branches of the paniculate inflorescence.

leaf forms, see Fig. 3.1.

maguey, an American Indian name for agave. It was picked up by the earliest Spaniards and appears to have been in general use in the Caribbean region and Mexico. Cortes in "Historia de Mexico" wrote, "miel de unas plantas, que llaman en las otras, y estas maguey, que es muy mejor que arrope; y de estas plantas hacen azucar, y vino, que es asi mismo venden."

mescal, an American Indian name applied to agave plants, to the cooked parts of the same, and to the distilled liquor made from the meristem. Used more in northern Mexico. Also applied loosely to *Manfreda, Tillandsia,* and other monocots.

metl, a Nahuatl name for agave, still in use among native tribal people in central Mexico: "papalo metl" (butterfly agave, *A. potatorum*), "tlaca metl" (*A. salmiana*).

monocarpic, a plant or rosette that flowers once and dies. Compare polycarpic.

multiannual, a plant that flowers once and dies, but requires several to many years to mature.

neck, apical portion of ovary between the ovarian cells and the base of the tube.

panicle, the branched inflorescence of the subgenus *Agave* with flowers borne in umbellate clusters on lateral branches.

paniculate, like a panicle.

patulous, standing open, spreading.

plane, applied to leaves having the upper surface flattened as compared with guttered or explanate.

polycarpic, a plant or rosette that flowers repeatedly, but not necessarily every year. Compare monocarpic.

proterandrous, the condition of a perfect flower when the anthers dehisce before the pistil is receptive. Hence, the flower cannot normally be self-fertilizing.

pruinose, having a waxy exudate on the surface of the leaf.

pulque, the fermented juice of the larger agaves. Word derived from Nahuatl "poliuhqui" or "ocli poliuhqui," but which was applied to soured or spoiled "ocli," fide Nuñez Ortega, the chronicler of Hernando Cortes.

raceme, an inflorescence in which the flowers are borne on pedicels along a central axis.

reclinate, reclining, applied to leaves pressing downward upon lower leaves or on the ground.

recurved, recurvate, curved backward or downward.

reflexed, bent sharply downward.

retrorse, directed towards the base.

rhizome, underground stem or shoot.

rosette, a closely spaced group of radiating leaves limited to a portion of the stem, usually at the base of the inflorescence.

sapogenin, a compound derived by hydrolysis from saponin. A large group of such compounds have been found in plant tissues and named according to their specific molecular configuration, e.g., diosgenin, smilogenin, hecogenin.

shaft, in *Agave* the central axis of the inflorescence including the peduncle and the central rachis of the flowering portion or branches.

spike, an inflorescence with the flowers more or less sissile along a common or single peduncle or shaft.

spine, terminal spine, in *Agave* the pungent indurated tip of the leaf.

spreading, of leaves extending outward less than 20° from horizontal.

subulate, awl-shaped, long tapering.

surculose, producing suckers or offsets.

sympatric, applied to related species or populations inhabiting the same geographic area.

teat, fleshy prominences under the teeth on the leaf margins.

teeth, the prickles along the leaf edge.

tepal, a combination of sepal and petal, applied when the segments of the perianth are not differentiated into two dissimilar ranks (common in monocotyledons).

trigonous, three-angled with plane faces.

tubular, tube-shaped.

umbel, a flat-topped or low-rounded flower cluster with the pedicels of unequal length from a common point, like an inverted umbrella.

umbellate, having the inflorescence in umbels or in similar form.

undulate, applied to leaf margins with low teats, wavy, as compared to straight margins.

urceolate, urn-shaped.

valleculate, little valley-shaped, as applied to agaves having folds in the leaf towards apex, plicate.

Index

Page numbers in boldface indicate taxon description and essay; taxa synonyms are marked with an asterisk.

[663]